Aarti Gupta Sharad Malik (Eds.)

Computer Aided Verification

20th International Conference, CAV 2008
Princeton, NJ, USA, July 7-14, 2008
Proceedings

Volume Editors

Aarti Gupta
NEC Laboratories America, Inc.
4 Independence Way, Suite 200
Princeton, NJ 08540, USA
E-mail: agupta@nec-labs.com

Sharad Malik
Princeton University
Department of Electrical Engineering
Princeton, NJ 08544-5263, USA
E-mail: malik@princeton.edu

Library of Congress Control Number: 2008930070

CR Subject Classification (1998): F.3, D.2.4, D.2.2, F.4.1, I.2.3, B.7.2, C.3

LNCS Sublibrary: SL 1 – Theoretical Computer Science and General Issues

ISSN 0302-9743
ISBN-10 3-540-70543-0 Springer Berlin Heidelberg New York
ISBN-13 978-3-540-70543-7 Springer Berlin Heidelberg New York

Springer is a part of Springer Science+Business Media

springer.com

Printed in Germany

Typesetting: Camera-ready by author, data conversion by Scientific Publishing Services, Chennai, India
Printed on acid-free paper SPIN: 12319992 06/3180 5 4 3 2 1 0

Preface

This volume contains the proceedings of the 20th International Conference on Computer Aided Verification (CAV) held in Princeton, New Jersey, USA, during July 7–14, 2008. CAV is dedicated to the advancement of the theory and practice of computer-aided formal analysis methods for hardware and software systems. Its scope ranges from theoretical results to concrete applications, with an emphasis on practical verification tools and the underlying algorithms and techniques.

Overall, 2008 has been a historical year for CAV.

- It marks the 20th anniversary of CAV, which has served as a forum for ideas whose impact is now clearly felt in research and practice.
- It celebrates the recognition received by Edmund M. Clarke, E. Allen Emerson and Joseph Sifakis as winners of the 2007 ACM Turing Award for their research in model checking. CAV is proud to have been the intellectual home for model checking over these 20 years.
- In recognition of the large body of contributions made to the field of computer-aided verification, the CAV Award was instituted this year with the first winner announced at the conference, and a citation to appear in the proceedings of the 21st CAV.

There were 131 paper submissions, divided into 104 regular and 27 tool papers. These went through an active review process, with each submission reviewed by at least 3, and on average 4, members of the Program Committee. We also sought external reviews from experts in certain areas. Authors had the opportunity to respond to the initial reviews during an author response period. All these inputs were used by the Program Committee in selecting a final program with 33 regular papers and 14 tool papers.

In addition to the reviewed papers, the program included:

- A session recognizing the Turing Award winners, with invited talks by them on the future of model checking.
- Two invited talks:
  - Edward Felten (Princeton University): *Coping with Outside-the-Box Attacks*
  - James Larus (Microsoft Research): *Singularity: Designing Better Software*
- Four invited tutorials:
  - Harry Foster (Mentor Graphics): *Assertion-based Verification*
  - John Harrison (Intel): *Theorem Proving for Verification*
  - Peter O' Hearn (University of London): *Tutorial on Separation Logic*
  - Reinhard Wilhelm (Saarland University): *Abstract Interpretation with Applications to Timing Validation*

- Reports on the results of two competitions:
  - The fourth *Satisfiability Modulo Theories Competition (SMT-Comp)*
  - The second *Hardware Model Checking Competition (HWMCC)*

CAV 2008 had seven affiliated workshops:

- Automated Formal Methods (AFM)
- Bit-Precise Reasoning (BPR)
- Exploiting Concurrency Efficiently and Correctly ($(\mathrm{EC})^2$)
- Formal verification of Analog Circuits (FAC)
- Heap Analysis and Verification (HAV)
- Numerical abstractions for Software Verification (NSV)
- Satisfiability Modulo Theories (SMT)

We gratefully acknowledge the support to CAV 2008 from Princeton University through facilities and other resources, the Institute for Advanced Study for hosting the conference banquet, and our corporate sponsors – Cadence Design Systems, IBM, Intel, Jasper Design Automation, Mentor Graphics, Microsoft Research, NEC Labs America, and Synopsys – for their financial contributions.

Many individuals were very generous with their time and expertise that went into putting the conference together. We thank the Program Committee and the external reviewers for their efforts in the assessment and evaluation needed to put together the technical program. We thank Alan Hu and Byron Cook for their help in organizing the tutorials and workshops, respectively. We thank the CAV 2007 Chairs and organizers – Werner Damm, Holger Hermanns, and Jüergen Niehaus – for their help and advice. We thank Avi Wigderson for his support for the Conference Banquet, held at the Institute for Advanced Study. We thank the CAV Steering Committee – Ed Clarke, Mike Gordon, Bob Kurshan and Amir Pnueli – for their help and advice.

We would especially like to thank Tara Zarillo (Conference Services, Princeton University), Stacey Weber Jackson (Electrical Engineering, Princeton University) and Susan Olson (Institute for Advanced Study) for their help with local arrangements; Maggie Westergaard (Communications Office, Princeton University) for designing the conference poster; and Nadia Papakonstantinou (NEC Labs America) for serving as the webmaster. We thank Andrei Voronkov for creating and supporting the invaluable EasyChair conference management system. Finally, we thank Springer for their help in publishing the 20th Anniversary DVD Compendium (with proceedings of CAV 1-20), and for providing a complimentary copy of the proceedings of the 25MC Workshop (held at FLoC 2006) to all CAV 2008 attendees.

July 2008

Aarti Gupta
Sharad Malik

Conference Organization

Program Chairs

Aarti Gupta (NEC Laboratories America, USA)
Sharad Malik (Princeton University, USA)

Program Committee

Rajeev Alur (University of Pennsylvania, USA)
Nina Amla (Cadence, USA)
Clark Barrett (New York University, USA)
Armin Biere (Johannes Kepler Universitat Linz, Austria)
Roderick Bloem (TU Graz, Austria)
Ahmed Bouajjani (University Paris 7, France)
Alessandro Cimatti (IRST Trento, Italy)
Werner Damm (Carl von Ossietzky Universität Oldenburg, Germany)
Steven German (IBM, USA)
Ganesh Gopalakrishnan (University of Utah, USA)
Michael Gordon (University of Cambridge, UK)
Orna Grumberg (Technion, Israel)
David Harel (Weizmann Institute, Israel)
John Harrison (Intel, USA)
Thomas Henzinger (EPFL, Switzerland)
Holger Hermanns (Universität des Saarlandes, Germany)
Pei Hsin Ho (Synopsys, USA)
Robert Jones (Intel, USA)
Daniel Kroening (Oxford University, UK)
Orna Kupferman (Hebrew University, Israel)
Shuvendu Lahiri (Microsoft Research, USA)
Rupak Majumdar (University of California – Los Angeles, USA)
Oded Maler (Verimag, France)
Kenneth McMillan (Cadence, USA)
Kedar Namjoshi (Alcatel-Lucent Bell Labs, USA)
Corina Pasareanu (NASA, USA)
Amir Pnueli (New York University)
Andreas Podelski (University of Freiburg, Germany)
Shaz Qadeer (Microsoft Research, USA)
Koushik Sen (University of California – Berkeley, USA)
Fabio Somenzi (University of Colorado, USA)
Ofer Strichman (Technion, Israel)
Karen Yorav (IBM Haifa, Israel)
Lenore Zuck (University of Illinois, USA)

Organizing Committee

Tutorials Chair: Alan J. Hu (University of British Columbia, Canada)
Workshops Chair: Byron Cook (Microsoft Research, UK)

Steering Committee

Edmund M. Clarke (Carnegie Mellon University, USA)
Michael Gordon (University of Cambridge, UK)
Robert P. Kurshan (Cadence, USA)
Amir Pnueli (New York University, USA)

Sponsors

Princeton University
Cadence Design Systems
IBM
Intel Corporation
Jasper Design Automation
Mentor Graphics
Microsoft Research
NEC Laboratories America
Synopsys

Reviewers

Yasmina Abdeddaim
Parosh Abdulla
Alfred Aho
Daphna Amit
Eli Arbel
Gilad Arnold
Tamarah Arons
Cyrille Valentin Artho
Eugene Asarin
Joanne Atlee
Domagoj Babic
Christel Baier
Gogul Balakrishnan
Gérard Basler
Gregory Batt
Jason Baumgartner
Dirk Beyer
Kunal Bindal
Jesse Bingham
Per Bjesse
Nicolas Blanc
Mihaela Gheorghiu Bobaru
Marius Bozga
Marco Bozzano
Aaron Bradley
Davide Bresolin
Angelo Brillout
James Brotherston
Robert Brummayer
Roberto Bruttomesso
Annette Bunker
Sebastian Burckhardt
Luis Caíres
Paul Caspi
Rohit Chadha
Yury Chebiryak

Xiaofang Chen
Ching-Tsun Chou
Christopher Conway
Scott Cotton
Cas Cremers
Vijay D'Silva
Dennis Dams
Thao Dang
Pallab Dasgupta
Alexandre David
Aldric Degorre
James Demmel
Henning Dierks
Alexandre Donzé
Laurent Doyen
Ashvin Dsouza
Stefan Edelkamp
Christian Eisentraut
Cindy Eisner
Michael Emmi
Görschwin Fey
Bernd Finkbeiner
Alain Finkel
Dana Fisman
Check Fleckenstein
Harry Foster
Anders Franzén
Goran Frehse
Martin Franzle
Oded Fuhrmann
Masahiro Fujita
Pierre Ganty
Yeting Ge
Amit Goel
Dan Goldwasser
Mark Greenstreet
Karin Greimel
Alberto Griggio
Jim Grundy
Colas Le Guernic
Sumit Gulwani
Anubhav Gupta
Arie Gurfinkel
Peter Habermehl
Georg Hofferek
Florian Horn
Hardi Hungar
Michael Huth
Franjo Ivančić
Christian Jacobi
Joxan Jaffar
Himanshu Jain
Geert Janssen
Susmit Jha
Ranjit Jhala
Barbara Jobstmann
Pallavi Joshi
Dejan Jovanovic
Vineet Kahlon
Gila Kamhi
Sharon Keidar-Barner
Uri Klein
Alfred Koelbl
Dmitry Korchemny
Sava Krstic
Ekaterina Kutsy
Akash Lal
Rom Langerak
Francois Laroussinie
Kim Guldstrand Larsen
Tal Lev-Ami
Jeremy Levitt
Guodong Li
Rhishikesh Limaye
Yoad Lustig
P. Madhusudan
Alexander Malkis
Freddy Mang
Panagiotis Manolios
Maria Mateescu
Paulo Mateus
Arie Matsliah
Bill McCloskey
Yael Meller
Shin-ichi Minato
In-Ho Moon
Leonardo de Moura
David Naumann
Uwe Nestmann
Ziv Nevo

Dejan Nickovic
Rotem Oshman
Joel Ouaknine
Chang-Seo Park
SeungJoon Park
Udo Payer
Dmitry Pidan
Ingo Pill
Carl Pixley
Alberto Policriti
Mitra Purandare
Zvonimir Rakamaric
Silvio Ranise
Jean-Francois Raskin
Kavita Ravi
Arend Rensink
Marco Roveri
Amitabha Roy
Pritam Roy
Abhik Roychoudhury
Mirron Rozanov
Theo Ruys
Michael Ryabtsev
Andrey Rybalchenko
Sriram Sankaranarayanan
Jun Sawada
Julien Schmaltz
Viktor Schuppan
Roberto Sebastiani
Peter-Michael Seidel
Sanjit A. Seshia
Subodh Sharma
Natasha Sharygina
Vitaly Shmatikov
Sharon Shoham
Stephen Siegel
Mihaela Sighireanu
Eli Singerman
Vasu Singh
Oleg Sokolsky
Scott Stoller
André Sülflow
Murali Talupur
Oliver Theel
Gregory Theoduloz
Ronald Tögl
Stefano Tonetta
Tayssir Touili
Stavros Tripakis
Viktor Vafeiadis
Martin Vechev
Helmut Veith
Michael Veksler
Miroslav Velev
Yakir Visel
Willem Visser
Anh Vo
Tomas Vojnar
Bjoern Wachter
Silke Wagner
Thomas Wahl
Ou Wei
Georg Weissenbacher
Bernd Westphal
Thomas Wies
Christoph M. Wintersteiger
Verena Wolf
Jin Yang
Yu Yang
Greta Yorsh

CAV Award

An annual award, called the CAV Award, has been established:

"For a specific fundamental contribution or a series of outstanding contributions to the field of computer-aided verification."

The cited contribution(s) must have been made not more recently than 5 years ago and not over 20 years ago. In addition, the contribution(s) should not yet have received recognition via a major award, such as the ACM Turing or Kanellakis Awards. (The nominee may have received such an award for other contributions.)

The award of $10,000 will be granted to an individual or group of individuals chosen by the Award Committee from a list of nominations. The Award Committee will select the nomination that most compellingly demonstrates a specific fundamental contribution or a series of outstanding contributions to the field of computer-aided verification, evidenced by its influence over the last 5–20 years, and ratified by a majority of the Award Committee. If the Award Committee does not so ratify any nomination, then no award shall be made in the given year.

The CAV Award shall be presented in an award ceremony at the Computer-Aided Verification Conference.

The Award Committee will provide a detailed citation that explains the basis of the award. This citation will be published together with selected papers from the conference in a forthcoming Special Issue of a Journal of Record. The present Journal of Record is the Springer journal *Formal Methods in System Design.*

Anyone, with the exception of members of the Award Committee, is eligible to receive the Award.

CAV Award Process

A description of the full process governing the administration of the CAV Award follows. This will also be published in the Journal of Record in which the citation for the first CAV Award is given.

Nominations

Anyone can submit a nomination. The Award Committee can originate a nomination.

A nomination must state clearly the contribution(s), explain why the contribution is fundamental or the series of contributions is outstanding, and be accompanied by supporting letters and other evidence of worthiness. Nominations should include a proposed citation (up to 25 words), a succinct (100-250 words) description of the contribution(s), and a detailed statement to justify the nomination.

A call for nominations will be part of the CAV call for papers, with the same deadline as for papers.

Nominations shall be sent to the Award Committee Chair.

Award Committee

The Award Committee consists of four individuals, each of whom shall have been an author of a paper accepted by CAV within the previous five years, unless this requirement is waived by the CAV Steering Committee. The members of the CAV Steering Committee are not eligible to serve on the Award Committee.

Two members of the Award Committee shall hold positions in the United States and two shall hold non-U.S. positions.

The four positions on the Award Committee are referred to as p_1, p_2, p_3 and p_4. The Steering Committee has appointed the first Award Committee to the respective positions.

The tenure of a member in position p_i will be i years.

In respective subsequent years, the member in position p_1 will retire, the member in position p_i ($i > 1$) will assume position p_$(i - 1)$, and the current Award Committee will select a member to fill position p_4 in the following year, by a majority vote of the current Award Committee.

The retiring member in position p_1 will not be selected for position p_4 in the following year, in order to assure turn-over in the Award Committee.

The member in position p_1 will serve as Chair of the Award Committee, with the responsibility of receiving nominations, distributing them to the other members of the Award Committee, and overseeing the selection processes for the Award and new member.

The Award Committee will take into account all individuals who have contributed to an awarded accomplishment, as well as independent discoveries of an awarded accomplishment, and assure that all individuals are treated fairly.

Awards will be for contributions not already honored by another major award.

The 2008 CAV Award Committee consisted of Thomas Henzinger (Chair), Randal Bryant, Orna Grumberg, and Moshe Vardi.

The 2009 CAV Award Committee consists of Randal Bryant (Chair), Orna Grumberg, Moshe Vardi, and Joseph Sifakis.

Rights of the Steering Committee

In the event of an unanticipated vacancy of a sitting Award Committee, the Steering Committee will assign someone to fill the vacated position.

Any circumstances that are unaccounted for through the above process will be resolved by the Steering Committee. The Steering Committee reserves the right to change the amount of the award, change the Journal of Record for citations or dissolve the Award Committee and cancel the CAV Award at any time.

The Steering Committee reserves the right to veto the selection by the Award Committee of a member for position p_4, in which case the Award Committee will select someone else.

Table of Contents

Invited Talks

Invited Tutorials

Session 1: Concurrency

Session 2: Memory Consistency

Session 3: Abstraction/Refinement

Session 4: Hybrid Systems

Session 5: Tools – Dynamic Verification

Session 6: Modeling and Specification Formalisms

Session 7: Decision Procedures

Session 8: Tools – Decision Procedures

Session 9: Program Verification

Session 10: Program and Shape Analysis

Session 11: Tools – Security and Program Analysis

Session 12: Hardware Verification I

Session 13: Hardware Verification II

Session 14: Model Checking

Session 15: Space Efficient Algorithms

Session 16: Tools – Model Checking

Singularity: Designing Better Software
(Invited Talk)

James R. Larus

Microsoft Research
One Microsoft Way
Redmond WA 98052
larus@microsoft.com
http://research.microsoft.com/~larus

Five years ago, frustrated by the never-ending process of finding bugs that developers had cleverly hidden throughout our software, I started a new project with Galen Hunt to rethink what software might look like if it was written, from scratch, with the explicit intent of producing more robust and reliable software artifacts. The Singularity project [1] in Microsoft Research pursued several novel strategies to this end. It has successfully encouraged researchers and product groups to think beyond the straightjacket of time-tested software architectures, to consider new solutions that cross the bounds of academic disciplines such as programming languages, operating systems, and tools.

Singularity built a new operating system using a new programming language, new software architecture, and new verification tools. The Singularity OS incorporates a system architecture based on software isolation of processes. Sing#, the programming language is an extension of C# that provides pre- and post-conditions; object invariants; verifiable, first-class support for OS communication primitives; and strong support for systems programming and code factoring.

From its start, the Singularity project was driven by the question of what would a software platform look like if it was designed with the primary goal of improving the reliability and robustness of software? To this end, we adopted three strategies. First, Singularity is almost entirely written in a safe, modern programming language, which eliminates many serious defects such as buffer overruns. Second, the system architecture limits the propagation of runtime errors by providing numerous, inexpensive, well-defined failure boundaries, thereby making it easier to achieve robust and correct system behavior, even in the presence of imperfect software. Finally, Singularity was designed from the start to facilitate the widespread use of sound program verification tools, with the belief that these tools could provide strong guarantees that entire classes of errors were eliminated.

The success of Singularity raises the possibility that it is time to rethink the traditional design, architecture, and construction practices for software in light of its increasingly central role in the world and the unprecedented threats to its security and integrity. It also poses interesting questions about today's balance

A. Gupta and S. Malik (Eds.): CAV 2008, LNCS 5123, pp. 1–2, 2008.

of effort between finding defects in existing software and developing the next generation of languages and tools, which could make a qualitative improvement in software robustness. The advent of parallel programming, occasioned by the Multicore revolution, makes these changes even more relevant, as this sea change opens the door for other radical changes in software.

References

1. Hunt, G., Larus, J.: Singularity: Rethinking the Software Stack. Operating System Review 41, 37–49 (2007)

Coping with Outside-the-Box Attacks

Edward W. Felten

Department of Computer Science
and
Woodrow Wilson School of Public and International Affairs
Princeton University

There is a long history of security attacks that succeed by violating the system designer's assumptions about how things work. Even if a designer does everything right—within the "obvious" model—such attacks can still succeed. How can we, as designers and verifiers of systems, cope with these "outside-the-box" attacks?

The classic examples of assumption-violating attacks are the timing attacks on cryptosystems first introduced by Kocher [1]. Cryptosystems are designed so that an attacker who has black-box access to an implementation (and does not know the secret key) cannot deduce the key. Extensive mathematical analysis of the input-output behavior of cryptographic functions led to the belief (though unfortunately not proof) that an attacker who can observe the input-output behavior of cryptosystems cannot feasibly find the secret key. Kocher showed that even if this is true, the running time of common cryptographic algorithms does depend on the secret key. Though the dependence of running time on the key is complex, Kocher showed how to use randomized experiments to extract enough signal to deduce the key, at least in principle. Brumley and Boneh later showed that such attacks are practical, even across a network[2].

Systems can be redesigned to resist Kocher-style timing attacks, and many systems were redesigned after the attacks became known. However, any new attack that breaks existing systems imposes significant cost and puts existing systems at risk. Designers began to worry that more outside-the-box attacks might be coming.

This fear turned out to be justified, as more attacks on cryptosystems were uncovered. Kocher, Jaffe, and Jun discovered that the power consumption of crypto implementations leaked information about their internal computations, allowing secret keys to be discovered. Fine-grained fluctuations in power consumption during a computation turned out to leak even more information [3]. Again, systems could be redesigned—though not so easily this time—but again existing systems were vulnerable.

Cryptosystems fell victim to other outside-the-box attacks. Boneh, DeMillo, and Lipton showed how to defeat public-key cryptosystems by inducing faults in the cryptographic computation. Remarkably, even if the attacker couldn't determine exactly where, when, and how the error occurred, just knowing there was an error and having access to the corrupted output was sufficient to deduce the secret key in most cases [4]. Biham and Shamir found a similar result for

A. Gupta and S. Malik (Eds.): CAV 2008, LNCS 5123, pp. 3–4, 2008.

symmetric cryptosystems [5]. More systems had to be redesigned, even though they were secure under (then-)existing models.

Similar attacks worked in non-crypto settings as well. Kuhn and Anderson showed attacks on satellite television smartcards [6]. Govindavajhala and Appel showed how to break Java virtual machine security by using heatlamps to generate thermal errors in PC execution [7]. Several researchers showed how to breach security barriers by observing the timing of execution. Halderman *et al.* defeated operating system memory protection by cooling DRAM chips then cutting system power [8].

The upshot of all this is that outside-the-box attacks seem to be a fact of life. We can expect to see more attacks new unexpected attack modes. Though we can't predict what the attacks will be, we know they are coming. How can we, as system designers and verifiers, cope?

We are often advised to think outside the box. While it is good to widen our horizons and think creatively from time to time, "think outside the box" is not really actionable advice. What we need instead is to find a better box—one whose boundaries are not so often broken.

How exactly to do this is not a question we can answer now. One goal of this presentation is to catalyze a discussion about how we might improve our models, or how we might design systems to be more resilient against failures of our models to match reality. Until we can answer this question, our systems will remain fragile, no matter how hard we work on verifying them.

References

1. Kocher, P.C.: Timing attacks on implementations of diffie-hellman, rsa, dss, and other systems, 104–113 (1996)
2. Brumley, D., Boneh, D.: Remote timing attacks are practical (1996)
3. Kocher, P., Jaffe, J., Jun, B.: Differential power analysis. In: Wiener, M.J. (ed.) CRYPTO 1999. LNCS, vol. 1666, pp. 388–397. Springer, Heidelberg (1999)
4. Boneh, D., DeMillo, R., Lipton, R.: On the importance of checking cryptographic protocols for faults. Journal of Cryptology 14(2), 101–119 (2001)
5. Biham, E., Shamir, A.: Differential fault analysis of secret key cryptosystems. In: Kaliski Jr., B.S. (ed.) CRYPTO 1997. LNCS, vol. 1294, pp. 513–525. Springer, Heidelberg (1997)
6. Anderson, R., Kuhn, M.: Tamper resistance - a cautionary note. In: Proc. USENIX Workshop on Electronic Commerce, pp. 1–11 (November 1996)
7. Govindavajhala, S., Appel, A.W.: Using memory errors to attack a virtual machine. In: Proc. IEEE Symposium on Security and Privacy (May 2003)
8. Halderman, J.A., Schoen, S.D., Heninger, N., Clarkson, W., Paul, W., Calandrino, J.A., Feldman, A.J., Appelbaum, J., Felten, E.W.: Lest we remember: Cold boot attacks on encryption keys. In: Proc. USENIX Security Symposium (2008)

Assertion-Based Verification: Industry Myths to Realities (Invited Tutorial)

Harry Foster

Mentor Graphics Corporation
Plano, Texas
Harry_Foster@mentor.com

Abstract. Debugging, on average, has grown to consume more than 60% of today's ASIC and SoC verification effort. Clearly, this is a topic the industry must address, and some organizations have done just that. Those that have adopted an assertion-based verification (ABV) methodology have seen significant reduction in simulation debugging time (as much as 50% [1]) due to improved observability. Furthermore, organizations that have embraced an ABV methodology are able to take advantage of more advanced verification techniques, such as formal verification, thus improving their overall verification quality and results. Nonetheless, even with multiple published industry case studies from various early adopters—each touting the benefits of applying ABV—the industry as a whole has resisted adopting assertion-based techniques. This tutorial provides an industry survey of today's ABV landscape, ranging from myths to realities. Emerging challenges and possible research opportunities are discussed. The following extended abstract provides a reference on which the tutorial builds.

Keywords: Assertion, Assertion-Based Verification, Debugging, Formal Verification, Functional Verification, Property Specification, Simulation.

1 Introduction

Ensuring functional correctness on RTL designs continues to pose one of the greatest challenges for today's ASIC and SoC design teams. Very few project managers would disagree with this statement. In fact, an often cited 2004 industry study by Collett International Research revealed that 35 percent of the total ASIC development effort was spent in verification [2]. In 2008, Far West Research published a study that indicated the verification effort has risen to 46 percent of the total ASIC development effort [3]. Furthermore, these industry studies reveal that debugging is the fastest-growing component of the verification effort, and that it consumes 60 percent of the total verification effort. Unfortunately, with this increase in verification effort, the industry has not experienced a measurable increase in quality of results. For example, the Collett International Research study focused on design closure and revealed that only 29 percent of projects developing ASICs were able to achieve first silicon

A. Gupta and S. Malik (Eds.): CAV 2008, LNCS 5123, pp. 5–10, 2008.

success. To make matters worse, the industry is witnessing increasing pressure to shorten the overall ASIC and SoC development cycle. Clearly, new design and verification techniques, combined with a focus on maturing functional verification process capabilities within an organization (and the industry as a whole), are required. Assertion-based verification (ABV), although certainly not an end-all to the verification challenge, does directly address today's debugging problem, while providing an integration path for more advanced forms of verification into the design flow (such as formal verification). This tutorial discussion provides a survey of today's ABV landscape, ranging from assertion language standardization efforts to industry case-studies, to common industry myths and objections that are impeding adoption, to emerging challenges and research opportunities.

2 Background

Alan Turing made the following observation over 50 years ago [4]: "How can one check a large routine in the sense of making sure that it's right? In order that the man who checks may not have too difficult a task, the programmer should make a number of definite assertions which can be checked individually, and from which the correctness of the whole program easily flows." In essence, this view is at the heart of ABV.

Informally, an assertion is a statement of design intent that can be used to specify design behavior. Assertions may specify internal design behaviors (such as a specific FIFO structure) or external design behavior (such as protocol rules or even higher-level, end-to-end behavior that spans multiple design blocks). One key characteristic of assertions is that they allow the user to specify *what* the design is supposed to do at a high level of abstraction, without having to describe the details of *how* the design intent is to be implemented. Thus, this abstract view of the design intent is ideal for the verification process—whether we are specifying high-level requirements or lower-level internal design behavior by means of assertions.

As a background for the ABV discussion, this tutorial traces events (from an industry perspective) that led to the emergence of assertion-based techniques.

3 The Road to Assertion Language Standards

Assertions are certainly not a new phenomenon in either software programming or hardware description languages. For example, languages such as Java and VHDL have contained simple assertion constructs for years —thus providing a convenient mechanism for ferreting out a class of bugs that can be identified by checking a Boolean condition (such as referencing a NULL pointer). Today's assertion languages, such as the IEEE Std 1850™-2005 Standard for Property Specification Language (PSL) and the assertion language contained within the IEEE Std 1800™-2005 SystemVerilog Verification and Hardware Description Language (SVA), not only allow the user to specify Boolean conditions, but also their relation over time using temporal logic and a generalized form of regular expressions.

The foundation for today's assertion language standards is built on the works of Amir Pnueli (linear time logic LTL [5]), and Ed Clarke and Allen Emerson (computation tree logic CTL [6]). Furthermore, the works of Moshe Vardi and Pierre Wolper provided significant contributions in that they helped improve expressiveness of LTL through the use of regular sequences of Boolean events [7, 8, 9]. Extending the expressiveness of CTL was later demonstrated by [10].

In the early 1990's, researchers at the IBM Haifa Research Laboratory developed the temporal language Sugar, which was a syntactic simplification (or sugaring) of CTL. The goal was to simplify the specification process for the RuleBase model checker. To improve usability and expressiveness, regular expressions were added to the language in the mid 1990's [11]. By the late 1990's, IBM had expanded its use of the Sugar assertion language for simulation [1].

With a similar motive, researchers at Intel Strategic CAD Labs developed the ForSpec property specification language, whose underlying logic is the ForSpec Temporal Logic (FTL) [12], which is based on LTL. Their decision to base FTL on LTL was driven by a desire to combine formal verification and dynamic validation techniques in a limited fashion. Furthermore, experience had demonstrated that mainstream verification engineers generally find branching time unintuitive—particularly since they are familiar with dynamic validation, which is inherently linear.

In 2000, both Sugar and ForSpec, in addition to the temporal property languages CBV from Motorola and Temporal ***e*** from Verisity, were donated to Accellera Formal Verification Technical Committee (FVTC) as candidate languages for standardization. The process within the committee was to establish a set of requirements for an assertion language and select a single language from four candidates. The final selection would then form the basis for the new standard, which then would undergo modification and enhancements dictated by the language requirements identified by the committee. For example, one of the committee's identified requirements was that its underlying semantics for the final standard should be based on linear time. This requirement influenced the IBM team to move Sugar from its branching-time semantics based on CTL to the linear-time semantics of LTL. In 2002, the FVTC selected Sugar as the base language, and it was approved by Accellera in 2004. Ultimately, the IEEE 1850™-2005 Property Specification Language PSL standard, based on the Accellera standard, was approved in October 2005 [13].

In 2002, work was underway in Accellera for the creation of a new version of Verilog, which would combine hardware description and hardware verification language capabilities into a single language. This effort resulted in the IEEE 1800™ 2005 SystemVerilog – Unified Hardware Design, Specification, and Verification Language standard, which was approved in November 2005. A major feature of this new language was the addition of temporal assertions, referred to as SystemVerilog Assertions (SVA). SVA has its roots in Open Vera Assertions (from Synopsys), ForSpec, and PSL. SVA provides direct links to control the verification environment by using action blocks associated with its cover and assertion directives. This capability allows the user to create reusable verification IP that can easily communicate with other verification components within the testbench, thus providing a separation between verification IP detection and action. In addition, the language provides a convenient mechanism for expressing a data integrity class of properties through the use of local variables. An SVA local variable provides the benefit of sampling and manipulating data in a property or sequence without requiring the property writer to define auxiliary state machines to model the intended behavior [14].

This tutorial compares and contrasts these two new industry standards, PSL and SVA, and then discusses future language directions for both.

3 Industry Challenges

A few industry surveys indicate that approximately 60 percent of the industry is currently employing assertion-based techniques [15]. However, these surveys are flawed in that they were conducted at conferences with a large attendance of engineers already using advanced verification techniques. From my own experience of engaging with a larger more diverse population of engineers in the industry, ranging from the extremely advanced to extremely basic, I would estimate that the figure is closer to 25 percent. Hence, it is a myth that ABV is a mainstream process. Increased adoption will only occur as organizations begin to invest in maturing their process capabilities.

In the early 1990's, the design community moved design up a level of abstraction from gate level to RT level. You will see evidence of this shift in Fig. 1, with the increase in the curve representing our ability to design larger blocks [3]. Yet even with today's synthesis breakthroughs in design productivity, designing and synthesizing RTL entirely from scratch cannot keep pace with what we are capable of fabricating. Hence, third-party IP that moves design to the transaction level will be necessary to increase design productivity.

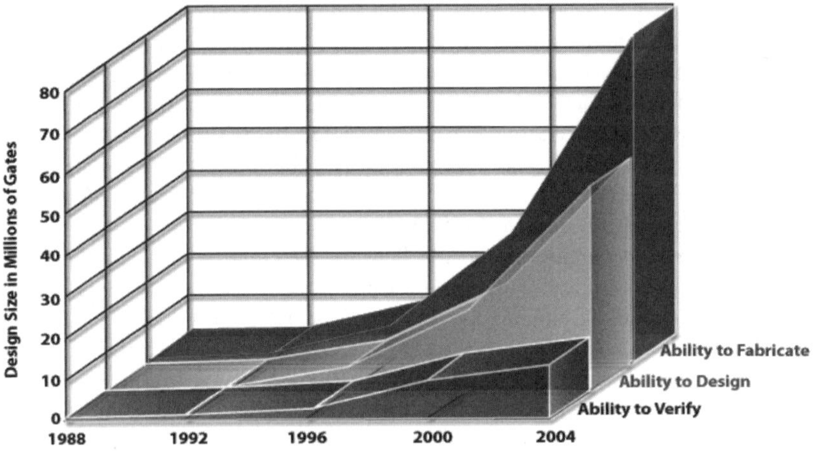

Fig. 1. Productivity gap, as reported by the Collett 2004 industry study [2]

Upon further examining Fig. 1, you might be drawn to the disparity between what we can design and what we are able to verify. Yet in many respects, the data in Fig. 1 seems to defy reality. Design teams actually do verify complex chips today, which is obvious from the myriad new electronic products available. In fact, today's verification gap is not due to a lack of innovation in verification technology. What differentiates a successful team from an unsuccessful team is process and adoption of new verification methods. Unsuccessful teams tend to approach development in an ad hoc fashion, while successful teams employ a more mature level of methodology that is systematic.

In this tutorial, I present multiple case studies illustrating successful integration of ABV by more advanced verification teams. In addition, I present case studies that illustrate multiple challenges faced by mainstream verification teams when attempting to adopt assertion-based techniques.

4 Future Direction and Research Opportunities

The industry is currently facing a design and verification productivity crisis, as illustrated by Fig. 1. Today's RTL-based flows cannot accommodate rapid iterations in design explorations, nor can they accommodate late stage changes in design features required by the growing consumer and wireless electronics market. Historically, increases in productivity have been achieved by raising the level of design and verification abstraction. Today, industry is just beginning to witness a shift in abstraction level from RTL to transaction level. While the increase in abstraction offers many advantages, there are a number of unanswered questions in terms of how to describe design intent (that is, assertions) on transaction-level models. These unanswered questions present opportunities for future research.

In this tutorial, I present a number of ABV research opportunities, which are based on discussions with multiple tool developers and industry experts currently applying assertion-based techniques.

References

1. Abarbanel, Y., Beer, I., Gluhovsky, L., Keidar, S., Wolfsthal, Y.: FoCs—Automatic Generation of Simulation Checkers from Formal Specifications. In: Proc. 12th International Conference Computer Aided Verification, pp. 414–427 (2000)
2. 2004 IC/ASIC Functional Verification Study, Industry report from Collett International Research, p. 34 (2004)
3. EDA Market Statistics Service Report, Far West Research (2008)
4. Turing, A.: In Report of a conference on high speed automatic calculating machines, pp. 67–69, Univ. Math. Laboratory, Cambridge (1949)
5. Pnueli, A.: The temporal logic of programs. In: Proc. 18th IEEE Symp. on Foundation of Computer Science, pp. 46–57 (1977)
6. Clarke, E.M., Emerson, E.A.: Design and synthesis of synchronization skeletons using branching time temporal logic. In: Kozen, D. (ed.) Logic of Programs 1981. LNCS, vol. 131, pp. 52–71. Springer, Heidelberg (1982)
7. Wolper, P.: Temporal logic can be more expressive. Information and Control 56(1/2), 72–99 (1983)
8. Vardi, M.Y., Wolper, P.: Reasoning about infinite computations. Information and Computation 115(1), 1–37 (1994)
9. Vardi, M.Y.: Branching vs. linear time: Final showdown. In: Margaria, T., Yi, W. (eds.) ETAPS 2001 and TACAS 2001. LNCS, vol. 2031, Springer, Heidelberg (2001)
10. Iwashita, H., Nakata, T.: Forward Model Checking Techniques Oriented to Buggy Designs. In: International Conference on Computer Aided Design, ICCAD (1997)
11. Beer, I., Ben-David, S., Landver, A.: On-the-fly model checking of RCTL formulas. In: Vardi, M.Y. (ed.) CAV 1998. LNCS, vol. 1427, pp. 184–194. Springer, Heidelberg (1998)

12. Armoni, R., Fix, L., Flaisher, A., Gerth, R., Ginsburg, B., Kanza, T., Landver, A., Mador-Haim, S., Singerman, E., Tiemeyer, A., Vardi, M.Y., Zbar, Y.: The ForSpec Temporal Logic: A New Temporal Property-Specification Language. In: Katoen, J.-P., Stevens, P. (eds.) ETAPS 2002 and TACAS 2002. LNCS, vol. 2280, pp. 296–311. Springer, Heidelberg (2002)
13. Eisner, C., Fisman, D.: A Practical Introduction to PSL. Springer, Heidelberg (2006)
14. Long, J., Seawright, A.: Synthesizing SVA Local Variables for Formal Verification. In: Proceedings of the 44th Design Automation Conference, DAC 2007, pp. 75–80 (2007)
15. Verification Census, extracted from the world-wide-web on April 16 (2008), http://www.deepchip.com/posts/dvcon07.html

Theorem Proving for Verification
(Invited Tutorial)

John Harrison

Intel Corporation, JF1-13
2111 NE 25th Avenue, Hillsboro OR 97124, USA
johnh@ichips.intel.com

1 The Scope of Automation

There are numerous verification techniques in active use. Traditional testing and simulation usually only provide a limited guarantee, since they can seldom exercise all possible situations. Methods based on abstraction consciously simplify the problem to make its complete analysis tractable, but still do not normally completely verify the ultimate target. We will confine ourselves here to full formal verification techniques that can be used to prove complete correctness of a (model of a) system with respect to a formal specification. Roughly speaking, these methods model the system and specification in a logical formalism and then apply general methods to determine whether the formal expressions are valid, indicating correctness of the model with respect to the specification. Typical formalisms include:

- Propositional logic, a.k.a. Boolean algebra
- Temporal logic (CTL, LTL etc.)
- Quantifier-free combinations of first-order theories
- Full first order logic
- Higher-order logic or first-order logic with arithmetic or set theory

This list is organized approximately in order of increasing logical generality, with formalisms later in the list often subsuming earlier ones. But there is a price to be paid for this generality: deciding validity in the formalisms becomes successively more difficult.

Testing validity (tautology) for propositional logic is just [the dual of] the well-studied propositional satisfiability problem (SAT), and even though the problem is known to be [co-]NP-complete [15], there are many reasonable algorithms [4,17,33,59] and some of these are implemented in practical SAT solvers with very good performance on typical problems, e.g. [22,26,46]. Many temporal logics, even quite rich ones, permit efficient 'model checking' algorithms based on explicit-state reachability analysis [13,52], its symbolic refinement [7] or automata-related techniques [68]. Algorithms for testing validity of formulas without quantifiers in combinations of common first-order theories like linear arithmetic and lists, originating in [47] and further studied in many recent papers [48,40], have been implemented in a series of tools now commonly known as SMT (Satisfiability Modulo Theories) solvers [2].

A. Gupta and S. Malik (Eds.): CAV 2008, LNCS 5123, pp. 11–18, 2008.

Once we reach full first-order logic, where we allow arbitrary use of the quantifiers 'for all' and 'there exists' over domain objects, the validity problem actually becomes undecidable [12,66]. However it is at least semidecidable, and there has been an extensive line of research in developing first-order provers, going right back to [18,24] and leading to several effective search algorithms incorporating unification, most famously resolution [54]. There are many good modern implementations of first-order proof search, and they have even occasionally answered significant open problems [43,44]. Yet the scope in principle of first-order logic is vast: any deductive inference from the accepted set-theoretic axioms of mathematics can in principle be found by first-order proof search. In contrast with this, the practical scope of these methods is tiny.

As soon as we reach higher-order logic, where quantification over functions or predicates is allowed, the problem is no longer even semidecidable [25,60]; in fact this holds even if we merely ask whether *first-order* formulas about arithmetic are *true*, as distinct from being deductive consequences of some decidable axioms. A particularly sharp form of this result is the undecidability of Hilbert's 10th problem [42]: there is no algorithm to decide whether a multivariate polynomial equation has a solution over the integers.

2 Interactive Theorem Proving

Traditionally, in the 'theorem proving' approach to verification, one works in a relatively expressive formalism such as higher-order logic, and accepts the fact that automation of the validity checking process is going to be impossible, or at least practically infeasible. Instead, one approaches the task in a more deductive way just like a traditional mathematical proof, verifying the result by machine, but applying a sequence of logical reasoning principles wholly or partly under human guidance. In fact, interactive theorem provers in this sense only appeared some time after the first experiments with fully automated provers, perhaps in disillusionment over their relatively limited scope. The first interactive provers in the modern sense were probably the SAM (semi-automated mathematics) series, whose manifesto stated [29]:

> Semi-automated mathematics is an approach to theorem-proving which seeks to combine automatic logic routines with ordinary proof procedures in such a manner that the resulting procedure is both efficient and subject to human intervention in the form of control and guidance. Because it makes the mathematician an essential factor in the quest to establish theorems, this approach is a departure from the usual theorem-proving attempts in which the computer *unaided* seeks to establish proofs.

Nowadays there is active research activity in both 'interactive' and 'automated' theorem proving. Not long after the SAM project, the AUTOMATH [19,20], Mizar [64,65], NQTHM [3] and LCF [28] proof checkers appeared, and each of them in its way has been profoundly influential. Many of the most successful interactive theorem provers around today are directly descended from one (or more) of these. As well as the automated methods we have already mentioned, such as SMT and first-order proving, there have been some successes for more specialized automated methods, e.g. in real algebra [14,61], polynomial ideals [5,39] and geometry [11,69].

It is desirable to make even 'interactive' provers as efficient as possible by incorporating powerful automated subsystems, so that the human can focus on the really difficult and creative parts of the proof. One approach is simply to combine the interactive provers with automated external systems [36,53,56]. However, most interactive theorem provers aim to prove theorems with a high degree of logical rigor, so relying on external tools is not uncontroversial. One way of combining efficiency and rigor is to use the external tool only to provide a certificate that the theorem prover can relatively easily check in a rigorous fashion [32]. Indeed, one can even have the external tool provide a complete logical *proof*, in the case of an automated theorem prover [34].

There are numerous interactive theorem provers in the world. The book [70] gives an instructive survey of some of the main interactive theorem provers: HOL, Mizar, PVS, Coq, Otter/IVY, Isabelle/Isar, Alfa/Agda, ACL2, PhoX, IMPS, Metamath, Theorema, Lego, Nuprl, Omega, B and Minlog. In each case, a proof of the irrationality of $\sqrt{2}$ is given, and some of the key features surveyed.

3 Why Theorem Proving?

Experience suggests, unsurprisingly, that highly automated techniques such as symbolic simulation [9], symbolic trajectory evaluation [57] and model checking [7] are much easier to learn, and often more productive to use, than methods like theorem proving that rely more on human interaction. They therefore tend to win acceptance much more easily, particularly in industrial practice but even in terms of academic interest. For example, the idea of using temporal logic for program verification goes at least back to [8,50]. Yet this attracted relatively little interest until the subsequent development of effective decision algorithms, particularly symbolic model checking. So with this in mind, why would one want to use theorem proving instead of model checking or other highly automated techniques? We can think of several reasons, which we rank in approximately decreasing order of perceived importance.

3.1 Beyond the Scope of Automated Methods

Some problems are simply not in the scope of any of the mainstream automated methods. Anything of interest about a finite-state transition system is decidable in principle, and thanks to model checkers, quite effectively in practice. But this trivial decidability breaks down as soon as we start to think about infinite-state systems, or even systems of an arbitrary finite size. For example, in a multiprocessor cache system with N nodes, we can usually model the essentials of cache coherence as a reachability problem in a finite state transition system for any *particular* value of N. But we might actually want a guarantee that such a parametrized system is correct for *any* (finite) value of the parameter N, and this is not a priori within the scope of automated methods. Although there is an extensive body of research on techniques for fully automated verification of parametrized systems [21,23,51], methods requiring at least some human guidance seem to be necessary in most practical applications.

For another example, consider verifying the correctness of floating-point arithmetic circuits. The desired specification in the IEEE Standard governing binary floating-point arithmetic [35] is in terms of real numbers, not bitstrings. Roughly speaking, one needs to prove that, for example, the result of a floating-point square root operation $\sqrt{x}$ is the closest floating-point number to the exact mathematical answer, which is in general an irrational number. It is not at all clear how to express this in limited formalisms where reasoning about arbitrary real numbers is impossible. One can come up with reasonably natural specifications for simple integer adders and multipliers in Boolean terms, but this becomes progressively more difficult when one considers division and square root, and seems quite impractical for transcendental functions. Thus, it is not surprising that one of the most popular and successful application areas of theorem proving to industrial verification is in the domain of floating-point arithmetic [30,37,38,45,49,55,58].

3.2 Verification of Underlying Theory

Even if some key properties of the system can be proved automatically, a global verification often demands deeper analysis of the underlying environment and background assumptions. Many program verification techniques simply rely on extracting verification conditions syntactically from an annotated program. Yet the connection between those verification conditions and the correct running of the program is often not formalized or verified. However, by starting from a formal semantics of the programming language in a general mathematical theorem prover, such properties can be rigorously proved in a unified way [27]. Also, some programs depend essentially on non-trivial mathematics. If one merely verifies the programming aspects, taking for granted the underlying theory, one risks either making a mistake in the theory itself or mis-applying it. For example, when verifying a program or circuit to perform elliptic curve cryptography, one could prove a specification of satisfying finality by formalizing also the underlying mathematics [63].

3.3 More Efficient

Even when something is within the scope of automated methods in principle, and despite the remarkable efficiency of many modern automated tools, larger systems can be difficult to verify in a practical amount of time. For example, going back to the example of a parametrized system, one may find that some such system is automatically verifiable for each specific N, but that the state space, and hence the runtime, increases so dramatically as N increases that one can't get past $N = 2$ or $N = 3$. In general, techniques like model checking that rely on state exploration tend to degrade in performance as the size of the system and/or specification increases. By contrast, the more deductive style of proof that theorem provers encourage often proceeds in a more structured way, e.g. using induction, and is largely independent of the size. Indeed, to describe techniques like model-checking as 'automatic' is in some ways a bit misleading. Very often, one needs to make significant modifications to or decompositions of the problem, or tweak parameters of the checker such as BDD variable ordering, in order to bring it within practical reach. Sometimes this gets so tedious and unproductive that one would be better off just settling on a deductive proof in the first place.

3.4 More Intellectually Stimulating

Theorem proving encourages a style of verification where the human uses a conceptual understanding of the system to construct a mathematical proof, rather than a 'push-button' approach of waiting for a yes/no answer from a black box. Although this may be, in a crude sense, much less productive, it can be valuable because it forces the human to articulate dimly perceived intuitions about the system, and so perhaps gain a significantly deeper conceptual understanding that may even help to improve the system. An example is reported in [31], which describes the formal verification of division algorithms. On formalizing one of the standard theorems [41], the author noticed that the full strength of one of the assumptions was never used, and a sharper theorem actually allowed the implementation of faster algorithms with the same behavior. Of course, it is not inconceivable that a sufficiently close reading would have led to the same revelation, but it is less likely without the level of logical rigor that a theorem prover imposes.

References

1. Aagaard, M., Harrison, J. (eds.): TPHOLs 2000. LNCS, vol. 1869. Springer, Heidelberg (2000)
2. Barrett, C., Sebastiani, R., Seshia, S., Tinelli, C.: Satisfiability modulo theories. In: Biere, A., van Maaren, H., Walsh, T. (eds.) Handbook of Satisfiability, vol. 4. IOS Press, Amsterdam (2008)
3. Boyer, R.S., Moore, J.S.: A Computational Logic. ACM Monograph Series. Academic Press, London (1979)
4. Bryant, R.E.: Graph-based algorithms for Boolean function manipulation. IEEE Transactions on Computers C-35, 677–691 (1986)
5. Buchberger, B.: Ein algorithmisches Kriterium fur die Lösbarkeit eines algebraischen Gleichungssystems. Aequationes Mathematicae 4, 374–383 (1970); English translation, An Algorithmical Criterion for the Solvability of Algebraic Systems of Equations. In: [6], pp. 535–545
6. Buchberger, B., Winkler, F. (eds.): Gröbner Bases and Applications. London Mathematical Society Lecture Note Series, vol. 251. Cambridge University Press, Cambridge (1998)
7. Burch, J.R., Clarke, E.M., McMillan, K.L., Dill, D.L., Hwang, L.J.: Symbolic model checking: 10^{20} states and beyond. Information and Computation 98, 142–170 (1992)
8. Burstall, R.M.: Program proving as hand simulation with a little induction. In: Information Processing 1974: Proceedings of IFIP Congress 1974, Stockholm, pp. 308–312. North-Holland, Amsterdam (1974)
9. Carter, W.C., Joyner, W.H., Brand, D.: Symbolic simulation for correct machine design. In: Proceedings of the 16th ACM/IEEE Design Automation Conference, pp. 280–286. IEEE Computer Society Press, Los Alamitos (1979)
10. Caviness, B.F., Johnson, J.R. (eds.): Quantifier Elimination and Cylindrical Algebraic Decomposition. Texts and monographs in symbolic computation. Springer, Heidelberg (1998)
11. Chou, S.-C.: An introduction to Wu's method for mechanical theorem proving in geometry. Journal of Automated Reasoning 4, 237–267 (1988)
12. Church, A.: An unsolvable problem of elementary number-theory. American Journal of Mathematics 58, 345–363 (1936)
13. Clarke, E.M., Emerson, E.A.: Design and synthesis of synchronization skeletons using branching-time temporal logic. In: Kozen, D. (ed.) Logic of Programs 1981. LNCS, vol. 131, pp. 52–71. Springer, Heidelberg (1982)

14. Collins, G.E.: Quantifier elimination for real closed fields by cylindrical algebraic decomposition. In: Brakhage, H. (ed.) GI-Fachtagung 1975. LNCS, vol. 33, pp. 134–183. Springer, Heidelberg (1975)
15. Cook, S.A.: The complexity of theorem-proving procedures. In: Proceedings of the 3rd ACM Symposium on the Theory of Computing, pp. 151–158 (1971)
16. Davis, M. (ed.): The Undecidable: Basic Papers on Undecidable Propositions, Unsolvable Problems and Computable Functions. Raven Press, NY (1965)
17. Davis, M., Logemann, G., Loveland, D.: A machine program for theorem proving. Communications of the ACM 5, 394–397 (1962)
18. Davis, M., Putnam, H.: A computing procedure for quantification theory. Journal of the ACM 7, 201–215 (1960)
19. de Bruijn, N.G.: The mathematical language AUTOMATH, its usage and some of its extensions. In: Laudet, M., Lacombe, D., Nolin, L., Schützenberger, M. (eds.) Symposium on Automatic Demonstration. Lecture Notes in Mathematics, vol. 125, pp. 29–61. Springer, Heidelberg (1970)
20. de Bruijn, N.G.: A survey of the project AUTOMATH. In: Seldin, J.P., Hindley, J.R. (eds.) To H. B. Curry: Essays in Combinatory Logic, Lambda Calculus, and Formalism, pp. 589–606. Academic Press, London (1980)
21. Delzanno, G.: Automatic verification of parameterized cache coherence protocols. In: Emerson, E.A., Sistla, A.P. (eds.) CAV 2000. LNCS, vol. 1855, pp. 53–68. Springer, Heidelberg (2000)
22. Een, N., Sörensson, N.: An extensible SAT-solver. In: Giunchiglia, E., Tacchella, A. (eds.) SAT 2003. LNCS, vol. 2919, pp. 502–518. Springer, Heidelberg (2004)
23. Fontaine, P.: Techniques for verification of concurrent systems with invariants. PhD thesis, Institut Montefiore, Université de Liège (2004)
24. Gilmore, P.C.: A proof method for quantification theory: Its justification and realization. IBM Journal of research and development 4, 28–35 (1960)
25. Gödel, K.: Über formal unentscheidbare Sätze der Principia Mathematica und verwandter Systeme, I. Monatshefte für Mathematik und Physik 38, 173–198 (1931); English translation, On Formally Undecidable Propositions of Principia Mathematica and Related Systems, I. In: [67], pp. 592–618 or [16], pp. 4–38
26. Goldberg, E., Novikov, Y.: BerkMin: a fast and robust Sat-solver. In: Kloos, C.D., Franca, J.D. (eds.) Design, Automation and Test in Europe Conference and Exhibition (DATE 2002), Paris, France, pp. 142–149. IEEE Computer Society Press, Los Alamitos (2002)
27. Gordon, M.J.C.: Mechanizing programming logics in higher order logic. In: Birtwistle, G., Subrahmanyam, P.A. (eds.) Current Trends in Hardware Verification and Automated Theorem Proving, pp. 387–439. Springer, Heidelberg (1989)
28. Gordon, M.J.C., Milner, R., Wadsworth, C.P.: Edinburgh LCF: A Mechanised Logic of Computation. LNCS, vol. 78. Springer, Heidelberg (1979)
29. Guard, J.R., Oglesby, F.C., Bennett, J.H., Settle, L.G.: Semi-automated mathematics. Journal of the ACM 16, 49–62 (1969)
30. Harrison, J.: Formal verification of floating point trigonometric functions. In: Johnson, S.D., Hunt Jr., W.A. (eds.) FMCAD 2000. LNCS, vol. 1954, pp. 217–233. Springer, Heidelberg (2000)
31. Harrison, J.: Formal verification of IA-64 division algorithms. In: Aagaard and Harrison [1], pp. 234–251
32. Harrison, J., Théry, L.: A sceptic's approach to combining HOL and Maple. Journal of Automated Reasoning 21, 279–294 (1998)
33. Hooker, J.N.: A quantitative approach to logical inference. Decision Support Systems 4, 45–69 (1988)

34. Hurd, J.: Integrating Gandalf and HOL. In: Bertot, Y., Dowek, G., Hirschowitz, A., Paulin, C., Théry, L. (eds.) TPHOLs 1999. LNCS, vol. 1690, pp. 311–321. Springer, Heidelberg (1999)
35. IEEE. Standard for binary floating point arithmetic. ANSI/IEEE Standard 754-1985, The Institute of Electrical and Electronic Engineers, Inc., 345 East 47th Street, New York, NY 10017, USA (1985)
36. Joyce, J.J., Seger, C.: The HOL-Voss system: Model-checking inside a general-purpose theorem-prover. In: Joyce, J.J., Seger, C. (eds.) HUG 1993. LNCS, vol. 780, pp. 185–198. Springer, Heidelberg (1994)
37. Kaivola, R., Aagaard, M.D.: Divider circuit verification with model checking and theorem proving. In: Aagaard and Harrison [1], pp. 338–355
38. Kaivola, R., Kohatsu, K.: Proof engineering in the large: Formal verification of the Pentium (R) 4 floating-point divider. In: Margaria, T., Melham, T.F. (eds.) CHARME 2001. LNCS, vol. 2144, pp. 196–211. Springer, Heidelberg (2001)
39. Kandri-Rody, A., Kapur, D.: Algorithms for computing Gröbner bases of polynomial ideals over various Euclidean rings. In: Fitch, J. (ed.) EUROSAM 1984 and ISSAC 1984. LNCS, vol. 174, pp. 195–206. Springer, Heidelberg (1984)
40. Krstic, S., Goel, A.: Architecting solvers for SAT modulo theories: Nelson-Oppen with DPLL. In: Konev, B., Wolter, F. (eds.) FroCos 2007. LNCS (LNAI), vol. 4720, pp. 1–27. Springer, Heidelberg (2007)
41. Markstein, P.W.: Computation of elementary functions on the IBM RISC System/6000 processor. IBM Journal of Research and Development 34, 111–119 (1990)
42. Matiyasevich, Y.V.: Enumerable sets are Diophantine. Soviet Mathematics Doklady 11, 354–358 (1970)
43. McCune, W.: Solution of the Robbins problem. Journal of Automated Reasoning 19, 263–276 (1997)
44. McCune, W., Padmanabhan, R.: Automated Deduction in Equational Logic and Cubic Curves. LNCS, vol. 1095. Springer, Heidelberg (1996)
45. Moore, J.S., Lynch, T., Kaufmann, M.: A mechanically checked proof of the correctness of the kernel of the $AMD5_K86$ floating-point division program. IEEE Transactions on Computers 47, 913–926 (1998)
46. Moskewicz, M.W., Madigan, C.F., Zhao, Y., Zhang, L., Malik, S.: Chaff: Engineering an efficient SAT solver. In: Proceedings of the 38th Design Automation Conference (DAC 2001), pp. 530–535. ACM Press, New York (2001)
47. Nelson, G., Oppen, D.C.: Simplification by cooperating decision procedures. ACM Transactions on Programming Languages and Systems 1, 245–257 (1979)
48. Nieuwenhuis, R., Oliveras, A., Tinelli, C.: Solving SAT and SAT modulo theories: from an abstract Davis-Putnam-Logemann-Loveland procedure to DPLL(T). Journal of the ACM 53, 937–977 (2006)
49. O'Leary, J., Zhao, X., Gerth, R., Seger, C.-J.H.: Formally verifying IEEE compliance of floating-point hardware. Intel Technology Journal 1999-Q1, 1–14 (1999), `http://developer.intel.com/technology/itj/q11999/articles/art_5.htm`
50. Pnueli, A.: The temporal logic of programs. In: Proceedings of the 18th IEEE Symposium on Foundations of Computer Science, pp. 46–67 (1977)
51. Pnueli, A., Ruah, S., Zuck, L.: Automatic Deductive Verification with Invisible Invariants. In: Margaria, T., Yi, W. (eds.) ETAPS 2001 and TACAS 2001. LNCS, vol. 2031. Springer, Heidelberg (2001)
52. Queille, J.P., Sifakis, J.: Specification and verification of concurrent programs in CESAR. In: Dezani-Ciancaglini, M., Montanari, U. (eds.) Programming 1982. LNCS, vol. 137, pp. 195–220. Springer, Heidelberg (1982)

53. Rajan, S., Shankar, N., Srivas, M.K.: An integration of model-checking with automated proof-checking. In: Wolper, P. (ed.) CAV 1995. LNCS, vol. 939, pp. 84–97. Springer, Heidelberg (1995)
54. Robinson, J.A.: A machine-oriented logic based on the resolution principle. Journal of the ACM 12, 23–41 (1965)
55. Rusinoff, D.: A mechanically checked proof of IEEE compliance of a register-transfer-level specification of the AMD-K7 floating-point multiplication, division, and square root instructions. LMS Journal of Computation and Mathematics 1, 148–200 (1998), `http://www.onr.com/user/russ/david/k7-div-sqrt.html`
56. Seger, C., Joyce, J.J. : A two-level formal verification methodology using HOL and COSMOS. Technical Report 91-10, Department of Computer Science, University of British Columbia, 2366 Main Mall, University of British Columbia, Vancouver, B.C, Canada V6T 1Z4 (1991)
57. Seger, C.-J.H., Bryant, R.E.: Formal verification by symbolic evaluation of partially-ordered trajectories. Formal Methods in System Design 6, 147–189 (1995)
58. Slobodová, A.: Challenges for Formal Verification in Industrial Setting. In: Brim, L., Haverkort, B.R., Leucker, M., van de Pol, J. (eds.) FMICS 2006 and PDMC 2006. LNCS, vol. 4346, pp. 1–22. Springer, Heidelberg (2007)
59. Stålmarck, G., Säflund, M.: Modeling and verifying systems and software in propositional logic. In: Daniels, B.K. (ed.) Safety of Computer Control Systems (SAFECOMP 1990), Gatwick, UK, pp. 31–36. Pergamon Press, Oxford (1990)
60. Tarski, A.: Der Wahrheitsbegriff in den formalisierten Sprachen. Studia Philosophica 1, 261–405 (1936); English translation, The Concept of Truth in Formalized Languages. In: [62], pp. 152–278
61. Tarski, A.: A Decision Method for Elementary Algebra and Geometry. University of California Press (1951); Previous version published as a technical report by the RAND Corporation (1948); prepared for publication by McKinsey, J.C.C. Reprinted In: [10], pp. 24–84
62. Tarski, A. (ed.): Logic, Semantics and Metamathematics. Clarendon Press (1956)
63. Théry, L., Hanrot, G.: Primality proving with elliptic curves. In: Schneider, K., Brandt, J. (eds.) TPHOLs 2007. LNCS, vol. 4732, pp. 319–333. Springer, Heidelberg (2007)
64. Trybulec, A.: The Mizar-QC/6000 logic information language. ALLC Bulletin (Association for Literary and Linguistic Computing) 6, 136–140 (1978)
65. Trybulec, A., Blair, H.A.: Computer aided reasoning. In: Parikh, R. (ed.) Logics of Programs, Brooklyn. LNCS, vol. 193, pp. 406–412. Springer, Heidelberg (1985)
66. Turing, A.M.: On computable numbers, with an application to the Entscheidungsproblem. Proceedings of the London Mathematical Society 42(2), 230–265 (1936)
67. van Heijenoort, J. (ed.): From Frege to Gödel: A Source Book in Mathematical Logic 1879–1931. Harvard University Press (1967)
68. Vardi, M.: An automata-theoretic approach to linear temporal logic. In: Moller, F., Birtwistle, G. (eds.) Logics for Concurrency. LNCS, vol. 1043, pp. 238–266. Springer, Heidelberg (1996)
69. Wen-tsün, W.: On the decision problem and the mechanization of theorem proving in elementary geometry. Scientia Sinica 21, 157–179 (1978)
70. Wiedijk, F.: The Seventeen Provers of the World. LNCS (LNAI), vol. 3600. Springer, Heidelberg (2006)

Tutorial on Separation Logic
(Invited Tutorial)

Peter O'Hearn*

Queen Mary, Univ. of London

Separation logic is an extension of Hoare's logic for reasoning about programs that manipulate pointers. Its assertion language extends classical logic with a separating conjunction operator $A * B$, which asserts that A and B hold for separate portions of memory.

In this tutorial I will first cover the basics of the logic, concentrating on highlights from the early work [1,2,3,4].

(i) The separating conjunction fits together with inductive definitions in a way that supports natural descriptions of mutable data structures [1].

(ii) Axiomatizations of pointer operations support *in-place reasoning*, where a portion of a formula is updated in place when passing from precondition to postcondition, mirroring the operational locality of heap update [1,2].

(iii) Notorious "dirty" features of low-level programming (pointer arithmetic, explicit deallocation) are dealt with smoothly, even embraced [2,3].

(iv) Frame axioms, which state what does not change, can be avoided when writing specifications [2,3].

These points together enable specifications and proofs of pointer programs that are dramatically simpler than was possible previously, in many cases approaching the simplicity associated with proofs of pure functional programs. I will describe how that is, and where rough edges lie (programs whose proofs are still more complex than we would like).

In describing these highlights I will outline how many of the points flow from Separation Logic's model theory, particularly an interaction between properties concerning the local way that imperative programs operate [5], and the abstract properties of its models, which it inherits from bunched logic [6,7] (a species of substructural logic related to linear and relevant logics, and Lambek's syntactic calculus). Using the model theoretic perspective, I will attempt to describe the extent to which Separation Logic's "benefits" do and do not depend on its language of assertions.

After the basic part, I will then discuss how these points (i)-(iv) feed into research on mechanized verification, both for interactive proof in proof assistants (e.g., [8,9,10,11,12]) and for automatic proof and abstract interpretation (e.g.

* Supported by the EPSRC and by a Royal Society Wolfson Research Merit Award.

A. Gupta and S. Malik (Eds.): CAV 2008, LNCS 5123, pp. 19–21, 2008.

[13,14,15,16,17,18,19,20,21,22,23]). Time permitting, I will close with more recent highlights, on concurrency, data abstraction, and object-oriented programming [24,25,26].

References

1. Reynolds, J.C.: Intuitionistic reasoning about shared mutable data structure. In: Davies, J., Roscoe, B., Woodcock, J. (eds.) Millennial Perspectives in Computer Science, Houndsmill, Hampshire, pp. 303–321. Palgrave (2000)
2. Isthiaq, S., O'Hearn, P.W.: BI as an assertion language for mutable data structures. In: 28th POPL, pp. 36–49 (2001)
3. O'Hearn, P., Reynolds, J., Yang, H.: Local reasoning about programs that alter data structures. In: Fribourg, L. (ed.) CSL 2001 and EACSL 2001. LNCS, vol. 2142, pp. 1–19. Springer, Heidelberg (2001)
4. Reynolds, J.C.: Separation logic: A logic for shared mutable data structures. In: 17th LICS, pp. 55–74 (2002)
5. Calcagno, C., O'Hearn, P., Yang, H.: Local action and abstract separation logic. In: 22nd LICS, pp. 366–378 (2007)
6. O'Hearn, P.W., Pym, D.J.: The logic of bunched implications. Bulletin of Symbolic Logic 5(2), 215–244 (1999)
7. Pym, D., O'Hearn, P., Yang, H.: Possible worlds and resources: the semantics of BI. Theoretical Computer Science 315(1), 257–305 (2004)
8. Yu, D., Hamid, N.A., Shao, Z.: Building certified libraries for PCC: Dynamic storage allocation. In: Degano, P. (ed.) ESOP 2003 and ETAPS 2003. LNCS, vol. 2618, pp. 363–379. Springer, Heidelberg (2003)
9. Marti, N., Affeldt, R., Yonezawa, A.: Formal verification of the heap manager of an operating system using separation logic. In: Liu, Z., He, J. (eds.) ICFEM 2006. LNCS, vol. 4260, pp. 400–419. Springer, Heidelberg (2006)
10. Tuch, H., Klein, G., Norrish, M.: Types, bytes, and separation logic. In: 34th POPL, pp. 97–108 (2007)
11. Myreen, M.O., Gordon, M.J.C.: Hoare logic for realistically modelled machine code. In: Grumberg, O., Huth, M. (eds.) TACAS 2007. LNCS, vol. 4424, pp. 568–582. Springer, Heidelberg (2007)
12. Varming, C., Birkedal, L.: Higher-order separation logic in Isabelle/HOLCF. In: 24th MFPS (2008)
13. Berdine, J., Calcagno, C., O'Hearn, P.W.: Smallfoot: Automatic modular assertion checking with separation logic. In: 4th FMCO, pp. 115–137 (2006)
14. Distefano, D., O'Hearn, P., Yang, H.: A Local Shape Analysis Based on Separation Logic. In: Hermanns, H., Palsberg, J. (eds.) TACAS 2006 and ETAPS 2006. LNCS, vol. 3920, pp. 287–302. Springer, Heidelberg (2006)
15. Magill, S., Nanevski, A., Clarke, E., Lee, P.: Inferring invariants in Separation Logic for imperative list-processing programs. In: 3rd SPACE Workshop (2006)
16. Berdine, J., Cook, B., Distefano, D., O'Hearn, P.: Automatic termination proofs for programs with shape-shifting heaps. In: Ball, T., Jones, R.B. (eds.) CAV 2006. LNCS, vol. 4144, pp. 386–400. Springer, Heidelberg (2006)
17. Gotsman, A., Berdine, J., Cook, B., Sagiv, M.: Thread-modular shape analysis. In: PLDI 2007 (2007)
18. Guo, B., Vachharajani, N., August, D.: Shape analysis with inductive recursion synthesis. In: PLDI (2007)

19. Chang, B., Rival, X., Necula, G.: Shape Analysis with Structural Invariant Checkers. In: Riis Nielson, H., Filé, G. (eds.) SAS 2007. LNCS, vol. 4634, pp. 384–401. Springer, Heidelberg (2007)
20. Berdine, J., Calcagno, C., Cook, B., Distefano, D., O'Hearn, P., Wies, T., Yang, H.: Shape analysis of composite data structures. In: Damm, W., Hermanns, H. (eds.) CAV 2007. LNCS, vol. 4590, pp. 178–192. Springer, Heidelberg (2007)
21. Nguyen, H.H., Chin, W.-N.: Enhancing program verification with lemmas. In: Gupta, A., Malik, S. (eds.) CAV 2008. LNCS, vol. 5123, pp. 355–369. Springer, Heidelberg (2008)
22. Magill, S., Tsai, M.-S., Lee, P., Tsay, Y.-K.: THOR: A tool for reasoning about shape and arithmetic. In: Gupta, A., Malik, S. (eds.) CAV 2008. LNCS, vol. 5123, pp. 428–432. Springer, Heidelberg (2008)
23. Yang, H., Lee, O., Berdine, J., Calcagno, C., Cook, B., Distefano, D., O'Hearn, P.: Scalable shape analysis for systems code. In: Gupta, A., Malik, S. (eds.) CAV 2008. LNCS, vol. 5123, pp. 385–398. Springer, Heidelberg (2008)
24. O'Hearn, P.W.: Resources, concurrency and local reasoning. Theoretical Computer Science 375(1-3), 271–307 (2007); Preliminary version appeared In: O'Hearn, P.W.: Resources, Concurrency and Local Reasoning. In: Gardner, P., Yoshida, N. (eds.) CONCUR 2004. LNCS, vol. 3170, pp. 49–67. Springer, Heidelberg (2004)
25. Parkinson, M., Bierman, G.: Separation logic and abstraction. In: 32nd POPL, pp. 59–70 (2005)
26. Biering, B., Birkedal, L., Torp-Smith, N.: BI-hyperdoctrines, higher-order separation logic, and abstraction. ACM TOPLAS 5(29) (2007)

Abstract Interpretation with Applications to Timing Validation*
Invited Tutorial

Reinhard Wilhelm and Björn Wachter

Universität des Saarlandes, Saarbrücken, Germany
{wilhelm,bwachter}@cs.uni-sb.de

Abstract. Abstract interpretation is one of the main verification technologies besides model checking and deductive verification.

Abstract interpretation has a rich theory of abstraction and strong support for the construction of abstract domains. It allows to express a precise relation to the (concrete) semantics of the programming language inducing a clear relation between the results of an abstract interpretation and the properties of the analyzed program. It permits trading efficiency against precision and offers means to enforce termination where this is not guaranteed.

We explain abstract interpretation using examples from a particular application domain: the determination of bounds on the execution times of programs. These bounds are used to show reliably that hard real-time systems satisfy their timing constraints.

The application domain requires a number of static analyses and domains with different characteristics. Most domains exhibit Galois connections, a few do not. Some analyses require widening to leap infinite ascending chains and ensure termination.

1 Introduction

Abstract interpretation, the theory behind static program analysis, has its roots in the compiler domain. From early on, compilers used static analysis to compute invariants at program points, which would imply the applicability conditions of optimizing transformations. First strong theoretical results about static program analysis were obtained in the 70s [1,2]. Exactly 30 years ago, Patrick Cousot submitted his PhD thesis [3], which contained the very rich theory of abstract interpretation and new static program analyses. He showed that all static analyses were abstractions of a suitable concrete semantics and hereby opened the way to analyses that could be proved correct or were even correct by construction.

* Work reported herein was partially supported by the European IST Project DAEDALUS, Validation of Critical Software by Static Analysis and Abstract Testing, the German Transregional Collaborative Research Centre AVACS (Automatic Verification and Analysis of Complex Systems) of the Deutsche Forschungsgemeinschaft, the European Networks of Excellence ARTIST2 and ARTIST DESIGN.

A. Gupta and S. Malik (Eds.): CAV 2008, LNCS 5123, pp. 22–36, 2008.

Since then, static program analysis has left the compiler domain and has become a verification method in its own right providing means to automatically prove safety properties of programs. Among the most spectacular applications of static program analysis to real systems probably are the analysis of the reasons for the failure of the Ariane5 rocket [4], the proof of the absence of runtime errors in safety-critical avionics code in the ASTRÉE project [5], and our method to determine reliable and precise execution-time bounds for hard real-time systems [6,7]. Today, static analysis tools based on abstract interpretation [8,9,10,11,12] are widely used in industry and abstract interpretation continues to make inroads into new application domains [13,14,15].

2 Timing Analysis - The Application Domain

Hard real-time systems are subject to stringent timing constraints which are dictated by the surrounding physical environment. We are concerned with the problem of guaranteeing that all the timing constraints of tasks when executed on a given processor architecture will be met ("timing validation").

Systems show a variability of execution times depending on

The input data: This has always been so and will remain so as it is a property of the algorithm,

The initial execution state: This is caused by modern architectural features such as caches, pipelines, and speculation, and

Interference from the environment: Preemptions and interrupts.

The unit-time (executing an instruction always takes exactly one time unit) or constant-time abstraction used in many approaches to timing validation is thus rendered obsolete by the advent of modern processors.

In general, the state space of input data and initial states is too large to exhaustively explore all possible executions and so determine the *exact* worst-case and best-case execution times. Some abstraction of the execution platform is necessary to make a timing analysis of the system feasible. These abstractions inevitably lose information, and yet must guarantee upper bounds for worst-case and lower bounds for best-case execution time, respectively.

The alternative to exhaustive end-to-end measurement, which was just argued to be infeasible, and non-exhaustive measurement, which is unsound in general because it may underestimate, will be explained in this paper. It consists in the computation of upper (and possibly lower) bounds on the execution times of instructions or basic blocks and the determination of a worst-case path through the program. However, the variability of execution times also appears on the instruction level, even on the level of individual memory accesses and arithmetic operations, but the problem to bound execution times is easier to solve for instructions than for whole programs.

We can look at the execution times of an instruction as an interval, from the best case, e.g. all memory accesses hit the cache, all needed pipeline units are free, no branch misprediction occurs, etc., to the worst case, where memory accesses miss the cache, pipeline units are occupied, buffers to fetch from are empty and buffers to write to are full, etc. Relative to the best case, we will call any increase in execution time during an instruction's execution a *timing accident* and the number of cycles by which it increases the execution time as compared to the fastest time the *timing penalty* for this accident. Timing penalties for an instruction can add up to several hundred processor cycles. Whether the execution of an instruction encounters a timing accident depends on the execution state, e.g., the contents of the cache(s), the occupancy of other resources, and thus on the execution history. It is therefore obvious that the attempt to predict or exclude timing accidents needs information about the execution history. Excluding timing accidents means decreasing the upper bounds.

The computation of worst-case bounds for a program is realized by first employing an abstract processor model to compute a cycle-level abstract semantics of the program and, in a second phase, mapping resulting time bounds for program portions to an Integer Linear Program (ILP) whose optimal solution yields the final bound. This tool architecture has been successfully used to determine precise upper bounds on the execution times of real-time programs running on processors used in embedded systems [6,7,16,17,18]. A commercially available tool, `aiT` by AbsInt, cf. `http://www.absint.de/wcet.htm`, was implemented and is used in the aeronautics and automotive industries.

In this tutorial, we deal with the more compute-intensive first phase. It has the following three constituents, which we will treat in more detail later:

1. Value analysis attempts to compute information about data accesses and control flow, in particular it tries to identify infeasible paths, syntactically possible paths that will never be taken because of contradictory conditions.
2. Cache-behavior prediction determines a safe and concise approximation of the contents of caches in order to classify memory accesses as definite cache hits or misses.
3. Pipeline-behavior prediction analyzes how instructions pass through the pipeline taking cache-hit or miss information into account. The cache-miss penalty is assumed for all cases where a cache hit cannot be guaranteed.
   At the end of simulating one instruction, a certain set of final states has been reached. The pipeline analysis starts the analysis of the next instruction in all those states.

Most powerful microprocessors have so-called *timing anomalies*. These are counter-intuitive influences of the (local) execution time of one instruction on the (global) execution time of the whole program. The interaction of several processor features can cause a locally faster execution of an instruction to lead to a globally longer execution time of the whole program. For example, a cache miss contributes the cache-miss penalty to the execution time of a program. It was, however, observed for the MCF 5307 [18] that a cache miss may actually speed up program execution if it prevents a costly branch misprediction. The existence of timing anomalies forces the analysis to consider a rather large search space since it has to follow not only the local worst-case transitions in the architecture.

3 Abstract Interpretation

This section describes crucial program analyses in the context of timing validation. In Subsection 3.1, we introduce the theoretical foundations before we describe constant propagation in 3.2, interval analysis in 3.3, cache analysis in 3.4 and pipeline analysis in 3.5.

3.1 The Theory

Program. A program is represented by a control flow graph which consists of a set of program points V, an initial location v_{in} (models program entry), and a set of labeled control flow edges $E \subseteq V \times Op \times V$ (the elements of Op model the operation that is executed when the edge is taken).

A program semantics consists of a (possibly infinite) set S of program states, a set of initial states $S_0 \subseteq S$ and a semantics function $[\![.]\!] : Op \rightarrow (S \rightarrow S)$ that assigns to each operation and thus to each control flow edge, a transfer function modeling its effect on the current program state.

Concerning the operations and the semantics, we observe that the execution-time bounds of a program cannot be determined from the source code of a high-level language like C: executable code has to be analyzed.

For readability we employ an imperative toy language rather than an assembly language to explain the first two example analyses. We will only regard assignment statements, $x \leftarrow e$ and the labels of the two outgoing edges of conditionals, true(e) and false(e), for a condition e. As a semantic domain, it uses states assigning integer values to variables, $\rho : Vars \rightarrow \mathbb{Z}$. A statement op transforms the state ρ. The semantics of the operations of the toy language is defined by:

$$\begin{array}{lll} [\![\mathsf{true}\,(e)]\!]\,\rho & = \rho & \text{if } [\![e]\!]\,\rho = \mathtt{tt} \\ [\![\mathsf{false}\,(e)]\!]\,\rho & = \rho & \text{if } [\![e]\!]\,\rho = \mathtt{ff} \\ [\![x \leftarrow e]\!]\,\rho & = \rho \oplus \{x \mapsto [\![e]\!]\,\rho\} & \end{array}$$

Compared to our toy language, executables have a different concept of "variables" as they employ registers defined in the instruction set architecture (ISA). Note that the ISA is still somewhat machine independent, e.g. the PowerPC architecture has many implementations for which the same value analysis can be used. Machine-dependent semantics for cache and pipeline behavior prediction are discussed in Subsection 3.4 and Subsection 3.5, respectively.

Collecting Semantics. The collecting semantics of a program assigns to each program point the set of states that may occur at it during some execution. The collecting semantics is expressible as the fixed point of a set of recursive equations and is, in general, not computable and, even in the finite-state case, not efficiently computable. To this end, the program analyses presented here compute a safe over-approximation of the collecting semantics of a program by computing a fixed point in a simpler domain.

The collecting semantics $\mathbb{S} : V \to 2^S$ is defined by the least fixpoint $lfp(F) = F^*(\lambda v.\emptyset)$ of the functional $F : (V \to 2^S) \to (V \to 2^S)$:

$$F(f)(v') = \begin{cases} S_0 & \text{if } v' = v_{in} \\ \bigcup\limits_{(v,op,v') \in E} [\![op]\!](f(v)) & \text{otherwise} \end{cases} .$$

Program Analysis. A program analysis $\mathbb{A} = (D, [\![.]\!]^\sharp)$ consists of an abstract domain D and an abstract semantics $[\![.]\!]^\sharp$.

An *abstract domain* $D = (S, A, \beta, \gamma)$ is defined by a complete semi-lattice $A = (A, \sqsubseteq, \bigsqcup, \bot, \top)$, a representation function $\beta : S \to A$, mapping concrete to abstract states, and a concretization $\gamma : A \to 2^S$, mapping abstract states to the set of concrete states they represent. Concretization and representation function are required to be monotone functions with respect to set inclusion and $\sqsubseteq$, respectively, and must be consistent to each other, i.e. the representation of a concrete state s must concretize to a set of states containing that concrete state, i.e. $s \in \gamma(\beta(s))$.

We define an abstraction function $\alpha : 2^S \to A$ by $\alpha(S') = \bigsqcup\{\beta(s) \mid s \in S'\}$. Given a lattice and a concretization, there may be a plethora of admissible representation functions with varying precision that lead to a domain, e.g. one could map some or all values to $\top$. To formalize the notion of optimal precision at the level of the domain, the concept of a *Galois connection* was introduced. If the concretization and the abstraction fulfill the condition $\alpha(X) \sqsubseteq a \Leftrightarrow X \subseteq \gamma(a)$, we shall call the pair (α, γ) a *Galois connection.*

The *abstract semantics* $[\![.]\!]^\sharp : Op \to (A \to A)$ assigns abstract transfer functions $[\![op]\!]^\sharp : A \to A$ to statements. We impose two requirements, first, the transfer functions are monotone with respect to $\sqsubseteq$ and, second, they approximate (or even equal) the best abstract transfer function $[\![op]\!]^\sharp{}_{\text{best}}(a) = \alpha([\![op]\!](\gamma(a)))$, i.e. $[\![op]\!]^\sharp{}_{\text{best}} \sqsubseteq [\![op]\!]^\sharp$ (or $[\![op]\!]^\sharp = [\![op]\!]^\sharp{}_{\text{best}}$).

The *program analysis problem* is to compute invariants $\mathbb{S}^\sharp : V \to A$ (in terms of the abstract domain) for all program points v such that $\mathbb{S}(v) \subseteq \gamma(\mathbb{S}^\sharp(v))$. This is solved by computing the fixpoint $lfp(F^\sharp) = F^{\sharp^*}(\lambda v.\bot)$ of the functional $F^\sharp : (V \to A) \to (V \to A)$:

$$F^\sharp(f)(v') = \begin{cases} l_0 & \text{if } v' = v_{in} \\ \bigsqcup\limits_{(v,op,v') \in E} [\![op]\!]^\sharp(f(v)) & \text{otherwise} \end{cases}$$

where the initial abstract state is chosen such that $\alpha(S_0) \sqsubseteq l_0$.

Termination. The transfer functions are required to be monotone, so that in each fixpoint iteration the values at the program points do not decrease with respect to $\sqsubseteq$. Nontermination can only occur if the lattice exhibits infinite ascending chains, i.e. sequences $a_1, a_2, a_3, \ldots$ of distinct elements with increasing order $a_1 \sqsubseteq a_2 \sqsubseteq a_3 \sqsubseteq \ldots$. Then widening is used [19,20] to enforce termination of fixpoint iteration. A widening operator accumulates (monotonely) increasing or decreasing values in such a way that each variable in the system of equations will only be changed finitely many times. This will guarantee termination albeit at the cost of a loss of precision. Widening for numerical domains has received a lot of attention, e.g. [21,22].

In this context, we discuss the interval domain (see 3.3), a simple and yet very useful numerical domain with infinite ascending chains.

3.2 Constant Propagation

Constant propagation attempts to find out for each program point which variables have which constant values whenever execution reaches that point. The resulting information can be used to *fold* (sub-)expressions and conditions, i.e., compute their values at compile time.

The abstract domain of constant propagation is constructed in two steps; we first define a partial order for the potential values of variables, the domain:

$$\mathbb{Z}^\top = \mathbb{Z} \cup \{\top\} \qquad \text{and} \quad x \sqsubseteq_{\mathbb{Z}^\top} y \quad \text{iff } y = \top \text{ or } x = y$$

where $\top$ is an extension of the set of integer values used to denote that the value of a variable is unknown.

The representation function maps integers to the corresponding element in $\mathbb{Z}^\top$, i.e. $\beta_{\mathbb{Z}^\top}(z) = z$. The abstraction function takes subsets $M \subseteq \mathbb{Z}$ of the integers as arguments; it maps singleton sets to their element and all other sets to unknown:

$$\alpha_{\mathbb{Z}^\top}(M) = \begin{cases} z & \text{if } M = \{z\} \\ \top & \text{otherwise} \end{cases} .$$

The concretization is defined by

$$\gamma_{\mathbb{Z}^\top}(\top) = \mathbb{Z} \qquad \text{and} \quad \gamma_{\mathbb{Z}^\top}(z) = \{z\} \quad \text{iff } z \neq \top .$$

In a second step, we lift the abstraction for values to an abstraction of *variable bindings* (states) and consider the complete lattice:

$$A = (\mathit{Vars} \to \mathbb{Z}^\top)_\bot = (\mathit{Vars} \to \mathbb{Z}^\top) \cup \{\bot\}$$

The new element $\bot$ denotes the fact that the analysis has not yet reached this program point. The partial order on this abstract domain, "$\sqsubseteq$", is defined as:

$$D_1 \sqsubseteq D_2 \qquad \text{iff} \qquad \bot = D_1 \quad \text{or} \quad D_1\, x \sqsubseteq_{\mathbb{Z}^\top} D_2\, x \quad \text{for all } x \in \mathit{Vars}$$

An abstract variable binding D_1 is more precise than a binding D_2 if D_1 binds all variables that D_2 also "knows" to the same values, but possibly knows some more values of variables. A together with this partial order is a complete lattice.

The concretization $\gamma(\bot)$ of the bottom element is the empty set of variable bindings, for all other abstract variable bindings D, it is the lifting of the concretization of $\mathbb{Z}^\top$:

$$\gamma(D) = \{s \mid \forall v \in Vars: \; s(v) \in \gamma_{\mathbb{Z}^\top}(D(v))\} .$$

The representation function is given by: $\beta(s)(v) = \beta_{\mathbb{Z}^\top}(s(v))$. The abstraction function maps the empty set to the bottom element, for non-empty sets we lift the abstraction function of $\mathbb{Z}^\top$:

$$\alpha(S')(v) = \alpha_{\mathbb{Z}^\top}(\{s(v) \mid s \in S'\})$$

The transfer functions of statements, $[\![op]\!]^\sharp : A \to A$, simulate the concrete evaluation function. They employ an abstract evaluation function for arithmetic expressions. For a binary operator $\Box$, it is defined by:

$$a \,\Box^\sharp\, b = \begin{cases} \top & \text{if} \quad a = \top \text{ or } b = \top \\ a \,\Box\, b & \text{otherwise} \end{cases}$$

The evaluation function is able to deal with unknown values of variables. It propagates this information; the result is $\top$ if one of the operands is unknown, i.e., is $\top$. The result is the same as in the concrete case for two known operands. The transfer functions of the abstract semantics for the toy language are given in Figure 1.

$$\begin{aligned} [\![x \leftarrow e]\!]^\sharp\, D &= D \oplus \{x \mapsto [\![e]\!]^\sharp\, D\} \\ [\![\mathsf{true}\,(e)]\!]^\sharp\, D &= \begin{cases} \bot \text{ if } & [\![e]\!]^\sharp\, D = \mathtt{ff} \\ D \text{ otherwise} \end{cases} \\ [\![\mathsf{false}\,(e)]\!]^\sharp\, D &= \begin{cases} \bot \text{ if } & [\![e]\!]^\sharp\, D = \mathtt{tt} \\ D \text{ otherwise} \end{cases} \end{aligned}$$

Fig. 1. The abstract semantics for constant propagation

If the condition can be definitely evaluated to ff, then the true branch is unreachable, and if it can be definitely evaluated to tt, then the false branch is unreachable. An assignment is analyzed by evaluating the right side in the abstract variable binding and over-writing the binding of the left side with the resulting value, which may be $\top$.

Constant propagation is useful for timing analysis since it transports statically available information to relevant places. The computed information is used by value analysis and control-flow analysis. Though of infinite size the domain is only of finite height, i.e. there are no infinite ascending chains. Further, the presented abstraction and concretization function form a Galois connection.

3.3 Interval Analysis

A static method for data-cache behavior prediction needs to know effective memory addresses of data, in order to determine where a memory access goes. However, effective addresses are only available at run time. Here interval analysis as described by Cousot and Cousot [19] comes into play. It can compute intervals for address-valued objects like registers and variables. An interval computed for such an object at some program point bounds the set of potential values the object may have when program execution reaches this program point. Such an analysis, as part of aiT's *value analysis*, has been shown to be very effective on disciplined code [7].

Interval analysis generalizes constant propagation by replacing the domain $\mathbb{Z}^\top$ for variables by that of intervals. The interval domain is given by

$$\mathbb{I} = \{[l, u] \mid l \in \mathbb{Z} \cup \{-\infty\}, u \in \mathbb{Z} \cup \{+\infty\}, l \leq u\}$$

Note that this definition admits only intervals that represent non-empty sets of integers. The set of intervals is ordered by "$\sqsubseteq$", defined by

$$[l_1, u_1] \sqsubseteq_{\mathbb{I}} [l_2, u_2] \qquad \text{iff} \qquad l_2 \leq l_1 \wedge u_1 \leq u_2$$

Least upper bound and greatest lower bound of two intervals are defined by:

$$\begin{aligned}
&[l_1, u_1] \sqcup [l_2, u_2] = [\mathsf{min}\{l_1, l_2\}, \mathsf{max}\{u_1, u_2\}] \\
&[l_1, u_1] \sqcap [l_2, u_2] = [\mathsf{max}\{l_1, l_2\}, \mathsf{min}\{u_1, u_2\}], \text{ if } \mathsf{max}\{l_1, l_2\} \leq \mathsf{min}\{u_1, u_2\}
\end{aligned}$$

The representation function maps an integer to a singleton interval: $\beta_{\mathbb{I}}(z) = [z, z]$ and the abstraction function maps a subset $M \subseteq \mathbb{Z}$ of the integers to an interval with the infimum and supremum of M as endpoints: $\alpha_{\mathbb{I}}(M) = [\inf_{z \in M} z, \sup_{z \in M} z]$. The concretization function relates concrete values and intervals:

$$\gamma_{\mathbb{I}}([l, u]) = \{z \in \mathbb{Z} \mid l \leq z \leq u\} \ .$$

Analogous to constant propagation, the numerical abstraction for variable values is lifted to an abstraction of variable bindings (states), i.e. we consider the complete lattice with elements $(Vars \to \mathbb{I})_\perp$.

To obtain an abstract semantics, some arithmetic on intervals is defined, first the sum of two intervals: $[l_1, u_1] +^\sharp [l_2, u_2] = [l_1 + l_2, u_1 + u_2]$ where $-\infty + \_ = -\infty$ and $+\infty + \_ = +\infty$ (the underscore _ stands for "any value"). For unary minus, we define: $-^\sharp [l, u] = [-u, -l]$. Multiplication on intervals is more involved and division yet more difficult. For this and comparisons on intervals, we refer to [23]. The abstract semantics is the same as the one for constant propagation (cf. Figure 1) *except* that it uses the expression evaluation function for intervals.

Achieving termination of interval analysis requires some extra work because the partially ordered domain $\mathbb{I}$, as opposed to $\mathbb{Z}^\top$, exhibits infinitely ascending chains, e.g.

$$[0, 0] \sqsubset [0, 1] \sqsubset [0, 2] \sqsubset [-1, 2] \sqsubset \ldots$$

and so does the lifted lattice $(Vars \to \mathbb{I})_\perp$. In order to enforce termination of fixpoint iteration, widening is used [19,20]. In our present interval analysis, any increasing upper bound of an interval will immediately be set to ∞, any decreasing lower bound of an interval to $-\infty$.

Interval analysis produces relevant information for cache analysis. The smaller the intervals bounding potential data addresses, the more precise are the results of cache analysis. The presented abstraction and concretization function form a Galois connection.

3.4 Cache Analysis

Abstract interpretation is also used to compute invariants about cache contents at all program points.

For brevity, we restrict our description to the semantics of fully associative caches with LRU replacement strategy. We refer to [24,17] for descriptions of how to deal with direct-mapped and A-way set associative caches.

In the following, we consider a (fully associative) cache as a set of cache lines $L = \{l_1, \ldots, l_n\}$ and the store as a set of memory blocks $M = \{m_1, \ldots, m_k\}$. To indicate the absence of any memory block in a cache line, we introduce a new element I; $M' = M \cup \{I\}$. A *(concrete) cache state* is a function $c : L \rightarrow M'$. C denotes the set of all concrete cache states. The initial cache state c_I maps all cache lines to I. If $c(l) = m_i$ for a concrete cache state c, then i is the relative age of the memory block.

The cache *update* function $\mathcal{U} : C \times M \rightarrow C$ determines the new cache state for a given cache state and a referenced memory block. The LRU (Least-Recently Used) strategy always makes the referenced memory block the youngest, i.e. the referenced memory block moves into l_1 if it was in the cache already (cache hit). All memory blocks in the cache that had been used more recently than the referenced block increase their relative age by one, i.e., they are shifted by one position to the next cache line. If the referenced memory block was not yet in the cache (cache miss), it is loaded into l_1 after all memory blocks in the cache have been shifted and the 'oldest', i.e., least recently used memory block, has been removed from the cache if the cache was full. This is depicted in Figure 2.

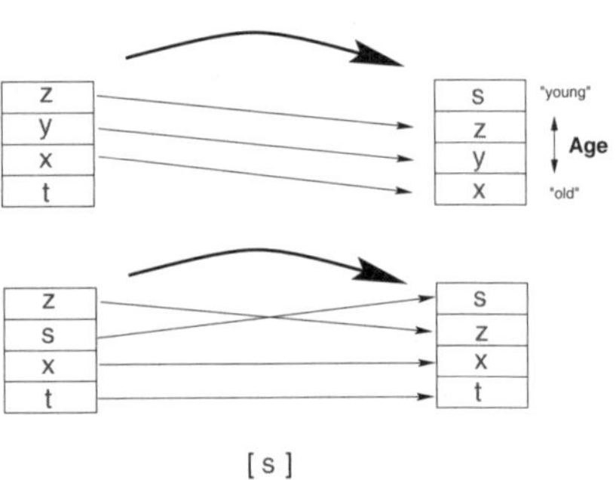

Fig. 2. Update of a concrete cache

In our present exposition, we assume that for each basic block, the sequence of references to memory is known, i.e., there exists a mapping from operations to sequences of memory blocks: $\mathcal{L} : Op \rightarrow M^*$. This is realistic for instruction caches. For data caches, only intervals may be available. The techniques described here are also routinely applied to data caches. The slight adaptions necessary to handle adress intervals can be found in [16].

We can describe the effect of such a sequence on a cache with the help of the update function $\mathcal{U}$. Therefore, we extend $\mathcal{U}$ to sequences of memory references by sequential composition: $\mathcal{U}(c, \langle m_{x_1}, \ldots, m_{x_y} \rangle) = \mathcal{U}(\ldots(\mathcal{U}(c, m_{x_1}))\ldots, m_{x_y})$. The cache semantics of an operation *op* at a control-flow edge is then $[\![op]\!] = \mathcal{U}(\cdot, \mathcal{L}_{op})$.

The collecting semantics would be computable, although often of enormous size. Therefore, another step abstracts it into a compact representation, so called abstract cache states. Note that every information drawn from the abstract cache states allows to safely deduce information about sets of concrete cache states, i.e., only precision may be reduced in this two step process. Correctness is guaranteed.

The abstraction consists in two analyses one computes an under- and the other an overapproximation of the cache content as follows: To classify definite cache hits, the *must analysis* determines a set of memory blocks that are in the cache at a given program point whenever execution reaches this point. To classify definite misses, a *may analysis*, not described in this paper, determines all memory blocks that may be in the cache at a given program point.

The domains for the must analysis (and also the may analysis) consist of *abstract cache states*: An *abstract cache state* $c^\sharp : L \rightarrow 2^M$ maps cache lines to sets of memory blocks. These sets are disjoint so that each memory block has unique position: it is either in one of the abstract cache lines or it is not in the cache. The position of a memory block in an abstract cache denotes, as in the case of concrete caches, the relative age of the corresponding memory blocks. As explained above, must analysis determines a set of memory blocks that are in the cache at a given program point whenever execution reaches this point. The positions of the memory blocks in the abstract cache state are thus the upper bounds of the *ages* of the memory blocks in the concrete caches occurring in the collecting cache semantics.

Good information, in the sense of being valuable for the prediction of cache hits, is the knowledge that a memory block is in the cache. The bigger the set the better. This is connected to the "age" of a memory block. Therefore, the partial order $\sqsubseteq$ is as define follows: Take an abstract cache state $c^\sharp$. Elements that are higher up with respect to $\sqsubseteq$ than $c^\sharp$ in the domain, i.e., less precise, are states where memory blocks from $c^\sharp$ are either missing or are older than in $c^\sharp$. Therefore, the $\bigsqcup$-operator applied to two abstract cache states $c_1^\sharp$ and $c_2^\sharp$ will produce a state $c^\sharp$ containing only those memory blocks contained in both, and will give them the maximum of their ages in $c_1^\sharp$ and $c_2^\sharp$ (see Figure 3). The *representation function* $\beta : C \rightarrow C^\sharp$ forms singleton sets from concrete cache states it is applied to, i.e., $\beta(c)(l_i) = \{m_x\}$ if $c(l_i) = c_x$. Concretization of an abstract cache state, $c^\sharp$, produces the set of all concrete cache states, which contain all the memory blocks contained in $c^\sharp$ with ages not older than in $c^\sharp$. Cache lines not filled by these are filled with other memory blocks. The *concretization function* $\gamma : C^\sharp \rightarrow 2^c$ is defined by $\gamma(c^\sharp) = \{c \mid \beta(c) \sqsubseteq c^\sharp\}$.

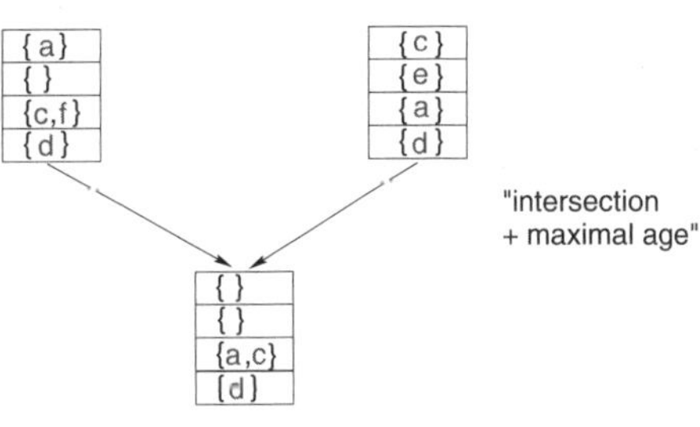

Fig. 3. Join

The abstract semantics is defined by *abstract cache update* functions, denoted $\mathcal{U}^\sharp$, which describe the effects of a control flow edge on an element of the abstract domain. An abstract cache update function (example depicted in Figure 4) is a lifted version of the corresponding concrete update function to sets, in that the referenced memory block goes to line l_1, all younger blocks age by one.

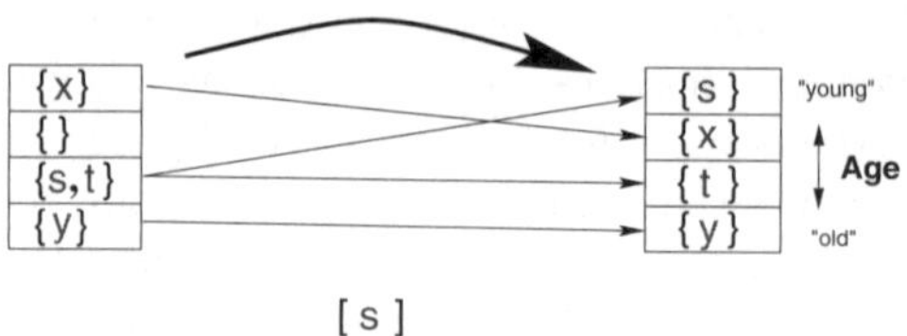

Fig. 4. Update of an abstract cache

The solution of the must analysis problem is interpreted as follows: Let $c^\sharp$ be an abstract cache state at some program point. If $m_x \in c^\sharp(l_i)$ for a cache line l_i then m_x will definitely be in the cache whenever execution reaches this program point. A reference to m_x is categorized as *always hit* (ah).

Termination. There are only a finite number of cache lines and for each program a finite number of memory blocks. This means, that the domain of abstract cache states $c : L \rightarrow 2^M$ is finite. Hence, every ascending chain is finite. Additionally, the abstract cache update functions, $\mathcal{U}^\sharp$, are monotonic. This guarantees that all the analyses will terminate.

The abstract cache-state domain is essential for the efficiency and therefore for the feasibility of cache analysis. The results are precise enough although this type of cache analysis loses information at merge points. More precise analyses are possible. However, experiments have shown that the corresponding analyses are too slow.

The domain for LRU caches fulfills the ascending chain condition and forms a Galois connection. For other cache domains, such as for the popular Pseudo-LRU caches, there does not exist an optimal abstraction function, and hence no Galois connection [18].

3.5 Pipeline Analysis

Most state-of-the-art processors employed in embedded systems today have an instruction pipeline, i.e. the execution of several instructions is overlapped. Instructions simultaneously pass through different pipeline stages: An instruction is first loaded from memory (fetch stage). The duration of this stage is determined by the contents of the instruction cache. The instruction is then ready to be dispatched: it is decoded and operands are fetched. In the execution stage, instructions compete for resources, such as execution units, buses and memory, producing complex interdependences. Depending on the internal state of the processor the time from fetch to completion of an instruction can vary by several orders of magnitude.

Pipeline analysis works on executable programs and is based on an abstract timing model for the specific processor. A timing model is a state machine whose transitions correspond to clock cycles of the modeled processor. Technically, pipeline analysis is a program analysis on the basic block graph[1] that computes

[1] A basic block is a maximal sequence of straight-line code in the program.

for each basic block an invariant on the machine states that can occur at it and an execution time bound for the number of cycles it takes to execute it whenever execution reaches that block. The abstract semantics of a basic block computes from the abstract processor states at entry to the block the set of processor states on exit of the block together with the bound. To this end, the analysis runs the abstract timing model of the processor cycle per cycle. Whenever abstraction produces uncertainty, e.g. inability to classify a cache access as a hit or a miss, the analysis follows all possibilities (both hit and miss case).

The structure of a timing model is determined by the different processor units and its memory system: the pipeline stages, a model of the processor chipset, the bus unit, the branch predictor, register files, and arithmetic units etc. Though structurally similar to the processor, the model concentrates on timing-relevant control components and data, e.g., it is not interested in what an arithmetic instruction computes, but in how many cycles the instruction takes.

Value analysis can be considered as factored-out arithemtic. The pipeline timing model imports the results of the value analsysis. The memory system is concisely abstracted by the chipset unit, bus unit and the cache domain described in the previous section and similarly factored out.

Cache and pipeline analysis are integrated to reflect the interdependences between the caches and the pipeline due to speculation and prefetching [25].

Abstract domain and transfer functions are determined by the timing model. The pipeline analysis for a state-of-the-art processor described in [25] uses the following domains:

- An abstract state of the timing model is a tuple $(p, c^\sharp)$ consisting of the pipeline state p and an abstract cache state $c^\sharp$. It represents a set of concrete states of the timing model.
- From the above domain of tuples, the disjunctive completion is taken. This results in a lattice whose elements are sets of states of the timing model. However, rather than taking set union as a join operator, a more sophisticated join is used that leverages the join operator of the cache domain. It takes a set of sets of processor states as input and produces a set of processor states that overapproximates the union of the input sets. Its number of elements equals the number of distinct pipeline states in the input sets. The join operator loses precision only on the cache side by joining abstract cache states where abstract pipeline state is identical. In the result set, each abstract pipeline state p is adjoined with the *join* of a set of abstract caches, namely the join of the abstract caches $c^\sharp$ such that $(p, c^\sharp)$ appears in one of the input sets:

$$\bigsqcup\{S_i \mid i = 1, ..., k\} = \{(p, c) \mid \exists i.(p, c') \in S_i \wedge c = \bigsqcup\{c'' \mid (p, c'') \in S_j\}\}$$

- The abstract domain results as the product of the semi-lattice of natural numbers with the maximum as join operator and the lattice of timing model states. The join operator is defined by:

$$\bigsqcup\{(n_i, S_i) \mid i = 1, ..., k\} = (\max_i n_i, \bigsqcup\{S_i \mid i = 1, ..., k\}$$

An abstract state is a tuple: the first component is a time bound (a number) and the second component a *set of states* of the timing model.

To evaluate the transfer function $[\![op]\!]^\sharp(a)$ for a control flow edge (b, op, b') and a tuple $(., S)$ (the first component is ignored), a finite transition system is computed. Its initial states are all those states $I \subseteq S$ that load at least one instruction of basic block b. The transitions are determined by the timing model. Its final states F are those in which all instruction within basic block b have been completed. The transition system is acyclic and finite. Let k be the maximal length of a path from initial to the final states, then $[\![op]\!]^\sharp(a) = (k, F)$.

4 Related Approaches

Timed automata [26] (or networks thereof) have been used to express timing constraints of real-time systems and require durations and time bounds. Timing analysis can deliver such bounds in the form of lower and upper bounds on the execution time for a realistic architecture.

Campos et al. [27,28,29] leverage finite-state BDD-based model checking for timing analysis. This work is not comparable with the approach proposed in this tutorial since results were only obtained for highly simplified architectures without typical features of modern processors such as caches and pipelining.

The works by Logothetis, Schuele and Schneider [30,31,32] describe timing analyses of assembler programs using symbolic simulation. This work remains at the machine-independent level and is based on the unit-time assumption.

Metzner [33] proposed to use BDD-based model checking for cache behavior prediction instead of abstract interpretation and reported some gain (1.5-5%) in precision over cache analysis by abstract interpretation [34] because joins at control-flow merge points are avoided. The experiments considered a very small cache and an extremely simple pipeline. Scalability of the analysis to industrial-scale benchmarks was not shown. Furthermore, the experimental results are limited to instruction caches for which cache analysis is easier than for data caches because the addresses are statically known and access patterns are more regular.

5 Conclusions

A short introduction into the theory of abstract interpretation was given, and several instances of abstract interpretations are described that are used in timing analysis. The different analyses have quite different characteristics. Constant propagation, interval analysis, and cache analysis live on the design of the right abstract domain. They represent sets of concrete values by single abstract values. Further, while cache and pipeline analysis employ a finite domain, constant propagation and interval analysis exhibit infinite domains, yet only the interval domain has infinite ascending chains and requires widening for termination.

For pipeline analysis, a suitable representation of sets of concrete pipeline states by single abstract states has not been found and is probably hard to find.

Pipeline analysis is not a typical static program analysis. It can rather be seen as a hybrid: it employs both state traversal of the pipeline evolution and join operations typical for static analysis.

References

1. Kildall, G.A.: A Unified Approach to Global Program Optimization. In: Proceedings of the ACM Symposium on Principles of Programming Languages, Boston, Massachusetts, pp. 194–206 (October 1973)
2. Kam, J., Ullman, J.D.: Monotone Data Flow Analysis Frameworks. Acta Informatica 7(3), 305–318 (1977)
3. Cousot, P.: Méthodes itératives de construction et d'approximation de point fixes d'opérateurs monotone sur un treilis, analyse sémantique des programmes. PhD thesis, Université de Grenoble (1978)
4. Lacan, J., Monfort, J.N., Ribal, V.Q., Deutsch, A., Gonthier, G.: The software reliability verification process: The Ariane 5 example. DAta Systems In Aerospace SP-422 (1998)
5. Blanchet, B., Cousot, P., Cousot, R., Feret, J., Mauborgne, L., Miné, A., Monniaux, D., Rival, X.: A static analyzer for large safety-critical software. In: Proceedings of the ACM SIGPLAN 2003 Conference on Programming Language Design and Implementation (PLDI 2003), San Diego, California, USA, pp. 196–207. ACM Press, New York (2003)
6. Ferdinand, C., Heckmann, R., Langenbach, M., Martin, F., Schmidt, M., Theiling, H., Thesing, S., Wilhelm, R.: Reliable and precise WCET determination for a real-life processor. In: Henzinger, T.A., Kirsch, C.M. (eds.) EMSOFT 2001. LNCS, vol. 2211, pp. 469–485. Springer, Heidelberg (2001)
7. Thesing, S., Souyris, J., Heckmann, R., Randimbivololona, F., Langenbach, M., Wilhelm, R., Ferdinand, C.: An abstract interpretation-based timing validation of hard real-time avionics software systems. In: Proceedings of the 2003 International Conference on Dependable Systems and Networks (DSN 2003), pp. 625–632. IEEE Computer Society, Los Alamitos (2003)
8. Mathworks: Polyspace, `http://www.polyspace.com`.
9. Coverity: Coverity Prevent, `http://www.coverity.com`
10. Fasoo.com, Seoul University: Sparrow, `http://spa-arrow.com:8000/index.htm`
11. GrammaTech Inc.: Codesurfer, `http://www.grammatech.com`
12. AbsInt Angewandte Informatik: aiT Worst-Case Execution Time Analyzers
13. Nielson, F., Nielson, H.R., da Rosa, D.S., Priami, C.: Static analysis for systems biology. In: Proc. of workshop on Systeomatics - dynamic biological systems informatics. Computer Science Press, Trinity College Dublin (2004)
14. Bodei, C., Buchholtz, M., Degano, P., Nielson, F., Nielson, H.R.: Static validation of security protocols. Journal of Computer Security 13(3), 347–390 (2005)
15. Chawdhary, A., Cook, B., Gulwani, S., Sagiv, M., Yang, H.: Ranking Abstractions. In: Proceedings of the 17th European Symposium on Programming (to appear, April 2008)
16. Alt, M., Ferdinand, C., Martin, F., Wilhelm, R.: Cache Behavior Prediction by Abstract Interpretation. In: Cousot, R., Schmidt, D.A. (eds.) SAS 1996. LNCS, vol. 1145, pp. 52–66. Springer, Heidelberg (1996)
17. Ferdinand, C., Martin, F., Wilhelm, R.: Cache Behavior Prediction by Abstract Interpretation. Science of Computer Programming 35, 163–189 (1999)

18. Heckmann, R., Langenbach, M., Thesing, S., Wilhelm, R.: The influence of processor architecture on the design and the results of WCET tools. IEEE Proceedings on Real-Time Systems 91(7), 1038–1054 (2003)
19. Cousot, P., Cousot, R.: Abstract Interpretation: A Unified Lattice Model for Static Analysis of Programs by Construction or Approximation of Fixpoints. In: Proceedings of the 4th ACM Symposium on Principles of Programming Languages, Los Angeles, California, pp. 238–252 (1977)
20. Cousot, P., Cousot, R.: Comparison of the Galois connection and widening/narrowing approaches to abstract interpretation. In: JTASPEFL 1991, Bordeaux. BIGRE, vol. 74, pp. 107–110 (1991)
21. Bagnara, R., Hill, P.M., Ricci, E., Zaffanella, E.: Precise widening operators for convex polyhedra. Sci. Comput. Program. 58(1-2), 28–56 (2005)
22. Miné, A.: The octagon abstract domain. Higher Order Symbol. Comput. 19(1), 31–100 (2006)
23. Wilhelm, R., Seidl, H.: Übersetzerbau: Analyse und Transformation. Springer, Heidelberg (2008)
24. Ferdinand, C.: Cache Behavior Prediction for Real-Time Systems. PhD Thesis, Universität des Saarlandes (September 1997)
25. Thesing, S.: Safe and Precise WCET Determination by Abstract Interpretation of Pipeline Models. PhD thesis, Saarland University (2004)
26. Alur, R., Dill, D.L.: A theory of timed automata. Theor. Comput. Sci. 126(2), 183–235 (1994)
27. Campos, S., Grumberg, O.: Selective quantitative analysis and interval model checking: Verifying different facets of a system. In: Alur, R., Henzinger, T.A. (eds.) CAV 1996. LNCS, vol. 1102, pp. 257–268. Springer, Heidelberg (1996)
28. Campos, S.V.A., Clarke, E.M.: Analysis and verification of real-time systems using quantitative symbolic algorithms. International Journal on Software Tools for Technology Transfer 2(3), 260–269 (1999)
29. Hartonas-Garmhausen, V., Campos, S.V.A., Cimatti, A., Clarke, E.M., Giunchiglia, F.: Verification of a safety-critical railway interlocking system with real-time constraints. Sci. Comput. Program. 36(1), 53–64 (2000)
30. Schüle, T., Schneider, K.: Exact runtime analysis using automata-based symbolic simulation. In: MEMOCODE, pp. 153–162 (2003)
31. Schüle, T., Schneider, K.: Abstraction of assembler programs for symbolic worst case execution time analysis. In: DAC, pp. 107–112 (2004)
32. Logothetis, G., Schneider, K.: Exact high level WCET analysis of synchronous programs by symbolic state space exploration. In: DATE, pp. 10196–10203 (2003)
33. Metzner, A.: Why model checking can improve WCET analysis. In: Alur, R., Peled, D.A. (eds.) CAV 2004. LNCS, vol. 3114, pp. 334–347. Springer, Heidelberg (2004)
34. Ferdinand, C., Wilhelm, R.: Fast and Efficient Cache Behavior Prediction for Real-Time Systems. Real-Time Systems 17, 131–181 (1999)

Reducing Concurrent Analysis Under a Context Bound to Sequential Analysis*

Akash Lal[1,**] and Thomas Reps[1,2]

[1] University of Wisconsin; Madison, WI; USA
{akash,reps}@cs.wisc.edu
[2] GrammaTech, Inc., Ithaca, NY, USA

Abstract. This paper addresses the analysis of concurrent programs with shared memory. Such an analysis is undecidable in the presence of multiple procedures. One approach used in recent work obtains decidability by providing only a partial guarantee of correctness: the approach bounds the number of context switches allowed in the concurrent program, and aims to prove safety, or find bugs, under the given bound. In this paper, we show how to obtain simple and efficient algorithms for the analysis of concurrent programs with a context bound. We give a general reduction from a *concurrent* program P, and a given context bound K, to a *sequential* program P_s^K such that the analysis of P_s^K can be used to prove properties about P. We give instances of the reduction for common program models used in model checking, such as Boolean programs and pushdown systems.

1 Introduction

The analysis of concurrent programs is a challenging problem. While in general the analysis of both concurrent and sequential programs is undecidable, what makes concurrency hard is the fact that even for simple program models, the presence of concurrency makes their analysis computationally very expensive. When the model of each thread is a finite-state automaton, the analysis of such systems is PSPACE-complete; when the model is a pushdown system, the analysis becomes undecidable [18]. This is unfortunate because it does not allow the advancements made on such models in the sequential setting, i.e., when the program has only one thread, to be applied in the presence of concurrency.

This paper addresses the problem of automatically extending analyses for sequential programs to analyses for concurrent programs under a bound on the number of context switches.[1] We refer to analysis of concurrent programs under a context bound as *context-bounded analysis* (CBA). Previous work has shown the value of CBA: KISS [17], a model checker for CBA with a fixed context

* Supported by NSF under grants CCF-0540955 and CCF-0524051 and by AFRL under contract FA8750-06-C-0249.

** Supported by a Microsoft Research Fellowship.

[1] A context switch occurs when execution control passes from one thread to another.

A. Gupta and S. Malik (Eds.): CAV 2008, LNCS 5123, pp. 37–51, 2008.

bound of 2, found numerous bugs in device drivers; a study with explicit-state model checkers [13] found more bugs with slightly higher context bounds. It also showed that the state space covered with each increment to the context-bound decreases as the context bound increases. Thus, even a small context bound is sufficient to cover many program behaviors, and proving safety under a context bound should provide confidence towards the reliability of the program. Unlike the above-mentioned work, this paper addresses CBA with any given context bound and with different program abstractions (for which explicit-state model checkers would not terminate).

The decidability of CBA, when each program thread is abstracted as a *push-down system* (PDS)—which serves as a general model for a recursive program with finite-state data—was shown in [16]. These results were extended to PDSs with bounded heaps in [3] and to weighted PDSs (WPDSs) in [10]. All of this work required devising new algorithms. Moreover, each of the algorithms have certain disadvantages that make them impractical to implement.

In the sequential setting, model checkers, such as those described in [1,21,5], use symbolic techniques in the form of BDDs for scalability. With the CBA algorithms of [16,3], it is not clear if symbolic techniques can be applied. Those algorithms require the enumeration of all reachable states of the shared memory at a context switch. This can potentially be very expensive. However, those algorithms have the nice property that they only consider those states that actually arise during valid (abstract) executions of the model. (We call this *lazy* exploration of the state space.)

Our recent paper [10] showed how to extend the algorithm of [16] to use symbolic techniques. However, the disadvantage there is that it requires computing auxiliary information for exploring the reachable state space. (We call this *eager* exploration of the state space.) The auxiliary information summarizes the effect of executing a thread from any control location to any other control location. Such summarizations may consider many more program behaviors than can actually occur (whence the term "eager").

This problem can also be illustrated by considering interprocedural analysis of sequential programs: for a procedure, it is possible to construct a summary for the procedure that describes the effect of executing it for any possible inputs to the procedure (eager computation of the summary). It is also possible to construct the summary lazily (also called partial transfer functions [12]) by only describing the effect of executing the procedure for input states under which it is called during the analysis of the program. The former (eager) approach has been successfully applied to Boolean programs[2] [1], but the latter (lazy) approach is often desirable in the presence of more complex abstractions, especially those that contain pointers (based on the intuition that only a few aliasing scenarios occur during abstract execution). The option of switching between eager and lazy exploration exists in the model checkers described in [1,20].

[2] Boolean programs are imperative programs with only the Boolean datatype (§3).

Contributions. This paper makes two main contributions. First, we show how to reduce a concurrent program to a sequential one that simulates all its executions for a given number of context switches. This has the following advantages:

- It allows one to obtain algorithms for CBA using different program abstractions. We specialize the reduction to Boolean programs (§3), PDSs (§4), and symbolic PDSs (see [8]). The former shows that the use of PDS-based technology, which seemed crucial in previous work, is not necessary: standard interprocedural algorithms [19,22,7] can also be used for CBA. Moreover, it allows one to carry over symbolic techniques designed for sequential programs for CBA.
- The reduction introduces symbolic constants and `assume` statements. Thus, any sequential analysis that can deal with these two additions can be extended to handle concurrent programs as well (under a context bound).
- For the case in which a PDS is used to model each thread, we obtain better asymptotic complexity than previous algorithms, just by using the standard PDS algorithms (§4).
- The reduction shows how to obtain algorithms that scale linearly with the number of threads (whereas previous algorithms scaled exponentially).

Second, we show how to obtain a lazy symbolic algorithm for CBA on Boolean programs (§5). This combines the best of previous algorithms: the algorithms of [16,3] are lazy but not symbolic, and the algorithm of [10] is symbolic but not lazy.

The rest of the paper is organized as follows: §2 gives a general reduction from concurrent to sequential programs; §3 specializes the reduction to Boolean programs; §4 specializes the reduction to PDSs; §5 gives a lazy symbolic algorithm for CBA on Boolean programs; §6 reports early results with our algorithms; §7 discusses related work. Additional details and proofs can be found in [8].

2 A General Reduction

This section gives a general reduction from concurrent programs to sequential programs under a given context bound. This reduction transforms the non-determinism in control, which arises because of concurrency, to non-determinism on data. (The motivation is that the latter problem is understood much better than the former one.)

The execution of a concurrent program proceeds in a sequence of *execution contexts*, defined as the time between consecutive context switches during which only a single thread has control. In this paper, we do not consider dynamic creation of threads, and assume that a concurrent program is given as a fixed set of threads, with one thread identified as the starting thread.

Suppose that a program has two threads, T_1 and T_2, and that the context bound is $2K - 1$. Then any execution of the program under this bound will have up to $2K$ execution contexts, with control alternating between the two threads, informally written as $T_1; T_2; T_1, \cdots$. Each thread has control for at most

K execution contexts. Consider three consecutive execution contexts $T_1; T_2; T_1$. When T_1 finishes executing the first of these, it gets swapped out and its local state, say l, is stored. Then T_2 gets to run, and when it is swapped out, T_1 has to resume execution from l (along with the global store produced by T_2).

The requirement of resuming from the same local state is one difficulty that makes analysis of concurrent programs hard—during the analysis of T_2, the local state of T_1 has to be remembered (even though it is unchanging). This forces one to consider the cross product of the local states of the threads, which causes exponential blowup when the local state space is finite, and undecidability when the local state includes a stack. An advantage of introducing a context bound is the reduced complexity with respect to the size $|L|$ of the local state space: the algorithms of [16,3] scale as $\mathcal{O}(|L|^5)$, and [10] scales as $\mathcal{O}(|L|^K)$. Our algorithm, for PDSs, is $\mathcal{O}(|L|)$. (Strictly speaking, in each of these, $|L|$ is the size of the local transition system.)

The key observation is the following: for analyzing $T_1; T_2; T_1$, we modify the threads so that we only have to analyze $T_1; T_1; T_2$, which eliminates the requirement of having to drag along the local state of T_1 during the analysis of T_2. For this, we *assume* the effect that T_2 might have on the shared memory, apply it while T_1 is executing, and then *check* our assumption after analyzing T_2.

Consider the general case when each of the two threads have K execution contexts. We refer to the state of shared memory as the *global state*. First, we guess $K-1$ (arbitrary) global states, say $s_1, s_2, \cdots, s_{K-1}$. We run T_1 so that it starts executing from the initial state s_0 of the shared memory. At a non-deterministically chosen time, we record the current global state s_1', change it to s_1, and resume execution of T_1. Again, at a non-deterministically chosen time, we record the current global state s_2', change it to s_2, and resume execution of T_1. This continues $K-1$ times. Implicitly, this implies that we assumed that the execution of T_2 will change the global state from s_i' to s_i in its ith execution context. Next, we repeat this for T_2: we start executing T_2 from s_1'. At a non-deterministically chosen time, we record the global state s_1'', we change it to s_2' and repeat $K-1$ times. Finally, we verify our assumption: we check that $s_i'' = s_{i+1}$ for all i between 1 and $K-1$. If these checks pass, we have the guarantee that T_2 can reach state s if and only if the concurrent program can have the global state s after K execution contexts per thread.

The fact that we do not alternate between T_1 and T_2 implies the linear scalability with respect to $|L|$. Because the above process has to be repeated for all valid guesses, our approach scales as $\mathcal{O}(|G|^K)$, where G is the global state space. In general, the exponential complexity with respect to K may not be avoidable because the problem is NP-complete when the input has K written in unary [9]. However, symbolic techniques can be used for a practical implementation.

We show how to reduce the above assume-guarantee process into one of analyzing a sequential program. We add more variables to the program, initialized with symbolic constants, to represent our guesses. The switch from one global state to another is made by switching the set of variables being accessed by the program. We verify the guesses by inserting `assume` statements at the end.

Program P^s	$\mathtt{st} \in T_i$	Checker

$L_1 : T_1^s$;
$L_2 : T_2^s$;
L_3 : Checker
	if $\mathtt{k} = 1$ **then**
	  $\tau(\mathtt{st}, 1)$;
	else if $\mathtt{k} = 2$ **then**
	  $\tau(\mathtt{st}, 2)$;
	  ...
	else if $\mathtt{k} = K$ **then**
	  $\tau(\mathtt{st}, K)$;
	end if
	if $\mathtt{k} \leq K$ and $*$ **then**
	  $\mathtt{k}$ ++
	end if
	if $\mathtt{k} = K + 1$ **then**
	  $\mathtt{k} = 1$
	  goto L_{i+1}
	end if
		for $i = 1$ to $K - 1$ **do**
		  for $j = 1$ to n **do**
		    assume $(\mathtt{x}_j^i = v_j^{i+1})$
		  end for
		end for

Fig. 1. The reduction for general concurrent programs under a context bound $2K - 1$. In the second column, $*$ stands for a nondeterministic Boolean value.

The reduction. Consider a concurrent program P with two threads T_1 and T_2 that only has scalar variables (i.e., no pointers, arrays, or heap).[3] We assume that the threads share their global variables, i.e., they have the same set of global variables. Let Var_G be the set of global variables of P. Let $2K - 1$ be the bound on the number of context switches.

The result of our reduction is a sequential program P^s. It has three parts, performed in sequence: the first part T_1^s is a reduction of T_1; the second part T_2^s is a reduction of T_2; and the third part, Checker, consists of multiple assume statements to verify that a correct interleaving was performed. Let L_i be the label preceding the ith part. P^s has the form shown in the first column of Fig. 1.

The global variables of P^s are K copies of Var_G. If $\text{Var}_G = \{\mathtt{x}_1, \cdots, \mathtt{x}_n\}$, then let $\text{Var}_G^i = \{\mathtt{x}_1^i, \cdots, \mathtt{x}_n^i\}$. The initial values of Var_G^i are a set of symbolic constants that represent the ith guess s_i. P^s has an additional global variable $\mathtt{k}$, which will take values between 1 and $K + 1$. It tracks the current execution context of a thread: at any time P^s can only read and write to variables in $\text{Var}_G^{\mathtt{k}}$. The local variables of T_i^s are the same as those of T_i.

Let $\tau(\mathtt{x}, i) = \mathtt{x}^i$. If $\mathtt{st}$ is a program statement in P, let $\tau(\mathtt{st}, i)$ be the statement in which each global variable $\mathtt{x}$ is replaced with $\tau(\mathtt{x}, i)$, and the local variables remain unchanged. The reduction constructs T_i^s from T_i by replacing each statement $\mathtt{st}$ by what is shown in the second column of Fig. 1. The third column shows Checker. Variables Var_G^1 are initialized to the same values as Var_G in P. Variable $\mathtt{x}_j^i$, when $i \neq 1$, is initialized to the symbolic constant v_j^i (which is later referenced inside Checker), and $\mathtt{k}$ is initialized to 1.

Because local variables are not replicated, a thread resumes execution from the same local state it was in when it was swapped out at a context switch.

The Checker enforces a correct interleaving of the threads. It checks that the values of global variables when T_1 starts its $i + 1^{\text{st}}$ execution context are the

[3] Such models are often used in model checking and numeric program analysis.

same as the values produced by T_2 when T_2 finished executing its ith execution context. (Because the execution of T_2^s happens after T_1^s, each execution context of T_2^s is guaranteed to use the global state produced by the corresponding execution context of T_1^s.)

The reduction ensures the following property: when P^s finishes execution, the variables VAR_G^K can have a valuation s if and only if the variables VAR_G in P can have the same valuation after $2K - 1$ context switches.

Symbolic constants. One way to deal with symbolic constants is to consider all possible values for them (eager computation). We show instances of this strategy for Boolean programs (§3) and for PDSs (§4). Another way is to lazily consider the set of values they may actually take during the (abstract) execution of the concurrent program, i.e., only consider those values that pass the `Checker`. We show an instance of this strategy for Boolean programs (§5).

Multiple threads. If there are n threads, $n > 2$, then a precise reasoning for K context switches would require one to consider all possible thread schedulings, e.g., $(T_1; T_2; T_1; T_3)$, $(T_1; T_3; T_2; T_3)$, etc. There are $\mathcal{O}((n-1)^K)$ such schedulings. Previous analyses [16,10,3] enumerate explicitly all these schedulings, and thus have $\mathcal{O}((n-1)^K)$ complexity even in the best case. We avoid this exponential factor as follows: we only consider the round-robin thread schedule $T_1; T_2; \cdots T_n; T_1; T_2; \cdots$ for CBA, and bound the length of this schedule instead of bounding the number of context switches. Because a thread is allowed to perform no steps during its execution context, CBA still considers other schedules. For example, when $n = 3$, the schedule $T_1; T_2; T_1; T_3$ will be considered by CBA only when $K = 5$ (in the round-robin schedule, T_3 does nothing in its first execution context, and T_2 does nothing in its second execution context).

Setting the bound on the length of the round-robin schedule to nK allows CBA to consider all thread schedulings with K context switches (as well as some schedulings with more than K context switches). Under such a bound, a schedule has K execution contexts per thread. The reduction for multiple threads proceeds in a similar way to the reduction for two threads. The global variables are copied K times. Each thread T_i is transformed to T_i^s, as shown in Fig. 1, and P^s calls the T_i^s in sequence, followed by `Checker`. `Checker` remains the same (it only has to check that the state after the execution of T_n^s agrees with the symbolic constants).

The advantages of this approach are as follows: (*i*) we avoid an explicit enumeration of $\mathcal{O}((n-1)^K)$ thread schedules, thus, allowing our analysis to be more efficient in the common case; (*ii*) we explore more of the program behavior with a round-robin bound of nK than with a context-switch bound of K; and (*iii*) the cost of analyzing the round-robin schedule of length nK is about the same (in fact, better) than what previous analyses take for exploring one schedule with a context bound of K (see §4). These advantages allow our analysis to scale much better in the presence of multiple threads than previous analyses.

In the rest of the paper, we only consider two threads because the extension to multiple threads is straightforward for round-robin scheduling.

Applicability of the reduction to different analyses. Certain analysis, like affine-relation analysis (ARA) over integers, as developed in [11], cannot make use of this reduction. The presence of `assume` statements makes the ARA problem undecidable. However, any abstraction prepared to deal with branching conditions can also handle `assume` statements.

It is harder to make a general claim as to whether most sequential analyses can handle symbolic values. One place where symbolic values are used in sequential analyses is to construct summaries for recursive procedures. Eager computation of a procedure summary is similar to analyzing the procedure while assuming symbolic values for the parameters of the procedure.

3 The Reduction for Boolean Programs

Boolean Programs. A Boolean program consists of a set of procedures, represented using their control-flow graphs (CFGs). The program has a set of global variables, and each procedure has a set of local variables, where each variable can only receive a Boolean value. Each edge in the CFG is labeled with a statement that can read from and write to variables in scope, or call a procedure. An example is shown in Fig. 2.

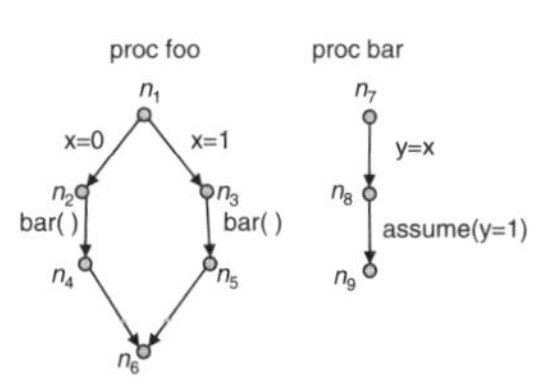

Fig. 2. A Boolean program

For ease of exposition, we assume that all procedures have the same number of local variables, and that they do not have any parameters. Furthermore, the global variables can have any value when program execution starts, and similarly for the local variables when a procedure is invoked.

Let G be the set of valuations of the global variables, and L be the set of valuations of the local variables. A program *data-state* is an element of $G \times L$. Each program statement `st` can be associated with a relation $[\![\texttt{st}]\!] \subseteq (G \times L) \times (G \times L)$ such that $(g_0, l_0, g_1, l_1) \in [\![\texttt{st}]\!]$ when the execution of `st` on the state (g_0, l_0) can lead to the state (g_1, l_1). For instance, in a procedure with one global variable $\mathtt{x}_1$ and one local variable $\mathtt{x}_2$, $[\![\mathtt{x}_1 = \mathtt{x}_2]\!] = \{(a, b, b, b) \mid a, b \in \{0, 1\}\}$ and $[\![\texttt{assume}(\mathtt{x}_1 = \mathtt{x}_2)]\!] = \{(a, a, a, a) \mid a \in \{0, 1\}\}$.

The goal of analyzing such programs is to compute the set of data-states that can reach a program node. This is done using the rules shown in Fig. 3 [1]. These rules follow standard interprocedural analyses [19,22]. Let entry(`f`) be the entry node of procedure `f`, proc(n) the procedure that contains node n, ep(n) = entry(proc(n)), and exitnode(n) is true when n is the exit node of its procedure. Let Pr be the set of procedures of the program, which includes a distinguished procedure `main`. The rules of Fig. 3 compute three types of relations: $H_n(g_0, l_0, g_1, l_1)$ denotes the fact that if (g_0, l_0) is the data state at entry(n), then the data state (g_1, l_1) can reach node n; $S_{\mathtt{f}}$ is the summary relation

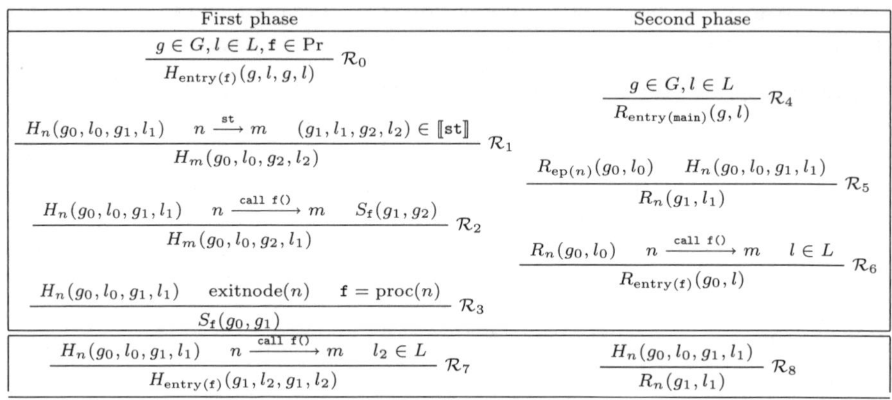

Fig. 3. Rules for the analysis of Boolean programs

for procedure f, which captures the net transformation that an invocation of the procedure can have on the global state; R_n is the set of data states that can reach node n. All relations are initialized to be empty.

Eager analysis. Rules $\mathcal{R}_0$ to $\mathcal{R}_6$ describe an eager analysis. The analysis proceeds in two phases. In the first phase, the rules $\mathcal{R}_0$ to $\mathcal{R}_3$ are used to saturate the relations H and S. In the next phase, this information is used to build the relation R using rules $\mathcal{R}_4$ to $\mathcal{R}_6$.

Lazy analysis. Let rule $\mathcal{R}'_0$ be the same as $\mathcal{R}_0$ but restricted to just the `main` procedure. Then the rules $\mathcal{R}'_0, \mathcal{R}_1, \mathcal{R}_2, \mathcal{R}_3, \mathcal{R}_7, \mathcal{R}_8$ describe a lazy analysis. The rule $\mathcal{R}_7$ restricts the analysis of a procedure to only those states it is called in. As a result, the second phase gets simplified and consists of only the rule $\mathcal{R}_8$.

Practical implementations [1,20] use BDDs to encode each of the relations H, S, and R and the rule applications are changed into BDD operations. For example, rule $\mathcal{R}_1$ is simply the relational composition of relations H_n and $[\![\mathtt{st}]\!]$, which can be implemented efficiently using BDDs.

Concurrent Boolean Programs. A concurrent Boolean program consists of one Boolean program per thread. The Boolean programs share their set of global variables. In this case, we can apply the reduction presented in §2 to obtain a single Boolean program by making the following changes to the reduction: (*i*) the variable `k` is modeled using a vector of $\log(K)$ Boolean variables, and the increment operation implemented using a simple Boolean circuit on these variables; (*ii*) the **if** conditions are modeled using `assume` statements; and (*iii*) the symbolic constants are modeled using additional global variables that are not modified in the program. Running any sequential analysis algorithm, and projecting out the values of the Kth set of global variables from R_n gives the precise set of reachable global states at node n in the concurrent program.

The worst-case complexity of analyzing a Boolean program P is bounded by $\mathcal{O}(|P||G|^3|L|^2)$, where $|P|$ is the number of program statements. Thus, using our approach, a concurrent Boolean program P_c with n threads, and K execution contexts per thread (with round-robin scheduling), can be analyzed in time $\mathcal{O}(K|P_c|(K|G|^K)^3|L|^2|G|^K)$: the size of the sequential program obtained from P_c is $K|P_c|$; it has the same number of local variables, and its global variables have $K|G|^K$ number of valuations. Additionally, the symbolic constants can take $|G|^K$ number of valuations, adding an extra multiplicative factor of $|G|^K$. The analysis scales linearly with the number of threads ($|P_c|$ is $\mathcal{O}(n)$).

This reduction actually applies to any model that works with finite-state data, which includes Boolean programs with references [2,14]. In such models, the heap is assumed to be bounded in size. The heap is included in the global state of the program, hence, our reduction would create multiple copies of the heap, initialized with symbolic values. Our experiments (§6) used such models.

Such a process of duplicating the heap can be expensive when the number of heap configurations that actually arise in the concurrent program is very small compared to the total number of heap configurations possible. The lazy version of our algorithm (§5) addresses this issue.

4 The Reduction for PDSs

PDSs are also popular models of programs. The motivation for presenting the reduction for PDSs is that it allows one to apply the numerous algorithms developed for PDSs for CBA. For instance, one can use backward analysis of PDSs to get a backward analysis on concurrent programs.

Definition 1. *A* ***pushdown system*** *is a triple $\mathcal{P} = (P, \Gamma, \Delta)$, where P is a set of states, Γ is a set of stack symbols, and $\Delta \subseteq P \times \Gamma \times P \times \Gamma^*$ is a finite set of rules. A* ***configuration*** *of $\mathcal{P}$ is a pair $\langle p, u\rangle$ where $p \in P$ and $u \in \Gamma^*$. A rule $r \in \Delta$ is written as $\langle p, \gamma\rangle \hookrightarrow \langle p', u\rangle$, where $p, p' \in P$, $\gamma \in \Gamma$ and $u \in \Gamma^*$. These rules define a transition relation $\Rightarrow_{\mathcal{P}}$ on configurations of $\mathcal{P}$ as follows: If $r = \langle p, \gamma\rangle \hookrightarrow \langle p', u'\rangle$ then $\langle p, \gamma u''\rangle \Rightarrow_{\mathcal{P}} \langle p', u'u''\rangle$ for all $u'' \in \Gamma^*$. The reflexive transitive closure of $\Rightarrow_{\mathcal{P}}$ is denoted by $\Rightarrow^*_{\mathcal{P}}$.*

Without loss of generality, we restrict the PDS rules to have at most two stack symbols on the right-hand side [21].

The standard way of modeling control-flow of programs using PDSs is as follows: the set P consists of a single state $\{p\}$; the set Γ consists of program nodes, and Δ has one rule per edge in the control-flow graph as follows: $\langle p, u\rangle \hookrightarrow \langle p, v\rangle$ for an intraprocedural edge (u, v); $\langle p, u\rangle \hookrightarrow \langle p, e\ v\rangle$ for a procedure call at node u that returns to v and calls the procedure starting at e; $\langle p, u\rangle \hookrightarrow \langle p, \varepsilon\rangle$ if u is the exit node of a procedure. Finite-state data is encoded by expanding P to be the set of global states, and expanding Γ by including valuations of local variables. Under such an encoding, a configuration $\langle p, \gamma_1\gamma_2\cdots\gamma_n\rangle$ represents the instantaneous state of the program: p is the valuation of global variables, γ_1 has

For each $\langle p, \gamma \rangle \hookrightarrow \langle p', u \rangle \in (\Delta_1 \cup \Delta_2)$ and for all $p_i \in P, k \in \{1, \cdots, K\}$:
$\langle (k, p_1, \cdots, p_{k-1}, p, p_{k+1}, \cdots, p_K), \gamma \rangle \hookrightarrow \langle (k, p_1, \cdots, p_{k-1}, p', p_{k+1}, \cdots, p_K), u \rangle$

For each $\gamma \in \Gamma_j$ and for all $p_i \in P, k \in \{1, \cdots, K\}$:
$\langle (k, p_1, \cdots, p_K), \gamma \rangle \hookrightarrow \langle (k+1, p_1, \cdots, p_K), \gamma \rangle$
$\langle (K+1, p_1, \cdots, p_K), \gamma \rangle \hookrightarrow \langle (1, p_1, \cdots, p_K), e_{j+1}\ \gamma \rangle$

Fig. 4. PDS rules for $\mathcal{P}_s$

the current program location and values of local variables in scope, and $\gamma_2 \cdots \gamma_n$ store the return addresses and values of local variables for unfinished calls.

A concurrent program with two threads is represented with two PDSs that share their global state: $\mathcal{P}_1 = (P, \Gamma_1, \Delta_1), \mathcal{P}_2 = (P, \Gamma_2, \Delta_2)$. A configuration of such a system is the triplet $\langle p, u_1, u_2 \rangle$ where $p \in P, u_1 \in \Gamma_1^*, u_2 \in \Gamma_2^*$. Define two transition systems: if $\langle p, u_i \rangle \Rightarrow_{\mathcal{P}_i} \langle p', u_i' \rangle$ then $\langle p, u_1, u \rangle \Rightarrow_1 \langle p', u_1', u \rangle$ and $\langle p, u, u_2 \rangle \Rightarrow_2 \langle p', u, u_2' \rangle$ for all u. The problem of interest with concurrent programs, under a context bound $2K - 1$, is to find the reachable states under the transition system $(\Rightarrow_1^*; \Rightarrow_2^*)^K$ (here the semicolon denotes relational composition, and exponentiation is repeated relational composition).

We reduce the concurrent program $(\mathcal{P}_1, \mathcal{P}_2)$ to a single PDS $\mathcal{P}_s = (P_s, \Gamma_s, \Delta_s)$. Let P_s be the set of all $K+1$ tuples whose first component is a number between 1 and K, and the rest are from the set P, i.e., $P_s = \{1, \cdots, K\} \times P \times P \times \cdots \times P$. This set relates to the reduction from §2 as follows: an element $(k, p_1, \cdots, p_K) \in P_s$ represents that the value of the variable `k` is k; and p_i encodes a valuation of the variables VAR_G^i. When $\mathcal{P}_s$ is in such a state, its rules would only modify p_k.

Let $e_i \in \Gamma_i$ be the starting node of the ith thread. Let Γ_s be the disjoint union of Γ_1, Γ_2 and an additional symbol $\{e_3\}$. $\mathcal{P}_s$ does not have an explicit checking phase. The rules Δ_s are defined in Fig. 4.

We deviate slightly from the reduction presented in §2 by changing the **goto** statement, which passes control from the first thread to the second, into a procedure call. This ensures that the stack of the first thread is left intact when control is passed to the next thread. Furthermore, we assume that the PDSs cannot empty their stacks, i.e., it is not possible that $\langle p, e_1 \rangle \Rightarrow_{\mathcal{P}_1}^* \langle p', \varepsilon \rangle$ or $\langle p, e_2 \rangle \Rightarrow_{\mathcal{P}_2}^* \langle p', \varepsilon \rangle$ for all $p, p' \in P$ (in other words, the `main` procedure should not return). This can be enforced for arbitrary PDSs [8].

Theorem 1. *Starting execution of the concurrent program* $(\mathcal{P}_1, \mathcal{P}_2)$ *from the state* $\langle p, e_1, e_2 \rangle$ *can lead to the state* $\langle p', c_1, c_2 \rangle$ *under the transition system* $(\Rightarrow_1^*; \Rightarrow_2^*)^K$ *if and only if there exist states* $p_2, \cdots, p_K \in P$ *such that* $\langle (1, p, p_2, \cdots, p_K), e_1 \rangle \Rightarrow_{\mathcal{P}_s} \langle (1, p_2, p_3, \cdots, p_K, p'), e_3\ c_2\ c_1 \rangle$.

Note that the checking phase is implicit in the statement of Thm. 1.

Complexity. Using our reduction, one can find the set of all reachable configurations of the concurrent program $(\mathcal{P}_1, \mathcal{P}_2)$ in time $\mathcal{O}(K^2 |P|^{2K} |Proc| |\Delta_1 + \Delta_2|)$, where $|Proc|$ is the number of procedures in the program[4] [8]. Using backward reachability algorithms, one can verify if a given configuration in reachable in time $\mathcal{O}(K^3|P|^{2K}|\Delta_1 + \Delta_2|)$. Both these complexities are asymptotically better than those of previous algorithms for PDSs [16,10], with the latter being linear in the program size $|\Delta_1 + \Delta_2|$.

A similar reduction works for multiple threads as well (under round-robin scheduling). Moreover, the complexity of finding all reachable states under a bound of nK with n threads, using a standard PDS reachability algorithm, is $\mathcal{O}(K^3|P|^{4K}|Proc||\Delta|)$, where $|\Delta| = \Sigma_{i=1}^{n}|\Delta_i|$ is the total number of rules in the concurrent program.

This reduction produces a large number of rules in $\mathcal{P}_s$, but we can leverage work on *symbolic PDSs* [21] to obtain symbolic implementations [8].

5 Lazy CBA of Concurrent Boolean Programs

In the reduction presented in §3, the analysis of the generated sequential program had to assume all possible values for the symbolic constants. The lazy analysis will have the property that at any time, if the analysis considers the K-tuple $(g_1, \cdots, g_K)$ of valuations of the symbolic constants, then there is a *single* valid execution of the concurrent program in which the global state is g_i at the end of the ith execution context of the first thread for all $1 \leq i \leq K$.

The idea is to iteratively build up the effect that each thread can have on the global state in their K execution contexts. Note that T_1^s (or T_2^s) does not need to know the values of Var_G^i when $\texttt{k} < i$. Hence, the analysis proceeds by making no assumptions on the values of Var_G^i when $i > \texttt{k}$. When $\texttt{k}$ is incremented to $k+1$ in the analysis of T_1^s, it consults a table E^2 that stores the effect that T_2^s can have in its first k execution contexts. Using that table, it figures out a valuation of Var_G^{k+1} to continue the analysis of T_1^s, and stores the effect that T_1^s can have in its first k execution contexts in table E^1. These tables are built iteratively. More technically, if the analysis can deduce that T_1^s, when started in state $(1, g_1, \cdots, g_k)$, can reach the state $(k, g_1', \cdots, g_k')$, and T_2^s, when started in state $(1, g_1', \cdots, g_k')$ can reach $(k, g_2, g_3, \cdots, g_k, g_{k+1})$, then an increment of $\texttt{k}$ in T_1^s produces the global state $s = (k+1, g_1', \cdots, g_k', g_{k+1})$. Moreover, s can be reached when T_1^s is started in state $(1, g_1, \cdots, g_{k+1})$ because T_1^s could not have touched Var_G^{k+1} before the increment that changed $\texttt{k}$ to $k+1$. The algorithm is shown in Fig. 5. The entities used in it have the following meanings:

- Let $\overline{G} = \cup_{i=1}^{K} G^i$, where G is the set of global states. An element from the set $\overline{G}$ is written as $\overline{g}$. Let L be the set of local states.
- The relation H_n^j is related to program node n of the jth thread. It is a subset of $\{1, \cdots, K\} \times \overline{G} \times \overline{G} \times L \times \overline{G} \times L$. If $H_n^j(k, \overline{g_0}, \overline{g_1}, l_1, \overline{g_2}, l_2)$ holds, then each of the $\overline{g_i}$ are an element of G^k (i.e., a k-tuple of global states), and the thread T_j is in its kth execution context. Moreover, if the valuation of

[4] The number of procedures of a PDS is defined as the number of symbols appearing as the first of the two stack symbols on the right-hand side of a call rule.

VAR^i_G, $1 \leq i \leq k$, was $\overline{g_0}$ when T^s_j (the reduction of T_j) started executing, and if the node ep(n) could be reached in data state $(\overline{g_1}, l_1)$, then n can be reached in data state $(\overline{g_2}, l_2)$, and the variables VAR^i_G, $i > k$ are not touched (hence, there is no need to know their values).
- The relation $S_{\mathtt{f}}$ captures the summary of procedure `f`.
- The relations E^j store the *effect* of executing a thread. If $E^j(k, \overline{g_0}, \overline{g_1})$ holds, then $\overline{g_0}, \overline{g_1} \in G^k$, and the execution of thread T^s_j, starting from $\overline{g_0}$ can lead to $\overline{g_1}$, without touching variables in VAR^i_G, $i > k$.
- The function $\mathtt{check}(k, (g_1, \cdots, g_k), (g'_1, \cdots, g'_k))$ returns g'_k if $g_{i+1} = g'_i$ for $1 \leq i \leq k-1$, and is undefined otherwise. This function checks for the correct transfer of the global state from T_2 to T_1 at a context switch.
- Let $[(g_1, \cdots, g_i), (g_{i+1}, \cdots g_j)] = (g_1, \cdots, g_j)$. We sometimes write g to mean (g), i.e., $[(g_1, \cdots, g_i), g] = (g_1, \cdots, g_i, g)$.

Understanding the rules. The rules $\mathcal{R}'_1, \mathcal{R}'_2, \mathcal{R}'_3$, and $\mathcal{R}'_7$ describe intra-thread computation, and are similar to the corresponding unprimed rules in Fig. 3. The rule $\mathcal{R}_{10}$ initializes the variables for the first execution context of T_1. The rule $\mathcal{R}_{12}$ initializes the variables for the first execution context of T_2. The rules $\mathcal{R}_8$ and $\mathcal{R}_9$ ensure proper hand off of the global state from one thread to another. These two are the only rules that change the value of k. For example, consider rule $\mathcal{R}_8$. It ensures that the global state at the end of kth execution context of T_2 is passed to the $(k+1)^{\text{th}}$ execution context of T_1, using the function `check`. The value g returned by this function represents a reachable valuation of the global variables when T_1 starts its $(k+1)^{\text{th}}$ execution context.

The following theorem shows that the relations E^1 and E^2 are built lazily, i.e., they only contain relevant information. A proof can be found in [8].

Theorem 2. *After running the algorithm described in Fig. 5, $E^1(k, (g_1, \cdots, g_k), (g'_1, \cdots, g'_k))$ and $E^2(k, (g'_1, \cdots, g'_k), (g_2, \cdots, g_k, g))$ hold if and only if there is an execution of the concurrent program with $2k-1$ context switches that starts in state g_1 and ends in state g, and the global state is g_i at the start of the ith execution context of T_1 and g'_i at the start of the ith execution context of T_2. The set of reachable global states of the program in $2K-1$ context switches are all $g \in G$ such that $E^2(K, \overline{g_1}, [\overline{g_2}, g])$ holds.*

6 Experiments

We did a proof-of-concept implementation of the eager algorithm for Boolean programs, presented in §3, using the model checker MOPED [20]. We took sequential programs and assumed that there were two copies of the program running concurrently (except for `BlueT`). The input programs are obtained from a variety of sources: `BlueT` is a model of a Bluetooth driver [17]; `Java*` are the result of abstracting Java programs [2]; `Reg*` are from the regression suite of MOPED; `Toy` is a toy program we wrote for checking correctness. Some programs, especially ones obtained from Java programs, have pointers and a bounded heap (which is

$$\frac{H^j_n(k, \overline{g_0}, \overline{g_1}, l_1, [\overline{g_2}, g_3], l_3) \quad n \xrightarrow{\texttt{st}} m \quad (g_3, l_3, g_4, l_4) \in [\![\texttt{st}]\!]}{H^j_m(k, \overline{g_0}, \overline{g_1}, l_1, [\overline{g_2}, g_4], l_4)} \; \mathcal{R}'_1$$

$$\frac{H^j_n(k, \overline{g_0}, \overline{g_1}, l_1, \overline{g_2}, l_2) \quad n \xrightarrow{\texttt{call f()}} m \quad S_{\texttt{f}}(k+i, [\overline{g_2}, \overline{g}], [\overline{g_3}, \overline{g'}])}{H^j_m(k+i, [\overline{g_0}, \overline{g}], [\overline{g_1}, \overline{g}], l_1, [\overline{g_3}, \overline{g'}], l_2)} \; \mathcal{R}'_2$$

$$\frac{H^j_n(k, \overline{g_0}, \overline{g_1}, l_1, \overline{g_2}, l_2) \quad \text{exitnode}(n) \quad \texttt{f} = \text{proc}(n)}{S_{\texttt{f}}(k, \overline{g_1}, \overline{g_2})} \; \mathcal{R}'_3 \qquad \frac{g \in G, l \in L, e = \text{entry}(\texttt{main})}{H^1_e(1, g, g, l, g, l)} \; \mathcal{R}_{10}$$

$$\frac{H^j_n(k, \overline{g_0}, \overline{g_1}, l_1, \overline{g_2}, l_2) \quad n \xrightarrow{\texttt{call f()}} m \quad l_3 \in L}{H^j_{\texttt{entry(f)}}(k, \overline{g_0}, \overline{g_2}, l_3, \overline{g_2}, l_3)} \; \mathcal{R}'_7 \qquad \frac{H^j_n(k, \overline{g_0}, \overline{g_1}, l_1, \overline{g_2}, l_2)}{E^j(k, \overline{g_0}, \overline{g_2})} \; \mathcal{R}_{11}$$

$$\frac{H^1_n(k, \overline{g_0}, \overline{g_1}, l_1, \overline{g_2}, l_2) \quad E^2(k, \overline{g_2}, \overline{g_3}) \quad g = \texttt{check}(\overline{g_0}, \overline{g_3})}{H^1_n(k+1, [\overline{g_0}, g], [\overline{g_1}, g], l_1, [\overline{g_2}, g], l_2)} \; \mathcal{R}_8 \qquad \frac{E^1(1, g_0, g_1), l \in L}{H^2_{e_2}(1, g_1, g_1, l, g_1, l)} \; \mathcal{R}_{12}$$

$$\frac{H^2_n(k, \overline{g_0}, \overline{g_1}, l_1, \overline{g_2}, l_2) \quad E^1(k+1, [g_3, \overline{g_2}], [\overline{g_0}, g_4])}{H^2_n(k+1, [\overline{g_0}, g_4], [\overline{g_1}, g_4], l_1, [\overline{g_2}, g_4], l_2)} \; \mathcal{R}_9$$

Fig. 5. Rules for lazy analysis of concurrent Boolean programs

accounted for in the number of variables). We verified if a certain program node was reachable by finding the set of reachable data-states at the node. In most cases, we modified the programs to have both positive and negative instances.

Prog	Inst	$2K$	Time (s)	|Prog|	#gvars	#lvars
Toy	pos	20	0.3	12	5	0
Reg-blast1	neg	20	3.9	19	7	21
Reg-blast1	pos	20	4.1	19	7	21
Reg-slam1	pos	20	19.6	19	1	10
BlueT	neg	20	7.2	30	10	1
BlueT	pos	10	7.6	30	10	1
JavaMeeting	neg	10	168.5	537	16	64
JavaMeeting	pos	10	361.3	537	16	64
JavaChange	neg	10	770.8	601	24	38
JavaChange	pos	10	1134.4	601	24	38

Fig. 6. Experiments on finite-data-state models

The results are shown in Fig. 6. The last three columns give the total number of CFG edges, the number of global variables, and the maximum number of local variables in a procedure, respectively. They show that our algorithm is practical—the data-state space of `JavaChange` has about 2^{158} possible states. The negative cases take less time than positive cases because of the way we implemented the BDD operations. In some cases, we can conclude that a set is empty, i.e., a node is not reachable, without applying all the required operations. For positive cases this never happens, and all the operations are applied.

7 Related Work

Most of the related work on CBA has been covered in the body of the paper. A reduction from concurrent programs to sequential programs was given in [17] for the case of two threads and two context switches (it has a restricted extension to multiple threads as well). In such a case, the only thread interleaving is $T_1; T_2; T_1$. The context switch from T_1 to T_2 is simulated by a procedure call. Then T_2 is executed on the program stack of T_1, and at the next context switch, the stack of T_2 is popped off to resume execution in T_1. Because the stack of T_2 is destroyed, the analysis cannot return to T_2 (hence the context bound of 2). Their algorithm cannot be generalized to an arbitrary context bound.

Analysis of message-passing concurrent systems, as opposed to ones having shared memory, has been considered in [4]. They bound the number of messages that can be communicated, similar to bound the number of contexts.

There has been a large body of work on verification of concurrent programs. Some recent work is [6,15]. However, CBA is different because it allows for precise analysis of complicated program models, including recursion. As future work, it would be interesting to explore CBA with the abstractions used in the aforementioned work.

References

1. Ball, T., Rajamani, S.: Bebop: A symbolic model checker for Boolean programs. In: SPIN (2000)
2. Berger, F., Schwoon, S., Suwimonteerabuth, D.: jMoped (2005), `http://www.informatik.uni-stuttgart.de/fmi/szs/tools/moped/jmoped/`
3. Bouajjani, A., Fratani, S., Qadeer, S.: Context-bounded analysis of multithreaded programs with dynamic linked structures. In: Damm, W., Hermanns, H. (eds.) CAV 2007. LNCS, vol. 4590, pp. 207–220. Springer, Heidelberg (2007)
4. Chaki, S., Clarke, E.M., Kidd, N., Reps, T.W., Touili, T.: Verifying concurrent message-passing C programs with recursive calls. In: Hermanns, H., Palsberg, J. (eds.) TACAS 2006 and ETAPS 2006. LNCS, vol. 3920, pp. 334–349. Springer, Heidelberg (2006)
5. Henzinger, T., Jhala, R., Majumdar, R., Sutre, G.: Lazy abstraction. In: POPL (2002)
6. Henzinger, T.A., Jhala, R., Majumdar, R.: Race checking by context inference. In: PLDI (2004)
7. Knoop, J., Steffen, B.: The interprocedural coincidence theorem. In: CC (1992)
8. Lal, A., Reps, T.: Reducing concurrent analysis under a context bound to sequential analysis. Technical Report 1629, University of Wisconsin (2008)
9. Lal, A., Touili, T., Kidd, N., Reps, T.: Interprocedural analysis of concurrent programs under a context bound. TR-1598, University of Wisconsin (July 2007)
10. Lal, A., Touili, T., Kidd, N., Reps, T.: Interprocedural analysis of concurrent programs under a context bound. In: TACAS (2008)
11. Müller-Olm, M., Seidl, H.: Precise interprocedural analysis through linear algebra. In: POPL (2004)
12. Murphy, B., Lam, M.: Program analysis with partial transfer functions. In: PEPM (2000)
13. Musuvathi, M., Qadeer, S.: Iterative context bounding for systematic testing of multithreaded programs. In: PLDI (2007)
14. Qadeer, S., Rajamani, S.: Deciding assertions in programs with references. Technical Report MSR-TR-2005-08, Microsoft Research, Redmond (January 2005)

15. Qadeer, S., Rajamani, S.K., Rehof, J.: Summarizing procedures in concurrent programs. In: POPL (2004)
16. Qadeer, S., Rehof, J.: Context-bounded model checking of concurrent software. In: Halbwachs, N., Zuck, L.D. (eds.) TACAS 2005. LNCS, vol. 3440, pp. 93–107. Springer, Heidelberg (2005)
17. Qadeer, S., Wu, D.: KISS: Keep it simple and sequential. In: PLDI (2004)
18. Ramalingam, G.: Context-sensitive synchronization-sensitive analysis is undecidable. In: TOPLAS (2000)
19. Reps, T., Horwitz, S., Sagiv, M.: Precise interprocedural dataflow analysis via graph reachability. In: POPL (1995)
20. Schwoon, S.: Moped, `http://www.fmi.uni-stuttgart.de/szs/tools/moped/`
21. Schwoon, S.: Model-Checking Pushdown Systems. PhD thesis, Technical Univ. of Munich, Munich, Germany (July 2002)
22. Sharir, M., Pnueli, A.: Two approaches to interprocedural data flow analysis. In: Program Flow Analysis: Theory and Applications, Prentice-Hall, Englewood Cliffs (1981)

Monitoring Atomicity in Concurrent Programs

Azadeh Farzan[1] and P. Madhusudan[2]

[1] Carnegie Mellon University
afarzan@cs.cmu.edu
[2] Univ. of Illinois at Urbana-Champaign
madhu@cs.uiuc.edu

Abstract. We study the problem of monitoring concurrent program runs for atomicity violations. Unearthing fundamental results behind scheduling algorithms in database control, we build space-efficient monitoring algorithms for checking atomicity that use space polynomial in the number of active threads and entities, and independent of the length of the run monitored. Second, by interpreting the monitoring algorithm as a *finite automaton*, we solve the model checking problem for atomicity of finite-state concurrent models. This establishes (for the first time) that model checking finite-state concurrent models for atomicity is decidable, and remedies incorrect proofs published in the literature. Finally, we exhibit experimental evidence that our atomicity monitoring algorithm gives substantial time and space benefits on benchmark applications.

1 Introduction

Correct concurrent programs are way harder to write than sequential ones. Sequential bug-free programs are already hard to write, maintain, and test, though the tremendous effort over the last 20 years in finding errors in programs has yielded certain tractable approaches and tools to assure correctness. The advent of multi-core technologies and the increasing use of threads and communicating modules in software design has brought all concurrency issues to the forefront. Consequently, one of the most important problems in software analysis is to understand concurrency idioms used in practice, and leverage the understanding to build testing and verification tools.

While programming for a multicore (shared-memory) architecture to exploit concurrency, a useful mechanism to have is the ability to parallelize tasks such that there is controlled interaction amongst them. For instance, the proposal of transactional memory (and software transactional memory [23]) introduces such an atomicity construct in a programming language. Programmers writing in current programming languages (such as Java or C with Pthreads) implicitly need such a construct, but since it is not available, implement their own concurrency control mechanism (say using locks) in order to mutually exclude their threads from accessing shared data. A large number of errors in these concurrent programs are due to mismanaged atomicity. For instance, a recent study on bug characteristics in real-world concurrent programs revealed that more than 68%

A. Gupta and S. Malik (Eds.): CAV 2008, LNCS 5123, pp. 52–65, 2008.

of concurrency bugs were due to atomicity violations (where blocks of code were intended to be atomic but the mechanisms did not ensure atomicity) [16].

The above intuition motivates a remarkable *generic specification* (a specification common across applications) for concurrent programs called *atomicity*. Consider a concurrent program where certain blocks of code are annotated as transaction blocks, capturing the *intention* of the programmer that they be atomically executed. We declare an interleaved run to be atomic if it is semantically equivalent to a *serial* run where the transaction blocks are scheduled one after another, without any interleaving. The idea then is that non-atomic runs violate programmer intentions and hence are likely to be unintended interactions that may be concurrency errors. This notion of atomicity stems from the concept of *serializability* studied in *database concurrency control*, and the idea of using the notion of atomicity as a generic specification for concurrent programs (running on non-TM platforms) was first proposed by Flanagan and Qadeer in 2003 [9].

The most well-studied, accepted and tractable notion of serializability is *conflict serializability* [20,4]. Intuitively, we declare events to be *dependent* if they cannot be commuted— so, two accesses to the same variable are dependent if one of them is a write. Two runs r and r' are equivalent if we can obtain one from the other by commuting independent events in the run. A run is conflict-serializable if it has an equivalent serial run. We study only conflict-serializability in this paper, and will henceforth refer to it as simply serializability. The terms serializability and atomicity are synonymous in this paper.

Notice that our methodology of finding errors in programs is parameterized with annotations of blocks of code intended to be atomic. While we can choose natural syntactic blocks of code (such as methods in a class) to be blocks intended to be atomic (as many papers in the literature have done), it is also possible to learn the intended atomic blocks from positive test runs [17] (see also [27]). Here, we are interested in building algorithms for identifying non-atomic runs, and hence we will assume that the annotations of transactional blocks as given.

The objective of this paper is to (a) study the algorithmics of monitoring atomicity of individual runs of concurrent programs, and (b) leverage the monitoring algorithm to solve the model checking problem for checking atomicity in concurrent Boolean programs.

A simple monitoring algorithm for serializability works by maintaining a *conflict graph*, which is a graph depicting the precedence order imposed by the run on the transactions (blocks of code). A run is serializable if, and only if, this graph remains acyclic. A tempting idea to minimize the conflict graph while monitoring a run is to remove *completed transactions* from the conflict-graph, replacing it with *transitive edges* that summarize its effect, with the intuition being that the completed transactions cannot play any role in the future. However, this intuition is *wrong*. In a paper by Alur, McMillan and Peled [2], automata-based algorithms were designed for checking serializability that overlooked this subtlety and deleted transactions, resulting in an *erroneous algorithm* (confirmed by one of its authors [1]).

Unearthing techniques from designs of database schedulers, we obtain simple space-efficient algorithms for monitoring runs for serializability. The main idea is not to remove completed transactions, but rather summarize their effects by throwing in transitive edges and absorbing their event content into active transactions. Such algorithms are present in the database literature (see [5]), and we have adapted these to the multithreaded software realm (especially to the setting of threads executing multiple transactions) to provide space-efficient monitoring. We refer the interested reader to two textbooks on database theory [20,4], a paper on the combinatorics of removing completed transactions and related deletion policies on the conflict-graph [12], and a volume on concurrency control by Casanova [5], from which our intuitions have been gained.

The monitoring algorithm we obtain is a streaming algorithm that reads runs of a program, and after reading a run r uses space $O(k^2 + k.n)$, where k is the number of active threads after r and n is the number of entities accessed by r. Furthermore, the time taken to update the information on reading an event is only linear in this graph. Note that there is no dependence on the *number* of events or transactions executed in r, and hence our algorithm scales well when working on long executions of programs.

The monitoring algorithm, surprisingly, paves the way to decision procedures to *model check* programs for atomicity violations. When there are a finite number of threads and entities, our monitoring algorithm uses a bounded amount of space, and hence can be seen as a *deterministic finite automaton*. Using this, we prove that concurrent Boolean programs without recursion (where each thread runs a program with a regular control structure and where all variables are interpreted over finite domains), the model checking problem for atomicity is decidable and is . PSPACE-complete. As far as we know, this is the first time that the decidability of model checking atomicity for finite-state programs has been established. Note that a similar claim appears in [2], but is incorrect due to the grave error we mentioned.

Turning to the experiments, we have implemented both the conflict-graph based monitoring algorithm and the new monitoring algorithm based on summarized conflict graphs. We evaluate these on a suite of benchmarks, and illustrate the significant space (and hence time) gains our algorithm provides in monitoring long runs of realistic concurrent benchmark programs.

Related Work: Atomicity is a new notion of correctness; the more classical notion is *race checking*: a race is a pair of accesses to the same variable by different threads, where one of the accesses is a write [19,22,6]. Data races also signal improper synchronization in code, and are routinely used to find errors in concurrent programs (see [22,6] for testing and runtime checking for data races and [18] for static data race analysis). It has been suggested [10,8,26,25] that *atomicity notions based on serializability* are more appropriate and yield fewer false positives; some practical tools built for serializability demonstrate this [27].

Most work in software verification for atomicity errors are based on approximations of the concept, including Lipton transactions (a *sufficient* but not necessary condition) that ensures serializability [15]. Type systems [10] and model checkers [13] for atomicity based on Lipton transactions have been developed. The work in [8] reports ways of exploring a run and a possible serialization of it simultaneously and checks whether they result in the same effect. In [7], we had proposed a slightly different notion of atomicity called *causal atomicity* which can be checked using partial-order methods. The work in [17] defines *access interleaving invariants* that are certain patterns of access interactions on variables, learns the intended specifications using tests, and monitors runs to find errors.

The closest work to ours is a series of papers by Wang and Stoller on runtime verification of atomicity [25,26]. In these papers, the authors consider a different and harder problem than what we tackle: they consider a run, project the run onto each thread, and ask whether they can be recombined in some way to produce a non-serializable run This is significantly harder: first, recombinations of runs may not be feasible in the original program in general, and though the authors handle locks accurately, the runs may still be infeasible (say due to data checks) and hence raise false alarms. Also, even abstracting the program to a set of reads and writes as they do, the problem of checking if a non-serializable run exists can be shown to be NP-hard. The authors provide approximate algorithms, that are neither sound nor complete, for the settings where locks are nested and transactions have no potential for deadlocks. The experimental results are reported for a small number of threads (3 in most cases). Our problem however is to simply check if the current run is itself non-serializable for which we show a scalable algorithm and where we have tested the benchmarks for a large number of threads (going up to 50; see Section 5 for more details).

A variant of dynamic two-phase locking algorithm [20] for detection of serializability violations is used in the atomicity tool developed in [27]. As discussed in [20], the set of runs that are detected as atomic by a two-phase locking algorithm are a strict subset of the set of conflict serializable runs.

2 Preliminaries

Modeling Runs of Concurrent Programs: We consider programs that run threads concurrently, with accesses to local and global data. We also assume that blocks of program code are marked as *transactions*, with each thread running a sequence of transactions on any run. We will check runs of programs for atomicity violations with respect to these blocks. We first define a general notion of a run of a concurrent program, where we assume the global accesses, the thread creations and termination, and the beginning and ending of transactions are observable.

Let us assume an infinite but countable set of *thread identifiers* $\mathcal{T}=\{T_1, T_2, ..\}$. Let us also assume a countable set of (global) *entity names* (or just entities) $\mathcal{X} = \{x_1, x_2, \ldots, \}$. The set of actions $\mathcal{A}$ over X, is defined as: $\mathcal{A} = \{rd(x), wr(x) \mid x \in \mathcal{X}\}$. The alphabet of events of a thread $T \in \mathcal{T}$ is

$$\Sigma_T = \{T{:}a \mid a \in \mathcal{A}\} \cup \{T{:}\rhd, T{:}\lhd\} \cup \{Beg_T, End_T\}.$$

The events $T{:}rd(x)$ and $T{:}wr(x)$ correspond to thread T reading and writing to entity x, respectively, and $T{:}\rhd$ and $T{:}\lhd$ correspond to transaction boundaries that begin and end blocks of code in thread T, while Beg_T and End_T denote the creation and termination of the thread T itself. Let $\Sigma = \bigcup_{T \in \mathcal{T}} \Sigma_T$ denote the set of all events.

Note that the above can model dynamic memory allocation as well, provided we observe the memory allocation/release actions. The only difference is that the set of actions $\mathcal{A}$ changes during the monitoring. We can assign a fresh name to every new piece of memory allocated (based on the desired granularity), and maintain aliasing information to observe accesses to the location.

For any alphabet A, $w \in A^*$, let $w[i]$ denote the i'th element of w, and $w[i,j]$ denote the substring from position i to position j in w. For $w \in A^*$ and $B \subseteq A$, let $w|_B$ denote the word w projected to the letters in B. For a word $\sigma \subseteq \Sigma^*$, let $\sigma|_T$ be a an abbreviation of $\sigma|_{\Sigma_T}$, which includes only actions of thread T.

The runs of concurrent programs, which we call *schedules*, are executions of the program where actions of threads are interleaved.

Definition 1. *A schedule is a word $\sigma \in \Sigma^*$ such that for each $T \in \mathcal{T}$, $\sigma|_T$ is a prefix of the word $Beg_T \cdot \left[(T{:}\triangleright) \cdot \{T{:}a \mid a \in \mathcal{A}\}^* \cdot (T{:}\triangleleft)\right]^* \cdot End_T$.*

In other words, the actions of thread T start with Beg_T and end with End_T, and the actions within are divided into a sequence of transactions, where each transaction begins with $T{:}\triangleright$, is followed by a set of reads and writes, and ends with $T{:}\triangleleft$. Let *Sched* denote the set of all schedules.

Notice that schedules do not observe synchronization mechanisms such as mutual exclusion using locks, semaphores, etc. as serializability of *one schedule* is independent of the synchronization mechanism.

A transaction *tr* of thread T is a word of the form $T{:}\triangleright \; w \; T{:}\triangleleft$, where $w \in \{T{:}a \mid a \in A\}^*$. Let $Tran_T$ denote the set of all transactions of thread T, and let *Tran* denote the set of all transactions.

When we refer to two particular events $\sigma[i]$ and $\sigma[j]$ in σ, we say they *belong* to the same transaction if they belong to the same transaction block: i.e. if there is some T such that $\sigma[i] = T{:}a$, $\sigma[j] = T{:}a'$, where $a, a' \in A$, and there is no i', $i < i' < j$ such that $\sigma[i'] = T{:}\triangleleft$. We refer to the transaction blocks freely and associate (arbitrary) names to them, using notations such as tr, tr_1, tr', etc.

Defining atomicity: We define atomicity through the notion of *conflict serializability.* The *dependency* relation D is a symmetric relation defined over the events in Σ, and captures the dependency between (a) two events accessing the same entity, one of them being a write, and (b) any two events of the same thread, i.e.,

$$\begin{aligned} D = \{(T_1{:}a_1, T_2{:}a_2) \mid & \; (T_1 = T_2 \wedge a_1, a_2 \in A \cup \{\triangleright, \triangleleft\}) \vee \\ & [\exists x \in \mathcal{X} \textit{ such that } (a_1 = rd(x) \wedge a_2 = wr(x)) \vee \\ & (a_1 = wr(x) \wedge a_2 = rd(x)) \vee (a_1 = wr(x) \wedge a_2 = wr(x))]\} \end{aligned}$$

Definition 2 (Equivalence of schedules). *The equivalence of schedules is defined as the* smallest *equivalence relation* $\sim \subseteq$ *Sched* $\times$ *Sched such that the following condition holds:*

if $\sigma = \rho e e' \rho', \sigma' = \rho e' e \rho' \in Sched$ *with* $(e, e') \notin D$, *then* $\sigma \sim \sigma'$.

It is easy to see that the above notion is a well-defined equivalence relation. Intuitively, two schedules are considered equivalent if we can derive one schedule from the other by iteratively swapping consecutive independent actions in the schedule. It is also clear (given that two actions accessing an entity are independent only if both are reads), that equivalent schedules produce the same valuation of all entities. Formally, assuming each thread has its view limited to the entities it has read and written, we can show that no matter what the domain of the entries are, and what functions the individual threads may compute, and no matter how many *local* variables a thread may have, the final values of both local variables and global entities remains unchanged when executing two equivalent schedules.

We call a schedule σ *serial* if all the transactions in it occur atomically: formally, for every i, if $\sigma[i] = T{:}a$ where $T \in \mathcal{T}$ and $a \in A$, then there is some $j < i$ such that $T[i] = T{:}\rhd$ and every $j < j' < i$ is such that $\sigma[j'] \in \Sigma_T$. In other words, the schedule is made up of a sequence of complete transactions from different threads, interleaved at boundaries only (the final transactions of a thread may be incomplete, but even then the actions in each incomplete transaction must occur together).

Definition 3. *A schedule is* serializable *if it has an equivalent serial schedule. That is, σ is a serializable schedule if there a* serial *schedule σ' such that $\sigma \sim \sigma'$.*

The conflict-graph characterization: A simple characterization of atomic (or conflict-serializable) schedules uses the notion of a conflict-graph, and is a classic characterization from the database literature (this notion is so common that many papers in database theory *define* conflict serializability using it).

If a transaction tr of thread T reads x and is followed by a transaction tr' of thread T' that writes x, then we must schedule the (entire) transaction T before the (entire) transaction T'. The conflict graph [11,20,4] is a graph that captures these constraints, and is made up of transactions as nodes, and edges capturing ordering constraints imposed by a schedule. A schedule is serializable iff its conflict graph represents a partial order (i.e. is acyclic).

Formally, for any schedule σ, let us give names to transactions in σ, say $tr_1, \ldots, tr_n$. The *conflict-graph* of σ is the graph $CG(\sigma) = (V, E, S)$ where $V = \{tr_1, \ldots, tr_n\}$, $S : V \longrightarrow 2^{\Sigma}$ is a labeling of vertices such that $S(tr_i)$ is precisely the set of events that have been scheduled in tr_i, and E contains an edge from tr to tr' iff there is some event e in transaction tr and some event e' in transaction tr' such that (1) the e-event occurs before e' in σ, and (2) eDe'.

Note that transactions of the same thread are always ordered in the order they occur (since all actions of a thread are dependent on each other to preserve sequential consistency).

Lemma 1. *A schedule σ is atomic iff the conflict graph of σ is acyclic.* □

The above lemma essentially follows from [11] (also [4,20]). The classic model is however a database model where transactions are independent; we require an extension of this lemma to our model where there are additional constraints imposed by the fact that some transactions are executed by the same thread. The above characterization actually happens to hold when the underlying alphabet has *any arbitrary* dependence relation, and hence holds in our threaded model since we have ensured that any two events of a thread are dependent.

If the conflict graph is acyclic, then it can be viewed as a partial order, and it is clear that *any* linearization of the partial order will define a serial schedule equivalent to σ. If the conflict-graph has a cycle, then it is easy to show that the cyclic dependency rules out the existence of any equivalent serial schedule.

The above characterization yields a simple algorithm for serializability that simply constructs the conflict-graph and checks for cycles. The algorithm can process the schedule event by event, updating the graph, and finally call a cycle-detection routine. Hence [20,4]:

Proposition 1. *The problem of checking whether a singe schedule σ is atomic is decidable in polynomial time.* □

3 Monitoring Atomicity

The goal of this section is to build a space-efficient monitoring algorithm for checking serializability violations. Our algorithm, after reading a run r, will take a space at most polynomial in the (maximum) number of *active threads* (at every moment) in r and the number of entities accessed in r; most importantly, this will have no dependence on the length of r itself.

Note that there is a simple algorithm that monitors serializability violations by keeping track of the conflict graph and checking it for cycles. However, since the conflict graph contains one node for every transaction that has ever happened in the system, it can grow arbitrarily large, and does not result in the monitoring algorithm we seek. We want to keep a reduced conflict graph when monitoring.

Let us look at an example to see how to reduce the conflict graph. Consider the following non-serializable schedule:

$$T_1{:}\rhd\ T_1{:}rd(x)\ T_2{:}\rhd T_2{:}wr(x)\ T_2{:}rd(z)$$
$$T_3{:}\rhd\ T_3{:}wr(z)\ T_2{:}\lhd\ T_1{:}wr(x)$$

The figure on the right demonstrates the conflict graph for the above schedule without the last event $T_1{:}wr(x)$ (t_1, t_2, and t_3 are transactions of threads T_1, T_2, and T_3).

Now, in this graph, since the transaction t_2 of T_2 has finished, it is very tempting to remove the node t_2, and summarize its effect by replacing it by an edge from t_1 to t_3. However, this is a *serious error:* for example, if we deleted t_2, then the next event $T_1{:}wr(x)$ *does not cause* a cycle. The reason is that though t_2 is a completed transaction, it may still participate in later cycles as *outgoing* edges from t_2 can always be introduced even after t_2 has completed.

The work by Alur, McMillan and Peled [2] considers precise algorithms for monitoring serializability violations (using automata that keep track of reduced conflict-graphs on the schedule it has read). Their paper has the grave error mentioned above, and the algorithms establishing upper bounds of checking serializability in the paper are *wrong*.

In fact, the problem of when completed transactions can be deleted is a well studied problem in database concurrency control, and there have been several approaches to finding safe deletion policies [4,5,12]. Next, we present a way to *summarize* the essential information of the conflict-graph using just a graph of nodes formed by active threads[1].

3.1 Summarized Conflict Graph for Serializability

The following notion of summarized conflict graphs is adapted from a conflict-serializable scheduling algorithm in [5]. Intuitively, we keep the conflict graph restricted to edges between the active transactions only (paths between active transactions summarized as edges), and also maintain for each active transaction/thread T a set C which denotes *the set of events that occurred in transactions that must be scheduled later than this transaction.* Moreover, when keeping track of events of completed threads, we erase the thread id from its description.

Recall that $\mathcal{A}$ is the set of actions of reads and writes to global entities, and Σ includes the set $\{T{:}a \mid T \in \mathcal{T}, a \in \mathcal{A}\}$ as well as begin and end events for transactions and threads. In the summarized graph, each node corresponding to an active thread T will be associated with two sets, S and C, where $S \subseteq \Sigma_T$ is the set of *scheduled events* of the current active transaction of thread T (as in the conflict graph), and $C \subseteq \Sigma \cup \mathcal{A}$ is the set of *conflicting events*, events that occurred in completed transactions that must be scheduled later than the current transaction of T.

Definition 4. *Let σ be a schedule and let $CG(\sigma)$ be its conflict graph. The* summarized conflict graph *of a schedule σ is a tuple $SCG_\sigma = (V, E, S, C)$, where (V, E) is a graph and S and C are two vertex-labeling functions $S : V \longrightarrow 2^\Sigma$, $C : V \longrightarrow (2^{\Sigma \cup \mathcal{A}})$, where*

- *V contains a node v_i for each active thread T_i;*
- *E contains an edge from v to v' (respectively associated to transactions tr and tr') if and only if there exists a path from the node corresponding to tr to that corresponding to tr' in $CG(\sigma)$ which does not contain any nodes corresponding to active transactions;*
- *For any $v_i \in V$, if v_i corresponds to active transaction tr, then $S(v_i)$ consists of precisely the label of tr in the conflict-graph of σ, and $C(v_i)$ contains:*
  - *the set of events $T{:}a \in \Sigma$ such that T is an active thread and there is some* completed *transaction tr', reachable from tr in $CG(\sigma)$, whose label contains $T{:}a$, and*
  - *the set of actions $a \in \mathcal{A}$ such that there is some* completed *transaction tr', reachable from tr in $CG(\sigma)$, whose label contains $T{:}a$, and where T is a thread that has already ended in σ.*

[1] An active thread T is a thread for where the $begin_T$ action has appeared in the schedule, and end_T has not appeared yet .

The above is a static definition of the summarized conflict graph, and it is easy to see that cycles in the conflict graph manifest themselves in the summarized conflict graph:

Lemma 2. *There is a cycle in the conflict graph of σ iff there is a cycle in the summarized cycle graph of σ' for some σ' that is a prefix of σ.* □

Notice that when a cycle gets formed in the summarized conflict graph, it may get removed later (for example if all the threads that form the transactions of the cycle end) However, since whenever a node is removed, we combine incoming edges and outgoing edges from this node with a transitive edge, it follows that unless all nodes in the cycle are removed, the cycle is preserved. When the cycle is finally removed, it will be in the form of a self-loop on a transaction node that is being deleted. Hence, when monitoring, it is sufficient to check for self-loops.

Maintaining the summarized conflict graph: Let us now turn to an algorithm for maintaining the summarized conflict graph. The following set of rules show how the summarized conflict graph can be constructed dynamically as the schedule σ progresses. The dynamic algorithm updates the graph based on these rules until a self-loop is created, and at which point reports a serializability violation. The algorithm maintains a set AT of the currently active threads.

– **(Rule 1):** If the next event in σ is $T_i{:}\rhd$, then create a new node v_i and set $S(v_i) = \emptyset$ and $C(v_i) = \emptyset$.
– **(Rule 2):** If the next event in σ is $T_i{:}\lhd$, then remove v_i by connecting all its (immediate) predecessors to all its (immediate) successors. Also for every (immediate) predecessor v_k of v_i, set $C(v_k) = C(v_k) \cup S(v_i) \cup C(v_i)$.
– **(Rule 3):** If the next event in σ is $T_i{:}a$, then set $S(v_i) := S(v_i) \cup \{T_i{:}a\}$. For all $v_k \neq v_i$, if there is an event $T_j{:}b \in S(v_k) \cup C(v_k)$ such that $(T_j{:}b, T_i{:}a) \in \mathcal{D}_\Sigma$, add an edge (v_k, v_i) to E. Also, for any action $b \in C(v_k)$ such that $(T_k{:}b, T_i{:}a) \in \mathcal{D}_\Sigma$, add an edge (v_k, v_i) to E.
– **(Rule 4):** If the next event in σ is Beg_{T_i} , then set $AT = AT \cup \{T_i\}$.
– **(Rule 5):** If the next event in σ is End_{T_i}, then set $AT = AT \setminus \{T_i\}$, and replace every $T_i{:}a$ in every conflict s et C by a.

The summarized conflict graph given here is derived from a similar one presented in [4,5], but adapted to handle threads. Furthermore, for terminated threads, we have adapted the algorithm to remove the thread id information, thereby bounding the information kept at each node to the product of the number of *active* threads and entities accessed only. In the case of dynamic memory allocation, when a location is freed, it can be removed from all the label sets.

The following theorem captures the correctness of the algorithm in maintaining the summarized conflict graph, and hence, by Lemma 2 and Lemma 1, is a streaming algorithm that detects serializability violations.

Theorem 1. *The algorithm presented streams events of a schedule σ, and maintains the summarized conflict graph. Hence, for schedules in which all started threads end, the algorithm detects a self-loop on some node at some point in time iff the schedule is non-serializable.* □

Note that if one is interested in checking a schedule that does not conform to the above, it is always easy to add transaction end and thread end actions to the end of the schedule to make it so.

Complexity of the algorithm: Our monitoring algorithm using the summarized conflict graph simply reads events of a schedule, maintains the summarized conflict graph, and checks if at any point a self-loop is introduced. Its space complexity is as follows:

Proposition 2. *For any schedule σ, the number of nodes of the summarized conflict graph is bounded by k and the combined sizes of the sets associated with the nodes is bounded by $k.n$, where k is the maximum number of active threads during σ and n is the number of entities accessed by it. The size of the summarized conflict graph is therefore of $O(k^2 + k \cdot n)$.* □

While scanning the schedule σ, the algorithm spends time $O(k \cdot \log n)$ when the next event is a read or a write action by some thread. If the next event is an end of a transaction, then the monitoring algorithm spends time $O(n \cdot k)$. The updates for other action are performed in constant time.

4 Model Checking Atomicity for Boolean Programs

In this section, we present the second main result of the paper: a solution for the problem of checking atomicity of concurrent Boolean programs and establishing its complexity to be precisely PSPACE-complete. The result we prove here was claimed in [2], but as we mentioned, the proof there was wrong.

We consider succinct representations of programs with Boolean variables with logical encodings of initial positions and the transition relation (very much akin to how systems are described in model checking based on Boolean decision diagrams such as NuSMV).

Let us fix a finite set of threads $\mathcal{T} = \{T_1, \ldots, T_k\}$ and a finite set of entities $\mathcal{X} = \{x_1, \ldots, x_n\}$. Recall the set of actions associated with these threads and entities:

$$\Sigma_T = \{T{:}a \mid a \in \mathcal{A}\} \cup \{T{:}\rhd, T{:}\lhd\} \cup \{Beg_T, End_T\}$$

and $\Sigma = \bigcup_{T \in \mathcal{T}} \Sigma_T$. Note that Σ is a finite set.

A Boolean program over threads $\mathcal{T}$ and $\mathcal{X}$ is defined over a finite set of Boolean variables $\mathcal{V}$, where the initial set of states is described using a Boolean formula $Init(V)$, and for each $e \in \Sigma$, we have a Boolean formula $Trans_e(\mathcal{V}, \mathcal{V}')$, where $\mathcal{V}' = \{v' \mid v \in \mathcal{V}\}$, which describes the transitions on the event e. The size of the program is defined as $|\mathcal{V}| + |Init(\mathcal{V})| + \sum_{e \in \Sigma} |Trans_e(\mathcal{V}, \mathcal{V}')|$. The semantics of the program is the natural one.

Imperative concurrent programs with common synchronization constructs can easily be described as a Boolean program provided there are only a finite number of threads and entities, and the data manipulated by the program has been abstracted into a finite domain. This model is particularly interesting in an *abstract-interpretation* framework where data domains are abstracted, say using predicate abstraction [3].

The key to model checking for atomicity violations of Boolean programs is to realize that the monitoring algorithm presented in this paper maintains a *bounded* graph to check atomicity, when the number of threads and entities are bounded. We can turn this monitoring algorithm into a deterministic automaton that checks for atomicity violations, and using the automata-theoretic method, reduce model checking to a decidable emptiness problem on automata. We build an automaton *Ser* that accepts the set of all serializable runs (of k threads and n entities). It is easy to see that size of *Ser* is exponential in n and k. By building an automaton B that accepts all the runs of the program (which will also be exponential in the size of the program), we can model check for atomicity by checking if $L(B) \subseteq L(Ser)$, which can be achieved in PSPACE by generating these automata on-the-fly.

Theorem 2. *The problem of checking if a Boolean program is serializable is* PSPACE*-complete.* □

5 Experimental Evaluation

We implemented two algorithms to monitor serializability of program runs: one was the classic algorithm based on conflict graphs, and the other of our new algorithm based on summarized conflict graphs. Comparing with the conflict graph algorithm is useful since existing methods [25] use structures that are similar to the conflict graph.

We evaluated the algorithms on a benchmark suite of 5 programs. These benchmarks include `sor` (successive over-relaxation), `lufact` (LU factorization), and `raytracer` from the Java Grande multithreaded benchmarks [14], and `elevator` and `tsp` from [21]. `sor` and `lufact` are (data-intensive) scientific computation programs which perform numerical computation on matrices, `elevator` simulates multiple lifts in a building, `tsp` solves the traveling sales man problem for a given input map, and `raytracer` renders a frame of an arrangement of spheres from a given view point.

We extracted runs by manually instrumenting programs to output the accesses to entities while executing. We have a simple automatic *escape analysis* unit that excludes from the run all accesses to thread-local entities. We then run the monitoring algorithms on these output files offline. We use the `glib` library to efficiently implement set and graph operations. In the case of summarized conflict graphs, we check for cycles by checking for self loops in the graph, and for conflict graphs, we check the graph for cycles *once* at the end.

Table 1 presents the results of our evaluation. We ran each benchmark with different input parameters such as number of threads, and input files. For each program, we report in the table the number of lines of code (LOC) (appears below the program names), number of threads used in the run and number of *truly shared* entities between threads, the length of the run (number of events). The table presents the results of running the two monitoring algorithms, conflict graph

Table 1. Monitoring Results (K=1000; M=1000000)

Application (LOC)	Spec	Threads; Entities	Length of the Run	CG (n, e)	CG (time)	SCG (n, e)	SCG (time)	ser-viol/ bug
sor (470)	100×100	3; 400	97M	600, 80K	118s	3, 0	0.5s	no/–
	100×100	10; 1800	97M	2K, 300K	8872s	6, 0	3.5s	no/–
	100×100	50; 10000	101M	—	> 8h	17, 0	30.4s	no/–
lufact (1234)	100×100	3; 10K	17M	894, 168K	1824s	3, 2	37s	no/–
	100×100	10; 10K	18M	3K, 617K	3940s	10, 9	55s	no/–
	100×100	50; 10K	22M	15K, 3M	11640s	50, 49	84s	no/–
elevator (566)	data	2; 32	7K	137, 6K	0.06s	3, 2	0	yes/no
	data2	4; 32	220K	416, 26K	0.97s	5, 2	0.01s	yes/no
	data3	4; 200	571K	258, 10K	6.3s	5, 3	0.1s	yes/no
tsp (794)	map4	3; 15	80	6, 5	0	1, 0	0	no/–
	map14	5; 40	1.4M	18, 109	2.2s	4, 3	1.0s	yes/no
raytracer (1537)	150×150	10; 1	66	10, 90	0.02s	10,9	0.02s	yes/yes
	200×200	10; 1	66	10, 79	0.02s	10,9	0.02s	yes/yes

(CG) and summarized conflict graph (SCG) on these runs, and we report the size of each graph (in number of nodes and edges), and the time (in seconds) consumed to monitor each run (an entry of 0 means the time was less than 0.01 seconds). Note that the times are for processing the run only, and not generating the run (as that part is common to both algorithms). Finally, we report whether we found a serializability violation (yes/no), and if yes, whether it pointed to a real bug in the program (yes/no), the latter determined manually. All experiments were performed on Linux PC with two 4GHz processors and 4GB of memory.

Our results clearly illustrate the tremendous impact of using our summarized conflict-graph algorithm, giving orders of magnitude speedup when compared to the classic conflict-graph algorithm. For example, for `sor` with 50 threads, the CG-algorithm did not finish even after 8 hours while SCG finishes in 30s. This is primarily due to space savings; for example, `lufact` with 50 threads gives a conflict graph of $15K$ nodes and $3M$ edges, while the SCG graph never uses more than 50 vertices (one for each thread) and 49 edges.

Note that the algorithm reported in [25] solves a harder problem, as it tries to find serializability violations in the current run as well as all other runs that can be inferred from this run. In fact, their technique does not scale well as they use a graph similar to the conflict graph; they reported to us that they get timed-out after 25 minutes while checking the `sor` benchmark for 50 threads while we check the run in 30 seconds [24].

All the runs of `sor` and `lufact` that we monitored were serializable. `tsp` and `elevator` generated non-serializable runs. But a closer investigation of the source of non-serializability of these runs shows that they do not correspond to real errors in the program. They both refer to interesting cases of non-trivial thread interactions which are intended by the programmer, and we believe a programmer would find such reports of interactions useful. The non-serializable runs of `raytracer`, however, are related to a real bug in that program. The benchmarks, the monitoring programs, and the precise runs monitored are available at `http://www.cs.uiuc.edu/~madhu/cav08/`.

Acknowledgement. We thank Liqiang Wang and Scott Stoller for useful discussions and clarifications both theoretical and experimental.

References

1. Alur, R.: Personal communication
2. Alur, R., McMillan, K.L., Peled, D.: Model-checking of correctness conditions for concurrent objects. Information and Computation 160(1-2), 167–188 (2000)
3. Ball, T., Rajamani, S.K.: The slam project: debugging system software via static analysis. In: POPL, pp. 1–3 (2002)
4. Bernstein, P.A., Goodman, N.: Concurrency control in distributed database systems. ACM Comput. Surv. 13(2), 185–221 (1981)
5. Casanova, M.A.: Concurrency Control Problem for Database Systems. Springer, New York (1981)
6. Engler, D.R., Ashcraft, K.: Racerx: effective, static detection of race conditions and deadlocks. In: SOSP, pp. 237–252 (2003)
7. Farzan, A., Madhusudan, P.: Causal Atomicity. In: Ball, T., Jones, R.B. (eds.) CAV 2006. LNCS, vol. 4144, pp. 315–328. Springer, Heidelberg (2006)
8. Flanagan, C., Freund, S.N.: Atomizer: a dynamic atomicity checker for multithreaded programs. In: POPL 2004, pp. 256–267 (2004)
9. Flanagan, C., Qadeer, S.: Types for atomicity. In: TLDI, pp. 1–12 (2003)
10. Flanagan, C., Qadeer, S.: A type and effect system for atomicity. In: ACM SIGPPLAN PLDI 2003, pp. 338–349. ACM Press, New York (2003)
11. Fle, M.P., Roucairol, G.: On serializability of iterated transactions. In: PODC 1982: Proc. of ACM SIGACT-SIGOPS PODC, pp. 194–200. ACM Press, New York (1982)
12. Hadzilacos, T., Yannakakis, M.: Deleting completed transactions. In: ACM SIGACT-SIGMOD PODS, pp. 43–46 (1986)
13. Hatcliff, J., Robby, Dwyer, M.: Verifying atomicity specifications for concurrent object-oriented software using model checking. In: Steffen, B., Levi, G. (eds.) VMCAI 2004. LNCS, vol. 2937, pp. 175–190. Springer, Heidelberg (2004)
14. Java Grand Benchmark Suite, `http://www.javagrande.org/`
15. Lipton, R.J.: Reduction: a method of proving properties of parallel programs. Commun. ACM 18(12), 717–721 (1975)
16. Lu, S., Park, S., Seo, E., Zhou, Y.: Learning from mistakes — a comprehensive study on real world concurrency bug characteristics. In: ASPLOS (2008)
17. Lu, S., Tucek, J., Qin, F., Zhou, Y.: Avio: Detecting atomicity violations via access-interleaving invariants. In: ASPLOS (2006)
18. Naik, M., Aiken, A., Whaley, J.: Effective static race detection for java. In: Schwartzbach, M.I., Ball, T. (eds.) PLDI, pp. 308–319. ACM, New York (2006)
19. Netzer, R.H.B., Miller, B.P.: What are race conditions? some issues and formalizations. LOPLAS 1(1), 74–88 (1992)

20. Papadimitriou, C.: The theory of database concurrency control. Computer Science Press, Inc., USA (1986)
21. von Praun, C., Gross, T.R.: Object race detection. SIGPLAN Not. 36(11), 70–82 (2001)
22. Savage, S., Burrows, M., Nelson, G., Sobalvarro, P., Anderson, T.: Eraser: A dynamic data race detector for multi-threaded programs. In: SOSP (1997)
23. Shavit, N., Touitou, D.: Software transactional memory. In: PODC, pp. 204–213 (1995)
24. Wang, L., Stoller, S.D.: Personal communication
25. Wang, L., Stoller, S.D.: Accurate and efficient runtime detection of atomicity errors in concurrent programs. In: PPoPP, pp. 137–146 (2006)
26. Wang, L., Stoller, S.D.: Runtime analysis of atomicity for multi-threaded programs. IEEE Transactions on Software Engineering 32, 93–110 (2006)
27. Xu, M., Bodík, R., Hill, M.D.: A serializability violation detector for shared-memory server programs. SIGPLAN Not. 40(6), 1–14 (2005)

Dynamic Verification of MPI Programs with Reductions in Presence of Split Operations and Relaxed Orderings*

Sarvani Vakkalanka, Ganesh Gopalakrishnan, and Robert M. Kirby

School of Computing, University of Utah, Salt Lake City UT 84112, USA
http://www.cs.utah.edu/formal_verification/cav08

Abstract. Dynamic verification methods are the natural choice for debugging real world programs when model extraction and maintenance are expensive. Message passing programs written using the MPI library fall under this category. Partial order reduction can be very effective for MPI programs because for each process, all its local computational steps, as well as many of its MPI calls, commute with the corresponding steps of all other processes. However, when dependencies arise among MPI calls, they are often a function of the runtime state. While this suggests the use of dynamic partial order reduction (DPOR), three aspects of MPI make previous DPOR algorithms inapplicable: (i) many MPI calls are allowed to complete out of program order; (ii) MPI has global synchronization operations (e.g., *barrier*) that have a special weak semantics; and (iii) the runtime of MPI cannot, without intrusive modifications, be forced to pursue a specific interleaving because of MPI's liberal message matching rules, especially pertaining to 'wildcard receives'. We describe our new dynamic verification algorithm 'POE' that exploits the out of order completion semantics of MPI by delaying the issuance of MPI calls, issuing them only according to the formation of *match-sets*, which are ample 'big-step' moves. POE guarantees to manifest any feasible interleaving by dynamically rewriting wildcard receives by specific-source receives. This is the first dynamic model-checking algorithm with reductions for (a large subset of) MPI that guarantees to catch all deadlocks and local assertion violations, and is found to work well in practice.

1 Introduction

MPI [1] programs are an important class of *concurrent* programs used for the distributed programming of virtually all high performance computing clusters in the world. MPI will also be widely used for programming peta-scale supercomputers under construction [2]. Typical MPI programs are C programs (or C++/Fortran programs) that create a fixed number of processes at inception. These processes then perform computations in their private stores, invoking various flavors of send and receive API functions in the MPI library to exchange

* Supported in part by NSF award CNS00509379, Microsoft HPC Institute Program, and SRC Contract 2005-TJ-1318.

A. Gupta and S. Malik (Eds.): CAV 2008, LNCS 5123, pp. 66–79, 2008.

data, and also invoke global synchronization operations in the MPI library. Most MPI programs create processes that eventually terminate.

MPI programs can contain many types of errors, including deadlocks, local assertion violations, resource leaks, and numerical inaccuracies. The primary goal of our work is to develop efficient methods to detect deadlocks and local assertion violations in MPI programs. Dynamic verification methods are the natural choice for verifying MPI programs because model extraction and model maintenance of MPI programs can be very expensive. This paper presents the first dynamic verification algorithm called POE (Partial Order reduction avoiding Elusive interleavings) for MPI that guarantees soundness (within the practical limits of runtime verification) and employs an effective partial order reduction algorithm. Of the many features of POE, the manner in which it guarantees coverage and implements reduction during dynamic verification are our main contributions. A good partial order reduction approach is crucial for verifying MPI programs because these programs mostly perform their computations in private stores, invoking MPI operations for message exchanges, where most (but not all) of these operations commute. Also, MPI calls occur with a high static and dynamic frequency, thanks to the many `for` loops in which MPI calls occur.

In this context, our verification tool ISP that uses the POE algorithm detects deadlocks missed by existing state-of-the-art tools. While the MPI 2.0 library itself supports over 300 MPI functions, ISP can handle 24 of the most commonly used MPI functions. In this paper, we describe the handling of five of these functions, namely `MPI_Isend`, `MPI_Irecv`, `MPI_Barrier`, `MPI_Wait`, and `MPI_Test`, and refer to them as 'send, receive, barrier, wait, and test.' Send and receive are, respectively, *non-blocking* operations, meaning that the issuing process can *start* the activity and proceed to execute later instructions while the send/receive proceeds in the background. The primary arguments of send are the destination process (this may not be a compile-time constant), the data being shipped, and a 'handle.' (Note: We do not detail some of the function arguments allowed by MPI calls, such as MPI 'tags' that affect message matching. Our implementation handles all allowed MPI arguments.) The issuing process may wait on the handle or test the handle. A wait blocks till the send operation finishes, while test returns false unless the send has finished, at which time it returns true. A send is deemed to have finished when the background process of copying the message out of the memory space of the sending process has finished. The arguments of receive are the source process ID (not necessarily a compile-time constant), the data receipt buffer, and a handle (with a semantics similar to the send handle). Instead of specifying a specific source process, receive can also mention '*', which is a *wildcard* receive that is open for receipt from any send that targets the receiving process. In effect, send and receive are *split operations*.

When an MPI process invokes a sequence of MPI calls, some of the calls may complete out of program order. For instance, if a process P0 invokes two consecutive non-blocking send operations targeting P1 and P2 respectively, the second send is allowed to finish before the first one (especially if the second send is shipping a much smaller amount of data). However, if both sends target the

```
P0:  MPI_ISend(to P1, data = 22); ...rest of P0...
P1:  MPI_Irecv(*, x); if (x==22) then error1 else ...rest of P1...;
P2:  MPI_ISend(to P1, data = 33); ...rest of P2...
```

Fig. 1. Simple MPI Example Illustrating Wildcard Receives

same process (say P1), then FIFO message ordering is required. This relaxed program order of MPI facilitates higher performance.

Dynamic verification with persistent set based reductions was introduced in [5]. The dynamic partial order reduction algorithm (DPOR) [6] allows these dependencies to be accurately computed based on runtime state. This algorithm works by generating one interleaving of the program (maintained as a stack trace) and generating its interleaving variants. It ensures that the set of transitions explored from a state s forms a persistent set as follows. Consider the transitions t_i and t_j of processes p_i and p_j respectively such that $i < j$ in the current interleaving (this means that in the current interleaving t_i is executed before transition t_j). If t_i and t_j are dependent (i.e., t_i can either enable or disable t_j and vice versa), and t_i and t_j are co-enabled, then p_j is added to the pre-state of t_i hoping to eventually execute $t_j \ldots t_i$. This approach does not work with MPI, as explained with the help of a short example (Figure 1).

In this example, MPI processes `P0` and `P2` are targeting `P1` which entertains a 'wildcard match,' *i.e.*, can receive from *any* process that has a concurrently enabled `ISend` targeting `P1`. As soon as one such send is chosen (say `P0`'s), the other send is not eligible to match with this receive of `P1` (it has to match another receive of `P1` coming later). This disabling behavior of the sends induces a dependency between them, as can be seen from the fact that the particular send that matches may or may not cause `error1` to be triggered. Consider some $i < j < k$, and a trace t where the ith action of t, namely t_i, is P2's send, and similarly t_j is P1's receive, and t_k is P0's send. In this trace, it is not necessary that P0's receive is matched with P2's send just because t_i is executed before t_k. MPI implements its own buffering mechanism that can cause one send to race ahead of the other send. Formally, unlike in DPOR, MPI's program order does not imply *happens-before* [7] in an MPI program's execution. Hence, it is possible that t_j is matched with t_k. There is no way in an MPI run-time (short of making intrusive modifications to the MPI library, which is often impossible because of the proprietary nature of the libraries) to force a match either way (both sends matching the receive in turn) by just changing the order of executing sends from P2 and P0.

Roadmap: Section 2 presents an overview of POE and discusses related work. Section 3 presents POE formally. Section 4 provides a summary of experimental results. Section 5 concludes the paper.

2 Overview of POE, and Related Work

Section 2.1 presents the barrier semantics of MPI, followed by the POE algorithm. Section 2.2 presents related work.

```
P0:  S0(to P1, h0) ; B0 ;                 W(h0) ;
P1:  R (*, h1)     ; B1 ;                 W(h1) ;
P2:                  B2 ; S2(to P1, h2); W(h2) ;
```

Fig. 2. Illustration of Barrier Semantics and the POE Algorithm

2.1 Barrier Semantics and Overview of the POE Algorithm

Barrier Semantics: No MPI process can issue an instruction past its barrier unless all other processes have issued their barrier calls. Therefore, an MPI program must be designed in such a way that when an MPI process reaches a barrier call, all other MPI processes also reach their barrier calls (in the MPI parlance, these are *collective operations*); a failure to do so deadlocks the execution. While these rules match the rules followed by other languages and libraries in supporting their barrier operations, in case of MPI, it is possible for a process P_i to have an operation OP_i before its barrier call, for another process P_j to have an operation OP_j after P_j's matching barrier call, and where OP_i can observe OP_j's execution. This means that OP_i can, in effect, complete after P_i's barrier has been invoked. This shows that the program ordering from an operation to a following barrier operation need not be obeyed during execution. This is allowed in MPI (to ensure higher performance), as shown by the example in Figure 2, and requires special considerations in the design of POE. In this example, one `MPI_Isend` issued by `P0`, shown as `S0`, and another issued by `P2`, shown as `S2`, target a wildcard receive issued by `P1`[1]. The following execution is possible: (i) `S0(to P1, h0)` is issued, (ii) R(*, h1) is issued, (iii) each process *fully* executes its own barrier, (`B0`, `B1`, or `B2`), and this "collective operation" finishes (all the `B`'s indeed form an atomic set of events), (iv) `S2(to P1, h2)` is issued, (v) now both sends and the receive are alive, and hence `S0` and `S2` become dependent, requiring a dynamic algorithm to pursue both matches. Notice that `S0` can finish after `B0` and `R` can finish after `B1`. (Note: Because of the placement of this barrier that is after `P0`'s send and `P1`'s receive, but *before* `P2`'s send, we sometimes refer to such barriers as 'crooked barriers.')

To recapitulate, MPI respects program ordering between any MPI operation $x \in \{\text{barrier}, \text{wait}, \text{test}\}$ and the MPI operation immediately following x in program order. A dynamic verification algorithm for MPI must therefore maintain a *completes-before* relation $\prec$ (defined in Section 3.2), and use it to determine, at runtime, all senders that can match a wildcard receive.[2]

POE Algorithm: We now present an overview of POE, as implemented by our verification scheduler (called the POE scheduler) that can intercept MPI calls and send them into the MPI run-time as and when needed:

---

[1] While not central to our current example, we also take the opportunity to illustrate how the handles `h0` through `h2`, and `MPI_Wait` (`W`) are used.

[2] Section 3 presents another detail of MPI which we refer to as 'trumping,' captured by another relation $\prec_c$.

- The POE scheduler executes C program statements along each process. All C statements are executed in program order. When the scheduler encounters an MPI operation, it simply records this operation, but does not execute it. This process continues till the scheduler arrives, within each process, at an MPI operation that is program ordered with respect to some previously collected (but not issued) MPI operation (we call these points *fences*).
- While at a fence point for all processes, since all senders that match a wildcard receive are known, *rewrite* the receives into specific receives. In our example, `R(*)` is rewritten into `R(from P0)` and `R(from P2)`.
- Form *match-sets*. Each match-set is either a single big-step move (as in operational semantics) or a set of big-step moves. Each big-step move is a set of actions that are issued collectively into the MPI run-time by the POE scheduler (we enclose them in $\langle\langle \ldots \rangle\rangle$). In our example, the match-sets are:
  - { ⟨⟨ `S0(to P1)`, `R(from P0)` ⟩⟩, ⟨⟨ `S2(to P1)`, `R(from P2)` ⟩⟩ }
  - ⟨⟨ `B0`, `B1`, `B2` ⟩⟩
- Execute the match-sets in priority order, with all big-step moves executed first. The execution of a big-step move consists of executing all its constituent MPI operations. When no more big-step moves are left, then for each remaining set of big-step moves, recursively explore (according to depth-first search) all the big-step moves contained in it. In our example, this results in the big-step move ⟨⟨ `B0`, `B1`, `B2` ⟩⟩ from being performed first. Subsequently, both the big-step moves in
  { ⟨⟨ `S0(to P1)`, `R(from P0)` ⟩⟩, ⟨⟨ `S2(to P1)`, `R(from P2)` ⟩⟩ }
  are pursued.

Thus, one can notice that POE never actually issues into the MPI run-time any wildcard receive operations it encounters. It always dynamically rewrites these operations into receives with specific sources, and pursues each specific receive paired with the corresponding matching send as a match-set in a depth-first manner.

Additional Points About Barriers: It must be observed that the code snippet in Figure 1 can be verified with DPOR if the technique of dynamic rewriting of the wildcard receives is employed. However, the code snippet in Figure 2 cannot be verified with DPOR even with dynamic rewriting of wildcard receives employed. Due to the presence of the barrier, the send `S2` can never be executed *before* the send `S0`, whereas in DPOR, we will need dependent actions to be replayable in both orders. In any interleaving of this example, however, `S0` will always be issued before `S2`. The POE algorithm overcomes this problem by executing the big-step move ⟨⟨ `B0`, `B1`, `B2` ⟩⟩, and *then* forming the match-set { ⟨⟨ `S0(to P1)`, `R(from P0)` ⟩⟩, ⟨⟨ `S2(to P1)`, `R(from P2)` ⟩⟩ }.

2.2 Related Work

In [8], it was observed that DPOR may offer a way to determine, at runtime, which sends and receives can match in MPI programs. However, since no dynamic verification tool was built, the issues discussed in Section 1 pertaining to the

P1	P2	P3
$B_{1,1}$	$B_{2,1}$	$B_{3,1}$
$R_{1,2}(*, \langle 1,2 \rangle)$	$B_{2,2}$	$S_{3,2}(1, \langle 3,2 \rangle)$
$B_{1,3}$	$S_{2,3}(1, \langle 2,3 \rangle)$	$B_{3,3}$
$R_{1,4}(*, \langle 1,4 \rangle)$	$W_{2,4}(\langle 2,3 \rangle)$	$W_{3,4}(\langle 3,2 \rangle)$
$W_{1,5}(\langle 1,2 \rangle)$	$B_{2,5}$	$B_{3,5}$
$W_{1,6}(\langle 1,4 \rangle)$		
$B_{1,7}$		

Fig. 3. An Example MPI Program

difficulties of forcing specific send/receive matches were not faced. In [9], nothing more than the standard DPOR of [6] was needed, as we handled only some of the shared memory features of MPI for which a DPOR-like approach works. In our 2-page tools paper [10], we actually implemented DPOR for many of MPI's communication commands, and in the process observed the unsoundness resulting from our inability to force specific send/receive matches. The POE algorithm takes advantage of our formal understanding of MPI (as captured in an extensive TLA+ model for MPI we are building [11]), precisely builds the *completes-before* relation $\prec$, uses it to discover potential send/receive matches precisely, and employs dynamic rewriting to force desired matches.

While MPI-SPIN [12,13,14], which is based on SPIN [15], can detect the kinds of errors that POE can detect, this approach inherently requires major effort on the part of users in building, by hand, verification models of their MPI programs in Promela [15]. Given the extensive number of C constructs and user-level library calls used in writing many MPI programs, this effort is impractical in those cases. MPI-SPIN does provide a reduction algorithm called the *Urgent Algorithm* that allows all MPI send/receive channels to be treated as rendezvous channels. However, this algorithm applies only to programs that do not use wildcard receives (which are extensively used by many MPI program types). In general, MPI-SPIN relies on SPIN's POR algorithm which, unfortunately, does not "understand" the commuting properties of MPI calls. In its favor, MPI-SPIN supports a symbolic execution facility to compare a sequential algorithm against an MPI implementation of the algorithm to detect numerical inaccuracies - a feature not supported by ISP.

Other works [16,17,18,19] do not seem to run into the problems we run into with MPI, including out-of-order completion, barriers, split operations, or run-time scheduling realities.

The plethora of concurrency libraries catering to 'multicore programming' suggests that dealing with complex APIs will become important. Yet, most tools in this area are based on the conventional 'testing' approach. ISP can now handle 24 MPI function types (detailed on our website). We have successfully handled all 69 examples in the Umpire [4] tool distribution. These are examples for which Umpire itself, and approaches such as Jitterbug [20] do not offer coverage guarantees (conventional verification tools for MPI that we surveyed [21] are unsound). Inserting randomized 'padding' delays to potentially perturb MPI's internal schedules (as done in ConTest [22], Jitterbug, Marmot [3], and Umpire) is highly unreliable, and slows down testing by adding delays into computational paths. For instance, for many of our examples containing wildcard receives provided on our website, Marmot missed generating many feasible schedules that actually contain deadlocks.

3 Formal Presentation of POE

3.1 Abstract Syntax

Let $Nat = \{0, 1, 2 \ldots\}$, $Bool = \{0, 1\}$, and $Bool_\bot = \{0, 1, \bot\}$. Given $P \in Nat$ MPI programs, their PID ("MPI rank" of each process) set is $\{1 \ldots P\}$, and PID_* is the set $\{1 \ldots P\} \cup \{*\}$ ($*$ is to model 'wildcard receives'; see below). Let $L \in PID \rightarrow Nat$ be the lengths of the given programs, each program being viewed as a sequence of instructions. For any function f, its application to any argument i, $f(i)$, is often written f_i for brevity; for example $L(1)$ (often written L_1) is the length of the first program. Also, a function f of two arguments can be applied to two arguments i and j, written $f_{i,j}$, or partially applied to one argument i, and that is written f_i (this partial application returns a function which later "expects" a j). Let $p \in PID \rightarrow Nat \rightarrow I$ (where I is the set of MPI instructions defined in this sequel) be the programs. Thus $p_1 \ldots p_P$ are the P programs, and the jth instruction of the ith program is $p_{i,j}$. Let $l \in PID \rightarrow Nat$ be the program counters (PC) $l_1 \ldots l_P$. Let $f[i \leftarrow e]$ be function update, i.e. $f[i \leftarrow e] = (f \setminus \{\langle i, f(i) \rangle\}) \cup \{\langle i, e \rangle\}$. Also, $map\ f\ lst = \{f(i) \mid i \in lst\}$. Let $\pi_1 \langle a, b \rangle = a$. For a set of pairs S, let $f[S]$ denote function update performed for every pair in S, i.e., $f[S] = (f \setminus \{\langle i, f(i) \rangle \mid i \in (map\ \pi_1 S)\}) \cup S$.

Let $h \in PID \rightarrow Nat \rightarrow Bool_\bot$ be the handles $h_1 \ldots h_P$. In our formal model, *every* instruction has a handle; it is only the case that W and T (MPI wait and test instructions defined in this sequel) happen to use this handle in a specific way. Handle $h_{i,j}$ is initially $\bot$. In our description of POE, we use the setting of $h_{i,j}$ to 0 to model POE *encountering* (collecting) instruction $any_{i,j}(\ldots)$ in program order, and the setting to 1 to model POE *issuing* (executing) this instruction. POE will (i) set $h_{i,j}$ to 1 out of program order (but still correctly so according to $\prec$), and (ii) dynamically rewrite the wildcards before forming match-sets and executing them. The **total system state** is $\langle l, h \rangle$ (we keep track of the PC values and the handle array status).

The set of MPI instructions I is the smallest set that include the following: *Barrier*, written $B_{i,j}$, *Send*, written $S_{i,j}(k, \langle i, j \rangle)$, where $k \in PID$ is the process targeted, and $\langle i, j \rangle$ is the handle used to track the progress of this *Send*, *Receive*, written $R_{i,j}(k, \langle i, j \rangle)$ where $k \in PID_*$ is the process from which the message is sourced ($*$ means 'wildcard receive,' i.e., the message is sourced from any process), and $\langle i, j \rangle$ is the handle (as with send) to track the progress of this *Receive*. We do not show the data payloads for sends S and receives R; when needed in discussions, they will be shown as a third argument. For S (send) and R (receive), their handle $\langle i, j \rangle$ is used by a following W instruction, or tested by a following T instruction (not required to exist by the MPI standard, and we also do not require the W/T to exist). I also includes *Wait*, written $W_{i,j}(\langle m,n\rangle)$ where $\langle m,n\rangle$ refers to a handle. $W_{i,j}(\langle m,n\rangle)$ blocks till $h_{m,n}$ is set to 1. This event occurs when the instruction which set $h_{m,n}$ to 0 finishes. (This earlier instruction is an S or R.) I also includes *Test*, written $T_{i,j}(\langle m,n\rangle, l)$ where $\langle m,n\rangle$ refers to a handle and l is a PC. $T_{i,j}(\langle m,n\rangle, l)$ blocks till $h_{m,n}$ is set to 1, and this occurs when the instruction that set $h_{m,n}$ to 0 (an earlier S or R) finishes, in which case the control transfers to the new PC l. Finally, I includes a conditional goto to model loops (space prevents further discussion of goto and T).

Figure 3 illustrates our syntax. Process P1 has seven sequential commands, and P2 and P3 each have five each. All proper MPI programs start with `MPI_INIT`, and *terminate* with `MPI_FINALIZE`, and both these essentially have the semantics of a barrier. Thus, the set $B_{1,1}$, $B_{2,1}$, and $B_{3,1}$ models `MPI_INIT`. Likewise, the set $B_{1,7}$, $B_{2,5}$, and $B_{3,5}$ models `MPI_FINALIZE`. The set $B_{1,3}$, $B_{2,2}$, and $B_{3,3}$ is a 'crooked barrier'. Thus, notice that the two sends $S_{2,3}(1, \langle 2,3\rangle)$ and $S_{3,2}(1, \langle 3,2\rangle)$ both target P1, and they can *both* potentially match with $R_{1,2}(*, \langle 1,2\rangle)$.

Illustration: In this example, if $R_{1,4}(*, \langle 1,4\rangle)$ is changed to $R_{1,4}(2, \langle 1,4\rangle)$, it is possible that $R_{1,2}(*, \langle 1,2\rangle)$ matches $S_{2,3}(1, \langle 2,3\rangle)$, and then $S_{3,2}(1, \langle 3,2\rangle)$ cannot match $R_{1,4}(2, \langle 1,4\rangle)$ (this receive expects a message from P2, not P3). This results in a *deadlock*. Such deadlocks cannot be detected through static analysis alone, because in MPI, send targets (i.e., the 1 in $S_{3,2}(1, \langle 3,2\rangle)$) and receive sources can be computed at runtime.

3.2 Completes-before Relation of MPI

MPI guarantees process-pair-wise message delivery ordering with respect to the *issue* orders of sends and receives. To illustrate this idea, consider two sends that are issued by process i both targeting process j, and two matching receives that are issued by process j, hoping to source from i. These sends and receives must be carried out in program order. It is only when send operations target receive operations in *different* processes, or receive operations source from different processes, that program order can be relaxed.

Specifically, suppose process i has a send $S_{i,m}(j, \langle i,m\rangle, d_1)$, and another send $S_{i,n}(j, \langle i,j\rangle, d_2)$, for $n > m$. Here, d_1 and d_2 are the data payloads. Suppose process j has a receive $R_{j,u}(i, \langle j,u\rangle, x_1)$, and another receive $R_{j,v}(i, \langle j,v\rangle, x_2)$, for $v > u$. Here, x_1 and x_2 are j's receive buffers, MPI guarantees FIFO message ordering and ensure that x_1 is bound to d_1 and x_2 to d_2 during execution. The POE algorithm must never issue these sends and receives out of order. In fact, the POE algorithm can 'fire and forget' these operations in program order, and be guaranteed that the MPI runtime will *match them in this appropriate order*.

Now consider a slightly different example where there are three processes i, j, and k in the system. The receives are $R_{j,u}(k, \langle j,u\rangle, x_1)$ and $R_{j,v}(*, \langle j,v\rangle, x_2)$, where $k \neq i$, and furthermore, let process k never issue a send to process j. In this case, the first receive (which cannot match any of the offers made by i) will be *trumped* by the second receive, which now goes ahead; the result will be that x_2 is bound to d_1. The POE algorithm has to be aware of this 'trumping rule.'

A third variant of our example is one where the sends are as above, the receives are $R_{j,u}(k, \langle j,u\rangle, x_1)$ and $R_{j,v}(*, \langle j,v\rangle, x_2)$, where $k \neq i$, but now there is a third process k which issues a send, $S_{k,l}(j, \langle k,l\rangle, d_3)$. Now, $R_{j,v}(*, \langle j,v\rangle, x_2)$ does not trump. The receive $R_{j,u}(k, \langle j,u\rangle, x_1)$ can indeed match the new send $S_{k,l}(j, \langle k,l\rangle, d_3)$, thus binding x_1 to d_3, and x_2 to d_1. POE has to be aware of this lack of trumping, as well. Thus, we note that when the sequence $R_{j,u}(k, \langle j,u\rangle, x_1)$; $\ldots$ $R_{j,v}(*, \langle j,v\rangle, x_2)$ appears in process j, the second receive can *conditionally complete* before the first one, in a manner that depends on the runtime state of the system.

We now define the **completes-before** relation, $\prec$. The POE algorithm presented in Section 3.4 will be based on $\prec$. A variant of $\prec$ called *conditionally completes* ($\prec_c$) is used to model the concept of trumping discussed earlier. We do not discuss $\prec_c$ any more in this paper, for the sake of simplicity (it is of course incorporated into our implementation of POE, in forming match-sets according to $\prec_c$).

$$
\begin{array}{l}
\forall i, j_1, j_2, k:\ j_1 < j_2 \Rightarrow S_{i,j_1}(k,\ldots) \prec S_{i,j_2}(k,\ldots)\\
\forall i, j_1, j_2, k:\ j_1 < j_2 \Rightarrow R_{i,j_1}(k,\ldots) \prec R_{i,j_2}(k,\ldots)\\
\forall i, j_1, j_2, k:\ j_1 < j_2 \Rightarrow R_{i,j_1}(*,\ldots) \prec R_{i,j_2}(k,\ldots)\\
\forall i, j_1, j_2:\ j_1 < j_2 \Rightarrow R_{i,j_1}(*,\ldots) \prec R_{i,j_2}(*,\ldots)\\
\forall i, j_1, j_2, k:\ j_1 < j_2 \Rightarrow S_{i,j_1}(k, \langle i, j_1\rangle) \prec W_{i,j_2}(\langle i, j_1\rangle)\\
\forall i, j_1, j_2, k:\ j_1 < j_2 \Rightarrow R_{i,j_1}(k, \langle i, j_1\rangle) \prec W_{i,j_2}(\langle i, j_1\rangle)\\
\forall i, j_1, j_2:\ j_1 < j_2 \Rightarrow B_{i,j_1} \prec any_{i,j_2}(\ldots)\\
\forall i, j_1, j_2:\ j_1 < j_2 \Rightarrow W_{i,j_1}(\ldots) \prec any_{i,j_2}(\ldots)
\end{array}
$$

FIFO Lemma: Any MPI program execution respecting $\prec^*$, the transitive closure of $\prec$, guarantees the required FIFO message orderings between MPI processes.

3.3 Match-Set Formation

Fence Instructions: For an instruction $j \in I$, $fence(j)$ holds exactly when for all succeeding instructions $k \in I$ in program order, $j \prec^* k$. Notice that 'wait' and 'barrier' act as fences, and depending on the MPI program, other instructions may attain a fence status.

Ancestor Relation: The *ancestor* of an instruction i is some instruction j where $j \prec^* i$. The set $ancestors(i)$ is the set of *indices* of i's ancestors. To exploit the FIFO Lemma, POE issues instruction i to the MPI system only after all its ancestors j have been issued. POE can issue any instruction not connected by $\prec^*$ out of order, as the MPI system itself considers such instructions semantically unordered (and hence may reorder them).

Match-set definitions: We now define all match-set types. The match-set type MS_R^* will be a set of big-step moves. The match-set type MS_B will be one big-step move containing all the matching barriers. Match-set type MS_R will contain exactly one send $S_{i,u}(j,\ldots)$, and its matching non wild-card receive $S_{j,v}(i,\ldots)$). Match-set type MS_W will be a big-step move of exactly one wait, and MS_T will be a big-step move of exactly one test. Consider the big-step moves $\langle\langle\ldots\rangle\rangle$ themselves to be sets.

The main difficulty in forming match-sets is to determine which sends can match a wildcard receive. To compute MS_R^*, we start with a set containing just the wildcard receive in question. We then seek the maximal number of additional sends that we can add to this set, without hitting a fence. Finally we break $*$ into specific instances of PIDs. We also must make sure that for the members of any MS, all its *ancestors* have been *issued* into the MPI system. Modeling this requires the state of the h array.

Formal Definition of $MS(l,h)$: We define match-sets as a function of l (the array of PCs) and h (the array of handles). In our definitions, we often refer to a "band" of past PC values where the MS might lie; this is what the function ρ used below denotes:

$$\begin{aligned}
MS_B(l,h) = &\ if\ \exists\rho \in PID \rightarrow Nat\ :\ \forall x \in PID : 1 \le \rho_x \le l_x \\
&\wedge \forall k \in PID:\ p_{k,\rho_k} = B_{k,\rho_k}\ \wedge\ \forall u \in ancestors(p_{k,\rho_k}) : h_{k,u} = 1 \\
&\wedge h_{k,\rho_k} = 0\ \ then\ \langle\langle B_{k,\rho_k}\ \mid\ k \in PID\rangle\rangle\ \ else\ \emptyset\ .
\end{aligned}$$

$$\begin{aligned}
MS_R^*(l,h) = &\ if\ \exists\rho \in PID \rightarrow Nat\ s.t.\ \forall x \in PID : 1 \le \rho_x \le l_x \\
&\wedge \exists i \in PID : p_{i,\rho_i} = R_{i,\rho_i}(*,\ldots)\ \wedge\ \forall u \in ancestors(p_{i,\rho_i}) :\ h_{i,u} = 1 \\
&\wedge h_{i,\rho_i} = 0 \\
&\wedge \forall k \in PID \setminus \{i\} : p_{k,\rho_k} = S_{k,\rho_k}(i,\ldots)\ \wedge\ \forall u \in ancestors(p_{k,\rho_k}) : h_{k,u} = 1 \\
&\wedge h_{k,\rho_k} = 0 \\
&then\ \{\ \langle\langle R_{i,\rho_i}(k,\ldots), S_{k,\rho_k}(i,\ldots)\rangle\rangle\ \mid\ k \in PID \setminus \{i\}\}\ \ else\ \{\emptyset\}\ .
\end{aligned}$$

$$\begin{aligned}
MS_R(l,h) = &\ if\ \exists\rho \in PID \rightarrow Nat\ s.t.\ \forall x \in PID : 1 \le \rho_x \le l_x \\
&\wedge \exists i,j \in PID : p_{i,\rho_i} = R_{i,\rho_i}(j,\ldots)\ \wedge\ p_{j,\rho_j} = S_{j,\rho_j}(i,\ldots) \\
&\wedge \forall u \in ancestors(p_{i,\rho_i}) : h_{i,u} = 1\ \wedge\ \forall u \in ancestors(p_{j,\rho_j}) : h_{j,u} = 1 \\
&\wedge h_{i,\rho_i} = h_{j,\rho_j} = 0 \\
&then\ \langle\langle R_{i,\rho_i}(j,\ldots), S_{j,\rho_j}(i,\ldots)\rangle\rangle\ else\ \emptyset\ .
\end{aligned}$$

$$\begin{aligned}
MS_W(l,h) = &\ if\ \exists i \in PID, 1 \le j,k \le l_i, k < j : p_{i,j} = W_{i,j}(\langle i,k\rangle) \\
&\wedge h_{i,j} = 0 \wedge h_{i,k} = 1 \wedge \forall u \in ancestors(p_{i,j}) : h_{i,u} = 1 \\
&then\ \langle\langle W_{i,j}(\langle i,k\rangle)\rangle\rangle\ else\ \emptyset\ .
\end{aligned}$$

Priority Scheme: Let $MS(l,h)$ be an abbreviation for invoking $MS_B(l,h)$, $MS_R(l,h)$, and $MS_W(l,h)$ in some order. If this invocation returns $\emptyset$, we will explicitly invoke $MS_R^*(l,h)$ and pursue the contents of this set, if any. The above is the **priority search** scheme that POE uses (postpone wildcard receives until all senders are discovered).

3.4 The POE Algorithm

We present the transition relation as an inference system which infers new states. Let $\langle l,h\rangle \in Rch$ mean that the state $\langle l,h\rangle$ has been reached. We invariantly maintain that $h_{i,l_i} = 0$. In the following, $h_{i,j}$ is set to 1 only by match-set moves. Non-MS moves are PC advances, and they result only in $h_{i,j}$ being set to 0 (the instruction is encountered but not issued). For a process i, a PC advance move is permitted if the instruction at its current PC is not a fence, or if the instruction has been issued (handle is set). The atomic transitions are the one of the $MS(l,h)$ moves, a PC move, or all the moves within $MS^*_R(l,h)$. Also $move(l,h,R)$ takes a system state $\langle l,h\rangle$, an atomic transition (set of instructions) R, sets the handle bits at the indices of the instruction. It does not advance the PC, as that will be done by the 'PC move' transition. Formally, let $\alpha \in I \rightarrow PID$ and $\beta \in I \rightarrow Nat$ be such that for instruction $r \in I$, $r = p_{\alpha(r),\beta(r)}$. Then, $move(l,h,R) = \langle l,\ h[\{\langle\langle\alpha(r),\beta(r)\rangle,1\rangle \mid r \in R\}]\rangle$.

Init: $\langle l_0,h_0\rangle \in Rch$, where $l_0 = \lambda i.1$ and $h = (\lambda ij.if\ j = 1\ then\ 0\ else\ \bot)$.

Step: for $\langle l,h\rangle \in Rch$
// All the deterministic **singleton ample-set** *moves*
if $MS(l,h) \neq \emptyset$ **then** $move(l,h,MS(l,h)) \in Rch$
// PC move which is also a **singleton ample-set** *move*
elseif $\exists i \in PID : \neg fence(p_{i,l_i}) \vee (h_{i,l_i} = 1)$
  then $\langle l[i \leftarrow l_i + 1], h_i[(l_i + 1) \leftarrow 0]\rangle \in Rch$
// Recursive exploration upon dependency. **Ample = enabled.**
elseif $MS^*_R(l,h) \neq \{\emptyset\}$ **then** $(map\ (\lambda r.move(l,h,r))\ (MS^*_R(l,h))) \subseteq Rch$
else *Deadlocked.*

Illustration of POE: POE will form match-sets (MS) from only those instructions that have a handle value of 0. In system state $\langle l,h\rangle$, if there exists a MS other than MS^*_R (will be a subset of I), POE picks any such set and invokes its operations (sets $h_{i,j}$ for that instruction to 1). MS^*_R is a set of subsets of I, and POE recursively invokes each member set in any order (in the implementation, these are backtrack points). If no MS can be built in the current system state, if possible, POE advances the PC l_i of some process i; else, the system is deadlocked. In our example (Figure 3), the first MS will be $\langle\langle B_{1,1}, B_{2,1}, B_{3,1}\rangle\rangle$, and these barrier calls are issued, setting $h_{1,1}, h_{2,1}$ and $h_{3,1}$ to 1. When $R_{1,2}(*, \langle 1,2\rangle)$ from P1 is encountered, $h_{1,2}$ is set to 0 (instruction encountered, but recorded for future issue). Likewise, from P3, we encounter $S_{3,2}(1, \langle 3,2\rangle)$, and set $h_{3,2} = 0$; we do not issue this send, as we have not carved out the maximal MS and we have not hit a fence. The system advances the PCs, finds the next MS $\langle\langle B_{1,3}, B_{2,2}, B_{3,3}\rangle\rangle$, and invokes it, setting the handle bits to 1. Following this, it will encounter $S_{2,3}(1, \langle 2,3\rangle)$, setting $h_{2,3} = 0$. At this point, further PC advancement will place P1's PC facing $R_{1,4}(*, \ldots)$, which is $\prec$ ordered after $R_{1,2}(*, \ldots)$, and hence serves as a fence within P1. Now P2 encounters fence $W_{2,4}$, and P3 encounters fence $W_{3,4}(\langle 3,2\rangle)$. At this point, the set $\langle\langle S_{2,3}(1, \langle 2,3\rangle), S_{3,2}(1, \langle 3,2\rangle), R_{1,2}(*, \langle 1,2\rangle)\rangle\rangle$ is *promoted to an MS status.* The dynamic rewriting process produces two MSs (actually a set containing two MS sets) $\langle\langle R_{1,2}(2, \langle 1,2\rangle), S_{2,3}(1, \langle 2,3\rangle)\rangle\rangle$ and $\langle\langle R_{1,2}(2, \langle 1,2\rangle), S_{3,2}(1, \langle 3,2\rangle)\rangle\rangle$, and recursively invokes POE with these MSs. When the last MS-B is encountered, this corresponds to `MPI_FINALIZE`. At this time, if any handle is still a 0, and no more MS remains, an invalid end-state error is reported. In this example, no deadlock is encountered.

Correctness of POE: The correctness of POE consists of two steps. First, we must ensure that we abide by the FIFO Lemma in all scheduling decisions. This follows from POE never issuing actions contrary to $\prec$. However, whenever

$\prec$ does not hold, POE may issue actions out of order. Second, we must ensure that we are executing according to conditions C0-C2 ([23]) of a correct partial order reduction algorithm (we do not require C3 owing to the acyclicity of MPI's state space). C2 is satisfied because local assertions only observe local process steps which are singleton ample. The priority scheme on Page 75 ensures that all singleton ample-sets contributed to by match-sets other than MS_R^* are exhausted. These preserve C1. Finally, the dependencies among the sends targeting a wildcard receive are correctly handled by doing a full recursive expansion of the constituents of MS_R^*, which also preserves C1.

4 Summary of Experimental Results

We have implemented the POE algorithm in our ISP runtime model-checker for MPI that is downloadable, along with our examples, from our website. A summary of our results is as follows:

- In all the 69 examples from the Umpire test suite, ISP produces the same theoretical number of interleavings required by our formal algorithm. This number is far smaller than the number of interleavings without reduction.
- Existing MPI program testing approaches (e.g., Umpire, Marmot) cannot detect deadlocks with assurance on many simple examples. In all these cases, the POE algorithm detects the deadlocks (see our webpage for the results).
- For some examples with several hundreds of lines of code that have no wildcard receives (where the code checks for local assertions), POE requires exactly *one* interleaving. Existing testing tools will wastefully explore multiple interleavings where the MPI operations have no dependencies.
- POE's setting of handle bits turns into collecting MPI operations without issuing them. These book-keeping steps of ISP have negligible overheads. The main overhead of ISP is that of restarting MPI for each replay. In [10], we provide techniques that can dramatically reduce this overhead. This technique will be integrated into our current ISP version.
- ISP supports 24 MPI functions, including many collective operations, MPI communicators, and non-deterministic wait functions such as `MPI_WAIT_ANY`. However, in a significant number of cases, we can allow an MPI program to issue operations even outside of this set. These extra functions (such as `MPI_TYPE_CREATE`) can still be issued into the MPI run-time without being trapped by the verification scheduler of POE.
- POE's scheduler is designed to be parallelized using MPI in future versions of ISP. Also a static analysis package to remove computations that do not affect control flow has been prototyped and will be integrated into ISP.

5 Concluding Remarks

We have described an algorithm for handling out of order execution and barrier semantics in verifying MPI programs for deadlocks and local assertions. We emphasize that POE works on unaltered MPI source programs. The verification tool implementing POE works well in practice, and is sound within the practical limits of runtime verification. An example of such a limitation is captured by the `MPI_Test` function. The outcome of `MPI_Test` (true or false) depends on the speed of computation of MPI processes. It is possible for a given MPI runtime to always produce the `true` outcome, for example. We are investigating the modification of the open source MPICH 2.0 library to overcome such limitations.

We have a formal TLA+ model of MPI 2.0 [11] (and we even have an execution framework that takes short MPI programs and runs them against this semantics [24]), we are in a position to rigorously prove the MPI semantics described in this paper.

Acknowledgements. The authors wish to thank Rajeev Thakur of Argonne National Labs and Bill Gropp of UIUC for their ideas and encouragement.

References

1. Snir, M., Otto, S.: MPI-The Complete Reference: The MPI Core. MIT Press, Cambridge (1998)
2. Invited Talk by Al Geist at EuroPVM/MPI 2007, Sustained Petascale: The Next MPI Challenge
3. Krammer, B., Bidmon, K., Müller, M.S., Resch, M.M.: Marmot: An MPI analysis and checking tool. In: Parallel Computing 2003 (September 2003)
4. Vetter, J.S., de Supinski, B.R.: Dynamic software testing of MPI applications with Umpire. In: Supercomputing, pp. 70–79 (2000)
5. Godefroid, P.: Model checking for programming languages using verisoft. In: POPL, pp. 174–186 (1997)
6. Flanagan, C., Godefroid, P.: Dynamic partial-order reduction for model checking software. In: Palsberg, J., Abadi, M. (eds.) POPL, pp. 110–121. ACM, New York (2005)
7. Lamport, L.: Time, clocks and the ordering of events in a distributed system. Communications of the ACM 21(7), 558–565 (1978)
8. Palmer, R., Gopalakrishnan, G., Kirby, R.M.: Semantics driven dynamic partial-order reduction of MPI-based parallel programs. In: Parallel and Distributed Systems - Testing and Debugging (PADTAD-V) (July 2007)
9. Pervez, S., Palmer, R., Gopalakrishnan, G., Kirby, R.M., Thakur, R., Gropp, W.: Practical model checking method for verifying correctness of MPI programs. In: EuroPVM/MPI, pp. 344–353 (2007)
10. Vakkalanka, S., Sharma, S.V., Gopalakrishnan, G., Kirby, R.M.: ISP: A tool for model checking MPI programs. In: Principles and Practices of Parallel Programming (PPoPP), pp. 285–286 (2008)
11. Li, G., DeLisi, M., Gopalakrishnan, G., Kirby, R.M.: Formal specification of the MPI-2.0 standard in TLA+. In: Principles and Practices of Parallel Programming (PPoPP), pp. 283–284 (2008)
12. Siegel, S.F.: Efficient Verification of Halting Properties for MPI Programs with Wildcard Receives. In: Cousot, R. (ed.) VMCAI 2005. LNCS, vol. 3385, pp. 413–429. Springer, Heidelberg (2005)

13. Siegel, S.F., Avrunin, G.S.: Modeling Wildcard-free MPI Programs for Verification. In: Proceedings of the ACM SIGPLAN Symposium on Principles and Practices of Parallel Programming (to appear, 2005)
14. Siegel, S.F., Avrunin, G.S.: Verification of MPI-based software for scientific computation. In: Proceedings of the 11th International SPIN Workshop on Model Checking Software, Barcelona, April 2004, pp. 286–303 (2004)
15. Holzmann, G.J.: The Spin Model Checker: Primer and Reference Manual. Addison-Wesley, Reading (2004)
16. Yang, Y., Chen, X., Gopalakrishnan, G., Kirby, R.M.: UUCS-07-008:Runtime Model Checking of Multithreaded C/C++ Programs. Technical report, University of Utah, School of Computing (2007), `http://www.cs.utah.edu/research/techreports/2007/ps/UUCS-07-008.ps`
17. Yang, Y., Chen, X., Gopalakrishnan, G., Kirby, R.M.: Distributed dynamic partial order reduction based verification of threaded software. In: Bošnački, D., Edelkamp, S. (eds.) SPIN 2007. LNCS, vol. 4595, pp. 58–75. Springer, Heidelberg (2007)
18. Musuvathi, M., Qadeer, S.: Iterative context bounding for systematic testing of multithreaded programs. In: Proceedings of the ACM SIGPLAN Conference on Programming Language Design and Implementation, pp. 446–455 (2007)
19. `http://research.microsoft.com/projects/CHESS/`
20. Vuduc, R., Schulz, M., Quinlan, D., de Supinski, B., Saebjornsen, A.: Improved distributed memory applications testing by message perturbation. In: Parallel and Distributed Systems: Testing and Debugging (PADTAD - IV) (2006)
21. Sharma, S.V., Gopalakrishnan, G., Kirby, R.M.: A survey of MPI related debuggers and tools. Technical Report UUCS-07-015, University of Utah, School of Computing (2007), `http://www.cs.utah.edu/research/techreports.shtml`
22. Edelstein, O., Farchi, E., Goldin, E., Nir, Y., Ratsaby, G., Ur, S.: Framework for testing multi-threaded Java programs. Concurrency and Computation: Practice and Experience 15(3-5), 485–499 (2003)
23. Clarke, E.M., Grumberg, O., Peled, D.A.: Model Checking. MIT Press, Cambridge (2000)
24. Palmer, R., Delisi, M., Gopalakrishnan, G., Kirby, R.M.: An approach to formalization and analysis of message passing libraries. In: Formal Methods for Industry Critical Systems (FMICS), Berlin (2007)
25. DeSouza, J., Kuhn, B., de Supinski, B.R., Samofalov, V., Zheltov, S., Bratanov, S.: Automated, scalable debugging of MPI programs with intel message checker. In: SE-HPCS 2005, pp. 78–82 (2005)

A Hybrid Type System for Lock-Freedom of Mobile Processes

Naoki Kobayashi[1] and Davide Sangiorgi[2]

[1] Tohoku University
[2] Università di Bologna

Abstract. We propose a type system for lock-freedom in the π-calculus, which guarantees that certain communications will eventually succeed. Distinguishing features of our type system are: it can verify lock-freedom of concurrent programs that have sophisticated recursive communication structures; it can be fully automated; it is hybrid, in that it combines a type system for lock-freedom with local reasoning about deadlock-freedom, termination, and confluence analyses. Moreover, the type system is parameterized by deadlock-freedom/termination/confluence analyses, so that any methods (e.g. type systems and model checking) can be used for those analyses. A lock-freedom analysis tool has been implemented based on the proposed type system, and tested for non-trivial programs.

1 Introduction

In this paper, we attack the problem of verifying concurrent programs that create threads and communication channels dynamically. More specifically, we choose the π-calculus [16] as the target language, and consider the problem of verifying the lock-freedom property, which intuitively means that certain communications (or synchronizations) will eventually succeed (possibly under some fairness assumption). Lock-freedom is important for communication-centric computation models like the π-calculus; indeed, in the pure π-calculus, most liveness properties can be turned into the lock-freedom property. For example, the following properties can be reduced to instances of lock-freedom: Will the request of accessing a resource be eventually granted? In a client-server system, will a client request be eventually received from the server? And if so, will the server eventually send back an answer to the client? In multi-threaded programs, can a thread eventually acquire a lock? And if so, will the thread eventually release the lock? The lock-freedom property has also applications to other verification problems and program transformation, such as information flow analysis and program slicing (dependency analysis in general). Verification of liveness properties such as lock-freedom is notoriously hard in concurrency. In formalisms for mobile processes, such as the π-calculus, it is even harder, because of dynamic creation of threads and first-class channels. In these formalisms, *type systems* have emerged as a powerful means for disciplining and controlling the behaviors of the processes.

A. Gupta and S. Malik (Eds.): CAV 2008, LNCS 5123, pp. 80–93, 2008.

Type systems for lock-freedom include [1,8,9,20,21]. An automatic verification tool, TyPiCal [10], has been implemented based on Kobayashi's system [9]. The expressive power of such type systems is, however, very limited. This shows up clearly, for instance, in the treatment of recursion. For example, even primitive recursive functions cannot be expressed in Kobayashi's lock-free type system, since it ignores value-dependent behaviors completely.

In this paper, we tackle lock-freedom by pursuing a different route. We overcome limitations of previous type systems by combining the lock-freedom analysis with two other analysis: *deadlock-freedom* and *termination*. The result, therefore, is not a "pure" type system, but one that is *parametric* in the techniques employed to ensure deadlock-freedom and termination. Such techniques may themselves be based on type systems (and indeed in the paper we indicate such type systems, or develop them when needed), but could also use other methods (model checking, theorem provers, etc.). The parameterization allows us to go beyond certain limits of type systems, by appealing to other methods. For instance, a type system, as a form of static analysis, may have difficulties in handling value-dependent behaviors (even very simple ones), which are more easily dealt with by other methods such as model checking.

Roughly, we use the deadlock-freedom analysis to ensure that a system can reduce if some of its expected communications have not yet occurred. We then apply a termination analysis to discharge the possibility of divergence and guarantee lock-freedom (i.e., the expected communication will indeed occur). The reasons for appealing to deadlock-freedom are that powerful type-based analyzers exist (notably Kobayashi's systems [11]), and that deadlock-freedom is a safety property, which is easier than liveness to verify in other verification methods such as model checking.

A major challenge was to be able to apply the deadlock and termination analysis *locally*, to subsystems of larger systems. The local reasoning is particularly important for termination. A result forcing a global termination analysis would not be very useful in practice: first, valid concurrent programs may not terminate (e.g., operating systems); second, even if a program is terminating, it can be extremely hard to verify it if the program is large, particularly in languages for mobile processes such as the π-calculus that subsume higher-order formalisms such as the λ-calculus.

Very approximately, our hybrid rule for local reasoning looks as follows:

$$\frac{\models_{\texttt{DF}} P \qquad \models_{\texttt{Ter}} P}{\Delta \vdash_{\mathrm{LT}} P} \qquad (*)$$

where $\models_{\texttt{DF}} P$ and $\models_{\texttt{Ter}} P$ indicate, respectively, that P is deadlock-free and terminating, and $\Delta \vdash_{\mathrm{LT}} P$ is a typing judgment for lock-freedom. The type environment Δ captures conditions, or "contracts", on the way P interacts with its environment, of the kind "P will eventually send a message on a" and "if P receives a message on a, then P is lock-free afterwards". Such contracts are necessary for the compositionality of the type system for lock-freedom (i.e., local reasoning on lock-freedom). We use Kobayashi's lock freedom types [9], which refine those of the simply-typed π-calculus with *channel usages*, to express the contracts. Therefore we add rule (∗), as an "axiom", to the rules of Kobayashi's lock freedom type system [9].

The contracts in Δ, however, are completely ignored—and are not guaranteed—in the premises of rule (∗). As a consequence, the resulting type system is unsound. In other words, knowing that P is deadlock-free and terminating is not sufficient to guarantee compositionality and local reasoning. As an example of missing information, P being terminating ensures that P itself has no infinite reductions; but it says nothing on the behaviour of P after it receives a message from other components in the system. (Indeed rule (∗) is only sound if applied globally, to the whole system.)

The first refinement we make for the soundness of rule (∗) is to replace deadlock-freedom and termination with more robust notions, which we call, respectively, *robust deadlock-freedom under* Δ, written $\Delta \models_{\mathtt{RD}} P$, and *robust termination*, written $\models_{\mathtt{RTer}} P$. These stronger notions approximately mean that P is deadlock-free or terminating after any substitution (P may be open, and therefore contain free variables), and any interaction with its environment; $\Delta \models_{\mathtt{RD}} P$ further ensures that P fulfills certain obligations in Δ. The problems of verifying robust deadlock-freedom and robust termination are more challenging than the ordinary ones, due to the additional requirements (e.g., quantifications over substitutions and transition sequences). Existing type systems for deadlock-freedom, notably [11], do meet however the extra conditions for robust deadlock-freedom. We also show how to tune type systems for ordinary termination in a generic manner so to guarantee the stronger property of robust termination. We should stress nevertheless that $\Delta \models_{\mathtt{RD}} P$ and $\models_{\mathtt{RTer}} P$ are semantic requirements: our type system is parametric on the verification methods that guarantee them—one need not employ type systems.

Even with the above refinement of the deadlock-freedom and termination conditions, the hybrid rule (∗) remains unsound. The reason is, roughly, the same as why assume-guarantee reasoning in concurrency often fails in the presence of circularity. In fact, the judgment $\Delta \vdash_{\mathrm{LT}} P$ can be considered a kind of assume-guarantee reasoning, where Δ expresses both assumptions on the environment and guarantees about P's behavior. To prevent circular reasoning, we add a condition $nocap(\Delta)$ that intuitively ensures us that P is independent of its environment, in the sense that P will fulfill its obligations (to perform certain input/output actions) without relying on its environment's behavior. (For example, suppose that there is an obligation to send a message on channel a. The process $\overline{a}[1]$, which sends 1 on a, is fine, since it fulfills the obligation with no assumption. On the other hand, the process $b(x).\,\overline{a}[x]$, which waits to receive a value on b before sending x on a, is not allowed since it fulfills the obligation only *on the assumption* that the environment will send a message on b.) This leads to the following hybrid rule:

$$\frac{\Delta \models_{\mathtt{RD}} P \qquad \models_{\mathtt{RTer}} P \qquad nocap(\Delta)}{\Delta \vdash_{\mathrm{LT}} P} \qquad \text{(LT-HYB)}$$

The resulting type system guarantees that any well-typed process P is *weakly lock-free*, in the sense that if an input/output action is declared in P as an action that should succeed, and if $P \longrightarrow^* Q$, then the action has already succeeded in $P \longrightarrow^* Q$ or there is a further reduction sequence from Q in which the action will succeed. This is similar to the way in which success of passing a test is defined in fair should/must testing [4] and bisimulation, (and also in accordance with other definitions of similar forms of liveness for π-calculus such as [20]).

We have also considered a stronger form of lock-freedom, guaranteeing that the marked actions will eventually succeed on the assumption that the scheduler is strongly fair. We show that our type system can be strengthened to guarantee the strong lock-freedom by adding a condition of partial confluence to rule LT-HYB above. Again, the partial confluence is only required locally; the whole program need not be confluent.

The verification framework outlined above for lock-freedom (including an automated robust termination analysis) has been implemented as an extension of TYPICAL program analysis tool (except the extension to strong lock-freedom; adding this on top of the present implementation would be tedious but not difficult). We have succeeded in automatically verifying several non-trivial programs, such as symbol tables and binary tree search. These examples are non-trivial because lists and trees are implented as networks of processes connected by channels, and they grow dynamically (both channels and processes are dynamically created and linked). Recursive structures of the kind illustrated in these examples are common in programming languages for mobile processes (the examples in fact, were taken or inspired from Pict programs [15]).

2 Target Language

Syntax. We write $\mathcal{L}$ for the set of *links* (also called *channels*), and $\mathcal{V}$ for the (disjoint) set of *variables*. We use meta-variables $a, b, c, \ldots$ and $x, y, z, \ldots$ for links and variables, respectively. We write $\mathcal{N}$ for the set $\mathcal{L} \cup \mathcal{V} \cup \{\texttt{true}, \texttt{false}\}$ of *names* (sometimes called *values*), where `true` and `false` are the usual boolean values. We use meta-variables u, v, w for names. The grammar is the following:

$$P ::= \mathbf{0} \mid \overline{v}^{\chi}[\widetilde{w}].\,P \mid v^{\chi}(\widetilde{y}).\,P \mid (P \,|\, Q) \mid *P \mid (\nu a)\,P \mid \mathbf{if}\ v\ \mathbf{then}\ P\ \mathbf{else}\ Q$$

Here, χ is either $\circ$ or $\bullet$, and $\widetilde{w}$ abbreviates a possibly empty sequence $w_1, \ldots, w_n$. The constructs are the standard ones of the polyadic π-calculus: nil, output and input prefixes, parallel composition, replication ($*P$ behaves like infinitely many copies of P running in parallel), restriction, and a conditional. The only difference is the annotation χ in prefixes, which indicates whether the action is expected to succeed (symbol $\circ$) or not (symbol $\bullet$). (In the type inference of TyPiCal these annotations are actually inferred, in the sense that if the analysis succeed then a set of prefixes that will eventually succeed is marked, see Section 5.) We call a prefix *marked* if its annotation is $\circ$. We usually omit the $\bullet$ annotation, thus for example $a(x).P$ stands for $a^{\bullet}(x).\,P$. As usual, restriction and input prefix are binders. A *closed* process has no free variables. We often omit trailing $\mathbf{0}$, and write $\overline{v}^{\chi}[\widetilde{w}]$ for $\overline{v}^{\chi}[\widetilde{w}].\,\mathbf{0}$. We also write $\overline{v}^{\chi}.P$ and $v^{\chi}.P$ for $\overline{v}^{\chi}[\,].\,P$ and $v^{\chi}(\,).\,P$ respectively. In examples, we use an extension of the above language with natural numbers, list, etc. as they are straightforward to accommodate.

Typing. The type systems that we will propose are defined on top of the simply-typed π-calculus (ST). The set of *simple types* is given by:

$$\mathtt{S} ::= \mathtt{Bool} \mid \sharp[\mathtt{S}_1, \ldots, \mathtt{S}_n]$$

$\sharp[\mathtt{S}_1, \ldots, \mathtt{S}_n]$ is the type of channels that are used for transmitting tuples consisting of values of types $\mathtt{S}_1, \ldots, \mathtt{S}_n$. A type judgment is of the form $\Gamma \vdash_{\mathtt{ST}} P$. A type environment Γ is a mapping from names to simple types, with the constraint that `true` and `false` are mapped to `Bool`, and that the links are mapped to channel types. $\Gamma, \widetilde{v}:\widetilde{\mathtt{S}}$ indicates the type environment obtained by extending Γ with the type assignments $\widetilde{v}:\widetilde{\mathtt{S}}$, with the understanding that for all v_i already defined in Γ it should be $\Gamma(v_i) = \mathtt{S}_i$. The standard typing rules are omitted.

Operational Semantics. We use the standard (early) labeled transition relation $P \xrightarrow{\eta} Q$ for the π-calculus. Here, η, called a transition label, is either a silent action τ (which represents an internal communication), an output action $(\nu\widetilde{c})\,\overline{a}[\widetilde{b}]$, or an input action $a[\widetilde{b}]$. We write $\xrightarrow{\tau}^{*}$ for the reflexive and transitive closure of $\xrightarrow{\tau}$; we write $P \xrightarrow{\tau}$ and $P \xrightarrow{\tau}^{*}$ if there is P' s.t. $P \xrightarrow{\tau} P'$ and $P \xrightarrow{\tau}^{*} P'$, respectively.

We extend the above transition relation to a *typed* transition relation, to show how a type environment evolves when a process performs a transition. $\Gamma \vdash_{\mathtt{ST}} P \xrightarrow{\eta} \Gamma' \vdash_{\mathtt{ST}} P'$ holds if: (1) $P \xrightarrow{\eta} P'$; (2) $\Gamma \vdash_{\mathtt{ST}} P$; and (3) if $\eta = \tau$ then $\Gamma = \Gamma'$; otherwise if η is an output $(\nu\widetilde{c})\,\overline{a}[\widetilde{b}]$ or an input $a[\widetilde{b}]$ and $\Gamma(a) = \sharp[\widetilde{\mathtt{S}}]$, then $\Gamma' = \Gamma, \widetilde{b} : \widetilde{\mathtt{S}}$. Note that $\Gamma \vdash_{\mathtt{ST}} P \xrightarrow{\eta} \Gamma' \vdash_{\mathtt{ST}} P'$ implies $\Gamma' \vdash_{\mathtt{ST}} P'$. We write $\Gamma_0 \vdash_{\mathtt{ST}} P_0 \xrightarrow{\eta_1} \cdots \xrightarrow{\eta_k} P_k$ to mean that $\Gamma_0 \vdash_{\mathtt{ST}} P_0$, and there are $\Gamma_1, ..., \Gamma_k$ s.t. for all $i < k$ it holds that $\Gamma_i \vdash_{\mathtt{ST}} P_i \xrightarrow{\eta_{i+1}} \Gamma_{i+1} \vdash_{\mathtt{ST}} P_{i+1}$.

Deadlock-Freedom and Lock-Freedom. A prefix is *at top level* if it is not underneath another input/output prefix or underneath a replication.

Definition 1 (deadlock-freedom). *P is* deadlock-free *if, whenever $P \xrightarrow{\tau}^{*} Q$ and Q has at least one marked prefix at top level, then $Q \xrightarrow{\tau}$.*

Deadlock-freedom indicates only the possibility for the system to evolve further; on the other hand, lock-freedom indicates the eventual success of marked actions at top-level. In the definition of lock-freedom, we track the success of a specific action (as several marked actions may simultaneously appear at top-level) by tagging it. We then demand success for all possible taggings. We call *tagged* a process in which exactly one unguarded and unreplicated prefix—the prefix that we wish to track—has the special annotation $\square$ (instead of $\circ$ as in the marked prefixes). Transitions of tagged processes are defined as for the untagged ones, except that the labels of transitions emanating from the tagged prefix are also tagged. We call a tagged τ-transition, written $P \xrightarrow{\tau^{\square}} P'$, a *success*.

Definition 2 ((weak) lock-freedom). A tagged process P is *successful* if whenever $P \xrightarrow{\tau}^{*} Q$ then $Q \xrightarrow{\tau}^{*} \xrightarrow{\tau^{\square}}$. Given an untagged process P, the *tagging* *of* P is the set of tagged processes obtained from P by replacing the annotation of a marked prefix at top level with $\Box$. Process P is *(weakly) lock-free* if whenever $P \xrightarrow{\tau}{}^{*} Q$ then all processes in the tagging of Q are successful.

A sequence of transitions $\xrightarrow{\tau}$ or $\xrightarrow{\tau^{\Box}}$ is *full* if it is finite and ends with an irreducible process, or if it is infinite. A sequence of transitions is *strongly fair* if, intuitively, any τ-action that is enabled infinitely often will eventually succeed (see [8, 3] for a formal definition of strong fairness in the π-calculus).

Definition 3 (strong lock-freedom). *P is* strongly lock-free *if whenever $P \xrightarrow{\tau}{}^{*} Q$ then every full and strongly fair transition sequence of each process in the tagging of Q contains the success transition $\xrightarrow{\tau^{\Box}}$.*

Experts in concurrency will easily recognize the difference between weak lock-freedom and strong lock-freedom: Weak lock-freedom combines safety and liveness guarantees, by requiring that a system never reaches a state where a marked action is at top-level, but there is no sequence of τ-actions in which the marked action is consumed. On other hand, strong lock-freedom is a purely liveness property that says that if a marked action is at top-level, the action will eventually be consumed. The example below shows the difference between weak lock-freedom and strong lock-freedom.

Example 1. Consider the following process P:

$$b^{\circ}(\,) \mid \overline{a}[b] \mid *a(y).\,(\nu c)\,(\overline{c}[y] \mid c(y).\,\overline{y}[\,] \mid c(y).\,\overline{a}[y])$$

The rightmost subprocess $(*a(y).\,\cdots)$ receives b on a and either sends a message on b or forwards b to itself non-deterministically. Since c is freshly created everytime b is received from a, the strong fairness does not guarantee that a message is eventually sent on b, and P is therefore *not* strongly lock-free. On the other hand, however, after any number of forwardings, there is a chance for a message to be sent on b; hence, P is weakly lock-free. See Example 3 for another example of a process that is weakly lock-free but not strongly lock-free.

3 Type System for Lock-Freedom

We introduce the type systems for weak/strong lock-freedom. They are obtained by augmenting Kobayashi's type system [9] with hybrid rules appealing to deadlock/termination/confluence analyses. For lack of space, precise definitions are often omitted; see the extended version [13].

3.1 Review of Previous Type System for Lock-Freedom

As mentioned in Section 1, to enable local reasoning about lock-freedom in terms of deadlock and termination analyses, we need to express some contracts between a process and its environment. We reuse the type judgments of Kobayashi's lock-freedom type system [9] to express the contracts. A type judgment is of the form

$\Delta \vdash_{\mathrm{LT}} P$, where Δ is a type environment, which expresses both requirements on the behavior of P, and assumptions on its environment. Ordinary channel types are extended with *usages*, which express how each communication channel is used. For example, $\sharp_{?.!}[\texttt{Bool}]$ describes a channel that should be first used for receiving a boolean once, and then for sending a boolean once. A channel of type $\sharp_{?}[\sharp_{!}[\texttt{Bool}]]$ should be first used for receiving a channel once, and then the received channel should be used once for sending a boolean. (! and ? express an output and an input respectively, and "." denotes the sequential composition;)

In order to express both assumptions on the environment (like, "a process can eventually receive a message from its environment") and guarantees by the process (like, "a process will certainly send a message"), each action (! or ?) in a usage is further annotated with *capability levels* and *obligation levels*, which range over the set of natural numbers extended with ∞. If a capability level of an action is finite, then that action is guaranteed to succeed (in other words, its co-action will be provided by the environment) if it becomes ready for execution (i.e., it is at top-level). If an obligation level of an action is finite, then that action must become ready for execution, only by relying on capabilities of smaller levels. For example, the type judgment $a : \sharp_{?^{\infty}_{0}}[\texttt{Bool}], b : \sharp_{!^{1}_{\infty}}[\texttt{Bool}] \vdash_{\mathrm{LT}} P$ means that P has a capability of level 0 to receive a boolean on channel a (but not an obligation to receive it) , and P has an obligation of level 1 to send a boolean on b. (Here, the superscript of ! or ? is the obligation level, and the subscript is the capability level.) Thus, P can be $\overline{b}[\texttt{true}]$ or $a(x).\,\overline{b}[x]$, but not $a(x).\,\mathbf{0}$. Thanks to the abstraction of process behavior by usages, the problem of checking lock-freedom of a process is reduced to that of checking whether the usage of each channel is consistent in the sense that, for each capability of level t, there is a corresponding obligation of level less than or equal to t.

To understand how this kind of judgment can be used for compositional reasoning about lock-freedom, consider the (deadlocked) process $a^{\circ}(x).\,\overline{b}[x] \mid b^{\circ}(x).\,\overline{a}[x]$. We have the following judgment for subprocesses:

$$a : \sharp_{?^{0}_{0}}[\texttt{Bool}], b : \sharp_{!^{1}_{\infty}}[\texttt{Bool}] \vdash_{\mathrm{LT}} a^{\circ}(x).\,\overline{b}[x]$$
$$a : \sharp_{!^{1}_{\infty}}[\texttt{Bool}], b : \sharp_{?^{0}_{0}}[\texttt{Bool}] \vdash_{\mathrm{LT}} b^{\circ}(x).\,\overline{a}[x]$$

For the entire process, we can simply combine both type environments by combining usages pointwise:

$$a : \sharp_{?^{0}_{0} \mid !^{1}_{\infty}}[\texttt{Bool}], b : \sharp_{!^{1}_{\infty} \mid ?^{0}_{0}}[\texttt{Bool}] \vdash_{\mathrm{LT}} a^{\circ}(x).\,\overline{b}[x] \mid b^{\circ}(x).\,\overline{a}[x]$$

Now, the capability level of the input on a (which is 0) is smaller than the obligation level of the corresponding output on a (which is 1); this indicates a failure of assume-guarantee reasoning (the assumption made by the left subprocess is not met by the guarantee by the right subprocess). Thus, we know the process may not be lock-free. On the other hand, if we replace the subprocess in the righthand side with $\overline{a}[\texttt{true}].\,b(x)$, then we get:

$$a : \sharp_{?^{0}_{0} \mid !^{0}_{0}}[\texttt{Bool}], b : \sharp_{!^{1}_{1} \mid ?^{1}_{1}}[\texttt{Bool}] \vdash_{\mathrm{LT}} a^{\circ}(x).\,\overline{b}[x] \mid \overline{a}[\texttt{true}].\,b^{\circ}(x)$$

The capability of each action is matched by the obligation of its co-action, which implies that the process is lock-free.

$$U \text{ (usages) } ::= \mathbf{0} \mid ?^{t_1}_{t_2}.U \mid !^{t_1}_{t_2}.U \mid (U_1 \mid U_2) \mid *U \qquad t \text{ (levels) } \in \mathbf{Nat} \cup \{\infty\}$$

$$\mathtt{L} \text{ (usage types) } ::= \mathtt{Bool} \mid \sharp_U[\widetilde{\mathtt{L}}] \qquad \Delta \text{ (type environments) } ::= v_1 : \mathtt{L}_1, \ldots, v_n : \mathtt{L}_n$$

$$\frac{\Delta_1 \vdash_{\mathtt{LT}} P \qquad t_c = \infty \Rightarrow \chi = \bullet}{v : \sharp_{!^0_{t_c}}[\widetilde{\mathtt{L}}]; (\Delta_1 \mid \widetilde{w} : \uparrow\widetilde{\mathtt{L}}) \vdash_{\mathtt{LT}} \overline{v}^{\chi}[\widetilde{w}].\, P} \qquad \frac{\Delta, \widetilde{y} : \widetilde{\mathtt{L}} \vdash_{\mathtt{LT}} P \qquad t_c = \infty \Rightarrow \chi = \bullet}{v : \sharp_{?^0_{t_c}}[\widetilde{\mathtt{L}}]; \Delta \vdash_{\mathtt{LT}} v^{\chi}(\widetilde{y}).\, P}$$

$$\frac{}{\emptyset \vdash_{\mathtt{LT}} \mathbf{0}} \qquad \frac{\Delta_1 \vdash_{\mathtt{LT}} P_1 \qquad \Delta_2 \vdash_{\mathtt{LT}} P_2}{\Delta_1 \mid \Delta_2 \vdash_{\mathtt{LT}} P_1 \mid P_2} \qquad \frac{\Delta' \vdash_{\mathtt{LT}} P \qquad \Delta \leq \Delta'}{\Delta \vdash_{\mathtt{LT}} P} \qquad \frac{\Delta \vdash_{\mathtt{LT}} P}{*\Delta \vdash_{\mathtt{LT}} *P}$$

$$\frac{\Delta, a : \sharp_U[\widetilde{\mathtt{L}}] \vdash_{\mathtt{LT}} P \qquad rel(U)}{\Delta \vdash_{\mathtt{LT}} (\nu a)\, P} \qquad \frac{\Delta \vdash_{\mathtt{LT}} P \qquad \Delta \vdash_{\mathtt{LT}} Q}{\Delta \mid (v : \mathtt{Bool}) \vdash_{\mathtt{LT}} \textbf{if } v \textbf{ then } P \textbf{ else } Q}$$

Fig. 1. Kobayashi's type system for lock-freedom [9]

Figure 1 summarizes the syntax of types, and typing rules of Kobayashi's lock-freedom type system [9]. A type inference algorithm for the type system, which serves as a lock-freedom verification algorithm, is discussed in [9].

3.2 Robust Deadlock-Freedom/Termination/Confluence

To enable local reasoning in the new type system for lock-freedom that we will present, we introduce a strengthening of the notions of deadlock-freedom, termination, and confluence.

A substitution $\sigma = [\widetilde{w}/\widetilde{x}]$ *respects* $\Gamma = \widetilde{v} : \widetilde{\mathtt{S}}$ if $\sigma\Gamma = \widetilde{\sigma v} : \widetilde{\mathtt{S}}$ is well-defined. A substitution σ is *closing for* Γ if σ respects Γ and $\sigma\Gamma$ has no variables. A process is robustly terminating if it cannot diverge, after any sequence of transition that conforms to the base type system ST.

Definition 4 (robust termination). *A process P is* terminating *if there is no infinite internal transition sequence $P \xrightarrow{\tau} P_1 \xrightarrow{\tau} P_2 \xrightarrow{\tau} \cdots$. An (open) process P is* robustly terminating *under Γ, written $\Gamma \models_{\mathtt{RTer}} P$, if $\Gamma \vdash_{ST} P$, and for every closing substitution σ for Γ and for any Q, k, and $\eta_1, \cdots \eta_k$ such that $\sigma\Gamma \vdash_{ST} \sigma P \xrightarrow{\eta_1} \cdots \xrightarrow{\eta_k} Q$, the derivative Q is terminating.*

We say that Δ is closed if $dom(\Delta) \cap \mathcal{V} = \emptyset$. We write $rel(\Delta)$ intuitively to mean that each capability in Δ is guaranteed by a corresponding obligation; and $ob_!(\mathtt{L})$ for the level of the obligation to send a message: again, precise definitions are in the extended version [13].

In the definition of robust deadlock-freedom below, the first condition say that P is deadlock-free when it is executed by itself, and that P either fulfills its obligations or reduces further. The other conditions say that the robust deadlock-freedom is preserved by substitutions and transitions. The relation $\Delta \xrightarrow{\eta} \Delta'$ (see [13] for the definition) expresses the increase/decrease of capabilities/obligations in Δ by the transition η. For example, $a : \sharp_{?^0_\infty}[\sharp_{!^1_\infty}[\mathtt{Bool}]] \xrightarrow{a[b]} a : \sharp_{\mathbf{0}}[\sharp_{!^1_\infty}[\mathtt{Bool}]], b : \sharp_{!^1_\infty}[\mathtt{Bool}]$ holds (where the usage $\mathbf{0}$ indicates that the channel cannot be used at all). Thus, $a : \sharp_{?^0_\infty}[\sharp_{!^1_\infty}[\texttt{Bool}]] \models_{\texttt{RD}} P$ means that P will eventually perform an input on a, and then send a boolean on the received channel, unless P at some point diverges.

Definition 5 (robust deadlock-freedom). *The relation $\Delta \models_{\texttt{RD}} P$ is the largest relation such that $\Delta \models_{\texttt{RD}} P$ implies all of the following conditions.*

1. *If Δ is closed and $rel(\Delta)$, then: (i) P is deadlock-free; (ii) If $ob_!(\Delta(a)) \neq \infty$, then either $P \xrightarrow{(\nu\widetilde{c})\,\overline{a}[\widetilde{b}]}$ or $P \xrightarrow{\tau}$; and (iii) If $ob_?(\Delta(a)) \neq \infty$ then either $P \xrightarrow{a[\widetilde{b}]}$ or $P \xrightarrow{\tau}$.*
2. *If $[v \mapsto a]\Delta$ is well-defined, then $[v \mapsto a]\Delta \models_{\texttt{RD}} [v \mapsto a]P$.*
3. *If $P \xrightarrow{\eta} P'$ and, furthermore, when η is an input, all names received are fresh, then $\Delta \xrightarrow{\eta} \Delta'$ and $\Delta' \models_{\texttt{RD}} P'$ for some Δ'.*

We say that P is robustly deadlock-free *under Δ if $\Delta \models_{\texttt{RD}} P$ holds.*

Partial confluence means that any τ-transition commutes with any other transitions. To define the partial confluence, we assume that each prefix is uniquely labeled (as in [3]), and extend the transition relation to $\xrightarrow{\eta,S}$ where S is the set of the labels of the prefixes involved in the transition: see [13]. Robust confluence indicates partial confluence after any sequence of transition that conforms to the base type system ST.

Definition 6 (robust confluence). *A process P is* partially confluent, *if whenever $P_1 \xleftarrow{\tau,S_1} P \xrightarrow{\eta,S_2} P_2$, either $\eta = \tau \wedge S_1 = S_2$, or $P_1 \xrightarrow{\eta,S_2}\equiv\xleftarrow{\tau,S_1} P_2$. A process P is* robustly confluent *under Γ, written $\Gamma \models_{\texttt{RConf}} P$, if $\Gamma \vdash_{ST} P$ and for any closing substitution σ that respects Γ and for any Q, k, and $\eta_1, \cdots \eta_k$ such that $\sigma\Gamma \vdash_{ST} \sigma P \xrightarrow{\eta_1} \cdots \xrightarrow{\eta_k} Q$, the derivative Q is partially confluent.*

While termination, deadlock-freedom, and confluence are frequently discussed in the literature, we are not aware of previous work that defines the robust counterparts above and verification methods for them.

We have proved that robust deadlock-freedom is guaranteed by Kobayashi's type system for deadlock-freedom [11]. In applications of robust deadlock-freedom, it is often the case that the environment Δ needed is of a restricted form, so that $\Delta \models_{\texttt{RD}} P$ then boils down to the verification of a few simple behavioral properties for which other type systems and model checkers can also be used. For example, if Δ is $a : \sharp_{!^0_\infty}[\texttt{Bool}]$, then $\Delta \models_{\texttt{RD}} P$ only means that P is deadlock-free and P will eventually send a boolean on a unless it diverges. Robust confluence is guaranteed, for instance, by types systems for linear channels [12] and race-freedom [18]; other static analysis methods such as model checking could also be used. Verification of robust termination is discussed in Section 4.

3.3 Hybrid Typing Rules

We now introduce the new rules LT-Hyb (for weak lock-freedom), and SLT-Hyb (for strong lock-freedom).

$$\frac{\Delta \models_{\mathtt{RD}} P \qquad Er(\Delta) \models_{\mathtt{RTer}} P \qquad nocap(\Delta)}{\Delta \vdash_{\mathtt{LT}} P} \quad \text{(LT-HYB)}$$

$$\frac{\Delta \models_{\mathtt{RD}} P \qquad Er(\Delta) \models_{\mathtt{RTer}} P \qquad Er(\Delta) \models_{\mathtt{RConf}} P \qquad nocap(\Delta)}{\Delta \vdash_{\mathtt{SLT}} P} \quad \text{(SLT-HYB)}$$

Here, $Er(\Delta)$ is the simple type environment obtained from Δ by removing all usage annotations. The condition $nocap(\Delta)$ holds if, intuitively, Δ describes a process that fulfills its obligations without relying on the environment. As mentioned in Section 1, this is used to avoid circular, unsound, assume-guarantee reasoning. The precise definition of $nocap(\Delta)$, given in [13], is subtle; for nested channel types, the nocap condition depends on whether a channel is used for input or output. For example, $nocap(\sharp_{?^0_\infty}[\sharp_{!^0_\infty}[\,]])$ holds but $nocap(\sharp_{!^0_\infty}[\sharp_{!^0_\infty}[\,]])$ does not. In the rule for strong lock-freedom, the robust confluence ensures that once a marked prefix is enabled, it cannot be disabled by any other transitions. See Example 3 for a non-trivial example, for which the rule LT-HYB fails to guarantee strong lock-freedom.

We write $\Delta \vdash_{\mathtt{LT}} P$ if it is derivable by using the typing rules in Section 3.1 and LT-HYB, and write $\Delta \vdash_{\mathtt{SLT}} P$ if it is derivable by using SLT-HYB instead of LT-HYB. The theorem below states the soundness of the type systems. Its proof is non-trivial because of the presence of the hybrid rules; for instance, conditions such as $nocap(\Delta)$ are not preserved by transitions, so in the proof we had to refine and extend the type systems. See the extended version [13].

Theorem 1 (lock-freedom). *If $\emptyset \vdash_{\mathtt{LT}} P$, then P is (weakly) lock-free. If $\emptyset \vdash_{\mathtt{SLT}} P$, then P is strongly lock-free.*

Example 2. Consider the following processes.

$$\begin{aligned} Clients &\stackrel{\mathrm{def}}{=} *(\nu r_1)\,(\overline{fact}^{\circ}[\mathtt{rnd}(), r_1] \mid r_1{}^{\circ}(x).\,\mathbf{0}) \\ Server &\stackrel{\mathrm{def}}{=} (\nu fact\_it)\,(*fact(n,r).\,\overline{fact\_it}[n,1,r] \\ &\qquad \mid *fact\_it(n,x,r).\,\mathbf{if}\ n=0\ \mathbf{then}\ \overline{r}[x]\ \mathbf{else}\ \overline{fact\_it}[n-1, x\times n, r]) \end{aligned}$$

The process *Server* creates an internal communication channel *fact_it* (used for computing factorial numbers in a tail-recursive manner), and waits on *fact* for a request $[n,r]$ on computing the factorial of n. Upon receiving a request, it returns the result on r. *Client* consists of infinitely many copies of the process that creates a fresh channel r_1 for receiving a reply, sends a request $[\mathtt{rnd}(), r_1]$ (where $\mathtt{rnd}()$ creates a random number) and then waits for the result on r_1.

Let Δ be $fact : \sharp_{*?^0_\infty}[\mathtt{Nat}, \sharp_{!^1_\infty}[\mathtt{Nat}]]$. Then, we have $\Delta \models_{\mathtt{RD}} Server$, $Er(\Delta) \models_{\mathtt{RTer}} Server$, and $Er(\Delta) \models_{\mathtt{RConf}} Server$ with $nocap(\Delta)$. Thus, by using SLT-HYB, we obtain $\Delta \vdash_{\mathtt{SLT}} Server$. From this judgment and $fact : \sharp_{*!^\infty_0}[\mathtt{Nat}, \sharp_{!^1_\infty}[\mathtt{Nat}]] \vdash_{\mathtt{SLT}} Clients$, we obtain: $\emptyset \vdash_{\mathtt{SLT}} (\nu fact)\,(Server \mid Clients)$. This means that all the clients can eventually receive replies. Note that the whole process diverges (since there are infinitely many clients), but we can derive strong lock-freedom by local reasoning based on SLT-HYB.

Example 3. This example shows a binary tree data structure, offering services for inserting and searching natural numbers. Each node of the tree is implemented as a process that has: a state, given by the integer stored in the node and pointers to the left and right subtree and that contain, respectively, smaller and greater integers; channels for the insert and search operations. In Figure 2, G is a generator of new nodes, which can then grow and originate a tree, and where: i and s will be the insertion and search channels; `state` stores the state of the node. Initially the node is a leaf. `TInit` is the initial tree, with an empty state and public channels `insert` and `search` to communicate with the environment. Once received a query for an integer n, the tree lets the request ripple down the nodes, following the order on the integers to find the right path, until either t is found in a node, or the end of the tree is reached. There is parallelism in the system: many requests can be rippling down the tree at the same time; in doing so, requests can even overtake each other.

Let Δ be $\mathtt{insert}:\sharp_{*?^0_\infty}[\mathtt{Nat},\sharp_{!^1_\infty}[\,]]$, $\mathtt{search}:\sharp_{*?^0_\infty}[\mathtt{Nat},\sharp_{!^1_\infty}[\mathtt{Bool}]]$. Then, we have:

$$\Delta \vdash_{\mathtt{RD}} \mathtt{TInit} \quad Er(\Delta) \models_{\mathtt{RTer}} \mathtt{TInit} \quad nocap(\Delta)$$

Thus, by using LT-HYB, we obtain $\Delta \vdash_{\mathtt{LT}} \mathtt{TInit}$. By applying rules for `LT` to the rest of the system, we get $\Delta \vdash_{\mathtt{LT}} \mathtt{Sys}$.

Note that SLT-HYB is not applicable since `TInit` is not robustly confluent (because, when multiple requests arrive simultaneously, there can be a race on the channel `state`). Indeed, the example is NOT strongly lock-free! A search request may never be replied if the request is overtaken by insertion requests so often that the tree grows faster than the search request goes down the tree. See [13] for a strongly lock-free version of binary trees.

4 Types for Robust Termination

For our analysis we need a refinement of the standard termination property, that we call robust termination. *Termination* of a term means that all its reduction sequences are of finite length. *Robust termination* guarantees that termination is maintained when the process interacts with its environment. Termination is strictly weaker than robust termination. Consider for instance the term $P \stackrel{\text{def}}{=} \overline{c}[b] \mid c(x).(\overline{x} \mid *a.\overline{x})$. The process P has one reduction only, and therefore it is terminating. It is indeed typable in the simplest of the type systems in [7]. However, P is not robustly terminating. It can interact with other processes via the input at c and, in doing so, it may receive a resulting in the non-terminating derivative $\overline{c}[b] \mid \overline{a} \mid *a.\overline{a}$.

A number of type systems for termination of mobile processes have appeared in the literature [6, 7, 17, 21]. We have isolated some some abstract conditions which allows us to turn a type system for termination into one for robust termination. For lack of space we refer the reader to [13] for the details.

$$
\begin{array}{rl}
G \stackrel{\mathrm{def}}{=} & *\mathtt{newtree}(i,s).(\nu\mathtt{state})\Big(\overline{\mathtt{state}}[\mathbf{leaf}] \\
& \mid *i(n,r).\mathtt{state}(x).\quad \texttt{/*** insertion ***/} \\
& \quad \mathbf{match}\ x\ \mathbf{with\ leaf}\ \rightarrow \\
& \qquad (\nu\mathtt{left\_i},\mathtt{left\_s},\mathtt{right\_i},\mathtt{right\_s}) \\
& \qquad \Big(\overline{\mathtt{newtree}}[\mathtt{left\_i},\mathtt{left\_s}] \mid \overline{\mathtt{newtree}}[\mathtt{right\_i},\mathtt{right\_s}] \\
& \qquad \mid \overline{\mathtt{state}}[\mathbf{node}(n,\mathtt{left\_i},\mathtt{left\_s},\mathtt{right\_i},\mathtt{right\_s})] \mid \overline{r}\Big) \\
& \quad \| \ \mathbf{node}(n_1,i_l,s_l,i_r,s_r)\ \rightarrow \\
& \quad \Big(\overline{\mathtt{state}}[x] \mid \mathbf{if}\ n=n_1\ \mathbf{then}\ \overline{r}[]\ \mathbf{else\ if}\ n<n_1\ \mathbf{then}\ \overline{i_l}[n,r]\ \mathbf{else}\ \overline{i_r}[n,r]\Big) \\
& \mid *s(n,r).\mathtt{state}(x).\Big(\overline{\mathtt{state}}[x] \quad \texttt{/*** search ***/} \\
& \quad \mid \mathbf{match}\ x\ \mathbf{with\ leaf}\ \rightarrow \overline{r}[\mathtt{false}] \\
& \quad \| \ \mathbf{node}(n_1,i_l,s_l,i_r,s_r) \rightarrow \\
& \qquad \mathbf{if}\ n_1=n\ \mathbf{then}\ \overline{r}[\mathtt{true}]\ \mathbf{else\ if}\ n<n_1\ \mathbf{then}\ \overline{s_l}[n,r]\ \mathbf{else}\ \overline{s_r}[n,r])\Big) \\
\mathtt{TInit} \stackrel{\mathrm{def}}{=} & (\nu\mathtt{newtree})\,(G \mid \overline{\mathtt{newtree}}[\mathtt{insert},\mathtt{search}]) \\
\mathtt{Sys} \stackrel{\mathrm{def}}{=} & (\nu\mathtt{insert},\mathtt{search}) \\
& (\mathtt{TInit} \mid *(\nu r_1)\,(\overline{\mathtt{insert}}^{\circ}[\mathtt{rnd}(),r_1] \mid r_1{}^{\circ}) \mid *(\nu r_2)\,(\overline{\mathtt{search}}^{\circ}[\mathtt{rnd}(),r_2] \mid r_2{}^{\circ}(x)))
\end{array}
$$

Fig. 2. A binary tree

5 Implementation

We have implemented the new weak lock-freedom analysis as a feature of TyPiCal Version 1.6.0 [10]. TyPiCal takes as an input a program written in the π-calculus, and marks all input/output prefixes that are guaranteed to succeed.

The original type system for lock-freedom (reviewed in Section 3.1) had been implemented already [11, 9]. A major challenge in the implementation of the new system was to automate verification of the robust termination property. We have modified the type systems of Deng and Sangiorgi [7], so that the resulting systems can guarantee robust termination, and also so to make them more suited for automatic verification (e.g., using heuristic and incomplete algorithms when the original ones were NP complete). We also integrated them with a termination analysis based on size-change graphs [2]. See the extended version for details.

We have applied the implementation to non-trivial programs (including the examples in Section 3), and verified them fully automatically (without any type annotations). According to benchmark results (shown in [13]), the new components (dealing with termination) run fast; most of the analysis time is spent by the other components (dealing with deadlock- and lock-freedom). For the binary tree (Example 3), the verification time was 5.47 sec., of which the time for robust termination analysis was only 0.02 sec.

6 Related Work

Several type systems for lock-freedom (sometimes referred to by different names) have been already proposed [8, 9, 20, 1, 19, 21]. Our type system substantially improves the expressiveness of previous type systems; for instance, it can handle non-trivial recursive structures (e.g., the binary trees as in Example 3), and value-dependent behaviors. This is possible through a parameterization that appeals to other analyzers, in particular those for deadlock freedom (so that more powerful analyzers make the lock-freedom type system more powerful too). Another important point is that none of the previous type systems for lock-freedom, except Kobayashi's one [9], has been implemented. In fact, most of the type systems classify channels into a few usage patterns, and prepare separate typing rules for each of the usage patterns. Thus, verification based on those type systems would not be possible without heavy program annotations.

Type systems for deadlock-freedom have been studied extensively. As already mentioned, deadlock-freedom is weaker than lock-freedom, so that those type systems alone cannot be used for lock-freedom analysis. For example, the divergent process obtained by replacing $\overline{\mathit{fact\_it}}[n-1, x \times n, r]$ in Example 2 with $\overline{\mathit{fact\_it}}[n, x \times n, r]$ is deadlock-free.

The idea of reducing verification of lock-freedom to verification of robust termination is a reminiscence of Cook et al.'s work on reducing verification of liveness properties to that of fair termination [5]. The target language of their work is a sequential, imperative language and is quite different from our language, which is concurrent and allows dynamic creation of communication channels and threads. The used techniques are also quite different; they use model checking while we use types.

There are a number of methods for proving termination of programs, and they have been extensively studied in the context of term rewriting systems and sequential programs. The point of parameterizing our type system for lock-freedom by the robust termination property was to reuse those techniques for termination verification, instead of developing a sophisticated type system that can reason about both termination and deadlock within the single type system.

Parameterized, or hybrid, type systems of this kind presented in this paper are fairly rare in the literature, mainly due to the difficulties in combining the analyses. For instance, in Leroy's modular module system [14] a type system for module is presented that is parametric on the type system used for the core language. This is quite different from ours, as the world on which the two type systems operate—modules and core languages—are stratified, hence clearly separated.

References

1. Acciai, L., Boreale, M.: Responsiveness in process calculi. In: Proc. of 11th Annual Asian Computing Science Conference (ASIAN 2006). LNCS, vol. 4435, pp. 136–150. Springer, Heidelberg (2006)
2. Ben-Amram, A.M., Lee, C.S.: Program termination analysis in polynomial time. ACM Trans. Prog. Lang. Syst. 29(1 (Article 5)) (2007)
3. Bidinger, P., Compagnoni, A.B.: Pict correctness revisited. In: Bonsangue, M.M., Johnsen, E.B. (eds.) FMOODS 2007. LNCS, vol. 4468, pp. 206–220. Springer, Heidelberg (2007)

4. Brinksma, E., Rensink, A., Volger, W.: Fair testing. In: Lee, I., Smolka, S.A. (eds.) CONCUR 1995. LNCS, vol. 962, pp. 313–327. Springer, Heidelberg (1995)
5. Cook, B., Gotsman, A., Podelski, A., Rybalchenko, A., Vardi, M.Y.: Proving that programs eventually do something good. In: Proc. of POPL, pp. 265–276 (2007)
6. Demangeon, R., Hirschkoff, D., Kobayashi, N., Sangiorgi, D.: On the complexity of termination inference for processes. In: Barthe, G., Fournet, C. (eds.) Proceedings of TGC 2007. LNCS, vol. 4912, pp. 140–155. Springer, Heidelberg (2008)
7. Deng, Y., Sangiorgi, D.: Ensuring termination by typability. Info. Comput. 204(7), 1045–1082 (2006)
8. Kobayashi, N.: A type system for lock-free processes. Info. Comput. 177, 122–159 (2002)
9. Kobayashi, N.: Type-based information flow analysis for the pi-calculus. Acta Informatica 42(4-5), 291–347 (2005)
10. Kobayashi, N.: TyPiCal: A type-based static analyzer for the pi-calculus, `http://www.kb.ecei.tohoku.ac.jp/~koba/typical/`
11. Kobayashi, N.: A new type system for deadlock-free processes. In: Baier, C., Hermanns, H. (eds.) CONCUR 2006. LNCS, vol. 4137, pp. 233–247. Springer, Heidelberg (2006)
12. Kobayashi, N., Pierce, B.C., Turner, D.N.: Linearity and the pi-calculus. ACM Trans. Prog. Lang. Syst. 21(5), 914–947 (1999)
13. Kobayashi, N., Sangiorgi, D.: A hybrid type system for lock-freedom of mobile processes. An extended version (2008), `http://www.kb.ecei.tohoku.ac.jp/~koba/papers/hybrid.pdf`
14. Leroy, X.: A modular module system. J. Funct. Program. 10(3), 269–303 (2000)
15. Pierce, B.C., Turner, D.N.: Pict: A programming language based on the pi-calculus. In: Plotkin, G., Stirling, C., Tofte, M. (eds.) Proof, Language and Interaction: Essays in Honour of Robin Milner, pp. 455–494. MIT Press, Cambridge (2000)
16. Sangiorgi, D., Walker, D.: The Pi-Calculus: A Theory of Mobile Processes. Cambridge University Press, Cambridge (2001)
17. Sangiorgi, D.: Termination of processes. Math. Struct. Comput. Sci. 16(1), 1–39 (2006)
18. Terauchi, T., Aiken, A.: A Capability Calculus for Concurrency and Determinism. In: Baier, C., Hermanns, H. (eds.) CONCUR 2006. LNCS, vol. 4137, pp. 218–232. Springer, Heidelberg (2006)
19. Sangiorgi, D.: The name discipline of uniform receptiveness. Theor. Comput. Sci. 221(1-2), 457–493 (1999)
20. Yoshida, N.: Type-based liveness guarantee in the presence of nontermination and nondeterminism. Technical Report 2002-20, MSC Technical Report, University of Leicester (April 2002)
21. Yoshida, N., Berger, M., Honda, K.: Strong normalisation in the pi-calculus. Info. Comput. 191(2), 145–202 (2004)

Implied Set Closure and Its Application to Memory Consistency Verification

Surender Baswana, Shashank K. Mehta, and Vishal Powar

Indian Institute of Technology, Kanpur - 208016, India
{sbaswana,skmehta,vishalp}@cse.iitk.ac.in

Abstract. Hangal et. al. [3] have developed a procedure to check if an instance of the execution of a shared memory multiprocessor program, is consistent with the Total Store Order (TSO) memory consistency model. They also devised an algorithm based on this procedure with time complexity $O(n^5)$, where n is the total number of instructions in the program. Roy et. al. [6] have improved the implementation of the procedure and achieved $O(n^4)$ time complexity.

We have identified the bottleneck in these algorithms as a graph problem of independent interest, called *implied-set closure* (ISC) problem. In this paper we propose an algorithm for ISC problem and show that using this algorithm, Hangal's consistency checking procedure can be implemented with $O(n^3)$ time complexity. We also experimentally show that the new algorithm is significantly faster than Roy's algorithm.

Keywords: Memory consistency model verification, Incremental transitive closure, Total store order, Shared memory multi-processor.

1 Introduction

Modern processors aggressively employ chip multiprocessing and simultaneous multi-threading to achieve high processor performance. Traditionally memory being the slower subsystem various techniques, such as hierarchical implementation of the memory, have been employed to reduce the bottleneck. With the fast multiprocessors, modern architectures exploit new techniques to improve the memory performance. Foremost of these techniques is executing the memory instructions out of program-order according to a predetermined consistency model.

A memory consistency model specifies the restriction on the order in which memory instructions may be executed. Commercial architectures support a variety of memory consistency models. The strictest of these is sequential-consistency (SC), which requires that all the instructions of one processors must be executed in their program-order. An execution is valid in this model if and only if the global order of the instructions result from interleaving the processor-programs. Relaxing the restrictions progressively lead to Total-Store-Order (TSO) and Release Consistency (RC), these along with further relaxed models are given in [3].

The problem of *memory consistency verification* is to ensure that a given architecture always executes every program in accordance to the memory consistency model adopted in its design. This problem is shown to be NP complete

A. Gupta and S. Malik (Eds.): CAV 2008, LNCS 5123, pp. 94–106, 2008.

as shown by Catlin et. al. [2]. With increasingly relaxed consistency model the verification of compliance becomes extremely difficult. Adve and Gharachorloo [1] discuss many issues related to these models and their implementation.

A more practical approach, adopted in industry, is to execute a test program and then verify from its execution trace that indeed it was executed in accordance with the consistency model. This approach can never prove that the design is error free but a large number of tests can give significant confidence. The test programs are multi-threaded programs of memory instructions: Store, Load, Memorybar etc. We call each thread a *processor-program.*

In this paper we address the problem of analyzing the execution trace. Hangal et. al. [3] have developed TSOtool for Sun Microsystems processors to analyze the outcome of test programs run under TSO model. Their algorithm has $O(n^5)$ time complexity and $O(n^2)$ space complexity, where n is the number of instructions in the test program. The basic procedure is general enough to be applicable to memory consistency models other than TSO. Manovit and Hangal [5] have exploited the total order among store instructions to reduce the search space. Their algorithm has time complexity $O(k.n^3)$ where k is the number of processors (number of threads in the program). But this algorithm crucially depend on the total order of store instructions and it cannot be applied to other consistency models.

Roy et. al. [6] have developed Intel MPRIT Tool for the same task by improving the general algorithm by Hangal, reducing the time complexity to $O(n^4)$. Once again, this algorithm can be applied to any consistency model without change in the time complexity.

Our contribution in this paper is to identify a graph problem called *implied-set-closure*(ISC) which is the abstraction of the bottleneck of the above mentioned general high-level algorithm. We present an efficient incremental algorithm for ISC problem based on the incremental transitive closure algorithm by Italiano [4]. The application of this algorithm reduces the time complexity of the general memory consistency verification algorithm to $O(n^3)$, which is a significant improvement over the $O(n^4)$ bound achieved by the previous best algorithm of Roy et al. [6]. The space requirement of the algorithm remains $\Theta(n^2)$. Other salient feature of our algorithm is its compact and simple description and no hidden constant, which makes it an ideal candidate to be a practical algorithm (a quality not possessed by many theoretically efficient algorithms). We also compared our algorithm with the algorithm of Roy et al. [6] experimentally in an identical computing environment. In MPRIT, the algorithm of Roy et. al. is implemented using vector instructions. For our experiment we implemented both algorithms without vector instructions. The results show that the new algorithm outperforms their algorithm in real time.

The remainder of the paper is organized as follows. In the following section, we reproduce the description of the memory consistency verification problem and its algorithm given in [3,6]. In Section 3 a graph problem, which we refer by the name *implied set closure* (ISC) problem, is formally defined and its relation to the memory consistency problem is established. In Section 4, we develop an algorithm for ISC, and also argue about the optimality of its theoretical time complexity. In Section 5, experimental results comparing the performance of the new algorithm to the previous best algorithm are presented.

2 Formal Description of TSO Model and Consistency Verification Algorithm

We reproduce here the formal description of TSO model and the high level consistency verification algorithm presented in [3]. The axiomatic description of the model is originally borrowed from Sindhu et. al. [7]. The model used by [6] is slightly generalized which will also be covered in the following discussion.

A Load instruction is considered executed when the issuing processor receives the data, while a Store instruction is considered executed when it is visible to all processors in the system. It is assumed that each instruction eventually gets executed, i.e., no instruction takes infinite time to complete. The notations used here are as follows.

L_a^i	a *Load* from location a by processor i
S_a^i	a *Store* to location a by processor i
$\lfloor L_a^i; S_a^i \rfloor$	an *Atomic operation* to location a by processor i
$Val[L_a^i]$	the value read by L_a^i
$Val[S_a^i]$	the value written by S_a^i
O_a^i	either a L_a^i or a S_a^i
$\mathcal{S}(L_a^i)$	the store instruction S_a^j s.t. $Val(S_a^j) = Val(L_a^i)$ (it is assumed that each Store instruction in the test program stores a unique value so $\mathcal{S}(L_a^i)$ is well defined.)

There are two partial-ordering relations defined over the set of all memory instructions: local ';' and global '$\leq$'. $x; y$ iff x and y are in the same processor-program with x before y; and $x \leq y$ iff it is required by TSO axioms or by the data dependency, that instruction x must be executed before y. In the latter ordering x and y may belong to different processor-programs. Now we present the axioms of total-store-order (TSO) memory consistency model.

Total Store Order. $\forall S_a^i, S_b^j : (S_b^j \leq S_a^i \vee S_a^i \leq S_b^j)$.

Atomic Operation. $\lfloor L_a^i; S_a^i \rfloor \Rightarrow (L_a^i \leq S_a^i) \wedge (\forall S_b^j : S_b^j \leq L_a^i \vee S_a^i \leq S_b^j)$.

Value Coherence. $Val[L_a^i] = Val[\max_{\leq}[\{S_a^j : S_a^j \leq L_a^i\} \cup \{S_a^i : S_a^i; L_a^i\}]]$.

Local Ordering. This tells when the local ordering must be preserved in the global ordering. We first present Roy et.al.'s[6] version of the axiom. If $O_a^i; O_b^i$ then subject to various criteria depending on the the type of the operations (load or store) and the type of the locations a, b (write-back, write-through, write protected, uncacheable, uncacheable speculative write combine), we require $O_a^i \leq O_b^i$. The criteria can be expressed by a function $f : (\{load, store\} \times \{location\text{-}types\})^2 \rightarrow \{0, 1\}$ and state the axiom as $(O_a^i; O_b^i) \wedge (f(type(O_a^i), type(a), type(O_b^i), type(b)) = 1) \Rightarrow (O_a^i \leq O_b^i)$.

This axiom is stated in a restricted way in [3] where a specific f is assumed, namely, $f(t_1, t_1', t_2, t_2') = 1$ iff $t_1 \neq$ store or $t_2 \neq$ load.

Axiomatic description of other memory consistency models are similarly described in [7].

2.1 Consistency Verification Problem

TSOtool developed by Hangal et. al. [3] generates a test program, for a multiprocessor system (consisting only of memory instructions), executes it on the system or a system simulator, and then analyzes the outcome of the run to check if it is consistent with the axioms of the TSO model. In order to carry out the analysis each Store instruction stores a unique value. It allows to determine $\mathcal{S}(L_a^i)$ for each L_a^i unambiguously. Their main contribution is the consistency checking algorithm. We will present the high level description of the algorithm in this section. The implementation of this algorithm has $O(n^5)$ time complexity, where n is the total number of (memory) instructions included in all processor-programs. Roy at. al. [6] improved the implementation leading to the time complexity $O(n^4)$. In the following sections we will show that it can be further improved to $O(n^3)$.

Algorithm 1 analyzes the program output by computing a directed graph (V, E) based on the outcome of the program. The nodes, V, of this graph are the instructions of the program. An edge is placed from node O_1 to O_2 if $O_1 \leq O_2$. If the graph contains a cycle, i.e., $\leq$ is not found to be a partial ordering, then we can conclude that the requirement of the consistency model must have been violated in the execution.

We reproduce here the rules of including the edges in the graph from [3,6] which incorporate the TSO axioms.

Static Edges: This rule is due to the local ordering axiom.

R_1: If $f(type(O_a^i), type(a), type(O_b^i), type(b)) = 1$, then include edge (O_a^i, O_b^i).

The edges due to this rule can be determined from the program itself and they are independent of the outcome of the run. Assuming that the function can be evaluated in $O(1)$ time, the static edges can be computed in $O(n^2)$ time.

Observed Edges: These rules are direct implication of the Value and the Total-store axioms.

R_2: If $(\mathcal{S}(L_a^i) = S_a^j) \wedge (i \neq j)$, then add the edge (S_a^j, L_a^i).

R_3: If $(\mathcal{S}(L_a^i) = S_a^j) \wedge (S_a'^i; L_a^i)$, then add the edge $(S_a'^i, S_a^j)$

The first part of these conditions is decided by finding S such that $Val(L) = Val(S)$ for a given L. This is because each Store instruction stores a unique value. Thus these edges can be computed only after the outcome of the run is known. The second part of the conditions only depend on the program. Hence it is easy to see that the observed edges can be computed in $O(n^2)$ time.

Inferred Edges: These rules are the indirect implications of Value Coherence axiom.

R_4: If $(\mathcal{S}(L_a^i) = S_a^j) \wedge (S_a'^k \leq L_a^i) \wedge (S_a'^k \neq S_a^j)$, then add the edge $(S_a'^k, S_a^j)$.

R_5: If $(\mathcal{S}(L_a^i) = S_a^j) \wedge (S_a^j \leq S_a'^k)$, then add the edge $(L_a^i, S_a'^k)$.

The conditions in this case, unlike in the earlier cases, depend on the structure (edges) of the graph. As the new inferred edges are added to E new pairs of vertices $S_a'^k, L_a^i$ or $S_a^j, S_a'^k$ may satisfy the respective precondition and be eligible for an edge between them. Thus inferred edges need to be computed iteratively. The computation of the inferred edges is the bottleneck in the performance of the algorithm, so the complexity of Algorithm 1 is determined by steps 9 and 12. In Sections 3 and 4 we describe an efficient solution for these steps.

2.2 Example

We present an example borrowed from [3] to show how above described rules detect inconsistency.

Let $S[X]\#n$ denote a store instruction which stores n in location X; and $L[X] = n$ denote a load instruction which loads value n from location X. In this example we assume that $f(t_1, t_1', t_2, t_2') = 1$ iff $t_1 \neq$ store or $t_2 \neq$ load. Consider the following 4-thread program along with the relevant information from an execution.

P_1	P_2	P_3	P_4
$S[B]\#91$	$S[A]\#2$	$S[B]\#92$	$L[B] = 92$
$S[A]\#1$		$L[A] = 2$	$L[B] = 91$
$L[A] = 2$		$L[B] = 92$	

In Figure 1 we show the directed graph generated using the rules. All edges due to rules R_1, R_2 and R_3 are easy to deduce from the rules. Here is the explanation for the two R_4 edges. Since $S[B]\#91 \leq S[A]\#1 \leq S[A]\#2 \leq L[A] = 2 \leq L[B] = 92$. So from R_4 there must be an edge from $S[B]\#92$ to $S[B]\#91$. We also have $S[B]\#92 \leq L[B] = 92 \leq L[B] = 91$. Again from R_4, there should be an edge from $S[B]\#91$ to $S[B]\#92$. These edges from a cycle so we conclude that the TSO model is violated in this execution.

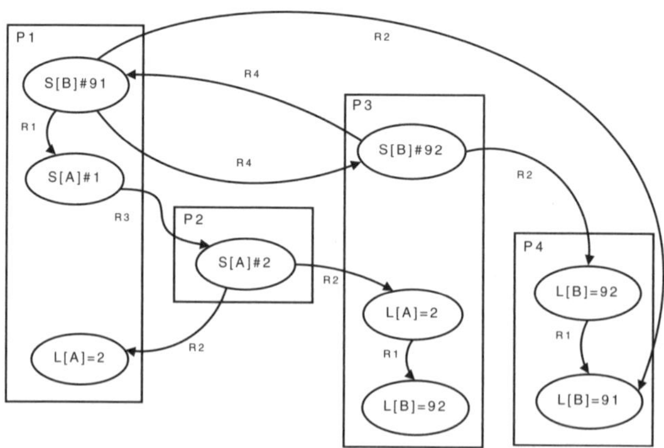

Fig. 1. Edges due to application of rules

2.3 Limitation of Algorithm 1

We have seen that a cycle in the graph implies that TSO model is violated but the converse is not true.

The TSO requires that all Stores must be totally ordered. There may be some cases where insufficient data dependence information is present to totally order all the Stores. To remain true to TSO, all total orders must be considered which are consistent with the partial order on stores determined by the data dependence. This would make the worst case complexity of the algorithm exponential [3]. The procedure chooses to ignore this lacunae in an attempt to trade off accuracy for reasonable analysis time.

All the three algorithms that implement this procedure are designed with this limitation. As explained above, the Store instructions do not get totally ordered in the global ordering (as required by TSO) only when there is insufficient data dependence. This situation arises when different threads do not have sufficient interaction. In practice this situation can easily be resolved by having Load instruction for each location in several threads. Therefore these algorithms are extremely useful in spite of the incompleteness.

3 Implied-Set Closure (ISC) Problem

Definition 1. *Let $G = (V, E_0)$ be a directed graph and $T : V \times V \to powerset$ $(V \times V)$. Then the* implied-set closure *of G is $G^{ISC} = (V, E_0^{ISC})$ where E_0^{ISC} is the smallest edge set such that*
(i) $E_0 \subset E_0^{ISC}$,
(ii) if there is a (directed) path from vertex a to vertex b in E_0^{ISC}, then $T(a,b) \subseteq E_0^{ISC}$.

If any two edge sets satisfy the above conditions, then their intersection also satisfies the same. Therefore there is a unique smallest set satisfying the conditions. Observe that if $T(a,b) = \{(a,b)\}$, then ISC problem reduces to transitive closure problem.

3.1 Relation to Memory Consistency Verification Problem

The computation of the edges generated by the rules R_4 and R_5 is an instance of the computation of implied-set closure problem as explained below.

Let E_0 be the edge set E in Algorithm 1 after step 8. Define (i) $T(S_a^k, L_a^i) = \{(S_a^k, \mathcal{S}(L_a^i))\}$ for every S_a^k and L_a^i such that $S_a^k \neq \mathcal{S}(L_a^i)$, (ii) $T(S_a^j, S_a^k) = \{(L_a^i, S_a^k) : \mathcal{S}(L_a^i) = S_a^j\}$, (iii) $T(x,y) = \emptyset$ for all the remaining ordered pairs (x,y). Then the final set E in Algorithm 1, computed after step 14, is E_0^{ISC} with respect to the given T-sets.

4 Algorithm for ISC Problem

We present an algorithm for implied-set closure problem which is based on incremental algorithm for transitive closure proposed by Italiano [4].

```
Data: Sequence of memory-instructions for each processor (Swap is considered
      both, a Load and a Store instruction), value associated with each
      Load/Store instruction. V denotes the set of all instruction
Result: It outputs true if the execution of the program obeys all TSO axioms,
        else outputs false
for each L ∈ V do
    if Val[L] = Val[S] then
        S(L) = S
    end
end
initialize graph (V, E) to (V, ∅);
/* R1:                                                              */
Add edges to E according to R1;
/* R2, R3:                                                          */
Add edges to E according to R2 and R3;
/* R4:                                                              */
while ∃L, S' such that S'_a ≠ S(L_a) = S_a and (S'_a, L_a) ∈ E do
    Add edge (S'_a, S_a) to E;
end
/* R5:                                                              */
while ∃L_a, S'_a such that S'_a ≠ S(L_a) = S_a and (S_a, S'_a) ∈ E do
    Add edge (L_a, S'_a) to E;
end
if E contains a cycle then
    declare that TSO violation found;
end
else
    declare no violation found;
end
```

Algorithm 1. Algorithm to analyze a program output for TSO violation

4.1 An Incremental Algorithm for ISC Problem

Algorithm 2 computes the implied-set closure of a set of directed-edges $E_0 \subseteq V \times V$. For convenience we shall denote the existence of a directed path from a to b in graph H, in the algorithm, by $a \rightsquigarrow b$ and the transitive closure of H by H^c.

The correctness of the algorithm can be established by observing that at the end of each iteration of the *while*-loop following four assertion are always true: (i) $E_0 \subseteq H \cup X$, (ii) $HC = H^c$, (iii) For each pair of vertices u, v, if $u \rightsquigarrow v$ in H, then $T(u,v) \subseteq H \cup X$, (iv) $H \cup X \subseteq E_0^{ISC}$, where E_0^{ISC} is the implied-set-closure of E_0. Condition (iii) is equivalent to: for each pair of vertices u, v, if $(u,v) \in HC$, then $T(u,v) \subseteq H \cup X$.

The algorithm terminates since H grows monotonically. On termination, X is empty so the loop invariant conditions imply that finally $H = E_0^{ISC}$.

In the next step we will show how to efficiently compute the incremental transitive closure of step 8 of Algorithm 2.

Data: vertex set V; edge set E_0; implied-edge-sets $T(a,b)$ for all pairs $(a,b) \in V \times V$

Result: $H = E_0^{ISC}$ and $HC = H^c$ (transitive closure of H)

```
H ← ∅;
HC ← ∅;
X ← E0;
while X ≠ ∅ do
    (x, y) ← Select(X);
    /* pick an arbitrary edge (a, b) from X and delete it from the set
       */
    H' ← H ∪ {(x, y)};
    if (x, y) ∉ HC then
        HC' ← (HC ∪ {(x, y)})^c;
    end
    else
        HC' = HC;
    end
    /* here superscript 'c' denotes transitive-closure                 */
    for each (u, v) ∈ (HC' − HC) do
        X ← X ∪ (T(u, v) − H');
    end
    H ← H';
    HC ← HC';
end
return H;
```

Algorithm 2. An incremental algorithm to compute implied-set closure of E_0

4.2 Improved Algorithm for ISC Problem

In order to update the transitive closure of the graph upon insertion of an edge, the algorithm uses following observation to minimize the computation required. Let (x, y) be an edge added to H. If y was already reachable from x, then the transitive closure of the graph will remain unchanged. Otherwise, transitive closure needs to be updated. In particular, we need to add the edges implied by the transitivity in context of all those source vertices w such that $w \rightsquigarrow x \wedge w \not\rightsquigarrow y$ prior to insertion of edge (x, y) because vertex reachable from y has subsequently become reachable from w too. In order to update the transitive closure for each such vertex w, a simple way is to scan all vertices reachable from y. This will require $O(n)$ work per vertex and leads to $O(n^4)$ time algorithm as designed by Roy et al. [6]. However, note that it would suffice if we can efficiently compute only those vertices which are reachable from y but not reachable from w (prior to the current edge insertion). Let us denote this set by $\mathcal{D}(w, y)$. The following Lemma would pave the way for its efficient computation, and hence updating the transitive closure.

Lemma 1. *For each vertex $v \in \mathcal{D}(w, y)$, and any path $\mathcal{P}$ from y to v, each vertex lying on $\mathcal{P}$ is also present in $\mathcal{D}(w, y)$.*

The set $\mathcal{D}(w, y)$ is a subset of the set of vertices reachable from y, and we know that the latter can be computed by performing DFS (or BFS) traversal in the graph starting from y. In order to compute $\mathcal{D}(w, y)$ efficiently, it would suffice to perform a *bounded* DFS traversal in the graph starting from y wherein we extend DFS recursively only along those vertices which were not reachable from w prior to insertion of the edge (x, y). This is because, as follows from Lemma 1, the DFS traversal pursued from a vertex already reachable from w won't lead to any vertex of set $\mathcal{D}(w, y)$, and so there is no point extending DFS traversal beyond such vertices.

Algorithm 3 is the bounded depth-first search based procedure to update transitive closure for a vertex w upon insertion of an edge. This algorithm implicitly computes $\mathcal{D}(w, y)$.

```
HC ← HC ∪ (w, y);
for each (y, z) ∈ H do
    if (w, z) ∉ HC then
        bDFS(w, z);
    end
end
```

Algorithm 3. $bDFS(w, y)$

Based on the above discussion, it follows that we can replace Step 8 in Algorithm 2 by the following step.

Step 8 for Algorithm 2:

for each $w \in V$ **do**

  if $((w, x) \in HC)\&((w, y) \notin HC)$ **then** $bDFS(w, y)$; **end**

end

Along with this modification we also absorb the for-loop at step 13 of Algorithm 2 in the bounded DFS routine. The final algorithm is given in Algorithm 4. Here H is stored in two data-structures, adjacency-list HL as well as adjacency-matrix HM. $HL[a]$ points to the list of vertices to which there are edges from a, and $HM[a, b] = 1$ iff $(a, b) \in H$. The transitive closure of H is stored as adjacency matrix, where $HC[a, b] = 1$ iff (a, b) belongs to H^c.

4.3 Time and Space Complexity

Analysis of running time: Let the number of vertices in V be n, number of edges in E_0 be m, and $\overline{m}$ denote the number of edges in the implied-set closure of E_0.

The **for**-loop in Algorithm 4 runs n times in each call and **while**-loops iterates $\overline{m}$ times so total time complexity of the algorithm, excluding the cost of $bDFS'$-calls is $O(\overline{m}n)$.

We bound the cost due to $bDFS'$-calls in two parts, one for each for-loop. Observe that the $bDFS'$-routine is never called again with the same arguments. Therefore the cumulative cost of all calls due to the second for-loop

```
Data: vertex set V; edge set E0; induced-edge-sets T(a,b) for all pairs
      (a,b) ∈ V × V
Result: HL is the adjacency list of the implied-set-closure, and HC stores its
        transitive closure
HL ← ∅;
HM ← ∅;
HC ← ∅;
X ← E0;
while X ≠ ∅ do
    (x,y) ← Select(S);
    Insert y in HL[x];
    HM[x,y] ← 1;
    if HC[x,y] = 0 then
        for each w ∈ V do
            if HC[w,x] = 1 and HC[w,y] = 0 then
                bDFS'(w,y)
            end
        end
    end
end
return HL, HC;
```

Algorithm 4. Algorithm to compute implied-set closure: Final version

```
HC[w,y] ← 1;
for each (u,v) ∈ T(w,y) do
    if HM[u,v] ≠ 1 then
        X ← X ∪ {(u,v)}
    end
end
for each z ∈ HL[y] do
    if HC[w,z] = 0 then
        bDFS'(w,z);
    end
end
```

Algorithm 5. subroutine $bDFS'(w, y)$

is $\sum_w \sum_y deg_H(y)$, where $deg_H(y)$ denotes the out-degree of y in H, which is $O(\overline{m}n)$. The contribution of the first for-loop can be estimated by observing that each $T(w, y)$ is scanned at most once and the membership test takes $O(1)$ time. So its cost is $O(\sum_{(w,y)\in HC} |T(w, y)|)$. The total time complexity of the algorithm is $O(\overline{m}n + \sum_{(w,y)\in HC} |T(w, y)|)$.

Note that the second component in the expression can't be got rid of by any algorithm of ISC problem because this is also the input size for the problem and each algorithm has to scan it at least once. The additional cost is $O(\overline{m}n)$. It seems quite difficult, if not infeasible, to beat this bound since the ordinary transitive closure problem is a special case of the ISC problem and there does not exist any combinatorial algorithm for transitive closure problem with running time $o(mn)$ where $m = |E_0|$.

Space requirement: The algorithm uses adjacency list as well as adjacency matrix representation for the graph H which amounts to $\Theta(\overline{m} + n^2) = \Theta(n^2)$ space. In addition we use $n \times n$ matrix HC. Thus the total space requirement of the algorithm is $\Theta(n^2)$. Hence, in addition to the input which consists of original edge set $E_)$ and the lists $T(a,b)$ for each $(a,b) \in V \times V$, the algorithm uses only $\Theta(n^2)$ additional space.

Theorem 1. *For a given graph on n vertices, a base set of edges E_0, and sets $T(a,b)$ $\forall(a,b) \in V \times V$ of implied edges, there exists an algorithm which solves the ISC problem in $O(\overline{m}n + \sum_{a,b} |T(a,b)|)$ time and $\Theta(n^2)$ space, in addition to the space used by input, where $\overline{m}$ is the number of edges in the implied-set closure of E.*

In Section 3 we have seen that computation of edges due to rules R_4 and R_5 in Algorithm 1, which dominates the time complexity, is the implied-set closure of the edge set resulting after applications of the first three rules. Therefore the time complexity for memory consistency verification problem is the same as that for computing the corresponding implied-set closure.

In this ISC problem the non-empty $T(a,b)$ sets are either of the form $T(S,S')$ or $T(S,L)$, so $\sum_{a,b} |T(a,b)| = \sum_{S,S'} |T(S,S')| + \sum_{S,L} |T(S,L)|$. From the definition it is clear that $T(S,S') \cap T(x,y) \neq \emptyset$ iff $S = x$ and $S' = y$. Hence $\sum_{S,S'} |T(S,S')| \leq |V \times V|$. Again from the definition $|T(S,L)| \leq 1$ so $\sum_{S,L} |T(S,L)| \leq \sum_{S,L} 1 \leq |V \times V|$. This gives $\sum_{a,b} |T(a,b)| < 2n^2$. We have the following corollary.

Corollary 1. *Memory consistency verification problem can be solved in $O(n^3)$ time and $O(n^2)$ space, where n is the total number of instructions in the test program.*

5 Experimental Results

Theoretically, the new algorithm achieves a speed-up by a factor of n over the worst case time complexity of the previous best algorithm of Roy et al. [6]. To show that there are no hidden large constants, we compared it with the algorithm of Roy et. al. [6] experimentally (see Table 1).

Our experiments are based on test programs with 2, 4, 6, and 8 program-threads. In each case the number of instructions per thread were kept the same and these varied from 100 instruction per thread to 500 instruction per thread. Each result reported here is the average of 100 programs in each category. The test programs were generated by Intel's MPRIT tool [6]. Here Algorithm A refers to the algorithm reported here and Algorithm B is that of Roy et. al. [6] .

To make a fair comparison, the algorithms were executed in an identical environment consisting of a single processor. The experiment used implementation of Roy et. al. [6] algorithm without vector instructions.

Table 1. Experimental results

	Instructions per thread									
	100		200		300		400		500	
#threads	algo A	algo B	algo A	algo B	algo A	algo B	algo A	algo B	algo A	algo B
2	119	82	230	240	591	626	1216	1350	2216	2476
4	149	163	614	953	1834	3289	4198	7913	8181	16071
6	225	332	1222	2997	3926	11023	9508	27914	18593	57671
8	328	688	2082	7216	7204	28183	16846	69851	32904	146350

In each experiment the number of instructions is fixed. So n is equal to the number of instructions-per-thread times the number of threads. As number of threads increase, n increases and consequently performance gap increases. Besides, we see that algorithm A performs significantly better compared to algorithm B in case of 100 instructions per thread with 8 thread, while the gap is insignificant when instruction per thread is 400 and threads are 2. In both the cases $n = 800$, but the reason for this difference is the number of threads. The number of inferred edges increases with more threads, and expensive computation involves the computation of these edges.

Intel's MPRIT Tool incorporates an implementation of the algorithm of Roy et al. [6] using vector instructions on SIMD processor which performs 128 bit operations per instruction. This results in a significant speedup (about a factor of 100). Yet asymptotically, for large values of n, the new algorithm will outperform even the vector-instruction aided implementation of the algorithm in [6].

5.1 Parallelization of the New Algorithm

Step 10 in Algorithm 4 involves processing for each $w \in V$. In each iteration of the *for*-loop all computations exclusively depend on the HC edges of the form $(w, *)$ and the sets $T(w, *)$. This implies that computations in different iterations are mutually independent. Therefore in a multi-processor environment different processor could be assigned different passes of this loop to be performed in parallel.

6 Conclusion

In this work we studied the memory consistency compliance algorithm of Hangal et.al. [3] which was improved by Roy et. al. [6]. The former had $O(n^5)$ complexity, while the latter was improved to $O(n^4)$. We identified the bottleneck in these algorithms and proposed it as a graph problem called induced-set closure problem. We proposed an efficient algorithm for this problem using Italiano's [4] incremental algorithm for transitive closure and showed that using this approach the memory consistency compliance algorithm can be implemented in $O(n^3)$ time. An efficient parallel implementation of our algorithm remains a task for the future research.

Acknowledgment

We thank Amitabha Roy for introducing the memory consistency compliance problem to us. We thank Mainak Chaudhuri for explaining various concepts in the area of memory architecture. We also thank Mayur Shardul for pointing out an error in the bound for $|\sum T(S, S')|$ in an earlier draft.

References

1. Adve, S.V., Gharachorloo, K.: Shared memory consistency models : A tutorial. In Digital Western Research Laboratory Technical Report (1995)
2. Cantin, J., Lipasti, M., Smith, J.: The complexity of verifying memory coherence. In: Proceedings of 15th Annual ACM Symposium on Parallel Algorithms and Architectures, pp. 254–255 (2003)
3. Hangal, S., Vahia, D., Manovit, C., Lu, J.-Y.J.: TSOtool: A program for verifying systems using the memory consistency model. In: Proceedings of the 31st annual international symposium on computer architecture (ISCA), pp. 114–123 (2004)
4. Italiano, G.F.: Amortized efficiency of a path retrieval data structure. Theoretical Computer Science 48, 273–281 (1986)
5. Manovit, C., Hangal, S.: Efficient algorithm for verifying memory consistency. In: Proceedings of the 17th annual ACM symposium on Parallelism in algorithms and architectures (SPAA 2005), pp. 245–252 (2005)
6. Roy, A., Zeisset, S., Fleckenstein, C.J., Huang, J.C.: Fast and generalized polynomial time memory consistency verification. In: Ball, T., Jones, R.B. (eds.) CAV 2006. LNCS, vol. 4144, pp. 503–516. Springer, Heidelberg (2006)
7. Sindhu, P.S., Frailong, J. M., Cekleov, M.: Formal specification of memory models. In Xerox PARC Technical Report (1991)

Effective Program Verification for Relaxed Memory Models

Sebastian Burckhardt and Madanlal Musuvathi

Microsoft Research

Abstract. Program verification for relaxed memory models is hard. The high degree of nondeterminism in such models challenges standard verification techniques. This paper proposes a new verification technique for the most common relaxation, store buffers. Crucial to this technique is the observation that all programmers, including those who use low-lock techniques for performance, expect their programs to be sequentially consistent. We first present a monitor algorithm that can detect the presence of program executions that are not sequentially consistent due to store buffers while *only* exploring sequentially consistent executions. Then, we combine this monitor with a stateless model checker that verifies that every sequentially consistent execution is correct. We have implemented this algorithm in a prototype tool called Sober and present experiments that demonstrate the precision and scalability of our method. We find relaxed memory model bugs in several programs, including two previously unknown bugs in a production-level concurrency library that would have been difficult to find by other means.

1 Introduction

Developers of performance-critical multi-threaded software often try to avoid the overhead of traditional locking by either making direct use of hardware primitives for atomic operations (such as interlocked exchange, or compare-and-swap), or by employing regular loads and stores for synchronization purposes. Unfortunately, such "low-lock" programs are notoriously hard to get right [4,20]. Subtle bugs can arise in these programs due to memory reordering caused by the relaxed memory model of the underlying hardware [1] . These errors are hard to find and debug as they most often show up only in specific thread interleavings and in particular hardware configurations. On the other hand, low-lock code is heavily used both in low-level libraries and in critical paths of a system. Because these parts are crucial to the reliability of the entire system, it is important to develop verification techniques.

In general, the same program may exhibit more executions on a relaxed model than on a sequentially consistent (SC) machine [18], as illustrated in Fig. 1. Let $\mathcal{T}_\pi^Y$ denote the set of executions of program π on memory model Y. Most existing program verification tools can not verify directly whether the executions in $\mathcal{T}_\pi^Y$ are correct (unless $Y = SC$). A few specialized memory model sensitive verification tools exist [4,13,22,25] but scalability and automation remain a challenge.

A. Gupta and S. Malik (Eds.): CAV 2008, LNCS 5123, pp. 107–120, 2008.

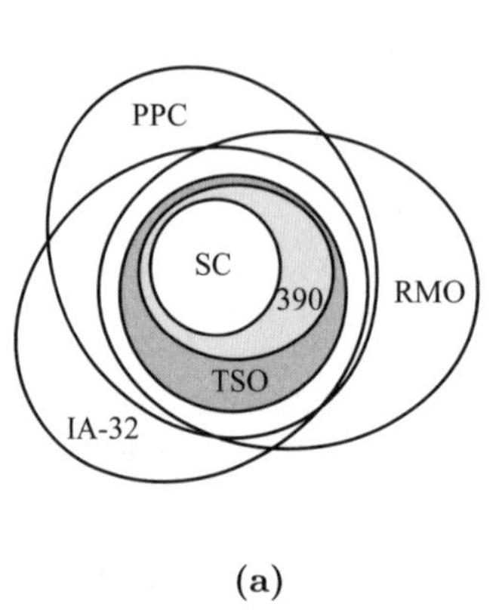

(a)

Initially: $X = Y = 0$

processor 1	processor 2
X = 1	Y = 1
r_1 = Y	r_2 = X

Eventually: $r_1 = 0$, $r_2 = 0$

When the two threads run on different processors, the stores to X and Y in the first line can possibly be delayed by the store buffers in these processors. Subsequent loads in the second line see the initial values of X and Y if the store buffers have not yet been committed.

(b)

Fig. 1. (a) A comparison of various memory models [6,9,14,15,24]. (b) An execution that is possible on *TSO* but not on *SC*.

A key observation of this paper is that programmers, even those writing low-lock code, *expect* their programs to be sequentially consistent. They design their programs to be correct for SC executions and insert memory ordering fences to counter relaxations where necessary. In particular, any program execution that is not SC is almost always an error, resulting either from an insufficient use of fences or a misunderstanding of the underlying memory model.

This observation suggests that we can sensibly verify the relaxed executions $\mathcal{T}_\pi^Y$ by solving the following two verification problems separately:

1. Use standard verification methodology for concurrent programs to show that the executions in $\mathcal{T}_\pi^{SC}$ are correct.
2. Use specialized methodology for *memory model safety* verification, showing that $\mathcal{T}_\pi^Y = \mathcal{T}_\pi^{SC}$. We say the program π is Y-*safe* if $\mathcal{T}_\pi^Y = \mathcal{T}_\pi^{SC}$.

In this paper, we focus on verifying memory model safety for the most common relaxation in modern multiprocessors, *store buffers* with store-load forwarding. The corresponding memory model is historically called *TSO* (total store order) [24], and we use the terms *TSO*-safety and store buffer safety interchangeably. Under *TSO*, processors may delay the effect of a store instruction in a processor-local FIFO buffer (to hide the memory latency). While the values of these store instructions are immediately visible to the local processor, other processors see these values only when the store buffer is committed at a later time. Fig. 1(b) shows a simple example. We provide a rigorous characterization of *TSO* in Section 2.

Apart from the fact that store buffers are so common (as apparent in Fig. 1(a), $\mathcal{T}_\pi^{TSO} \subseteq \mathcal{T}_\pi^Y$ for almost all models Y), our motivation for focusing on *TSO* largely arises from the need to prepare the huge body of legacy code heavily optimized to run on x86 machines for future multicore chip generations. These processors are likely to make increased use of store buffers but are otherwise fairly conservative as far as the memory model is concerned [16].

The main contribution of this paper is a technique for checking the store buffer safety of a program while *only* exploring its sequentially consistent executions, which lets us perform the steps 1 and 2 above *simultaneously*. Our technique relies on a notion of *borderline* execution, which is an SC execution that can be extended into an execution in $\mathcal{T}_\pi^{TSO} \setminus \mathcal{T}_\pi^{SC}$. We establish that a program is store buffer safe *exactly* if there are no borderline executions (Theorem 1). Then we present an efficient, precise monitor for detecting borderline executions, using a novel *generalized* vector clock algorithm.

We have implemented these ideas in a prototype tool called Sober. Sober combines our store buffer safety monitor with the stateless model checker CHESS [21] which systematically enumerates the *SC* executions of a bounded concurrent test program and checks them for errors such as null pointers or assertion violations. In principle, Sober terminates with one of three possible outputs. First, Sober may detect a regular program error and output an erroneous execution. Second, Sober may report that the program is not store buffer safe. Finally, Sober may terminate without finding an error, proving that all *TSO* executions of the program are correct. In practice, exhaustive verification is too time-consuming for most programs and we resort to iterative context-bounding [21], which provides verification guarantees up to a specific preemption bound.

Section 4 describes our initial experiments. Using Sober we found and fixed store buffer issues in several programs, including Dekker's mutual exclusion protocol [2] and the Bakery protocol [17]. We got our greatest success so far when we applied Sober to a component of a concurrency library at Microsoft. This component implements a low-lock datastructure. Sober demonstrated two store buffer problems that the developer immediately agreed were real errors. These bugs were never detected during the extensive code-review and testing the component underwent.

Related Work. Prior work has addressed the verification of programs for relaxed memory models using explicit state enumeration [7,13,22] and using constraint solving [3,4,11,26]. Our work improves upon them in scalability. To our knowledge, this paper is the first to demonstrate the possibility of program verification without exploring the additional nondeterminism of memory-model relaxation. See the experiments in Huynh and Roychoudhury [13] for the state space explosion caused by this nondeterminism even for simple programs. This paper is definitely not the first to observe that sequential consistency is the most natural memory model for programmers [1,12,18]. The Java Memory Model [19] guarantees sequential consistency for a broad class of programs, namely those which are data-race free. In contrast, our characterization of memory model safety *precisely* captures those programs which behave sequentially consistent in a memory model. In particular, a program with data-races might still be memory-model safe. Specialized algorithms to automatically insert fences based on static analysis [8,23] can guarantee memory-safety in principle. However, doubts remain about their precision in the presence of aliasing, loops, and conditionals and the performance implication of conservative fence insertion. Also, the memory models considered in these algorithms assume atomic memory and cannot model store buffers, the main emphasis of this paper.

2 Problem Formulation

We represent the relevant aspects of a program executions by a *memory trace*, or just trace. A trace is a collection of events, each representing a memory access (either a store, a load, or an interlocked operation[1]) by a specific processor to a specific address. Each event has an *issue index*, which is a sequence number relative to all events by the same processor. Furthermore, each event has a *coherence index*, which is the sequence number of the value that is read or written by the event, relative to the entire value sequence written to the targeted memory location during the execution.

Formally, let $Op = \{st, ld, il\}$, let $\mathbb{N}$ be the set of natural numbers, let $Proc = \{1, \ldots, N\}$ be a finite set of processor identifiers for some fixed bound $N \in \mathbb{N}$, let Adr be a finite set of memory addresses, and let $\mathbb{N}_0 \subseteq \mathbb{Z}$ be the set of nonnegative integers. Then we define the set of events as $Evt = Op \times Proc \times \mathbb{N} \times Adr \times \mathbb{N}_0$, and we denote elements $e \in Evt$ using the syntax $o(p, i, a, c)$, where $o \in Op$, $p \in Proc$, $i \in \mathbb{N}$ is the issue index, $a \in Adr$, and $c \in \mathbb{N}_0$ is the coherence index. We use corresponding projection functions $o(e), p(e), i(e), a(e), c(e)$ for an event e. Given a set $E \subseteq Evt$ of events, we define the following subsets for notational convenience:

$$
\begin{aligned}
\text{(commands issued by processor } p\text{)}\ E(p) &= \{e \in E \mid p(e) = p\} \\
\text{(load events)}\ L(E) &= \{e \in E \mid o(e) = ld\} \\
\text{(store events)}\ S(E) &= \{e \in E \mid o(e) = st\} \\
\text{(events that write to memory)}\ W(E) &= \{e \in E \mid o(e) \in \{st, il\}\} \\
\text{(events that read from memory)}\ R(E) &= \{e \in E \mid o(e) \in \{ld, il\}\} \\
\text{(events that write location } a\text{)}\ W(E, a) &= \{e \in W(E) \mid a(e) = a\}
\end{aligned}
$$

We call a function $f : Evt \to \mathbb{N}$ an *index* function for a subset $S' \subseteq Evt$ if $f(S') = \{1, \ldots, |S'|\}$ (including the special case where S' is empty).

Definition 1 (Traces). *A* trace *is a subset $E \subseteq Evt$ satisfying*

(E1) *For all $p \in Proc$, i is an index function for $E(p)$.*
(E2) *For all $a \in Adr$, c is an index function for $W(E, a)$.*
(E3) *For all $l \in L(E)$, either $c(l) = 0$, or there exists a $w \in W(E, a(l))$ such that $c(l) = c(w)$.*

Define $\mathcal{T} \subseteq \mathcal{P}(Evt)$ to be the set of all traces. We say a trace E is a prefix *of a trace E' if $E \subseteq E'$.*

---

[1] We do not need to include memory fence operations because a full fence is semantically equivalent to an interlocked operation to a location that is not accessed anywhere else.

To reason about traces, we introduce binary relations $\rightarrow_p$ and $\rightarrow_c$:

- We use the program order $\rightarrow_p \subseteq Evt \times Evt$ to describe the relative order of events by the same processor. Specifically, we define $e \rightarrow_p e'$ if and only if $p(e) = p(e')$ and $i(e) < i(e')$. For any trace E, $\rightarrow_p$ is a partial order on E and a total order on $E(p)$ for all $p \in Proc$.
- We use the conflict order $\rightarrow_c \subseteq Evt \times Evt$ to describe the relative order of conflicting accesses (where we call two accesses $e, e' \in Evt$ *conflicting* if $a(e) = a(e')$ and $\{e, e'\} \cap W(Evt) \neq \emptyset$). Specifically, we define: $e \rightarrow_c e'$ if and only if $a(e) = a(e')$ and either (1) $o(e') \in W(Evt)$ and $c(e) < c(e')$, or (2) $(e, e') \in W(Evt) \times L(Evt)$ and $c(e) \leq c(e')$. The conflict order is not actually an 'order' in the mathematical sense because it is not transitive.

We now proceed to define the memory models *SC* (sequential consistency) and *TSO* (total store order) using an axiomatic style. To state the definitions concisely, we define the binary relation $\rightarrow_{hb}$, called *happens-before* relation, to be the union of the program and conflict orders: $\rightarrow_{hb} = (\rightarrow_p \cup \rightarrow_c)$. Note that this definition does not make $\rightarrow_{hb}$ implicitly transitive; we will take the transitive closure $\rightarrow_{hb}^*$ explicitly if required by the context.

Definition 2 (*SC*). *Define the set $\mathcal{T}^{SC} \subseteq \mathcal{T}$ of* sequentially consistent *traces to consist of all traces E that satisfy the following condition:*

(SC1) *The relation $\rightarrow_{hb}$ is acyclic on E.*

To define *TSO* for any given event set E, we first define the *relaxed happens-before relation* $\rightarrow_{rhb}$:

$$\rightarrow_{rhb} \;=\; \rightarrow_{hb} \setminus \{ (e, e') \mid e \rightarrow_p e' \wedge o(e) = st \wedge o(e') = ld \}$$

Thus the $\rightarrow_{rhb}$ relation does not put a happens-before edge between a store and a subsequent load of the same processor (even if they have the same address). This reflects the existence of a store buffer: a store may globally commit after subsequent loads by the same processor, and thus not globally appear as 'happening before the load'.

Definition 3 (*TSO*). *Define the set $\mathcal{T}^{TSO} \subseteq \mathcal{T}$ of* totally-store-ordered *traces to consist of all traces E that satisfy the following conditions:*

(TSO1) *The relation $\rightarrow_{rhb}$ is acyclic on E.*
(TSO2) *never $(e \rightarrow_p e' \wedge e' \rightarrow_c e)$ for any $e, e' \in E$*

The axiom (TSO2) is required to guarantee that loads correctly "snoop" the store buffer: the coherence index of a load may not be less than that of a previous store to the same address by the same processor. For a detailed proof that Definitions 2 and 3 are equivalent to more intuitive operational descriptions, we refer to our technical report [5].

We now formally define the set of traces $\mathcal{T}_\pi^Y$ that a program π may exhibit on a memory model $Y \in \{SC, TSO\}$. To keep our formalization light, we represent a program π abstractly by a function $next_\pi : \mathcal{T} \times Proc \rightarrow \mathcal{P}(Op \times Adr)$. The set $next_\pi(E, p)$ describes what instructions (combinations of operation and address) may possibly be issued by processor p next, after having executed E. For a trace E, let $last(E, p)$ be the element $e \in E(p)$ such that $i(e)$ is maximal, or undefined if $E(p) = \emptyset$. We say that a program π is *locally deterministic* if for all $(E, p) \in \mathrm{dom}\, next_\pi$, we have (1) $|next_\pi(E, p)| \leq 1$, and (2) for all prefixes $E' \subseteq E$ such that $last(E', p) = last(E, p)$, we have $next_\pi(E, p) = next_\pi(E', p)$. In the following, we will assume without further mention that all programs are locally deterministic. For a trace $E \in \mathcal{T}$, define the set of possible successor events under program π as

$$succ_\pi(E) = \{e \in (\mathrm{Evt} \setminus E) \mid (E \cup \{e\} \in \mathcal{T}) \text{ and } next_\pi(E, p(e)) = (o(e), a(e))\}.$$

Definition 4 (Program Traces). *For a program π and memory model $Y \in \{SC, TSO\}$, define the set of traces $\mathcal{T}_\pi^Y$ inductively as the smallest set satisfying (i) $\emptyset \in \mathcal{T}_\pi^Y$, and (ii) for all $E \in \mathcal{T}_\pi^Y$ and $e \in succ_\pi(E)$ such that $E \cup \{e\} \in \mathcal{T}^Y$, we have $E \cup \{e\} \in \mathcal{T}_\pi^Y$.*

Definition 5 (Store Buffer Safety). *The program π is called* store buffer safe *if and only if $\mathcal{T}_\pi^{TSO} = \mathcal{T}_\pi^{SC}$.*

3 Solution

We now describe how we can check store buffer safety by exploring $\mathcal{T}_\pi^{SC}$ only. The idea is to look for *borderline traces* which are defined as follows.

Definition 6 (Borderline Trace). *A sequentially consistent trace $E \in \mathcal{T}_\pi^{SC}$ of a program π is called a* borderline trace *if there exists an $e \in succ_\pi(E)$ such that $E \cup \{e\} \in (\mathcal{T}_\pi^{TSO} \setminus \mathcal{T}_\pi^{SC})$.*

Theorem 1. *A program π is store buffer safe if and only if it has no borderline traces.*

Proof. If $E \in \mathcal{T}_\pi^{SC}$ is a borderline trace, then there exists a trace $E \cup \{e\} \in (\mathcal{T}_\pi^{TSO} \setminus \mathcal{T}_\pi^{SC})$ implying $\mathcal{T}_\pi^{SC} \neq \mathcal{T}_\pi^{TSO}$. Conversely, assume $\mathcal{T}_\pi^{SC} \neq \mathcal{T}_\pi^{TSO}$. Because $\mathcal{T}_\pi^{SC} \subseteq \mathcal{T}_\pi^{TSO}$, there must exist $E \in (\mathcal{T}_\pi^{TSO} \setminus \mathcal{T}_\pi^{SC})$. By construction of $\mathcal{T}_\pi^{TSO}$, there exist traces $E_0, \ldots, E_n \in \mathcal{T}_\pi^{TSO}$ and events $e_1, \ldots, e_n$ such that $E_0 = \emptyset$, $\{e_k\} = E_k \setminus E_{k-1}$, and $E_n = E$. Because $E_n \notin \mathcal{T}_\pi^{SC}$ but $E_0 \in \mathcal{T}_\pi^{SC}$, there exists a minimal k such that $E_k \notin \mathcal{T}_\pi^{SC}$. This implies that $E_{k-1} \in \mathcal{T}_\pi^{SC}$ and E_{k-1} is a borderline trace (because $E_{k-1} \cup \{e_k\} \in (\mathcal{T}_\pi^{TSO} \setminus \mathcal{T}_\pi^{SC})$).

The following cycle characterization lemma provides an efficient method to detect borderline traces. For a trace E, let $lastR(E, p)$ be the element $e \in E(p) \cap R(E)$ such that $i(e)$ is maximal, or be undefined if $(E(p) \cap R(E)) = \emptyset$; and let $write(E, a, c)$ denote the element $e \in W(E, a)$ such that $c(e) = c$ if it exists, or be undefined otherwise.

Lemma 1 (Cycle Characterization). *Let $E \in \mathcal{T}_\pi^{SC}$ be a sequentially consistent trace of π, and let $e = o(p,i,a,c) \in succ_\pi(E)$. Let $E' = E \cup \{e\}$. Then:*

(1) $E' \notin \mathcal{T}_\pi^{SC}$ if and only if $o = ld$ and $write(E,a,c+1) \rightarrow_{hb}^* last(E,p)$.
(2) $E' \notin \mathcal{T}_\pi^{TSO}$ if and only if $o = ld$ and either
(i) $write(E,a,c+1) \rightarrow_{rhb}^* lastR(E,p)$, or
(ii) there exists $c' > c$ such that $p(write(E,a,c')) = p$.

Proof. **(1⇐).** If $o = ld$ and $write(E,a,c+1) \rightarrow_{hb}^* last(E,p)$, then

$$e \rightarrow_c write(E,a,c+1) \rightarrow_{hb}^* last(E,p) \rightarrow_p e$$

which forms a $\rightarrow_{hb}$-cycle, implying $E' \notin \mathcal{T}^{SC}$ by (SC1), and thus $E' \notin \mathcal{T}_\pi^{SC}$. **(2⇐).** either (i) or (ii) must hold; if (i) holds, we proceed as in case (1⇐): we use $e \rightarrow_c write(E,a,c+1)$ and $lastR(E,p) \rightarrow_p e$ to construct a cycle (this time, a $\rightarrow_{rhb}$-cycle) which implies $E' \notin \mathcal{T}^{TSO}$ by (TSO1), and thus $E' \notin \mathcal{T}_\pi^{TSO}$. If (ii) holds, then either $write(E,a,c') \rightarrow_p e$ or $e \rightarrow_p write(E,a,c')$; but the latter is impossible because both E and E' are traces (specifically, because i is an index function on both $E(p)$ and $E'(p)$). Therefore, $write(E,a,c') \rightarrow_p e$. Along with $e \rightarrow_c write(E,a,c')$ we conclude $E' \notin \mathcal{T}^{TSO}$ by (TSO2), and thus $E' \notin \mathcal{T}_\pi^{TSO}$. **(1⇒).** Assume $E' \notin \mathcal{T}_\pi^{SC}$. Then $E' \notin \mathcal{T}^{SC}$ (by Def. 4(ii)), which means (SC1) does not hold: specifically, $E \cup \{e\}$ has a $\rightarrow_{hb}$-cycle. Because $\rightarrow_{hb}$ is acyclic on E (because $E \in \mathcal{T}^{SC}$), it must be of the form $e \rightarrow_{hb} e_1 \rightarrow_{hb} \ldots \rightarrow_{hb} e_n \rightarrow_{hb} e$ where all $e_k \in E$ and $n \geq 1$. Now, $e \rightarrow_{hb} e_1$ by definition implies that either $e \rightarrow_p e_1$ or $e \rightarrow_c e_1$. As reasoned earlier, it can not be the case that $e \rightarrow_p e_1$ (because E and E' are both traces), thus $e \rightarrow_c e_1$. This implies that $o = ld$ (because c is an index function on both $W(E,a)$ and $W(E',a)$). Because e is a load and $e \rightarrow_c e_1$, we know $o(e_1) \in \{st, il\}$, $a(e_1) = a$ and $c(e_1) > c$, and thus either $write(E,a,c{+}1) = e_1$ or $write(E,a,c{+}1) \rightarrow_c e_1$. Therefore $write(E,a,c{+}1) \rightarrow_{hb}^* e_n$. Now, it can not be the case that $e_n \rightarrow_c e$ (otherwise $e_n \rightarrow_c^* e_1$ which creates a $\rightarrow_{hb}$-cycle within E, contradicting $E \in \mathcal{T}_\pi^{SC}$), thus $e_n \rightarrow_p e$. Therefore, either $e_n = last(E,p)$ or $e_n \rightarrow_p last(E,p)$. We can thus conclude that $write(E,a,c+1) \rightarrow_{hb}^* last(E,p)$ as desired. **(2⇒).** If $E' \notin \mathcal{T}_\pi^{TSO}$ then $E' \notin \mathcal{T}^{TSO}$ (by Def. 4(ii)). Thus either (TSO1) or (TSO2) must be violated. First, assume that E' does not satisfy (TSO1). Just as in (1⇒) (but using the relation $\rightarrow_{rhb} \subseteq \rightarrow_{hb}$), we conclude that there exists a cycle of the form $e \rightarrow_{rhb} e_1 \rightarrow_{rhb} \ldots \rightarrow_{rhb} e_n \rightarrow_{rhb} e$, that $e \rightarrow_c e_1$, that $o = ld$, that $write(E,a,c+1) \rightarrow_{rhb}^* e_n$, and that $e_n \rightarrow_p e$. The latter implies that $o(e_n) \neq st$ (otherwise not $e_n \rightarrow_{rhb} e$), and therefore either $e_n = lastR(E,p)$ or $e_n \rightarrow_{rhb} lastR(E,p)$. Thus condition (i) is satisfied. Next, assume that E' does not satisfy (TSO2). Because E does, and because we know that not $e \rightarrow_p e'$ for any $e' \in E$ (because E and E' are both traces), there must exist an $e' \in E$ such that $e' \rightarrow_p e$ and $e \rightarrow_c e'$. This implies $o(e) = ld$ (because c is an index function on both $W(E,a)$ and $W(E',a)$). Because e is a load and $e \rightarrow_c e'$, we know $o(e') \in \{st, il\}$, $a(e') = a$ and $c(e') > c$. Thus, condition (ii) is satisfied with $c' = c(e')$.

```
1  function is_store_buffer_safe(e1e2...en) returns boolean {
2    var k,p,a,c : ℕ; var E : 𝒯;
3    E := ∅;
4    for (k := 1; k <= n; k++) {
5      if (o(ek) = ld) {
6        p := p(ek); a := a(ek); c := c(ek);
7        while (c > 0) {
8          if (p = i(write(E,a,c)))
9            break;
10         if (write(E,a,c) →*rhb lastR(E,p))
11           break;
12         if (write(E,a,c) →*hb last(E,p))
13           return false;
14         c := c - 1;
15       }
16     }
17     E := E ∪ ek;
18   }
19   return true;
20 }
```

Fig. 2. Our algorithm to monitor store buffer safety in a given interleaving

3.1 Monitor Algorithm

Fig. 2 shows our implementation of a monitor that can monitor store buffer safety in any interleaved execution of the program. It processes the events in the sequence in order (and can thus be used online or offline) and reports any detected borderline traces. We now qualify the soundness and completeness of this monitor. For a sequence $w = e_1 \dots e_n \in Evt^*$ of events, let $E_w = \{e_1, \dots e_n\}$. The sequence w is called an *interleaving* of a program π if (1) the e_k are pairwise distinct, (2) $E_w \in \mathcal{T}_\pi^{SC}$, (3) $e_x \rightarrow_{hb} e_y \implies x < y$, and (4) $next_\pi(E_w, p) = \emptyset$ for all $p \in Proc$.

Theorem 2 (Soundness). *If an an interleaving w of program π is reported unsafe by our monitor, then π is not store buffer safe.*

Proof. Assume `is_store_buffer_safe(w)` returns false for $w = e_1 \dots e_n$. Let E, k, p, i, a and c' be the values of the program variables `E`, `k`, `p`, `i`, `a`, and `c` at the time of the return, respectively. Then $E = \{e_1, \dots, e_{k-1}\}$, and $e_k = ld(p, i, a, c)$ for some c. Let $e = e_k$, and let $e' = ld(p, i, a, c' - 1)$. We now argue that $E' = E \cup \{e'\} \in (\mathcal{T}_\pi^{TSO} \setminus \mathcal{T}_\pi^{SC})$, which implies that E is a borderline trace and thus $\mathcal{T}_\pi^{SC} \neq \mathcal{T}_\pi^{TSO}$ by Theorem 1 as desired. First, note that $e' \in succ_\pi(E)$ because $E \cup \{e\} \in \mathcal{T}_\pi^{SC}$ implies $E \cup \{e'\} \in \mathcal{T}$ and $(o, a) \in next_\pi(E, p)$ (using that π is locally deterministic). We can thus enlist the help of Lemma 1 to show $E' \in (\mathcal{T}_\pi^{TSO} \setminus \mathcal{T}_\pi^{SC})$. First, because the program returned at line 13, we know $write(E, a, c') \rightarrow_{hb}^* last(E, p)$, which implies $E' \notin \mathcal{T}_\pi^{SC}$ by Lemma 1, part (1). Second, because the program did not break at line 11 right before returning on line 13, we know that not $(write(E, a, c') \rightarrow_{rhb}^* lastR(E, p))$. Moreover, because the while loop was not broken at line 9, we know that $p(write(E, a, c'')) \neq p$ for all $c'' \geq c'$. By Lemma 1, part (2) we conclude that $E' \in \mathcal{T}_\pi^{TSO}$.

As for completeness, we clearly cannot detect all borderline traces by looking at a single interleaving w only. However, it is possible to detect them reliably by checking a sufficient set of interleavings. Specifically, we call a set of interleavings $I \subseteq Evt^*$ *representative* for program π if for all $E \in \mathcal{T}_\pi^{SC}$ there exists an interleaving $w \in I$ such that $E \subseteq E_w$ and there are no $\rightarrow_{hb}$-edges from $E_w \setminus E$ into E.

Theorem 3 (Completeness). *Let I be a representative set of interleavings of a program π. Then, if π is not store buffer safe, our monitor will detect it on some interleaving $w \in I$.*

Proof. By Theorem 1, we know that $\mathcal{T}_\pi^{SC} \neq \mathcal{T}_\pi^{TSO}$ implies that there exists a borderline trace $E \in \mathcal{T}_\pi^{SC}$. Thus there exists an element $e = o(p, i, a, c) \in Evt$ such that $E' = (E \cup \{e\}) \in \mathcal{T}_\pi^{TSO} \setminus \mathcal{T}_\pi^{SC}$. Because I is representative, it must contain an interleaving $w = e_1 \dots e_n$ such that $E \subseteq E_w$ is a prefix. Because $(o(e), a(e)) \in next_\pi(E, p)$, there must be a k such that $p(e_k) = p$ and $i(e_k) = i$ (otherwise $last(E_w, p) = last(E, p)$ and thus $next_\pi(E, p) = next_\pi(E_w, p)$, contradicting $next_\pi(E_w, p) = \emptyset$). We now claim that if the algorithm reaches the k-th iteration, it must return false (if it returns prior to that, it also returns false and we are satisfied). Let $E_k = \{e_1, \dots, e_{k-1}\}$. By Lemma 1, part (1), we know that $write(E, a, c+1) \rightarrow_{hb}^* last(E, p)$ within E. Now, by the choice of k, we know $E(p) = E_k(p)$, thus $last(E, p) = last(E_k, p)$, and because w is an interleaving (respects $\rightarrow_{hb}$), this implies $write(E_k, a, c+1) \rightarrow_{hb}^* last(E_k, p)$ within E_k. Moreover, we know that $c(e_k) \geq (c+1)$ because w is an interleaving and $write(E_k, a, c+1)$ appears before e_k in w. Thus, the while loop (which assigns $c(e_k)$ to the variable c initially, and then keeps decrementing it) must eventually return true at line 13 unless it is broken at either line 9 or line 11. But that is not possible, for the following reasons. First, suppose line 9 breaks. Let c' be the value of the variable c at that time; then $c+1 \leq c' \leq c(e_k)$ and $p(write(E_k, a, c')) = p$. Now, because $E(p) = E_k(p)$, we know $write(E_k, a, c') \in E$. Thus, $write(E, a, c') = write(E_k, a, c')$, implying $p(write(E, a, c')) = p$ which in turn implies $E' \notin \mathcal{T}_\pi^{TSO}$ by Lemma 1, part (2ii), contradicting the assumption. Next, suppose line 11 breaks. Let c' be the value of the variable c at that time; then $c+1 \leq c' \leq c(e_k)$ and $write(E_k, a, c') \rightarrow_{rhb}^* lastR(E_k, p)$ within E_k. Now, because $E(p) = E_k(p)$, $lastR(E_k, p) = lastR(E, p)$. Because there are no $\rightarrow_{hb}$-edges (and thus no $\rightarrow_{rhb}$-edges) from E_w into E, this implies that $write(E, a, c') \rightarrow_{rhb}^* lastR(E, p)$. Because $c + 1 \leq c'$, this implies $write(E, a, c) \rightarrow_{rhb}^* lastR(E_k, p)$, which in turn implies $E' \notin \mathcal{T}_\pi^{TSO}$ by Lemma 1, part (2i), contradicting the assumption.

A stateless model checker (such as Verisoft [10] or CHESS[21]) can provide us with a representative set of interleavings if the program is bounded (we call a program *bounded* if there exists a number $M \in \mathbb{N}$ such that $|E| < M$ for all $E \in \mathcal{T}_\pi^{SC}$). The following theorem (proved in [5]) clarifies that this is true even if partial order reduction is employed. We call a set of interleavings $I \subseteq Evt^*$

```
type       timestamp: array[2*N] of ℕ0;
var        lc: array[Proc] of timestamp;
           sc: array[Proc] of timestamp;
           mc1: array[Proc][Adr] of timestamp;
           mc2: array[Adr] of timestamp;
initially  lc[*][*] = sc[*][*] = mc1[*][*][*] = mc2[*][*] = 0;
function merge(ts1, ... tsn : timestamp) returns timestamp {
  return (maxi(tsi[1]), ... , maxi(tsi[N*2]));
}
function process_event(e : Evt) returns timestamp {
  match e with
    ld(p,i,a,c) ->
      ts := merge(lc[p], mc1[p][a]);
      ts[2*p] := ts[2*p] + 1;     // advance load count for p
      lc[p] := merge(lc[p], ts);
      mc2[a] := merge(mc2[a], ts);
    st(p,i,a,c) ->
      ts := merge(sc[p], lc[p], mc2[a]);
      ts[2*p+1] := ts[2*p+1] + 1; // advance store count for p
      forall q ≠ p do
        mc1[q][a] := merge(mc1[q][a], ts);
      mc2[a] := merge(mc2[a], ts);
      sc[p] := merge(sc[p], ts);
    il(p,i,a,c) ->
      ts := merge(sc[p], lc[p], mc2[a]);
      ts[2*p] := ts[2*p] + 1;     // advance load count for p
      ts[2*p+1] := ts[2*p+1] + 1; // advance store count for p
      forall q ∈ Proc do
        mc1[q][a] := merge(mc1[q][a], ts);
      mc2[a] := merge(mc2[a], ts);
      lc[p] := merge(lc[p], ts);
      sc[p] := merge(sc[p], ts);
  return ts;
}
```

Fig. 3. A vector clock for tracking the transitive closure $\rightarrow^*_{rhb}$

a *partial-order-complete set* for program π if for all interleavings w of π, there exists a w' in I such that $E_w = E_{w'}$.

Theorem 4. *If I is a partial-order-complete set of interleavings for a bounded program π, then it is representative for π.*

3.2 Vector Clocks

The pseudocode in Fig. 2 does not detail how to decide the conditions on lines 10 and 12. While it is well known how to use vector clocks to compute the transitive closure $\rightarrow^*_{hb}$ for a given interleaving of length n in time $O(nN)$, it is not immediately clear how to do the same for $\rightarrow^*_{rhb}$. We solved this problem by generalizing vector clocks (Def. 7 below) and by engineering a vector clock instance (Fig. 3) that can compute the transitive closure $\rightarrow^*_{rhb}$ in time $O(nN^2)$.

Theorem 5. *Let $w = e_0 \dots e_n$ be an interleaving of some program π, and let $t_1, \dots, t_n$ be the timestamps returned by the corresponding sequence of calls to* `process_event` *(Fig. 3). Then $e_i \rightarrow^*_{rhb} e_j$ if and only if $i \leq j$ and $t_i[k] \leq t_j[k]$ for all $k \in \{1, \dots, 2N\}$.*

We now describe informally how this vector clock works (for a detailed proof of the theorem see [5]). Our vector clock uses timestamps of a fixed width (here $2N$, where N is the maximal number of processors) and maintains a number of clocks (defined as global variables in Fig. 3). The computation of each timestamp follows the following pattern: (1) some of the clocks are read and merged, (2) some positions of the resulting vector are incremented to form the timestamp, and (3) the timestamp is merged back into some of the clocks. The following definition clarifies the conditions that underly this general mechanism ($in(e)$ and $out(e)$ represent the clock sets in step (1) and (3), respectively, and $gps(e)$ represents the set of positions in step (2)).

Definition 7 (General Vector Clock). *Let Σ be a set of events, and let $\rightarrow$ be a binary relation on Σ. A* general vector clock *for $(\Sigma, \rightarrow)$ is a tuple (C, G, in, out, gps) where C is a set of clocks, G is a set of groups, in, out are functions $\Sigma \rightarrow \mathcal{P}(C)$, and gps is a function $\Sigma \rightarrow \mathcal{P}(G)$ such that the following conditions are satisfied:*

(VC1) for all $\sigma \in \Sigma$, $gps(\sigma) \neq \emptyset$.
(VC2) for all $g \in G$, $\rightarrow$ is a total order on $\{\sigma \in \Sigma \mid g \in gps(\sigma)\}$.
(VC3) for all $\sigma, \sigma' \in \Sigma$, we have $(out(\sigma) \cap in(\sigma') \neq \emptyset) \Leftrightarrow (\sigma \rightarrow \sigma')$.

4 Experiments

We present experimental results for four C# programs (Fig. 4(a)). The largest one (takequeue) implements a low-lock datastructure and is part of a concurrency library at Microsoft. For all programs, Sober (1) *falsified* the original version (found that it is not store buffer safe), and (2) *verified* a fixed version (which we obtained by adding more memory fences whenever Sober showed us a borderline trace) up to some bound on the number of preemptions [21] (column 2).

We make two observations. First, a large percentage of interleavings trip the monitor (columns 3,4). Therefore, a violation is found quickly (column 5). This indicates that our monitor may be useful for falsification even in a plain testing setup (without doing exhaustive space exploration). Second, when verifying a correct program, the number of interleavings and the verification time increase dramatically with the context bound as usual [21]; however, the overhead by the store buffer safety monitor is fairly low in practice (columns 6,7), indicating that it makes sense to turn it on by default within the CHESS tool.

Figure 4(b) describes a memory model bug that we found in a production level concurrency library at Microsoft [5]. The program uses two flags `isIdling` and `hasWork` as well as a condition variable to synchronize between consumers and producers. An idle consumer waits on the condition variable if `hasWork` is

program name	context bound	# interleavings total	borderline	time [s]	ver. time [s] SoBeR	CHESS
Fig. 1(b)	∞	10	4	< 0.1	< 0.2	< 0.2
dekker	1	5	4	< 0.1	< 0.2	< 0.2
(2 threads,	2	36	23	< 0.1	0.39	0.37
2 crit-sec)	3	183	50	< 0.1	1.9	1.8
(loc 82)	4	1,219	124	< 0.1	13.2	13.0
	5	8,472	349	< 0.1	106.0	100.6
bakery	0	1	1	< 0.1	< 0.2	< 0.2
(2 threads,	1	25	20	< 0.1	0.47	0.43
3 crit-sec)	2	742	533	< 0.1	10.3	9.8
(loc 122)	3	12,436	8,599	< 0.1	189.0	181.0
takequeue	0	3	0	n.a.	< 0.3	< 0.3
(2 threads,	1	47	14	0.34	0.72	0.69
6 ops)	2	402	189	0.43	5.2	4.9
(loc 374)	3	2,318	1,197	0.74	28.9	27.8
	4	9,147	5,321	0.84	125.5	118.9
	5	29,821	17,922	0.86	481.5	461.6

(a)

```
volatile bool isIdling;
volatile bool hasWork;

//Consumer thread
void BlockOnIdle(){
   lock (condVariable){
      isIdling = true;
      if (!hasWork)
          Wait(condVariable);
      isIdling = false;
   }
}

//Producer thread
void NotifyPotentialWork(){
   hasWork = true;
   if (isIdling)
     lock (condVariable) {
        Pulse(condVariable);
     }
}
```

(b)

Fig. 4. (a) Experiments on a 2.2GHz Intel Core Duo laptop running Windows Vista. **(b)** An example of a store buffer safety bug we found in a production-level C# program.

false, but only after setting `isIdling` to true. To optimize for the common case in which there are no idle consumers, the producer acquires the lock only when `isIdling` is true. Also, to account for a possible race on the `isIdling` flag, the producer sets `hasWork` to true before checking the `isIdling` flag. We can see that in all sequentially consistent executions the producer correctly wakes up the idle consumer, if any. However, in the presence of store buffers, a store can be delayed past a subsequent load.[2] In particular, the consumer can read `hasWork` before its write to `isIdling` is visible to the producer. Thus, the producer may erroneously believe that no consumer is idling, not perform a signal, and leave the consumer waiting forever.

5 Conclusions and Future Work

We have presented a novel method to verify store buffer safety using a non-intrusive monitor that is run alongside sequentially consistent executions of the program. We have demonstrated that this method is scalable, automatic and precise enough to find store-buffer-related bugs in realistic low-lock code, such as concurrency libraries.

As future work, we consider including memory model relaxations other than store buffers, and we plan to apply our monitor to larger execution traces.

[2] Note that unlike Java, the C# memory model does not guarantee a sequentially consistent ordering of volatile accesses.

References

1. Adve, S., Gharachorloo, K.: Shared memory consistency models: a tutorial. Computer 29(12), 66–76 (1996)
2. Ben-Ari, M.: Principles of Concurrent Programming. Prentice Hall, Englewood Cliffs (1982)
3. Burckhardt, S., Alur, R., Martin, M.: Bounded verification of concurrent data types on relaxed memory models: A case study. In: Ball, T., Jones, R.B. (eds.) CAV 2006. LNCS, vol. 4144, pp. 489–502. Springer, Heidelberg (2006)
4. Burckhardt, S., Alur, R., Martin, M.: CheckFence: Checking consistency of concurrent data types on relaxed memory models. In: PLDI, pp. 12–21 (2007)
5. Burckhardt, S., Musuvathi, M.: Effective program verification for relaxed memory models. Technical Report MSR-TR-2008-12, Microsoft Research (2008)
6. Compaq Computer Corporation. Alpha Architecture Reference Manual, 4th edn. (January 2002)
7. Dill, D., Park, S., Nowatzyk, A.: Formal specification of abstract memory models. In: Symposium on Research on Integrated Systems, pp. 38–52. MIT Press, Cambridge (1993)
8. Fang, X., Lee, J., Midkiff, S.: Automatic fence insertion for shared memory multiprocessing. In: ICS, pp. 285–294 (2003)
9. Frey, B.: PowerPC Architecture Book v2.02. IBM Corporation (2005)
10. Godefroid, P.: Model checking for programming languages using Verisoft. In: POPL 1997: Principles of Programming Languages, pp. 174–186 (1997)
11. Gopalakrishnan, G., Yang, Y., Sivaraj, H.: QB or not QB: An efficient execution verification tool for memory orderings. In: Alur, R., Peled, D.A. (eds.) CAV 2004. LNCS, vol. 3114, pp. 401–413. Springer, Heidelberg (2004)
12. Hill, M.: Multiprocessors should support simple memory-consistency models. IEEE Computer 31(8), 28–34 (1998)
13. Huynh, T., Roychoudhury, A.: A Memory Model Sensitive Checker for C#. In: Misra, J., Nipkow, T., Sekerinski, E. (eds.) FM 2006. LNCS, vol. 4085, pp. 476–491. Springer, Heidelberg (2006)
14. IBM Corporation. z/Architecture Principles of Operation, 1st edn. (2000)
15. Intel Corporation. Intel 64 and IA-32 Architectures Software Developer's Manual, vol. 3A (November 2006)
16. Intel Corporation. Intel 64 Architecture Memory Ordering White Paper (August 2007)
17. Lamport, L.: A new solution of dijkstra's concurrent programming problem. Communications of the ACM 17(8), 453–455 (1974)
18. Lamport, L.: How to make a multiprocessor computer that correctly executes multiprocess programs. IEEE Trans. Comp. C-28(9), 690–691 (1979)
19. Manson, J., Pugh, W., Adve, S.: The Java memory model. In: Principles of Programming Languages (POPL), pp. 378–391 (2005)
20. Morrison, V.: Understand the impact of low-lock techniques in multithreaded apps. MSDN Magazine 20(10) (October 2005)
21. Musuvathi, M., Qadeer, S.: Iterative context bounding for systematic testing of multithreaded programs. In: PLDI, pp. 446–455 (2007)
22. Park, S., Dill, D.L.: An executable specification, analyzer and verifier for RMO (relaxed memory order). In: SPAA, pp. 34–41 (1995)

23. Shasha, D., Snir, M.: Efficient and correct execution of parallel programs that share memory. ACM Trans. Program. Lang. Syst. 10(2), 282–312 (1988)
24. Weaver, D., Germond, T.: The SPARC Architecture Manual Version 9. PTR Prentice Hall, Englewood Cliffs (1994)
25. Yang, Y., Gopalakrishnan, G., Lindstrom, G.: Memory-model-sensitive data race analysis. In: Davies, J., Schulte, W., Barnett, M. (eds.) ICFEM 2004. LNCS, vol. 3308, pp. 30–45. Springer, Heidelberg (2004)
26. Yang, Y., Gopalakrishnan, G., Lindstrom, G., Slind, K.: Nemos: A framework for axiomatic and executable specifications of memory consistency models. In: IPDPS (2004)

Mechanical Verification of Transactional Memories with Non-transactional Memory Accesses*

Ariel Cohen[1], Amir Pnueli[1], and Lenore D. Zuck[2]

[1] New York University,
{arielc,amir}@cs.nyu.edu
[2] University of Illinois at Chicago,
lenore@cs.uic.edu

Abstract. *Transactional memory* is a programming abstraction intended to simplify the synchronization of conflicting memory accesses (by concurrent threads) without the difficulties associated with locks. In a previous work we presented a formal framework for proving that a transactional memory implementation satisfies its specifications and provided with model checking verification of some using small instantiations. This paper extends the previous work to capture non-transactional accesses to memory, which occurs, for example, when using legacy code. We provide a mechanical proof of the soundness of the verification method, as well as a mechanical verification of a version of the popular TCC implementation that includes non-transactional memory accesses. The verification is performed by the deductive temporal checker TLPVS.

1 Introduction

Transactional Memory [5] is a simple solution for coordinating and synchronizing concurrent threads that access the same memory locations. It transfers the burden of concurrency management from the programmers to the system designers and enables a safe composition of scalable applications, we well as efficiently utilizes the multiple cores. Multicore and many-core processors, which require concurrent programs in order to gain a full advantage of the multiple number of processors, has become the mainstream architecture for microprocessor chips and thus many new transactional memory implementations have been proposed recently (see [9] for an excellent survey).

A transactional memory (TM) receives requests from clients and issues responses. The requests are usually part of a *transaction* that is a sequence of operations starting with a request to *open* a transaction, followed by a sequence of read/write requests, followed by a request to *commit* (or *abort*). The TM responds to requests. When a transaction requests a successful "commit," all of its effects are stored in the memory. If a transaction is aborted (by either issuing an abort request or when TM detects that it should be aborted) all of its effects are removed. Thus, a transaction is a sequence of *atomic* operations, either all complete successfully and all its write operations update the memory, or none completes and its write operations do not alter the memory. In

* This research was supported in part by ONR grant N00014-99-1-0131 and NSF grants CCF 0742686 and CNS-0720525.

A. Gupta and S. Malik (Eds.): CAV 2008, LNCS 5123, pp. 121–134, 2008.

addition, committed transaction should be serializable – the sequence of operations belonging to successful transactions should be such that it can be reordered (preserving the order of operations in each transaction) so that the operation of each transaction appear consecutive, and a "read" from any memory location returns the value of the last "write" to that memory location.

TMs are often parameterized by their properties. These may include the conflicts they are to avoid, when are the conflicts detected, how they are resolved, when is the memory updated, whether transactions can be nested, etc. (see [9] for a list of such properties). Each set of parameters defines a unique set of sequences of events that can occur is a TM so to guarantee atomicity and serializability. We refer to the set of sequences of events allowed by a TM as its s*specifications*. A particular implementation does not necessarily generate all allowed sequences, but should only generate allowed sequences. The topic of this paper (as well as [3]) is to formally verify that a TM implementation satisfies its specification that is uniquely defined by its parameters.

Such parameters were given in [12]'s widely-cited paper, which was the first to characterize transactional memory in a way that captured and clarified the many semantic distinctions among the most popular implementations of TMs. Scott's ([12]) approach is to begin with classical notions of transactional histories and sequential specifications, and to introduce two important notions. The first is a *conflict function* which specifies when two overlapping (concurrent) transactions cannot both succeed (a safety condition). The second is an *arbitration function* which specifies which of two transactions must fail (a liveness condition). Scott's work went a long way towards clarification of the semantics of TMs, but did not facilitate mechanical verification of implementations.

The work in [3] (co-authored by the authors of this paper) took a first step towards modeling TMs, accordingly to [12]'s parameters, so to as allow for mechanical verification of their implementations. There, a specification of a TM is represented by a fair state machine that is parameterized by a set of *admissible* interchanges — a set of rules specifying when a pair of consecutive operations in a sequence of transactional operations can be safely swapped without introducing or removing a conflict. All the conflicts described in [12] can be cast as admissible sets. The specification machine takes a stream of transactional requests as inputs, and outputs a serializable sequence of the input requests and their responses. The fairness is used to guarantee that each transaction is eventually closed (committed or aborted) and, if committed, appears in the output. Some proof rules are given to show that a TM implementation satisfies its specification. The applicability of the approach is demonstrated on several well-known TM implementations. Small instantiations of each of the case study were shown to specify their specification using the model checker TLC [8].

This paper extends the work of [3] in two directions. The first is to add another parameter to the system — *non-transactional* memory accesses. Unlike their transactional counterparts, non-transactional accesses cannot be aborted. While atomicity and serializability requirements remain, where a non-transaction operation is cast as a singleton, successfully committed, transaction. The second direction is a framework that allows for a mechanical formal verification that TM implementations satisfy their specifications. The tool we use is TLPVS [11], which embeds temporal logic and its deductive frame-work within the theorem prover PVS [10]. Using TLPVS entailed some changes to the [3] proof rules that establish that an implementation indeed refines its specification. In fact, the rule presented here is more general than its predecessor. Using TLPVS also entailed restricting to interchange rules that can be described by temporal logics (which still covers all of [12]'s conflicts). For simplicity, we chose to restrict to interchanges whose temporal description uses only past temporal operators (i.e., depend only on the history leading to the interchange), which rules out [12]'s mixed invalidation conflict.

We make here a strong assumption on non-transactional accesses, namely, that the transactional memory is aware that non-transactional accesses, as soon as they occur. While the TM cannot abort such accesses, it may use them in order to abort transactions that are under its control. It is only with such or similar assumption that total consistency or coherence can be maintained.

We demonstrate the new framework by presenting TLPVS proofs that some TM implementations with non-transactional accesses satisfy their specifications, given an admissible interchange.

To the best of our knowledge, the work presented here is the first to employ a theorem prover for verifying correctness of transactional memories and the first to formally verify an implementation that handles non-transactional memory accesses.

The rest of the paper is organized as follows: Section 2 provides preliminary definitions related to transactional memory, and defines the concept of admissible interchanges. Section 3 provides a specification model of a transactional memory. Section 4 discusses a proof rule for verifying implementations. Section 5 presents a simple implementation of transactional memory that handles non-transactional memory accesses. Section 6 shows how to apply deductive verification using TLPVS to verify this implementation. Section 7 provides some conclusions and open problems.

2 Transactional Sequences and Interchanges

We extend the [3] to support non-transactional memory accesses and separate each action into a request/response pair, as well as give a temporal definition for interchanges.

2.1 Transactional Sequences

Assume n *clients* that direct requests to a *memory system*, denoted by *memory*. For every client p, let the set of *non-transactional invocations by client* p consists of:

- $\iota R_p^{nt}(x)$ – A non-transactional request to read from address $x \in \mathbb{N}$.
- $\iota W_p^{nt}(y, v)$ – A non-transactional request to write value $v \in \mathbb{N}$ to address $y \in \mathbb{N}$.

Let the set of *transactional invocations by client* p consists of:

- $\iota \blacktriangleleft_p$ – An open transaction request.
- $\iota R_p^t(x)$ – A transactional read request from address $x \in \mathbb{N}$.
- $\iota W_p^t(y, v)$ – A transactional request to write the value $v \in \mathbb{N}$ to address $y \in \mathbb{N}$.
- $\iota \blacktriangleright_p$ – A commit transaction request.
- $\iota \not\blacktriangleright_p$ – An abort transaction request.

The memory provides a response for each invocation. Erroneous invocations (e.g., a $\iota \blacktriangleleft_p$ while client p has a pending transaction) are responded by the memory returning an error flag err. Non-erroneous invocations, except for ιR^t and ιR^{nt} are responded by the memory returning an acknowledgment ack. Finally, for non-erroneous $\iota R^t_p(x)$ and $\iota R^{nt}_p(x)$ the memory returns the (natural) value of the memory at location x. We assume that invocations and responses occur atomically and consecutively, i.e., there are no other operation that interleave an invocation and its response.

Let $E^{nt}_p\colon \{R^{nt}_p(x,u), W^{nt}_p(x,v)\}$ be the set of *non-transactional observable events*, $E^t_p\colon \{\blacktriangleleft_p, R^t_p(x,u), W^t_p(x,v), \blacktriangleright_p, \not\blacktriangleright_p\}$ be the set of *transactional observable events* and $E_p = E^{nt} \cup E^t$, i.e. all events associated with client p. We consider as observable events only requests that are accepted, and abbreviate the pair (*invocation, non-err response*) by omitting the ι-prefix of the invocation. Thus, $W^t_p(x,v)$ abbreviates $\iota W^t_p(x,v), ack_p$. For read actions, we include the value read, that is, $R^t_p(x,u)$ abbreviates $\iota R^t(x), \rho R(u)$. When the value written/read has no relevance, we write the above as $W^t_p(x)$ and $R^t_p(x)$. When both values and addresses are of no importance, we omit the addresses, thus obtaining W^t_p and R^t_p (symmetric abbreviations and shortcuts are used for the non-transactional observable events). The output of each action is its relevant observable event when the invocation is accepted, and undefined otherwise. Let E be the set of all observable events over all clients, i.e., $E = \bigcup_{p=1}^{n} E_p$ (similarly define E^{nt} and E^t to be the set of all non-transactional and the set of all transactional observable events, respectively).

Let $\sigma\colon e_0, e_1, \ldots, e_k$ be a finite sequence of observable E-events. We say that the sequence $\hat{\sigma}$ over E^t is *σ's transactional sequence*, where $\hat{\sigma}$ is obtained from σ by replacing each R^{nt}_p and W^{nt}_p by $\blacktriangleleft_p\ R^t_p\ \blacktriangleright_p$ and $\blacktriangleleft_p\ W^t_p\ \blacktriangleright_p$, respectively. That is, each non-transactional event of σ is transformed into a singleton committed transaction in $\hat{\sigma}$. The sequence σ is called a *well-formed transactional sequence* (TS for short) if the following all hold:

1. For every client p, let $\hat{\sigma}|_p$ be the sequence obtained by projecting $\hat{\sigma}$ onto E^t_p. Then $\hat{\sigma}|_p$ satisfies the regular expression T^*_p, where T_p is the regular expression $\blacktriangleleft_p\ (R^t_p + W^t_p)^*(\blacktriangleright_p + \not\blacktriangleright_p)$. For each occurrence of T_p in $\hat{\sigma}|_p$, we refer to its first and last elements as *matching*. The notion of matching is lifted to $\hat{\sigma}$ itself, where $\blacktriangleleft_p$ and $\blacktriangleright_p$ (or $\not\blacktriangleright_p$) are matching if they are matching in $\hat{\sigma}|_p$;
2. The sequence $\hat{\sigma}$ is *locally read-write consistent*: for any subsequence of $\hat{\sigma}$ of the form $\langle W^t_p(x,v)\,\eta\, R^t_p(x,u)\rangle$ where η contains no $\blacktriangleright_p, \not\blacktriangleright_p$, or $W^t_p(x)$ events, $u = v$.

We denote by $\mathcal{T}$ the set of all well-formed transactional sequences, and by *pref*$(\mathcal{T})$ the set of $\mathcal{T}$'s prefixes. Note that the requirement of local read-write consistency can be enforced by each client locally. To build on this observation, we assume that, within a single transaction, there is no $R^t_p(x)$ following a $W^t_p(x)$, and there are no two reads or two writes to the same address. With these assumptions, the requirement of local read-write consistency is always (vacuously) satisfied. A TS σ is *atomic* if:

1. $\hat{\sigma}$ satisfies the regular expression $(T_1 + \cdots + T_n)^*$. That is, there is no overlap between any two transactions;
2. $\hat{\sigma}$ is *globally read-write consistent*: namely, for any subsequence $W^t_p(x,v)\eta$ $R^t_q(x,u)$ in $\hat{\sigma}$, where η contains $\blacktriangleright_p$, which is not preceded by $\not\blacktriangleright_p$, and contains no event $W^t_k(x)$ followed by event $\blacktriangleright_k$, it is the case that $u = v$.

2.2 Interchanging Events

The notion of a correct implementation is that every TS can be transformed into an atomic TS by a sequence of interchanges which swap two consecutive events. This definition is parameterized by the set $\mathcal{A}$ of *admissible interchanges* which may be used in the process of serialization. Rather than attempt to characterize $\mathcal{A}$, we choose to characterize its complement $\mathcal{F}$, the set of *forbidden interchanges*. The definition here differs from the one in [3] in two aspects: There, in order to characterize $\mathcal{F}$, we allowed arbitrary predicates over the TS, here, we restrict to temporal logic formulae. Also, while [3] allowed swaps that depend on future events, here we restrict to swaps whose soundness depends only on the history leading to them. This restriction simplifies the verification process, and is the one used in all TM systems we are aware of. Note that it does not allow to express [12]'s mixed invalidation conflict. In all our discussions, we assume *strict serializability* which implies that while serializing a TS, the order of committed transactions has to be preserved.

Consider a temporal logic over E using the past operators $\ominus$ (previously), $\diamondminus$ (sometimes in the past), and $\mathcal{S}$ (since). Let σ be a prefix of a well-formed TS over E^t (i.e., $\sigma = \hat{\sigma}$). We define a satisfiability relation $\models$ between σ and a temporal logic formula φ so that $\sigma \models \varphi$ if at the end of σ, φ holds. (The more standard notation is $(\sigma, |\sigma| - 1) \models \varphi$, but since we always interpret formulae at the end of sequences we chose the simplified notation.)

Some of the restrictions we place in $\mathcal{F}$ are structural. For example, the formula $p \neq q \wedge \blacktriangleright_p \wedge \ominus \blacktriangleright_q$ forbids the interchange of closures of transactions belonging to different clients. This guarantees the strictness of the serializability process. Similarly, the restriction $u_p \wedge \ominus v_p$, where $u_p, v_p \in E_p$, forbids the interchanges of two events belonging to the same client. Other formulas may guarantee the absence of certain conflicts. For example, following [12], a *lazy invalidation* conflict occurs when committing one transaction may invalidate a read of another, i.e., if for some transactions T_p and T_q and somc mcmory address x, we have $R_p(x), W_q(x) \prec \blacktriangleright_q \prec \blacktriangleright_p$ (where "$e_i \prec e_j$" denotes that e_i precedes e_j). Formally, the last two events in σ *cannot* be interchanged when for some $p \neq q$,

$$\sigma \models \blacktriangleright_q \wedge \ominus(R_p(x) \wedge (\neg \blacktriangleright_q) \mathcal{S}\, W_q(x)) \tag{1}$$

Similarly, we express conflicts by TL formulae that determine, for any prefix of a TS (that includes only E^t events), whether the two last events in the sequence can be safely interchanged without removing the conflict. For a conflict c, the formula that forbids interchanges that may remove instances of this conflict is called the *maintaining formula for c* and is denoted by m_c. Thus, Formula 1 is the maintaining formula for the conflict lazy invalidation. See [2] for a list of the maintaining formulae for each[12]'s conflicts (expect for mixed invalidation that requires future operators).

Let $\mathcal{F}$ be a set of forbidden formulae characterizing all the forbidden interchanges, and let $\mathcal{A}$ denote the set of interchanges which do not satisfy any of the formulas in $\mathcal{F}$. Assume that $\sigma = a_0, \ldots, a_k$. Let σ' be obtained from σ by interchanging two elements, say a_{i-1} and a_i. We then say that σ' is *1-derivable from σ with respect to $\mathcal{A}$* if $(a_0, \ldots, a_i) \not\models \bigvee \mathcal{F}$. Similarly, we say that σ' is *derivable from σ with respect to $\mathcal{A}$* if there exist $\sigma = \sigma_0, \ldots, \sigma_\ell = \sigma'$ such that for every $i < \ell$, σ_{i+1} is 1-derivable from σ_i with respect to $\mathcal{A}$.

A TS is *serializable with respect to* $\mathcal{A}$ if there exists an atomic TS that is derivable from it with respect to $\mathcal{A}$. The sequence $\breve{\sigma}$ is called the *purified version* of TS σ if $\breve{\sigma}$ is obtained by removing from $\hat{\sigma}$ all aborted transactions, i.e., removing the opening and closing events for such a transaction and all the read-write events by the same client that occurred between the opening and closing events. When we specify the correctness of a transactional memory implementation, only the purified versions of the implementation's transaction sequences will have to be serializable.

3 Specification and Implementation

Let $\mathcal{A}$ be a set of admissible interchanges which we fix for the remainder of this section. We next describe $Spec_{\mathcal{A}}$, a specification of transactional memory that generates all sequences whose corresponding TSs are serializable with respect to $\mathcal{A}$. The process $Spec_{\mathcal{A}}$ is described as a fair transition system. In every step, it outputs an element in $E_{\perp} = E \cup \{\perp\}$. The sequence of outputs it generates, once the $\perp$ elements are projected away, is the set of TSs that are admissible with respect to $\mathcal{A}$. $Spec_{\mathcal{A}}$ uses the following data structures:

- *spec_mem*: **array** $\mathbb{N} \mapsto \mathbb{N}$ — A persistent memory. Initially, $\mathit{spec\_mem}[i] = 0$ for all $i \in \mathbb{N}$;
- $\mathcal{Q}$: **list over** $E^t \cup \bigcup_p \{mark_p\}$ — A queue-like structure, to which elements are appended, interchanged, deleted, and removed. The sequence of elements removed from this queue-like structure defines an atomic TS that can be obtained by serialization of $Spec_{\mathcal{A}}$'s output with respect to $\mathcal{A}$. For each client p, it is assumed that $mark_p \notin E_p$ is a new symbol. Initially, $\mathcal{Q}$ is empty;
- *spec_out*: **scalar in** $E_{\perp} = E \cup \{\perp\}$ — An output variable, initially $\perp$;
- *spec_doomed*: **array** $[1..n] \mapsto$ **bool** — An array recording which pending transactions are doomed to be aborted. Initially $\mathit{spec\_doomed}[p] = \mathrm{F}$ for every p.

Fig. 1 summarized the steps taken by $Spec_{\mathcal{A}}$. The first column describes the value of *spec_out* with each step; it is assumed that every step produces an output. The second column describes the effects of the step on the other variables. The third column describes the conditions under which the step can be taken. The following abbreviations are used in Fig. 1:

- A client p is *pending* if $\mathit{spec\_doomed}[p] = \mathrm{T}$ or if $\mathcal{Q}|_p$ is not empty and does not terminate with $\blacktriangleright_p$;
- a client p is *unmarked* if $\mathcal{Q}|_p$ does not terminate with $mark_p$;
- a p-action a is *locally consistent with* $\mathcal{Q}$ if $\mathcal{Q}|_p, a$ is a prefix of some locally consistent p-transaction;
- a transaction T is *consistent with* spec_mem if every $R^t(x, v)$ in T is either preceded by some $W^t(x, v)$, or else $v = \mathit{spec\_mem}[x]$;
- the *update of* spec_mem *by a transaction* T is $\mathit{spec\_mem}'$ where for every location x for which T has no $W^t(x, v)$ actions, $\mathit{spec\_mem}'[x] = \mathit{spec\_mem}[x]$, and for every memory location x such that T has some $W^t(x, v)$ actions, $\mathit{spec\_mem}'[x]$ is the value written in the last such action in T;

spec_out	other updates	conditions
$\blacktriangleleft_p$	append $\blacktriangleleft_p$ to $\mathcal{Q}$	p is not pending
$R^t_p(x, v)$	append $R^t_p(x, v)$ to $\mathcal{Q}$	p is pending, unmarked and *spec_doomed*$[p]$ = F; $R(x, v)$ is locally consistent with $\mathcal{Q}$
$R^t_p(x, v)$	none	p is pending, unmarked and *spec_doomed*$[p]$ = T
$W^t_p(x, v)$	append $W^t_p(x, v)$ to $\mathcal{Q}$	p is pending, unmarked and *spec_doomed*$[p]$ = F
$W^t_p(x, v)$	none	p is pending, unmarked and *spec_doomed*$[p]$ = T
$\not\blacktriangleright_p$	delete p's pending transaction from $\mathcal{Q}$; set *spec_doomed*$[p]$ to F	p is pending
$\blacktriangleright_p$	update *spec_mem* by p's pending transaction; remove p-pending transaction from $\mathcal{Q}$	p has a consistent transaction at the front of $\mathcal{Q}$ that ends with $mark_p$ (p is pending and marked)
$R^{nt}_p(x, v)$	append $\blacktriangleleft_p$, $R^t_p(x, v)$, $\blacktriangleright_p$ to $\mathcal{Q}$	p is not pending
$W^{nt}_p(x, v)$	append $\blacktriangleleft_p$, $W^t_p(x, v)$, $\blacktriangleright_p$ to $\mathcal{Q}$	p is not pending
$\perp$	set *spec_doomed*$[p]$ to T; delete all pending p-events from $\mathcal{Q}$	p is pending and *spec_doomed*$[p]$ = F
$\perp$	apply a $\mathcal{A}$-valid transformation to $\mathcal{Q}$	none
$\perp$	append $mark_p$ to $\mathcal{Q}$	p is pending and unmarked
$\perp$	none	none

Fig. 1. Steps of *Spec*$_{\mathcal{A}}$

- an $\mathcal{A}$*-valid transformation to* $\mathcal{Q}$ is a sequence of interchanges of $\mathcal{Q}$'s entries that is consistent with $\mathcal{A}$. To apply the transformations, each $mark_p$ is treated as if it is $\blacktriangleright_p$.

The role of *spec_doomed* is to allow *Spec*$_{\mathcal{A}}$ to be implemented with various arbitration policies. It can, at will, schedule a pending transaction to be aborted by setting *spec_doomed*$[p]$, by so "dooming" p's pending transaction to be aborted. The variable *spec_doomed*$[p]$ is reset once the transaction is actually aborted (when *Spec*$_{\mathcal{A}}$ outputs $\not\blacktriangleright_p$). Note that actions of doomed transactions are not recorded on $\mathcal{Q}$.

We assume a fairness requirement, namely, that for every client $p = 1, \ldots, n$, there are infinitely many states of *Spec*$_{\mathcal{A}}$ where $\mathcal{Q}|_p$ is empty and *spec_doomed*$[p]$ = F. This implies that every transaction eventually terminates (commits or aborts). It also guarantees that the sequence of outputs is indeed serializable. Note that unlike the specification described in [3] where progress can always be guaranteed by aborting transactions, here, because of the non-transactional accesses, there are cases where $\mathcal{Q}$ cannot be emptied.

While *Spec*$_{\mathcal{A}}$ resembles its counterpart in [3], the treatment of non-transactional accesses entailed numerous changes: Roughly speaking, each transactional access is appended to the queue, and removed from it when the transaction commits, aborts, or is doomed to abort. When a transaction attempts to commit, a special marker $mark_p$ is appended to the queue, and, if there are admissible interchanges that move the whole transaction into the head of the queue, its events are removed from the queue and it commits. Thus, the queue never contains $\blacktriangleright$-events, and it contains at most one $mark$-event. A non-transactional p-access a^{nt}_p (that can only be accepted when p-has no pending transaction) is treated by appending $\blacktriangleleft_p$, a^t_p, $\blacktriangleright_p$ to the queue. (Note that the non-transactional event is replaced by its transactional counterpart.) Hence, here the queue may have $\blacktriangleright$-events. The transactions (or, rather, non-transaction) corresponding to them *cannot* be "doomed to abort" since such a transaction is, by definition (see below), not pending. The liveness properties require that all transaction are eventually removed from the queue. As a consequence, unlike its [3] version, *Spec*$_{\mathcal{A}}$ does not support "eager version management" that eagerly updates the memory with every W^t-action (that doesn't conflict any pending transaction) which is reasonable since eager version management and the requirement to commit each non-transactional access are contradictory.

A sequence σ over E is *compatible with* Spec$_\mathcal{A}$ if it can be obtained by the sequence of *spec_out* that *Spec*$_\mathcal{A}$ outputs once all the $\bot$'s are removed. We then have:

Claim. For every sequence σ over E, σ is compatible with *Spec*$_\mathcal{A}$ iff $\hat{\sigma}$ is serializable with respect to $\mathcal{A}$.

An *implementation TM*: $(\mathit{read}, \mathit{commit})$ of a transactional memory consists of a pair of functions *read*: $\mathit{pref}(TS) \times [1..n] \times \mathbb{N} \to \mathbb{N}$ and *commit*: $\mathit{pref}(TS) \times [1..n] \to \{\mathit{ack}, \mathit{err}\}$ For a prefix σ of a TS, $\mathit{read}(\sigma, p, x)$ is the response (value) of the memory to an accepted $\iota R_p^{nt}(x)/\iota R_p^t(x)$ request immediately following σ, and $\mathit{commit}(\sigma, p)$ is the response (*ack* or *err*) of the memory to a $\iota \blacktriangleright_p$ request immediately following σ.

A TS σ is said to be *compatible* with the memory *TM* if:

1. For every prefix $\eta R_p^{nt}(x, u)$ or $\eta R_p^t(x, u)$ of σ, $\mathit{read}(\eta, p, x) = u$.
2. For every prefix $\eta \blacktriangleright_p$ of σ, $\mathit{commit}(\eta, p) = \mathit{ack}$.

An implementation *TM*: $(\mathit{read}, \mathit{commit})$ is a *correct implementation of a transactional memory with respect to* $\mathcal{A}$ if every TS compatible with *TM* is also compatible with *Spec*$_\mathcal{A}$.

4 Verifying Implementation Correctness

We present a proof rule for verifying that an implementation satisfies the specification *Spec*. The rule is adapted [7], which, in turn is based on [1]'s *abstraction mapping*. In addition to being case in a different formal framework, and ignoring compassion (which is rarely, if ever, used in TMs), the rule generalizes on the two given in [3] by allowing for general stuttering equivalence.

To apply the underlying theory, we assume that both the implementation and the specifications are represented as OPFSs (see [2] for details). In the current application, we prefer to adopt an *event-based* view of reactive systems, by which the observed behavior of a system is a (potentially infinite) set of events. Technically, this implies that one of the system variables O is designated as an *output variable*. The observation function is then defined by $\mathcal{O}(s) = s[O]$. It is also required that the observation domain always includes the value $\bot$, implying no observable event. In our case, the observation domain of the output variable is $E_\bot = E \cup \{\bot\}$.

Let η: $e_0, e_1, \ldots$ be an infinite sequence of $E_\bot$-values. The $E_\bot$-sequence $\widetilde{\eta}$ is called a *stuttering variant* of the sequence η if it can be obtained by removing or inserting finite strings over $\{\bot\}$ at (potentially infinitely many) different positions within η.

Let σ: $s_0, s_1, \ldots$ be a computation of OPFS $\mathcal{D}$, that is, a sequence of states where s_0 satisfies the initial condition, each state is a successor of the previous one, and for every justice (weak fairness) requirement, σ has infinitely many states that satisfy the requirement. The *observation* corresponding to σ (i.e., $\mathcal{O}(\sigma)$) is the $E_\bot$ sequence $s_0[O], s_1[O], \ldots$ obtained by listing the values of the output variable O in each of the states. We denote by $\mathit{Obs}(\mathcal{D})$ the set of all observations of system $\mathcal{D}$.

Let $\mathcal{D}_C$ be a *concrete* system whose set of states is Σ_C, set of observations is $E_\perp$, observation function is $\mathcal{O}_C$, initial condition is Θ_C, transition relation is ρ_C, and justice requirements are $\cup_{f\in\mathcal{F}_C}\mathcal{J}(f)$. Similarly, let $\mathcal{D}_A$ be an *abstract* system whose set of states, set of observations, observation function, initial condition, transition relation, and justice requirements are Σ_A, $E_\perp$, $\mathcal{O}_A$, Θ_A, ρ_A, and $\cup_{f\in\mathcal{F}_A}\mathcal{J}(f)$ respectively. (For simplicity, we assume that neither system contains compassion requirements.) Note that we assume that both systems share the same observations domain $E_\perp$. We say that system $\mathcal{D}_A$ *abstracts* system $\mathcal{D}_C$ (equivalently $\mathcal{D}_C$ *refines* $\mathcal{D}_A$), denoted $\mathcal{D}_C \sqsubseteq \mathcal{D}_A$ if, for every observation $\eta \in Obs(\mathcal{D}_C)$, there exists $\widetilde{\eta} \in Obs(\mathcal{D}_A)$, such that $\widetilde{\eta}$ is a stuttering variant of η. In other words, modulo stuttering, $Obs(\mathcal{D}_C)$ is a subset of $Obs(\mathcal{D}_A)$. We denote by s and S the states of $\mathcal{D}_C$ and $\mathcal{D}_A$, respectively.

Rule ABS-REL in Fig. 2 is a proof rule to establish that $\mathcal{D}_C \sqsubseteq \mathcal{D}_A$. The method advocated by the rule assumes the identification of an *abstraction relation* $R(s,S) \subseteq \Sigma_C \times \Sigma_A$. If the relation $R(s,S)$ holds, we say that the abstract state S is an R-image of the concrete state s.

$$
\begin{array}{lll}
\textbf{R1}.\ \Theta_C(s) \rightarrow \exists S : R(s,S)\ \wedge\ \Theta_A(S) & & \\
\textbf{R2}.\ \mathcal{D}_C \quad \models \Box(R(s,S)\ \wedge\ \rho_C(s,s') \quad \rightarrow \quad \exists S' : R(s',S')\ \wedge\ \rho_A(S,S')) & & \\
\textbf{R3}.\ \mathcal{D}_C \quad \models \Box(R(s,S)\ \rightarrow\ \mathcal{O}_C(s) = \mathcal{O}_A(S)) & & \\
\textbf{R4}.\ \mathcal{D}_C \quad \models \Box\Diamond(\forall S : R(s,S) \rightarrow \mathcal{J}(f)(S)), & & \text{for every } f \in \mathcal{F}_A \\
\hline
\qquad\qquad \mathcal{D}_C \sqsubseteq \mathcal{D}_A & &
\end{array}
$$

Fig. 2. Rule ABS-REL

Premise R1 of the rule states that for every initial concrete state s, it is possible to find an initial abstract state $S \models \Theta_A$, such that $R(s,S) = \text{T}$.

Premise R2 states that for every pair of concrete states, s and s', such that s' is a ρ_C-successor of s, and an abstract state S which is a R-related to s, there exists an abstract state S' such that S' is R-related to s' and is also a ρ_A-successor of S. Together, R1 and R2 guarantee that, for every run $s_0, s_1, \ldots$ of $\mathcal{D}_C$ there exists a run $S_0, S_1, \ldots,$ of $\mathcal{D}_A$, such that for every $j \geq 0$, S_j is R-related to s_j. Premise R3 states that if abstract state S is R-related to the concrete state s, then the two states agree on the values of their observables. Together with the previous premises, we conclude that for every σ a run of $\mathcal{D}_C$ there exists a corresponding run of $\mathcal{D}_A$ which has the same observation as σ. Premise R4 ensures that the abstract justice requirements hold in any abstract state sequence which is a (point-wise) R-related to a concrete computation. Here, $\Box$ is the (linear time) temporal operator for "henceforth", $\Diamond$ the temporal operator for "eventually", thus, $\Box\Diamond$ means "infinitely often." It follows that every sequence of abstract states which is R-related to a concrete computation σ and is obtained by applications of premises R1 and R2 is an abstract computation whose observables match the observables of σ. This leads to the following claim which was proved using TLPVS (see Section 6):

Claim. If the premises of rule ABS-REL are valid for some choice of R, then $\mathcal{D}_A$ is an abstraction of $\mathcal{D}_C$.

5 An Example: TCC with Non-transactional Accesses

We demonstrate our proof method by verifying a TM implementation which is essentially TCC [4] augmented with non-transactional accesses. Its specifications is given by $Spec_{\mathcal{A}}$ where $\mathcal{A}$ is the admissible set of events corresponding to the lazy invalidation conflict described in Subsection 2.2.

In the implementation, transactions execute speculatively in the clients' caches. When a transaction commits, all pending transactions that contain some read events from addresses written to by the committed transaction are "doomed." Similarly, non-transactional writes cause pending transactions that read from the same location to be "doomed." A doomed transactions may execute new read and write events in its cache, but it must eventually abort.

Here we present the implementation, and in Section 6 explain how we can verify that it refines its specification using the proof rule ABS-REL in TLPVS. We refer to the implementation as *TM*. It uses the following data structures:

- *imp_mem*: $\mathbb{N} \rightarrow \mathbb{N}$ — A persistent memory. Initially, for all $i \in \mathbb{N}$, $imp\_mem[i] = 0$;
- *cache*: **array**$[1..n]$ **of list of** E^t — Caches of clients. For each $p \in [1..n]$, $cache[p]$, initially empty, is a sequence over E^t_p that records the actions of p's pending transaction;
- *imp_out*: **scalar in** $E_\perp = E \cup \{\perp\}$ — an output variable recording responses to clients, initially $\perp$;
- *imp_doomed*: **array** $[1..n]$ **of booleans** — An array recording which transactions are doomed to be aborted. Initially $imp\_doomed[p] = \mathrm{F}$ for every p.

TM receives requests from clients, to each it updates its state, including updating the output variable *imp_out*, and issues a response to the requesting client. The responses are either a value in $\mathbb{N}$ (for a ιR^t or ιR^{nt} requests), an error err (for $\iota\blacktriangleright$ requests that cannot be performed), or an acknowledgment ack for all other cases. Fig. 3 describes the actions of *TM*, where for each request we describe the new output value, the other updates to *TM*'s state, the conditions under which the updates occur, and the response to the client that issues the request. For now, ignore the comments in the square brackets under the "conditions" column. The last line represents the idle step where no actions occurs and the output is $\perp$.

Comment: For simplicity of exposition, we assume that clients only issue reads for locations they had not written to in the pending transaction.

The specification of Section 3 specifies not only the behavior of the Transactional Memory but also the combined behavior of the memory when coupled with a typical clients module. A generic clients module, *Clients*(n), may, at any step, invoke the next request for client p, $p \in [1..n]$, provided the sequence of E_p-events issued so far (including the current one) forms a prefix of a well-formed sequence. The justice requirement of *Clients*(n) is that eventually, every pending transaction issues an ack-ed $\iota\blacktriangleright$ or an $\iota\not\blacktriangleright_p$.

Combining modules *TM* and *Clients*(n) we obtain the complete implementation, defined by:

$$Imp: \quad TM \;|||\; Clients(n)$$

Request	*imp_out*	Other Updates	Conditions	Response
$\iota \blacktriangleleft_p$	$\blacktriangleleft_p$	append $\blacktriangleleft_p$ to *cache*$[p]$	[*cache*$[p]$ is empty]	*ack*
$\iota R^t_p(x)$	$R^t_p(x, v)$	append $R_p(x, v)$ to *cache*$[p]$	$v =$ *imp_mem*$[x]$; [*cache*$[p]$ is empty] (see comment)	*imp_mem*$[x]$
$\iota W^t_p(x, v)$	$W^t_p(x, v)$	append $W_p(x, v)$ to *cache*$[p]$	[*cache*$[p]$ is not empty]	*ack*
$\iota \not\blacktriangleright_p$	$\not\blacktriangleright_p$	set *cache*$[p]$ to empty; set *imp_doomed*$[p]$ to F	[*cache*$[p]$ is not empty]	*ack*
$\iota \blacktriangleright_p$	$\blacktriangleright_p$	set *cache*$[p]$ to empty; for every x and $q \neq p$ such that $W^t_p(x) \in$ *cache*$[p]$ and $R^t_p(x) \in$ *cache*$[q]$ set *spec_doomed*$[q]$ to T; update *imp_mem* by *cache*$[p]$	*imp_doomed*$[p] =$ F; [*cache*$[p]$ is not empty]	*ack*
$\iota \blacktriangleright_p$	$\perp$	none	*imp_doomed*$[p] =$ T; [*cache*$[p]$ is not empty]	*err*
$\iota R^{nt}_p(x)$	$R^{nt}_p(x, v)$	none	$v =$ *imp_mem*$[x]$; [*cache*$[p]$ is empty]	*imp_mem*$[x]$
$\iota W^{nt}_p(x, v)$	$W^{nt}_p(x, v)$	set *imp_mem*$[x]$ to v; for every q such that $R^t(x) \in$ *cache*$[q]$ set *imp_doomed*$[q]$ to T	[*cache*$[p]$ is empty]	*ack*
none	$\perp$	none	none	none

Fig. 3. The actions of *TM*

where ||| denote the *synchronous* composition operator defined in [6]; This composition in combines several of the actions of each of the modules into one.

The actions of *Imp* can be described similarly to the one given by Fig. 3, where the first and last column are ignored, the conditions in the brackets are added. The justice requirements of *Clients*(n), together with the observation that both $\iota \not\blacktriangleright$ and an *ack*-ed $\iota \blacktriangleright$ cause the cache of the issuing client to be emptied, imply that *Imp*'s justice requirement is that for every $p = 1, \ldots, n$, *cache*$[p]$ is empty infinitely many times.

The application of rule ABS-REL requires the identification of a relation R which holds between concrete and abstract states. In [3], we used the relation defined by:

$$\begin{aligned} \mathit{spec\_out} = \mathit{imp\_out}\ \wedge\ \mathit{spec\_mem} = \mathit{imp\_mem} \\ \wedge\ \mathit{spec\_doomed} = \mathit{imp\_doomed} \\ \wedge\ \textstyle\bigwedge_{p=1}^{n} \mathit{imp\_doomed}[\mathrm{p}] \longrightarrow (\mathcal{Q}|_p = \mathit{cache}[p]) \end{aligned}$$

however, there the implementation did not support non-transactional accesses. In Section 6 we provide the relation that was applied when proving the augmented implementation using TLPVS.

6 Deductive Verification in TLPVS

In this section we describe how we used TLPVS [11] to verify the correctness of the implementation provided in Section 5. TLPVS was developed to reduce the substantial manual effort required to complete deductive temporal proofs of reactive systems. It embeds temporal logic and its deductive framework within the high-order theorem prover, PVS [10]. It includes a set of theories defining linear temporal logic (LTL), proof rules for proving soundness and response properties, and strategies which aid in conducting the proofs. In particular, it has a special framework for verifying unbounded systems and theories. See [10] and [11] for thorough discussions for proving with PVS and TLPVS, respectively. In [3] we described the verification of three known transactional memory implementations with the (explicit-state) model checker TLC. This verification involved TLA^+ [8] modules for both the specification and implementation, and abstraction *mapping* associating each of the specification's variables with an expression over the implementation's variables.

This effort has several drawbacks: The mapping does not allow for *abstraction* relations between *states*, but rather for *mappings* between *variables*. We therefore used a proof rule that is weaker than ABS-REL and auxiliary structures. For example, for $\mathcal{Q}|_p = \mathit{cache}[p]$, which cannot be expressed in TLA^+, we used an auxiliary queue that can be mapped into $\mathcal{Q}$ and that records the order in which events are invoked in the implementation. And, like any other model checking took, TLC can only be used to verify small instantiations, rather than the general case. A full deductive verification requires a theorem prover.

Our tool of choice is TLPVS. Since, however, TLPVS only supports the model of PFS, we formulate OPFS in the PVS specification language. We then defined a new theory that uses two OPFSs, one for the abstract system (specification) and another for the concrete system (implementation), and proved, in a rather straightforward manner, that the rule ABS-REL is sound.

We then defined a theory for the queue-like structures used in both specification and implementation. This theory required, in addition to the regular queue operations, the definition of the projection ($|$) and deletion of projected elements, which, in turn, required the proofs of several auxiliary lemmas.

Next all the components of both OPFS's defining the abstract and concrete systems were defined. To simplify the TLPVS proofs, some of the abstract steps were combined. For example, when a $\mathit{Spec}_{\mathcal{A}}$ commits a transaction, we combined the steps of interchanging events, removing them from $\mathcal{Q}$, and setting $\widetilde{\mathit{spec\_out}}$ to ▶. This restricts the set of $\mathit{Spec}_{\mathcal{A}}$'s runs but retains soundness. Formally, $\mathit{TM} \sqsubseteq \widetilde{\mathit{Spec}}_{\phi_{li}}$ implies that $\mathit{TM} \sqsubseteq \mathit{Spec}_{\phi_{li}}$, where $\widetilde{\mathit{Spec}}_{\phi_{li}}$ is the restricted specification

The abstraction relation R between concrete and abstract states was defined by:

```
rel: RELATION = (LAMBDA s_c, s_a:
  s_c'out = s_a'out  AND  s_c'mem = s_a'mem  AND
  s_c'doomed = s_a'doomed AND
  FORALL(id: ID): (NOT s_c'doomed(id)) IMPLIES
                         project(id,s_a'Q) = s_c'caches(id)  AND
  FORALL(id: ID): (empty(s_c'caches(id))) IMPLIES
                         empty(project(id,s_a'Q)))
```

Here, s_c is a concrete state and s_a is an abstract state. The relation R equates the values of out, mem and doomed in the two systems. It also states that if the transaction of a client is not doomed, then its projection on the abstract $\mathcal{Q}$ equals to the concrete client's cache, and if the concrete cache is empty then so is the projection of the abstract $\mathcal{Q}$ on the client's current transaction. An additional invariant, stating that each value read by a non-doomed client is consistent with the memory was also added.

In order to prove that *TM* $\sqsubseteq$ $Spec_{\phi_{li}}$, $\mathcal{D}_C$ and $\mathcal{D}_A$ of ABS-REL were instantiated with *TM* and $Spec_{\phi_{li}}$, respectively, and all the premises were verified. The proofs are in $http://cs.nyu.edu/acsys/tlpvs/tm.html$.

7 Conclusion and Future Work

We extended our previous work on a verification framework for transactional memory implementations against their specifications, to accommodate non-transactional memory accesses. We also developed a methodology for verifying transactional memory implementations based on the theorem prover TLPVS that provides a framework for verifying parameterized systems against temporal specifications. We obtained mechanical verifications of both the soundness of the method and the correctness of an implementation which is based on TCC augmented with non-transactional accesses.

Our extension for supporting non-transactional accesses is based on the assumption that an implementation can detect such accesses. We are currently working on weakening this assumption. We are also planning to study liveness properties of transactional memory.

References

1. Abadi, M., Lamport, L.: The existence of refinement mappings. Theoretical Computer Science 82(2), 253–284 (1991)
2. Cohen, A., Pnueli, A., Zuck, L.D.: Verification of transactional memories that support non-transactional memory accesses. In: TRANSACT 2008 (February 2008)
3. Cohen, A., O'Leary, J.W., Pnueli, A., Tuttle, M.R., Zuck, L.D.: Verifying correctness of transactional memories. In: Proceedings of the 7th International Conference on Formal Methods in Computer-Aided Design (FMCAD), November 2007, pp. 37–44 (2007)
4. Hammond, L., Wong, W., Chen, M., Carlstrom, B.D., Davis, J.D., Herzberg, B., Prabhu, M.K., Wijaya, H., Kozyrakis, C., Olukotun, K.: Transactional memory coherence and consistency. In: Proc. 31st annu. Int. Symp. on COmputer Architecture, p. 102. IEEE Computer Society, Los Alamitos (2004)
5. Herlihy, M., Moss, J.E.B.: Transactional memory: architectural support for lock-free data structures. In: ISCA 1993: Proceedings of the 20th annual international symposium on Computer architecture, San Diego, California, United States, pp. 289–300. ACM Press, New York (1993)
6. Kesten, Y., Pnueli, A.: Verification by augmented finitary abstraction. Inf. Comput. 163(1), 203–243 (2000)
7. Kesten, Y., Pnueli, A., Shahar, E., Zuck, L.D.: Network invariants in action. In: Brim, L., Jančar, P., Křetínský, M., Kucera, A. (eds.) CONCUR 2002. LNCS, vol. 2421, pp. 101–115. Springer, Heidelberg (2002)
8. Lamport, L.: Specifying Systems: The TLA^+ Language and Tools for Hardware and Software Engineers. Addison-Wesley, Reading (2002)

9. Larus, J.R., Rajwar, R.: Transactional Memory. Morgan & Claypool Publishers (2007)
10. Owre, S., Shankar, N., Rushby, J.M., Stringer-Calvert, D.W.J.: PVS System Guide. Computer Science Laboratory, SRI International, Menlo Park, CA (September 1999)
11. Pnueli, A., Arons, T.: Tlpvs: A pvs-based ltl verification system. In: Dershowitz, N. (ed.) Verification: Theory and Practice. LNCS, vol. 2772, pp. 596–625. Springer, Heidelberg (2004)
12. Scott, M.L.: Sequential specification of transactional memory semantics. In: Proc. TRANSACT the First ACM SIGPLAN Workshop on Languages, Compiler, and Hardware Suppport for Transactional Computing, Ottawa (2006)

Automated Assume-Guarantee Reasoning by Abstraction Refinement

Mihaela Gheorghiu Bobaru[1,2], Corina S. Păsăreanu[1], and Dimitra Giannakopoulou[1]

[1] PSGS and RIACS, NASA Ames Research Center, Moffett Field, CA 94035, USA
[2] Department of Computer Science, University of Toronto, Toronto, ON M5S 3G4, Canada
mg@cs.toronto.edu,
{corina.s.pasareanu,dimitra.giannakopoulou}@nasa.gov

Abstract. Current automated approaches for compositional model checking in the assume-guarantee style are based on learning of assumptions as deterministic automata. We propose an alternative approach based on abstraction refinement. Our new method computes the assumptions for the assume-guarantee rules as conservative and not necessarily deterministic abstractions of some of the components, and refines those abstractions using counterexamples obtained from model checking them together with the other components. Our approach also exploits the alphabets of the interfaces between components and performs iterative refinement of those alphabets as well as of the abstractions. We show experimentally that our preliminary implementation of the proposed alternative achieves similar or better performance than a previous learning-based implementation.

1 Introduction

Despite impressive recent progress in the application of model checking to the verification of realistic systems, the essential challenge in model checking remains the well-known state-space explosion problem [8]. Compositional techniques attempt to tame this problem by applying verification to individual components and merging the results without analyzing the whole system. In checking components individually, it is often necessary to incorporate some knowledge of the context in which each component is expected to operate correctly. Assume-guarantee reasoning [13,15] addresses this issue by using *assumptions* that capture the expectations that a component makes about its environment. Assumptions have traditionally been developed manually, which has limited the practical impact of assume-guarantee reasoning.

In recent work, automation has been achieved through learning-based techniques [10]. The L* learning algorithm [2] is used to generate the assumptions needed for the assume-guarantee rules. The simplest such rule checks if a system composed of components M_1 and M_2 satisfies a property P by checking that M_1 under assumption A satisfies P (*Premise 1*) and discharging A on the environment M_2 (*Premise 2*). For safety properties, *Premise 2* amounts to checking that A is a conservative abstraction of M_2, *i.e.*, an abstraction that preserves all of M_2's execution paths. This rule is also represented as follows, where the notation is described in more detail in Section 2.

A. Gupta and S. Malik (Eds.): CAV 2008, LNCS 5123, pp. 135–148, 2008.

$$\frac{\begin{array}{l}\textit{(Premise 1)}\ \langle A\rangle\ M_1\ \langle P\rangle\\ \textit{(Premise 2)}\ \langle \textit{true}\rangle\ M_2\ \langle A\rangle\end{array}}{\langle \textit{true}\rangle\ M_1 \parallel M_2\ \langle P\rangle} \tag{1}$$

Learning-based assume-guarantee verification is an iterative process, during which L* makes conjectures in the form of automata that represent intermediate assumptions. Each conjectured assumption A is used to check the two premises of Rule 1. The process ends if A passes both premises of the rule, in which case the property holds in the system, or if it uncovers a real violation. Otherwise, a counterexample is returned and L* modifies the conjecture. Similar approaches are proposed in [1,4,17]; the work in [12] uses sampling rather than L* to learn the assumptions in a similar way.

In this paper we propose an alternative approach, AGAR (Assume-Guarantee Abstraction Refinement), that automates assume-guarantee reasoning by iteratively computing assumptions as conservative abstractions of the interface behavior of M_2, *i.e.*, the behavior that concerns the interaction with M_1. In each iteration, the computed assumption A satisfies *Premise 2* of the Rule 1 by construction and it is only checked for *Premise 1*. If the check is successful, we conclude that $M_1 \parallel M_2$ satisfies the property; if the check fails, we get a counterexample trace that we analyze to see if it corresponds to a real error in $M_1 \parallel M_2$ or it is spurious due to the over-approximation in the abstraction. If it is spurious, we used it to refine A and then repeat the entire process. Unlike learning-based assumption generation, AGAR does not constrain assumptions to be *deterministic*. Therefore the assumptions constructed with AGAR can be (potentially) exponentially smaller than those obtained with learning, resulting in smaller verification problems.

To reduce the assumption sizes even further, we also combine the abstraction refinement with an orthogonal technique, *interface alphabet refinement*, which extends AGAR so that it starts the construction of A with a small subset of the interface alphabet and adds actions to the alphabet as necessary until the required property is shown to hold or to be violated in the system. Actions to be added are discovered also by counterexample analysis. We introduced alphabet refinement in [11] for learning-based assume-guarantee reasoning; we adapt it here for AGAR[1]. We have implemented AGAR with alphabet refinement in the explicit state model checker LTSA [14] and performed a series of experiments that demonstrate that it can achieve better performance than L* for Rule 1 above.

Related work. AGAR is a variant of the well-known CEGAR (Counter Example-Guided Abstraction Refinement) [7] with the notable differences that the computed abstractions keep information only about the interface behavior of M_2 that concerns the interaction with M_1 while it abstracts away its internal behavior, and that the counterexamples used for the refinement of M_2's abstractions are obtained in an assume-guarantee style by model checking the other component, M_1.

CEGAR has been used before in compositional reasoning in [5]). In that work, a conservative abstraction of every component is constructed and then all the resulting abstractions are composed and checked. If the check passes, the verification concludes successfully, otherwise the resulting abstract counterexample is analyzed on every

[1] Note that [6] introduced a related alphabet minimization technique for L* as well.

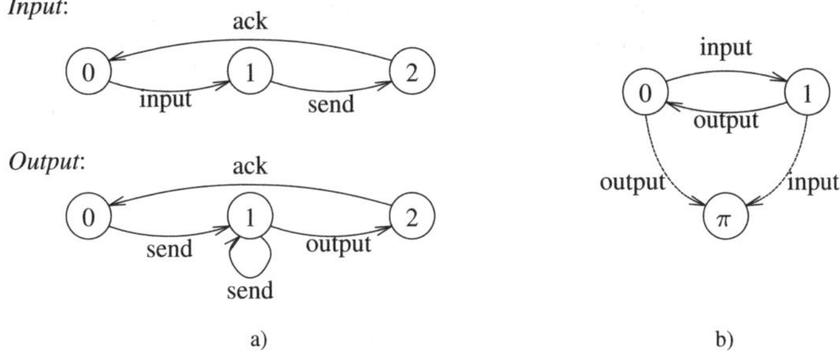

Fig. 1. (a) Example LTSs; (b) *Order* property

abstraction that is refined if needed. The work does not use assume-guarantee reasoning, it does not address the reduction of the interface alphabets and it has not been compared with learning-based techniques.

A comparison of learning and CEGAR-based techniques has been performed in [3] but for a different problem: the "interface synthesis" for a single component whose environment is unknown. In our context, this would mean generating an assumption that passes *Premise 1*, in the absence of a second component against which to check *Premise 2*. The interface being synthesized by the CEGAR-based algorithm in [3] is built as an abstraction of M_1. The work does not apply reduction to interface alphabets, nor does it address the verification of the generated interfaces against other components, *i.e.*, completing the assume-guarantee reasoning.

2 Preliminaries

Labeled Transition Systems (LTSs). We model components as finite-state *labeled transition systems* (LTSs), as considered by LTSA. Let $\mathbf{U}$ be the universal set of observable actions and let τ denote a special action that is unobservable.

An LTS M is a tuple $\langle Q, \Sigma, \delta, q_0 \rangle$, where: Q is a finite non-empty set of states; $\Sigma \subseteq \mathbf{U}$ is the *alphabet* of M; $\delta \subseteq Q \times (\Sigma \cup \{\tau\}) \times Q$ is a transition relation, and q_0 is the initial state. We write $(q, a, q') \in \delta$ as $q \xrightarrow{a} q'$. An LTS M is *non-deterministic* if it contains τ-transitions or if $\exists (q, a, q'), (q, a, q'') \in \delta$ such that $q' \neq q''$. Otherwise, M is *deterministic*. π denotes an *error state* with no outgoing transitions, and Π denotes the LTS $\langle \{\pi\}, \mathbf{U}, \emptyset, \pi \rangle$. Let $M = \langle Q, \Sigma, \delta, q_0 \rangle$ and $M' = \langle Q', \Sigma', \delta', q_0' \rangle$; M *transits* into M' with action a, denoted $M \xrightarrow{a} M'$, if $(q_0, a, q_0') \in \delta$ and either $Q = Q'$, $\Sigma = \Sigma'$, and $\delta = \delta'$ for $q_0' \neq \pi$, or, in the special case where $q_0' = \pi$, $M' = \Pi$.

Parallel Composition. Parallel composition "$\|$" is a commutative and associative operator such that: given LTSs $M_1 = \langle Q_1, \Sigma_1, \delta^1, q_0^1 \rangle$ and $M_2 = \langle Q_2, \Sigma_2, \delta^2, q_0^2 \rangle$, $M_1 \parallel M_2$ is Π if either one of M_1, M_2 is Π. Otherwise, $M_1 \parallel M_2$ is an LTS $M = \langle Q, \Sigma, \delta, q_0 \rangle$ where $Q = Q_1 \times Q_2, q_0 = (q_0^1, q_0^2), \Sigma = \Sigma_1 \cup \Sigma_2$, and δ is defined as follows (the symmetric version also applies): $M_1 \parallel M_2 \xrightarrow{a} M_1' \parallel M_2$ if $M_1 \xrightarrow{a} M_1', a \notin \Sigma_2$, and $M_1 \parallel M_2 \xrightarrow{a} M_1' \parallel M_2'$ if $M_1 \xrightarrow{a} M_1', M_2 \xrightarrow{a} M_2', a \neq \tau$.

As an example [10], consider a simple communication channel that consists of two components whose LTSs are shown in Fig. 1(a).

Paths and traces. A *path* in an LTSs $M = \langle Q, \Sigma, \delta, q_0 \rangle$ is a sequence p of alternating states and (observable or unobservable actions) of $M, p = q_{i_0}, a_0, q_{i_1}, a_1, \ldots, a_{n-1}, q_{i_n}$ such that for every $k \in \{0, \ldots, n-1\}$ we have $(q_{i_k}, a_k, q_{i_{k+1}}) \in \delta$.

The *trace of path* p, denoted $\sigma(p)$ is the sequence $b_0, b_1, \ldots, b_l$ of actions along p, obtained by removing all τ from $a_0, \ldots, a_{n-1}$. A state q *reaches* a state q' in M with a sequence of actions t, denoted $q \stackrel{t}{\Rightarrow} q'$, if there exists a path p from q to q' in M whose trace is t, *i.e.*, $\sigma(p) = t$. A *trace of* M is the trace of a path in M starting from q_0. The set of all traces of M forms the *language* of M, denoted $\mathcal{L}(M)$. For any trace $t = a_0, a_1, \ldots, a_{n-1}$, a *trace LTS* can be constructed whose only transitions are $q_0 \stackrel{a_0}{\rightarrow} q_1 \stackrel{a_1}{\rightarrow} q_2 \ldots \stackrel{a_{n-1}}{\rightarrow} q_n$. We sometimes abuse the notation and denote by t both a trace and its trace LTS. The meaning should be clear from the context. For $\Sigma' \subseteq \Sigma$, $t\downarrow_{\Sigma'}$ is the trace obtained by removing from t all actions $a \notin \Sigma$. Similarly, $M\downarrow_{\Sigma'}$ is an LTS over Σ obtained from M by renaming to τ all the action labels not in Σ. Let t_1, t_2 be two traces. Let Σ_1, Σ_2 be the sets of actions occurring in t_1, t_2, respectively. By the *symmetric difference* of t_1 and t_2 we mean the symmetric difference of sets Σ_1 and Σ_2.

Safety properties. A *safety LTS* is a deterministic LTS not containing π. A safety property P is a *safety LTS* whose language $\mathcal{L}(P)$ defines the acceptable behaviors over Σ_P.

An LTS $M = \langle Q, \Sigma, \delta, q_0 \rangle$ satisfies $P = \langle Q_P, \Sigma_P, \delta_P, q_0^P \rangle$, denoted $M \models P$, iff $\forall t \in \mathcal{L}(M) \cdot t\downarrow_{\Sigma_P} \in \mathcal{L}(P)$. For checking a property P, its safety LTS is *completed* by adding error state π and transitions on all the missing outgoing actions from all states into π so that the resulting transition relation is (left-)total (when seen as in $(Q \times (\Sigma \cup \{\tau\})) \times Q)$ and deterministic; the resulting LTS is denoted by P_{err}. LTSA checks $M \models P$ by computing $M \parallel P_{err}$ and checking if π is reachable in the resulting LTS.

For example, the *Order* property in Fig. 1(b) states that inputs and outputs come in matched pairs, with the input always preceding the output. The dashed arrows represent transitions to the error state that were added to obtain *Order$_{err}$*.

Assume-guarantee triples. An assume-guarantee triple $\langle A \rangle M \langle P \rangle$ is true if whenever component M is part of a system satisfying assumption A, the system must also guarantee property P. In LTSA, this reduces to checking whether $A \parallel M \models P$.

Learning assumptions with L*. Previous work [10] uses the L* algorithm [2] to iteratively learn the assumption A for Rule 1, as a *deterministic* finite state automaton. L* needs to interact with a *teacher* that answers *queries* and validates *conjectures*. For membership queries on string s, the teacher uses LTSA to check $\langle s \rangle\ M_1\ \langle P \rangle$; if true, then $s \in \mathcal{L}(A)$ and the Teacher returns "true". Otherwise, the answer to the query is "false". The conjectures returned by L* are intermediate assumptions; the teacher implements two *oracles* to validate these conjectures: *Oracle 1* guides L* towards a conjecture that makes $\langle A \rangle\ M_1\ \langle P \rangle$ true and then *Oracle 2* is invoked to discharge A on M_2. If this is also true, then the assume guarantee rule ensures that P holds on $M_1 \parallel M_2$; the teacher returns "true" and the computed assumption A. If model checking returns "false", the returned counterexample is analyzed to determine if P is indeed violated in $M_1 \parallel M_2$ or if A is imprecise due to learning, in which case A is modified and the process repeats. If A has n states, L* makes at most $n-1$ incorrect conjectures. The number of membership queries made by L* is $O(kn^2 + n \log m)$, where k is the size of A's alphabet and m is the length of the longest counterexample returned when a conjecture is made.

Interface alphabet. When reasoning in an assume-guarantee style, there is a natural notion of the complete *interface* between M_1 and M_2, when property P is checked. Let $M_1 = \langle Q_1, \Sigma_1, \delta^1, q_0^1 \rangle$ and $M_2 = \langle Q_2, \Sigma_2, \delta^2, q_0^2 \rangle$ be LTSs modeling two components and let $P = \langle Q_P, \Sigma_P, \delta^P, q_0^P \rangle$ be a safety property. The interface alphabet Σ_I is defined as $\Sigma_I = (\Sigma_1 \cup \Sigma_P) \cap \Sigma_2$.

3 Motivating Example

We motivate our approach using the input-output example from Section 2. We show that even on this simple example AGAR leads to smaller assumptions in fewer iterations than the learning approach, and therefore it potentially leads faster to smaller verification problems.

Let $M_1 =$ *Input*, $M_2 =$ *Output*, and $P =$ *Order*. As mentioned, we aim to automatically compute an assumption according to Rule 1. Instead of "guessing" an assumption and then checking both premises of the rule, as in the learning approaches, we *build* an abstraction that satisfies *Premise 2* by construction. Therefore, all that needs to be checked is *Premise 1*.

The initial abstraction A of *Output* is illustrated in Figure 2(a). Its alphabet consists of the interface between *Input* and the *Order* property on one side, and *Output* on the other, *i.e.*, the alphabet of A is $\Sigma_I = \{(\Sigma_{Input} \cup \Sigma_{Order}) \cap \Sigma_{Output}$. The LTS A is constructed simply by mapping all concrete states in *Output* to the same abstract state 0 which has a self-loop on every action in Σ_I and no other transitions. By construction, A is an overapproximation of M_2, *i.e.*, $\mathcal{L}(M_2{\downarrow_{\Sigma_I}}) \subseteq \mathcal{L}(A)$, and therefore *Premise 2* $\langle true \rangle$ M_2 $\langle A \rangle$ holds. Checking *Premise 1* of the assume-guarantee rule using A as the assumption fails, with abstract counterexample: 0, *output*, 0. We simulate this counterexample on M_2 and find that it is spurious (*i.e.*, it does not correspond to a trace in M_2), therefore A needs to be refined so that the refined abstraction no longer contains this trace. We split abstract state 0 into two new abstract states: abstract state 0, representing concrete states 0 and 2 that do not have an outgoing *output* action, and abstract state 1, representing concrete state 1 that has an outgoing *output* action, and adjust the transitions accordingly. The refined abstraction A', shown in Figure 2(a), is checked again for *Premise 1* and this time it passes, therefore AGAR terminates and reports that the property holds.

The sequence of assumptions learned with L* is shown in Figure 2(b). The assumption computed by AGAR thus has two states fewer than that obtained from learning and is computed in two fewer iterations.

4 Assume-Guarantee Abstraction Refinement (AGAR)

The abstraction refinement presented here is an adaptation of the CEGAR framework of [7], with the following notable differences: 1) abstraction refinement is performed

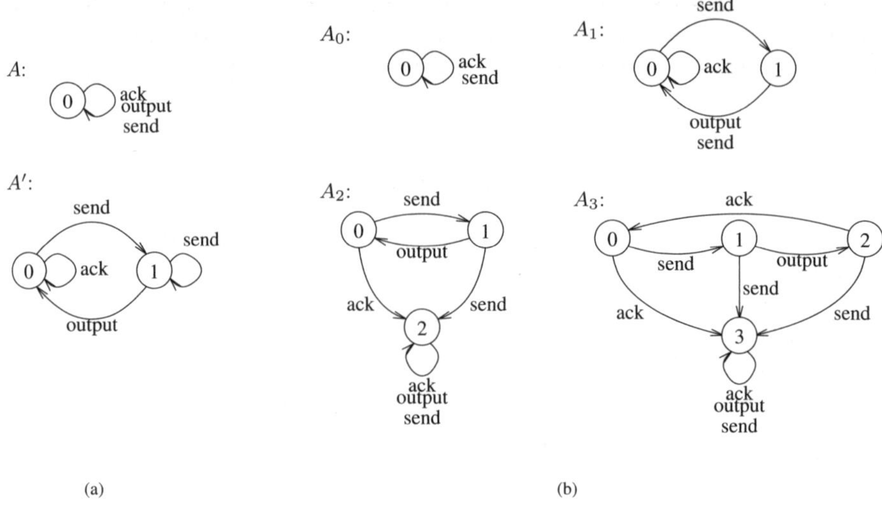

Fig. 2. Assumptions computed (a) with our algorithm and (b) with L*

in the context of LTSs; abstract transitions for LTSs are computed using *closure* with respect to actions that are not in their interface alphabet, and 2) counterexample analysis is performed in an assume-guarantee style: a counterexample obtained from model checking one component is used to refine abstractions of a different component.

In this section, we start by describing, independently of the assume-guarantee rule, abstraction refinement as applied to LTSs. We then describe how we use this abstraction refinement in an iterative algorithm (AGAR) that computes assumptions for Rule 1. Later on, we combine AGAR with an orthogonal algorithm that performs iterative refinement of the interface alphabet between the analyzed components.

4.1 Abstraction Refinement for LTSs

Abstraction. Let $C = \langle Q_C, \Sigma_C, \delta^C, q_0^C \rangle$ be an LTS that we refer to as *concrete*. Let alphabet Σ_A be such that $\Sigma_A \subseteq \Sigma_C$. An *abstraction* A of C is an LTS $\langle Q_A, \Sigma_A, \delta^A, q_0^A \rangle$ such that there exists a surjection $\alpha : Q_C \rightarrow Q_A$, called the *abstraction* function, that maps each *concrete state* $q^C \in Q_C$ to an *abstract state* $q^A \in Q_A$; q_0^A must be such that $\alpha(q_0^C) = q_0^A$. The *concretization* function $\gamma : Q_A \rightarrow 2^{Q_C}$ is defined for any $q^A \in Q_A$ as $\gamma(q^A) = \{q^C \in Q_C \mid \alpha(q^C) = q^A\}$. Note that γ induces a partition on Q_C, namely $\{\gamma(q^A) \mid q^A \in Q_A\}$.

To define the abstract transition relation δ^A, we first introduce the notion of reachability with respect to a subset alphabet. For $q^C \in C, a \in \Sigma_C$, we define the set $Reachable_C(q^C, a, \Sigma_A)$ of concrete states q_i^C reachable from q^C on action a, under the transitive closure of δ^C over actions in $(\Sigma_C \setminus \Sigma_A) \cup \{\tau\}$:

$$Reachable_C(q^C, a, \Sigma_A) = \{q_i^C \in C | \exists t, t' \in ((\Sigma_C \setminus \Sigma_A) \cup \{\tau\})^* \cdot q^C \stackrel{t}{\Rightarrow} q_i^C \text{ or } q^C \stackrel{t,b,t'}{\Rightarrow} q_i^C\}.$$

Algorithm 1. CEGAR for LTSs with respect to subset alphabets

Inputs: Concrete LTS C, its abstraction A, and an abstract counterexample $p = q_0^A, a_1, q_1^A, a_2, \ldots, a_n, q_n^A$ in A.

Outputs: a concrete counterexample t, if p is not spurious, or a refined abstraction A' without path p, if p is spurious.

1: $i \leftarrow 0$
2: $S_0 \leftarrow \{q_0^C\}$
3: **while** $S_i \neq \emptyset \wedge i \leq n-1$ **do**
4: $\quad i \leftarrow i + 1$
5: $\quad S_i \leftarrow \gamma(q_i^A) \cap \mathit{Reachable}_C(S_{i-1}, a_i, \Sigma_A)$
6: **end while**
7: **if** $S_i = \emptyset$ **then**
8: $\quad$ split q_{i-1}^A into two new abstract states x_{i-1}^A, z_{i-1}^A s.t. $\gamma(x_{i-1}^A) = \gamma(q_{i-1}^A) \cap \{q^C \mid \mathit{Reachable}_C(q^C, a_i, \Sigma_A) \cap q_i^A \neq \emptyset\}$, $\gamma(z_{i-1}^A) = \gamma(q_{i-1}^A) \setminus \gamma(x_{i-1}^A)$
9: $\quad$ build new abstraction A' with $Q_{A'} = Q_A \setminus \{q_{i-1}^A\} \cup \{x_{i-1}^A, z_{i-1}^A\}$
10: $\quad$ change only incoming and outgoing transitions for q_{i-1}^A in A to/from $\{x_{i-1}^A, z_{i-1}^A\}$ in refined abstraction A', according to Definition 2
11: $\quad$ **return** A'
12: **else**
13: $\quad$ **return** concrete trace $t \leftarrow \sigma(p)$
14: **end if**

We define the abstraction to be *existential*, but using $\mathit{Reachable}_C$ instead of the usual transition relation of C [7]: $\exists (q_i^A, a, q_j^A) \in \delta^A$ iff

$$\exists q_i^C, q_j^C \in C \cdot \alpha(q_i^C) = q_i^A, \alpha(q_j^C) = q_j^A, \text{ and } q_j^C \in \mathit{Reachable}_C(q_i^C, a, \Sigma_A) \quad (2)$$

From the above definition and that of weak simulation [16], it follows that the abstraction defines a weak simulation relation between $C{\downarrow}_{\Sigma_A}$ and A. It is known that weak simulation implies trace inclusion [16]. We therefore have the following:

Proposition 1. *Given concrete LTS C and and its abstraction A defined as above, $\mathcal{L}(C{\downarrow}_{\Sigma_A}) \subseteq \mathcal{L}(A)$, and consequently $\langle true \rangle$ C $\langle A \rangle$ hold.*

The CEGAR algorithm for LTSs is defined by Algorithm 1. It takes as inputs a concrete system C, an abstraction A (as defined above), and an abstract counterexample path p (in A). The algorithm analyzes the counterexample (lines 1–6) to see if it is real, in which case it is returned (line 13) or spurious, in which case it is used to refine the abstraction (lines 7–11). The refined abstraction A' is such that it no longer contains p. We discuss Algorithm 1 in more detail below.

Analysis of abstract counterexamples. Suppose we have obtained an *abstract counterexample* in the form of a path $p = q_0^A, a_1, q_1^A, a_2, \ldots, a_n, q_n^A$ in the abstraction A of C. We want to determine if it corresponds to a concrete path in C. For this we need to "play" (i.e. symbolically simulate) p in C from the initial state q_0^C. We do so considering that $\Sigma_A \subseteq \Sigma_C$ and thus we use $\mathit{Reachable}_C$ again.

We first extend $\mathit{Reachable}_C$ to sets: for $S \subseteq Q_C$, $\mathit{Reachable}_C(S, a, \Sigma_A) = \{q_j^C \in C \mid \exists q_i^C \in S . q_j^C \in \mathit{Reachable}(q_i^C, a, \Sigma_A)\}$. We play the abstract counterexample p

following [7]. We start at step 0 with the set $S_0 = \{q_0^C\}$ of concrete states, and the first transition $q_0^A \xrightarrow{a_1} q_1^A$ from p. Note that $S_0 = \{q_0^C\} \cap \gamma(q_0^A)$. At each step $i \in \{1, \ldots, n\}$, we compute the set $S_i = \gamma(q_i^A) \cap \mathit{Reachable}_C(S_{i-1}, a_i, \Sigma_A)$. If, for some $i \leq n$, S_i is empty, the abstract counterexample is spurious and we need to refine the abstraction to eliminate it. Otherwise, the counterexample corresponds to a concrete path.

Abstraction refinement. The abstraction refinement is performed in lines 8–10 of Algorithm 1: p is spurious because abstract state q_{i-1}^A does not distinguish between two disjoint, non-empty sets of concrete states [7]: (i) those that reach, with action a_i, states in the concretization of q_i^A (these are the states defined as $\gamma(x_{i-1}^A)$ in line 8) and (ii) those reached so far from q_0^C with the prefix $a_1, a_2, \ldots, a_{i-1}$, *i.e.*, the states in S_{i-1}.

To eliminate the spurious abstract path, we need to refine A by splitting its state q_{i-1}^A into (at least) two new abstract states that separate the (concrete) states of types (i) and (ii) (line 9). We split q_{i-1}^A into x_{i-1}^A where $\gamma(x_{i-1}^A)$ contains the set of states in (i) and z_{i-1}^A where $\gamma(z_{i-1}^A)$ contains the set of states in (ii) and any remaining states in $\gamma(q_{i-1}^A)$. Note that this results in a finer partition of the concrete states. After the splitting, we update the abstract transitions in line 10. The refined abstraction A' has the same transitions as A except for those incoming or outgoing for the split state q_{i-1}^A: they are readjusted to point to or from the states x_{i-1}^A, z_{i-1}^A according to condition 2. We therefore can conclude that:

Lemma 1. *If a counterexample p input to Algorithm 1 is spurious, the returned abstraction A' results in a strictly finer partition than A and does not contain p.*

4.2 The AGAR Algorithm

The pseudocode that combines Algorithm 1 with Rule 1 is given in Algorithm 2. Recall that Σ_I denotes the alphabet $(\Sigma_{M_1} \cup \Sigma_P) \cap \Sigma_{M_2}$ of the interface between M_1 and M_2, with respect to P. The algorithm checks that $M_1 \parallel M_2$ satisfies P using Rule 1. It builds abstractions A of M_2 in an iterative fashion (while loop at lines 2–15); these abstractions are used to check *Premise 1* of the assume guarantee rule using model checking (lines 3–5). If the check is successful, then, according to the rule (and since A satisfies *Premise* 2 by construction), P indeed holds in $M_1 \parallel M_2$ and the algorithm returns "true". Otherwise, a counterexample p is obtained from model checking *Premise 1* (line 7) and Algorithm 1 is invoked to check if p corresponds to a real path in M_2 (in which case it means p is a real error in $M_1 \parallel M_2$ and this is reported to the user in line 11). If p is spurious, Algorithm 1 returns a refined abstraction A' for which we repeat the whole process starting from checking *Premise 1*.

Obtaining an abstract counterexample. As mentioned, we use counterexamples from failed checks of Premise 1 (that involves checking component M_1) to refine abstractions of M_2. Obtaining an abstract counterexample involves several steps (lines 7–9). First, a counterexample from line 4 is a path $o = q_0, b_1, q_1, b_2, \ldots, b_l, q_l$ in $A \parallel M_1 \parallel P_{err}$. Thus, for every $i \in \{0, l\}$, q_i is a triple of states (q_i^A, q_i^1, p_i) from $A \times M_1 \times P_{err}$. We first project every triple on A to obtain the sequence $o' = q_0^A, b_1, q_1^A, b_2, q_2^A, \ldots, b_l q_l^A$; o' is not yet a path in A as it may contain actions from M_1 and P_{err} that are not observable to A; those actions have to be between the same consecutive abstract states in the

Algorithm 2. AGAR: assume-guarantee verification by abstraction-refinement

Inputs: Component LTSs M_1, M_2, safety property LTS P, and alphabet $\Sigma_A = \Sigma_I$.
Outputs: **true** if $M_1 \parallel M_2$ satisfies P, **false** with a counterexample, otherwise.
Uses: Algorithm 1

1: Compute initial abstraction A of M_2, with a single state q_0^A having self-loops on all actions in Σ_A
2: **while true do**
3: Check *Premise 1*: $\langle A \rangle\ M_1\ \langle P \rangle$
4: **if** successful **then**
5: **return true**
6: **else**
7: Get counterexample $o = q_0, b_1, q_1, b_2, \ldots, b_l, q_l$ from line 3, where each $q_i = (q_i^A, q_i^1, p_i)$
8: Project o on A to get $o' = q_0^A, b_1, q_1^A, b_2, q_2^A, \ldots b_l, q_l^A$
9: Project o' on Σ_A to get abstract counterexample $p = q_0^A, a_1, q_1^A, \ldots, a_n, q_n^A$ in A.
10: **end if**
11: Call Algorithm 1 with inputs: M_2, A, p
12: **if** Algorithm 1 returned real counterexample t **then**
13: **return false** with counterexample t
14: **else**
15: $A = A'$
16: **end if**
17: **end while**

sequence, since they do not change the state of A; we eliminate from o' those actions and the duplicate abstract states that they connect, and finally obtain p that we pass to Algorithm 1.

Theorem 1. *Our algorithm (AGAR) computes a sequence of increasingly refined abstractions of M_2 until both premises of Rule 1 are satisfied, and we conclude that the property is satisfied by $M_1 \parallel M_2$, or a real counterexample is found that shows the violation of the property on $M_1 \parallel M_2$.*

Proof. Correctness The algorithm terminates when *Premise 1* is satisfied by the current abstraction or when a real counterexample is returned by Algorithm 1. In the former case, since the abstraction satisfies *Premise 2* by construction (Proposition 1), Rule 1 ensures that $M_1 \parallel M_2$ indeed satisfies P, so AGAR correctly returns answer "true". In the latter case, the counterexample returned by Algorithm 1 is a common trace of M_1 and of M_2 that leads to error in P_{err}. This shows that property P is violated on $M_1 \parallel M_2$ and in this case again AGAR correctly returns answer "false".

Termination. AGAR continues to refine the abstraction until a real counterexample is reported or the property holds. Refining the abstraction always results in a finer partition of its states (Lemma 1), and is thus guaranteed to terminate since in the worst case it converges to M_2 which is finite-state. □

If M_2 has n states, AGAR makes at most n refinement iterations, and in each iteration, counterexample analysis performs at most m closure operations, each of cost $O(n^3)$,

Algorithm 3. AGAR with alphabet refinement

Inputs: Component LTSs M_1, M_2, safety property LTS P, and alphabet $\Sigma_A \subseteq \Sigma_I$.
Outputs: **true** if $M_1 \parallel M_2$ satisfies P, **false** with a counterexample, otherwise.
Uses: Algorithm 2

1: **while true do**
2: Call Algorithm 2 with M_1, M_2, P, Σ_A.
3: **if** Algorithm 2 returned **true then**
4: **return true**
5: **else**
6: Obtain counterexample $t = a_1, \ldots, a_n$ from Algorithm 2 **and** trace $s = \sigma(o')$ from line 8 of Algorithm 2.
7: Check if error reachable in $s^{err}\downarrow_{\Sigma_I} \parallel M_2$ where $s^{err}\downarrow_{\Sigma_I}$ is the trace-LTS ending with an extra transition into error state π
8: **if** error reached **then**
9: **return false** with counterexample $s\downarrow_{\Sigma_I}$
10: **else**
11: Compare t to $s\downarrow_{\Sigma_I}$ to find difference action set Σ
12: $\Sigma_A \leftarrow \Sigma_A \cup \Sigma$
13: **end if**
14: **end if**
15: **end while**

where m is the length of the longest counterexample analyzed. This bound is not very tight as the closure steps are done on-the-fly to seldom exhibit worst-case behavior, and actually involve only parts of M_2's transition relation as needed.

4.3 AGAR with Interface Alphabet Refinement

In [11] we introduced an *alphabet refinement* technique to reduce the alphabet of the assumptions learned with L*. This technique improved significantly the performance of compositional verification. We show here how alphabet refinement can be similarly introduced in AGAR. Instead of the full interface alphabet Σ_I, we start AGAR from a small subset $\Sigma_A \subseteq \Sigma_I$. A good strategy is to start from those actions in Σ_I that appear in the property to be verified, since the verification should depend on them. We then run Algorithm 2 with this small Σ_A. Alphabet refinement introduces an extra layer of approximation, due to the smaller alphabet being used.

The pseudocode is in Algorithm 3. This algorithm adds an outer loop to AGAR (lines 1–15). At each iteration, it invokes AGAR (line 2) for the current alphabet Σ_A. If AGAR returns "true", it means that alphabet Σ_A is enough for proving the property (and "true" is returned to the user). Otherwise, the returned counterexample needs to be further analyzed (lines 5–13) to see if it corresponds to a real error (which is returned to the user in line 9) or it is spurious due to the approximation introduced by the smaller interface alphabet, in which case it is used to refine this alphabet (lines 11–12).

Additional counterexample analysis. As explained in [11], when $\Sigma_A \subset \Sigma_I$, the counterexamples obtained by applying Rule 1 may be spurious, in which case Σ_A needs to be extended. Intuitively, a counterexample is real if it is still a counterexample when

Table 1. Comparison of AGAR and learning for 2 components, with and without alphabet refinement

Case	k	No alpha. ref.						With alpha. ref.						Sizes	
		AGAR			Learning			AGAR			Learning				
		$\|A\|$	Mem.	Time	$\|A\|$	Mem.	Time	$\|A\|$	Mem.	Time	$\|A\|$	Mem.	Time	$\|M_1 \| P_{err}\|$	$\|M_2\|$
Gas Station	3	**16**	**4.11**	**3.33**	177	42.83	–	**5**	**2.99**	**2.09**	8	3.28	3.40	1960	643
	4	**19**	**37.43**	**23.12**	195	100.17	–	**5**	**22.79**	**12.80**	8	25.21	19.46	16464	1623
	5	**22**	**359.53**	**278.63**	45	206.61	–	**5**	216.07	**83.34**	8	**207.29**	188.98	134456	3447
Chiron, Property 2	2	10	**1.30**	**0.92**	9	**1.30**	1.69	10	1.30	**1.56**	**8**	**1.22**	5.17	237	102
	3	36	**2.59**	**5.94**	**21**	5.59	7.08	36	**2.44**	**10.23**	**20**	6.00	30.75	449	1122
	4	160	**8.71**	152.34	**39**	27.1	**32.05**	160	**8.22**	252.06	**38**	41.50	**180.82**	804	5559
	5	4	55.14	–	**111**	**569.23**	**676.02**	3	58.71	–	110	–	386.6	2030	129228
Chiron, Property 3	2	**4**	**1.07**	**0.50**	9	1.14	1.57	4	1.23	**0.62**	**3**	**1.06**	0.91	258	102
	3	**8**	**1.84**	**1.60**	25 n jmj	4.45	7.72	8	**2.00**	3.65	**3**	2.28	**1.12**	482	1122
	4	**16**	**4.01**	**18.75**	45	25.49	36.33	16	**5.08**	107.50	**3**	**7.30**	1.95	846	5559
	5	4	52.53	–	**122**	**134.21**	**271.30**	1	81.89	–	**3**	**163.45**	**19.43**	2084	129228
MER	2	**34**	**1.42**	11.38	40	6.75	**9.89**	**5**	**1.42**	5.02	6	1.89	**1.28**	143	1270
	3	**67**	**8.10**	**247.73**	335	133.34	–	9	11.09	180.13	**8**	**8.78**	**12.56**	6683	7138
	4	58	341.49	–	38	377.21	–	9	532.49	–	**10**	**489.51**	**1220.62**	307623	22886
Rover Exec.	2	**10**	**4.07**	**1.80**	11	2.70	2.35	**3**	2.62	**2.07**	4	2.46	3.30	544	41

considered with Σ_I. For counterexample analysis, we modify Algorithm 2 to also output the trace $s = \sigma(o')$ of actions along the intermediate path o' obtained at its line 8. Since p is a path obtained from o' by eliminating transitions labeled with actions from $\Sigma_I \setminus \Sigma_A$ (See Section 4.2) and $t = \sigma(p)$, it follows that s is an "extension" of t to Σ_I.

We check whether $s{\downarrow_{\Sigma_I}}$ is a trace of M_2 by making it into a trace LTS ending with the error state π, and whose alphabet is Σ_I (line 7). Since M_2 does not contain π, the only way to reach error is if $s{\downarrow_{\Sigma_I}}$ is a trace of M_2; if we reach error, the counterexample t is real. If $s{\downarrow_{\Sigma_I}}$ is not a trace of M_2, since t is, we need to refine the current alphabet Σ_A. At this point we have two traces, $s{\downarrow_{\Sigma_I}}$ and t that agree with respect to Σ_A and only differ on the actions from $\Sigma_I \setminus \Sigma_A$; since one trace is in M_2 and the other is not, we are guaranteed to find in their symmetric difference at least an action that we can add to Σ_A to eliminate the spurious counterexample t. We include the new action(s) and then repeat AGAR with the new alphabet. Termination follows from the fact that the interface alphabet is finite.

5 Evaluation

We implemented AGAR with alphabet refinement for Rule 1 in the LTSA tool. We compared AGAR with learning based assume guarantee reasoning, using a similar experimental setup as in [11]. The case studies are: *Gas Station* (with 3 . . . 5 customers), *Chiron* – a model of a GUI (with 2 . . . 5 event handlers), and two NASA models: *MER* resource arbiter (with 2 . . . 4 threads competing for a common resource) and *Rover*, with an executive and an event monitoring component. We first used the same two-way decompositions of these models as described in [11]. For *Gas Station* and *Chiron*, these decompositions were demonstrated to be the best for the performance of learning (without alphabet refinement) among all possible two-way decompositions [9].

Table 2. Comparison of AGAR and learning for balanced decompositions

Case	k	No alpha. ref.						With alpha. ref.						Sizes	
		AGAR			Learning			AGAR			Learning				
		$\|A\|$	Mem.	Time	$\|A\|$	Mem.	Time	$\|A\|$	Mem.	Time	$\|A\|$	Mem.	Time	$\|M_1 \parallel P\|$	$\|M_2\|$
Gas Station	3	**10**	**3.35**	**3.36**	294	367.13	–	**5**	**2.16**	**3.06**	59	11.14	81.19	1692	1942
	4	269	174.03	–	433	188.94	–	10	15.57	191.96	**5**	**9.25**	**4.73**	4608	6324
	5	**7**	**47.91**	**184.64**	113	82.59	–	2	47.48	–	**15**	**52.41**	**71.29**	31411	32768
Chiron, Property 2	2	**41**	**2.45**	**5.46**	140	118.59	395.56	**9**	**1.91**	**3.89**	17	2.73	13.09	906	924
	3	**261**	**81.24**	**710.1**	391	134.57	–	**79**	**39.94**	**663.53**	217	36.12	–	6104	6026
	4	**54**	**7.11**	**37.91**	354	383.93	–	**45**	**9.55**	**121.66**	586	213.78	–	1308	1513
	5	402	73.74	–	112	90.22	–	**33**	**19.66**	**157.35**	46	30.05	686.37	11157	11748
Chiron, Property 3	2	**2**	**0.98**	**0.37**	40	5.21	8.30	**2**	**1.02**	**0.49**	3	1.04	0.91	168	176
	3	**88**	**15.45**	**102.93**	184	284.83	–	46	41.40	115.77	**3**	**5.97**	**2.26**	4240	4186
	4	**2**	**5.60**	**2.65**	408	222.54	–	**2**	**6.14**	11.90	20	9.33	**7.44**	4156	4142
	5	**79**	**44.16**	405.03	179	104.25	–	42	42.04	430.47	**3**	**21.94**	**7.00**	16431	16840
MER	4	9	27.62	–	311	104.72	–	2	27.60	–	**10**	**65.42**	**35.78**	10045	66230

All experiments were performed on a Dell PC with a 2.8 GHz Intel Pentium 4 CPU and a 1.0 GB RAM running Linux Fedora Core 4 and Sun's Java SDK version 1.5. We report the maximum assumption size (*i.e.*, number of states) reached ("$|A|$"), the memory consumed ("Mem.") in MB, the time ("Time") in seconds, and the numbers of states on each side of the two-way decomposition: "$|M_1 \parallel P_{err}|$" and "$|M_2|$". A " " indicates that the limit of 1G of memory or 30 minutes has been exceeded. For those cases, the other quantities are shown as they were when the limit was reached. We also highlight in bold font the best results.

The results for the first set of experiments are shown in Table 1. Overall, AGAR shows similar or better results than learning in more than half of the cases. From the results, we noticed that the relative sizes of $M_1 \parallel P_{err}$ and M_2 seem to influence the performance of the two algorithms; *e.g.*, for *Gas Station*, where M_2 is consistently smaller, AGAR is consistently better, while for *Chiron*, as the size of M_2 becomes much larger, the performance of AGAR seems to degrade. Furthermore, we observed that the learning runs exercise more the first component, whereas AGAR exercises both. We therefore considered a second set of experiments were we tried to compare the relative performance of the two approaches for two-way system decompositions that are more balanced in terms of number of states.

We generated off-line all the possible two-way decompositions and chose those minimizing the difference in number of states between $M_1 \parallel P_{err}$ and M_2. The rest of the setup remained the same. The results for these new decompositions are in Table 2 (for MER, in only one case we found a more balanced partition than previously). These results show that with these new decompositions AGAR is consistently better in terms of time (14/21 cases), memory (16/21 cases) and assumption size (16/21 cases)[2]. The results also indicate that the benefits of alphabet refinement are more pronounced for learning. The results are somewhat non-uniform as k increases because for each larger value of k we re-computed balanced decompositions independently of those for smaller values. This is why we even found smaller components for larger parameter, as for Chiron, Property 2, $k = 3$ vs. $k = 4$.

[2] We did not count the cases when both algorithms ran out of limits.

6 Conclusions and Future Work

We have introduced an assume-guarantee abstraction-refinement technique (AGAR) as an alternative to learning-based approaches. Our preliminary results clearly indicate that the alternative is feasible. We are currently extending AGAR with the following rule (for reasoning about n components).

$$\frac{\begin{array}{l}\textit{(Premise 1)}\ \langle A_1\rangle\ M_1\ \langle P\rangle \\ \textit{(Premise 2)}\ \langle A_2\rangle\ M_2\ \langle A_1\rangle \\ \quad \ldots \\ \textit{(Premise n)}\ \langle \textit{true}\rangle\ M_n\ \langle A_{n-1}\rangle\end{array}}{\langle \textit{true}\rangle\ M_1 \parallel M_2 \parallel \ldots \parallel M_n\ \langle P\rangle} \tag{3}$$

In previous work [11], learning with this rule overcame the intermediate state explosion related to two-way decompositions (*i.e.*, when components are larger than the entire system). That helped us demonstrate better scalability of compositional vs. non-compositional verification which we believe to be the ultimate test of any compositional technique. We expect to similarly achieve better scalability for AGAR.

The implementation of AGAR for Rule 3 involves the creation of $n-1$ instances AR_i of our abstraction-refinement code for computing each A_i as an abstraction of $M_{i+1} \parallel A_{i+1}$, except for A_{n-1} which abstracts M_n . Counterexamples obtained from *(Premise 1)* are used to refine the intermediate abstractions $A_1, \ldots, A_{n-1}$. When A_i is refined, all the abstractions $A_1, \ldots, A_{i-1}$ are refined as well to eliminate the spurious trace. In the future, we also plan to explore extensions of AGAR to liveness properties.

Acknowledgements. We thank Moshe Vardi and Orna Grumberg for helpful suggestions and the CAV reviewers for their comments.

References

1. Alur, R., Madhusudan, P., Nam, W.: Symbolic Compositional Verification by Learning Assumptions. In: Etessami, K., Rajamani, S.K. (eds.) CAV 2005. LNCS, vol. 3576, pp. 548–562. Springer, Heidelberg (2005)
2. Angluin, D.: Learning regular sets from queries and counterexamples. Inf. and Comp. 75(2), 87–106 (1987)
3. Beyer, D., Henzinger, T.A., Singh, V.: Algorithms for Interface Synthesis. In: Damm, W., Hermanns, H. (eds.) CAV 2007. LNCS, vol. 4590, pp. 4–19. Springer, Heidelberg (2007)
4. Chaki, S., Clarke, E.M., Sinha, N., Thati, P.: Automated Assume-Guarantee Reasoning for Simulation Conformance. In: Etessami, K., Rajamani, S.K. (eds.) CAV 2005. LNCS, vol. 3576, pp. 534–547. Springer, Heidelberg (2005)
5. Chaki, S., Ouaknine, J., Yorav, K., Clarke, E.: Automated Compositional Abstraction Refinement for Concurrent C Programs: A Two-Level Approach. ENTCS 89(3) (2003)
6. Chaki, S., Strichman, O.: Optimized L*-Based Assume-Guarantee Reasoning. In: Grumberg, O., Huth, M. (eds.) TACAS 2007. LNCS, vol. 4424, pp. 276–291. Springer, Heidelberg (2007)
7. Clarke, E., Grumberg, O., Jha, S., Lu, Y., Veith, H.: Counterexample-Guided Abstraction Refinement. In: Emerson, E.A., Sistla, A.P. (eds.) CAV 2000. LNCS, vol. 1855, pp. 154–169. Springer, Heidelberg (2000)

8. Clarke, E.M., Grumberg, O., Peled, D.: Model Checking. MIT, Cambridge (2000)
9. Cobleigh, J.M., Avrunin, G.S., Clarke, L.A.: Breaking Up is Hard to Do: An Investigation of Decomposition for Assume-Guarantee Reasoning. In: Proc. of ISSTA 2006, pp. 97–108. ACM, New York (2006)
10. Cobleigh, J.M., Giannakopoulou, D., Pasareanu, C.S.: Learning Assumptions for Compositional Verification. In: ETAPS 2003 and TACAS 2003. LNCS, vol. 2619, pp. 331–346. Springer, Heidelberg (2003)
11. Gheorghiu, M., Giannakopoulou, D., Pasareanu, C.S.: Refining Interface Alphabets for Compositional Verification. In: Grumberg, O., Huth, M. (eds.) TACAS 2007. LNCS, vol. 4424, pp. 292–307. Springer, Heidelberg (2007)
12. Gupta, A., McMillan, K.L., Fu, Z.: Automated Assumption Generation for Compositional Verification. In: Damm, W., Hermanns, H. (eds.) CAV 2007. LNCS, vol. 4590, pp. 420–432. Springer, Heidelberg (2007)
13. Jones, C.B.: Specification and Design of (Parallel) Programs. In: Inf. Proc. 1983: Proc. of IFIP 9th World Congress, pp. 321–332. North Holland, Amsterdam (1983)
14. Magee, J., Kramer, J.: Concurrency: State Models & Java Programs. John Wiley & Sons, Chichester (1999)
15. Pnueli, A.: In Transition from Global to Modular Temporal Reasoning about Programs. Logic and Models of Conc. Sys. 13, 123–144 (1984)
16. Milner, R.: Communication and Concurrency. Prentice-Hall, New York (1989)
17. Sinha, N., Clarke, E.M.: SAT-Based Compositional Verification Using Lazy Learning. In: Damm, W., Hermanns, H. (eds.) CAV 2007. LNCS, vol. 4590, pp. 39–54. Springer, Heidelberg (2007)

Local Proofs for Linear-Time Properties of Concurrent Programs

Ariel Cohen[1] and Kedar S. Namjoshi[2]

[1] New York University
arielc@cs.nyu.edu
[2] Bell Labs, Alcatel-Lucent
kedar@research.bell-labs.com

Abstract. This paper develops a local reasoning method to check linear-time temporal properties of concurrent programs. In practice, it is often infeasible to model check over the product state space of a concurrent program. The method developed in this paper replaces such global reasoning with checks of (abstracted) individual processes. An automatic refinement step gradually exposes local state if necessary, ensuring that the method is complete. Experiments show that local reasoning can hold a significant advantage over global reasoning.

1 Introduction

Model Checking [5,34] has been singularly successful at automating (in)correctness proofs of programs. On the other hand, the standard model checking method suffers from a serious *state explosion* problem [6]. For concurrent programs, state explosion is caused by an exponential growth of the global state space with increasing number of components. It is often necessary to use abstraction and compositional reasoning methods to break up a model checking question into a series of local questions.

This paper develops just such a *local reasoning* method, for analyzing linear temporal properties of asynchronously composed, concurrent programs. In the first step, a *split invariant* (a vector of local, per-process assertions whose conjunction is a program invariant) is calculated, as shown in [30,9]. The next step is to construct abstractions of individual processes, based on the split invariant. Finally, liveness properties are checked individually on each abstraction. A contribution of this work is the *derivation* of this method from a deductive reasoning rule similar to that of Owicki and Gries [31]. The deductive rule is based on user-supplied rank functions: the derivation shows how to replace these with model checking.

Local reasoning is inherently incomplete: informally speaking, each process abstraction can view only part of the behavior of the other processes, which may not suffice to establish a property. Both Owicki and Gries, and Lamport [28] propose ways to resolve this problem. Owicki and Gries suggest introducing auxiliary (typically unbounded, "history") variables, while Lamport suggests

A. Gupta and S. Malik (Eds.): CAV 2008, LNCS 5123, pp. 149–161, 2008.

making some local state globally visible. In essence, both methods expose information about the internal behavior of the component processes.

History variables, which are often valuable for a deductive proof, complicate model checking, as the generic construction turns a finite-state problem into an infinite-state one. Lamport's proposal retains the finite-state nature of the problem, but it is not clear how to choose the local state to be exposed. (Exposing *all* local state results in a global model checking problem.) A second contribution of this work is a systematic refinement scheme, which is based on an analysis of counter-examples produced for the abstracted processes.

This combination of local reasoning with refinement is sound for all programs, and is shown to be complete for finite-state programs. I.e., for a finite-state program, a property is eventually proved or disproved. The combined method is simple to implement in terms of BDD manipulations. Experiments with an implementation based on TLV [33] show that local reasoning can have a significant advantage over global reasoning. Full proofs and more experiments are in [10].

2 Motivating Example

Figure 1(a) shows a simple mutual exclusion protocol. It satisfies mutual exclusion (a safety property), and the weak progress property: "infinitely often $x = 0$", i.e., infinitely often, some process is in its critical section. It does not satisfy the stronger individual progress property "every waiting process eventually enters its critical section". In the following, consider a 2-process instance.

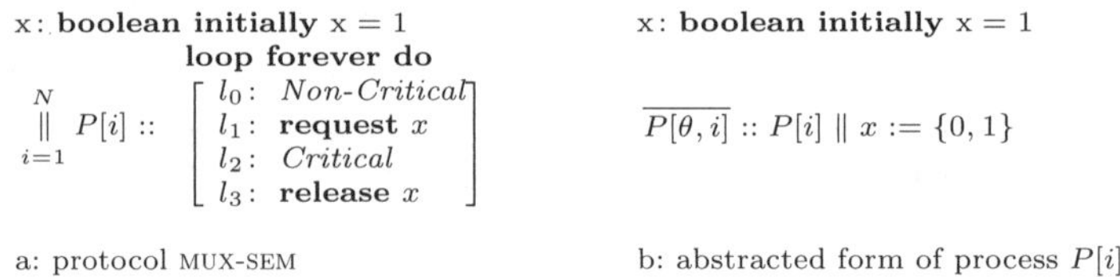

Fig. 1. Example Illustrating the Local Reasoning Method

The first step of the local method is to compute the strongest *split invariant*, θ, which is the strongest vector of local, per-process assertions whose conjunction is an inductive program invariant. (The reachable states form the strongest *global* inductive invariant.) This computation, by the fixpoint method in [30], results in $\theta_i = true$, for all i.

The second step uses θ to compute an over-approximation $\overline{P[\theta, i]}$ of each process P_i. For process i, the transition relation of $\overline{P[\theta, i]}$ is the disjoint union of (a) the transition relation, T_i, of P_i and (b) for every other process, a transition relation which summarizes its effect on the *shared state*, under the constraint θ. The contribution for process P_j ($j \neq i$) is computed by quantifying out the local variables of process P_j (both current and next-state) from the formula $\theta_j \wedge T_j$.

For this protocol, the summary for P_j allows x to change arbitrarily, leading to the abstract process shown in Figure 1(b).

The final step is to check using a standard model checker whether, for *all* i, the abstraction $\overline{P[\theta, i]}$ satisfies a weakening of the original progress property: that there is no execution on which a Büchi automaton for the negated original property accepts infinitely often from T_i transitions. This check succeeds for both abstract processes, which proves the original property for the original program.

It is interesting to note that the abstract form of process P_1 does fail the *original* property, as the summary transition for process P_2 can ensure that x is "stuck" at 1 forever. However, in this execution, a T_1 transition does not appear infinitely often, so it does not represent a failure for the weaker property defined above.

The particular form of these checks arises from the derivation that follows. In broad brush, however, the steps can be seen as: (1) constraining the globally reachable states with a split invariant, (2) computing an over-approximation of the behavior of each component process, restricted by the split invariant and (3) checking a weakened property on these over-approximations. In each step, the method avoids computing with the global product state space.

Although it is not the case for this protocol, it is possible that the abstraction of step (2) is too weak, resulting in false error reports in step (3). Exposing more local information strengthens the split invariant at the next iteration and, consequently, tightens the process summaries computed in step (2). This eventually results in the elimination of false errors.

Finally, for this protocol, the local proof for the 2-process instance suffices to show that the property holds for *all* instances (i.e., in a parameterized sense). This is true as both the split invariant $\theta = \mathit{true}$, and the abstract process $\overline{P[\theta, 1]}$ are unchanged for larger instances, and $\overline{P[\theta, j]}$ is identical to $\overline{P[\theta, 1]}$ by symmetry. We do not explore such parameterized proofs further in this paper.

3 The Local Reasoning Method

This section defines the computational model, and presents the derivation of the local reasoning method. Some of the preliminary definitions are taken from [30], and are repeated here for convenience.

3.1 Basic Definitions

Programs and Composition. A *program* is defined as a tuple (V, I, T), where V is a set of (typed) variables, $I(V)$ is an initial condition, and $T(V, V')$ is a transition condition, where V' is a fresh set of variables in 1-1 correspondence with V.

Programs need not be finite-state. The local reasoning method is sound in general, but we show completeness only for finite-state programs.

The semantics of a program is given by a *transition system*, which is defined by a triple (S, S_0, R), where S is the state domain defined by the Cartesian product of the domains of variables in V, $S_0 = \{s : I(s)\}$, and $R = \{(s, t) : T(s, t)\}$.

We assume that T is left-total, i.e., every state has a successor. A *state predicate* (or *assertion*) is a Boolean expression over the program variables. The value of a variable w at state s is denoted by $w(s)$. The truth value of a predicate p at a state s, denoted by $p(s)$, is defined as usual by induction on formula structure.

Given a non-empty, finite, set of programs, $\{P_i\}$, their *asynchronous composition*, written as $(\|i :: P_i)$, is the program $P = (V, I, T)$, with components defined as follows. Let $V = (\bigcup i :: V_i)$ and $I = (\bigwedge i :: I_i)$. The *shared variables*, denoted X, are those that belong to $V_i \cap V_j$, for a distinct pair (i, j). For simplicity, we assume that $X \subseteq V_i$, for all i. The *local variables* of P_i, denoted L_i, are the non-shared variables in V_i (i.e., $L_i = V_i \setminus X$). We assume that the local variables of distinct processes are disjoint. The set of local variables is $L = (\bigcup i :: L_i)$. The transition condition T_i of program P_i is constrained to leave local variables of other processes unchanged. Define $\hat{T}_i$ as $T_i(V_i, V_i') \wedge unch(L \setminus L_i)$, where $unch(W)$, for a set of variables W, is defined as $(\forall w : w \in W : w' = w)$. Then T is defined simply as $(\bigvee i :: \hat{T}_i)$.

Inductiveness and Invariance. A state predicate φ is an *invariant* of program $P = (V, I, T)$ if it holds at all reachable states of P. A state assertion ξ is an *inductive invariant* for P if it is initial (1) and inductive (2).

$$[I \Rightarrow \xi] \tag{1}$$

$$[\xi \Rightarrow wlp(T, \xi)] \tag{2}$$

Notation: *wlp* is the *weakest liberal precondition* transformer introduced by Dijkstra in [14]. The notation $[\psi]$, from [15], is read as "ψ is valid". For $P = (\|i :: P_i)$, it is the case that $[wlp(T, \varphi) \equiv (\bigwedge i :: wlp(\hat{T}_i, \varphi))]$.

Split Invariance. A *local assertion*, θ_i, is defined over the variables of P_i. Thus, θ_i is defined over L_i and X, but *does not* refer to other local variables. A *split assertion* for a composition $(\|i :: P_i)$ is a vector of local assertions, one for each process. A split assertion, θ, is a *split invariant* if the conjunction $(\wedge i :: \theta_i)$ is an inductive invariant for the composition. To simplify notation, θ refers indifferently either to the vector or to the conjunction of its components, with the interpretation clear from the context. In [30], it is shown that the *strongest* split invariant can be computed as the (simultaneous) least fixpoint of the set of equations below, one for each i.

$$[\theta_i \equiv (\exists L \setminus L_i :: I \vee (\vee j :: sp(\hat{T}_j, \theta)))]$$

Here, *sp* is the strongest post-condition operator (also known as "post"). The expression takes successors of θ (i.e., of $(\wedge i :: \theta_i)$) for each $\hat{T}_j$, adds the initial states, and quantifies out non-P_i local variables, to ensure that the result is a local assertion for P_i. For finite-state programs, these calculations are easily implemented with standard BDD operations.

3.2 Background: Proofs of Linear-Time Properties

The automata-theoretic approach to model checking [37] is followed here. Linear-time properties are specified by a finite-state automaton over infinite words for the *complemented* property. The derivations below rely on a few assumptions.

1. *Programs are deadlock-free.* As the transition relation of each component is left-total, the program as a whole never gets stuck. As in the UNITY model [3], however, "deadlock" may be defined as a state where the only transition is a self-loop. Deadlock-freedom is thus a safety property, which can be checked locally using the method from [9].
2. *The property is defined by a non-deterministic Büchi automaton for its complement, and refers only to the shared variables.* This enables the automaton transitions to be inserted into the program, synchronously with each component transition. The predicate *accept*, on the shared state, indicates that the automaton is in an accepting state.
3. *Fairness constraints are enforced by the automaton.* Liveness properties often depend on fairness assumptions about execution schedules. For simplicity, we assume that any fairness conditions are part of the automaton; i.e., the automaton accepts an execution if it is fair but fails the property.

Under these assumptions, the following non-compositional proof rule can be used to prove a linear-time property for a program $P = (V, I, T)$. The proof rule requires an assertion θ, and a *rank function* ρ, a partial map from states to a well-founded domain, $(W, \prec)$, which satisfy the conditions below.

$$[I \Rightarrow \theta] \tag{3}$$

$$[\theta \Rightarrow wlp(T, \theta)] \tag{4}$$

$$[\theta \Rightarrow domain(\rho)] \tag{5}$$

$$\forall k : k \in W : [\theta \wedge (\rho = k) \Rightarrow wlp(T, \rho \preceq k)] \tag{6}$$

$$\forall k : k \in W : [\theta \wedge accept \wedge (\rho = k) \Rightarrow wlp(T, \rho \prec k)] \tag{7}$$

Theorem 1. *The proof rule is sound and relatively complete.*

3.3 Localizing the Proof Rule

Consider now the case where P is a composition $(||i : P_i)$. The goal is to *localize* the reasoning rule given previously. To this end, a first change is to make θ a *split invariant.* A second change is to let ρ be a *vector* of local functions, with ρ_i defined over the variables V_i of P_i, with a well-founded domain, $(W_i, \prec_i)$, as its range. The local proof rule is as follows.

$$[I \Rightarrow \theta] \tag{8}$$

$$[\theta \Rightarrow wlp(T, \theta)] \tag{9}$$

$$\forall i : [\theta \Rightarrow domain(\rho_i)] \tag{10}$$

$$\forall i, k : k \in W_i : [\theta \wedge (\rho_i = k) \Rightarrow wlp(T, \rho_i \preceq_i k)] \tag{11}$$

$$\forall i, k : k \in W_i : [\theta \wedge accept \wedge (\rho_i = k) \Rightarrow wlp(\ddot{T}_i, \rho_i \prec_i k)] \tag{12}$$

Theorem 2. *The local proof rule is sound.*

Proof Sketch. The first two conditions ensure that θ is a (split) inductive invariant. Define a global rank function ρ by $\rho(s) = (vec\ i :: \rho_i(s_i))$, where s_i is s restricted to V_i. Global rank vectors are compared point-wise. From (10)-(12), it follows that the pair (θ, ρ) satisfies the hypotheses of the previous proof rule, ensuring soundness by Theorem 1. □

Condition (11) ensures that ρ_i is not adversely affected by any transition of P. This is one of the "non-interference" properties defined by Owicki and Gries in [31]. The local formulation has the following interesting consequence.

Theorem 3. *If the local proof rule is applicable, it can be applied with θ being the* strongest *split invariant.*

Proof. Suppose that the local proof rule is applicable for some θ. Let θ^* represent the strongest split invariant. Conditions (8) and (9) are satisfied by θ^* by definition. The other conditions are anti-monotone in θ; as θ^* is stronger than θ, they hold also for θ^*. □

This theorem provides the first hint for mechanizing the local proof rule, as it eliminates one part of the guesswork: one can let θ be the strongest split invariant. The next section shows how to replace the rank function requirements with model checking.

3.4 Guessing Ranks through Model Checking

In this section, we consider a *fixed* split invariant, θ. The goal is to replace the reasoning about rank functions with a local model checking procedure. First, note that by the conjunctivity of *wlp* for asynchronous composition, conditions (11) and (12) are equivalent to saying that, for each i, each j, and $k \in W_i$,

$$[\theta \wedge (\rho_i = k) \Rightarrow wlp(\hat{T}_j, \rho_i \preceq_i k)] \tag{13}$$

$$[\theta \wedge accept \wedge (\rho_i = k) \Rightarrow wlp(\hat{T}_i, \rho_i \prec_i k)] \tag{14}$$

These conditions are rewritten below, exploiting locality. In these derivations, we do not explicitly write the variable dependencies, to avoid clutter. For reference, they are: $\theta(X, L)$, $\theta_i(X, L_i)$, $\rho_i(X, L_i)$, $\hat{T}_i(X, L, X', L')$, and $T_i(X, L_i, X', L'_i)$. We write ρ'_i to refer to $\rho_i(X', L'_i)$. The calculations make extensive use of the following fact: $[p(x, y) \Rightarrow q(y)]$ is equivalent to $[(\exists x :: p(x, y)) \Rightarrow q(y)]$.

For each i, each j, and $k \in W_i$,

$$
\begin{array}{ll}
 & [\theta \wedge (\rho_i = k) \Rightarrow wlp(\hat{T}_j, \rho_i \preceq_i k)] \\
\equiv & \quad \{ \text{ definition of } wlp \ \} \\
 & [\theta \wedge (\rho_i = k) \wedge \hat{T}_j \Rightarrow \rho'_i \preceq_i k] \\
\equiv & \quad \{ \text{ by locality, as the consequent is independent of } L \setminus L_i \text{ and } L' \setminus L'_i \ \} \\
 & [(\exists L \setminus L_i, L' \setminus L'_i :: \theta \wedge (\rho_i = k) \wedge \hat{T}_j) \Rightarrow \rho'_i \preceq_i k]
\end{array}
$$

$$\equiv \quad \{ \text{ pushing quantifiers inwards } \}$$
$$[(\exists L \setminus L_i :: \theta \wedge (\exists L' \setminus L'_i :: \hat{T}_j)) \wedge (\rho_i = k) \Rightarrow \rho'_i \preceq_i k]$$

For $j \neq i$, the term $(\exists L' \setminus L'_i :: \hat{T}_j)$ simplifies to $(\exists L'_j :: T_j) \wedge unch(L_i)$. As θ is really $(\wedge i :: \theta_i)$, the final implication simplifies to

$$[(\exists L \setminus L_i :: \theta) \wedge (\exists L_j : \theta_j \wedge (\exists L'_j :: T_j)) \wedge unch(L_i) \wedge (\rho_i = k) \Rightarrow \rho'_i \preceq_i k] \quad (15)$$

This has a shape similar to that of the condition (6) from the non-local rule, with $(\exists L \setminus L_i :: \theta)$ playing the role of the invariant assertion, and the term "$(\exists L_j : \theta_j \wedge (\exists L'_j :: T_j)) \wedge unch(L_i)$" playing the role of a transition relation. This observation leads to the following definition.

Definition 1. *Define $\overline{T_j[\theta, i]}$ as $(\exists L_j :: \theta_j \wedge (\exists L'_j :: T_j)) \wedge unch(L_i)$, for $j \neq i$, and as T_i, for $j = i$. With free variables (X, L_i, X', L'_i), this is a transition term for P_i.*

Similarly transforming the other conditions, one obtains for any i, and any j,

$$[(\exists L \setminus L_i :: I) \Rightarrow (\exists L \setminus L_i :: \theta)] \quad (16)$$
$$[(\exists L \setminus L_i :: \theta) \Rightarrow wlp(\overline{T_j[\theta, i]}, (\exists L \setminus L_i :: \theta))] \quad (17)$$
$$[(\exists L \setminus L_i :: \theta) \Rightarrow domain(\rho_i)] \quad (18)$$
$$\forall k : k \in W_i : [(\exists L \setminus L_i :: \theta) \wedge (\rho_i = k) \Rightarrow wlp(\overline{T_j[\theta, i]}, \rho_i \preceq_i k)] \quad (19)$$
$$\forall k : k \in W_i : [(\exists L \setminus L_i :: \theta) \wedge accept \wedge (\rho_i = k) \Rightarrow wlp(T_i, \rho_i \prec_i k)] \quad (20)$$

The implications (16)-(20) suggests the definition of an abstract process.

Definition 2. *The abstraction of process P_i relative to θ is a process denoted $\overline{P[\theta, i]}$, with variables V_i, initial condition $(\exists L \setminus L_i :: I)$, and transition relation formed by the terms T_i and $\overline{T_j[\theta, i]}$, for $j : j \neq i$, combined disjunctively.*

Conditions (19) and (20) lead to the following theorem.

Theorem 4. *For fixed θ: if there is a rank function vector which satisfies the local proof conditions then, for any i, $\overline{P[\theta, i]}$ satisfies the property "for all executions, it is not the case that T_i occurs infinitely often from states satisfying accept".*

The contrapositive of this theorem implies that, for a given θ, if the check fails for one of the abstract processes, there is *no* rank function vector which can satisfy the local proof rule (for the same θ). This forces a refinement of the split invariant in order to rule out false errors, as described in the next section. On the other hand, if the check succeeds for *all* of the abstract processes, the property must hold of the original program.

Theorem 5. *For any split invariant θ: if, for every i, $\overline{P[\theta, i]}$ satisfies the property "for all executions, it is not the case that T_i occurs infinitely often from states satisfying accept", then the original property is true of the composition $(\|i :: P_i)$.*

3.5 The Local Reasoning Algorithm

The algorithm is now easily stated. Given $P = (\|i :: P_i)$, with an embedded Büchi automaton for a negated property, and acceptance condition *accept*.

1. Compute a split invariant θ of P, ideally the strongest split invariant.
2. For each i, define the abstract program $\overline{P[\theta, i]}$ (Def. 2). Form the product of the abstract program with the property automaton. Check the property stated in Theorem 4.
3. If each check succeeds, by Theorem 5, the property holds of P.

These operations have complexity polynomial in the number of processes, the size of each process, and the size of the automaton. (Compare this with the exponential complexity in the number of processes for a global model check.)

What if the check fails for some i? By Theorem 3, if the strongest split invariant was used, it is either the case that the property is false, or that a refinement step is needed to expose more of the local state and rule out a false counter-example.

3.6 Modifications

Quantified Properties. For parameterized protocols, it is common to have quantified liveness properties (e.g., "every waiting process eventually enters its critical section"). Fortunately, such protocols typically have a high degree of symmetry. Under symmetry, a composition $(\|i :: P_i)$ satisfies a quantified property $(\forall i :: \varphi(i))$ if, and only if, it satisfies $\varphi(1)$ [16]. Hence, making the local variables of P_1 part of the shared state suffices to meet the requirement that the property is defined over the shared variables. Symmetry can also be exploited to reduce the computations for the split invariance calculation, and to reduce the number of checks needed in step 2 above to the abstract processes for P_1 and P_2.

Fairness. For an unconditional fairness assumption, it suffices to annotate each transition with the index of the process making the transition. These indices are carried over to the abstract processes. To express stronger fairness assumptions, it is necessary to shadow the local predicates mentioned in the fairness assumption with auxiliary shared variables.

4 A Refinement Strategy

The local reasoning algorithm defined above is necessarily incomplete. If the property cannot be proved, a local proof requires exposing more of the local state. We describe a simple, yet effective, strategy to choose the portions of the local state to be exposed. This strategy is based on examining counter-example executions for those abstract processes which fail to model-check.

A failure for $\overline{P[\theta, i]}$ implies that there is a finite, "lasso" shaped counter-example: a path ending in a cycle which contains at least one T_i transition from an *accept* state. By construction, in $\overline{P[\theta, i]}$, the T_i transitions are exact, while the transitions of other processes may be approximate. Recall that the definition of $\overline{T_j[\theta, i]}$ (Def. 1) has an $\exists L_j \exists L'_j$ form. In terms of language used in branching-time abstraction methods [12], this is a *may*-transition. It is a *must*-transition if, for *every* value of L_j that satisfies θ_j, there is a T_j transition from (X, L_j) to (X', L_j). The distinction between may- and must-transitions is useful in determining whether a counterexample lasso represents a real execution.

Theorem 6. *Let π be a counterexample for $\overline{P[\theta, i]}$. If every abstract transition along π is also a must-transition, there is an induced global counterexample for the full program P.*

This theorem leads to the refinement procedure below.

1. If, for some abstract program $\overline{P[\theta, i]}$, the transitions on a counterexample path, π, meet the condition of Theorem 6, HALT("the property is false"), and provide the induced global path as a counterexample.
2. Otherwise, let $t = (u, \overline{T_k[\theta, i]}, v)$ be a non-must transition on π. Define p_k by

$$p_k(L_k) := \theta_k(X(u), L_k) \wedge \neg(\exists L'_k :: T_k(X(u), L_k, X(v), L'_k))$$

For local state a of P_k, $p_k(a)$ is true if a is an *obstacle* to forming a must-transition, since there is no transition to $X(v)$ from $(X(u), a)$ in P_k. Predicate p_k is non-trivial: it is not valid, as t is a may transition; neither is it unsatisfiable, as t is not a must transition. As required for local reasoning, p_k is a local assertion for P_k. A *shared* boolean variable, b_k, is added to the program, such that $b_k \equiv p_k(L_k)$ is an invariant, and the local reasoning algorithm is repeated for the augmented program. The initial value of b_k is the initial value of p_k; the constraint $(b'_k \equiv p_k(L'_k))$ is conjoined to T_k, and $(b'_k \equiv b_k)$ to T_j, for $j \neq k$.

Theorem 7. *For finite-state programs, this procedure eventually terminates.*

Proof. First, we show that each refinement step discovers at least one new predicate. Existing predicates are preserved by the split invariance calculation (Lemma 1 of [9]). Thus, any existing predicates have the same value for all local states a of P_k that satisfy $\theta_k(X(u), a)$; and this is not true of p_k by its definition. As there are a bounded number of predicates, the refinement process cannot continue forever. □

Theorem 8. *The combination of local reasoning with refinement is complete for finite-state programs.*

Proof. Consider first the case where the property holds. Thus, any counterexamples are not real, so the hypothesis of Theorem 6 does not apply, and the procedure will not terminate with an incorrect answer. As termination is guaranteed by Theorem 7, the procedure must terminate with success.

Next, consider the case where the property does not hold. By the contrapositive of Theorem 5, at every stage, at least one of the abstract processes fails the check. Thus, the procedure cannot terminate with success. As termination is guaranteed by Theorem 7, the procedure must terminate with failure. □

While the termination argument relies on exhausting the set of available predicates, the hope is that, in most cases, termination occurs before the problem is transformed back into a global model checking question. It is important to note that the procedure is *sound*—but not necessarily terminating—for all programs.

5 Experiments

We implemented our method in TLV [33], a BDD-based model checker. The experiments use parameterized protocols, as the global state space can be varied simply by altering the number of processes. We do not use symmetry to optimize the calculations, as the intent is to compare local with global reasoning, as represented by algorithm TEMP_ENTAIL, based on [27]. The experiments show that local reasoning can have significantly better performance than global reasoning.

Table 1. Test results for the property $\Box\Diamond x = 0$

Method	Processes	BDDs	Time (sec)	BDDs [variant]	Time (sec) [variant]
Local Reasoning	2	433	0	7.6k	0
Global Reasoning	2	440	0	6.2k	0
Local Reasoning	10	10k	0.1	19k	1.3
Global Reasoning	10	10k	0.05	248k	294
Local Reasoning	20	15k	1.28	62k	4.6
Global Reasoning	20	23k	1.53	-	>2hrs
Local Reasoning	50	88k	21.7	330k	46.8
Global Reasoning	50	141k	53.5	-	>2hrs

We checked two different properties for MUX-SEM, the motivating example of Figure 1(a). For the first property, $\Box\Diamond x = 0$ ("infinitely often $x = 0$"), our method does not require any refinement step. Compared to TEMP_ENTAIL, it runs significantly faster and requires nearly half the amount of BDDs (Table 1). A variant of the protocol models the situation where there is some state irrelevant to the property (in this case, a counter in each process). The locality of the analysis results in this excess state being eliminated from the abstract processes. The effect is shown in the two final columns of Table 1.

The second property is $\Box(P[1].at\_l_1 \rightarrow \Diamond P[1].at\_l_2)$ ("if process $P[1]$ is at location 1 it eventually enters its critical section"), which is not satisfied by MUX-SEM under unconditional fairness. Expressing the property requires exposing the location variable of $P[1]$. The method detects a counter example after one

Table 2. Test results for the property $\Box(P[1].at\_l_1 \rightarrow \Diamond P[1].at\_l_2)$.

Method	Processes	BDDs	Time (sec)	Refinements	New Variables
Local Reasoning	3	2.6k	0.02	1	1
Global Reasoning	3	1.6k	0.01	-	-
Local Reasoning	10	10k	0.22	1	8
Global Reasoning	10	16k	0.13	-	-
Local Reasoning	20	37k	1.2	1	18
Global Reasoning	20	66k	1.46	-	-
Local Reasoning	50	215k	13.5	1	48
Global Reasoning	50	414k	30.6	-	-
Local Reasoning	100	852k	382	1	98
Global Reasoning	100	1.6M	586	-	-

refinement step, during which one bit of information per process (whether the process is waiting) is exposed. The results are provided in Table 2.

6 Related Work and Conclusions

Compositional reasoning about concurrency has a long history, going back 30 years to the seminal papers of Owicki and Gries, and Lamport. Early work focuses on deductive proof methods for safety [4,26] and liveness [32,13]. Tools such as Cadence SMV support guided compositional proofs [29,25]. "Thread-modular" reasoning[18,19,24,23] uses per-process transition relations to prove safety, but the method is incomplete. The split invariance method was introduced in [30], with completeness obtained by a refinement method [9].

The new contribution here is the mechanization of an Owicki-Gries style proof rule for liveness properties, coupled with a refinement procedure. The procedure is fully automatic, and complete for finite-state processes. It has a simple implementation, and the experimental results support the hypothesis that local reasoning is often significantly faster than global reasoning. To the best of our knowledge, this is the first fully automated and complete method of its type for checking linear-time properties of asynchronous programs.

Recent work [22,11] has shown that local reasoning can be effective for proving termination properties. However, the algorithms do not include a mechanism to expose additional local state, which is necessary for completeness.

In [35], Shoham and Grumberg propose a complete compositional method, coupled with refinement, for proving mu-calculus properties. Our methods have several points of commonality, including the use of the may-must distinction for refinement, but also some important differences: the method in [35] operates on *synchronous* compositions, rather than the asynchronous compositions considered here, and has as a key step the global analysis of a composition of abstract processes, which differs from the separate analysis of individual abstract processes in our method. Earlier work [1] uses per-process invariants to constrain abstractions (in the synchronous setting) as is done in Definitions 1 and 2.

Automata learning algorithms have been used for compositional analysis of safety properties [21,36,20,2], and recently extended to liveness properties [17]. The algorithms are complete, but can be expensive in practice [8]. Our completion procedure is based on a form of counter-example guided refinement [7], which may be viewed as a process of learning from failure.

Acknowledgements. This research was supported in part by the National Science Foundation under grant CCR-0341658.

References

1. Alur, R., de Alfaro, L., Henzinger, T.A., Mang, F.Y.C.: Automating Modular Verification. In: Baeten, J.C.M., Mauw, S. (eds.) CONCUR 1999. LNCS, vol. 1664, pp. 82–97. Springer, Heidelberg (1999)
2. Chaki, S., Clarke, E.M., Sinha, N., Thati, P.: Automated assume-guarantee reasoning for simulation conformance. In: Etessami, K., Rajamani, S.K. (eds.) CAV 2005. LNCS, vol. 3576, pp. 534–547. Springer, Heidelberg (2005)
3. Mani Chandy, K., Misra, J.: Parallel Program Design: A Foundation. Addison-Wesley, Reading (1988)
4. Chandy, K.M., Misra, J.: Proofs of networks of processes. IEEE Transactions on Software Engineering 7(4) (1981)
5. Clarke, E.M., Emerson, E.A.: Design and synthesis of synchronization skeletons using branching time temporal logic. In: Kozen, D. (ed.) Logic of Programs 1981. LNCS, vol. 131. Springer, Heidelberg (1982)
6. Clarke, E.M., Grumberg, O.: Avoiding the state explosion problem in temporal logic model checking. In: PODC, pp. 294–303 (1987)
7. Clarke, E.M., Grumberg, O., Jha, S., Lu, Y., Veith, H.: Counterexample-guided abstraction refinement for symbolic model checking. J. ACM 50(5), 752–794 (2003)
8. Cobleigh, J.M., Avrunin, G.S., Clarke, L.A.: Breaking up is hard to do: an investigation of decomposition for assume-guarantee reasoning. In: ISSTA, pp. 97–108 (2006)
9. Cohen, A., Namjoshi, K.S.: Local proofs for global safety properties. In: Damm, W., Hermanns, H. (eds.) CAV 2007. LNCS, vol. 4590, pp. 55–67. Springer, Heidelberg (2007)
10. Cohen, A., Namjoshi, K.S.: Local proofs for linear-time temporal properties of concurrent programs. Technical report, Bell Labs (2008), Available at: `http://www.cs.bell-labs.com/who/kedar`
11. Cook, B., Podelski, A., Rybalchenko, A.: Proving thread termination. In: PLDI, pp. 320–330. ACM, New York (2007)
12. Dams, D., Grumberg, O., Gerth, R.: Abstract interpretation of reactive systems. TOPLAS 19(2) (1997)
13. de Roever, W.-P., de Boer, F., Hannemann, U., Hooman, J., Lakhnech, Y., Poel, M., Zwiers, J.: Concurrency Verification: Introduction to Compositional and Non-compositional Proof Methods. Cambridge University Press, Cambridge (2001)
14. Dijkstra, E.W.: Guarded commands, nondeterminacy, and formal derivation of programs. CACM 18(8) (1975)
15. Dijkstra, E.W., Scholten, C.S.: Predicate Calculus and Program Semantics. Springer, Heidelberg (1990)
16. Emerson, E.A., Sistla, A.P.: Symmetry and model checking. In: Courcoubetis, C. (ed.) CAV 1993. LNCS, vol. 697. Springer, Heidelberg (1993)
17. Farzan, A., Chen, Y., Clarke, E.M., Tsan, Y., Wang, B.: Extending automated compositional verification to the full class of omega-regular languages. In: Ramakrishnan, C.R., Rehof, J. (eds.) TACAS 2008. LNCS, vol. 4963, pp. 2–17. Springer, Heidelberg (2008)

18. Flanagan, C., Freund, S.N., Qadeer, S., Seshia, S.A.: Modular verification of multithreaded programs. Theor. Comput. Sci. 338(1-3), 153–183 (2005)
19. Flanagan, C., Qadeer, S.: Thread-modular model checking. In: Ball, T., Rajamani, S.K. (eds.) SPIN 2003. LNCS, vol. 2648, pp. 213–224. Springer, Heidelberg (2003)
20. Giannakopoulou, D., Pasareanu, C.S.: Learning-based assume-guarantee verification (tool paper). In: Godefroid, P. (ed.) SPIN 2005. LNCS, vol. 3639, pp. 282–287. Springer, Heidelberg (2005)
21. Giannakopoulou, D., Pasareanu, C.S., Barringer, H.: Assumption generation for software component verification. In: ASE, pp. 3–12 (2002)
22. Gotsman, A., Berdine, J., Cook, B., Sagiv, M.: Thread-modular shape analysis. In: PLDI, pp. 266–277. ACM, New York (2007)
23. Henzinger, T.A., Jhala, R., Majumdar, R.: Race checking by context inference. In: PLDI, pp. 1–13 (2004)
24. Henzinger, T.A., Jhala, R., Majumdar, R., Qadeer, S.: Thread-modular abstraction refinement. In: Hunt Jr., W.A., Somenzi, F. (eds.) CAV 2003. LNCS, vol. 2725, pp. 262–274. Springer, Heidelberg (2003)
25. Jhala, R., McMillan, K.L.: Microarchitecture verification by compositional model checking. In: Berry, G., Comon, H., Finkel, A. (eds.) CAV 2001. LNCS, vol. 2102, pp. 396–410. Springer, Heidelberg (2001)
26. Jones, C.B.: Development methods for computer programs including a notion of interference. PhD thesis, Oxford University (1981)
27. Kesten, Y., Pnueli, A., Raviv, L., Shahar, E.: Model checking with strong fairness. Formal Methods in System Design 28(1), 57–84 (2006)
28. Lamport, L.: Proving the correctness of multiprocess programs. IEEE Trans. Software Eng. 3(2) (1977)
29. McMillan, K.L.: A compositional rule for hardware design refinement. In: Grumberg, O. (ed.) CAV 1997. LNCS, vol. 1254, Springer, Heidelberg (1997)
30. Namjoshi, K.S.: Symmetry and completeness in the analysis of parameterized systems. In: Cook, B., Podelski, A. (eds.) VMCAI 2007. LNCS, vol. 4349, pp. 299–313. Springer, Heidelberg (2007)
31. Owicki, S.S., Gries, D.: Verifying properties of parallel programs: An axiomatic approach. Commun. ACM 19(5), 279–285 (1976)
32. Pnueli, A.: In transition from global to modular reasoning about programs. In: Logics and Models of Concurrent Systems, NATO ASI Series (1985)
33. Pnueli, A., Shahar, E.: A platform for combining deductive with algorithmic verification. In: Alur, R., Henzinger, T.A. (eds.) CAV 1996. LNCS, vol. 1102, pp. 184–195. Springer, Heidelberg (1996), `www.cs.nyu.edu/acsys/tlv`
34. Queille, J.P., Sifakis, J.: Specification and verification of concurrent systems in CESAR. In: Dezani-Ciancaglini, M., Montanari, U. (eds.) Programming 1982. LNCS, vol. 137, Springer, Heidelberg (1982)
35. Shoham, S., Grumberg, O.: Compositional verification and 3-valued abstractions join forces. In: Riis Nielson, H., Filé, G. (eds.) SAS 2007. LNCS, vol. 4634, pp. 69–86. Springer, Heidelberg (2007)
36. Tkachuk, O., Dwyer, M.B., Pasareanu, C.S.: Automated environment generation for software model checking. In: ASE, pp. 116–129 (2003)
37. Vardi, M., Wolper, P.: An automata-theoretic approach to automatic program verification. In: IEEE Symposium on Logic in Computer Science (1986)

Probabilistic CEGAR*

Holger Hermanns, Björn Wachter, and Lijun Zhang

Universität des Saarlandes, Saarbrücken, Germany
{hermanns,bwachter,zhang}@cs.uni-sb.de

Abstract. Counterexample-guided abstraction refinement (CEGAR) has been *en vogue* for the automatic verification of very large systems in the past years. When trying to apply CEGAR to the verification of probabilistic systems, various foundational questions arise. This paper explores them in the context of predicate abstraction.

1 Introduction

Probabilistic behavioral descriptions are widely used to analyze and verify systems that exhibit "quantified uncertainty", such as embedded, networked, or biological systems. The semantic model for such systems often are Markov chains or Markov decision processes. We here consider homogeneous discrete-time Markov chains (MCs) and Markov decision processes (MDPs). Properties of these systems can be specified by formulas in temporal logics such as PCTL [1], where for instance quantitative probabilistic reachability ("the probability to reach a set of bad states is at most 3%") is expressible. Model checking algorithms for such logics have been devised mainly for finite-state MCs [1] and MDPs [2], and effective tool support is provided by probabilistic model checkers such as PRISM [3] or MRMC [4]. Despite its remarkable versatility, the approach is limited by the state explosion problem, aggravated by the cost of numerical computation compared to Boolean CTL model checking.

Predicate abstraction [5] is a method for creating finite abstract models of non-probabilistic systems where symbolic expressions, so-called predicates, induce a partitioning of its (potentially infinite) state space into a finite number of regions. For automation, it is typically coupled [6,7] with *counterexample-guided abstraction refinement* (CEGAR) [8] where an initially very coarse abstraction is refined using diagnostic information (predicates) derived from abstract counterexamples, until either the property is proved or refuted.

In this paper, we discuss how counterexample-guided abstraction refinement can be developed in a probabilistic setting. Predicate abstraction without abstraction refinement has been presented in [9] for a guarded command language whose concrete semantics maps to MDPs – more precisely, to probabilistic automata [10]. This is the natural basis for our present work. We restrict our

* This work is supported by DFG as part of the Transregional Collaborative Research Center SFB/TR 14 AVACS and by the NWO-DFG bilateral project VOSS.

A. Gupta and S. Malik (Eds.): CAV 2008, LNCS 5123, pp. 162–175, 2008.

treatment to probabilistic reachability and aim to determine if the probability to reach a set of bad states exceeds a given threshold.

The core challenge of developing probabilistic CEGAR lies in the notion and analysis of counterexamples. In the traditional setting, an abstract counterexample is a single finite path (to some bad state) and counterexample analysis consists in checking if the concrete model exhibits a corresponding error path. In contrast, a counterexample to a probabilistic reachability property can be viewed as a finite, but generally cyclic, Markov chain [11]. Due to these cycles, probabilistic counterexample analysis is not directly amenable to conventional methods. We circumvent this problem, by preprocessing the abstract counterexample using the strongest evidence idea of [12]: We generate a finite set *afp* of abstract finite paths that together carry enough abstract probability mass, and formulate the problem of computing the realizable probability mass of *afp* in terms of a weighted MAX-SMT problem [13]. The set *afp* is built incrementally in an *on-the-fly* manner, until either enough probability is realizable, or *afp* cannot be enriched with sufficient probability mass to make the probability threshold realizable, in which case the counterexample is spurious.

These ingredients result in a theory for probabilistic CEGAR. We have evaluated the approach on various case studies. Indeed, CEGAR entirely mechanizes the verification process: predicates are added mechanically on demand based on counterexample analysis. To this end, our implementation employs interpolation [14] to generate new predicates from spurious paths.

Related Work. Aljazzar & Hermanns [15] and Katoen & Han [12] introduced concepts and algorithms to deal with probabilistic counterexamples. However, both papers do not consider counterexamples in the context of abstraction.

We are not aware of previous work that combines abstraction refinement with predicate abstraction for probabilistic systems. Up to now different abstraction refinement approaches for finite-state probabilistic models have been proposed: among those [16,17,18,19] do not exploit counterexamples, while Chatterjee et. al [11] apply CEGAR to finite probabilistic two-player game structures. Although finite MDPs are a special case of these game structures, our work differs from [11] since it considers infinite-state models, stays more in the classical predicate abstraction realm, and is supplemented with a running implementation.

Outline. We briefly review our previous work on predicate abstraction for probabilistic programs in Section 2. The contributions of this paper, counterexample analysis and abstraction refinement for probabilistic programs, are presented in Section 3. Experimental results are presented in Section 4.

2 Preliminaries

Probabilistic Programs. We consider probabilistic programs in a guarded command language [9] which is inspired by the PRISM [3] input language, but supports infinite data domains. We fix a finite set of program variables X and a finite set of actions *Act*. Variables are typed in a definition such as `i : int`. We denote the set of expressions over the set of variables V by $Expr_V$ and the set of Boolean expression over V by $BExpr_V$. An *assignment* is a total function $E : \mathtt{X} \to Expr_{\mathtt{X}}$ from variables $\mathtt{x} \in \mathtt{X}$ to expressions $E(\mathtt{x})$. Given an expression $\mathtt{e} \in Expr_{\mathtt{X}}$ and an assignment E, we denote by $\mathtt{e}[\mathtt{X}/E(\mathtt{X})]$ the expression obtained from $\mathtt{e}$ by substituting each occurrence of a variable $\mathtt{x}$ with $E(\mathtt{x})$.

A *guarded command* $\mathtt{c}$ consists of an action a, a guard $\mathtt{g} \in BExpr_{\mathtt{X}}$ and assignments $E_{\mathtt{u}_1}, ..., E_{\mathtt{u}_k}$ weighted with probabilities $p_1, ..., p_k$ where $\sum_{i=1}^{k} p_i = 1$. We denote by $\mathtt{X}' = E$ the simultaneous update E of variables $\mathtt{X}$. With the i-th update of $\mathtt{c}$, we associate a unique update label $\mathtt{u}_i \in \mathtt{U}$. Updates are syntactically separated by a "+": $[a]\ \mathtt{g} \to \mathtt{p}_1 : \mathtt{X}'\text{=}E_{\mathtt{u}_1} + \ldots + \mathtt{p}_k : \mathtt{X}'\text{=}E_{\mathtt{u}_k}$. If the guard is satisfied, the i-th update will be executed with probability p_i. For $\mathtt{c}$, we write $a_{\mathtt{c}}$ for its action, $\mathtt{g}_{\mathtt{c}}$ for its guard. If $\mathtt{c}$ is clear from the context, we write $a, \mathtt{g}$ and $\mathtt{u}_i$ instead. We define the weakest liberal precondition of an expression with respect to an update as follows: $\mathtt{WP}_{E_{\mathtt{u}}}(\mathtt{e}) = \mathtt{e}[\mathtt{X}/E_{\mathtt{u}}(\mathtt{X})]$.

A *program* $\mathtt{P} = (\mathtt{X}, \mathtt{I}, \mathtt{C})$ consists of a Boolean expression $\mathtt{I} \in BExpr_{\mathtt{X}}$ that defines the set of initial states and a set of guarded commands $\mathtt{C}$. The program has distinctly labeled guarded commands – two different commands have distinct actions and update labels.

Probabilistic Automata. The semantics of a probabilistic program is a probabilistic automaton. To enable reconstruction of commands and updates from the semantics, automata are decorated with labels from two alphabets: an action alphabet *Act* for commands, and an update alphabet $\mathtt{U}$ for probabilistic choices. A *simple distribution* π over Σ is a function $\pi : \Sigma \to [0,1]$ such that $\sum_{s \in \Sigma} \pi(s) = 1$. Let $Distr_{\Sigma}$ denote the set of all simple distributions over Σ.

We decorate the distribution with the alphabet $\mathtt{U}$ as follows: an *(update-labeled) distribution* π over $\mathtt{U} \times \Sigma$ is a distribution $\pi : \mathtt{U} \times \Sigma \to [0,1]$ such that (i) the update alphabet is right-unique, i.e., $\pi(\mathtt{u}, s) > 0$ and $\pi(\mathtt{u}, s')$ implies that $s = s'$ and (ii) $\sum_{\mathtt{u} \in \mathtt{U}} \sum_{s \in \Sigma} \pi(\mathtt{u}, s) = 1$. Let $Distr_{(\mathtt{U},\Sigma)}$ denote the set of update-labeled distributions over $\mathtt{U} \times \Sigma$. For $\pi \in Distr_{(\mathtt{U},\Sigma)}$, we call the set of states $Supp(\pi) = \{(\mathtt{u}, s) \mid \pi(\mathtt{u}, s) > 0\}$ the support of π.

A *probabilistic update automaton* $\mathcal{M}$ is a tuple $(\Sigma, I, Act, \mathtt{U}, R)$ where Σ is a set of states, $I \subseteq \Sigma$ is a set of initial states, *Act* is the action alphabet, $\mathtt{U}$ is the update alphabet, and $R \subseteq \Sigma \times Act \times Distr_{(\mathtt{U},\Sigma)}$ the probabilistic transition relation. $\mathcal{M}$ is called finite if Σ is finite. A finite path is a finite sequence $(s_0, a_0, \mathtt{u}_0, \pi_0), (s_1, a_1, \mathtt{u}_1, \pi_1), \ldots s_n$ such that $s_0 \in I$, $(s_i, a_i, \pi_i) \in R$, and $(\mathtt{u}_i, s_{i+1}) \in Supp(\pi_i)$ for all $i = 0, \ldots, n-1$. Let $Path_{fin}(\mathcal{M})$ denote the set of all finite paths over $\mathcal{M}$. We write $\sigma \leq \sigma'$, if the finite path σ is a prefix of σ'. A finite path σ is maximal if $\sigma \leq \sigma'$ implies that $\sigma = \sigma'$. An infinite path σ is an infinite sequence $(s_0, a_0, \mathtt{u}_0, \pi_0), (s_1, a_1, \mathtt{u}_1, \pi_1), \ldots$ starting with an initial state $s_0 \in I$, $(s_i, a_i, \pi_i) \in R$, and $(\mathtt{u}_i, s_{i+1}) \in Supp(\pi_i)$ for all $i = 0, 1, \ldots$. Let $Path(\mathcal{M})$ denote the set of all infinite or maximal paths over $\mathcal{M}$. For $\sigma \in Path_{fin}$, let $C(\sigma) = \{\sigma' \in Path(\mathcal{M}) \mid \sigma \leq \sigma'\}$ denote the *cylinder set* for σ. For $\sigma \in Path$, let $\sigma[i] = s_i$ denote the $i+1$-th state of σ.

Our definition of probabilistic update automata adds labels at probabilistic choices to the probabilistic automata of [10]. This does not give additional modeling power, it rather allows us to develop the CEGAR approach succinctly. Dropping the labels of the distributions, $\mathcal{M}$ induces a probabilistic automaton $ind(\mathcal{M})$ in the style of [10] as follows: replace every update-labeled distribution π by its induced distribution $ind(\pi)$, defined by $ind(\pi)(s) = \sum_{\mathtt{u} \in \mathtt{U}} \pi(\mathtt{u}, s)$. If the context is clear, we use $\mathcal{M}$ and $ind(\mathcal{M})$ interchangeably.

An adversary is a resolution of non-determinism. In general, an adversary A of an automaton $\mathcal{M}$ is a function from paths to pairs of actions and distributions. We let $\mathcal{D}_\delta$ denote the Dirac distribution defined by: $\mathcal{D}_\delta(\delta) = 1$ where δ is a special symbol for termination. An adversary A is called *simple* if it only looks at the last state in a path, i.e. if it is a function $A : \Sigma \to (Act \times Distr_\Sigma) \cup \{\mathcal{D}_\delta\}$. Note that if $A(s) = \mathcal{D}_\delta$, the adversary A decides to stop at state s. For a given state $s \in \Sigma$ and an adversary A, let P_s^A denote the corresponding probability measure [20] over $Path(\mathcal{M})$. Given a probabilistic automaton $\mathcal{M}$, a simple adversary A induces an MC $\mathcal{M}_A = (\Sigma, I, Act, R_A)$ where $R_A = \{(s, a, \pi) \in R \mid A(s) = (a, \pi)\}$. Note s has no outgoing transitions if $A(s) = \mathcal{D}_\delta$.

MCs and MDPs are special cases of probabilistic automata. An MC is a deterministic probabilistic automaton, i.e. an automaton where for every state s there is at most one transition $(s, a, \pi) \in R$. An MDP is an action-deterministic probabilistic automaton, i.e. an automaton where for every pair $s \in \Sigma$ and $a \in Act$, there exists at most one π with $(s, a, \pi) \in R$.

Program Semantics. A *state* over variables $\mathtt{X}$ is a type-consistent total function from variables in $\mathtt{X}$ to their semantic domains. We denote the set of states by $\Sigma(\mathtt{X})$ or Σ for short and a single state by s. For an expression $\mathsf{e} \in Expr_{\mathtt{X}}$, we denote by $[\![\mathsf{e}]\!]_s$ its valuation in state s. The valuation of a Boolean expression e is a value $[\![\mathsf{e}]\!]_s \in \{0, 1\}$ (0 for "false", 1 for "true"). For a Boolean expression e and a state s, we write $s \models \mathsf{e}$ iff $[\![\mathsf{e}]\!]_s = 1$. Semantic brackets around a Boolean expression e without a subscript denote the set of states fulfilling e, i.e. $[\![\mathsf{e}]\!] = \{s \in \Sigma \mid s \models \mathsf{e}\}$.

The semantics of a program $\mathtt{P} = (\mathtt{X}, \mathtt{I}, \mathtt{C})$ is the probabilistic update automaton $\mathcal{M} = (\Sigma, I, Act, R)$ with set of states $\Sigma = \Sigma(\mathtt{X})$, set of initial states $I = [\![\mathtt{I}]\!]$, set of actions $Act = \{a_{\mathsf{c}} \mid \mathsf{c} \in \mathtt{C}\}$, and transitions induced by the guarded commands $R = \bigcup_{\mathsf{c} \in \mathtt{C}} [\![\mathsf{c}]\!]$ where $(s, a, \pi) \in [\![\mathsf{c}]\!]$ if $s \models \mathsf{g}$ and π such that $\pi(\mathtt{u}_i, s') = p_i$ if $s'(\mathtt{x}) = [\![E_{\mathtt{u}_i}(\mathtt{x})]\!]_s$ for all $\mathtt{x} \in \mathtt{X}$.

Properties. In this paper, we consider probabilistic reachability properties which we write as $Reach_{\leq p}(\mathsf{e})$ where $p \in [0, 1]$ is a probability value, and the Boolean expression $\mathsf{e} \in BExpr_{\mathtt{X}}$ describes the states to be reached. For a state s and an adversary A, let $p_s^A(\rightsquigarrow\mathsf{e}) = P_s^A(\{\sigma \in Path(\mathcal{M}) \mid \exists i \in \mathbb{N}\ \sigma[i] \models \mathsf{e}\})$ be the probability of set of paths reaching an e-state. Then, $Reach_{\leq p}(\mathsf{e})$ is satisfied by s if $p_s^A(\rightsquigarrow\mathsf{e}) \leq p$ for all adversary A, and it is satisfied by the model if it is satisfied by all initial states. Algorithmically, it is sufficient to only consider simple adversaries [2], as extremal probabilities are already attained among them.

Predicate Abstraction. Predicates are Boolean expressions over the program variables. A predicate φ stands for the set of states satisfying it, namely $[\![\varphi]\!]$. We fix a set of predicates $\mathcal{P} = \{\varphi_1, ..., \varphi_n\}$. The set $\mathcal{P}$ partitions the states into disjoint sets characterized by which predicates hold and which not. An equivalence class can therefore be represented by a bit vector of length n. We call such a bit-vector an *abstract state* and denote the set of abstract states by $\Sigma^\sharp$. We define a state-abstraction function by: $h_\mathcal{P}(s) = ([\![\varphi_1]\!]_s, ..., [\![\varphi_n]\!]_s)$. For a given $s^\sharp \in \Sigma^\sharp$, we call the corresponding equivalence class the concretization of $s^\sharp$ and denote it by $\gamma(s^\sharp)$. The concretization of $s^\sharp$ is characterized by a Boolean expression $F(s^\sharp)$ such that $\gamma(s^\sharp) = [\![F(s^\sharp)]\!]$. $F(s^\sharp)$ is exactly the conjunction containing satisfied predicates as positive literals and unsatisfied ones as negated literals.

We recall predicate abstraction of probabilistic programs [9]: The state abstraction $h_\mathcal{P}$ induces a quotient automaton, denoted by $\mathcal{M}^\sharp = (\Sigma^\sharp, I^\sharp, Act, \mathtt{U}, R^\sharp)$ with the set of initial states $I^\sharp = \{h(s) \mid s \in I\}$, and transitions $R^\sharp = \{(h(s), a, h(\pi)) \mid (s, a, \pi) \in R\}$ where $h(\pi) = \{(\mathtt{u}, h(s)) : p \mid \pi(\mathtt{u}, s) = p\}$. If the reachability property is satisfied by the quotient automaton $\mathcal{M}^\sharp$, we can safely conclude that it holds for the original model $\mathcal{M}$ as well. The soundness follows from the fact that the quotient automaton $\mathcal{M}^\sharp$ simulates $\mathcal{M}$.

3 Refinement

In this section, we present a novel refinement scheme for probabilistic programs based on CEGAR (counterexample-guided abstraction refinement). The plain CEGAR approach is the obvious strategy also to follow in the probabilistic case: start with a coarse abstraction and successively refine it using predicates learned from spurious counterexamples until either a realizable counterexample is found or the abstract model is precise enough to establish the property. However, in order to put refinement to work for probabilistic models, several questions of both principal and practical nature need to be answered. We *(i)* need to identify what an abstract counterexample constitutes, *(ii)* lift it to the concrete system, *(iii)* decide if it is spurious, and *(iv)* identify appropriate predicates to refine the abstract quotient automaton.

Counterexamples for Quotient Automata. Intuitively, a counterexample is a pair of an initial state and an adversary that violates the property to be checked. This pair induces an MC in the abstract setting. In the sequel, we fix the probabilistic update automaton $\mathcal{M}$ and the reachability property $Reach_{\leq p}(\mathsf{e})$. Let $\mathcal{M}^\sharp$ be the quotient automaton of $\mathcal{M}$.

Definition 1. *A counterexample for* $Reach_{\leq p}(\mathsf{e})$ *is a pair* $(s^\sharp, A^\sharp)$ *where* $s^\sharp \in I^\sharp$ *is an initial state and* $A^\sharp$ *is an adversary such that* $P_{s^\sharp}^{A^\sharp}(\rightsquigarrow \mathsf{e}) > p$. *In this case,* $A^\sharp$ *is called a* counter-adversary.

Spurious Counterexamples. In the non-probabilistic setting, a counterexample is a path, which is called spurious if there does not exist a corresponding concrete path. We now introduce the notion of spurious counterexamples for probabilistic automata based on the concretization of an abstract counter-adversary and an abstract counterexample.

Concretization of Counter-adversaries. For a counter-adversary $A^\sharp$ in the quotient $\mathcal{M}^\sharp$, its concretization $\gamma(A^\sharp)$ is an adversary in M defined by: $\gamma(A^\sharp)(s)$ equals (a_c, π) if $A^\sharp(s^\sharp) = (a_\mathsf{c}, \pi^\sharp)$ with $s^\sharp = h(s)$ and $\pi^\sharp = h(\pi)$, otherwise $\mathcal{D}_\delta$.

In case of $\gamma(A^\sharp)(s) = \mathcal{D}_\delta$, the adversary $A^\sharp$ has chosen $(a_\mathsf{c}, \pi^\sharp)$ from $s^\sharp$, however s does not satisfy the guard g_c associated with c. Thus, we let $\gamma(A^\sharp)$ stop at state s, as no corresponding concretization exists. Recall that the program has distinctly labeled guarded commands, thus we can choose at most one corresponding outgoing concrete transition. For illustration, consider the fragment of a probabilistic automaton and its corresponding quotient automaton in the figure below. If the adversary $A^\sharp$ chooses a_{c_2} at state $s_2^\sharp$, the concretization $\gamma(A^\sharp)$ chooses also action a_{c_2} at state s.

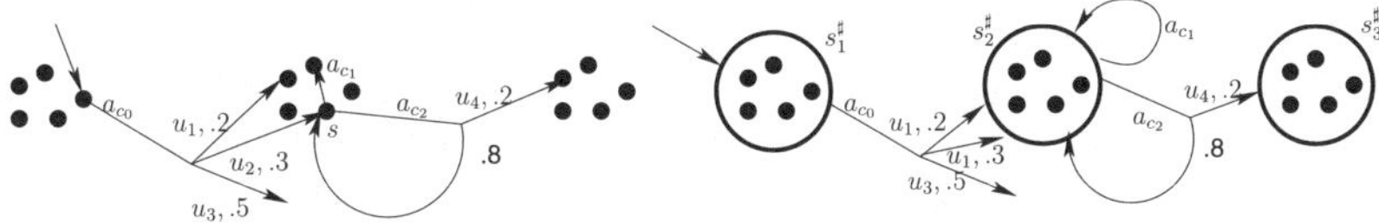

Concretization of Counterexamples. Now consider a counterexample $(s^\sharp, A^\sharp)$. Its concretization, denoted $\gamma(s^\sharp, A^\sharp)$, is the set: $\{(s, A) \mid A = \gamma(A^\sharp) \wedge s \in (I \cap \gamma(s^\sharp))\}$. Directly linked to the cardinality of the initial state set, the concretization can contain many (even infinitely many) elements, and thus induce many MCs. The reachability probability $P(\rightsquigarrow \mathsf{e})$ may differ from element to element. A counterexample $(s^\sharp, A^\sharp)$ is spurious if its concretization does not contain a pair (s, A) such that the probability threshold is exceeded. In other words, a spurious counterexample does not induce any concrete MC for which the probability measure of reaching concrete e-states exceeds the specified threshold.

Definition 2. *Let $(s^\sharp, A^\sharp)$ be a counterexample for $Reach_{\leq p}(\mathsf{e})$ in $\mathcal{M}^\sharp$. Then, $(s^\sharp, A^\sharp)$ is called* realizable *if there exists $(s, A) \in \gamma(s^\sharp, A^\sharp)$ such that $P_s^A(\rightsquigarrow \mathsf{e}) > p$. Otherwise we say that the counterexample is* spurious.

Checking Counterexamples. Checking realizability of counterexamples is a key element of the refinement procedure: If a counterexample turns out to be realizable, the property is refuted with A playing the role of a counter-adversary in the concrete model, which can be used for debugging purposes. Otherwise, the abstract model is too coarse and additional predicates will need to be added to eliminate the false negative.

Overall Idea. In the non-probabilistic setting, an abstract counterexample is a single finite abstract path $\sigma^\sharp$ starting in an abstract initial state. Its concretization is a set of corresponding paths in the concrete model each of which starts in some concrete initial state and respects the concrete transition relation. This set might potentially be infinite. If it is empty, the counterexample is spurious. It is common practice to check emptiness of the concretization by expressing the behavior enforced on the concrete program by the abstract path implicitly by a formula and checking the satisfiability of that formula [6,7]. If the formula is satisfied, then the concretization is non-empty, and we have found a concrete counterexample violating the property. In this case, the counterexample is realizable. Otherwise, it is spurious, and additional predicates can be extracted from the path $\sigma^\sharp$ for refinement.

In the probabilistic setting, however, the situation is much more involved. What makes the counterexample $(s^\sharp, A^\sharp)$ realizable is a concrete initial state $s \in (I \cap \gamma(s^\sharp))$ and adversary A such that the probability of reaching an e-state in the thus induced concrete MC exceeds the given threshold p. All candidates (s, A) are contained in $\gamma(s^\sharp, A^\sharp)$ but this set might be infinite. We preprocess the abstract counterexample using the strongest evidence idea of Katoen & Han [12]. They have devised a method that, for a given MC, can be used to construct the smallest set of paths reaching e-states with an accumulated probability measure above p. This fits well to our needs.

As the abstract counterexample $(s^\sharp, A^\sharp)$ induces an abstract MC, we can apply the algorithm from [12] yielding *a finite* set of *finite paths* in the quotient automaton starting from state $s^\sharp$, such that the probability measure exceeds p. To check if the counterexample is spurious, our goal is then to compute how much measure out of this set of paths can be reproduced in $\mathcal{M}$ with respect to *any* $(s, A) \in \gamma(s^\sharp, A^\sharp)$. If that is indeed larger than the threshold p for some (s, A), we have found a realizable counterexample. Otherwise we may be able to conclude that it is spurious, or conclude that more work is needed, as we will explain below.

Spuriousness of Abstract Paths. Before coming to adversaries, we first explain how to check if a single abstract path is realizable or spurious. Let $(s^\sharp, A^\sharp)$ be a counterexample and let $\sigma^\sharp = (s_0^\sharp, a_0, \mathtt{u}_0, \pi_0^\sharp)\ (s_1^\sharp, a_1, \mathtt{u}_1, \pi_1^\sharp) \dots s_k^\sharp$ be a path in $\mathcal{M}^\sharp_{A^\sharp}$ where $s_0^\sharp = s^\sharp$ and $s_k^\sharp$ satisfies e. The concretization $\gamma(\sigma^\sharp)$ of an abstract path $\sigma^\sharp$ is a set of finite paths in $\mathcal{M}$ with consistent states, and the same update and action labels, i.e. $\gamma(\sigma^\sharp) = \{(s_0, a_0, \mathtt{u}_0, \pi_0) \dots s_k \mid (s_0, \dots, s_k) \models \mathit{TF}(\sigma^\sharp)\}$ where $\mathit{TF}(\sigma^\sharp)$ is the trace formula which is defined by:

$$\mathit{TF}(\sigma^\sharp) = \mathtt{I}(\mathtt{X}_0) \wedge \bigwedge_{i=0}^{k} F(s_i^\sharp)(X_i) \wedge \bigwedge_{i=0}^{k-1} \left(\mathtt{g}_{\mathtt{c}_i}(\mathtt{X}_i) \wedge \mathtt{X}_{i+1} = E_{u_i}(\mathtt{X}_i)\right) \wedge \mathtt{e}(\mathtt{X}_k) \ .$$

The measure of $\sigma^\sharp$ under $(s^\sharp, A^\sharp)$ is $\prod_{i=0}^{k-1} \pi_i^\sharp(\mathtt{u}_i, s_{i+1}^\sharp)$. Note that the paths in the concretization of $\sigma^\sharp$ share the same measure. The path $\sigma^\sharp$ is called *realizable* if its concretization is non-empty, $\gamma(\sigma^\sharp) \neq \emptyset$, otherwise it is called *spurious*. As for the non-probabilistic setting [7,6], an abstract path is realizable if its trace formula is satisfiable or, equivalently, its weakest precondition. The weakest precondition of an abstract path is formalized in Figure 1 as the repeated application of the standard *syntactic weakest precondition* $\mathtt{WP}_E(e)$ where $\mathtt{WP}_E(e) := e[\mathtt{X}/E(\mathtt{X})]$ for an expression e and an update $\mathtt{X}'{=}E$.

WP($\sigma^\sharp = (s_0^\sharp, a_{\mathtt{c}_0}, \mathtt{u}_0, \pi_0^\sharp) \dots s_k^\sharp$)

```
exp_σ♯ ← F(s♯_k) ∧ e
for (j = k .. 0) do
    exp_σ♯ ← g_{c_j} ∧ F(s♯_j) ∧ WP_{u_j}(exp_σ♯)
end for
return exp_σ♯ ∧ I
```

Fig. 1. WP of an abstract path $\sigma^\sharp$

Lemma 1. *For an abstract path $\sigma^\sharp$, the following statements are equivalent: (i) The weakest precondition* $\mathrm{WP}(\sigma^\sharp)$ *of path $\sigma^\sharp$ is satisfiable. (ii) The trace formula $TF(\sigma^\sharp)$ of $\sigma^\sharp$ is satisfiable. (iii) The path $\sigma^\sharp$ is realizable, i.e. $\gamma(\sigma^\sharp) \neq \emptyset$.*

Checking Spuriousness. The counterexample $(s^\sharp, A^\sharp)$ is guaranteed to be realizable if it has a concretization with sufficiently high measure. We assume a nonempty set *afp* of abstract paths respecting $(s^\sharp, A^\sharp)$. Note that corresponding concrete paths may start in different initial states, so that the probability in the concrete model is possibly lower. Let us consider an abstract path $\sigma^\sharp$. For all $\sigma \in \gamma(\sigma^\sharp)$ with $\sigma = (s_0, a_0, \mathrm{u}_0, \pi_0) \ldots s_k$, the measure of the cylinder set $C(\sigma)$ under $(s, A) \in \gamma(s^\sharp, A^\sharp)$ is given by $\prod_{i=0}^{k-1} \pi_i(\mathrm{u}_i, s_{i+1})$ if $s = s_0$, which is the same as $\prod_{i=0}^{k-1} \pi_i^\sharp(\mathrm{u}_i, s_{i+1}^\sharp)$. For a set *afp* of abstract path we let $\gamma(\mathit{afp}) = \bigcup_{\sigma^\sharp \in \mathit{afp}} \gamma(\sigma^\sharp)$ denote the union of the concretizations. Now an interesting issue arises: what is the maximal probability measure of the set $\gamma(\mathit{afp})$ under some concretization of $\gamma(s^\sharp, A^\sharp)$. For illustration, consider the figure below where *afp* consists of two disjoint abstract paths $\sigma_1^\sharp, \sigma_2^\sharp$, but the intersection is empty: $exp_{\sigma_1^\sharp} \wedge exp_{\sigma_2^\sharp} = \emptyset$, hence only the maximum of both can be achieved.

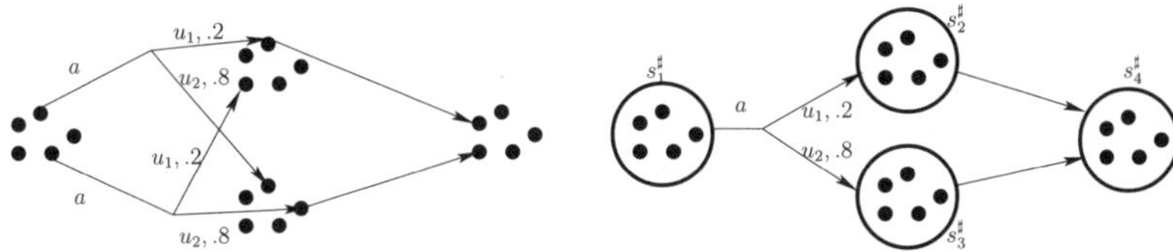

We resolve this problem by using weakest preconditions of abstract paths. Given an abstract path $\sigma^\sharp$, the backwards algorithm in Figure 1 computes its weakest precondition, i.e. those initial states in which a path from the concretization of $\sigma^\sharp$ starts. We use these weakest preconditions to obtain subsets of the given set of abstract paths sharing a common concrete initial state. The subset with maximal probability gives us the actual measure in the concrete model.

For $\mathit{afp} = \{\sigma_1^\sharp, \ldots, \sigma_n^\sharp\}$, let $exp_1, \ldots, exp_n$ denote the weakest preconditions returned by $\mathrm{WP}(\sigma_i^\sharp)$. Moreover, for each of them the probability measure of path $\sigma_i^\sharp$ is given as a weight, denoted by p_i, which corresponds to the probability of the set $\gamma(\sigma_i^\sharp)$ starting from some initial state in exp_i. We now formulate the problem of computing the realizable probability mass of a set of abstract paths in terms of a weighted MAX-SMT [13] problem, which consists in finding an assignment of $\mathtt{X}$ such that the total weight of the satisfied expression is maximal. Formally, it is defined by: $\textsc{MaxSmt}(exp_1, \ldots, exp_n) = \max\left\{\sum_{i=1}^{n} [\![exp_i]\!]_s \cdot p_i \mid s \in [\![\mathtt{I} \wedge F(s^\sharp)]\!]\right\}$.

Lemma 2. *Let $(s^\sharp, A^\sharp)$ be a counterexample for $Reach_{\leq p}(\mathsf{e})$, and let $\mathit{afp} = \{\sigma_1^\sharp, \ldots, \sigma_n^\sharp\}$ be a set of abstract paths with measure greater than p. It holds:*

(i) $\textsc{MaxSmt}(exp_1, \ldots, exp_n) > p$ *implies that $(s^\sharp, A^\sharp)$ is realizable,*
(ii) $\textsc{MaxSmt}(exp_1, \ldots, exp_n) + P_{s^\sharp}^{A^\sharp}(\rightsquigarrow \mathsf{e}) - P_{s^\sharp}^{A^\sharp}(\mathit{afp}) \leq p$ *implies that the counterexample $(s^\sharp, A^\sharp)$ is spurious.*

Let $\varepsilon = P_{s^\sharp}^{A^\sharp}(\leadsto \mathsf{e}) - P_{s^\sharp}^{A^\sharp}(afp)$ denote the probability of the set of abstract paths which violate the property $Reach_{\leq p}(\mathsf{e})$, but are not part of the set *afp*. The lemma indicates that the decision algorithm is only partial: if the value $\textsc{MaxSmt}(exp_1, \ldots, exp_n)$ lies in the interval $(p - \epsilon, p]$, we are not sure whether the counterexample $(s^\sharp, A^\sharp)$ is spurious or realizable. By enlarging the set *afp*, the ε can be made arbitrarily small. We will see later how this is exploited for the CEGAR algorithm.

Obtaining Predicates. There are two sources of potential imprecision: spurious abstract paths or a too coarse abstraction of the initial states.

Predicates to Remove Spurious Paths. Let $(s^\sharp, A^\sharp)$ be a counterexample in $\mathcal{M}^\sharp_{A^\sharp}$. Let $\sigma^\sharp = (s_0^\sharp, a_0, \mathrm{u}_0, \pi_0^\sharp)(s_1^\sharp, a_1, \mathrm{u}_1, \pi_1^\sharp) \ldots s_k^\sharp$ be a path such that $s_0^\sharp = s^\sharp$ and $\sigma^\sharp$ satisfies $\leadsto \mathsf{e}$. Assume that $\sigma^\sharp$ is spurious. Our goal is to find predicates to eliminate the spurious abstract path. The abstract path resolves both nondeterministic choice between different commands, and probabilistic choice between different updates. That enables us to use standard techniques developed in the non-probabilistic setting to find predicates. Here we employ interpolation and apply it to the trace formula of the abstract path along the lines of [21].

Predicates to Separate Initial States. Observe the case where no path in *afp* is spurious but the realizable probability of the paths is lower than the probability threshold p, i.e., $\textsc{MaxSmt}(exp_1, \ldots, exp_n) \leq p$. In this case, the initial state $s^\sharp$ may be too coarse. To this end, we choose the maximal solution obtained from $\textsc{MaxSmt}$. Let ψ^- denote the conjunction of non-satisfied exp_i, and ψ^+ denote the conjunction of satisfied exp_i. Obviously, $\psi^- \wedge \psi^+$ is not satisfiable. Hence, interpolants can be found to refine the abstraction of the initial states. Note that this is a heuristic choice and does not guarantee removal of the abstract counterexample.

CEGAR Algorithm. At the start of each iteration of the CEGAR loop, a quotient automaton $\mathcal{M}^\sharp$ is built using the current set of predicates. We submit the quotient automaton and the property to a probabilistic model checker. Due to soundness of the abstraction, we can safely report success if the property is satisfied in $\mathcal{M}^\sharp$. Otherwise the model checker produces an abstract counterexample $(s^\sharp, A^\sharp)$ which is passed to the counterexample analysis phase.

Counterexample analysis constitutes the next phase: Along the ideas of strongest evidence [12], we maintain a sequence $\eta = \langle \sigma_1, \sigma_2, ..., \sigma_n \rangle$ of abstract paths reaching an e-state in the MC induced by $(s^\sharp, A^\sharp)$, an additional set $afp \subseteq \eta$ contains realizable paths in η. As illustrated in the diagram below, sequence η

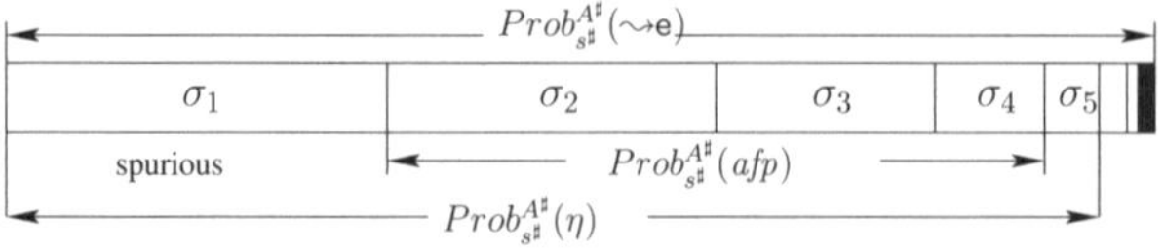

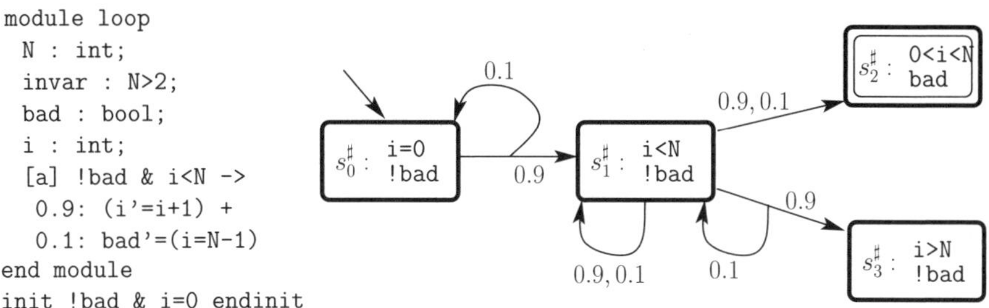

Fig. 2. Cycle program and the quotient automaton with respect to `i=0,bad,i<N`

is ordered by decreasing probability mass – a longer bar means higher probability measure of the path; η is computed incrementally by a variant of best first search [22] in a weighted graph obtained from the MC. Initially η contains only the path with the highest probability, path σ_1.

First we check if path σ_1 is realizable using Lemma 1 (in the diagram we assume σ_1 is spurious), and, if so, we add σ_1 to the set of confirmed paths *afp*. If enough "alleged" probability mass has already accumulated in *afp* to exceed the threshold, i.e. $P_{s^\sharp}^{A^\sharp}(afp) > p$, we check how much of that probability is actually realizable using Lemma 2. If the realizable probability mass exceeds the threshold, the property is refuted, since we can report a realizable counterexample. Otherwise we repeat the process with path σ_2 that has the second highest probability: we add it to η, and check if it is realizable. If realizable, we add σ_2 to *afp*. We continue in this way until either we can refute the property or $n - |afp| = C$, in which case we proceed to phase three. C is a verification parameter set by the user, in the diagram we have $C = 2$.

In the third phase predicates are generated from spurious paths or from weakest preconditions. Then the next iteration of the algorithm commences.

Toy example. Consider the program `Cycle` shown in the left part of Figure 2. The right part shows the quotient automaton with respect to predicates `i=0,bad,i<N` where we omitted the actions and updates. Assume we want to check $Reach_{\leq 0.3}(bad)$. In the quotient automaton, the probability of reaching the `bad` state is 1.0. Let $\mathtt{u}_0$ denote the update `i'=i+1`. We start with the abstract path with (highest) probability 0.81 (distributions are omitted): $\sigma^\sharp = (s_0^\sharp, a, \mathtt{u}_0)(s_1^\sharp, a, \mathtt{u}_0)s_2^\sharp$. Obviously, this path is not realizable as witnessed by the unsatisfiability of its trace formula ψ (see Lemma 1). Taking $C = 1$, we apply Lemma 2 from which we conclude that we have a spurious counterexample. To remove the spurious path $\sigma^\sharp$, we apply interpolation to the trace formula ψ, i.e. we compute a simplification of its prefix $\psi_1^- := N > 2 \wedge i_0 = 0 \wedge \neg bad_0 \wedge i_1 = i_0 + 1 \wedge bad_1 = bad_0 \wedge i_1 < N \wedge \neg bad_1$ that is disjoint with its postfix $\psi_1^+ := i_2 = i_1 + 1 \wedge bad_2 = bad_1 \wedge i_2 \geq N \wedge bad_2$. As an interpolant we obtain $i < N - 1$, add it as a fresh predicate and restart. In the ensuing iteration, the property is established.

4 Experimental Results

We have implemented a prototype of probabilistic CEGAR within the predicate abstraction tool PASS [9]. It is is written in C++ and interfaces to the SMT solver Yices [23] which also supports MAX-SMT. PASS uses CEGAR to obtain predicates based on the interpolant-generating theorem prover FOCI [24]. Experiments were run on a Pentium^{TM} IV 2.6 GHz with 1.5 GB RAM.

Table 1. Statistics. Shown on the left are model parameters and properties studied. On the right we display, apart from reachable state numbers, number of transitions (non-zero entries in transition matrix), and computation time, the number of iterations of the CEGAR loop (refs), of predicates generated (preds), and of abstract paths analyzed (paths). The number of states and transitions are given in thousands, i.e. 34K means 34,000.

Case study (parameters)		Property	Conventional			Abstraction					
			states	trans	time	states	trans	refs	preds	paths	time
WLAN (BOFF T)	5 315	k=3	5,195K	11,377K	93	34K	36K	9	120	604	72
	6 315	k=3	12,616K	28,137K	302	34K	42K	9	116	604	88
	6 315	k=6	12,616K	28,137K	2024	771K	113K	9	182	582	306
	6 9500	k=6	–	–	**TO**	771K	113K	9	182	582	311
CSMA/CD (BOFF)	3	p1	41K	52K	10	1K	2K	8	58	28	9
	4	p1	124K	161K	56	6K	9K	14	100	56	38
	3	p2	41K	52K	10	0.5K	0.9K	12	41	28	10
	4	p2	124K	161K	21	0.5K	1.5K	12	41	44	11
BRP (N MAX)	16 3	p1	2K	3K	5.4	2K	3K	9	46	41	9
	32 5	p1	5K	7K	12	5K	7K	9	64	111	21
	64 5	p1	10K	14K	26	10K	14K	8	95	585	91
	>16 3	p4	∞	–	–	0.5K	0.9K	7	26	17	3
	>16 4	p4	∞	–	–	0.6K	1K	7	27	17	4
	>16 5	p4	∞	–	–	0.7K	1K	8	28	18	5
SW		goodput	∞	–	–	5K	11K	3	40	7	87
		timeout	∞	–	–	27K	44K	3	49	6	89

We consider a selection of case studies summarized in Table 1. Time is measured in seconds. The timeout limit was set to two hours (timeout is indicated by **TO**). We have analyzed the finite-state models with PRISM 3.1.1 (column "Conventional") in comparison with PASS (column "Abstraction"). PRISM is the leading finite-state probabilistic model checker. All models except for the Sliding Window protocol are taken from the PRISM web repository and can be scaled via different parameters. Unconstrained parameters yield infinite-state models (denoted by "∞"). Surprisingly, although our method only guarantees *upper bounds* on probabilities in general, probabilities obtained for all case studies are *tight* upper bounds: they agree with those of PRISM for finite models.

IEEE 802.11 Wireless LAN Protocol (WLAN). The protocol is parameterized with an *exponential* back-off counter limit BOFF and a maximal package send time of T μs. We checked the property: "The maximum probability that either station's back-off counter reaches k" for k=3 and k=6. As shown in Table 1, increasing BOFF from 5 to 6 leads to an exponential increase in model size and running time in PRISM, while in PASS the row is identical for BOFF=5 and BOFF=6. This is because states with back-off counter higher than three can reach a goal state (via a reset), however they do not lie on paths with maximal probability. Hence refinement never splits abstract states with respect to back-offs beyond three. Similarly, for fixed value of BOFF, PASS scales much better in comparison to PRISM with respect to different values of T.

IEEE 802.3 CSMA/CD Protocol (CSMA/CD). Similar to the WLAN protocol, CSMA/CD is parameterized with an *exponential* back-off counter limit BOFF. We analyzed the properties: (p1): "The maximum probability that both stations deliver", and (p2): "The message of any station eventually delivered before 1 backoff". For both properties, as shown in the table, the abstract state space is significantly smaller. Consider property p2. Similar to the WLAN protocol, the size of the abstraction does not change with respect to the size of BOFF. However, the number of paths explored increases with BOFF. The reason is that for greater values of BOFF, there is more branching in the probabilistic model, thus in the abstraction there are more abstract paths being explored.

Bounded Retransmission Protocol (BRP). The BRP protocol has two parameters: N denotes the length of the file to be transmitted, and MAX denotes the maximal number of retransmissions. We have studied "Property 1" and "Property 4" (p1 and p4 in the table). On p1, PRISM outperforms PASS. It appears that this is due to little opportunity for abstraction as, seemingly, a lot of model detail is relevant and has to be discovered by refinement. On the other hand, p4 can be analyzed for an infinite parameter range with PASS, since it is an invariant property with respect to the file length N. Thus the constraint $N > 16$ allows us to verify the property for any possible file length greater than 16.

Sliding Window (SW). This is the standard protocol with lossy channels over an unbounded domain of sequence numbers. Thus the model is infinite and hence we have no comparison to PRISM. We checked goodput properties which consider the difference between the number of sent and received packages. We want to know the probability that the number of sent packages exceeds the number of received packages by a particular constant. PASS checked that, at any time, the probability of the difference exceeding three is at most three percent for windows size four. The second property concerns the probability of a protocol timeout.

Initial Predicates. Probabilistic CEGAR leaves the choice of the initial set of predicates as a parameter. Predicates appearing in the property under study are the minimal option and generally a good one (BRP, CSMA). Further, control locations of the program might also be part of the initial abstraction to avoid non-determinism between commands in the abstraction. Thus adding predicates from the initial condition (WLAN) or additionally from guards (SW) can improve running times. PASS features several automatic modes; Table 1 contains data obtained via the respective best mode.

Discussion. To compete with PRISM on finite models, the benefit of state space reduction has to offset the cost of repeating the CEGAR loop. On infinite or very large models only PASS can be used. We observe that abstraction for

WLAN and CSMA/CD, PASS is superior to the conventional approach, due to the significantly smaller abstract state space. Notably the analysis of BRP for $N > 16$ can be considered a parametric analysis: p4 is proven for any such N.

5 Conclusion

This paper explores fundamental questions and pragmatic issues of probabilistic abstraction refinement. The main contribution lies in our treatment of abstract counterexamples which are finite Markov chains, instead of finite paths. Spurious counterexamples are analyzed with interpolation-based predicate inference, leading to a refined model which closes the CEGAR loop. The resulting theory and tool work smoothly, as shown by our experimental evaluation. Our next goal is to enable model checking of full PCTL.

Acknowledgements. Thanks to E. Moritz Hahn for helping us with the implementation, and to the anonymous reviewers for their valuable comments.

References

1. Hansson, H., Jonsson, B.: A Logic for Reasoning about Time and Reliability. Formal Asp. Comput. 6, 512–535 (1994)
2. Bianco, A., de Alfaro, L.: Model checking of probabilistic and nondeterministic systems. In: Thiagarajan, P.S. (ed.) FSTTCS 1995. LNCS, vol. 1026, pp. 499–513. Springer, Heidelberg (1995)
3. Hinton, A., Kwiatkowska, M.Z., Norman, G., Parker, D.: PRISM: A Tool for Automatic Verification of Probabilistic Systems. In: Hermanns, H., Palsberg, J. (eds.) TACAS 2006 and ETAPS 2006. LNCS, vol. 3920, pp. 441–444. Springer, Heidelberg (2006)
4. Katoen, J.P., Khattri, M., Zapreev, I.S.: A Markov reward model checker. In: QEST, pp. 243–244 (2005)
5. Graf, S., Saídi, H.: Construction of Abstract State Graphs with PVS. In: Grumberg, O. (ed.) CAV 1997. LNCS, vol. 1254, pp. 72–83. Springer, Heidelberg (1997)
6. Ball, T., Rajamani, S.K.: The SLAM project: debugging system software via static analysis. In: POPL, pp. 1–3 (2002)
7. Henzinger, T.A., Jhala, R., Majumdar, R., Sutre, G.: Lazy abstraction. In: POPL, pp. 58–70 (2002)
8. Clarke, E.M., Grumberg, O., Jha, S., Lu, Y., Veith, H.: Counterexample-Guided Abstraction Refinement. In: Emerson, E.A., Sistla, A.P. (eds.) CAV 2000. LNCS, vol. 1855, pp. 154–169. Springer, Heidelberg (2000)
9. Wachter, B., Zhang, L., Hermanns, H.: Probabilistic Model Checking Modulo Theories. In: QEST, pp. 129–138 (2007)
10. Segala, R., Lynch, N.A.: Probabilistic simulations for probabilistic processes. Nord. J. Comput. 2, 250–273 (1995)
11. Chatterjee, K., Henzinger, T.A., Majumdar, R.: Counterexample-Guided Planning. In: UAI (2005)
12. Han, T., Katoen, J.P.: Counterexamples in probabilistic model checking. In: Grumberg, O., Huth, M. (eds.) TACAS 2007. LNCS, vol. 4424, pp. 72–86. Springer, Heidelberg (2007)

13. Papadimitriou, C.H., Yannakakis, M.: Optimization, approximation, and complexity classes. J. Comput. Syst. Sci. 43, 425–440 (1991)
14. McMillan, K.L.: Lazy abstraction with interpolants. In: Ball, T., Jones, R.B. (eds.) CAV 2006. LNCS, vol. 4144, pp. 123–136. Springer, Heidelberg (2006)
15. Aljazzar, H., Hermanns, H., Leue, S.: Counterexamples for timed probabilistic reachability. In: Pettersson, P., Yi, W. (eds.) FORMATS 2005. LNCS, vol. 3829, pp. 177–195. Springer, Heidelberg (2005)
16. D'Argenio, P.R., Jeannet, B., Jensen, H.E., Larsen, K.G.: Reduction and Refinement Strategies for Probabilistic Analysis. In: Hermanns, H., Segala, R. (eds.) PROBMIV 2002, PAPM-PROBMIV 2002, and PAPM 2002. LNCS, vol. 2399, pp. 57–76. Springer, Heidelberg (2002)
17. de Alfaro, L., Roy, P.: Magnifying-lens abstraction for markov decision processes. In: Damm, W., Hermanns, H. (eds.) CAV 2007. LNCS, vol. 4590, pp. 325–338. Springer, Heidelberg (2007)
18. Kwiatkowska, M., Norman, G., Parker, D.: Game-based Abstraction for Markov Decision Processes. In: QEST, pp. 157–166 (2006)
19. Fecher, H., Leucker, M., Wolf, V.: Don't Know. In: Valmari, A. (ed.) SPIN 2006. LNCS, vol. 3925, pp. 71–88. Springer, Heidelberg (2006)
20. Puterman, M.L.: Markov Decision Processes: Discrete Stochastic Dynamic Programming. John Wiley & Sons, Chichester (1994)
21. Henzinger, T.A., Jhala, R., Majumdar, R., McMillan, K.L.: Abstractions from proofs. In: POPL, pp. 232–244 (2004)
22. Dechter, R., Pearl, J.: Generalized best-first search strategies and the optimality af A*. J. ACM 32, 505–536 (1985)
23. Dutertre, B., de Moura, L.M.: A Fast Linear-Arithmetic Solver for DPLL(T). In: Ball, T., Jones, R.B. (eds.) CAV 2006. LNCS, vol. 4144, pp. 81–94. Springer, Heidelberg (2006)
24. McMillan, K.L.: An interpolating theorem prover. Theor. Comput. Sci. 345, 101–121 (2005)

Computing Differential Invariants of Hybrid Systems as Fixedpoints

André Platzer[1] and Edmund M. Clarke[2]

[1] University of Oldenburg, Department of Computing Science, Germany
[2] Carnegie Mellon University, Computer Science Department, Pittsburgh, PA
{aplatzer,emc}@cs.cmu.edu

Abstract. We introduce a fixedpoint algorithm for verifying safety properties of hybrid systems with differential equations whose right-hand sides are polynomials in the state variables. In order to verify nontrivial systems without solving their differential equations and without numerical errors, we use a continuous generalization of induction, for which our algorithm computes the required *differential invariants*. As a means for combining local differential invariants into global system invariants in a sound way, our fixedpoint algorithm works with a compositional verification logic for hybrid systems. To improve the verification power, we further introduce a *saturation procedure* that refines the system dynamics successively with differential invariants until safety becomes provable. By complementing our symbolic verification algorithm with a robust version of numerical falsification, we obtain a fast and sound verification procedure. We verify roundabout maneuvers in air traffic management and collision avoidance in train control.

Keywords: verification of hybrid systems, differential invariants, verification logic, fixedpoint engine.

1 Introduction

Reachability questions for systems with complex continuous dynamics are among the most challenging problems in verifying embedded systems. Hybrid systems [1, 2, 3, 4] are models for these systems with interacting discrete and continuous transitions, with the latter being governed by differential equations. For simple systems whose differential equations have solutions that are polynomials in the state variables, quantifier elimination [5] can be used for verification [3,6,7,8,9]. Unfortunately, this symbolic approach does not scale to systems with complicated differential equations whose solutions do not support quantifier elimination (e.g., when they are transcendental functions) or cannot be given in closed form.

Numerical or approximation approaches [10, 11, 12] can deal with more general dynamics. However, numerical or approximation errors need to be handled carefully as they easily cause unsoundness [11]. More specifically, we have shown previously that even single image computations of fairly restricted classes of hybrid systems are undecidable by numerical computation [11]. Thus, numerical approaches can be used for falsification but not (ultimately) for verification.

A. Gupta and S. Malik (Eds.): CAV 2008, LNCS 5123, pp. 176–189, 2008.

In this paper, we present an approach that combines the soundness of symbolic approaches [3, 7, 8, 9] with support for nontrivial dynamics that is classically more dominant in numerical approaches [10, 11, 12]. During continuous transitions, the system follows a solution of its differential equation. But for nontrivial dynamics, these solutions are much more complicated than the original equations. Solutions quickly become transcendental even if the differential equations are linear. To overcome this, we handle continuous transitions based on their vector fields, which are described by their differential equations. We use *differential induction* [13], a continuous generalization of induction that works with the differential equations themselves instead of their solutions. For the induction step, we use a condition that can be checked easily based on *differential invariants* [13], i.e., properties whose derivative holds true in the direction of the vector field of the differential equation. The derivative is a directional derivative in the direction of (the vector field generated by) the differential equation, and we generalize derivatives from functions to formulas appropriately. For this to work in practice, the most crucial steps are to find sufficiently strong local differential invariants for differential equations and compatible global invariants for the hybrid system.

To this end, we introduce a *sound* verification algorithm for hybrid systems that computes the differential invariants and system invariants in a fixedpoint loop. We follow the invariants as fixedpoints paradigm [14] using a verification logic that is generalized to hybrid systems accordingly [8, 9]. For combining multiple local differential invariants into a global invariant in a sound way, we exploit the closure properties of the underlying verification logic [8, 9] by forming appropriate logical combinations of multiple safety statements. In addition, we introduce a *differential saturation process* that refines the hybrid dynamics successively with auxiliary differential invariants until the safety statement becomes an invariant of the refined system. Finally, each fixedpoint iteration of our algorithm can be combined with numerical falsification to accelerate the overall symbolic verification in a sound way [15]. We validate our algorithm by verifying *aircraft roundabout maneuvers* [16, 11] and train control applications [17].

The major contribution in this work is the fixedpoint algorithm for computing differential invariants coupled with a differential saturation process. We show that it can verify realistic applications that were out of scope for related invariant approaches [18,19,20] or [1,3,6], both for theoretical reasons [9,13] and scalability.

2 Hybrid Programs and Differential Dynamic Logic

As operational models for hybrid systems, we use *hybrid programs* (HP), a program notation for hybrid automata (HA) [1]. HP can be decomposed syntactically into *fragments*: subprograms which correspond to partial executions of only a part of the full HP (programs are easier to split structurally into parts than graphs, because handling dangling edges between graph fragments is complicated). This is important as our verification algorithm recursively decomposes an HP into fragments $\alpha_1, \ldots, \alpha_n$ (e.g., to find local invariants for each α_i) and recombines corresponding correctness statements about these fragments α_i later.

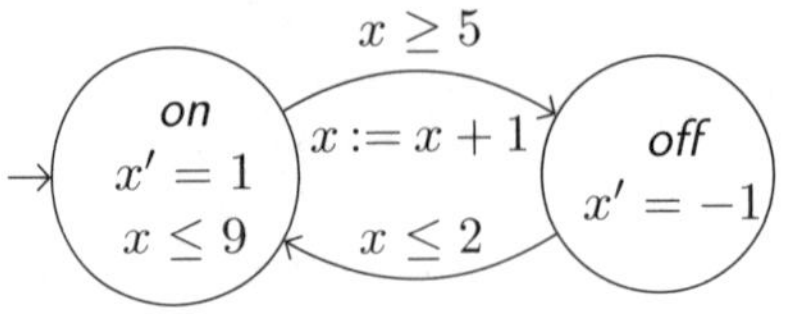

$q := on;$ /* *initial location is on* */
$(\ \ (?q = on;\ \ x' = 1 \wedge x \leq 9)$
$\cup\ (?q = on \wedge x \geq 5;\ \ x := x + 1;\ \ q := off)$
$\cup\ (?q = off;\ \ x' = -1)$
$\cup\ (?q = off \wedge x \leq 2;\ \ q := on;\ \ ?x \leq 9))^*$

Fig. 1. Natural hybrid program rendition of hybrid automaton (simple water tank)

Hybrid Programs. In order to represent HA [1] textually as an HP, we represent each discrete and continuous transition as a sequence of statements, with a nondeterministic choice ($\cup$) between these transitions. For instance, the second line in Fig. 1 represents a continuous transition. It tests (denoted by $?q = on$) if the current location q is *on*, and then follows a differential equation restricted to invariant region $x \leq 9$ (i.e., the conjunction $x' = 1 \wedge x \leq 9$). The third line tests the guard $x \geq 5$ when in state *on*, resets x by a discrete assignment, and then changes location q to *off*. The * at the end indicates that the transitions of a HA repeat indefinitely. Alternatively, the resulting HP in Fig. 1 can be considered as the essential part of a program exported from Stateflow/Simulink enriched with differential equations for the continuous dynamics. Every safety property that this HP satisfies is fulfilled for *all* deterministic implementation refinements.

Formally, let V be a set of state variables of the system and auxiliary variables. As *terms* we allow polynomials over $\mathbb{Q}$ with variables in V. To make a structural decomposition of HP into fragments possible, each operation of a HP only has a single effect. There are separate classes of program statements with purely discrete effect, purely continuous effect, and statements for regulating their interaction. *Hybrid programs (HP)* are built with the statements in Tab. 1. The effect of $x := \theta$ is an instantaneous discrete jump assigning θ to x. Instead, $x := random$ randomly assigns *any* real value to x by a nondeterministic choice. During a continuous evolution $x_1' = \theta_1 \wedge \cdots \wedge x_n' = \theta_n \wedge H$, all conjuncts need to hold. Its effect is a continuous transition controlled by the differential equation $x_1' = \theta_1, \ldots, x_n' = \theta_n$ that always satisfies the arithmetic constraint H (thus remains in the region described by H). This directly corresponds to a continuous evolution mode of a HA. The effect of state check $?H$ is a *skip* (i.e., no change) if H is true in the current state and that of *abort*, otherwise.

Table 1. Statements and (informal) effects of hybrid programs (HP)

notation	statement	effect
$x := \theta$	discrete assignment	assigns term θ to variable $x \in V$
$x := random$	nondet. assignment	assigns any real value to $x \in V$
$x_1' = \theta_1 \wedge \ldots$ $\cdots \wedge x_n' = \theta_n \wedge H$	continuous evolution	diff. equations for $x_i \in V$ and terms θ_i, with arithmetic constraint H (domain)
$?H$	state check	test formula H at current state
$\alpha;\ \beta$	seq. composition	HP β starts after HP α finishes
$\alpha \cup \beta$	nondet. choice	choice between alternatives HP α or β
α^*	nondet. repetition	repeats HP α n-times for any $n \in \mathbb{N}$

Non-deterministic choice $\alpha \cup \beta$ expresses alternatives in the behavior of the hybrid system. Sequential composition $\alpha; \beta$ expresses a behavior in which β starts after α finishes (as usual, β never starts if α continues indefinitely). Non-deterministic repetition α^*, repeats α an arbitrary number of times, possibly zero.

Formulas of d$\mathcal{L}$. Our verification algorithm repeatedly decomposes and recombines HP. As a logical framework where these operations are sound, we use a logic in which simultaneous correctness properties about multiple subsystems are expressible. The *differential dynamic logic* d$\mathcal{L}$ [8,9] is an extension of first-order logic over the reals with modal formulas like $[\alpha]\phi$, which is true iff all states reachable by following the transitions of HP α satisfy property ϕ (*safety*).

Definition 1 (d$\mathcal{L}$ formulas). *The* formulas of d$\mathcal{L}$ *are defined by the following grammar (where θ_1, θ_2 are terms, $\sim \in \{=, \leq, <, \geq, >\}$, ϕ, ψ are formulas, $x \in V$, and α is an HP built from the statements in Tab. 1):*

$$\text{Fml} ::= \theta_1 \sim \theta_2 \mid \neg\phi \mid \phi \wedge \psi \mid \phi \vee \psi \mid \phi \rightarrow \psi \mid \forall x\, \phi \mid \exists x\, \phi \mid [\alpha]\phi \ .$$

A Hoare-triple $\{\psi\}\alpha\{\phi\}$ can be expressed as $\psi \rightarrow [\alpha]\phi$, which is true iff all states reachable by HP α satisfy ϕ when starting from an initial state that satisfies ψ. Unlike Hoare-logics, dynamic logics are closed under logical connectives [21]. Hence, we can express simultaneous correctness statements about multiple fragments α_i using conjuncts $[\alpha_1]\phi_1 \wedge [\alpha_2]\phi_2$. With this, a proof for $[\alpha]\phi$ can be decomposed soundly into $[\alpha_1]\phi_1 \wedge [\alpha_2]\phi_2$, when $[\alpha]\phi$ and $[\alpha_1]\phi_1 \wedge [\alpha_2]\phi_2$ are equivalent for appropriate fragments α_i of α and subproperties ϕ_i of ϕ. In turn, if the verification algorithm with input $[\alpha_i]\phi_i$ yields $\tilde{\phi}_i$, these can be recombined soundly to the verification result $\tilde{\phi}_1 \wedge \tilde{\phi}_2$ for $[\alpha]\phi$. By the semantics of d$\mathcal{L}$, this process gives a *sound* way of combining local invariants required in the respective subgoals $[\alpha_i]\phi_i$ to a global system invariant. Finally, d$\mathcal{L}$ and its proof techniques are closed under quantification, which we use to quantify over parameter choices of local invariants. For example, $\exists p\,([\alpha_1]\phi_1 \wedge [\alpha_2]\phi_2)$ can be used to determine if there is a common choice for parameter p that makes both subgoals $[\alpha_i]\phi_i$ true. The *semantics* of d$\mathcal{L}$ and HP is a Kripke semantics [8,9].

3 Inductive Verification by Combining Local Fixedpoints

For verifying safety properties of hybrid systems without having to solve their differential equations, we use a continuous form of induction. In the induction step, we use a condition on directional derivatives in the direction of the vector field generated by the differential equation. The resulting properties are invariants of the differential equation (whence called *differential invariants* [13]). The crucial step for verifying discrete systems by induction is to find sufficiently strong invariants (e.g., for loops α^*). Similarly, the crucial step for verifying dynamical systems (which correspond to a single continuous mode of a hybrid system) by induction is to find sufficiently strong invariant properties of the differential equation. Consequently, for verifying hybrid systems inductively, local invariants need to be found for each differential equation and a global system invariant needs to be found that is compatible with all local invariants.

To compute the required invariants and differential invariants, we combine the invariants as fixedpoints approach from [14] with the lifting of verification logics to hybrid systems from [8,9]. We introduce a verification algorithm that computes invariants of a system as fixedpoints of safety constraints on subsystems. We exploit the fact that HP can be decomposed into subsystems and that d$\mathcal{L}$ can combine safety statements about multiple subsystems simultaneously.

A *safety statement* corresponds to a d$\mathcal{L}$ formula $\psi \rightarrow [\alpha]\phi$ with an HP α, a safety property ϕ about its reachable states, and an arithmetic formula ψ that characterizes the set of initial states symbolically. *Validity* of formula $\psi \rightarrow [\alpha]\phi$ (i.e., truth in all states) corresponds to ϕ being true in all states reachable by HP α from initial states that satisfy ψ [9]. Our verification algorithm defines the function *prove*($\psi \rightarrow [\alpha]\phi$) for verifying this safety statement recursively.

3.1 Verification by Symbolic Decomposition

The cases of *prove* where d$\mathcal{L}$ enables us to verify a property of an HP directly by decomposing it into a property of its parts [9] are shown in Fig. 2. For a concise presentation, the case in line 1 introduces an auxiliary variable $\hat{x}$ to handle discrete assignments by substituting $\hat{x}$ for x in $\phi_x^{\hat{x}}$: E.g., $x \geq 2 \rightarrow [x := x - 1]x \geq 0$ is shown by proving $x \geq 2 \wedge \hat{x} = x - 1 \rightarrow \hat{x} \geq 0$. Our implementation uses optimizations to avoid auxiliary variables [9]. State checks $?H$ are shown by assuming the test succeeds, i.e., H holds true (line 3), nondeterministic choices split into their alternatives (line 4), sequential compositions are proven using nested modalities (line 6), and random assignments by universal quantification (line 7).

The base case in line 8, where ϕ is a formula of first-order real arithmetic, can be proven by real quantifier elimination [5]. Despite its complexity, this can remain feasible, because the formulas resulting from our algorithm do not depend on the solutions of differential equations but only their right-hand sides. Using a temporary form of Skolemization together with Deskolemization, quantifier elimination can be lifted to eliminate quantifiers from d$\mathcal{L}$ formulas [9].

The algorithm in Fig. 2 recursively reduces safety of HP to properties of continuous evolutions or of repetitions, which we verify in the next sections.

3.2 Discrete and Differential Induction, Differential Invariants

In the sequel, we present algorithms for verifying loops by discrete induction and continuous evolutions by differential induction, which is a continuous form of induction. In either case, we prove that an invariant F holds initially (in the states characterized symbolically by ψ, thus $\psi \rightarrow F$ is valid) and finally entails the postcondition ϕ (i.e., $F \rightarrow \phi$). The cases differ in their induction step.

Definition 2 (Discrete induction). *Formula F is a* (discrete) invariant *of $\psi \rightarrow [\alpha^*]\phi$ iff the following formulas are valid: $\psi \rightarrow F$ (induction start), and $F \rightarrow [\alpha]F$ (induction step). An invariant is* sufficiently strong *if $F \rightarrow \phi$ is valid.*

```
function prove(ψ → [x := θ]φ):
  return prove(ψ ∧ x̂ = θ → φ^x̂_x) where x̂ is a new auxiliary variable
function prove(ψ → [?H]φ): return prove(ψ ∧ H → φ)
function prove(ψ → [α ∪ β]φ):
  return prove(ψ → [α]φ) and prove(ψ → [β]φ) /* thus ψ → [α]φ ∧ [β]φ */
function prove(ψ → [α; β]φ): return prove(ψ → [α][β]φ)
function prove(ψ → [x := random]φ): return prove(ψ → ∀x φ)
function prove(ψ → φ) where isFirstOrder(φ):
  return QuantifierElimination(ψ → φ)
```

Fig. 2. d$\mathcal{L}$-based verification by symbolic decomposition

Definition 3 (Continuous invariants). *Let $\mathcal{D}$ be a differential equation. Formula F is a* continuous invariant *of $\psi \to [\mathcal{D} \wedge H]\phi$ iff the following formulas are valid: $\psi \wedge H \to F$ (induction start), and $F \to [\mathcal{D} \wedge H]F$ (induction step). Again, a continuous invariant is* sufficiently strong *if $F \to \phi$ is valid.*

To prove that F is a continuous invariant, it is sufficient to check a condition on the directional derivatives of all terms of the formula, which expresses that no atomic subformula of F changes its truth-value along the dynamics of the differential equation [13]. This condition is much easier to check than a reachability property ($F \to [\mathcal{D} \wedge H]F$) of a differential equation. Applications like aircraft maneuvers need invariants with mixed equations and inequalities. Thus, we generalize directional derivatives from functions to logical formulas.

Definition 4 (Differential induction). *Let the differential equation system $\mathcal{D}$ be $x_1' = \theta_1 \wedge \cdots \wedge x_n' = \theta_n$. Formula F is a* differential invariant *of $\psi \to [\mathcal{D} \wedge H]\phi$ iff the following formulas are valid: $\psi \wedge H \to F$ and $H \to \nabla_{\mathcal{D}}F$, where $\nabla_{\mathcal{D}}F$ is defined as the conjunction of all directional derivatives of atomic formulas in F in the direction of the vector field of $\mathcal{D}$ (the partial derivative of b by x_i is $\frac{\partial b}{\partial x_i}$):*

$$\nabla_{\mathcal{D}}F \equiv \bigwedge_{(b \sim c) \in F} \left(\left(\sum_{i=1}^{n} \frac{\partial b}{\partial x_i}\theta_i \right) \sim \left(\sum_{i=1}^{n} \frac{\partial c}{\partial x_i}\theta_i \right) \right) \quad \text{for } \sim \in \{=, \geq, >, \leq, <\}.$$

Proposition 1 (Principle of differential induction [13]). *All differential invariants are continuous invariants.*

See [13] for the theory of differential invariants and [15] for specific proofs. The region corresponding to a differential invariant F is illustrated in Fig. 3. Formula $\nabla_{\mathcal{D}}F$ is a directional derivative of F in the direction of the dynamics of $\mathcal{D}$. Intuitively, formula $\nabla_{\mathcal{D}}F$ is true if the gradient arrows are pointing inside the (possibly unbounded) region consisting of the points where F is true. In Sections 3.4–3.6, we present algorithms for finding differential invariants for differential equations, and for finding global invariants for repetitions.

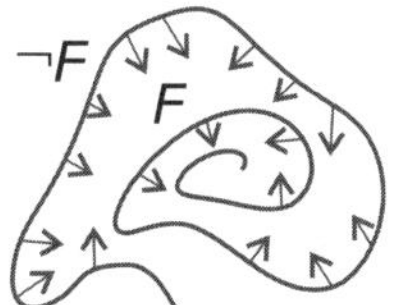

Fig. 3. Differential invariant F

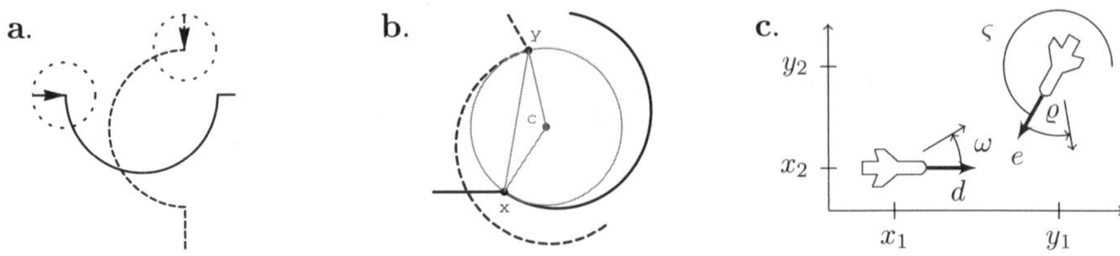

Fig. 4. Roundabout maneuvers for air traffic collision avoidance

3.3 Example: Flight Dynamics in Air Traffic Collision Avoidance

Aircraft collision avoidance maneuvers resolve conflicting flight paths, e.g., by roundabout maneuvers [16], see Fig. 4a–b. Their nontrivial dynamics makes safe separation of aircraft difficult to verify [16, 22, 23, 24, 11, 25]. The parameters of two aircraft at (planar) position $x = (x_1, x_2) \in \mathbb{R}^2$ and $y = (y_1, y_2)$ with angular orientation ϑ and ς are illustrated in Fig. 4c (with $\vartheta = 0$). Their dynamics is determined by their linear speeds $v, u \in \mathbb{R}$ and angular speeds $\omega, \varrho \in \mathbb{R}$, see [16]:

$$x_1' = v\cos\vartheta \quad x_2' = v\sin\vartheta \quad \vartheta' = \omega \quad y_1' = u\cos\varsigma \quad y_2' = u\sin\varsigma \quad \varsigma' = \varrho \tag{1}$$

In safe flight configurations, aircraft are separated by at least distance p:

$$(x_1 - y_1)^2 + (x_2 - y_2)^2 \geq p^2 \tag{2}$$

To handle the transcendental functions in (1), we axiomatize sin and cos by differential equations and reparametrize the system using a linear velocity vector $d = (d_1, d_2) := (v\cos\vartheta, v\sin\vartheta) \in \mathbb{R}^2$, which describes both the linear velocity $\|d\| := \sqrt{d_1^2 + d_2^2} = v$ and orientation of the aircraft in space, see Fig. 4c:

$$\begin{bmatrix} x_1' = d_1 & x_2' = d_2 & d_1' = -\omega d_2 & d_2' = \omega d_1 & t' = 1 \\ y_1' = e_1 & y_2' = e_2 & e_1' = -\varrho e_2 & e_2' = \varrho e_1 & s' = 1 \end{bmatrix} \tag{$\mathcal{F}$}$$

Equations ($\mathcal{F}$) and (1) are equivalent up to reparameterization [13]. We add clock variables t, s that we need for synchronizing collision avoidance maneuvers [15]. By a simple computation, $d_1^2 + d_2^2 \geq a^2$ is a differential invariant of ($\mathcal{F}$):

$$\begin{aligned} \nabla_{\mathcal{F}}(d_1^2 + d_2^2 \geq a^2) &\equiv \nabla_{(d_1' = -\omega d_2 \wedge d_2' = \omega d_1)}(d_1^2 + d_2^2 \geq a^2) \\ \equiv \frac{\partial(d_1^2 + d_2^2)}{\partial d_1}(-\omega d_2) &+ \frac{\partial(d_1^2 + d_2^2)}{\partial d_2}\omega d_1 \geq \frac{\partial a^2}{\partial d_1}(-\omega d_2) + \frac{\partial a^2}{\partial d_2}\omega d_1 \\ \equiv 2d_1(-\omega d_2) &+ 2d_2\omega d_1 \geq 0 \ . \end{aligned}$$

3.4 Local Fixedpoint Computation for Differential Invariants

Fig. 5 depicts the fixedpoint algorithm for constructing differential invariants for each continuous evolution $\mathcal{D} \wedge H$ with a differential equation system $\mathcal{D}$. The algorithm in Fig. 5 (called *Differential Saturation*) successively refines the domain H by differential invariants until saturation, i.e., H accumulates enough

```
1 function prove(ψ → [D ∧ H]φ):
2   if prove(∀cl(H → φ)) then return true /* property proven */
3   for each F ∈Candidates(ψ → [D ∧ H]φ, H) do
4     if prove(ψ ∧ H → F) and prove(∀cl(H → ∇D F)) then
5       H := H ∧ F          /* refine by differential invariant */
6       goto 2;             /* repeat fixedpoint loop */
7   end for
8   return "not provable using candidates"
```

Fig. 5. Fixedpoint algorithm for differential invariants (*Differential Saturation*)

information to become a strong invariant that implies postcondition ϕ (line 2). If domain H already entails ϕ, then $\psi \rightarrow [\mathcal{D} \wedge H]\phi$ is proven (line 2). Otherwise, the algorithm considers candidates F for augmenting H (line 3). If F is a differential invariant (line 4), then H can soundly be refined to $H \wedge F$ (line 5) without affecting the states reachable by $\mathcal{D} \wedge H$ (Proposition 2 below). Then, the fixedpoint loop repeats (line 6). At each iteration of this fixedpoint loop, the previous invariant H can be used to prove the next level of refinement $H \wedge F$ (line 4). The refinement of the dynamics at line 5 is correct by the following proposition, using that the conditions in line 4 imply that F is a differential invariant and, thus, a continuous invariant by Proposition 1, see proofs [15,13].

Proposition 2 (Differential saturation). *If F is a continuous invariant of $\psi \rightarrow [\mathcal{D} \wedge H]\phi$, then $\psi \rightarrow [\mathcal{D} \wedge H]\phi$ and $\psi \rightarrow [\mathcal{D} \wedge H \wedge F]\phi$ are equivalent.*

This progressive differential saturation turns out to be crucial in practice. For instance, the aircraft separation property (2) cannot be proven until ($\mathcal{F}$) has been refined by invariants for d and e, because these determine x' and y'.

Function *Candidates* determines candidates for induction (line 3) depending on transitive differential dependencies, as will be explained in Section 3.5. When these are insufficient for proving $\psi \rightarrow [\mathcal{D} \wedge H]\phi$, the algorithm fails (line 8, with improvements in subsequent sections). Finally, $\forall_{cl}\phi$ denotes the *universal closure* of ϕ. It is required in lines 2 and 4, because the respective formulas need to hold in *all* states (that satisfy H), see [15] for improvements.

3.5 Dependency-Directed Induction Candidates

In this section, we construct likely candidates for differential induction (function *Candidates*). Later, we use the same procedure for finding global loop invariants. We construct two kinds of candidates in an order induced by differential dependencies. Our algorithm enriches ψ successively with more precise information about the symbolic prestate as obtained by the symbolic decompositions and proof steps in Fig. 2 and 5. We first look for invariant symbolic state information in ψ and ϕ by selecting subformulas that are not yet

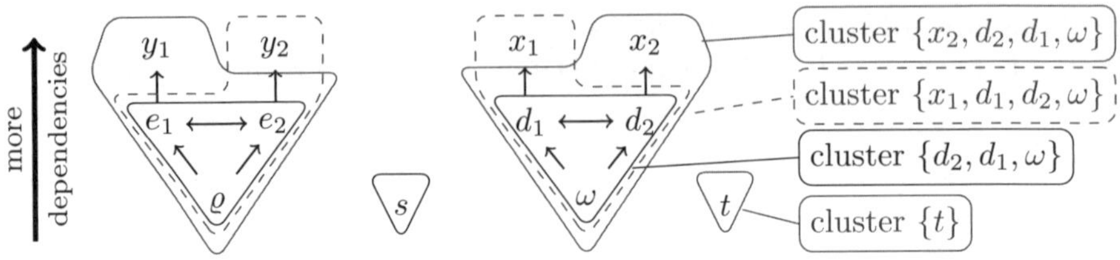

Fig. 6. Differential dependencies (arrows) and (triangular) variable clusters of $(\mathcal{F})$

contained in H. In practice, this gives good candidates for highly parametric hybrid systems.

Secondly, we generate parametric invariants. Let $V = \{x_1, \ldots, x_n\}$ be a set of variables. We choose fresh names $a^{(l)}_{i_1,\ldots,i_n}$ for *formal parameters* of the invariant candidates and build polynomials $p_1, \ldots, p_k$ of degree d with variables V using formal parameters as symbolic coefficients: $p_l := \sum_{i_1+\cdots+i_n \le d} a^{(l)}_{i_1,\ldots,i_n} x_1^{i_1} \ldots x_n^{i_n}$ for $1 \le l \le k$. We define the set of *parametric candidates* (operator $\vee$ is similarly):

$$ParaForm(k, d, V) := \left\{ \bigwedge_{l=1}^{i} p_l \ge 0 \wedge \bigwedge_{l=i+1}^{k} p_l = 0 \mid 0 \le i \le k \right\} .$$

For instance, the parametric candidate $a_{0,0} + a_{1,0}d_1 + a_{0,1}x_2 = 0$ yields a differential invariant of $(\mathcal{F})$ for the choice $a_{0,0} = 0$, $a_{1,0} = 1$, $a_{0,1} = \omega$. By simple combinatorics, *ParaForm* contains $k+1$ candidates with $k\binom{n+d}{d}$ formal parameters $a^{(l)}_{i_1,\ldots,i_n}$, which are existentially quantified. Existence of a common satisfying instantiation for these parameters can be expressed by adding $\exists a^{(l)}_{i_1,\ldots,i_n}$ to the resulting d$\mathcal{L}$ formulas. For this to be feasible, the number of parameters is crucial, which we minimize by respecting (differential) dependencies.

To accelerate the differential saturation process in Section 3.4, it is crucial to explore candidates in a promising order from simple to complex, because the algorithm in Fig. 5 uses successful differential invariants to refine the dynamics, thereby simplifying subsequent proofs: E.g., (2) is only provable after the dynamics has been refined with invariants for d and e. We construct candidates in a natural order based on variable occurrence that is consistent with the differential dependencies of the differential equations. For a differential equation $\mathcal{D}$, variable x depends on variable y according to the differential equation system $\mathcal{D}$ if y occurs on the right-hand side for x' (or transitively so). The resulting set *depend*($\mathcal{D}$) of dependencies is the transitive closure of $\{(x, y) \mid (x' = \theta) \in \mathcal{D}$ and y occurs in $\theta\}$. From the differential equation system $(\mathcal{F})$, we determine the differential dependencies indicated as arrows (pointing to the dependent variables x) in Fig. 6.

From these dependencies we determine an order on candidates. The idea is that, as the value of x_1 depends on that of d_1, it makes sense to look for invariant expressions of d_1 first, because refinements with these help differential saturation in proving invariant expressions involving also x_1. Thus, we order variables by differential dependencies, which resembles the back substitution order in Gaussian elimination (if, in triangular form, x_1 depends on d_1 then equations for d_1 must be solved first). Now we call a set V of variables a *cluster* of the differential equation $\mathcal{D}$ iff V is closed with respect to *depend*($\mathcal{D}$), i.e., variables of V only depend on variables in V. The resulting variable clusters for system ($\mathcal{F}$) are marked as triangular shapes in Fig. 6. Finally, we choose candidates from ψ and *ParaForm*(k, d, V) starting with candidates whose variables lie in small clusters V. Thus, the differential invariant $d_1^2 + d_2^2 \geq a^2$ of Section 3.3 within cluster $\{d_2, d_1, \omega\}$ can be discovered before invariants like $d_1 = -\omega x_2$ that involve x_2, because x_2 depends on d_2.

3.6 Global Fixedpoint Computation for Loop Invariants

With the uniform setup of d$\mathcal{L}$, we can adapt the algorithm in Fig. 5 easily to obtain a fixedpoint algorithm for loops ($\psi \to [\alpha^*]\phi$) in place of continuous evolutions ($\psi \to [\mathcal{D} \wedge H]\phi$): In line 4 of Fig. 5, we replace the induction step from Def. 4 by the step for loops (Def. 2). As an optimization, invariants H of previous iterations can be exploited as refinements of the hybrid system dynamics:

Proposition 3 (Loop saturation). *If H is a discrete invariant of $\psi \to [\alpha^*]\phi$, $H \wedge F$ is a discrete invariant iff $\psi \to F$ and $H \wedge F \to [\alpha](H \to F)$ are valid.*

See [15] for a proof. The induction step from Proposition 3 can generally be proven faster, because it is a weaker property than that of Def. 2.

To adapt our approach from Section 3.5 to loops, we use discrete data-flow and control-flow dependencies of α. There is a direct *data-flow dependency* with the value of x depending on y, if $x := \theta$ or $x' = \theta$ occurs in α with a term θ that contains y. Accordingly, there is a direct *control-flow dependency*, if, for any term θ, $x := \theta$ or $x' = \theta$ occurs in α after a $?H$ containing y.

3.7 Interplay of Local and Global Fixedpoint Loops

The local and global fixedpoint algorithms jointly verify correctness properties of HP. Their interplay needs to be coordinated with fairness. If the local fixedpoint algorithm in Fig. 5 does not converge, stronger invariants may need to be found by the global fixedpoint algorithm which result in stronger preconditions ψ for the local algorithm. Thus, the local fixedpoint algorithm should stop when it cannot prove its postcondition, either because of a counterexample or because it runs out of candidates for differential invariants. As in the work of Prajna [20], the degrees of parametric invariants, therefore, need to be bounded and increased iteratively. As in [20], there is no natural measure for how these degrees should be increased. Instead, here, we exploit the fact that the candidates of *Candidates* are independent and we explore them in parallel with fair time interleaving.

Table 2. Experimental results

Case study	Time(s)	Memory(MB)	Proof steps	Dimension
tangential roundabout (2 aircraft)	14	8	117	13
tangential roundabout (3 aircraft)	387	42	182	18
tangential roundabout (4 aircraft)	730	39	234	23
tangential roundabout (5 aircraft)	1964	88	317	28
bounded speed roundabout entry	20	34	28	12
flyable roundabout entry (simplified)	6	10	98	8
ETCS-kernel safety	41	28	53	9
ETCS safety	183	87	169	15
ETCS train controllability	1	6	17	5
ETCS RBC controllability	1	7	45	16

3.8 Soundness

Theorem 1 (Soundness). *The verification algorithm in Section 3 is sound, i.e., whenever* `prove`$(\psi \to [\alpha]\phi)$ *returns* "`true`", *the* d$\mathcal{L}$ *formula* $\psi \to [\alpha]\phi$ *is true in all states, i.e., all states reachable by* α *from states satisfying* ψ *satisfy* ϕ.

See [15] for a proof. Since reachability of hybrid systems is undecidable, our algorithm must be incomplete. It can fail to converge when the required invariants are not expressible in first-order logic (yet, they are always expressible in d$\mathcal{L}$ [9]).

4 Experimental Results: Aircraft Roundabout Maneuver

As an example with nontrivial dynamics, we analyze aircraft roundabout maneuvers [16]. Curved flight as in roundabouts is challenging for verification, because of its transcendental solutions. The maneuver in Fig. 4a from [16] and the maneuver in Fig. 4b from [11, 13] are not flyable, because they still involve a few instant turns. A flyable roundabout maneuver without instant turns is depicted in Fig. 7. We verify safety properties for most (but not yet all) phases of Fig. 7 and provide verification results in Tab. 2, see [15]. Finally, note that the required invariants for the roundabout maneuver cannot even be found from Differential Gröbner Bases [26].

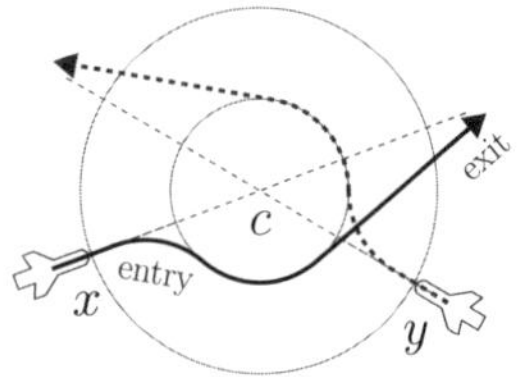

Fig. 7. Flyable aircraft roundabout

Verification results for roundabout aircraft maneuvers [16, 24, 11, 13, 15] and the European Train Control System (ETCS) [17] are in Tab. 2. Results are from a 2.6GHz AMD Opteron with 4GB memory. Memory consumption of quantifier elimination is shown in Tab. 2, excluding the front-end. The results are only slightly worse on a 1.7GHz Pentium M laptop with 1GB. We handle *all* variables symbolically. The dimension of the continuous state space is indicated.

5 Related Work

Other authors [18,19,20] already argued that invariant techniques scale to more general dynamics than explicit reach-set computations or techniques that require solutions for differential equations [3,6,8]. However, they cannot handle hybrid systems with inequalities in initial sets or switching surfaces [18,19], which occur in most real applications like aircraft maneuvers. *Barrier certificates* [20] only work for inequalities, but invariants of roundabout maneuvers require mixed equations and inequalities [13]. Prajna et al. [20] search for barrier certificates of a fixed degree by global optimization over the set of all proof attempts for the whole system at once, which is infeasible: Even with degree bound 2, it already requires solving a 5848-dimensional optimization problem for ETCS [17] and a 10005-dimensional problem for roundabouts with 5 aircraft.

Tomlin et al. [16] derive saddle solutions for aircraft maneuver games using Hamilton-Jacobi-Isaacs partial differential equations and propose roundabout maneuvers. Their exponential state space discretizations for PDEs, however, do not scale to larger dimensions (they consider dimension 3) and can be unsound [11]. Differential invariants, instead, work for 28-dimensional systems.

Straight-line aircraft maneuvers have been analyzed by geometrical meta-level reasoning [23,25]. We directly verify the hybrid flight dynamics, including curved roundabout maneuvers instead of straight-line maneuvers with non-flyable instant turns. A few approaches [22,24] have been undertaken to Model Check if there are *orthogonal* collisions in discretizations of roundabout maneuvers. However, the counterexamples found by our model checker in previous work [11] show that *non-orthogonal* collisions can happen in these maneuvers.

Tools like HyTech, PHAVer, CheckMate, or other approaches [1,3,6] cannot handle our applications with nonlinear switching, nonlinear discrete and continuous dynamics, and high-dimensional state spaces.

6 Conclusions and Future Work

We have presented a *sound* algorithm for verifying hybrid systems with nontrivial dynamics. It handles differential equations using differential invariants instead of requiring solutions of the differential equations, because the latter quickly yield undecidable arithmetic. We compute differential invariants as fixedpoints using a verification logic for hybrid systems. In the logic we can decompose the system for computing local invariants and we obtain sound recombinations into global invariants. Moreover, we introduce a differential saturation procedure that verifies more complicated properties by refining the system dynamics successively in a sound way. We validate our algorithm on challenging *roundabout collision avoidance maneuvers* for aircraft and on collision avoidance protocols for trains.

Our algorithm works particularly good for highly parametric hybrid systems, because their parameter constraints can be combined faster to find invariants than for systems with a single initial state, where simulation is more appropriate. Our decompositional approach exploits locality in system designs. Thus, it probably performs worse for systems that violate locality principles. We want to validate this in further experiments and analyze scalability.

References

1. Henzinger, T.A.: The theory of hybrid automata. In: LICS, pp. 278–292. IEEE, Los Alamitos (1996)
2. Davoren, J.M., Nerode, A.: Logics for hybrid systems. Proc. IEEE 88(7) (2000)
3. Fränzle, M.: Analysis of hybrid systems. In: Flum, J., Rodríguez-Artalejo, M. (eds.) CSL 1999. LNCS, vol. 1683, pp. 126–140. Springer, Heidelberg (1999)
4. Alur, R., Pappas, G.J. (eds.): HSCC 2004. LNCS, vol. 2993. Springer, Heidelberg (2004)
5. Collins, G.E., Hong, H.: Partial cylindrical algebraic decomposition for quantifier elimination. J. Symb. Comput. 12(3), 299–328 (1991)
6. Piazza, C., Antoniotti, M., Mysore, V., Policriti, A., Winkler, F., Mishra, B.: Algorithmic algebraic model checking I: Challenges from systems biology. In: Etessami, K., Rajamani, S.K. (eds.) CAV 2005. LNCS, vol. 3576, pp. 5–19. Springer, Heidelberg (2005)
7. Anai, H., Weispfenning, V.: Reach set computations using real quantifier elimination. In: Di Benedetto, M.D., Sangiovanni-Vincentelli, A.L. (eds.) HSCC 2001. LNCS, vol. 2034, pp. 63–76. Springer, Heidelberg (2001)
8. Platzer, A.: Differential dynamic logic for verifying parametric hybrid systems. In: Olivetti, N. (ed.) TABLEAUX 2007. LNCS (LNAI), vol. 4548, pp. 216–232. Springer, Heidelberg (2007)
9. Platzer, A.: Differential dynamic logic for hybrid systems. J. Autom. Reasoning (2008)
10. Asarin, E., Dang, T., Girard, A.: Reachability analysis of nonlinear systems using conservative approximation. In: Maler, O., Pnueli, A. (eds.) HSCC 2003. LNCS, vol. 2623, pp. 20–35. Springer, Heidelberg (2003)
11. Platzer, A., Clarke, E.M.: The image computation problem in hybrid systems model checking. In: [27], pp. 473–486
12. Donzé, A., Maler, O.: Systematic simulation using sensitivity analysis. In: [27], pp. 174–189
13. Platzer, A.: Differential algebraic dynamic logic for differential algebraic programs (submitted, 2007)
14. Clarke, E.M.: Program invariants as fixedpoints. Computing 21(4), 273–294 (1979)
15. Platzer, A., Clarke, E.M.: Computing differential invariants of hybrid systems as fixedpoints. Technical Report CMU-CS-08-103, Carnegie Mellon University (2008)
16. Tomlin, C., Pappas, G.J., Sastry, S.: Conflict resolution for air traffic management: a study in multi-agent hybrid systems. IEEE T. Automat. Contr. 43(4) (1998)
17. Platzer, A., Quesel, J.D.: Logical verification and systematic parametric analysis in train control. In: Egerstedt, M., Mishra, B. (eds.) HSCC. LNCS, vol. 4981, pp. 646–649. Springer, Heidelberg (2008)
18. Sankaranarayanan, S., Sipma, H., Manna, Z.: Constructing invariants for hybrid systems. In: [4], pp. 539–554
19. Rodríguez-Carbonell, E., Tiwari, A.: Generating polynomial invariants for hybrid systems. In: Morari, M., Thiele, L. (eds.) HSCC 2005. LNCS, vol. 3414, pp. 590–605. Springer, Heidelberg (2005)

20. Prajna, S., Jadbabaie, A., Pappas, G.J.: A framework for worst-case and stochastic safety verification using barrier certificates. IEEE T. Automat. Contr. 52(8) (2007)
21. Harel, D., Kozen, D., Tiuryn, J.: Dynamic logic. MIT Press, Cambridge (2000)
22. Massink, M., Francesco, N.D.: Modelling free flight with collision avoidance. In: ICECCS, pp. 270–280. IEEE Computer Society, Los Alamitos (2001)
23. Dowek, G., Muñoz, C., Carreño, V.A.: Provably safe coordinated strategy for distributed conflict resolution. In: AIAA Conference Proc. AIAA-2005-6047 (2005)
24. Damm, W., Pinto, G., Ratschan, S.: Guaranteed termination in the verification of LTL properties of non-linear robust discrete time hybrid systems. In: Peled, D.A., Tsay, Y.-K. (eds.) ATVA 2005. LNCS, vol. 3707, pp. 99–113. Springer, Heidelberg (2005)
25. Hwang, I., Kim, J., Tomlin, C.: Protocol-based conflict resolution for air traffic control. Air Traffic Control Quarterly 15(1) (2007)
26. Mansfield, E.L.: Differential Gröbner Bases. PhD thesis, University Sydney (1991)
27. Bemporad, A., Bicchi, A., Buttazzo, G. (eds.): HSCC 2007. LNCS, vol. 4416. Springer, Heidelberg (2007)

Constraint-Based Approach for Analysis of Hybrid Systems*

Sumit Gulwani[1] and Ashish Tiwari[2]

[1] Microsoft Research, Redmond, WA 98052
sumitg@microsoft.com
[2] SRI International, Menlo Park, CA 94025
tiwari@csl.sri.com

Abstract. This paper presents a constraint-based technique for discovering a rich class of inductive invariants (boolean combinations of polynomial inequalities of bounded degree) for verification of hybrid systems. The key idea is to introduce a template for the unknown invariants and then translate the verification condition into an $\exists\forall$ constraint, where the template unknowns are existentially quantified and state variables are universally quantified. The verification condition for continuous dynamics encodes that the system does not exit the invariant set from any point on the boundary of the invariant set. The $\exists\forall$ constraint is transformed into $\exists$ constraint using Farkas lemma. The $\exists$ constraint is solved using a bit-vector decision procedure. We present preliminary experimental results that demonstrate the feasibility of our approach of solving the $\exists\forall$ constraints generated from models of real-world hybrid systems.

1 Introduction

The model checking problem seeks to determine if a given system satisfies a given property. For several interesting classes of systems (and properties), the model checking problem is theoretically intractable. As a result, techniques have been developed that are relatively complete for either verification or falsification. Predicate abstraction and abstract interpretation are examples of the former, while bounded model checking (BMC) is an example of the latter. An attractive feature of BMC is that it reduces the search for (bounded) falsification to a single constraint that can be solved using powerful satisfiability modulo theories (SMT) solvers. One analog of BMC for verification is k-induction. The other analog, which we pursue in this paper in the context of hybrid systems, is an approach based on using templates to search for inductive invariants.

Inductive invariants are at the core of any general approach for verification. In the case of hybrid systems, initial work on discovering inductive invariants was based on using iterative fixed-point computation based approaches like abstract

* Research supported in part by the National Science Foundation under grant CNS-0720721 and by NASA under Grant NNX08AB95A.

A. Gupta and S. Malik (Eds.): CAV 2008, LNCS 5123, pp. 190–203, 2008.

interpretation or model checking [6, 11, 2]. Recently, constraint-based approaches have been proposed that search for invariants of some given form by reducing the problem to constraint solving over the unknowns in the templates [22, 17]. Constraint-based techniques offer two main advantages over fixed-point computation based techniques. First, they are goal-directed and hence have the potential to be more efficient. Second, they do not require the use of widening heuristics that lead to an uncontrollable loss of precision in fixed-point based techniques. Furthermore, constraint-based techniques can search for "deep" invariants of a known form, whereas the other techniques are more suited for "simple" invariants of (relatively speaking) an unknown form. Since hybrid systems typically have "deep" invariants of (a small number of) known simple forms, constraint-based technique are quite appealing. Even though they have demonstrated some success in the form of discovering *equational* invariants [22] and *conservatively* discovering conjunctions of polynomial inequalities [17], constraint-based techniques have not yet achieved their full potential in verification of hybrid systems.

In this paper, we develop the constraint-based approach further and show that it can be applied to discovering a rich class of inductive invariants for verification of hybrid systems. In particular, our constraint-based technique can be used for discovering invariants that involve disjunctions of polynomial inequalities. One part of the challenge here is in formulating the inductiveness requirement—if I holds in the current state $\boldsymbol{x}$ and there is a transition from $\boldsymbol{x}$ to $\boldsymbol{x}'$, then I holds in the state $\boldsymbol{x}'$—for the *continuous* dynamics. The key insight here is that this requirement can be captured precisely as a universally quantified formula, just as it can be done for *discrete* transitions. In the continuous case, inductiveness is equivalent to requiring that the vector field points "inwards" on the "boundary" of the invariant set I.

The key steps of our constraint-based approach for verification are

(1) introduce a template for the unknown inductive invariant and express the verification conditions as satisfiability of a $\exists\forall$ formula over the reals, where the existential quantification is over the template variables and the universal quantification is over the state variables (Section 3);
(2) use a generalization of Farkas' Lemma to eliminate the $\forall$ quantifiers and convert the $\exists\forall$ formula to an $\exists$ formula (over the reals) (Section 4.1); and
(3) use the bit-vector theory in SMT solvers to search for solutions of the $\exists$ formula in a bounded range (Section 4.2).

We start by defining continuous dynamical systems and hybrid systems in Section 2. We then show that the problem of discovering invariants and verifying safety can be reduced to solving $\exists\forall$ constraints over the reals (Section 3). We present our approach for solving these constraints in Section 4. We present several nontrivial examples of continuous dynamical systems and hybrid systems that were successfully analyzed using our approach (Section 5). We compare with related work in Section 6 before concluding.

2 Continuous Dynamical and Hybrid Systems

A *continuous dynamical system* is a tuple $\langle X, \texttt{Init}, \texttt{Inv}, f\rangle$ where X is a finite set of variables interpreted over the reals $\mathbb{R}$, $\mathbf{X} = \mathbb{R}^X$ is the set of all valuations of the variables X, $\texttt{Init} \subseteq \mathbf{X}$ is the set of initial states, $\texttt{Inv} \subseteq \mathbf{X}$ is the state invariant, and $f : \mathbf{X} \mapsto \mathbf{X}$ is a vector field that specifies the continuous dynamics (as $\dot{\boldsymbol{x}} = f(\boldsymbol{x})$). We assume that f satisfies the standard assumptions for existence and uniqueness of solutions to ordinary differential equations. The set Inv specifies the domain where the system is defined. The semantics of a continuous dynamical system are standard.

Example 1. Consider the following adaptive cruise controller where a car is following a leading car maintaining a safe distance [9, 19]. Let d, v_f, v, and a respectively represent the gap between the two cars, the velocity of the leading car, and the velocity and acceleration of the rear car. The system dynamics are given by the following differential equations [19]:

$$\dot{v} = a, \quad \dot{a} = -3a - 3(v - v_f) + (d - (v + 10)), \quad \dot{d} = v_f - v, \quad \dot{v_f} = a_f$$

where a_f is the acceleration of the leading car and an input in this model. Formally, we have a linear continuous dynamical system $\langle X, \texttt{Init}, \texttt{Inv}, f\rangle$ where $X = \{v, v_f, a, d, a_f\}$, f is defined by the right-hand sides of the above differential equations, $\texttt{Inv} = \{v \geq 0,\ v_f \geq 0,\ -2 \leq a \leq 5,\ -2 \leq a_f \leq 5\}$, $\texttt{Init} = \{d = 5, v = v_f, a = 0\}$. The invariant Inv captures the physical constraints that the cars do not move backwards and that the acceleration of the two cars is bounded from above and below. The initial states indicate when the above control law may be invoked. The problem is to verify that the rear car would never collide with the car in front, i.e., always $d > 0$. We note that reachability is decidable for certain classes of linear dynamical systems [13], but this example does not fall in these decidable classes. □

A *hybrid system* $\texttt{HS} = (\mathbf{Q}, X, \texttt{Init}, \texttt{Inv}, t, f)$ consists of a finite set of modes $\mathbf{Q}$, a finite set X of variables — that together define the state space $\mathbf{Q} \times \mathbf{X} := \mathbf{Q} \times \mathbb{R}^X$ of the system — a mapping $\texttt{Init} : \mathbf{Q} \mapsto 2^{\mathbf{X}}$ that defines the initial states (in each mode), a mapping $\texttt{Inv} : \mathbf{Q} \mapsto 2^{\mathbf{X}}$ that defines the state invariant of each mode, a mapping $f : \mathbf{Q} \mapsto (\mathbf{X} \mapsto \mathbf{X})$ that specifies the continuous dynamics in each mode, and a mapping $t : \mathbf{Q} \times \mathbf{Q} \mapsto 2^{\mathbf{X}}$ that specifies the discrete transitions. Specifically, for any two modes $q, q' \in \mathbf{Q}$, the system can jump from a state $(\mathbf{q}, \mathbf{x})$ to any state $(\mathbf{q}', \mathbf{x})$ if $\mathbf{x} \in t(\mathbf{q}, \mathbf{q}')$. Note that, for simplicity of presentation, we are forcing the discrete transitions to have identity reset maps (that is, $\boldsymbol{x}$ is not updated), but our method works in the other case as well. Hence, $t(\mathbf{q}, \mathbf{q}')$ is just the *guard*, or *switching condition*, for going from mode $\mathbf{q}$ to mode $\mathbf{q}'$. We assume that the semantics of hybrid systems and the set of *reachable states* are defined in the standard way, see [1].

Example 2. We consider a model of adaptive cruise control coupled with transmission from [25]. The hybrid system here is described by $\langle \mathbf{Q}, X, \texttt{Init}, \texttt{Inv}, t, f\rangle$

where $\mathbf{Q} := \{\mathit{1st}, \mathit{2nd}, \mathit{3rd}, \mathit{4th}\} \times \{cc, acc\} \times \{normal, maxbrake, maxacc\}$ and $X := \{d, v, v_f, a_f\}$. Thus, the hybrid system has 24 modes depending on the gear of the rear car (1st, 2nd, 3rd, 4th), its cruise control mode (regular cruise control *cc*, or adaptive cruise control *acc*), and its mode of operation (normal, max-braking, or max-acceleration). The dynamics in the 24 modes of the adaptive cruise control model is defined as follows:

$\dot{d} = v_f - v$, in all modes
$\dot{v} = -3.5$, in all maxbraking modes
$\dot{v} = 6 - i$, in all max-acceleration and i-th gear modes
$\dot{v} = 0.9(v_{des} - v)$, in all normal, regular cruise control (cc) modes, and
$\dot{v} = -0.66v + 0.08d - 0.4 + 0.26v_f$, in all normal, adaptive (acc) modes

where v_{des} is a parameter set to the desired velocity in the cruise control mode. The set $\mathtt{Inv}(q)$ is the conjunction of all the following applicable facts:

$-3.5 \geq -0.66v + 0.08d - 0.4 + 0.26v_f$	maxbrake, acc, all gears
$-3.5 \leq -0.66v + 0.08d - 0.4 + 0.26v_f \leq 6 - i$	normal, acc, i-th gear
$-0.66v + 0.08d - 0.4 + 0.26v_f \geq 6 - i$	maxacc, acc, i-th gear
$-3.5 \geq 0.9(v_{des} - v)$	maxbrake, cc, all gears
$-3.5 \leq 0.9(v_{des} - v) \leq 6 - i$	normal, cc, i-th gear
$0.9(v_{des} - v) \geq 6 - i$	maxacc, cc, i-th gear
$0 \leq v_f \leq 60, -3.5 \leq a_f \leq 5, d \leq 40$	acc
$d \geq 38$	cc
$0 \leq v \leq 6.7$	1st gear
$6.7 \leq v \leq 14.2$	2nd gear
$14.2 \leq v \leq 29.8$	3rd gear
$29.8 \leq v \leq 60$	4th gear

All discrete transitions have identity reset maps (that is, the continuous variables do not change values on discrete transitions). The guard $t(q, q')$ of a discrete transition from mode q to q' is given by $\mathtt{Inv}(q) \cap \mathtt{Inv}(q')$. For example, there is a transition from *normal, acc, 1st-gear* to *normal, acc, 2nd-gear* if $v = 6.7 \wedge -3.5 \leq -0.66v + 0.08d - 0.4 + 0.26v_f \leq 5$. For more details, see [25]. $\square$

The notation $K[X]$ denotes the set of polynomials with coefficients in K and variables in X. We use $\mathbb{Q}$ and $\mathbb{Z}$ ($\mathbb{Z}^+$) to denote the set of rationals and (positive) integers respectively.

3 Verification of Hybrid Systems

Given a *hybrid system* $\mathtt{HS} = (\mathbf{Q}, X, \mathtt{Init}, \mathtt{Inv}, t, f)$, and a safety property $S : \mathbf{Q} \to 2^{\mathbf{X}}$, the problem of hybrid system verification is to determine if the set of reachable states of the hybrid system in each mode $q \in \mathbf{Q}$ is a subset of $S(q)$.

The classical approach for solving the verification problem involves finding an inductive invariant map $I : \mathbf{Q} \to 2^{\mathbf{X}}$ such that the following constraints, referred to as the *verification condition*, hold for each mode $q \in \mathbf{Q}$.

A1. (Initial Constraint) $\texttt{Init}(q) \subseteq I(q)$.
A2. (Transition Constraint) For all modes $q' \in \mathbf{Q}$, $I(q) \cap t(q, q') \subseteq I(q')$.
A3. (Safety Constraint) $I(q) \subseteq S(q)$.
A4. (Inductive Constraint) If the system is in a state from the set $I(q) \cap \texttt{Inv}(q)$, then it continues to be in the set $I(q)$, provided it also remains in $\texttt{Inv}(q)$, at any time in the future as per the dynamics $f(q)$.[1]

In this section, we present a constraint-based technique for discovering an inductive invariant map that maps different modes to *closed semi-algebraic* invariants of the form $\bigwedge_i \bigvee_j p_{ij} \geq 0$, where $p_{ij} \in \mathbb{Q}[X]$ are polynomials of bounded degrees over X. We further assume that the initial conditions $\texttt{Init}(q)$, the safety conditions $S(q)$, and the transition conditions $t(q, q')$ are semi-algebraic, and that the flow f is given by polynomials. This class of *polynomial hybrid systems* is very general and covers a wide variety of examples.

The key idea of our technique is to translate the verification condition into a $\exists\forall$ constraint over real variables. (Section 4 then describes how to solve such formulas using Farkas lemma.) This is achieved by choosing a template, $\mathcal{I} : \mathbf{Q} \mapsto 2^{\mathbf{U},\mathbf{X}}$, for the inductive invariant I, where U is a finite set of new template parameters and $\mathcal{I}(q) := \bigwedge_i \bigvee_j p'_{ij} \geq 0$ with $p'_{ij} \in \mathbb{Q}[U, X]$. The first three constraints in the verification condition can be easily translated into a $\exists\forall$ constraint over real variables by simply substituting the invariant template $\mathcal{I}(q)$ in place of $I(q)$ and replacing $\subseteq$ relation by $\Rightarrow$ relation. (This is because the existence of I gets translated to existence of the unknown parameters U.) The challenge is to do this for the *invariant constraint* (A4). For that, we make use of continuity to obtain the following critical (necessary and sufficient) *verification condition for continuous dynamical systems.*

Proposition 1. *Let $\langle X, \mathit{Init}, \mathit{Inv}, f\rangle$ be a continuous dynamical system and I be a set such that $\bigwedge_{i=1}^{n}(\bigvee_{j=1}^{m} p_{ij} \geq 0)$ is the conjunctive normal form (CNF) of $\mathit{Inv} \Rightarrow I$. The set I satisfies Constraint A4 for the continuous dynamical system iff for all $i \in \{1, \ldots, n\}$ and all non-empty subsets $J \subseteq \{1, \ldots, m\}$:*

$$I(\boldsymbol{x}) \wedge \bigwedge_{j\in J}(p_{ij}(\boldsymbol{x}) = 0) \wedge \bigwedge_{j\notin J}(p_{ij}(\boldsymbol{x}) < 0) \wedge \mathit{Inv}(\boldsymbol{x}) \Rightarrow \bigvee_{j\in J}(\frac{dp_{ij}(\boldsymbol{x})}{dt} \geq 0) \quad \text{(A4')}$$

Here $\frac{dp}{dt}$ denotes the time derivative of p, also called the Lie derivative *of p, in the vector field defined by f; that is, $\frac{dp}{dt} := \sum_{x\in X} \frac{\partial p}{\partial x}\frac{dx}{dt} := \sum_{x\in X} \frac{\partial p}{\partial x} f_x$.*

Proposition 1 essentially says that the vector field should point "inwards" on the boundary of the set $\overline{\texttt{Inv}} \cup I$. The boundary of $\bigvee_{j=1}^{m} p_{ij} \geq 0$ is contained in the union (over all subsets J) of the sets $\bigwedge_{j\in J} p_{ij} = 0 \wedge \bigwedge_{j\notin J} p_{ij} < 0$. For each set in this disjoint union, we have a formula in Constraint A4' stating that the vector field is pointing inwards. For instance, consider the set $p_1 \geq 0 \vee p_2 \geq 0$. Choosing $J = \{1\}$, we get the boundary points $p_1 = 0 \wedge p_2 < 0$. On these boundary points, the vector field points inwards iff the Lie derivative, $\frac{dp_1}{dt}$, of p_1 is non-negative.

[1] If $I(q)$ is a *positive invariant set* [5], then it also satisfies our condition. In general, our condition is weaker than that of positive invariant sets.

```
HS2ExistsForall(HS,S,I) =
  // Inputs: HS := (Q, X, Init, Inv, t, f), Safety property S : Q → 2^X,
        Template I : Q → 2^{U,X}, where I(q) := ∧_{i=1}^{n} ∨_{j=1}^{m} p_ij ≥ 0, p_ij ∈ ℚ[U,X]
  ans := true
  for all q ∈ Q do
    ans := ans ∧ (Init(q) ⇒ I(q)) ∧ (I(q) ∧ Inv(q) ⇒ S(q))
    for all q' ∈ Q do ans := ans ∧ (I(q) ∧ t(q,q') ⇒ I(q'))
    for all i ∈ {1,...,n}, j ∈ {1,...,m} do
      Lp := Σ_{x∈X} (∂p_ij/∂x) f_x(q)
      ans := ans ∧ (I(q) ∧ Inv(q) ∧ p_ij = 0 ∧ ∧_{k=1,k≠j}^{k=m} p_ik ≤ 0 ⇒ Lp ≥ 0)
  return(∃U∀X ans)
```

Fig. 1. Translating safety verification to satisfiability of an ∃∀ formula

The Lie derivatives, $\frac{dp}{dt}$, in Equation A4' get reduced to polynomials as $\frac{dp}{dt} := \sum_{x \in X} \frac{\partial p}{\partial x}\frac{dx}{dt}$, and this summation simplifies into a polynomial if p is a polynomial and $\frac{d\boldsymbol{x}}{dt} := f(\boldsymbol{x})$ contains only polynomials. Since Constraints A1, A2, A3, and A4' are expressible in the first-order theory of reals, it follows from the decidability of this theory [26] that the problem of discovering bounded-size inductive invariants for polynomial hybrid systems over the class of positive boolean combination of (non-strict) polynomial inequalities of bounded degree is decidable. However, our interest is in obtaining a more *practical* technique for generating invariants. Figure 1 shows the construction of the ∃∀ formula over real variables using the Constraints A1, A2, and A3, and Constraint A4'. Since the number of constraints in A4' is exponential in m, Figure 1 uses a stronger variant of Constraint A4' that contains only a linear number of constraints, but that is usually sufficient in practice. In the next section we present our approach for solving these constraints.

Example 3 (Verification to ∃∀ Constraint). Consider the dynamical system from Example 1 and the verification problem stated therein. Let us assume a template that searches for linear invariants:

$$\mathcal{I} \;:=\; (\alpha v_f + \beta v + \gamma a + \delta d \geq \epsilon) \wedge (d \geq 0)$$

where $U := \{\alpha, \beta, \gamma, \delta, \epsilon\}$. Since Constraint A3 is trivially satisfied, the safety of the adaptive cruise control law reduces to the satisfiability of the following ∃∀ constraint which essentially says that $\mathcal{I}$ is an inductive invariant.

$$\begin{array}{ll}\exists U \forall X : ((d = 5 \wedge v = v_f \wedge a = 0 \Rightarrow \mathcal{I}) & \text{(A1)}\\ \quad \wedge(\mathcal{I} \;\wedge\; \texttt{Inv} \;\wedge\; \alpha v_f + \beta v + \gamma a + \delta d = \epsilon \Rightarrow p \geq 0) & \text{(A4')}\\ \quad \wedge(\mathcal{I} \;\wedge\; \texttt{Inv} \;\wedge\; d = 0 \Rightarrow v_f - v \geq 0)) & \text{(A4')}\end{array}$$

where p is the Lie derivative of $\alpha v_f + \beta v + \gamma a + \delta d - \epsilon$ and is equal to $\alpha \dot{v_f} + \beta \dot{v} + \gamma \dot{a} + \delta \dot{d} = \alpha a_f + \beta a - 3\gamma a - 3\gamma v + 3\gamma v_f + \gamma d - \gamma v - 10\gamma + \delta v_f - \delta v$. Similarly, $v_f - v$ is the Lie derivative of d. Note that X, Inv are defined in Example 1. □

```
ExistsForall2Exists(φ) =
  // Input: φ := ∃U∀X ⋀_{i=1}^n (⋁_{j=1}^m p_ij ≤ 0 ∨ ⋁_{k=1}^l p'_ik < 0), where p_ij, p'_ik ∈ ℚ[U, X]
  V := ∅, ans := true
  for i = 1 to n do
    V := V ∪ {μ_i} ∪ {μ_ij : j = 1,...,m} ∪ {ν_ik : k = 1,...,l}
    ans := ans∧ ElimX(μ_i + Σ_{j=1}^m μ_ij p_ij + Σ_{k=1}^l ν_ik p'_ik = 0) ∧ (μ_i > 0 ∨ ⋁_{j=1}^m μ_ij > 0)
  return(∃U∃V ans)

ElimX(p = 0) = // Input p ∈ ℚ[U, X]
  Let p := Σ_α p_α X^α, where p_α ∈ ℚ[U] are the coefficients of p in (ℚ[U])[X]
  return(⋀_α p_α = 0)
```

Fig. 2. Translating ∃∀ formula to an ∃ formula

4 Solving ∃∀ Formulas

We check for satisfiability of the ∃∀ formula in two steps. First we eliminate the inner universal quantifier and next we check for satisfiability of the resulting existential formula over a finite domain using a satisfiability modulo theories (SMT) solver.

4.1 Step 1: Eliminating Universal Quantification

The inner universal quantifier from the ∃∀ formula is eliminated using the following variant of Farkas Lemma.

Lemma 1. *For polynomials $p_j, r_k \in \mathbb{Q}[X]$, the formula $\bigwedge_{j\in J} p_j > 0 \wedge \bigwedge_{k\in K} r_k \geq 0$ is unsatisfiable (over the reals) if there exist non-negative constants μ, μ_j $(j \in J)$, and ν_k, $(k \in K)$ such that $\mu + \sum_{j\in J} \mu_j p_j + \sum_{k\in K} \nu_k r_k = 0$ and at least one of μ_j, μ is strictly positive.*

If polynomials p_j, r_k are linear (more generally, linear only in the universal variables; for example, see the constraint in Example 3), then the condition above is both necessary and sufficient. However, the condition is not necessary for unsatisfiability when p_j, r_k are arbitrary nonlinear polynomials.[2] After applying Lemma 1, the universal variables can be eliminated by just equating the coefficients of each of the power products in the following expression to zero.

$$\mu + \sum_{j\in J} \mu_j p_j + \sum_{k\in K} \nu_k r_k$$

We can convert *any* universally quantified arithmetic formula $\forall X : \phi$ into an existentially quantified formula using the above lemma, as shown in Figure 2.[3] We illustrate this on our running example below.

[2] There is a generalization of Farkas Lemma for arbitrary polynomials, called Positivstellensatz [15], obtained by replacing the multipliers μ_j, ν_k by sum of squares of polynomials, but we did not use it in our experiments.

[3] This is achieved by converting ϕ into conjunctive normal form $\bigwedge_i(\bigvee_j p_{ij} \geq 0 \vee \bigvee_k r_{ik} > 0)$ and noting that $\forall X : \phi \equiv \bigwedge_i \forall X(\bigvee_j p_{ij} \geq 0 \vee \bigvee_k r_{ik} > 0) \equiv \bigwedge_i(\neg(\bigvee_j p_{ij} \geq 0 \vee \bigvee_k r_{ik} > 0)$ is unsatisfiable$) \equiv \bigwedge_i((\bigwedge_j -p_{ij} > 0 \wedge \bigwedge_k -r_{ik} \geq 0)$ is unsatisfiable$)$. We can now use Lemma 1 on each conjunct.

Example 4. Consider the $\exists\forall$ formula in Example 3. To avoid clutter, we illustrate the $\forall$ elimination on a simpler subformula from that formula:

$$\exists U \forall X : (\alpha v_f + \beta v + \gamma a + \delta d \geq \epsilon \ \wedge \ d = 0 \ \wedge \ 2a \geq -7 \Rightarrow v_f - v \geq 0)$$

Using Lemma 1, the validity of the above formula is equivalent to the existence of constants $V := \{\nu_1, \lambda, \nu_2, \mu_1, \mu_2\}$ such that

$$\nu_1(\alpha v_f + \beta v + \gamma a + \delta d - \epsilon) + \lambda d + \nu_2(2a + 7) + \mu_1(v - v_f) + \mu_2 = 0,$$

and $\nu_1, \mu_1, \mu_2 \geq 0$ and at least one of the μ's is strictly positive. By equating the coefficients to 0, we get the following existentially quantified formula,

$$\begin{aligned} \exists U \exists V : \nu_1\alpha - \mu_1 = 0 \ \wedge \ \nu_1\beta + \mu_1 = 0 \ \wedge \ \nu_1\gamma + 2\nu_2 = 0 \ \wedge \ \nu_1\delta + \lambda = 0 \ \wedge \\ \mu_2 - \nu_1\epsilon + 7\nu_2 = 0 \ \wedge \ \textstyle\bigwedge_i \mu_i \geq 0 \ \wedge \ \bigwedge_i \nu_i \geq 0 \ \wedge \ (\mu_1 > 0 \vee \mu_2 > 0) \end{aligned}$$

A possible solution is

$$\nu_1 = 2, \lambda = -2, \nu_2 = 1, \mu_1 = 2, \mu_2 = 1, \alpha = 1, \beta = -1, \gamma = -1, \delta = 1, \epsilon = 4.$$

Note that μ_1 is strictly positive. This corresponds to the inductive invariant $v_f - v - a + d \geq 4$. We remark here that the full example contains additional constraints, but the above solution for U continues to be a solution and it is the solution computed by our tool. □

We now state the correctness of our approach as follows.

Theorem 1. *Let HS be a hybrid system and S a safety property. For any template $\mathcal{I}$, if the constraint* `ExistsForall2Exists(HS2ExistsForall(HS,` $S, \mathcal{I}$`))` *is satisfiable, then for every reachable state $(q, \boldsymbol{x})$ of HS, it is the case that $\boldsymbol{x} \subset S(q)$.*

4.2 Step 2: Solving the ∃ Constraint Using an SMT Solver

We have reduced the verification problem to the satisfiability of some (existentially quantified) nonlinear constraints. The important point to note here is that we are interested in finding solutions, rather than showing unsatisfiability, of the generated existential formula.

We search for solutions of the nonlinear constraints using the bit-vector decision procedure of an SMT solver. The translation of $\exists Y : \phi$ to bit-vectors is obtained in several steps. First polynomials in $\mathbb{Q}[Y]$ that occur in ϕ are converted to polynomials in $\mathbb{Z}[Y]$ by multiplying suitably by positive integer constants. Next we pick an integer lower bound $\boldsymbol{l}$ and an integer upper-bound $\boldsymbol{u}$ for the variables Y. Finally, we search for integer solutions for Y in the chosen finite range by searching for the bit-level representation. We choose a size for the bit-vectors by conservatively estimating the number of bits that would be required to evaluate the polynomials in ϕ over the range $\boldsymbol{l} \leq Y \leq \boldsymbol{u}$. The pseudo-code for the translator is given in Figure 3.

```
Exists2BitVector(∃Y : φ, l, u) =
  // Inputs:Y := {y_1,...,y_n}, φ := ∧_i ∨_j p_ij ~_ij 0, p_ij ∈ Z[Y], ~_ij ∈ {≥,=,>}
            l,u ∈ Z^n given lower- and upper-bounds for Y
  forall i,j: m_ij := estimate max #bits reqd to evaluate p_ij when l ≤ Y ≤ u
  Let m be the maximum of m_ij's
  ans := declare each y_i to be a bit-vector of size m
  return(ans, ∧_i ∨_j E2BVA(p_ij ~_ij 0) ∧ E2BVA(Y ≥ l) ∧ E2BVA(u ≥ Y))

E2BVA(p_1 ~ p_2) = // p_1,p_2 ∈ Z^+[Y]
  return(p'_1 ~' p'_2) where p'_1 ~' p'_2 is obtained by replacing *,+,≥,>,= by
                          corresponding bit-vector operations in p_1 ~ p_2
```

Fig. 3. Converting satisfiability checking to bit-vector satisfiability problem. The bit-vector instance searches for all bounded integer solutions for Y in the range $\boldsymbol{l} \leq Y \leq \boldsymbol{u}$ that satisfy ϕ.

4.3 Discussion

Comparing the overall approach to bounded model checking, we note that both approaches translate the analysis problem into a constraint satisfiability problem. In the case of BMC, the generated constraint encodes existence of a counterexample, whereas here the generated constraint encodes existence of a proof.

When verifying a large hybrid system, we start by applying our technique to a small component of the system using linear and quadratic templates. If we find an invariant for the smaller subsystem, we use it as a starting point to construct refined templates for the full system (Example 6).

Our constraint-based technique for verification can be used for solving instances of the *synthesis* problem as well. The technique uniformly treats the entities of the verification condition, which includes both the inductive invariants and the description of the system. It does not matter whether the invariants are unknown or parts of the system are unknown or both of them are unknown. As long as there is sufficient information in the system description, the constraint-based approach can potentially find a solution for the unknown quantities.

Example 5. Consider the classical thermostat example, which is a hybrid system with two modes: in the "on" mode, temperature x increases as $dx/dt = 100 - x$, and in the "off" mode, it decreases as $dx/dt = -x$. We want to synthesize the control logic that determines when to switch modes. Assume initially mode is "off" and $x = 78$. The goal is to ensure $75 \leq x \leq 80$ always. For simplicity, assume that the specified safety property, $75 \leq x \leq 80$, is also an inductive invariant (and we do not guess a template for the invariant). Assume that we guess that the guard for the transition from heater-on to heater-off mode is of the form $x \geq \alpha$ and that the guard for the reverse transition is $x \leq \beta$. We can now write the verification conditions as follows:

$$\begin{aligned}&\exists \alpha, \beta : \forall x : (x = 75 \wedge x > \beta \;\Rightarrow\; -x \geq 0) \wedge (x = 80 \wedge x > \beta \;\Rightarrow\; -x \leq 0) \wedge \\ &\quad (x = 75 \wedge x < \alpha \;\Rightarrow\; 100 - x \geq 0) \wedge (x = 80 \wedge x < \alpha \;\Rightarrow\; 100 - x \leq 0) \wedge \\ &\quad (x = 78 \;\Rightarrow\; x > \beta)\end{aligned}$$

One solution returned by our constraint solver was $\alpha = \beta = 76$. However, this solution leads to zeno behavior. We can add additional constraints (not described in this paper) that capture the requirement that the switching logic be most liberal, in which case we get $\beta = 75$ and $\alpha = 79$. □

5 Experimental Results

The approach described in this paper has been partially implemented in the form of two separate components. The first component takes an $\exists\forall$ formula (over arbitrary nonlinear polynomials) and returns an $\exists$ formula. The second component takes the $\exists$ formula and creates a Yices [8] formula over bit-vectors. The implementation is in Lisp. The bit-vector decision procedure of Yices is used to finally search for solutions. The front-end step of generating the $\exists\forall$ formula from a hybrid system description has not been automated yet.

All examples presented in this paper were analyzed automatically using the above tools. Some results are reported in Table 1.

Example 6 (Adaptive Cruise Control with Transmission). Consider the cruise control model from Example 2. The safety property to establish is that inter-vehicle separation remains positive; specifically, $d \geq 5$. We assume that initially the rear car is in the mode *normal, acc, 4-th gear* satisfying $v = v_f \ \wedge\ -3.5 \leq -0.66v + 0.08d - 0.4 + 0.26v_f \leq 2 \ \wedge\ 29.8 \leq v \leq 60$. We want to prove the safety property assuming that the velocity v_f of the leading car remains bounded between $30 \leq v_f \leq 60$.

Our tools prove the safety by generating the following invariant for each of the acc modes:

Invariant	Modes
$2d - 2v + v_f - 2 \geq 0,\ d \geq 5$	normal, acc, all gears
$-350 \leq -66v + 8d - 40 + 26v_f$	normal, acc, all gears
false	max-braking, acc, all gears
$2d - 2v + v_f - 2 \geq 0,\ d \geq 5$	max-acceleration, acc, all gears

Note that the max-braking mode is not reachable from the chosen initial states. We did not generate the invariants for all modes in one step. We first generated invariants for single modes and that gave us an idea of the form of invariants and helped refine our template. Using a refined template, we generated invariants for all the acc-modes simultaneously. □

Example 7 (Human Blood Glucose Metabolism). We consider the model of insulin metabolism in the body of a Type-I diabetic patient [24, 14]. For purposes of modeling insulin concentration in the human body, the body is divided into six compartments – brain (B), heart (H), gut (G), lungs (L), kidney (K), and periphery (P) – and each state variable represents the insulin concentration in one such compartment (there are two variables for the "periphery" compartment).

Table 1. Experimental Results. We report the number of continuous variables (Dim) in the example, the size of the Yices formulas generated by the example in terms of the number of variables (Vars), the size of bit-vectors (Bits), and number of assertions (Assertions), and the time (Time) taken by Yices to find a model on a 64-bit Pentium 3.4GHz cpu with 2MB cache.

Example	Dim	Vars	Bits	Assertions	Time
disjunction Ex. 9	2	14	6	50	7ms
delta-notch	4	34	8	120	30ms
plankton Ex. 8	3	31	8	110	56ms
thermostat	1	29	20	126	.45s
thermostat synthesis Ex. 5	1	21	20	75	1.2s
ACC Ex. 1	5	28	12	95	1.3s
acc-transmission Ex. 2	4	35	24	122	4.7s
insulin Ex. 7	7	66	18	180	18s

The dynamics of the system are given as follows, see also [24]:

$$
\begin{aligned}
dI_B/dt &= -45/26I_B + 45/26I_H \\
dI_H/dt &= 45/99I_B - 312/99I_H + 90/99I_L + 72/99I_K + 105/99I_{PV} + u \\
dI_G/dt &= 72/94I_H - 72/94I_G \\
dI_L/dt &= 18/114I_H - 720/10000I_H + 72/118I_G - 2880/10000I_G - 90/118I_L \\
dI_K/dt &= 72/51I_H - 72/51I_K - 2160/10000I_K \\
dI_{PV}/dt &= 105/74I_H - 105/74I_{PV} - 674/1480I_{PV} + 674/1480I_{PI} \\
dI_{PI}/dt &= 1/20I_{PV} - 1/20I_{PI} - 21231/51580I_{PI}
\end{aligned}
$$

The control input u in this case is the insulin injected into the body by an external insulin pump. Since we assume a Type-I diabetic, there is no pancreatic insulin release and hence no feedback from the glucose metabolism model. Assuming that the input u is bounded between 20 and 25, we can compute bounds or ranges for insulin concentrations in different body compartments. As remarked earlier, we can easily invert the analysis and ask for acceptable bounds on insulin injection rate that will ensure bounded insulin concentration levels in the body. □

Example 8. Consider the following Phytoplankton Growth Model (see [3] and references therein): $\dot{x}_1 = 1 - x_1 - \frac{x_1 x_2}{4}$, $\dot{x}_2 = (2x_3 - 1)x_2$, $\dot{x}_3 = \frac{x_1}{4} - 2x_3^2$, where x_1 denotes the substrate, x_2 the phytoplankton biomass, and x_3 the intracellular nutrient per biomass. For this nonlinear dynamical system, we can immediately generate the following invariant: $0 \leq x_1 \leq 2$, $0 \leq x_2 \leq 1$, $0 \leq x_3 \leq 1/2$. □

Example 9 (Disjunctive Invariants). Our technique can be used to generate disjunctive invariants. Consider the system $dx/dt = -y$, $dy/dt = -x$ with initial states given by $x \geq 3$. Using the template $x \geq \alpha \ \vee \ y \geq \beta$, we can generate the invariant $x \geq 0 \vee y \geq 0$. □

6 Related Work

The approach of using templates and generating invariants of a specific form for hybrid systems was introduced simultaneously by Sankaranarayanan et. al. [22] and Prajna et. al. [16, 17, 18]. In all such approaches, an $\exists\forall$ formula is generated, although this may not be explicitly stated. The various approaches differ in the form of the invariants considered, the technique used to generate the $\exists\forall$ formula, and the approach for solving it. Templates are restricted to polynomial equations in [22] and Proposition 1 is not required there. The approach for solving the $\exists\forall$ constraints is based on Gröbner basis computation. Polynomial inequality templates are used in [17], but a much weaker variant of Proposition 1 is used there. The constraint solving method in [17] is based on convex optimization and sum-of-squares computation and is, in essence, a slightly more general form of Lemma 1 inspired by Positivstellensatz. We build upon these works and explore a more precise translation into $\exists\forall$ constraints and the use of SMT solvers as the backend engine.

Tiwari [27] generated linear inductive invariants for linear systems. Rodriguez-Carbonell and Tiwari [20] showed that the best (strongest) possible polynomial equational invariant was computable for hybrid systems with linear dynamics in each mode. Pappas et al. have also considered the problem of computing invariants, but only for linear systems, using interesting techniques [28, 29].

In software program analysis, constraint based techniques have been successfully applied for discovering linear arithmetic invariants [7, 21, 23, 10], non-linear polynomial invariants [12] and invariants in the combined theory of linear arithmetic and uninterpreted functions [4].

7 Conclusion

The verification technique based on guessing the form of inductive invariant and searching for invariants of that form using SMT solvers is a potent approach for verifying hybrid systems. Its extension to solving the synthesis problem is left for future work. Using efficient nonlinear constraint solvers directly could also significantly improve the performance of our approach and remains to be explored.

References

[1] Alur, R., Courcoubetis, C., Halbwachs, N., Henzinger, T.A., Ho, P.-H., Nicollin, X., Olivero, A., Sifakis, J., Yovine, S.: The algorithmic analysis of hybrid systems. Theoretical Computer Science 138(3), 3–34 (1995)

[2] Alur, R., Henzinger, T., Lafferriere, G., Pappas, G.J.: Discrete abstractions of hybrid systems. Proceedings of the IEEE 88(2), 971–984 (2000)

[3] Bernard, O., Gouze, J.-L.: Global qualitative description of a class of nonlinear dynamical systems. Artificial Intelligence 136, 29–59 (2002)

[4] Beyer, D., Henzinger, T., Majumdar, R., Rybalchenko, A.: Invariant Synthesis for Combined Theories. In: Cook, B., Podelski, A. (eds.) VMCAI 2007. LNCS, vol. 4349, pp. 378–394. Springer, Heidelberg (2007)
[5] Blanchini, F.: Set invariance in control. Automatica 35, 1747–1767 (1999)
[6] Chutinan, A., Krogh, B.H.: Verification of Polyhedral-Invariant Hybrid Automata Using Polygonal Flow Pipe Approximations. In: Vaandrager, F.W., van Schuppen, J.H. (eds.) HSCC 1999. LNCS, vol. 1569. Springer, Heidelberg (1999)
[7] Colón, M., Sankaranarayanan, S., Sipma, H.: Linear Invariant Generation Using Non-linear Constraint Solving. In: Hunt Jr., W.A., Somenzi, F. (eds.) CAV 2003. LNCS, vol. 2725, pp. 420–432. Springer, Heidelberg (2003)
[8] Dutertre, B., de Moura, L.: A fast linear-arithmetic solver for dpll(t). In: Ball, T., Jones, R.B. (eds.) CAV 2006. LNCS, vol. 4144, pp. 81–94. Springer, Heidelberg (2006), http://yices.csl.sri.com/
[9] Godbole, D., Lygeros, J.: Longitudinal control of the lead car of a platoon. IEEE Transactions on Vehicular Technology 43(4), 1125–1135 (1994)
[10] Gulwani, S., Srivastava, S., Venkatesan, R.: Program analysis as constraint solving. In: Proc. PLDI (2008)
[11] Henzinger, T.A., Ho, P.-H., Wong-Toi, H.: HyTech: A model checker for hybrid systems. Software Tools for Technology Transfer 1, 110–122 (1997)
[12] Kapur, D.: Automatically generating loop invariants using quantifier elimination. In: Deduction and Applications (2005)
[13] Lafferriere, G., Pappas, G.J., Yovine, S.: Symbolic reachability computations for families of linear vector fields. J. Symbolic Computation 32(3), 231–253 (2001)
[14] Parker, R.S., Doyle, F.J., Peppas, N.A.: A model-based algorithm for blood glucose control in type I diabetes patients. IEEE Trans BioMed Eng. 46(2) (1999)
[15] Parrilo, P.A.: Structured semidefinite programs and semialgebraic geometric methods in robustness and optimization. PhD thesis, California Inst. of Tech. (2000)
[16] Prajna, S.: Barrier certificates for nonlinear model validation. In: Proc. IEEE Conference on Decision and Control (2003)
[17] Prajna, S., Jadbabaie, A.: Safety verification of hybrid systems using barrier certificates. In: Alur, R., Pappas, G.J. (eds.) HSCC 2004. LNCS, vol. 2993, pp. 477–492. Springer, Heidelberg (2004)
[18] Prajna, S., Jadbabaie, A., Pappas, G.: A framework for worst-case and stochastic safety verification using barrier certificates. IEEE Trans. Aut. Control (2005)
[19] Puri, A., Varaiya, P.: Driving safely in smart cars. In: Proceedings of the 1995 American Control Conference (1995)
[20] Rodriguez-Carbonell, E., Tiwari, A.: Generating polynomial invariants for hybrid systems. In: Morari, M., Thiele, L. (eds.) HSCC 2005. LNCS, vol. 3414, pp. 590–605. Springer, Heidelberg (2005)
[21] Sankaranarayanan, S., Sipma, H., Manna, Z.: Non-linear loop invariant generation using gröbner bases. In: POPL 2004 (2004)
[22] Sankaranarayanan, S., Sipma, H., Manna, Z.: Constructing invariants for hybrid systems. In: Alur, R., Pappas, G.J. (eds.) HSCC 2004. LNCS, vol. 2993, pp. 539–554. Springer, Heidelberg (2004)
[23] Sankaranarayanan, S., Sipma, H.B., Manna, Z.: Scalable Analysis of Linear Systems Using Mathematical Programming. In: Cousot, R. (ed.) VMCAI 2005. LNCS, vol. 3385, pp. 25–41. Springer, Heidelberg (2005)
[24] Sorensen, J.T.: A physiologic model of glucose metabolism in man and its use to design and assess improved insulin therapies for diabetes. PhD thesis, Dept. Chem. Eng., Massachusetts Inst. Technology (MIT), Cambridge (1985)

[25] Stursberg, O., Fehnker, A., Han, Z., Krogh, B.H.: Verification of a cruise control system using counterexample-guided search. Control Engineering Practice 12(10), 1269–1278 (2004)
[26] Tarski, A.: A Decision Method for Elementary Algebra and Geometry, 2nd edn. University of California Press (1948)
[27] Tiwari, A.: Approximate reachability for linear systems. In: Maler, O., Pnueli, A. (eds.) HSCC 2003. LNCS, vol. 2623, pp. 514–525. Springer, Heidelberg (2003)
[28] Yazarel, H., Pappas, G.J.: Geometric programming relaxations for linear system reachability. In: Proc. 2004 American Control Conference (2004)
[29] Yazarel, H., Prajna, S., Pappas, G.J.: S.O.S. for safety. In: Proc. 43rd IEEE Conference on Decision and Control (2004)

AutoMOTGen: Automatic Model Oriented Test Generator for Embedded Control Systems*
Tool Paper

Ambar A. Gadkari, Anand Yeolekar, J. Suresh, S. Ramesh, Swarup Mohalik, and K.C. Shashidhar

General Motors R&D - India Science Lab, Bangalore
{ambar.gadkari,anand.yeolekar,suresh.jeyaraman, ramesh.s,swarup.mohalik,shashidhar.kc}@gm.com

1 Introduction

We present AutoMOTGen, a tool for automatic test case generation (ATG) from MATLAB Simulink/Stateflow (SL/SF) models [6] for testing automotive controllers. Our methodology is based on model checking [2]. The main highlights of the tool are:

1. Enhanced coverage of the model elements as well as high-level requirements.
2. A modular design for *plug-and-play* of different model checkers, test data generators and coverage analysis tools for enhancing the test suite quality.
3. Implements sampling time abstraction to generate tests with *lesser* number of (discrete) steps in the intermediate model.
4. Implements coverage dependent instrumentation of the model for the structural coverage criteria.
5. Capability to handle SL/SF blocks commonly used in automotive controllers (including blocks such as integrator, delay, multiplication/division, look-up tables, triggered subsystems and hierarchical and parallel charts).

The current implementation of AutoMOTGen uses SAL [8] as an intermediate representation and uses associated tools such as `sal-atg`, `sal-bmc` and `sal-smc` for generation of test data and proving the unreachability of some of the coverage goals. AutoMOTGen is implemented in Java and C++ (.NET framework) and uses MATLAB scripting language for extracting the relevant information from SL/SF models required for the purpose of test generation.

2 Motivation

Model checking, besides formal verification, has also been shown to provide an efficient technique to automatically derive test sequences from transition system models [1,4,5]. This approach for ATG relies on capabilities of the model

* The opinions expressed in this article are those of the authors, and do not necessarily reflect the opinions or positions of their employers, or other organizations.

A. Gupta and S. Malik (Eds.): CAV 2008, LNCS 5123, pp. 204–208, 2008.

checkers to generate traces for counter-examples of properties that do not hold in the model. Test suites are usually derived to satisfy certain coverage criteria of a model. The coverage criteria are mostly based on structural coverage of the transition system model such as state and transition coverage. The structural elements (states and transitions) are typically associated with Boolean variables called *trap* variables. Structural coverage of a model element is then reduced to model checking the reachability of the state where the associated trap variable is true. Model checking based ATG strives to find the most efficient test suite using directed search techniques. The main advantage of this approach is that one can achieve a systematic coverage of undischarged goals by using various model checking engines employing techniques such as explicit model checking, bounded model checking (based on SAT/SMT solvers), symbolic model checking and others in combination with various model slicing and reduction techniques for covering the deeper goals. Also, whenever certain goals cannot be covered the model checking engines can be invoked to prove the unreachability of those goals. Various other model abstraction techniques such as counter-example guided abstraction refinement, predicate-based abstraction and others can be explored to enhance the coverage, efficiency and scalability. Compared to other techniques such as random generation or guided-simulation used by most of the current commercial ATG tools [9,7] used in automotive controller development, the model checking based approach for ATG holds a greater promise in covering the deep-rooted coverage goals. This motivated the work on development of AutoMOTGen two years back. It serves as an experimental testbed for evaluating various technologies for test generation using industrial case studies. Recently, The Mathworks has introduced a toolbox, Simulink Design Verifier [6] which has ATG capability based on Prover's SAT-solver technology. We believe that our tool with its unique capability to provide an integrated environment for generation of test cases with plug-n-play of diverse tools and combining various techniques can help in addressing the needs of industrial scale designs. Results from our case studies have been encouraging in this regard.

3 Overview of Test Generation Flow

We describe here the generic flow of AutoMOTGen as shown in Figure 1. There are three inputs, namely, SL/SF models, high-level requirements and test specifications including different coverage goals. The output is a test suite of timed input-output sequences that can be used for testing the implementation. The test specification includes different coverage goals based on various structural criteria defined over SL/SF models. The test specification and high-level requirements are translated into formal properties using a subset of Linear Temporal Logic (LTL) by the property generator module. The SL/SF model is translated into a formal language which can be fed to the model checking engine. The model checking engine then verifies the formal model. The generated counter-example traces are converted into test cases consisting of timed input-output sequences.

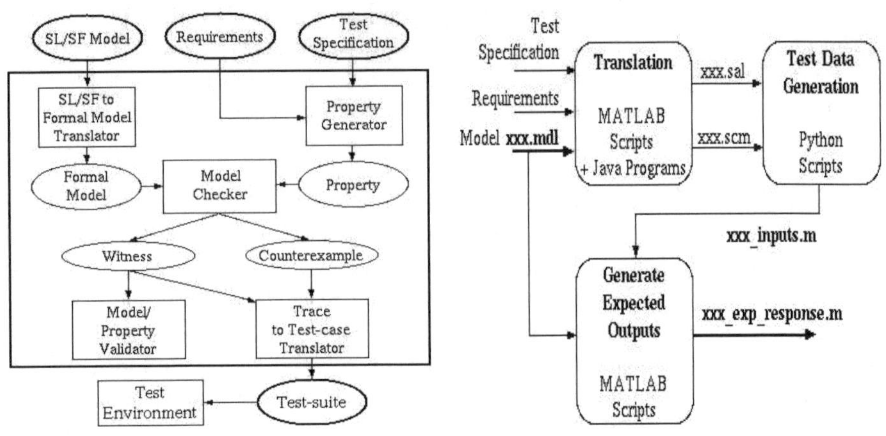

Fig. 1. AutoMOTGen architecture

Fig. 2. AutoMOTGen back-end flow

4 AutoMOTGen Implementation

The current implementation of AutoMOTGen uses SAL as an intermediate language. This enables use of associated tools such as sal-atg, sal-bmc, sal-smc, etc. The back-end flow for model translation and test case generation is shown in Figure 2. The translator extracts the relevant information from SL/SF models through scripts built using MATLAB APIs. The SL/SF models are simulated using the generated test data and the corresponding outputs are stored as reference for testing the implementation. During simulation the model coverage information is obtained to assess the completeness of the generated test data. The translation of SL/SF to SAL is non-trivial and involves various steps such as time discretization, type abstractions and captures the simulation semantics of SL/SF. The SAL model is structured such that it retains the hierarchy information and allows the mapping of structural coverage of SL/SF model to the coverage goals. Additionally, monitors are inserted to cover the high-level requirements specified in the form of temporal logic properties. The continuous blocks in Simulink are approximated using linear interpolation. The step-size used for sampling is taken as a user input during the translation step. The user can modify the step-size depending on the coverage information. The trap variables are selectively introduced into the SAL model based upon the user selected coverage options. The use of model-checker such as sal-bmc requires that all the variables should be of type bounded integers. The model-checker sal-inf-bmc can be used in cases where real datatypes are present in the model, however, it puts restrictions on the arithmetic operations which result in nonlinear constraints. In these cases the real variables are approximated by use of look-up-tables for arithmetic operations. The tool provides a capability to easily modify the variable types and their ranges before selecting the appropriate model-checking engine. The uncovered goals are checked for proving unreachability and the results are reported in the test generation logs. The unreachability of conditions or states in SAL model does not always imply unreachability in the SL/SF model. The tool provides a simple and intuitive GUI.

5 Case Studies

Our methodology has been evaluated using automotive controller case studies viz., Automatic Transmission Controller (ATC) and Adaptive Cruise Controller (ACC). The test results were compared with those obtained from a commercial tool that uses random test data generation techniques. The tests using our method were found to be more efficient in terms of providing model coverage with less number of input injections thus significantly reducing the test execution time, by almost factor of 10 in some cases. Results from these case studies based on a very preliminary implementation with a semi-automated flow are presented in [3]. In AutoMOTGen the entire end-to-end methodology has been fully automated. We have been able to handle various medium-sized controller models (corresponding RTW generated C code ranging between 2000-3000 lines) from real automotive subsystems. Recently, a larger case study has been initiated using controller modules from StabiliTrak™ project (Electronic Stability Control system).

6 Conclusion

We have presented AutoMOTGen for automatic test case generation from SL/SF models of automotive controllers. It uses model checking for efficient generation of test data. It is designed to be modular to enable plug-and-play of different model checkers, test data generators and coverage analysis tools for obtaining efficient test suites. The tool has been evaluated using automotive controller examples and the comparative results with respect to other commercially available tools have been encouraging. Various enhancements are being carried out in the tool such as counter-example based abstraction refinement and other techniques to address scalability issues arising in larger industrial designs. HybridSAL is also being explored as one of the approaches for discretization.

References

1. Ammann, P., Black, P.E., Majurski, W.: Using model checking to generate tests from specifications. In: ICFEM, p. 46 (1998)
2. Clarke, E.M., Grumberg, O., Peled, D.A.: Model Checking. The MIT Press, Cambridge (2000)
3. Gadkari, A., Mohalik, S.K., Shashidhar, K.C., Yeolekar, A., Suresh, J., Ramesh, S.: Automatic generation of test-cases using model checking for SL/SF models. In: 4th International Workshop on Model Driven Engineering, Verification and Validation (MoDeVVa 2007) (2007)

4. Gargantini, A., Heitmeyer, C.L.: Using model checking to generate tests from requirements specifications. In: ESEC / SIGSOFT FSE, pp. 146–162 (1999)
5. Hamon, G., deMoura, L., Rushby, J.: Generating efficient test sets with a model checker. In: 2nd International Conference on Software Engineering and Formal Methods, Beijing, China, September 2004, pp. 261–270. IEEE Computer Society, Los Alamitos (2004)
6. The Mathworks, Inc., http://www.mathworks.com
7. Reactis, Reactive Systems, Inc., http://www.reactive-systems.com
8. SAL homepage, http://sal.csl.sri.com/
9. Safety Test Builder, TNI-Software, http://www.tni-software.com

FSHELL: Systematic Test Case Generation for Dynamic Analysis and Measurement*

Tool Paper

Andreas Holzer, Christian Schallhart, Michael Tautschnig, and Helmut Veith

Technische Universität Darmstadt, Germany

Abstract. Although the principal analogy between counterexample generation and white box testing has been repeatedly addressed, the usage patterns and performance requirements for software testing are quite different from formal verification. Our tool FSHELL provides a versatile testing environment for C programs which supports both interactive explorative use and a rich scripting language. More than a frontend for software model checkers, FSHELL is designed as a database engine which dispatches queries about the program to program analysis tools. We report on the integration of CBMC into FSHELL and describe architectural modifications which support efficient test case generation.

1 Introduction

This paper introduces our prototype tool FSHELL which supports both interactive and scripted test case generation for real-world C code. We have consciously designed FSHELL in analogy to a database engine; FSHELL uses a query language tailored for program analysis, and dispatches queries about program paths to software analysis tools. The query language is built around the concept of a *test job*, i.e., a specification of program paths along with coverage requirements and other parameters, and provides the system engineer with convenient primitives for test job management and test job execution.

In the first iteration of the project, we have integrated CBMC [1] as model checking backend. We are using SAT enumeration techniques to generate families of test cases subject to coverage criteria in a *single run* of the model checker. We note that the program-as-database metaphor has been previously used in the BLAST project [2]. The query language of BLAST [3,4], however, is tailored towards model checking, and testing activities focus on basic block coverage only [5]. Other tools using model checkers for test case generation are Java PathFinder [6] and SAL2 [7]. Unlike FSHELL, neither of them supports full C semantics. Conversely, the CUTE toolkit implements concolic testing [8], which results in a tool that uses directed testing to form a model checker.

2 Features of the FSHELL Environment

The FSHELL environment assists the user to create, manage and execute test jobs. With each source file, the tool automatically associates a *generic test job* which the user

* Supported by DFG grant FORTAS – Formal Timing Analysis Suite for Real Time Programs (VE 455/1-1).

A. Gupta and S. Malik (Eds.): CAV 2008, LNCS 5123, pp. 209–213, 2008.

annotates and modifies for the purpose at hand: On the one hand, *structural constraints* restrict the test job to program paths matching a regular expression over program locations annotated by program predicates. To this end, FSHELL provides convenient primitives for structural notions of C programs such as function headers, labels etc. Internally, FSHELL represents structural constraints as *path automata*. On the other hand, *quality constraints* require the *set of test cases* generated by a test job to meet minimal requirements such as basic block coverage, condition coverage or predicate coverage (cf. [9]). Importantly, different quality constraints can be enforced locally in the program. An example of the representation of these requirements is given in Sec. 3.

Additional commands for *test job management* enable the user to maintain and develop multiple test jobs for complex applications. Thus, test jobs are viewed as objects which are loaded, stored, duplicated, merged, etc. When a test job is prepared, the user invokes *test case generation* commands to compute a *test suite* which satisfies the constraints expressed by the test job. FSHELL also provides support for *test case execution* to observe non-functional properties not only on desktop machines, but also on embedded devices. Fig. 1 presents an example session of FSHELL. We show the query used to generate a test suite with predicate coverage for `bubble.c` for an array of size 20.

```
void bubble(int a[], int N) {
  int i, j, t;
  for (i = N - 1; i >= 0; i--)
    for (j = 1; j <= i; j++)
      if (a[j - 1] > a[j])
        SWAP(a[j - 1], a[j]);
}
void main() {int a[20]; bubble(a, 20);}
```

```
> ADD_SOURCE(bubble.c);
> GENERATE_TC(COVERAGE(ENTRY(main),
  EXIT(main), PREDICATE)) AS tc1;
> SHOW(tc1);
IN: a={0,1128267777,...,1465438334}
IN: a={0,-2139095040,...,1709483866}
```

Fig. 1. Source code of `bubble.c` and corresponding FSHELL session

3 Tool Architecture

We chose CBMC 2.4 as the first backend for FSHELL because (1) it supports full C syntax and semantics, (2) BMC is conceptually closer to testing than an abstraction/refinement approach, (3) the source code is available, and (4) it is well engineered and offers a very clean design and a stable code base. FSHELL is, like CBMC, implemented in C++ and accounts for 13k lines of code.[1]

The design of FSHELL is based on three main layers. The *frontend* handles user interactions with a command line interface. There, job control commands, such as loading source files into the test job, and constraint specifications are entered by the user. The *management layer* implements control commands and redirects queries to the appropriate *backend*. The backend performs the actual path feasibility analysis and test case

[1] Visit `http://code.forsyte.cs.tu-darmstadt.de/fshell` to follow development.

generation. Both the management and the backend access a single shared cache. This cache stores all queries and their respective results. Fig. 2 gives an overview of the collaboration of components. Dashed rectangles represent parts reused from CBMC. Both path feasibility analysis and test case generation follow a bounded model checking workflow. In path analysis, the CBMC modules in use are unmodified and only invoked from within our code. Test case generation, however, requires an additional control flow graph analysis. To this end, we collect conditions along possible control flow paths and identify the relevant literals used in the SAT encoding. Further, to allow for highly efficient test case generation of large test suites, the SAT solver is invoked in an incremental fashion, retaining the conflict clause database. In this process, we use both blocking clauses and assumptions in a way that guarantees the conflict clause database to remain consistent. By the design of our procedure, the resulting SAT solver invocation returns UNSAT only if there is no further matching test case, i.e., all coverage criteria are satisfied.

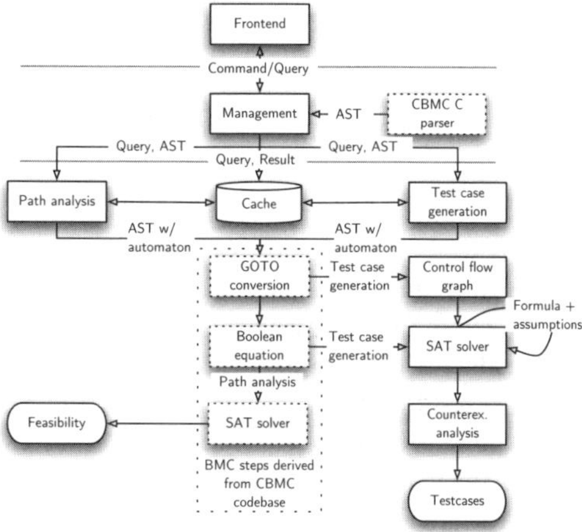

Fig. 2. Main components of FSHELL

As an example, consider decision coverage. In the test job the user specifies a segment of the program where all reachable branches have to be covered. From the AST and the CNF formula, FSHELL obtains a set B_0 of Boolean variables which correspond to the branches to be taken, and thus characterize the coverage criterion. Initially, we add B_0 as a clause, thus requiring that at least one of the variables is set to true, i.e., at least one of the branches is taken within the critical segment of code. Subsequently we compute clause B_{i+1} from B_i by removing already satisfied variables from B_i, and add the respective clause to the SAT instance. This incremental SAT solving process ends if either $B_{i+1} = \emptyset$ or the instance is found to be unsatisfiable. In this case, B_{i+1} marks the set of infeasible branches.

4 Experimental Results

In our experiments, we first analyzed an industrial engine controller which was autogenerated from a MATLAB/Simulink model. The resulting C source code of 2033 LLOC[2] was, without applying any abstraction, tested for basic block coverage. FSHELL achieved coverage with only five test cases, taking 18.18 seconds on a 3.2 GHz Intel P4 equipped with 3 GB of RAM.

[2] logical lines of code, i.e., number of occurrences of ';'

Table 1. Results on device drivers

Source file	LLOC	BLAST #cases	BLAST Time[s]	FSHELL #cases	FSHELL Time[s]	Speed-up
kbfiltr.i	4879	39	300	66	17	17.9
floppy.i	6435	111	1500	288	1305	1.1
cdaudio.i	8022	85	1500	159	748	2.0
parport.i	20698	213	5460	312	1999	2.7
parclass.i	45283	219	2520	716	1511	1.7

As the concepts implemented in FSHELL are related to both directed testing and model checking, we chose BLAST as benchmarking reference. Still, it should be noted that BLAST is a full fledged model checker and is not as optimized towards testing as FSHELL is. Their set of benchmarks, presented in [5], is well documented and all source files are publicly available. To achieve equivalent test goals, we generated test suites with full basic block coverage. Apart from `parclass.i`, in the tests of Table 1 FSHELL was run on the P4 system mentioned above. The file `parclass.i` required a cleanup of conflicting `typedef`'s and more than the addressable 2 GB of main memory. On a 3.0 GHz AMD64 system we succeeded with a memory usage of 2.3 GB. The results for BLAST are taken literally from [5], because the version of BLAST performing test case generation is currently unavailable. The hardware used, however, is very similar. We observe that FSHELL typically returns a higher number of test cases to achieve basic block coverage, but it takes less time to do so. We believe that these performance improvements outweigh the larger test sets. Nevertheless we plan to include minimization strategies in FSHELL.

Additionally, we generated test suites for sorting algorithms literally taken from [10]. The experiments are parameterized by the size of the array to be sorted. In Table 2, we present the speedup achieved by generating covering test suites in a single test job, compared to naïve iterative invocations of the model checker. Note that in the latter case, coverage constraints are not even considered. For each algorithm and each array size, we show the speedup for basic block- and condition coverage.

Table 2. Speedup for basic block/condition coverage

Source File	5	10	15	20
bubble.c	1.88/1.98	1.86/1.89	1.96/1.58	1.74/1.26
insertion.c	1.63/1.95	1.56/2.19	0.95/1.39	0.98/2.13
selection.c	2.13/1.76	1.41/2.01	1.38/1.97	0.99/2.27

Conclusion. We presented FSHELL as an environment to facilitate white-box testing of C programs. By design, FSHELL treats a C program as a database to be queried by the user. FSHELL serves as a framework which integrates multiple program analysis backends. Our experimental results confirm the practical feasibility and relative efficiency of our approach.

Acknowledgments. We are grateful to Raimund Kirner, Sven Bünte, Ingomar Wenzel, and Michael Zolda for discussions on the topic of this paper. Further we thank Dirk Beyer for his help with BLAST.

References

1. Clarke, E.M., Kroening, D., Lerda, F.: A Tool for Checking ANSI-C Programs. In: Jensen, K., Podelski, A. (eds.) TACAS 2004. LNCS, vol. 2988, pp. 168–176. Springer, Heidelberg (2004)
2. Henzinger, T.A., Jhala, R., Majumdar, R., Sutre, G.: Software Verification with BLAST. In: Ball, T., Rajamani, S.K. (eds.) SPIN 2003. LNCS, vol. 2648, pp. 235–239. Springer, Heidelberg (2003)
3. Beyer, D., Chlipala, A.J., Henzinger, T.A., Jhala, R., Majumdar, R.: The BLAST Query Language for Software Verification. In: Giacobazzi, R. (ed.) SAS 2004. LNCS, vol. 3148, pp. 2–18. Springer, Heidelberg (2004)
4. Beyer, D., Noack, A., Lewerentz, C.: Simple and Efficient Relational Querying of Software Structures. In: WCRE, pp. 216–225 (2003)
5. Beyer, D., Chlipala, A.J., Henzinger, T.A., Jhala, R., Majumdar, R.: Generating Tests from Counterexamples. In: ICSE, pp. 326–335 (2004)
6. Visser, W., Pasareanu, C.S., Khurshid, S.: Test input generation with java PathFinder. In: ISSTA, pp. 97–107 (2004)
7. Hamon, G., de Moura, L.M., Rushby, J.M.: Generating Efficient Test Sets with a Model Checker. In: SEFM, pp. 261–270 (2004)
8. Sen, K., Marinov, D., Agha, G.: CUTE: a concolic unit testing engine for C. In: FSE, pp. 263–272 (2005)
9. Ntafos, S.C.: A comparison of some structural testing strategies. IEEE Trans. Software Eng. 14(6), 868–874 (1988)
10. Sedgewick, R.: Algorithms in C. Addison-Wesley Publishing Company, Inc., Reading (1990)

Applying the Graph Minor Theorem to the Verification of Graph Transformation Systems*

Salil Joshi[1] and Barbara König[2]

[1] Indian Institute of Technology, Delhi, India
[2] Abteilung für Informatik und Angewandte Kognitionswissenschaft, Universität Duisburg-Essen, Germany

Abstract. We show how to view certain subclasses of (single-pushout) graph transformation systems as well-structured transition systems, which leads to decidability of the covering problem via a backward analysis. As the well-quasi order required for a well-structured transition system we use the graph minor ordering. We give an explicit construction of the backward step and apply our theory in order to show the correctness of a leader election protocol.

1 Introduction

In a series of seminal papers Robertson and Seymour have shown that graphs are well-quasi-ordered with respect to the minor ordering [7,8]: in any (infinite) sequence of graphs $G_0, G_1, G_2, \ldots$ there are always two indices $i < j$ such that G_i is a minor of G_j. This means that G_i can be obtained from G_j by deleting and contracting edges and by deleting isolated nodes.

The theorem has far-reaching consequences. It guarantees that every set of graphs that is upward-closed with respect to the minor ordering can be represented by a finite number of minimal graphs. Similarly, any downward-closed set of graphs (e.g., planar graphs, forests, graphs embeddable in a torus) can be characterized by a finite set of forbidden minors. A well-known special case are (undirected) planar graphs which are characterized by two forbidden minors: the complete graph with five nodes (K_5) and the complete bipartite graph with six nodes ($K_{3,3}$), a fact which is known as Kuratowski's theorem.

Well-quasi-orders (wqo's) also play a fundamental role in the analysis of a class of (infinite-state) transition systems, so called well-structured transition systems (WSTS) [4]. States in a WSTS are well-quasi-ordered and the standard analysis method shows whether some state in an upward-closed set is reachable from an initial state by performing backward analysis. The well-quasi-ordering guarantees that upward-closed sets are finitely representable, that the set of predecessors is also upward-closed and that the technique terminates after finitely many steps.

One important example for WSTS are Petri net transition system, where a marking m_1 is considered larger than or equal to m_2 if it contains at least as

* Research partially supported by the DFG project SANDS.

A. Gupta and S. Malik (Eds.): CAV 2008, LNCS 5123, pp. 214–226, 2008.

many tokens in every place. Other examples are string rewrite systems, basic process algebra and communicating finite state machines. A transition system that can not be naturally viewed as a WSTS can often be turned into one by introducing some notion of "lossiness". For instance an unreliable channel may lose messages and a suitable wqo considers the content c_1 of a channel as greater than c_2 if c_2 can be obtained from c_1 by dropping some messages.

The graph minor ordering fits well with this intuition of "lossiness" and seems to be applicable to networks where edges (connections or processes) may disappear—possibly due to faults—and where edges can be contracted. The latter phenomenon appears if a process leaves a network by connecting its predecessor and successor, something which typically happens in rings.

Here we show how to view certain graph transformation systems (GTS) as WSTS with respect to the minor ordering. GTS are an intuitive formalism, well-suited to model concurrent and distributed systems. In general GTS are Turing-complete and due to undecidability issues it is hard to imagine a useful wqo for the general case. However, if the GTS exhibits features as described above it can be successfully verified.

GTS are typically defined by means of category theory, which makes the definition of rewriting steps less tedious. Graph rewriting is defined via pushouts in a suitable category of graph morphisms and in the rest of this paper we will exploit certain well-known properties of pushouts. The relation of a graph G to its minor H can be represented by a partial graph morphism with specific properties. Since the theory requires the handling of partial morphisms, we have decided to work in the single-pushout approach (SPO) which uses partial morphisms [5,3].

The paper is organized as follows: Section 2 introduces the basic definitions. In Section 3 we consider classes of GTS that can be seen as WSTS, and introduce the techniques for their analysis. In Section 4 we will look at a leader election protocol and show how the analysis method works in practice.

2 Preliminaries

Here we introduce some of the basic notions needed in the paper, especially well-quasi-orders, well-structured transition systems, graph transformation systems and minors.

2.1 Well-Quasi-Order

Definition 1 (wqo). *A* well-quasi-order *(wqo) is any quasi-ordering*[1] $\leq$ *(over some set X) such that, for any infinite sequence $x_0, x_1, x_2, \ldots$ in X, there exist indices $i < j$ with $x_i \leq x_j$.*

An upward-closed set *is any set $I \subseteq X$ such that $y \geq x$ and $x \in I$ entail $y \in I$. A* downward-closed set *can be analogously defined.*

For an element $x \in I$, we define $\uparrow x = \{y \mid y \geq x\}$. Then, a basis *of an upward-closed set I is a set I^b such that $I = \bigcup_{x \in I^b} \uparrow x$.*

[1] Note that a quasi-order is the same as a preorder.

Lemma 2

1. *If $\leq$ is a well-quasi-ordering then any upward-closed I has a finite basis.*
2. *If $\leq$ is a wqo and $I_0 \subseteq I_1 \subseteq I_2 \subseteq \ldots$ is an infinite increasing sequence of upward-closed sets, then there exists an index $k \in \mathbb{N}$ such that $I_k = I_{k+1} = I_{k+2} = \ldots$*

2.2 Well-Structured Transition Systems

Definition 3 (WSTS). *A* well-structured transition system *(WSTS) is a transition system $T = (S, \Rightarrow, \leq)$, where S is a set of states and $\Rightarrow \subseteq S \times S$, such that the following conditions hold:*

1. ***Well quasi ordering:*** *$\leq$ is a well-quasi-ordering on S.*
2. ***Compatibility:*** *For all $s_1 \leq t_1$ and a transition $s_1 \Rightarrow s_2$, there exists a sequence $t_1 \Rightarrow^* t_2$ of transitions such that $s_2 \leq t_2$.*

$$\begin{array}{ccc} t_1 & \overset{*}{\Longrightarrow} & t_2 \\ \text{\rotatebox{90}{$\leq$}} & & \text{\rotatebox{90}{$\leq$}} \\ s_1 & \Longrightarrow & s_2 \end{array}$$

Given a set $I \subseteq S$ of states we denote by $\mathit{Pred}(I)$ the set of direct predecessors of I, i.e., $\mathit{Pred}(I) = \{s \in S \mid \exists s' \in I : s \Rightarrow s'\}$. Furthermore $\mathit{Pred}^*(I)$ is the set of all predecessors.

Let $(S, \Rightarrow, \leq)$ be a WSTS. Consider a set of states $I \subseteq S$. Backward reachability analysis involves the computation of $\mathit{Pred}^*(I)$ as the limit of the sequence $I_0 \subseteq I_1 \subseteq I_2 \subseteq \ldots$ where $I_0 = I$ and $I_{n+1} = I_n \cup \mathit{Pred}(I_n)$. However, in general this may not terminate. For WSTS, if I is upward-closed then it can be shown that $\mathit{Pred}^*(I)$ is also upward-closed (compatibility condition) and that termination is guaranteed (Lemma 2).

Definition 4 (Effective pred-basis). *A WSTS has an* effective pred-basis *if there exists an algorithm accepting any state $s \in S$ and returning $pb(s)$, a finite basis of $\uparrow \mathit{Pred}(\uparrow s)$.*

Now assume that T is a WSTS with effective pred-basis. Pick a finite basis I^b of I and define a sequence $K_0, K_1, K_2, \ldots$ of sets with $K_0 = I^b$ and $K_{n+1} = K_n \cup pb(K_n)$. Let m be the first index such that $\uparrow K_m = \uparrow K_{m+1}$. Such an m must exist by Lemma 2 and we have $\uparrow K_m = \mathit{Pred}^*(I)$. Finally, note that due to Lemma 2 every set K_n can be represented by a finite basis.

The *covering problem* is to decide, given two states s and t, whether starting from a state s it is possible to cover t, i.e. to reach a state t' such that $t' \geq t$. From the previous argument follows the decidability of the covering problem.

Theorem 5 (Covering problem). *The covering problem is decidable for a WSTS with an effective pred-basis and a decidable wqo $\leq$.*

Thus, if T is a WSTS and the "error states" can be represented as an upward-closed set I, then it is decidable whether any element of I is reachable from the start state.

2.3 Graphs and Graph Transformation

Definition 6 (Hypergraph). *Let Λ be a finite set of labels. A (Λ-)*hypergraph *is a tuple (V_G, E_G, c_G, l_G) where V_G is a finite set of nodes, E_G is a finite set of edges, $c_G: E_G \to V_G^*$ is a connection function and $l_G: E_G \to \Lambda$ is the labelling function for edges.*

Directed labelled graphs are a special case of hypergraphs where every sequence $c_G(e)$ is of length two.

Definition 7 (Partial hypergraph morphism). *Let G, G' be (Λ-)hypergraphs. A* partial hypergraph morphism *(or simply* morphism*) $\varphi: G \rightharpoonup G'$ consists of a pair of partial functions $(\varphi_V : V_G \rightharpoonup V_{G'}, \varphi_E : E_G \rightharpoonup E_{G'})$ such that for every $e \in E_G$ it holds that $l_G(e) = l_{G'}(\varphi_E(e))$ and $\varphi_V(c_G(e)) = c_{G'}(\varphi_E(e))$ whenever $\varphi_E(e)$ is defined. Furthermore if a morphism is defined on an edge, it must be defined on all nodes adjacent to it. (This condition need not hold in the other direction.)*

Total morphisms *are denoted by an arrow of the form $\to$.*

In the following we will drop the subscripts and write φ instead of φ_V and φ_E.

Gluing of graphs along a common subgraph is done via pushouts in the category of partial graph morphisms.

Definition 8 (Pushout)

Let $\varphi: G_0 \rightharpoonup G_1$ and $\psi: G_0 \rightharpoonup G_2$ be two partial graph morphisms. The pushout *of φ and ψ consists of a graph G_3 and two graph morphisms $\psi': G_1 \rightharpoonup G_3$, $\varphi': G_2 \rightharpoonup G_3$ such that $\psi' \circ \varphi = \varphi' \circ \psi$ and for every other pair of morphisms $\psi'': G_1 \rightharpoonup G_3'$, $\varphi'': G_2 \rightharpoonup G_3'$ such that $\psi'' \circ \varphi = \varphi'' \circ \psi$ there exists a unique morphism $\eta: G_3 \rightharpoonup G_3'$ with $\eta \circ \psi' = \psi''$ and $\eta \circ \varphi' = \varphi''$.*

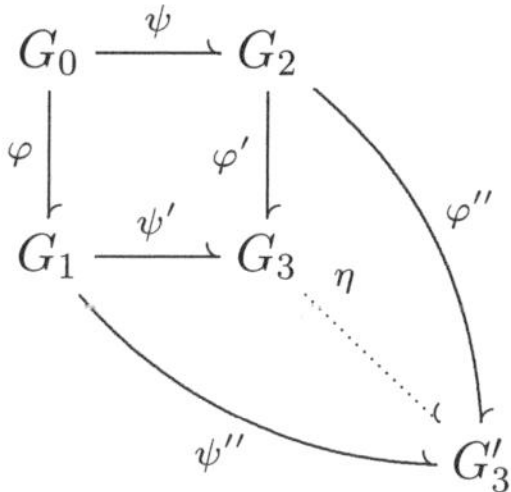

It is known that pushouts of partial graph morphisms always exist, that they are unique up to isomorphism and that they can be constructed as follows. The intuition behind the construction is that G_1, G_2 are glued together along a common interface G_0 and that an element is deleted if it is deleted by either φ or ψ.

Proposition 9 (Construction of pushouts). *Let $\varphi: G_0 \rightharpoonup G_1$, $\psi: G_0 \rightharpoonup G_2$ be partial hypergraph morphisms. Furthermore let $\equiv_V$ be the smallest equivalence on $V_{G_1} \cup V_{G_2}$ and $\equiv_E$ the smallest equivalence on $E_{G_1} \cup E_{G_2}$ such that $\varphi(x) \equiv \psi(x)$ for every element x of G_0.*

An equivalence class of nodes is called valid *if it does not contain the image of a node x for which $\varphi(x)$ or $\psi(x)$ are undefined. Similarly a class of edges is* valid *if the analogous condition holds and furthermore all nodes adjacent to these edges are contained in valid equivalence classes.*

Then the pushout G_3 of φ and ψ consists of all valid equivalence classes $[x]_\equiv$ as nodes and edges, where $l_{G_3}([e]_\equiv) = l_{G_i}(e)$ and $c_{G_3}([e]_\equiv) = [v_1]_\equiv \dots [v_k]_\equiv$ if $e \in E_{G_i}$ and $c_{G_i}(e) = v_1 \dots v_k$.

It can be seen that the pushout of two total morphisms (in the category of partial morphisms) always results in two total morphisms. Furthermore it is equal to their pushout in the category of total morphisms. However φ total and ψ partial does not necessarily imply that φ' is total. This is due to so-called *deletion/preservation conflicts* where two elements x_0, x_0' of G_0 are mapped to the same element of G_1, i.e., $\varphi(x_0) = \varphi(x_0')$, while $\psi(x_0)$ is defined, whereas $\psi(x_0')$ is undefined. The construction above suggests that then $\varphi'(\psi(x_0))$ must be undefined, i.e., φ' is not total. If no such elements x_0, x_0' can be found, then φ is said to be *conflict-free* with respect to ψ and in this case φ' is always total.

Definition 10 (Graph rewriting). *A* rewriting rule *is a partial morphism $r: L \rightharpoonup R$, where L is called left-hand side and R right-hand side.*

A match *(of r) is a total morphism $m: L \to G$ which is conflict-free wrt. r.*

Given a rule and a match, a rewriting step *or an application of the rule to the graph G, resulting in H, is a pushout diagram as shown in Fig. 1 on the left. In this case we write $G \Rightarrow H$.*

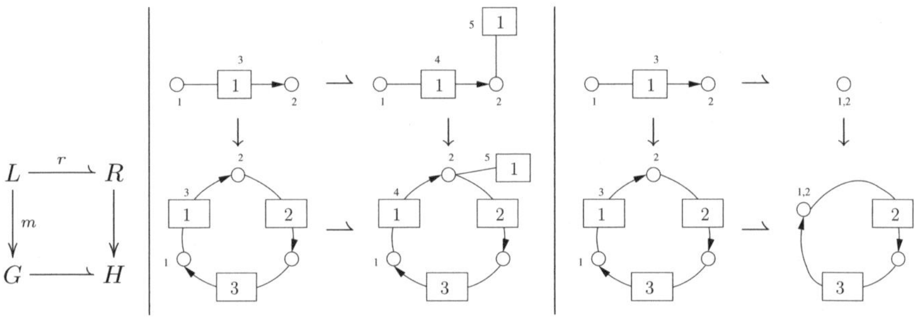

Fig. 1. Single-pushout graph rewriting (pushout diagram and example rewriting steps)

Intuitively, we can think of this as follows: L is a subgraph of G, all items of L whose image is undefined under r are deleted, the new items of R are added and connected as specified by r. Note that whenever a node is deleted, all adjacent edges will be deleted as well.

Fig. 1 shows two examples for graph rewriting steps. In the middle pushout a binary hyperedge generates another (unary) hyperedge, whereas in the right pushout an edge is contracted. The way in which the morphisms map nodes and edges is indicated by the small numbers next to the edges. These specific rewriting rules will also play a role in our application (see Section 4).

In the context of this paper a *graph transformation system (GTS)* consists of a finite set $\mathcal{R}$ of rewriting rules. Sometimes we will fix an *initial graph* or *start graph*.

2.4 Minors and Minor Morphisms

We will now review the notion of a graph minor.

Definition 11 (Minor). *A graph $\hat{G}$ is a* minor *of a graph G, if $\hat{G}$ can be obtained from G by (repeatedly) performing the following operations on G:*

1. *Deletion of an edge.*
2. *Contraction of an edge, thereby merging all nodes adjacent to the edge.*
3. *Deletion of an isolated node.*

The *Robertson-Seymour Theorem* [7] says that the minor order is a well-quasi-order. In fact, this theorem is true even if the edges and vertices of the graphs are labelled from a well-quasi-ordered set, and also for hypergraphs and directed graphs (see [8]).

Now, if we could show that a GTS satisfies the compatibility condition of Definition 3 (with respect to the minor ordering), we could analyze it using the theory of WSTS. But before we characterize such GTS we first need the definition of minor morphisms and their properties. A *minor morphism* is a partial morphism that identifies a minor of a graph.

Definition 12 (Minor morphism). *A partial morphism $\mu : G \rightharpoonup \hat{G}$ is a minor morphism (written $\mu : G \mapsto \hat{G}$) if*

1. *it is surjective,*
2. *it is injective on edges* and
3. *whenever $\mu(v) = \mu(w) = z$ for some $v, w \in V_G$ and $z \in V_{\hat{G}}$, there exists a path between v and w in G. If e is an edge on this path then $\mu(e)$ is undefined, and all nodes in $c_G(e)$ are mapped to z.*

In [8] a different way to characterize minors is proposed: a function, going in the opposite direction, mapping nodes of $\hat{G}$ to subgraphs of G. This however can not be seen as a morphism in the sense of Definition 7 and we would have problems integrating it properly into the theory of graph rewriting.

One can show the following facts about minor morphisms.

Lemma 13. *$\hat{G}$ is a minor of G iff there exists a minor morphism $\mu : G \mapsto \hat{G}$.*

Lemma 14. *Pushouts preserve minor morphisms in the following sense: If $f : G_0 \mapsto G_1$ is a minor morphism and $g : G_0 \to G_2$ is total, then the morphism f' in the pushout diagram below is a minor morphism.*

$$\begin{array}{ccc} G_0 & \xrightarrow{f} & G_1 \\ {\scriptstyle g}\downarrow & & \downarrow{\scriptstyle g'} \\ G_2 & \xrightarrow{f'} & G_3 \end{array}$$

3 GTS as WSTS!

As observed earlier, a GTS can be seen as a WSTS with the minor relation as the well-quasi-ordering, provided the GTS satisfies the compatibility condition introduced in Definition 3.

3.1 Characterization

We will first give a sufficient condition that allows us to view a GTS as a WSTS. Note that the fundamental problem is that whenever a minor of G contains a left-hand side, then G might contain a "disconnected" copy of the left-hand side.

Proposition 15 (GTS as WSTS). *Let $\mathcal{R}$ be a GTS that satisfies the following condition: For every rule $(r: L \rightharpoonup R) \in \mathcal{R}$, every minor morphism $\mu: G \mapsto \hat{G}$ and every match $m: L \to \hat{G}$ (see diagram on the left) there exists a graph G' such that $G \Rightarrow^* G'$, there is a minor morphism $\mu': G' \mapsto \hat{G}$ and there exists a match $m': L \to G'$ such that $m = \mu' \circ m'$ (see commuting diagram below on the right). Then $\mathcal{R}$ is a WSTS.*

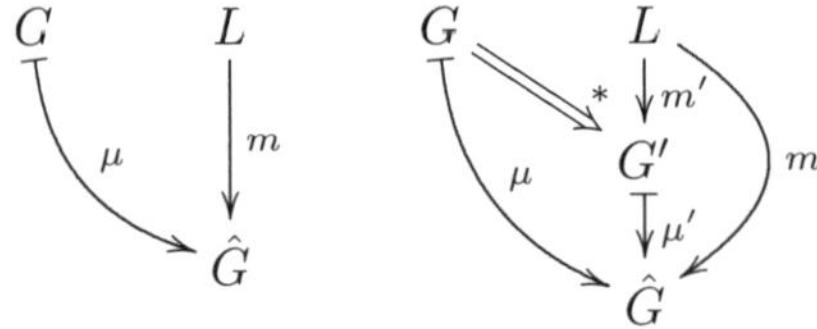

With this characterization we can now identify suitable types of GTS that are WSTS:

- Context-free graph grammars, where the left-hand side of every rule consists of a single hyperedge. Here G must always contain a match of L that makes the above diagram commute and no intermediate graph G' is needed.
- GTS where the left-hand sides of the rules consist of disconnected edges. The argument is analogous to the case above.
- Any arbitrary GTS can be transformed into a WSTS with the addition of an edge contraction rule for every edge label. Now, if $\hat{G}$ contains a subgraph which is isomorphic to a left-hand side, the pre-image of this subgraph under μ is present in G, but it might possibly be disconnected. The minor morphism μ makes the elements of L adjacent by contracting paths and the same can be done by applying the additional edge contraction rules.

3.2 Backward Analysis

Let $\mathcal{R}$ be a set of graph transformation rules which satisfies the compatibility condition. Now we consider the question of performing a backward reachability analysis on $\mathcal{R}$ which requires a method for computing an effective pred-basis $pb(S)$ for a given graph S.

Our method will involve the backwards application of an SPO rewriting rule. This requires the completion of a diagram of the form $L \rightharpoonup R \to H$ by a graph G and morphisms $L \to G \rightharpoonup H$ such that the square is a pushout. Then G is a so-called *pushout complement*. Pushout complements are well-studied for total morphisms since they are an essential ingredient in double-pushout rewriting. For partial morphisms they have been studied to a lesser extent.

We will first demonstrate some issues that can arise with pushout complements: for instance, the two total morphisms $L \rightharpoonup R \to H$ shown in Fig. 2 (left) (edges and nodes are unlabelled, morphisms are indicated by numbers $1, 2$) have five different pushout complements. Note also that each pair of total morphisms has only finitely many pushout complements (up to isomorphism).

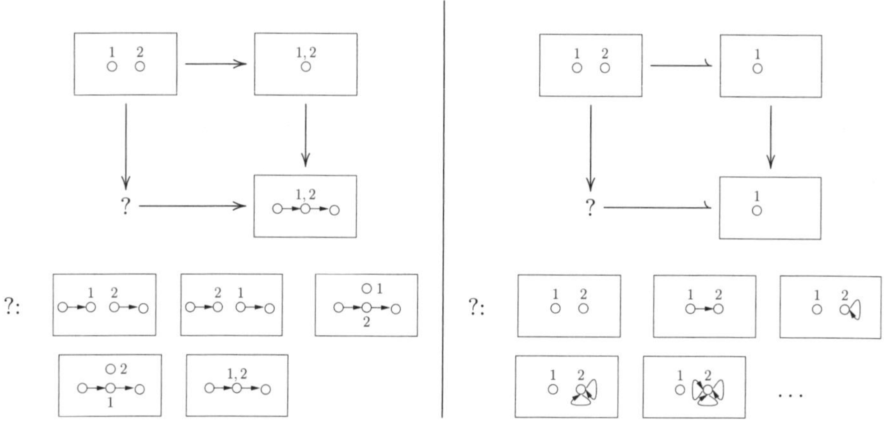

Fig. 2. Left: Two total morphisms with five pushout complements. Right: A partial and a total morphism with infinitely many pushout complements.

While the existence of multiple pushout complements is a feature that will be needed to determine the pred-basis, the situation for partial morphisms is more involved. Consider the diagram in Fig. 2 (right) where the morphism from L to R is partial. Here we have infinitely many pushout complements. Note however that the first graph is a minor of all other pushout complements. This suggests that only the computation of minimal pushout complements is needed.

Now we will give a high-level description of the procedure for computing $pb(S)$ for a given graph S. A more detailed account will be given in Section 3.3 where we will also argue that the procedure is indeed effective.

1. For each rule $(r : L \rightharpoonup R) \in \mathcal{R}$, let $\mathcal{M}_R$ be the (finite) set of all minor morphisms with source R.
2. For each $(\mu\colon R \mapsto M) \in \mathcal{M}_R$ consider the rule $\mu \circ r\colon L \rightharpoonup M$.
3. For each total match $m'\colon M \to S$ compute all minimal[2] pushout complements X such that $m\colon L \to X$ below is total and conflict-free wrt. r.

[2] "Minimimal" means "minimal wrt. the well-quasi ordering $\leq$".

$$\begin{array}{ccc} L & \xrightarrow{\mu \circ r} & M \\ {\scriptstyle m}\downarrow & & \downarrow{\scriptstyle m'} \\ X & \longrightarrow & S \end{array}$$

4. The set $pb(S)$ contains all graphs X obtained in this way.

That is, we use all minors of R as right-hand sides for the backward step. This is needed since S represents an upward-closed set and not all items of R must be present in S itself. We can now show the correctness of the procedure $pb(S)$, where the proof depends crucially on Lemma 14.

Theorem 16. *The procedure $pb(S)$ computes a finite subset of $Pred(\uparrow S)$.*

In order to prove that $pb(s)$ generates every member of the pred-basis, we first prove a general result in the category of graphs and partial morphisms.

Lemma 17. *Let $\psi_1 : L \to G$ be total and conflict-free wrt. ψ_2. If the diagram below on the left is a pushout and $\mu : H \mapsto S$ a minor morphism, then there exist minors M and X of R and G respectively, such that*

1. *the diagram below on the right commutes and the outer square is a pushout.*
2. *the morphisms $\mu_G \circ \psi_1 : L \to X$ and $\varphi_1 : M \to S$ are total and $\mu_G \circ \psi_1$ is conflict-free wrt. ψ_2.*

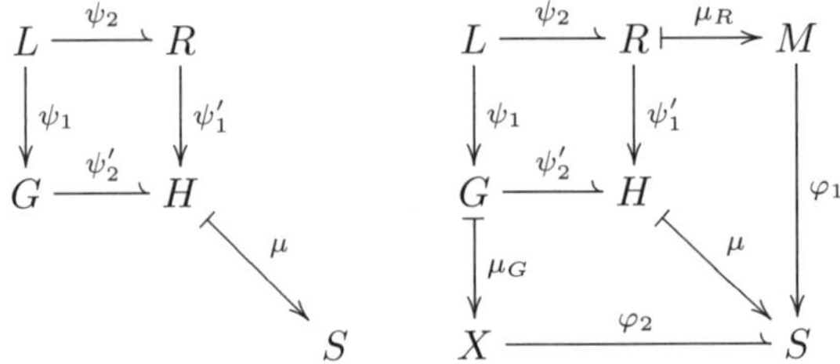

The lemma above says that whenever S is a minor of H and G is a predecessor of H, then we can make a backwards step for S and obtain X, a minor of G. Using this lemma we can now state the completeness of the procedure $pb(S)$.

Theorem 18. *The set generated by $pb(S)$ is a pred-basis of S.*

3.3 Computing Minimal Pushout Complements

Now we consider the question of how to construct pushout complements when some (but not all) of the morphisms involved may be partial. Hence consider a diagram $L \xrightarrow{\varphi} \tilde{L} \to \tilde{X}$. The idea is to split $L \xrightarrow{\varphi} \tilde{L} = L \rightharpoonup dom(\varphi) \to \tilde{L}$ where $dom(\varphi) \to \tilde{L}$ is total and $L \rightharpoonup dom(\varphi)$ is an inverse injection, i.e., a morphism which is injective, surjective, but not necessarily total. Now the task of computing pushout complements can be divided into two subtasks.

Lemma 19. *Let L and $\tilde{L}$ be graphs, $\varphi_1 : L \rightharpoonup \tilde{L}$ be an inverse injection, and $\psi_2 : \tilde{L} \to \tilde{X}$ be a total morphism. Now construct a specific pushout complement X' with morphisms $\psi_1' \colon L \to X'$, $\varphi_2' \colon X' \rightharpoonup \tilde{X}$ as follows:*

1. *Take a copy of the graph $\tilde{X}$, and let ψ_1' be $\psi_2 \circ \varphi_1$. The morphism φ_2' is the identity.*
2. *Let Y be the set of elements of L the image of which is undefined under φ_1. Add a copy of Y to this copy of $\tilde{X}$, and extend ψ_1' by mapping Y into this set. Furthermore φ_2' is undefined on all elements of the copy of Y.*
3. *Now merge these new elements (originally contained in Y) in all possible combinations, i.e., factor through all appropriate[3] equivalences. The morphisms ψ_1' and φ_2' are modified accordingly.*

$$\begin{array}{ccc} L & \xrightarrow{\varphi_1} & \tilde{L} \\ \downarrow{\scriptstyle \psi_1} & & \downarrow{\scriptstyle \psi_2} \\ X & \xrightarrow{\varphi_2} & \tilde{X} \end{array}$$

The set of graphs obtained in this way is denoted by $\mathcal{P}$. Each element X' of $\mathcal{P}$ is a pushout complement of φ_1, ψ_2 and the corresponding morphisms $\psi_1' \colon L \to X'$ are total. Any other pushout complement X where $\psi_1 : L \to X$ is total (see diagram on the right) has some graph $X' \in \mathcal{P}$ as a minor.

Finally, if $\psi_1 \colon L \to X$ is conflict-free wrt. to a rule $r : L \rightharpoonup R$, then there exists a pushout complement $X' \in \mathcal{P}$ with $\psi_1' \colon L \to X'$ conflict-free wrt. r, such that $X' \leq X$.

In order to do backwards application of rules in order to obtain $pb(s)$, we construct pushout complements (with total conflict-free matches) as follows:

Proposition 20. *Let $r \colon L \rightharpoonup R$ be a fixed rule. Furthermore let L, M and S be graphs, with a partial morphism $\varphi_1 : L \rightharpoonup M$ and a total morphism $\psi_2 : M \to S$. Then, if we apply the following procedure we only construct pushout complements X' of φ_1, ψ_2 and any other pushout complement X (with $\psi_1 \colon L \to X$ where ψ_1 is total and conflict-free wrt. r) has one of them as a minor.*

1. *Split φ_1 into two morphisms as follows: let $\varphi_1' : L \rightharpoonup dom(\varphi_1)$ be an inverse injection and let $\varphi_1'' : dom(\varphi_1) \to M$ be total.*
2. *Now consider the total morphisms $\varphi_1'' : dom(\varphi_1) \to M$, and $\psi_2 : M \to S$. Construct all their pushout complements as usual for total morphisms.[4]*
3. *Let $\tilde{X}$ be any such pushout complement with $\eta \colon dom(\varphi_1) \to \tilde{X}$.*
4. *For φ_1', η use the construction of Lemma 19 in order to obtain the minimal pushout complements X' (with total and conflict-free ψ_1').*
5. *Finally, from all such pushout complements X' take the minimal ones.*

The situation is depicted in the diagram below.

$$\begin{array}{ccccc} L & \xrightarrow{\varphi_1'} & dom(\varphi_1) & \xrightarrow{\varphi_1''} & M \\ \downarrow{\scriptstyle \psi_1'} & & \downarrow{\scriptstyle \eta} & & \downarrow{\scriptstyle \psi_2} \\ X' & \longrightarrow & \tilde{X} & \longrightarrow & S \end{array}$$

---

[3] Here "appropriate" means that whenever two edges are in the equivalence relation, all their adjacent nodes must be pairwise equivalent.

[4] We do not describe this construction here, but it is well-known that there are only finitely many such pushout complements and that they can be constructed effectively.

4 Example: Leader Election

As an example, we shall apply this technique to a typical leader election protocol, to verify its correctness. The rules for this leader election protocol are shown in Fig. 3. We start with a ring containing processes, each with a unique natural number as ID. These processes can generate messages containing their ID, which are forwarded whenever the ID of the message is smaller than the ID of the process which receives it. A process becomes the leader if it receives a message containing its own ID. Non-leader processes may also choose to leave the system at any time, connecting its predecessor and successor. We will prove that such a system can never create two leaders in the ring.

It can be seen that these rules satisfy the compatibility condition. The rule for edge contraction can be interpreted as a process leaving the system. Note that we do not need to add a rule for contracting messages (since messages are unary hyperedges), or for edge deletion in order to ensure compatibility.

All forbidden minors (which we computed manually) are shown in Fig. 4. We start with the first of these as the error state, and performing the backward analysis we obtain the rest of the forbidden minors. We consider natural numbers up to a certain bound, in order to keep the label and rule sets finite. Here, i, j or k as a label indicates "any number" (except where a constraint is indicated). Thus, the entire process has been fully parametrized, so that these forbidden minors are valid for a start graph with an arbitrarily large number of processes in the ring. Since the given start graph does not have any of these forbidden graphs as a minor, we can conclude that the leader election protocol is correct, i.e., it can never create two leaders.

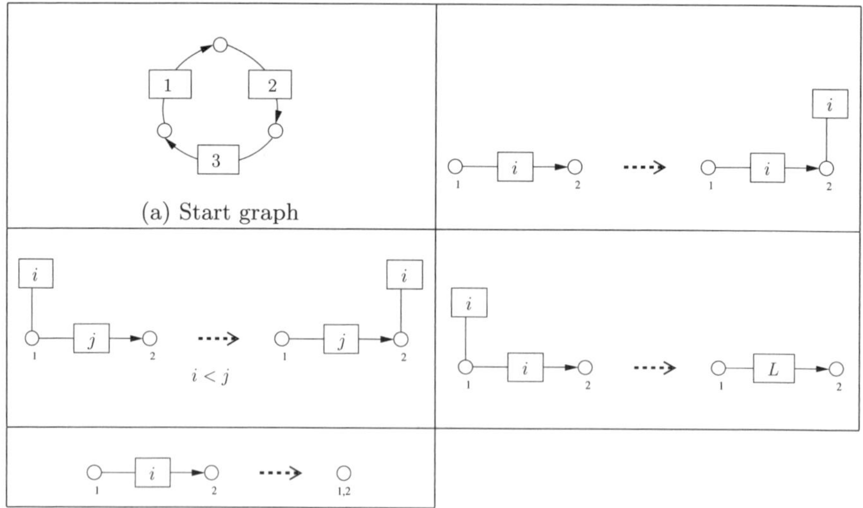

Fig. 3. Leader election (start graph and rewriting rules)

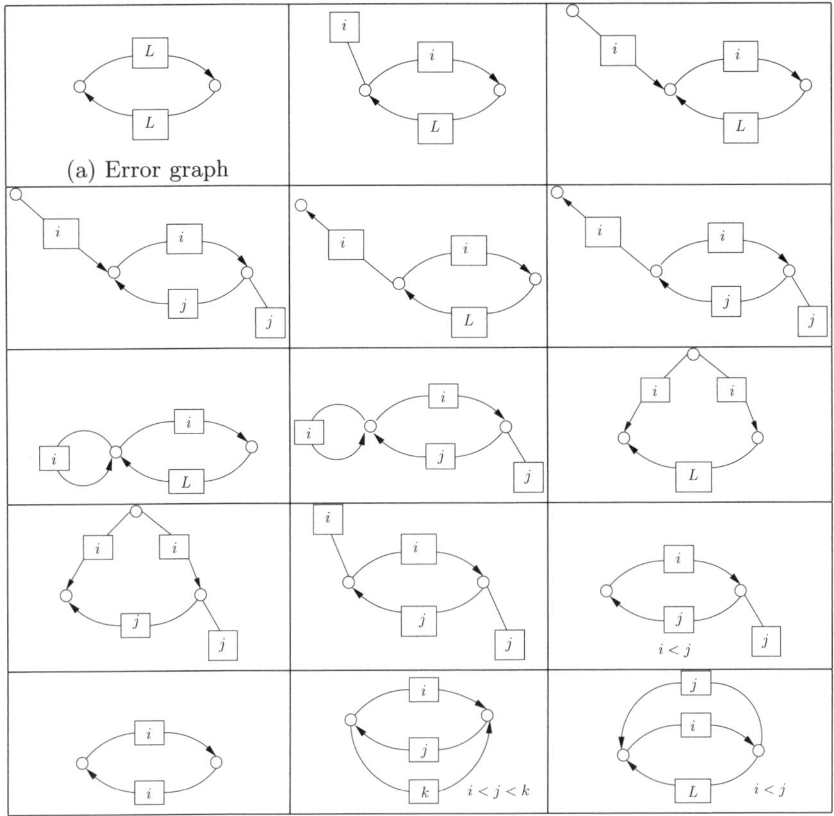

Fig. 4. Leader election (forbidden minors)

Note that since our technique can handle infinite state spaces, we could use the expressive power of graph transformation to extend the example in such a way that the ring is extended by new processes during runtime.

5 Conclusion

We have shown how to view subclasses of graph transformation systems as WSTS which gives us a decision algorithm for the covering problem. Currently we are working on an implementation which will help us to get a better insight into efficiency issues. Specifically it will help us to answer how many backward steps usually have to be taken and how many forbidden minors are generated. Although the worst case behaviour of this technique will certainly be bad, it might be feasible for many practical applications. We are also working on a more extended case study involving a termination detection protocol.

Another issue is the treatment of negative application conditions that have so far posed many problems in the analysis of GTSs. As already observed in [9] backward techniques seem to have fewer problems with negative application conditions than forward techniques which have so far mainly been studied. We also believe that such application conditions can be integrated with our technique.

Additional future work will be the investigation of partial order techniques (as in [1]) and the combination with (approximative) forward techniques (as described in [2,6]) in order to eliminate states which are not reachable from the start graph early on. In addition we work on a related technique which allows to show whether certain invariants (represented by forbidden minors) are preserved by graph transformation rules.

Acknowledgements. We would like to thank Javier Esparza for his suggestion to explore the relation between WSTS and graph transformation.

References

1. Abdulla, P.A., Jonsson, B., Kindahl, M., Peled, D.: A general approach to partial order reductions in symbolic verification. In: Y. Vardi, M. (ed.) CAV 1998. LNCS, vol. 1427, pp. 379–390. Springer, Heidelberg (1998)
2. Baldan, P., Corradini, A., König, B.: A static analysis technique for graph transformation systems. In: Larsen, K.G., Nielsen, M. (eds.) CONCUR 2001. LNCS, vol. 2154, pp. 381–395. Springer, Heidelberg (2001)
3. Ehrig, H., Heckel, R., Korff, M., Löwe, M., Ribeiro, L., Wagner, A., Corradini, A.: Algebraic approaches to graph transformation—part II: Single pushout approach and comparison with double pushout approach. In: Rozenberg, G. (ed.) Handbook of Graph Grammars and Computing by Graph Transformation, ch. 4, vol. 1: Foundations, World Scientific, Singapore (1997)
4. Finkel, A., Schnoebelen, P.: Well-structured transition systems everywhere! Theoretical Computer Science 256(1–2), 63–92 (2001)
5. Löwe, M.: Algebraic approach to single-pushout graph transformation. Theoretical Computer Science 109, 181–224 (1993)
6. Rensink, A., Distefano, D.: Abstract graph transformation. In: Proc. of SVV 2005 (3rd International Workshop on Software Verification and Validation). ENTCS, vol. 157.1, pp. 39–59 (2005)
7. Robertson, N., Seymour, P.: Graph minors. XX. Wagner's conjecture. Journal of Combinatorial Theory, Series B 92(2), 325–357 (2004)
8. Robertson, N., Seymour, P.: Graph minors. XXIII. Nash-Williams' immersion conjecture (2006) (submitted for publication), `http://www.math.princeton.edu/~pds/papers/GM23/GM23.pdf`
9. Saksena, M., Wibling, O., Jonsson, B.: Graph grammar modeling and verification of ad hoc routing protocols. In: Proc. of TACAS 2008. LNCS, vol. 4963, pp. 18–32. Springer, Heidelberg (2008)

Conflict-Tolerant Features*

Deepak D'Souza and Madhu Gopinathan

Indian Institute of Science,
Bangalore, India
{deepakd,gmadhu}@csa.iisc.ernet.in

Abstract. We consider systems composed of a base system with multiple "features" or "controllers", each of which independently advise the system on how to react to input events so as to conform to their individual specifications. We propose a methodology for developing such systems in a way that guarantees the "maximal" use of each feature. The methodology is based on the notion of "conflict-tolerant" features that are designed to continue offering advice even when their advice has been overridden in the past. We give a simple priority-based composition scheme for such features, which ensures that each feature is maximally utilized. We also provide a formal framework for specifying, verifying, and synthesizing such features. In particular we obtain a compositional technique for verifying systems developed in this framework.

1 Introduction

In this paper we consider systems that are composed of a base system along with multiple "features" or "controllers," each of which are meant to advise the system on how to adhere to their individual feature specifications. Such system models are common in software intensive domains such as telecommunications and automotive electronic control. One of the problems faced in integrating various features in such systems is that the system may reach a point of "conflict" between two (or more) features, where the features do not agree on a common action for the system to perform [1,2]. Such conflicts can be resolved by redesigning one of the features to satisfy a relaxed specification. However redesigning existing features is often not feasible in practice [3]. Redesign would also defeat the purpose of software product lines [4], which aim to improve software reuse by composing features to derive multiple products from a family of features. Another resolution technique is to suspend the feature with lower priority, and continue with the advice of the higher priority feature. However the issue now is how and when to "resume" the suspended feature so as to maximize its use.

In this paper we propose a formal framework for developing such systems in a way that overcomes some of these problems. The framework is based on a notion of "conflict-tolerance", which simply requires features to be "resilient" or "tolerant" with regard to violations of their advice due to conflicts with other features.

* This research was partially supported by a grant from India Science Lab, General Motors Corporation.

A. Gupta and S. Malik (Eds.): CAV 2008, LNCS 5123, pp. 227–239, 2008.

Thus, unlike a classical feature, a conflict-tolerant feature observes that its advice has been overridden, takes into account the violating event, and proceeds to offer advice for *subsequent* behavior of the system. Our methodology includes a way of specifying features, as well as a compositional verification technique for checking whether a feature implementation satisfies its specification.

The starting point in this framework is the notion of a *conflict-tolerant specification* of a feature. A classical specification (which we will assume in this paper to be safety specifications of linear-time behavior) can be viewed as a prefix-closed language of finite words containing all the system behaviors which are considered "safe". This can be pictured as a safety "cone" in the tree representing all possible behaviors, as shown in Fig. 1 (a). A conflict-tolerant specification on the other hand can be viewed as an "advice function" that specifies for *each* behavior w of the base system, a safety cone comprising all future behaviors that are considered safe, *after* the system has exhibited behavior w (Fig. 1 (b)).

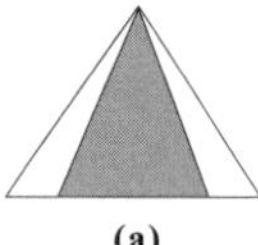

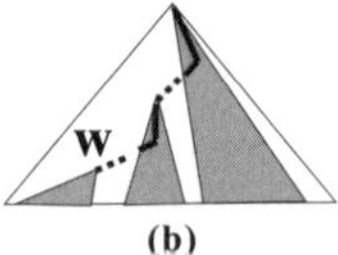

Fig. 1. The conflict-tolerant specification on the right advises on how to extend w even though its advice has been overridden (dashed line) in the past when generating w

To illustrate how a conflict-tolerant specification can capture a specifier's intent more richly than a classical specification, consider a feature that is required to release (by the event "**rel**") a single unit of medication in response to a "*timer*" event from the environment. A classical specification for this feature may be given by the transition system shown in Fig. 2 (a). The transition systems shown in Fig. 2 (b) and (c) denote conflict-tolerant specifications that both induce the same classical specification shown in (a). The dashed transitions are to be read as "not-advised", and their role is to keep track of events that violate the advice at a given state. Thus in specification (b), after an initial *timer* event, the advised action is **rel**; however, if the event **no-rel** occurs against its advice, the specification moves to the lightly shaded copy where on receiving another *timer* event it advises the action **rel-double**. The second specification thus attempts to maintain a unit average in every window of two cycles (it may, for example, be releasing oxygen), while the third specification does not attempt to do this (it may be releasing doses of insulin).

A conflict-tolerant feature implementation can be viewed as a transition system with transitions annotated as "advised" and "not-advised", similar to the conflict-tolerant specifications described above. A feature implementation is now said to satisfy a conflict-tolerant specification (with respect to a given base system), if after every possible behavior w of the base system, the behaviors of the base system that are according to the advice of the feature implementation, are all contained in the safety cone prescribed by the specification for w.

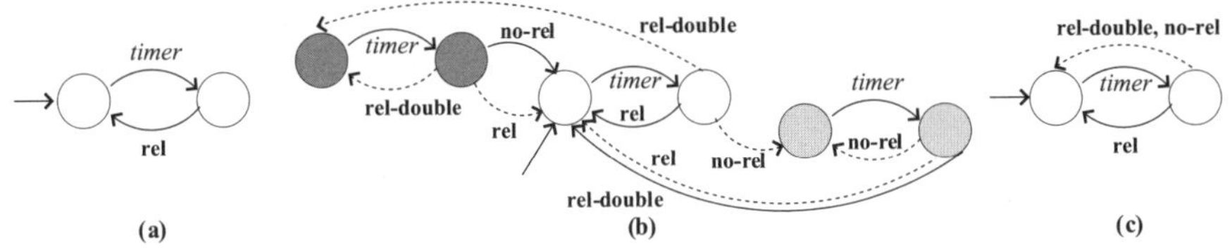

Fig. 2. The specification (a) is a classical specification whereas (b) and (c) are conflict-tolerant specifications. If "release double" (event **rel-double**) occurs against the advice of specification (b), it moves to the darkly shaded copy and advises "no release" (event **no-rel**) in the next cycle; if **no-rel** occurs against its advice, it moves to the lightly shaded copy and advises **rel-double** in the next cycle.

We give decision procedures to solve the natural synthesis and verification problems in this setting. In particular, we give a procedure to check whether a finite-state implementation satisfies a given finite-state conflict-tolerant specification, with respect to a given base system.

An important aspect of our framework is the fact that conflict-tolerant features admit a simple and effective composition scheme based on a prioritization of the features being composed, which can also be viewed as a conflict resolution technique. The composition scheme ensures that the resulting system always satisfies the specification of the highest priority feature. Additionally, it follows the advice of all other features F, *except* at points where each action in the advice of F conflicts with the advice of a higher priority feature. It is in this sense that each feature is "maximally" utilized. Together with our verification procedure, this gives us a compositional way of verifying the composed system, since once the individual features have each been verified to conform to their specifications, the composed system is guaranteed to be correct (in the sense above) "by construction."

Related Work. In [5], it is argued that system verification must be decomposed by features as every feature naturally has an associated property to be verified. There are several approaches in the literature where features are specified as state machines and a conflict is detected by checking whether a state, in which the features advise conflicting system actions, is reached. For a survey, see [1]. The problem of conflict detection is addressed in [6], where features are specified using temporal logic and conflict is detected automatically at the specification stage.

We now focus on previous work addressing conflict resolution. Our approach of viewing features as discrete event controllers [7] follows that of [8,9]. In both these works, the main issue addressed is that of resuming the advice of a controller once it has been overridden due to conflict with a higher priority controller. In [8] (see also [10]), when the lower priority controller (say $\mathcal{C}_2$) is suspended, the behavior of the base system is masked from $\mathcal{C}_2$. The resolution mechanism resumes $\mathcal{C}_2$ when it determines that the base system has reached a state (based on language equivalence) from which it is safe to accept the advice of $\mathcal{C}_2$. The drawback of this scheme is that the base system may never reach a state from which the advice of $\mathcal{C}_2$ can be accepted, and even if such a state was reached eventually, the utility of the controller is lost during the period of suspension.

The work of [9] is closer to ours, in that the specifications are designed to anticipate conflict by having two kinds of states: *in-spec* and *out-of-spec*. When a controller's specification is violated, it transitions to an out-of-spec state from where it passively observes the system behavior, till it sees a specified event that brings it back to an in-spec state. Thus, unlike our controllers, these controllers are designed to work with only certain anticipated conflicts, and moreover do not offer any useful advice in out-of-spec states.

In [11], a rule-based feature model and composition operators for resolving conflicts based on prioritization is presented. Their work is closest to ours in that it implicitly contains the notion of conflict-tolerance with a similar resolution mechanism. However the notion of a conflict-tolerant specification (as against the feature implementation itself) is absent in their work, while it is central in ours. In the absence of a specification one cannot address the important problems of verification and synthesis. With respect to their notion of "weak" (in a sense conflict-tolerant) invariants of features, the feature model is not obliged to offer advice on how to restore the invariant in case of violations.

The rest of the paper is structured as follows: After preliminary definitions, in Sect. 3 we view features as controllers and illustrate conflict between features. We then introduce the notion of conflict-tolerance in Sect. 4 and address the synthesis and verification problems in Sect. 5. Finally in Sect. 6, we describe our composition scheme and provide a precise formulation of the claim that the controllers are maximally utilized.

2 Preliminaries

Let Σ be a finite alphabet of events and let Σ^* denote the set of finite words over Σ. We denote the empty word by ϵ. A *language* over Σ is a subset of Σ^*. We write $v \cdot w$ (or simply vw) to denote the concatenation of two words v and w. Let L be a language over Σ and let v be a word over Σ. We define the set of *extensions* of v in L to be $ext_v(L) = \{w \in \Sigma^* \mid v \cdot w \in L\}$.

A *transition system* $\mathcal{T}$ over Σ is a tuple $(Q, s, \rightarrow)$, where Q is a set of states, $s \in Q$ is the start state, and $\rightarrow \subseteq Q \times \Sigma \times Q$ is a Σ-labeled transition relation. A *run* of $\mathcal{T}$ on a word $w = a_0 a_1 \cdots a_n$ starting from a state q, is a sequence $q_0, q_1, \ldots, q_{n+1}$ of states in Q such that $q_0 = q$, and for each $i \in \{0, \ldots, n\}$, we have $q_i \xrightarrow{a_i} q_{i+1}$. The language generated by $\mathcal{T}$, denoted $L(\mathcal{T})$, is the set of all words w on which $\mathcal{T}$ has a run starting from s. The language generated by $\mathcal{T}$ starting from a state $q \in Q$, denoted by $L_q(\mathcal{T})$, is the set of all words on which $\mathcal{T}$ has a run starting from q. We say the transition system $\mathcal{T}$ is *complete* (respectively *deterministic*) if for each $q \in Q$ and $a \in \Sigma$, there exists (respectively, exists at most one) q' such that $q \xrightarrow{a} q'$. For a deterministic transition system $\mathcal{T}$ and a word w on which $\mathcal{T}$ has a run, let q be the unique state reached in $\mathcal{T}$ after generating w. Then, we define $L_w(\mathcal{T}) = L_q(\mathcal{T})$.

Finally, for transition systems $\mathcal{T}_1 = (Q_1, s_1, \rightarrow_1)$ and $\mathcal{T}_2 = (Q_2, s_2, \rightarrow_2)$ over Σ, we define the *synchronized product* of $\mathcal{T}_1$ and $\mathcal{T}_2$, denoted $\mathcal{T}_1 \| \mathcal{T}_2$, to be the transition system $(Q_1 \times Q_2, (s_1, s_2), \rightarrow)$ over Σ, where $(p_1, p_2) \xrightarrow{a} (q_1, q_2)$ iff $p_1 \xrightarrow{a}_1 q_1$ and $p_2 \xrightarrow{a}_2 q_2$.

3 Features as Controllers

In this section we elaborate on the view of features as modular discrete event controllers [7], as proposed in [9,8].

In this paper, we focus on "safety" specifications. A *safety* specification over an alphabet Σ is a prefix-closed language over Σ. A safety specification can also be viewed as an "advice function" as defined below. This view will be useful when we introduce the notion of conflict tolerance in Sect. 4.

Definition 1 (Advice Function). *An advice function over Σ is a function $f : \Sigma^* \rightarrow 2^{\Sigma^*}$ such that $f(\epsilon)$ is prefix-closed language, and is* ***consistent*** *in the sense that for all $vw \in f(\epsilon)$ we have $f(vw) = ext_w(f(v))$.*

A safety specification L over Σ induces an advice function f_L given by $f_L(v) = ext_v(L)$. Conversely, an advice function f induces a safety language L_f given by $L_f = f(\epsilon)$. An advice function f induces in a natural way an *immediate* advice function $f^i : \Sigma^* \rightarrow \Sigma$ given by

$$f^i(v) = \{a \in \Sigma \mid \exists w \in \Sigma^* : aw \in f(v)\}.$$

We say a finite word w is *according to* an immediate advice f^i if for each prefix va of w, we have $a \in f^i(v)$. A deterministic transition system $\mathcal{T}$ over Σ induces an advice function $f_{\mathcal{T}}$ given by $f_{\mathcal{T}}(w) = L_w(\mathcal{T})$ for all $w \in L(\mathcal{T})$, and ϵ otherwise. We will say a safety specification L over Σ is *regular* if it is given by a deterministic finite-state transition system $\mathcal{S}$ over Σ.

We now define the notion of a base system. Let Σ be an alphabet which is partitioned into "environment events" Σ_e and "system events" Σ_s. We model systems over Σ by viewing their executions as a repeated cycle in which an environment event is sampled and a system event is performed in response to it. For simplicity we assume that exactly one environment action is sampled in each cycle.

Definition 2 (Base System). *A base system (or plant) over Σ is a deterministic finite-state transition system $\mathcal{B}$ over Σ, which is*

- ***alternating*** *in that $L(\mathcal{B}) \subseteq (\Sigma_e \cdot \Sigma_s)^* \cup ((\Sigma_e \cdot \Sigma_s)^* \cdot \Sigma_e)$.*
- ***non-blocking*** *in that whenever $w \in L(\mathcal{B})$ there exists $c \in \Sigma$ such that $wc \in L(\mathcal{B})$.*

Definition 3 (Controller). *Let $\mathcal{B}$ be a base system over Σ. A controller (or feature implementation) for $\mathcal{B}$ is a deterministic transition system over Σ. A controller $\mathcal{C}$ for $\mathcal{B}$ is valid if*

- $\mathcal{C}$ *is* ***non-restricting:*** *If* $w \in L(\mathcal{B}\|\mathcal{C})$ *and* $w \cdot e \in L(\mathcal{B})$ *for some environment event* $e \in \Sigma_e$, *then* $w \cdot e \in L(\mathcal{C})$. *Thus the controller must not restrict any environment event* e *enabled in the base system after any controlled behavior* w.
- $\mathcal{C}$ *is* ***non-blocking:*** *If* $w \in L(\mathcal{B}\|\mathcal{C})$, *then* $wc \in L(\mathcal{B}\|\mathcal{C})$ *for some* $c \in \Sigma$. *Thus the controller must not block the base system after any controlled behavior* w.

We carry over the notions of advice function $f_{\mathcal{C}}$ and corresponding immediate advice function $f^i_{\mathcal{C}}$ for a controller $\mathcal{C}$.

Let $\mathcal{B}$ be a base system and L a safety specification over Σ. We say a controller $\mathcal{C}$ for $\mathcal{B}$ *satisfies* L if $L(\mathcal{B}\|\mathcal{C}) \subseteq L$.

As a running example, we consider a lift system (in a building of only two floors – floor 0 and floor 1) and two of its features, adapted from a case study in [12]. Figure 3 shows the base system model. We denote the environment events in italics and system events in bold. In order to avoid clutter, we use a Statechart-like notation. An arrow from a box (dashed rectangle) to another box represents arrows between states with the same label in the two boxes. For example, the arrow labeled **open** represents two arrows: one between states labeled 1, and another between states labeled 0 in the boxes C and O. We use the convention that self-loops on states are labeled with all events in Σ, excluding those on which there is an outgoing transition from the state, and those events a for which the self-loop has a label $\sim a$. Thus the state 0 in box C has a transition to itself on the system event $\mathbf{nop_s}$ and on all environment events except *ttfull* and *ttnotfull*. The initial state is shown by an incoming arrow. To keep the figure simple we have not shown the base system as generating alternating environment and system events. However, we consider only the alternating behaviors.

The event fpress_i occurs when a user presses the button on floor i and the event cpress_i occurs when a user presses the car button for floor i inside the lift. The event *ttfull* indicates that the lift is two-thirds full and the event *ttnotfull* indicates that the lift is not two-thirds full. The events nop_e and $\mathbf{nop_s}$ respectively denote that the environment and the system has not performed any action.

The base system is typically run with several controllers including a "vanilla" controller which would keep track of the current direction of travel and require the base system to service all pending requests from floors and from within the lift along that direction before changing direction. However, we will focus on a controller for "executive floor" feature which requires that the requests from the "executive floor" (say, floor 1) must be serviced before other floor requests. Figure 4 shows a possible specification $\mathcal{S}_E$ for this requirement. The transition system simply keeps track of the current floor in its state. When it receives a fpress_1 event, it transitions to the same state in the box P indicating that an executive floor request is pending. It prohibits **open** on floor 0 and **down** on floor 1 so that the executive floor request is serviced before other floors. We take $\mathcal{S}_E$, with the additional constraint that $\mathbf{nop_s}$ is not allowed from the states in box P, as a valid controller $\mathcal{C}_E$ for the base system $\mathcal{B}$.

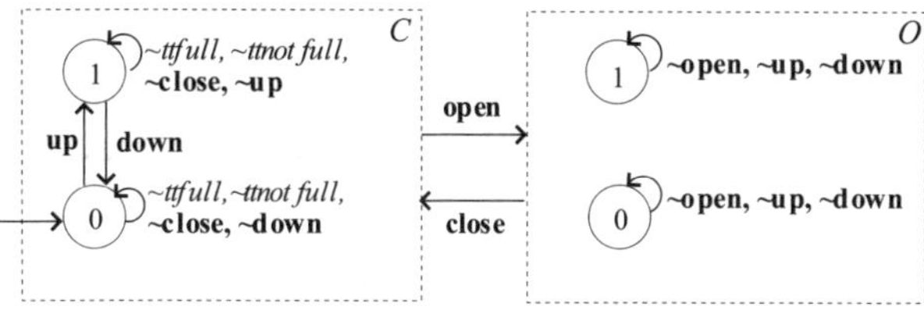

Fig. 3. Lift base system $\mathcal{B}$. The set Σ_e is $\{fpress_i, cpress_i, ttfull, ttnotfull, nop_e\}$. The events *ttfull* and *ttnotfull* can occur only when the lift door is open (in box O). The set Σ_s is $\{\mathbf{open}, \mathbf{close}, \mathbf{up}, \mathbf{down}, \mathbf{nop_s}\}$. The lift can move up or down only when its door is closed.

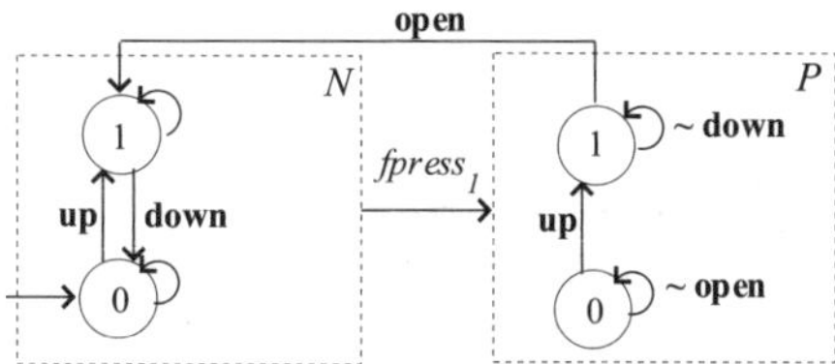

Fig. 4. Executive Floor Specification $\mathcal{S}_E$. Controller $\mathcal{C}_E$ is the same as $\mathcal{S}_E$ except that it does not advise $\mathbf{nop_s}$ from the states in box P.

Consider adding a feature called "two-thirds-full" which requires that the requests from within the lift should be serviced before requests from floors when the lift is two-thirds full. Thus, when the system receives a *ttfull* event, the lift should not service requests from floors as long as there are pending car requests. When the *ttnotfull* event occurs, the lift can go back to its normal functioning. Figure 5 shows a possible specification $\mathcal{S}_T$ for this feature. We take $\mathcal{S}_T$, with the additional constraint that $\mathbf{nop_s}$ is not allowed when the lift is two-thirds full and a car request is pending, as a valid controller $\mathcal{C}_T$ for $\mathcal{B}$. Note that a controller could impose additional constraints such as choosing **open** over *up* at floor 0 when car requests from both floors are pending.

We now illustrate the notion of conflict between controllers.

Definition 4 (Conflict). *Let $\mathcal{C}_1$ and $\mathcal{C}_2$ be valid controllers for a base system $\mathcal{B}$. The controllers $\mathcal{C}_1$ and $\mathcal{C}_2$ are in* conflict *with respect to $\mathcal{B}$, if there exists a behavior w in $L(\mathcal{B}\|\mathcal{C}_1\|\mathcal{C}_2)$ such that $ext_w(L(\mathcal{B}\|\mathcal{C}_1\|\mathcal{C}_2))$ is empty. In other words, there exists a behavior $w \in L(\mathcal{B})$ which is according to both $\mathcal{C}_1$ and $\mathcal{C}_2$, but $f^i_{\mathcal{B}}(w) \cap f^i_{\mathcal{C}_1}(w) \cap f^i_{\mathcal{C}_2}(w) = \emptyset$.*

Thus $\mathcal{C}_1$ and $\mathcal{C}_2$ are in conflict with respect to $\mathcal{B}$ if $\mathcal{C}_1\|\mathcal{C}_2$ is blocking with respect to $\mathcal{B}$. Consider the behavior

$$fpress_1 \cdot \mathbf{up} \cdot nop_e \cdot \mathbf{open} \cdot ttfull \cdot \mathbf{close} \cdot cpress_0 \cdot \mathbf{down} \cdot fpress_1$$

from the initial state of $\mathcal{B}$. The system events allowed by $\mathcal{B}$ are **up**, **open** and $\mathbf{nop_s}$. However, the controlled system is blocked as $\mathcal{C}_E$ does not advise **open** and $\mathbf{nop_s}$ while $\mathcal{C}_T$ does not advise **up** and $\mathbf{nop_s}$.

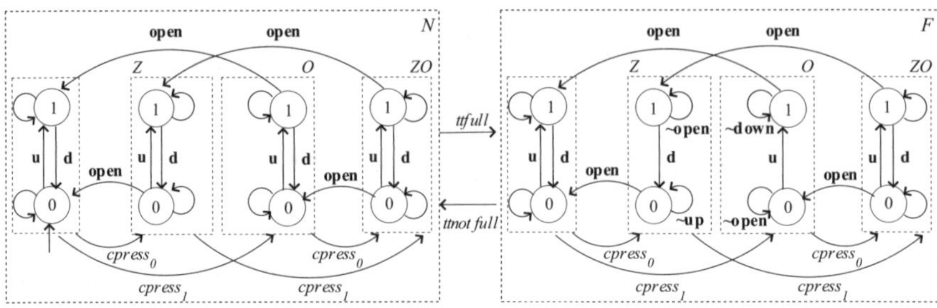

Fig. 5. Two-Thirds Full Specification $\mathcal{S}_T$. In box $Z(O)$, car request for floor 0 (respectively 1) is pending and in box ZO, car requests for both floors are pending. Controller $\mathcal{C}_T$ is the same as $\mathcal{S}_T$ except that it does not advise $\mathbf{nop_s}$ when the lift is two-thirds full (box F) and a car request is pending.

4 Conflict-Tolerant Controllers

In this section we introduce our notion of conflict-tolerance. Analogous to the notion of specification as an advice function given in Sect. 3, a *conflict-tolerant* safety specification over an alphabet Σ is a *conflict-tolerant* advice function in the following sense:

Definition 5 (Conflict-Tolerant Advice Function). *A conflict-tolerant advice function over an alphabet Σ is a function $f : \Sigma^* \rightarrow 2^{\Sigma^*}$ which assigns a prefix-closed language $f(v)$ to every finite word $v \in \Sigma^*$, and is consistent in the sense that for* **all** *$vw \in \Sigma^*$ with $w \in f(v)$, we have $f(vw) = ext_w(f(v))$.*

A *conflict-tolerant* transition system (or CTTS) over Σ is a tuple $\mathcal{T}' = (\mathcal{T}, N)$, where $\mathcal{T} = (Q, s, \rightarrow)$ is a deterministic transition system over Σ and $N \subseteq \rightarrow$ is a subset of transitions designated as "not-advised." The language generated by $\mathcal{T}'$ starting from a state q in Q, denoted $L_q(\mathcal{T}')$, is defined to be simply $L_q(\mathcal{T})$. The *constrained* language generated by $\mathcal{T}'$, denoted $L^c_q(\mathcal{T}')$, is defined to be $L_q(\widehat{\mathcal{T}})$ where $\widehat{\mathcal{T}}$ is the transition system obtained from $\mathcal{T}$ by deleting all not-advised transitions (i.e. transitions in N). Let $w \in L(\mathcal{T})$, and let q be the unique state reached by $\mathcal{T}$ on w. Then by $L^c_w(\mathcal{T}')$ we mean $L^c_q(\mathcal{T}')$. We say $\mathcal{T}'$ is *complete* with respect to a language $L \subseteq \Sigma^*$ if $L \subseteq L_\epsilon(\mathcal{T}')$.

The CTTS $\mathcal{T}'$ induces a natural conflict-tolerant advice function $f_{\mathcal{T}'}$ given by, for all $w \in \Sigma^*$, $f(w) = L^c_w(\mathcal{T}')$. We say a conflict-tolerant advice function is *regular* if it is given by a finite-state CTTS over Σ.

We define the synchronized product of a transition system $\mathcal{T}_1$ and a CTTS $\mathcal{T}'_2 = (\mathcal{T}_2, N_2)$ to be the CTTS $(\mathcal{T}_1 \| \mathcal{T}_2, N'_2)$ which is complete with respect to $L(\mathcal{T}_1)$ and N'_2 is the set of joint transitions where the $\mathcal{T}_2$ transitions are not-advised (thus $\mathcal{T}_1 \| \mathcal{T}'_2$ inherits the not-advised transitions of $\mathcal{T}'_2$).

In the definitions below let $\mathcal{B}$ be a base system over a partitioned alphabet Σ.

Definition 6 (Conflict-Tolerant Controller). *A conflict-tolerant controller for $\mathcal{B}$ is a CTTS over Σ that is complete with respect to $L(\mathcal{B})$. The controller $\mathcal{C}'$ for $\mathcal{B}$ is* valid *if*

- $\mathcal{C}'$ *is* ***non-restricting****: If* $w \cdot e \in L(\mathcal{B})$ *for some environment event* $e \in \Sigma_e$*, then* $e \in L^c_w(\mathcal{C}')$ *(or equivalently* $e \in f^i_{\mathcal{C}'}(w)$*). Thus the controller must not restrict any environment event e enabled in the base system after* ***any*** *system behavior w.*
- $\mathcal{C}'$ *is* ***non-blocking****: If* $w \in L(\mathcal{B})$*, then* $L^c_w(\mathcal{B}\|\mathcal{C}') \neq \emptyset$ *(equivalently* $f^i_{\mathcal{C}'}(w) \cap f^i_{\mathcal{B}}(w) \neq \emptyset$*). Thus the controller must not block the system after* ***any*** *system behavior w.*

Definition 7 ($\mathcal{C}'$ satisfies f). *Let f be a conflict-tolerant specification over Σ. A conflict-tolerant controller $\mathcal{C}'$ for $\mathcal{B}$ satisfies f if for each $w \in L(\mathcal{B})$, $L^c_w(\mathcal{B}\|\mathcal{C}') \subseteq f(w)$. Thus after any system behavior w, if the base system follows the advice of $\mathcal{C}'$, the resulting behavior conforms to the safety language $f(w)$.*

We now illustrate these definitions with our running example. Figure 6 shows a conflict-tolerant specification $\mathcal{S}'_E$ for the "executive floor" feature. The not-advised transitions are shown using dashed transitions. By ignoring the dashed transitions, we get back the conventional specification in Fig. 4. The dashed transitions in a conflict-tolerant specification indicate the obligation on a controller when the specification is overridden to meet the requirements of a higher priority specification. In $\mathcal{S}'_E$, **open** from state 0 of box P and **down** from state 1 of box P are not advised. However, even if the door opens at floor 0 when a request from floor 1 is pending, $\mathcal{S}'_E$ requires the controller's subsequent advice to be such that the request from floor 1 is serviced. Figure 7 shows a conflict-tolerant specification for the "two-thirds-full" feature.

5 Synthesis and Verification

In this section we address the natural synthesis and verification problems for conflict-tolerant controllers. Let Σ be a partitioned alphabet.

Theorem 1 (Synthesis). *Given a base system $\mathcal{B}$ over Σ, and a regular conflict-tolerant specification $\mathcal{S}'$ over Σ, we can check if there exists a conflict-tolerant controller for $\mathcal{B}$ that satisfies $\mathcal{S}'$, and if so, synthesize a finite-state one.*

Proof. We claim that there exists a controller for $\mathcal{B}$ satisfying $\mathcal{S}'$ iff in the synchronized product $\mathcal{B}\|\mathcal{S}'$ there does not exist a state (b, q) which is reachable from the start state and satisfies one of the conditions

1. (b, q) is "restricting" in the sense that there is an environment event e enabled at b in $\mathcal{B}$, but is not advised at q in $\mathcal{S}'$.
2. (b, q) is "blocking" in that there is no event advised at (b, q) in $\mathcal{B}\|\mathcal{S}'$.

If no such (b, q) exists in $\mathcal{B}\|\mathcal{S}'$, then clearly $\mathcal{S}'$ itself is a valid finite-state controller for $\mathcal{B}$ that satisfies $\mathcal{S}'$. Conversely, if there is a valid controller $\mathcal{C}'$ for $\mathcal{B}$ satisfying $\mathcal{S}'$, then again it is easy to see that no such (b, q) must exist in $\mathcal{B}\|\mathcal{S}'$ (recall that the base system is always non-blocking). These conditions can be checked in time linear in the product of the sizes of $\mathcal{B}$ and $\mathcal{S}'$. □

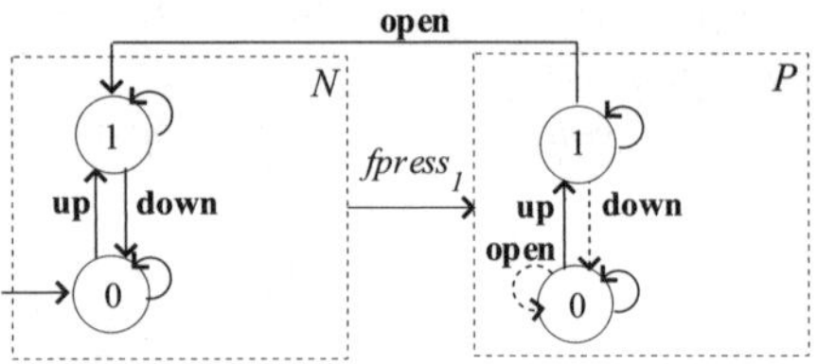

Fig. 6. Tolerant Executive Floor Specification $\mathcal{S}'_E$. Tolerant Controller $\mathcal{C}'_E$ is the same as $\mathcal{S}'_E$ except that $\mathbf{nop_s}$ is not advised from states in box P.

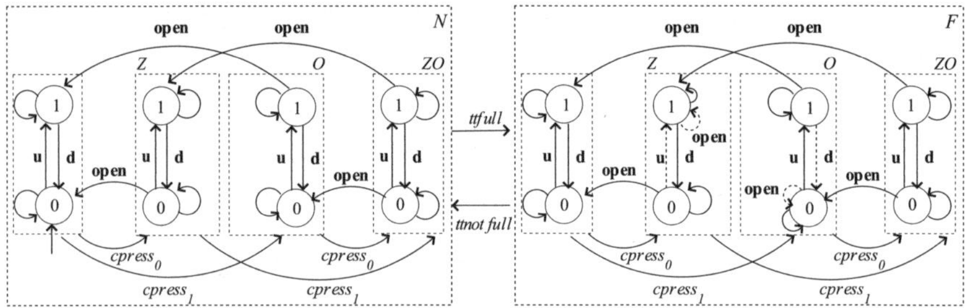

Fig. 7. Tolerant Two-Thirds Full Specification $\mathcal{S}'_T$. Tolerant Controller $\mathcal{C}'_T$ is the same as $\mathcal{S}'_T$ except that it does not advise $\mathbf{nop_s}$ when the lift is two-thirds full and a car request is pending.

Note that even if a state (b, q) as above exists, a classical controller [7] may still exist if it has a strategy to avoid reaching such a state.

Theorem 2 (Verification). *Given a base system $\mathcal{B}$ over Σ, a regular conflict-tolerant specification $\mathcal{S}'$, and a finite-state conflict-tolerant controller $\mathcal{C}'$, we can check whether $\mathcal{C}'$ is a valid conflict-tolerant controller for $\mathcal{B}$, that satisfies $\mathcal{S}'$.*

Proof. It is easy to see that a necessary and sufficient condition for $\mathcal{C}'$ to be a valid controller for $\mathcal{B}$ and satisfying $\mathcal{S}'$, is to check that in the synchronized product $\mathcal{B}||\mathcal{C}'||\mathcal{S}'$ there does *not* exist a state (b, p, q) which is reachable from the initial state and satisfies one of the following conditions:

1. ($\mathcal{C}'$ is restricting) there exists an event $e \in \Sigma_e$ enabled at b in $\mathcal{B}$, but is not advised at p in $\mathcal{C}'$.
2. ($\mathcal{C}'$ is blocking) there is no event c which is both enabled at b in $\mathcal{B}$ and advised at p in $\mathcal{C}'$.
3. ($\mathcal{C}'$ does not satisfy $\mathcal{S}'$) there is an event c which is both enabled at b in $\mathcal{B}$ and advised at p in $\mathcal{C}'$, but not advised at q in $\mathcal{S}'$.

This check can be carried out in time linear in the product of the sizes of $\mathcal{B}$, $\mathcal{C}'$, and $\mathcal{S}'$. □

6 Composition

We now give a way of composing conflict-tolerant controllers based on a prioritization of the controllers. The composition guarantees that the advice of each controller is used in a "best possible" way.

Let $\mathcal{B} = (B, r_0, \rightarrow)$ be a base system over an alphabet Σ. Let $\mathcal{C}'_1 = (Q_1, s_1, \rightarrow_1, N_1)$ and $\mathcal{C}'_2 = (Q_2, s_2, \rightarrow_2, N_2)$ be valid conflict-tolerant controllers for $\mathcal{B}$. Let P be a priority ordering between $\mathcal{C}'_1$ and $\mathcal{C}'_2$, and say P assigns a higher priority to $\mathcal{C}'_1$, denoted by $\mathcal{C}'_2 <_P \mathcal{C}'_1$. Then:

Definition 8 (Prioritized Composition). *The* P-prioritized composition *of the controllers $\mathcal{C}'_1$ and $\mathcal{C}'_2$ w.r.t. $\mathcal{B}$, is the conflict-tolerant transition system $\mathcal{C}'$, denoted by $\|_{P,\mathcal{B}}(\mathcal{C}'_1, \mathcal{C}'_2)$, and defined as*

$$\mathcal{C}' = (Q_1 \times Q_2 \times B, (s_1, s_2, r_0), \Rightarrow, N)$$

where $\Rightarrow$ is given by $(p_1, p_2, r) \stackrel{a}{\Rightarrow} (q_1, q_2, t)$ iff $p_1 \stackrel{a}{\rightarrow}_1 q_1$, $p_2 \stackrel{a}{\rightarrow}_2 q_2$, and $r \stackrel{a}{\rightarrow} t$; and the set of not-advised transitions N is defined as follows. With each transition $u = (p_1, p_2, r) \stackrel{a}{\Rightarrow} (q_1, q_2, t)$, we associate a bit-string (in this case of length 2) which denotes which of the controllers has advised this transition. Thus the associated bit-string for the transition above is $b_1 b_2$ where b_i is 1 iff $p_i \stackrel{a}{\rightarrow}_i q_i \notin N_i$. Let the "rank" of u be the number represented by this string in binary notation. Then the transition u is advised (i.e. not in N) iff there is no transition of higher rank going out of the state (p_1, p_2, r).

Figure 8 show how the ranks and "advised" status of transitions are calculated (assuming that the base system allows all the events shown). Thus from state (p, p') only the transition on d is advised, while all others are not advised.

Lemma 1. *The CTTS $\mathcal{C}' = \|_{P,\mathcal{B}}(\mathcal{C}'_1, \mathcal{C}'_2)$ defined above is a* valid *(i.e. non-restricting and non-blocking) conflict-tolerant controller for $\mathcal{B}$. In addition, the immediate advice function $f^i_{\mathcal{C}'}$ it induces is given as follows. For each $w \in L(\mathcal{B})$:*

$$f^i_{\mathcal{C}'}(w) = \begin{cases} f^i_{\mathcal{B}}(w) \cap f^i_{\mathcal{C}'_1}(w) \cap f^i_{\mathcal{C}'_2}(w) & \text{if } f^i_{\mathcal{B}}(w) \cap f^i_{\mathcal{C}'_1}(w) \cap f^i_{\mathcal{C}'_2}(w) \neq \emptyset \\ f^i_{\mathcal{B}}(w) \cap f^i_{\mathcal{C}'_1}(w) & \text{otherwise.} \end{cases}$$

□

We can now generalize this prioritized composition to any number of controllers. Let $\mathcal{C}'_1, \ldots, \mathcal{C}'_n$ be valid conflict-tolerant controllers for a base system $\mathcal{B}$. Let P be a priority that induces a total ordering $<_P$ on the controllers. Then the P-prioritized composition of $\mathcal{C}'_1, \ldots, \mathcal{C}'_n$ (wrt $\mathcal{B}$) is denoted $\|_{P,\mathcal{B}}(\mathcal{C}'_1, \ldots, \mathcal{C}'_n)$ and defined similarly as above. Let $\mathcal{S}'_1, \ldots, \mathcal{S}'_n$ be conflict-tolerant specifications, and suppose each $\mathcal{C}'_j$ individually satisfies the specification $\mathcal{S}'_j$ w.r.t. $\mathcal{B}$.

Theorem 3. *The conflict-tolerant transition system $\mathcal{C}' = \|_{P,\mathcal{B}}(\mathcal{C}'_1, \ldots, \mathcal{C}'_n)$ is a valid conflict-tolerant controller for $\mathcal{B}$. Further, $\mathcal{C}'$ satisfies each of the specifications $\mathcal{S}'_1, \ldots, \mathcal{S}'_n$ in the following "maximal" sense: Each $w \in L(\mathcal{B} \| \mathcal{C}')$ is always*

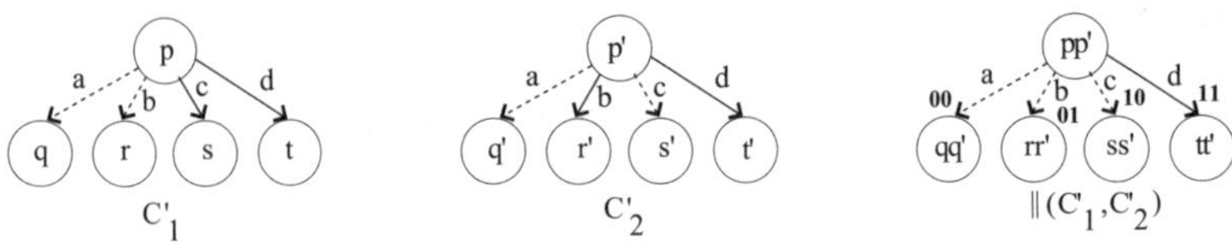

Fig. 8. Computation of ranks in the prioritized composition

according to the immediate advice of S'_j, except *at the points where* C'_j *is in conflict with the advice of higher-priority controllers, in that for each* a *in* $f^i_{C'_j}(w)$, *there is a controller* C'_k *such that* $C'_k >_P C'_j$ *and* $a \notin f^i_{C'_k}(w) \cap f^i_{\mathcal{B}}(w)$. *In particular, the highest priority specification* S'_1 *is always satisfied.* □

Consider the base system behavior

$$\mathit{fpress}_1 \cdot \mathbf{up} \cdot \mathit{nop}_e \cdot \mathbf{open} \cdot \mathit{ttfull} \cdot \mathbf{close} \cdot \mathit{cpress}_0 \cdot \mathbf{down} \cdot \mathit{fpress}_1$$

which we used to illustrate conflict in Sect. 3. With the priority order $C'_T > C'_E$, the conflict is resolved such that one possible extension is

$$\mathbf{open} \cdot \mathit{ttnotfull} \cdot \mathbf{close} \cdot \mathit{nop}_e \cdot \mathbf{up} \cdot \mathit{nop}_e \cdot \mathbf{open}$$

– i.e. the door is opened at floor 0 violating the advice of C'_E. As passengers go out at floor 0, the lift is not two-thirds full and the system immediately follows the advice of C'_E to service the executive floor. We emphasise that the same controllers can be composed with the priority $C'_E > C'_T$, to obtain a system in which conflicts will be resolved in favor of C'_E, while maximally utilizing the advice of C'_T.

7 Discussion

We note that conflict-tolerant specifications are somewhat stronger than classical specifications, and may not always admit a conflict-tolerant controller even when the induced classical specifications admit a classical controller. See [13] for an example. Nonetheless, whenever the conflict-tolerant specifications are realizable, our framework provides a flexible way of composing the controllers to obtain systems with guarantees on the usage of each controller.

We have considered extensions of this framework to include combinations of safety and liveness specifications [13], as well as real-time features [14] with similar results.

Our framework is also amenable to more flexible priority schemes like according priority dynamically based on history of events. Gößler and Sifakis [15] consider transition systems with priorities specified by predicates under which one action is prioritized over another. They provide conditions under which the predicates are consistent in that the prioritized system is non-blocking. Our composition scheme can be thought of as synthesizing consistent priority predicates from possibly inconsistent predicates obtained from individual controllers.

References

1. Keck, D.O., Kühn, P.J.: The feature and service interaction problem in telecommunications systems. a survey. IEEE Trans. Software Eng. 24(10), 779–796 (1998)
2. Hall, R.J.: Feature interactions in electronic mail. In: FIW, pp. 67–82 (2000)
3. Jackson, M., Zave, P.: Distributed feature composition: A virtual architecture for telecommunications services. IEEE Trans. Software Eng. 24(10), 831–847 (1998)
4. Software Engineering Institute: Software product lines, http://www.sei.cmu.edu/productlines
5. Fisler, K., Krishnamurthi, S.: Decomposing verification by features. In: IFIP Working Conference on Verified Software: Theories, Tools, Experiments (2006)
6. Felty, A.P., Namjoshi, K.S.: Feature specification and automated conflict detection. ACM Trans. Softw. Eng. Methodol. 12(1), 3–27 (2003)
7. Ramadge, P.J.G., Wonham, W.M.: The control of discrete event systems. Proc. of the IEEE 77, 81–98 (1989)
8. Wong, K.C., Thistle, J.G., Hoang, H.H., Malhamé, R.P.: Supervisory control of distributed systems: Conflict resolution. In: Conf. on Decision and Control, pp. 416–421. IEEE, Los Alamitos (1995)
9. Chen, Y.L., Lafortune, S., Lin, F.: Modular supervisory control with priorities for discrete event systems. In: Conf. on Decision and Control, pp. 409–415. IEEE Computer Society Press, Los Alamitos (1995)
10. Wong, K.C., Thistle, J.G., Hoang, H.H., Malhamé, R.P.: Supervisory control of distributed systems: Conflict resolution. In: Conf. on Decision and Control, pp. 3275–3280. IEEE, Los Alamitos (1998)
11. Hay, J.D., Atlee, J.M.: Composing features and resolving interactions. In: SIGSOFT Found. of Softw. Engg., pp. 110–119 (2000)
12. Plath, M., Ryan, M.: Feature integration using a feature construct. Sci. Comput. Program 41(1), 53–84 (2001)
13. D'Souza, D., Gopinathan, M.: Conflict-tolerant features. Technical Report IISc-CSA-TR-2007-11, Computer Science and Automation, Indian Institute of Science, India (2007), http://archive.csa.iisc.ernet.in/TR/2007/11/
14. D'Souza, D., Gopinathan, M., Ramesh, S., Sampath, P.: Conflict-detection and resolution for real-time features (manuscript in preparation)
15. Gößler, G., Sifakis, J.: Priority Systems. In: de Boer, F.S., Bonsangue, M.M., Graf, S., de Roever, W.-P. (eds.) FMCO 2003. LNCS, vol. 3188, pp. 314–329. Springer, Heidelberg (2004)

Ranking Automata and Games for Prioritized Requirements

Rajeev Alur, Aditya Kanade, and Gera Weiss

University of Pennsylvania

Abstract. Requirements of reactive systems are usually specified by classifying system executions as desirable and undesirable. To specify *prioritized* requirements, we propose to associate a rank with each execution. This leads to optimization analogs of verification and synthesis problems in which we compute the "best" requirement that can be satisfied or enforced from a given state. The classical definitions of acceptance criteria for automata can be generalized to ranking conditions. In particular, given a mapping of states to colors, the *Büchi ranking* condition maps an execution to the highest color visited infinitely often by the execution, and the *cyclic ranking* condition with cycle k maps an execution to the modulo-k value of the highest color repeating infinitely often. The well-studied parity acceptance condition is a special case of cyclic ranking with cycle 2, and we show that the cyclic ranking condition can specify all ω-regular ranking functions. We show that the classical characterizations of acceptance conditions by fixpoints over sets generalize to characterizations of ranking conditions by fixpoints over an appropriately chosen lattice of coloring functions. This immediately leads to symbolic algorithms for solving verification and synthesis problems. Furthermore, the precise complexity of a decision problem for ranking conditions is no more than the corresponding acceptance version, and in particular, we show how to solve Büchi ranking games in quadratic time.

1 Introduction

A requirement φ of a reactive system M can be formally described as a set L_φ of finite or infinite words over system states (or observations) [14]. Verification of the system M with respect to the requirement φ corresponds to checking whether there exists an execution of M that does not belong to L_φ. When the system has only finitely many states, and the requirement can be captured as an ω-regular language, the verification problem can be solved algorithmically using decision procedures for ω-automata [18]. When the choices within the system M are partitioned into controllable and uncontrollable, then synthesis with respect to the requirement φ corresponds to checking whether there exists a strategy to resolve the controllable choices to ensure that the resulting execution belongs to L_φ. For the finite-state case, the synthesis question can be solved by algorithms for solving games with ω-regular winning conditions [15,17].

In this paper, we propose a framework for specification, verification, and synthesis of *prioritized* requirements. Given a sequence $\varphi_0, \ldots \varphi_k$ of requirements,

A. Gupta and S. Malik (Eds.): CAV 2008, LNCS 5123, pp. 240–253, 2008.

listed in increasing order of importance, we want to find the *best* requirement that can be satisfied. For example, in Section 2, we describe a scheduling case study which employs several identical buffers to ensure uninterrupted flow of input messages, and using large buffers is expensive. We can specify prioritized requirements $\varphi_0, \ldots \varphi_k$, where φ_i states that the maximal buffer size required infinitely often during the execution is i or less. There are other potential scenarios where words can be naturally associated with ranks. For example, given multiple requirements, the priority of a word can be the number of requirements it satisfies. For software verification, we can associate different costs with different types of bugs, and have the analysis tool compute a classification of program statements grouped according to the worst bug that can manifest from them.

The optimization questions concerning prioritized requirements $\varphi_0, \ldots \varphi_k$ can, of course, be answered by solving k verification (or synthesis) questions separately, one for each requirement φ_i, using known techniques. However, as we establish in this paper, there is a better way of formulating and solving such questions. We formalize a prioritized requirement as a *ranking function* r that maps each word to a rank in an ordered set. We will focus only on *ω-regular ranking functions*, namely, functions that use only finitely many ranks such that the set of words with the same rank is an ω-regular language. For the case of two ranks, these notions coincide with the classical definitions of ω-regular sets.

Automata with different types of acceptance conditions (such as final-state, Büchi, Rabin, parity) are commonly used to specify ω-regular sets. To generalize acceptance to ranking, we consider deterministic automata in which each state is assigned a color. Then, given a run ρ of the automaton over a word w, the *reachability ranking* condition maps w to the highest color appearing in ρ. It is well-known that the set of states from which a target set can be reached can be computed as a least-fixpoint computation over sets of states starting with the target set. We show that this computation can be naturally generalized to fixpoints over functions that assign a color to each state (coloring functions). If the number of states is n and the number of colors (which captures the number of disjoint prioritized reachability requirements) is k, then, even though the number of iterations of the fixpoint computation is nk, we show that existing algorithm for solving reachability games (see [5]) can be adopted to solve the reachability ranking games in linear time.

Given a mapping of states to colors, the *Büchi ranking* condition maps a word to the highest color that repeats infinitely often in the corresponding run. The classical nested fixpoint characterization of Büchi acceptance [6] can now be generalized to an analogous fixpoint over coloring functions, and this allows us to compute the value of the corresponding verification/synthesis question at every state. We show that the number of iterations of the outer greatest-fixpoint loop is *independent* of the number of ranks. This gives us a quadratic-time algorithm for solving games with respect to Büchi ranking conditions.

Our final set of results concerns generalizing *parity* acceptance to *cyclic ranking* condition. Given an assignment of colors to states, the cyclic ranking condition with cycle k maps a word to the *modulo-k* value of the highest color that repeats infinitely often in the corresponding run. The well-studied parity acceptance is a cyclic ranking condition with cycle 2. It is known that parity acceptance can specify all ω-regular sets. We prove an analogous result: every ω-regular ranking condition can be captured by a cyclic ranking condition. We also show that the winning strategy in games with cyclic ranking condition is memoryless, and the corresponding decision problem is in NP $\cap$ coNP.

Related Work. In scheduling literature, priorities are usually assigned with tasks, and are used to make local decisions [3]. Our definition allows associating priorities with global executions, and can potentially be used to capture high-level quality-of-service goals. In verification literature, various quantitative generalizations of verification and synthesis questions have been studied, typically involving real-time and/or probabilities, and are orthogonal to our notion of rankings. Parametric temporal logic allows capturing some versions of optimization problems, but the corresponding model checking problem has very different technical flavor [1]. Lattice automata [12,13] generalize the notions of initialization, transitions, and acceptance from the Boolean case to a lattice, and can be used to associate ranks with words. However, it does not consider the problem of computing the optimum values of games, central to our motivation. While the decision procedures studied in [12,13] can be used for optimization, we propose a faster algorithm. Similar to our result for reachability ranking games, [4] proposes a linear-time algorithm for weak parity games. Finally, the games literature considers models with costs associated with each state or transition. The most widely studied ranking function for runs is the *mean-payoff* cost [20], which is not ω-regular, and does not capture prioritization of requirements. The work in [9] considers real valued ranking functions for stochastic games.

2 A Motivating Example

Our example is based on a case study in which scheduling policies are determined for a signal processing board [8,19]. We show that the problem of synthesizing an optimal scheduler is naturally modeled as a ranking question.

A block diagram of the signal processing board is depicted in Fig. 1. The board processes the two input streams shown at the top left and produces the two output streams shown at the bottom left. As depicted in the figure, data is stored in memory and is brought back when needed. The memory can only be accessed via a shared bus that transports data in quantities of 128 bits. Nine identical buffers are designed to allow uninterrupted flow of data when the bus is not available. The main objective of the case study is to analyze and to design an arbiter that schedules the use of the shared bus. A valid schedule must guarantee that, after a finite initialization phase, none of the data streams are interrupted. The optimal schedule is the one which minimizes the buffer size (the number of bits per buffer).

A schedule for bus arbitration is represented by a word over the alphabet $\Sigma = \{0, 1, \ldots, 9\}$. If the ith letter of the word is B, the bus is used to transfer

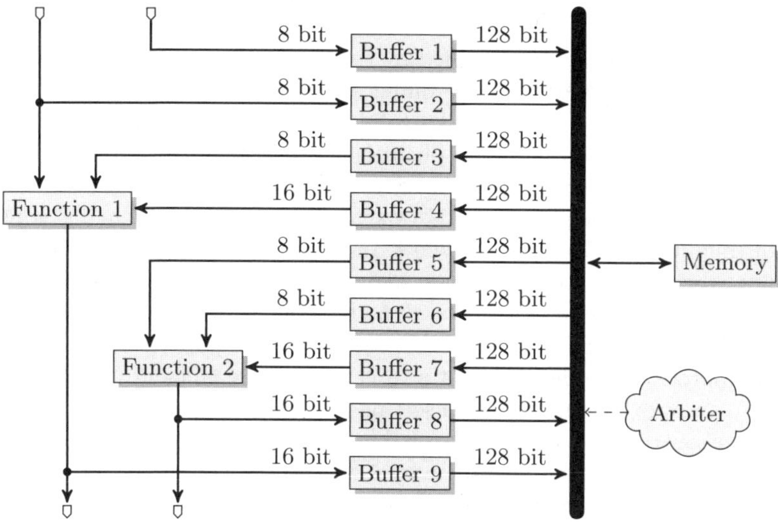

Fig. 1. A signal processing board

data to or from the Bth buffer at the ith clock cycle. Zero at the ith letter of the word means that the bus is idle at the ith clock cycle. Our goal is to form an automaton that accepts a word if and only if it represents a valid schedule (no overflows or underflows). The automaton will also be used to assign ranks (payoffs) to schedules and analyzed to determine the optimal schedule.

The automaton for the language of valid schedules is constructed as a composition of nine automata, one for each buffer. The automaton for a buffer B is as follows. Let M be an upper bound on the size of the buffer and $r \in \{-16, -8, 8, 16\}$ denote the inflow to the buffer (negative if B buffers a stream from memory to a functional block). These parameters can be read from Fig. 1. The states of the automaton are the numbers $Q(B) = \{0, |r|, 2|r|, \ldots, m\}$ where $m \leq M$ and $m + |r| > M$ for the quantum of flow r associated with buffer B.

We define two types of transitions. First, if the buffer is not scheduled to use the bus, it gets or loses a quantum of flow. This is modeled by $\delta(q, b) = q + r$ if $q + r \in Q(B)$ and $b \in \Sigma \setminus \{B\}$. Second, if the buffer is scheduled to use the bus, it loses or gets 128 bits. This is modeled by $\delta(q, B) = q - 128\,\mathrm{sgn}(r) + r$ if $q - 128\,\mathrm{sgn}(r) + r \in Q(B)$ (where $\mathrm{sgn}(r) := r/|r|$). All other transitions go to an implicit sink which is the only non-accepting state. The initial state is $q_0 = 0$.

Now, we use the product construction to get an automaton for the intersection of the languages of the nine automata. Clearly, a word is accepted by the resulting automaton if and only if it is a valid schedule. Note that the intersection may be empty, in which case we would take a larger M and recompute.

To analyze the schedules, we assign with a state $q = (q_1, \ldots, q_9) \in Q(1) \times \cdots \times Q(9)$ a number $c(q) = \max\{q_1, \ldots, q_9\}$. We call this number the color of the state and c the coloring function. Let the *rank* of an infinite word $w \in \Sigma^\omega$ be the maximal color visited infinitely often when the automaton reads the word. The rank of a word is thus the maximal buffer size needed for valid execution of the bus arbitration schedule identified by the word. The optimal buffer size thus identified could be smaller than the buffer size required during the initialization phase of the board (a finite prefix of a valid schedule). The potential loss of data (overflow) during the initialization phase is acceptable in the case study.

To get an optimal schedule, we need to search for a word with the minimal rank. A naïve approach is to solve this (quantitative) problem by solving a series of (qualitative) decision problems. In particular, let L_i be the set of valid schedules where no state with color higher than i is visited infinitely often. Then, one can check non-emptiness of L_i to identify whether the buffer size i is sufficient; until the minimal bound is found. In this paper we first study this iterative approach and then present a direct and more efficient algorithm.

We also analyze the more general setting of games. In games, nondeterminism is split into controllable and uncontrollable choices. The synthesis of an optimal schedule for the signal processing board, in its full generality, requires game based analysis. The memory needs to be refreshed periodically and this is not in the control of the arbiter. The chip decides the refresh timings and, during a memory refresh, the memory is not available. Specifically, the time between consecutive refreshes varies nondeterministically between 100 and 200 clocks and each refresh takes 10 clocks. Using the algorithms presented in this paper, we can find an optimal schedule even in the presence of memory refreshes, by considering the refresh mechanism as adversarial and finding the best strategy against its worst-case behavior. Note that games are essential here since the refresh mechanism is nondeterministic and the worst possible refresh (from the perspective of the arbiter) cannot be identified independently.

3 Ranking Functions and Games

We introduce a generalization of ω-regular languages called ω-regular ranking functions. Intuitively, languages specify qualitative properties of words by giving the set of accepted words whereas rankings specify quantitative properties by assigning numbers to words.

Definition 1. *An ω-regular ranking function over a finite alphabet Σ is a function $r\colon \Sigma^\omega \to \mathbb{N}$ with a finite range and $\{w\colon r(w) = n\}$ is an ω-regular language for every n in the range.*

Note that the characteristic function of an ω-regular language is an ω-regular ranking function with range $\{0, 1\}$.

A *(deterministic) finite state automaton* is a tuple (Q, Σ, δ, q_0) where Q is a set of states, Σ is an alphabet, $\delta\colon Q \times \Sigma \to Q$ is a transition function, and $q_0 \in Q$ is an initial state. Define $\delta^*\colon \Sigma^* \to Q$, inductively, by $\delta^*(\sigma_1 \cdots \sigma_l) = \delta(\delta^*(\sigma_1 \cdots \sigma_{l-1}), \sigma_l)$ and $\delta^*(\epsilon) = q_0$ (where ϵ is the empty word). For $w \in \Sigma^\omega$, define $\mathrm{Inf}_{\mathcal{A}}(w) := \{q \in Q\colon \delta^*(w') = q$ for infinitely many prefixes w' of $w\}$.

We introduce cyclic ranking conditions. A cyclic ranking condition with cycle k maps a word to the modulo-k value of the highest color repeating infinitely often. A parity acceptance condition is a cyclic ranking condition with $k = 2$.

Definition 2. *A cyclic ranking condition for a finite automaton $\mathcal{A} = (Q, \Sigma, \delta, q_0)$ is a pair (c, k) where $c\colon Q \to \mathbb{N}$ is a coloring function and $k \in \mathbb{N}$ is the number of ranks. The corresponding ranking function is given by* $\mathrm{Cyclic}_{\mathcal{A},c,k}(w) := \max\{c(q)\colon q \in \mathrm{Inf}_{\mathcal{A}}(w)\}$ modulo k.

Analogous to the known result that parity acceptance conditions can capture all ω-regular languages, we show that cyclic ranking conditions can capture all ω-regular ranking functions.

Proposition 3. *[Expressive completeness of cyclic ranking conditions] A ranking function $r\colon \Sigma^\omega \to \mathbb{N}$ is an ω-regular ranking function if and only if there is a deterministic finite automaton $\mathcal{A}$ and a cyclic ranking condition (c, k) such that* $\mathrm{Cyclic}_{\mathcal{A},c,k}(w) = r(w)$, *for all $w \in \Sigma^\omega$.*

Proof. The if direction is straightforward. For the only-if direction, let the ω-regular language $\{w\colon r(w) = s\}$ be specified by a deterministic Muller automaton $\mathcal{A}_s = (Q_s, \Sigma, \delta_s, q_{0_s}, \mathcal{F}_s)$ where $s \in S$ and S is the range of r. For simplicity, let $S = \{0, \ldots, k-1\}$. Let $\mathcal{A}_M = (Q_M, \Sigma, \delta_M, q_{0_M})$ be the product of the automata $\mathcal{A}_s$, $s \in S$. For $F \subseteq Q_M$, let $\mathrm{proj}_s(F) = \{q_s \in Q_s\colon \exists q = \langle \ldots, q_s, \ldots \rangle \in F\}$. Consider a Muller ranking function $\mathrm{Mul}_{\mathcal{A}_M}\colon 2^{Q_M} \to \mathbb{N}$ defined as $\mathrm{Mul}_{\mathcal{A}_M}(F) := \max(\{s\colon \mathrm{proj}_s(F) \in \mathcal{F}_s\} \cup \{0\})$.

Consider a ranking function $r_M\colon \Sigma^\omega \to \mathbb{N}$ where $r_M(w) = \mathrm{Mul}_{\mathcal{A}_M}(\mathrm{Inf}_{\mathcal{A}_M}(w))$. Suppose for an infinite word $w \in \Sigma^\omega$, $r(w) = i$. Clearly, $\mathrm{Inf}_{\mathcal{A}_i}(w) \in \mathcal{F}_i$ and for $j \neq i$, $\mathrm{Inf}_{\mathcal{A}_j}(w) \notin \mathcal{F}_j$. Thus $r_M(w) = \mathrm{Mul}_{\mathcal{A}_M}(\mathrm{Inf}_{\mathcal{A}_M}(w)) = i$.

Using the latest appearance record (LAR) [10,11] we construct an automaton $\mathcal{A} = (Q, \Sigma, \delta, q_0)$ which simulates $\mathcal{A}_M$ as follows:

$Q = Q_M! \times |Q_M|$ where $Q_M!$ is the set of all permutations of Q_M

$q_0 = ((p_1, \ldots, p_n), 1)$ for some permutation $(p_1, \ldots, p_n) \in Q_M!$ for $p_1 = q_{0_M}$

$\delta(((p_1, \ldots, p_n), h), \sigma) = ((\delta_M(p_1, \sigma), p_1, \ldots, p_{h'-1}, p_{h'+1}, \ldots, p_n), h')$

with h' as the index for $(p_1, \ldots, p_n)$ called *hit* position s.t. $\delta_M(p_1, \sigma) = p_{h'}$, if $\delta_M(p_1, \sigma)$ is defined.

Consider a cyclic ranking condition (c, k) where $k = |S|$ and a coloring function $c\colon Q \to \mathbb{N}$ defined as follows:

$$c(((p_1, \ldots, p_n), h)) := \begin{cases} kh & \text{if } \{p_1, \ldots, p_h\} \in \mathrm{Mul}^{-1}_{\mathcal{A}_M}(0) \\ kh + 1 & \text{if } \{p_1, \ldots, p_h\} \in \mathrm{Mul}^{-1}_{\mathcal{A}_M}(1) \\ \vdots & \\ kh + k - 1 & \text{if } \{p_1, \ldots, p_h\} \in \mathrm{Mul}^{-1}_{\mathcal{A}_M}(k-1) \end{cases}$$

Let h_{max} be the maximal hit position occurring infinitely often on the run of $\mathcal{A}$ on w. Eventually for any state $((p_1, \ldots, p_n), h)$, $h \leq h_{max}$ and $\{p_1, \ldots,$

$p_{h_{max}}\} = \mathrm{Inf}_{\mathcal{A}_M}(w)$. If $r_M(w) = i$ then $\{p_1, \ldots, p_{h_{max}}\} \in \mathrm{Mul}^{-1}_{\mathcal{A}_M}(i)$. Thus the maximal color occurring infinitely often on the run of $\mathcal{A}$ on w is $kh_{max} + i$. Hence $\mathrm{Cyclic}_{\mathcal{A},c,k}(w) = i = r_M(w) = r(w)$. □

Similarly, we generalize reachability, Büchi, and coBüchi acceptance conditions to reachability, Büchi, and coBüchi ranking conditions as follows.

Definition 4. *Reachability, Büchi, and coBüchi ranking conditions for a finite automaton* $\mathcal{A} = (Q, \Sigma, \delta, q_0)$ *are expressed by coloring functions* $c\colon Q \to \mathbb{N}$. *The corresponding ranking functions are defined as follows:*

$$\begin{aligned}\mathrm{Reach}_{\mathcal{A},c}(w) &:= \max\{c(\delta^*(w'))\colon\ w' \text{ is a prefix of } w\}\\ \mathrm{Buchi}_{\mathcal{A},c}(w) &:= \max\{c(q)\colon q \in \mathrm{Inf}_{\mathcal{A}}(w)\}\\ \mathrm{coBuchi}_{\mathcal{A},c}(w) &:= \min\{c(q)\colon q \in \mathrm{Inf}_{\mathcal{A}}(w)\}\end{aligned}$$

Next, we extend ω-win-lose games to ω-ranking games. A *game automaton* is a triplet $(\mathcal{A}, Q_0, Q_1)$ where $\mathcal{A} = (Q, \Sigma, \delta, q_0)$ is a finite state automaton and (Q_0, Q_1) is a partition of Q into Player 0 and Player 1 states, respectively. In figures, following the usual drawing convention, we use ○ to denote states in Q_0 and □ to denote states in Q_1.

Definition 5. *An* ω*-regular ranking game is a pair* $\mathcal{G} = (\mathcal{A}, r)$ *where* $\mathcal{A}$ *is a game automaton and* $r\colon \Sigma^\omega \to \mathbb{N}$ *is an* ω*-regular ranking function (rewards for Player 0, penalties for Player 1).*

A strategy for a player $p \in \{0, 1\}$ in an ω-regular ranking game is a function $s_p\colon \{w \in \Sigma^*\colon \delta^*(w) \in Q_p\} \to \Sigma$. The letter $s_p(w)$ models the move of player p after observing w. Let $\mathcal{S}_p(\mathcal{G})$ be the set of all strategies for player $p \in \{0, 1\}$.

A pair of strategies $(s_0, s_1) \in \mathcal{S}_0(\mathcal{G}) \times \mathcal{S}_1(\mathcal{G})$ induces an infinite word w whose $(i+1)$th letter is given by $w_{i+1} := s_0(w_{1..i})$ if $\delta^*(w_{1..i}) \in Q_0$ and $s_1(w_{1..i})$ if $\delta^*(w_{1..i}) \in Q_1$, where $w_{1..i}$ is the prefix of length i of w and $w_{1..0} = \epsilon$. We denote this word by $w_{\mathcal{G}}(s_0, s_1)$. The outcome of a pair of strategies is defined to be the rank of this word and the value of the game is defined accordingly, as follows.

Definition 6. *The value of a strategy* $s_0 \in \mathcal{S}_0(\mathcal{G})$ *in a game* $\mathcal{G} = (\mathcal{A}, r)$ *is defined by* $\mathrm{val}_{\mathcal{G}}(s_0) = \min\{r(w_{\mathcal{G}}(s_0, s_1))\colon s_1 \in \mathcal{S}_1(\mathcal{G})\}$. *The value of the game is defined by* $\mathrm{val}(\mathcal{G}) = \max\{\mathrm{val}_{\mathcal{G}}(s_0)\colon s_0 \in \mathcal{S}_0(\mathcal{G})\}$.

If we think of Player 0 as the system and Player 1 as the environment, the above definition captures the objective of Player 0 which is to maximize the outcome against the worst-case behavior of Player 1.

4 Algorithms for Ranking Games

We analyze algorithms for solving ω-regular ranking games. In our context, solving means for each state of the automaton, determining the value of the game starting at it and synthesizing a strategy that achieves these values.

4.1 Solving Ranking Games as a Series of Win-Lose Games

In this section we propose a simple scheme for solving ranking games as a series of appropriately defined win-lose games. In lattice games (cf. [2,13]), a similar decomposition to join-irreducible elements is used but it only gives a sufficiency condition. We show that for Büchi games such decomposition is also necessary. We focus on Büchi games for simplicity. Generalization to other games is straightforward.

For a game automaton $\mathcal{A}$ and a set of states B, let Buchi($\mathcal{A}, B$) be an algorithm for win-lose Büchi games [11] which computes the states from which Player 0 can force visiting B infinitely often (the winning region for Player 0).

Consider a game automaton whose set of states is Q, and a coloring function $c\colon Q \to \mathbb{N}$ that maps the states to colors. Algorithm 1 computes the function $r\colon Q \to \mathbb{N}$ where $r(q)$ is the maximal color such that Player 0 can force infinitely many visits to states with color $r(q)$, starting at q. In the first iteration, the algorithm computes all the states from which Player 0 can force infinitely many visits to the highest color. Then, it removes these nodes from the graph and proceeds to the second highest color and the process is repeated.

Algorithm 1. IteratedMaxBuchi($\mathcal{A}, c$)

foreach $q \in Q$ **do** mark $r(q)$ as undefined
foreach $\gamma \in \{c(q)\colon q \in Q\}$ *in decreasing order* **do**
  $Q_\gamma := \{q \in Q\colon r(q)$ is undefined$\}$
  $\mathcal{A}_\gamma :=$ the automaton $\mathcal{A}$ restricted to the states in Q_γ
  $B_\gamma := \{q \in Q_\gamma\colon c(q) \geq \gamma\}$
  foreach $q \in \text{Buchi}(\mathcal{A}_\gamma, B_\gamma)$ **do** $r(q) := \gamma$
return r

Remark 7. *For a color γ, it is necessary to also include the states (if any) with color $> \gamma$ from the set Q_γ in the Büchi set B_γ. For example, consider the following game automaton:*

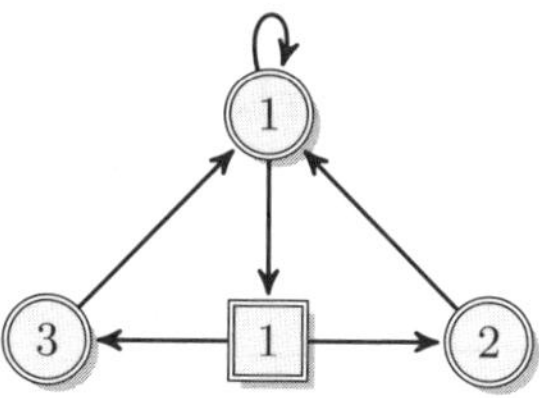

Player 0 cannot force infinite visits to color 3. Player 0 also cannot force infinite visits to color 2. However, the objective of Player 1 is to minimize the value and hence Player 1 will not choose color 3 over color 2. Hence the value that Player 0 can achieve starting from any state is 2. In terms of Büchi win-lose games, the winning set corresponding to color 2 is identified correctly only if the Büchi set, B_2, contains states with colors 2 as well as 3.

Complexity analysis. For $k, n \in \mathbb{N}$, consider the game automaton:

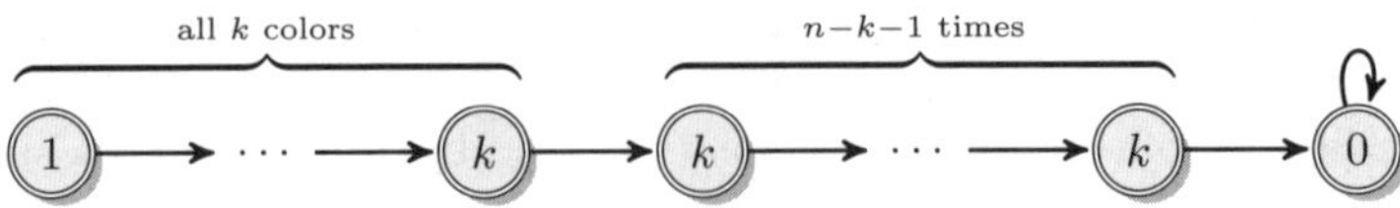

As the value for each state is 0, Algorithm 1 takes k iterations. Assuming the standard implementation of Buchi, it can be verified that each iteration runs in time $O(|B_\gamma| \cdot |\delta|)$ (Buchi$(\mathcal{A}_\gamma, B_\gamma)$ starts with B_γ and, in its ith iteration, removes the states from which Player 0 cannot force at least i visits to B_γ; each iteration takes $O(|\delta|)$). Further, from Remark 7, $|B_\gamma| = \sum_{i=\gamma}^{k} |c^{-1}(i)|$ and $\sum_{\gamma=k}^{1} |B_\gamma|$ is $O(k \cdot |Q|)$. Hence the worst-case execution time of Algorithm 1 is $O(k \cdot |Q| \cdot |\delta|)$ which is k times the worst-case execution time for solving a Büchi win-lose game.

Remark 8. *Denote the leftmost state in the above automaton by q_0. If we only need to compute the value of q_0, we can use binary search as follows. Check if $q_0 \in$ Buchi$(\mathcal{A}_{k/2}, B_{k/2})$ then if $q_0 \in$ Buchi$(\mathcal{A}_{k/4}, B_{k/4})$ and so on until we find that $q_0 \in$ Buchi$(\mathcal{A}_0, B_0)$. As only the last query is answered positively, the algorithm takes $\log k$ steps. The total execution time is $O(\log k \cdot |Q| \cdot |\delta|)$.*

Remark 7 shows that in case of Büchi games, the decomposition to join-irreducible elements (with respect to the order $1 \leq \ldots \leq k$) is necessary but the complexity of the resulting algorithm is not optimal as shown by the example above. In Section 4.4, we generalize the algorithm for solving Büchi win-lose games to solve Büchi ranking games without any increase in complexity. The advantage of the above algorithm, however, is that existing algorithms for Büchi games can be directly used for solving Büchi ranking games.

4.2 Fixpoints over Coloring Functions

The solutions of win-lose games are typically defined as fixpoints of functions over the lattice of the power-set of the set of states of the game automaton, ordered by inclusion. For ranking games, we use coloring functions as a generalization of sets of states and define the following lattice.

Definition 9. *For a set Q, $\mathbb{N}^Q$ is the set of all functions that map elements of Q to natural numbers (coloring functions). Consider the lattice $(\mathbb{N}^Q, \sqsubseteq)$ where $r_1 \sqsubseteq r_2$ if $r_1(q) \leq r_2(q)$ for all $q \in Q$. The join and meet operations of the lattice are $(r_1 \sqcup r_2)(q) = \max\{r_1(q), r_2(q)\}$ and $(r_1 \sqcap r_2)(q) = \min\{r_1(q), r_2(q)\}$, respectively.*

The lattice of coloring functions is infinite. However, the range of a coloring function used for defining a ranking condition is a finite subset of the set of natural numbers. We therefore identify a finite sub-lattice defined below.

Definition 10. *For a coloring function $c\colon Q \to \mathbb{N}$, let $\mathrm{Lat}(c)$ be the lattice $(\mathcal{R}(c)^Q, \sqsubseteq)$ where $\mathcal{R}(c) = \{c(q)\colon q \in Q\}$. It is easy to verify that $\mathrm{Lat}(c)$ is closed under join and meet, so it is a complete finite sub-lattice of $(\mathbb{N}^Q, \sqsubseteq)$. The bottom of the lattice is $\bot = Q \times \{\min \mathcal{R}(c)\}$ and the top is $\top = Q \times \{\max \mathcal{R}(c)\}$.*

In the following subsections, we give fixpoint characterizations for ranking games. It can be easily verified that the functions whose extremal fixpoints determine the solutions of the games are monotonic and closed over the lattice defined above. By finiteness of the lattice and by the Knaster–Tarski fixpoint theorem, we know that the extremal fixpoints of the functions can be computed.

4.3 Solving Reachability Ranking Games in Linear Time

For a game automaton $\mathcal{A}$ and a coloring function c, let the solution of the reachability ranking game be given by the function $r^{\mathcal{A},c}_{\text{Reach}}$ that maps each state q to the maximal color i such that Player 0 can force a visit to a state in $\{q' \in Q\colon c(q') \geq i\}$, starting from q. Let $\text{succ}(q) := \{\delta(q,\sigma)\colon \sigma \in \Sigma\}$ and $\text{pred}(q) := \{q'\colon q \in \text{succ}(q')\}$.

The fixpoint formulation of $r^{\mathcal{A},c}_{\text{Reach}}$ is given in Proposition 11 as the least fixpoint (LFP) of the function f which assigns to a state q the highest color that Player 0 can force in one step (or less) from q. Let $\text{apred}(r)(q) := \max\{r(q')\colon q' \in \text{succ}(q)\}$ for $q \in Q_0$ and $\text{apred}(r)(q) := \min\{r(q')\colon q' \in \text{succ}(q)\}$ for $q \in Q_1$.

Proposition 11. $r^{\mathcal{A},c}_{\text{Reach}} = \text{LFP}(f)$ *where* $f\colon \text{Lat}(c) \to \text{Lat}(c)$ *is given by* $f(r) := r \sqcup \text{apred}(r) \sqcup c$.

The function apred can be computed in time $O(|\delta|)$, where δ is the transition function of the game automaton. Hence, the function f can also be computed in time $O(|\delta|)$. Further, f is defined such that if, starting at a state q, Player 0 can force a visit to a color $\geq \gamma$ (with respect to a coloring function r) within i steps, it can do so within $\max\{0, i-1\}$ steps with respect to the coloring function $f(r)$. Since the length of an acyclic path in $\mathcal{A}$ can be at most $|Q|-1$, the effective height of the lattice $\text{Lat}(c)$ for f is $|Q|-1$. The overall complexity of computing the fixpoint appears to be quadratic. However, we now specify a suitable traversal of the automaton and also propose to keep a record of the transitions already processed. This ensures that any transition of the automaton is processed only once and we get a linear-time algorithm.

With each state $q \in Q_1$ associate a number $\text{count}(q) = |\,\text{succ}(q)|$. Also let all states be marked as not visited. Consider the following evaluation order: Starting with the highest color, for each color γ, perform a preorder backwards traversal starting with the states with color γ that are not marked as visited. Let q be the state being processed. If $q \in Q_0$ and not marked as visited then its color is set to γ and is marked as visited. Let $q \in Q_1$. If $\text{count}(q) = 1$ then all other outgoing transitions of q have been explored during processing of colors $> \gamma$. Hence the minimal color that Player 1 can force is γ. Set the color of q to γ and mark it as visited. Otherwise, set $\text{count}(q)$ to $\text{count}(q) - 1$.

Theorem 12. *Reachability ranking games can be solved in* $O(|\delta|)$ *time, where* δ *is the transition function of the game automaton.*

Remark 13. *The memoryless optimal winning strategy for Player 0 can be identified during the computation of values described above. If* $q \in Q_0$ *has color*

γ and is not marked as visited until the processing of color γ then the value at q is γ, that is, its own color. Thus the strategy for Player 0 at q is to select the label of any outgoing transition. Otherwise, the strategy is to select the label of the outgoing transition of q that lead to q being marked as visited (note that the game automaton is deterministic).

4.4 An Efficient Quadratic-Time Algorithm for Büchi Ranking Games

For a game automaton $\mathcal{A}$ and a coloring function c, the solution to the Büchi ranking problem is a function $r^{\mathcal{A},c}_{\text{Buchi}}$ that maps each state q to the maximal color i such that Player 0 can force infinitely many visits to $\{q' \in Q \colon c(q') \geq i\}$, starting from q. We present a fixpoint formulation of the solution function and show that it can be computed in quadratic-time. Its complexity is independent of the number of colors, as opposed to Algorithm 1.

The fixpoint formulation of $r^{\mathcal{A},c}_{\text{Buchi}}$ is given in Proposition 14. The function g identifies for each state q the maximal color that Player 0 can force to visit at least once, starting at q. The greatest fixpoint (GFP) of g computes for each state q the maximal color (less than or equal to $c(q)$) that Player 0 can force to visit infinitely many times. Finally, the solution of the reachability ranking for the coloring function given by GFP(g) determines the solution $r^{\mathcal{A},c}_{\text{Buchi}}$.

Proposition 14. $r^{\mathcal{A},c}_{\text{Buchi}} = r^{\mathcal{A},\text{GFP}(g)}_{\text{Reach}}$ *where* $g\colon \text{Lat}(c) \to \text{Lat}(c)$ *is defined by* $g(r) = r \sqcap c \sqcap \text{LFP}(f_r)$ *and* $f_r\colon \text{Lat}(r) \to \text{Lat}(r)$ *is defined by* $f_r(r') = r' \sqcup \text{apred}(r' \sqcup r)$.

Note that the function f_r is defined analogously to the function f given in Proposition 11 but considers states reachable in one or more steps instead of zero or more steps. From Theorem 12, we can deduce that LFP(f_r) can be computed in time $O(|\delta|)$. Consequently, the function g can be computed in time $O(|\delta|)$.

Complexity analysis. We now show that in each iteration of the fixpoint computation GFP(g), at least one additional state gets its final color. Thus, the fixpoint computation GFP(g) takes no more than $|Q|$ steps. This gives quadratic-time complexity for solving Büchi ranking games.

Let $r_0 = c$ and $r_0 \sqsupseteq r_1 \sqsupseteq \cdots \sqsupseteq r_l$ be the sequence of functions computed in the fixpoint computation GFP(g). Let $W_j(r_i) := \{q \in Q \colon r_i(q) = j > r_l(q)\}$ be the set of states whose color in the ith iteration (given by coloring function r_i) is j which is larger than their final color given by coloring function r_l.

In Lemma 15, we show that if $W_j(r_i) \neq \emptyset$ then, in the next iteration, at least one state from $\bigcup_{j'=j}^{k} W_{j'}(r_i)$ gets a color smaller than j or is assigned to its final color; and this is true for each color j.

Lemma 15. *If* $W_j(r_i) \neq \emptyset$ *then* $\bigcup_{j'=j}^{k} W_{j'}(r_i) \setminus \bigcup_{j'=j}^{k} W_{j'}(r_{i+1}) \neq \emptyset$.

Proof. Since $W_j(r_i) \neq \emptyset$ there exists a state q such that $r_i(q) = j > r_l(q)$. Because $r_l(q)$ is the final color of q and $r_0 = c$, Player 0 cannot force infinitely many visits to $\{q' : r_0(q') > r_l(q)\}$ starting from q. Since $r_0 \sqsupseteq r_i$ and $r_i(q) > r_l(q)$, we have $r_0(q) \geq r_i(q) > r_l(q)$. Thus, Player 0 cannot force infinitely many visits to $S := \{q' : r_i(q') \geq r_i(q)\}$, starting at q. Because $q \in S$, there exists at least one 'exit' state $q' \in S$ from which Player 1 cannot be forced to visit S again. In particular, Player 0 cannot force visiting S starting at q' which means that $\mathrm{LFP}(f_{r_i})(q') < j \leq r_i(q')$. Since $r_{i+1}(q') = \min\{r_i(q'), \mathrm{LFP}(f_{r_i})(q')\}$ we get that $r_{i+1}(q') < j \leq r_i(q')$. □

In Lemma 16, we show that if $W_j(r_i) = \emptyset$ then in the next iteration, no state gets color j unless it is its final color.

Lemma 16. *If $W_j(r_i) = \emptyset$ then $W_j(r_{i+1}) = \emptyset$.*

Proof. Assume that $W_j(r_{i+1}) \neq \emptyset$, that is there exists a state q such that $r_{i+1}(q) = j > r_l(q)$. Let $S := \{q' \in Q : r_i(q') = r_{i+1}(q)\}$. By the definition of $\mathrm{LFP}(f_{r_i})$, there is $S' \subseteq S$ that Player 0 can force visiting, starting from q. Consider also the set $S'' = \{q' \in S' : r_l(q') = r_{i+1}(q)\}$. If $S'' = S'$, starting at q, Player 0 can force infinite visits to S' which contradicts our assumption that $r_{i+1}(q) > r_l(q)$. Therefore, there exists $q' \in S' \setminus S''$. As $q' \in \{q'' : r_i(q'') > r_l(q'') = j\}$, we can infer that $W_j(r_i) \neq \emptyset$. □

Theorem 17. *Büchi ranking games can be solved in $O(|Q| \cdot |\delta|)$ time, where Q is the set of game states and δ is the transition function.*

Proof. The function g can be computed in time $O(|\delta|)$. We show that the number of iterations of the fixpoint computation $\mathrm{GFP}(g)$ is bounded by $|Q|$ by proving that in each iteration at least one additional state gets its final color.

For an iteration i, let $j = \min\{j' : W_{j'}(r_i) \neq \emptyset\}$. By Lemma 15, $\bigcup_{j'=j}^{k} W_{j'}(r_i) \setminus \bigcup_{j'=j}^{k} W_{j'}(r_{i+1}) \neq \emptyset$. By minimality of j, $\bigcup_{j'=0}^{j-1} W_{j'}(r_i) = \emptyset$. By Lemma 16, $\bigcup_{j'=0}^{j-1} W_{j'}(r_{i+1}) = \emptyset$. Therefore we have $\bigcup_{j'=j}^{k} W_{j'}(r_i) \setminus \bigcup_{j'=0}^{k} W_{j'}(r_{i+1}) \neq \emptyset$. This means that at least one state whose color in r_i was not its final color gets its final color in r_{i+1}. □

Remark 18. *The memoryless optimal winning strategy for Player 0 can be identified during the computation of values described above. In the final reachability computation with the coloring function as* $\mathrm{GFP}(g)$*, if $q \in Q_0$ has color γ and is not marked as visited until the processing of color γ then the strategy for Player 0 at q is same as the strategy determined in the last* $\mathrm{LFP}(f_r)$ *step (a one or more step reachability computation). Otherwise, the strategy is to select the label of the outgoing transition of q that lead to q being marked as visited in the reachability computation for the coloring function* $\mathrm{GFP}(g)$.

4.5 Cyclic Ranking Games

For a cyclic ranking game, consider the following decision problem. Given a number $i \in \mathbb{N}$, determine if Player 0 can force a word with rank $\geq i$.

Proposition 19. *The decision problem is in* NP ∩ coNP.

Proof. Assume that the game is defined by the automaton $\mathcal{A}$ and the pair (c, k) where $c\colon Q \to \mathbb{N}$ is a coloring function and $k \in \mathbb{N}$ is the number of ranks. Consider a parity game over $\mathcal{A}$ using the coloring function

$$c'(q) = 2 \cdot \lfloor c(q)/k \rfloor + \begin{cases} 1 \text{ if } c(q) \geq i (\text{modulo } k) \\ 0 \text{ if } c(q) < i (\text{modulo } k). \end{cases}$$

Player 0 wins if the maximal color appearing infinitely often is odd. Clearly, Player 0 can force a win in this game if and only if the answer to the decision problem is affirmative. The proof follows from the known result that deciding the winner of a parity game is in NP ∩ coNP [7]. □

Using the above reduction, we can determine the value of a game with a cyclic ranking condition by repeated queries. Once we know the value, any strategy that wins the parity game also assures the value in the cyclic game. Since parity games have memoryless optimal strategies, we get the following proposition.

Proposition 20. *Cyclic ranking games have memoryless optimal strategies.*

5 Conclusions

We have proposed a framework for specifying prioritized requirements by associating ranks with executions and shown how to generalize classical automata-theoretic notions of acceptance to rankings. The resulting optimization analogs of verification and synthesis problems can naturally be solved by adopting symbolic fixpoint algorithms to an appropriately chosen lattice of coloring functions. In particular, we have identified the cyclic ranking condition as a means of specifying all ω-regular ranking functions, and shown that Büchi ranking games can be solved in quadratic time. Implementation using binary decision diagrams (BDDs) and algebraic decision diagrams (ADDs) [16] is planned for future work.

Acknowledgments. This research was partially supported by NSF grants 0541149 and 0524059, and by General Motors India Science Lab.

References

1. Alur, R., Etessami, K., Torre, S.L., Peled, D.: Parametric temporal logic for "model measuring". ACM Trans. Comput. Log 2(3), 388–407 (2001)
2. Bruns, G., Godefroid, P.: Model Checking with Multi-valued Logics. In: Díaz, J., Karhumäki, J., Lepistö, A., Sannella, D. (eds.) ICALP 2004. LNCS, vol. 3142, pp. 281–293. Springer, Heidelberg (2004)
3. Buttazzo, G.: Hard Real-Time Computing Systems: Predictable Scheduling Algorithms and Applications. Kluwer Academic Publishers, Boston, USA (2000)
4. Chatterjee, K.: A linear-time algorithm for weak parity games. Technical Report UCB/EECS-2006-153, University of California, Berkeley (2006)

5. Cleaveland, R., Steffen, B.: A linear-time model-checking algorithm for the alternation-free modal mu-calculus. In: Proc. 3rd Conference on Computer Aided Verification, pp. 48–58 (1991)
6. Emerson, E.: Temporal and modal logic. In: van Leeuwen, J. (ed.) Handbook of Theoretical Computer Science, vol. B, pp. 995–1072. Elsevier Science Publishers, Amsterdam (1990)
7. Emerson, E.A., Jutla, C.S., Sistla, A.P.: On model checking for the μ-calculus and its fragments. Theor. Comput. Sci 258(1-2), 491–522 (2001)
8. Ernits, J.P.: Memory arbiter synthesis and verification for a radar memory interface card. Nord. J. Comput 12(2), 68–88 (2005)
9. Gimbert, H., Zielonka, W.: Perfect information stochastic priority games. In: Arge, L., Cachin, C., Jurdziński, T., Tarlecki, A. (eds.) ICALP 2007. LNCS, vol. 4596, pp. 850–861. Springer, Heidelberg (2007)
10. Gurevich, Y., Harrington, L.: Trees, automata, and games. In: Proc. 14th Annual ACM Symposium on Theory of Computing, pp. 60–65 (1982)
11. Büchi, J.R.: State-strategies for games in $F_{\sigma\delta} \cap G_{\sigma\delta}$. Journal of Symbolic Logic 48, 1171–1198 (1983)
12. Kupferman, O., Lustig, Y.: Lattice automata. In: Proc. 8th Int. Conf. on Verification, Model Checking, and Abstract Interpretation, pp. 199–213 (2007)
13. Kupferman, O., Lustig, Y.: Latticed simulation relations and games. In: Proc. 5th symp. on Aut. Technology for Verification and Analysis, pp. 316–330 (2007)
14. Pnueli, A.: The temporal logic of programs. In: Proc. 18th IEEE Symposium on Foundations of Computer Science, pp. 46–77 (1977)
15. Pnueli, A., Rosner, R.: On the synthesis of a reactive module. In: Proc. 16th ACM Symposium on Principles of Programming Languages (1989)
16. Bahar, R.I., Frohm, E.A., Gaona, C.M., Hachtel, G.D., Macii, E., Pardo, A., Somenzi, F.: Algebraic Decision Diagrams and Their Applications. In: Proc. 9th International Conference on CAD, pp. 188–191 (1993)
17. Thomas, W.: On the synthesis of strategies in infinite games. In: Proc. 12th Symp. on Theoretical Aspects of Computer Science, pp. 1–13 (1995)
18. Vardi, M.Y., Wolper, P.: Reasoning about infinite computations. Inf. Comput. 115(1), 1–37 (1994)
19. Weiss, G.: Optimal scheduler for a memory card. Research report, Dep. of Computer Science and Applied Mathematics, The Weizmann Institue of Science (2002)
20. Zwick, U., Paterson, M.: The complexity of mean payoff games on graphs. Theor. Comput. Sci. 158(1-2), 343–359 (1996)

Efficient Craig Interpolation for Linear Diophantine (Dis)Equations and Linear Modular Equations*

Himanshu Jain[1], Edmund Clarke[1], and Orna Grumberg[2]

[1] School of Computer Science, Carnegie Mellon University
[2] Department of Computer Science, Technion - Israel Institute of Technology

Abstract. The use of Craig interpolants has enabled the development of powerful hardware and software model checking techniques. Efficient algorithms are known for computing interpolants in rational and real linear arithmetic. We focus on subsets of integer linear arithmetic. Our main results are polynomial time algorithms for obtaining interpolants for conjunctions of linear diophantine equations, linear modular equations (linear congruences), and linear diophantine disequations. We show the utility of the proposed interpolation algorithms for discovering *modular/divisibility* predicates in a counterexample guided abstraction refinement (CEGAR) framework. This has enabled verification of simple programs that cannot be checked using existing CEGAR based model checkers.

1 Introduction

The use of Craig interpolation [8] has led to powerful hardware [14] and software [9] model checking techniques. In [14] the idea of interpolation is used for obtaining over-approximations of the reachable set of states without using the costly image computation (existential quantification) operations. In [9,11] interpolants are used for finding the *right* set of predicates in order to rule out *spurious counterexamples*. An interpolating theorem prover performs the task of finding the interpolants. Such provers are available for various theories such as propositional logic, rational and real linear arithmetic, and equality with uninterpreted functions [6,11,12,13,15,19,21].

Efficient algorithms are known for computing interpolants in rational and real linear arithmetic [6,15,19]. Linear arithmetic formulas where all variables are constrained to be integers are said to be formulas in *(pure) integer linear arithmetic* or $LA(\mathbb{Z})$, where $\mathbb{Z}$ is the set of integers. There are no known efficient algorithms for computing interpolants for formulas in $LA(\mathbb{Z})$. This is expected because checking the satisfiability of conjunctions of atomic formulas in $LA(\mathbb{Z})$ is itself NP-hard. We show that for various *subsets* of $LA(\mathbb{Z})$ one can compute interpolants efficiently.

Informally, a linear equation where all variables are integer variables is said to be a *linear diophantine equation (LDE)*. A *linear modular equation (LME)* or a *linear congruence* over integer variables is a type of linear equation that expresses divisibility relationships. A *system* of LDEs (LMEs) denotes conjunctions of LDEs (LMEs).

* This research was sponsored by the Gigascale Systems Research Center (GSRC), the Semiconductor Research Corporation (SRC), the Office of Naval Research (ONR), the Naval Research Laboratory (NRL), the Army Research Office (ARO), and the General Motors Lab at CMU.

A. Gupta and S. Malik (Eds.): CAV 2008, LNCS 5123, pp. 254–267, 2008.

Both LDEs and LMEs arise naturally in program verification when modeling assignments and conditional statements as logical formulas. These subsets of $LA(\mathbb{Z})$ are also known to be tractable, that is, polynomial time algorithms are known for deciding systems of LDEs and LMEs. We study the interpolation problem for LDEs and LMEs.

Given formulas F, G such that $F \wedge G$ is unsatisfiable, an interpolant for the pair (F, G) is a formula $I(F, G)$ with the following properties: 1) F implies $I(F, G)$, 2) $I(F, G) \wedge G$ is unsatisfiable, and 3) $I(F, G)$ refers only to the common variables of F and G. This paper presents the following new results.

- F, G denote a system of LDEs: We show that $I(F, G)$ can be obtained in polynomial time by using a proof of unsatisfiability of $F \wedge G$. The interpolant can be either a LDE or a LME. This is because in some cases there is no $I(F, G)$ that is a LDE. In these cases, however, there is always an $I(F, G)$ in the form of a LME. (Sec. 3)
- F, G denote a system of LMEs: We obtain $I(F, G)$ in polynomial time by using a proof of unsatisfiability of $F \wedge G$. We can ensure that $I(F, G)$ is a LME. (Sec. 4)
- Let S denote an unsatisfiable system of LDEs. The proof of unsatisfiability of S can be obtained in polynomial time by using the *Hermite Normal Form* of S (represented in matrix form) [20]. A system of LMEs R can be reduced to an equi-satisfiable system of LDEs R'. The proof of unsatisfiability for R is easily obtained from the proof of unsatisfiability of R'. (Sec. 5)
- Let S denote a system of LDEs. We show that if S has an integral solution, then every LDE that is implied by S, can be obtained by a linear combination of equations in S. We show that S is *convex* [17], that is, if S implies a disjunction of LDEs, then it implies one of the equations in the disjunction. In contrast, conjunctions of atomic formulas in $LA(\mathbb{Z})$ are not convex due to inequalities [17]. These results help in efficiently dealing with *linear diophantine disequations (LDDs)*. (Sec. 6)
- Let $S = S_1 \wedge S_2$, where S_1 is a system of LDEs, while S_2 is a system of LDDs. We say that S is a system of LDEs+LDDs. We show that S has no integral solution if and only if $S_1 \wedge S_2$ has no rational solution or S_1 has no integral solution. This gives a polynomial time decision procedure for checking if S has an integral solution. If S has no integral solution, then the proof of unsatisfiability of S can be obtained in polynomial time. (Sec. 6)
- F, G denote a system of LDEs+LDDs: We show $I(F, G)$ can be obtained in polynomial time. The interpolant can be a LDE, a LDD, or a LME. (Sec. 6)
- We show the utility of our interpolation algorithms in counterexample guided abstraction refinement (CEGAR) based verification [7]. Our interpolation algorithm is effective at discovering *modular/divisibility predicates*, such as $3x + y + 2z \equiv 1 \ (mod\ 4)$, from spurious counterexamples. This has allowed us to verify programs that cannot be verified by existing CEGAR based model checkers.

Polynomial time algorithms are known for solving (deciding) a system of LDEs [20,5] and LMEs (by reduction to LDEs) over integers. We do not give any new algorithms for solving a system of LDEs or LMEs. Instead we focus on obtaining proofs of unsatisfiability and interpolants for systems of LDEs, LMEs, LDEs+LDDs. We only consider conjunctions of LDEs, LMEs, LDEs+LDDs. Interpolants for any (unsatisfiable) Boolean combinations of LDEs can also be obtained by calling the interpolation algorithm for conjunctions of LDEs+LDDs multiple times in a satisfiability modulo theory (SMT) framework [6]. However, computing interpolants for Boolean combinations of LMEs is difficult. This is due to linear modular disequations (LMDs). We can show that even the decision problem for conjunctions of LMDs is NP-hard. The extended version of the paper [10] contains all proofs.

Related Work. It is known that Presburger arithmetic (PA) augmented with modulus operator (divisibility predicates) allows quantifier elimination. Kapur et al. [12] show that a recursively enumerable theory allows quantifier-free interpolants if and only if it allows quantifier elimination. The systems of LDEs, LMEs, LDEs+LDDs are subsets of PA. Thus, the existence of quantifier-free interpolants for these systems follows from [12]. However, quantifier elimination for PA has exponential complexity and does not immediately yield efficient algorithms for computing interpolants. We give polynomial time algorithms for computing interpolants for systems of LDEs, LMEs, LDEs+LDDs.

Let S_1, S_2 denote conjunctions of atomic formulas in $LA(\mathbb{Z})$. Suppose $S_1 \wedge S_2$ is unsatisfiable. Pudlak [18] shows how to compute an interpolant for (S_1, S_2) by using a *cutting-plane* (CP) proof of unsatisfiability. The CP proof system is a sound and complete way of proving unsatisfiability of conjunctions of atomic formulas in $LA(\mathbb{Z})$. However, a CP proof for a formula can be exponential in the size of the formula. Pudlak does not provide any guarantee on the size of CP proofs for a system of LDEs or LMEs. Our results show that polynomially sized proofs of unsatisfiability and interpolants can be obtained for systems of LDEs, LMEs and LDEs+LDDs.

McMillan [15] shows how to compute interpolants in the combined theory of rational linear arithmetic $LA(\mathbb{Q})$ and equality with uninterpreted functions $\mathcal{EUF}$ by using proofs of unsatisfiability. Rybalchenko and Sofronie-Stokkermans [19] show how to compute interpolants in combined $LA(\mathbb{Q})$, $\mathcal{EUF}$ and real linear arithmetic $LA(\mathbb{R})$ by using linear programming solvers in a black-box fashion. The key idea in [19] is to use an extension of Farkas lemma [20] to reduce the interpolation problem to constraint solving in $LA(\mathbb{Q})$ and $LA(\mathbb{R})$. Cimatti et al. [6] show how to compute interpolants in a satisfiability modulo theory (SMT) framework for $LA(\mathbb{Q})$, rational difference logic fragment and $\mathcal{EUF}$. By making use of state-of-the-art SMT algorithms they obtain significant improvements over existing interpolation tools for $LA(\mathbb{Q})$ and $\mathcal{EUF}$. Yorsh and Musuvathi [21] give a Nelson-Oppen [17] style method for generating interpolants in a combined theory by using the interpolation procedures for individual theories. Kroening and Weissenbacher [13] show how a bit-level proof can be lifted to a word-level proof of unsatisfiability (and interpolants) for equality logic.

To the best of our knowledge the work in [15,21,19,13,6] is not complete for computing interpolants in $LA(\mathbb{Z})$ or its subsets such as LDEs, LMEs, LDEs+LDDs. That is, the work in [15,21,19,13,6] cannot compute interpolants for formulas that are satisfiable over rationals but unsatisfiable over integers. Such formulas can arise in both hardware and software verification. We give sound and complete polynomial time algorithms for computing interpolants for conjunctions of LDEs, LMEs, LDEs+LDDs.

2 Notation and Preliminaries

We use capital letters $A, B, C, X, Y, Z, \ldots$ to denote matrices and formulas. A matrix M is *integral (rational)* iff all elements of M are integers (rationals). For a matrix M with m rows and n columns we say that the size of M is $m \times n$. A *row vector* is a matrix with a single row. A *column vector* is a matrix with a single column. We sometimes identify a matrix M of size 1×1 by its only element. If A, B are matrices, then AB denotes matrix multiplication. We assume that all matrix operations are well defined in this paper. For example, when we write AB without specifying the sizes of matrices A, B, it is assumed that the number of columns in A equals the number of rows in B.

For any rational numbers α and β, $\alpha|\beta$ if and only if, α divides β, that is, if and only if $\beta = \lambda\alpha$ for some integer λ. We say that α is equivalent to β *modulo* γ written as $\alpha \equiv \beta \ (mod\ \gamma)$ if and only if $\gamma|(\alpha - \beta)$. We say γ is the *modulus* of the equation $\alpha \equiv \beta \ (mod\ \gamma)$. We allow α, β, γ to be rational numbers. If $\alpha_1, \ldots, \alpha_n$ are rational numbers, not all equal to 0, then the largest rational number γ dividing each of $\alpha_1, \ldots, \alpha_n$ exists [20], and is called the *greatest common divisor*, or *gcd* of $\alpha_1, \ldots, \alpha_n$ denoted by $gcd(\alpha_1, \ldots, \alpha_n)$. We assume that gcd is always positive.

Basic Properties of Modular Arithmetic: Let a, b, c, d, m be rational numbers.
P1. $a \equiv a \ (mod\ m)$ (reflexivity).
P2. $a \equiv b \ (mod\ m)$ implies $b \equiv a \ (mod\ m)$ (symmetry).
P3. $a \equiv b \ (mod\ m)$ and $b \equiv c \ (mod\ m)$ imply $a \equiv c \ (mod\ m)$ (transitivity).
P4. If $a \equiv b \ (mod\ m)$, $c \equiv d \ (mod\ m)$, and x, y are integers, then $ax + cy \equiv bx + dy \ (mod\ m)$ (integer linear combination).
P5. If $c > 0$ then $a \equiv b \ (mod\ m)$ if, and only if, $ac \equiv bc \ (mod\ mc)$.
P6. If $a = b$, then $a \equiv b \ (mod\ m)$ for any m.

Example 1. Observe that $x \equiv 0 \ (mod\ 1)$ for any integer x. Also observe from P5 (with $c = 2$) that $\frac{1}{2}x \equiv 0 \ (mod\ 1)$ if and only if $x \equiv 0 \ (mod\ 2)$.

A *linear diophantine equation (LDE)* is a linear equation $c_1x_1 + \ldots + c_nx_n = c_0$, where $x_1, \ldots, x_n$ are integer variables and $c_0, \ldots, c_n$ are rational numbers. A variable x_i is said to *occur* in the LDE if $c_i \neq 0$. We denote a system of m LDEs in a matrix form as $CX = D$, where C denotes an $m \times n$ matrix of rationals, X denotes a column vector of n integer variables and D denotes a column vector of m rationals. When we write a (single) LDE in the form $CX = D$, it is implicitly assumed that the sizes of C, X, D are of the form $1 \times n, n \times 1, 1 \times 1$, respectively. A variable is said to *occur* in a system of LDEs if it occurs in at least one of the LDEs in the given system of LDEs.

A *linear modular equation (LME)* has the form $c_1x_1 + \ldots + c_nx_n \equiv c_0 \ (mod\ l)$, where $x_1, \ldots, x_n$ are integer variables, $c_0, \ldots, c_n$ are rational numbers, and l is a rational number. We call l the modulus of the LME. Allowing l to be a rational number leads to simpler proofs and covers the case when l is an integer. We abbreviate a LME $t \equiv c \ (mod\ l)$ by $t \equiv_l c$. A variable x_i is said to *occur* in a LME if l does not divide c_i.

A *system* of LDEs (LMEs) denotes conjunctions of LDEs (LMEs). If F, G are a system of LDEs (LMEs), then $F \wedge G$ is also a system of LDEs (LMEs).

2.1 Craig Interpolants

Given two logical formulas F and G in a theory $\mathcal{T}$ such that $F \wedge G$ is unsatisfiable in $\mathcal{T}$, an interpolant I for the ordered pair (F, G) is a formula such that

(1) $F \Rightarrow I$ in $\mathcal{T}$
(2) $I \wedge G$ is unsatisfiable in $\mathcal{T}$
(3) I refers to only the common variables of A and B.

The interpolant I can contain symbols that are interpreted by $\mathcal{T}$. In this paper such symbols will be one of the following: addition $(+)$, equality $(=)$, modular equality for some rational number m $(\equiv_m)$, disequality $(\neq)$, and multiplication by a rational number $(\times)$. The exact set of interpreted symbols in the interpolant depends on $\mathcal{T}$.

3 System of Linear Diophantine Equations (LDEs)

In this section we discuss proofs of unsatisfiability and interpolation algorithm for LDEs. The following theorem from [20] gives a necessary and sufficient condition for a system of LDEs to have an integral solution.

Theorem 1. *(Corollary 4.1(a) in Schrijver [20]) A system of LDEs $CX = D$ has no integral solution for X, if and only if there exists a rational row vector R such that RC is integral and RD is not an integer.*

Definition 1. *We say a system of LDEs $CX = D$ is* **unsatisfiable** *if it has no integral solution for X. For a system of LDEs $CX = D$ a* **proof of unsatisfiability** *is a rational row vector R such that RC is integral and RD is not an integer.*

Example 2. Consider the system of LDEs $CX = D$ and a proof of unsatisfiability R:

$$CX = D := \begin{bmatrix} 1 & 1 & 0 \\ 1 & -1 & 0 \\ 0 & 2 & 2 \end{bmatrix} \begin{bmatrix} x \\ y \\ z \end{bmatrix} = \begin{bmatrix} 1 \\ 1 \\ 3 \end{bmatrix} \qquad \begin{aligned} R &= [\tfrac{1}{2}, -\tfrac{1}{2}, \tfrac{1}{2}] \\ RC &= [0, 2, 1] \\ RD &= \tfrac{3}{2} \end{aligned}$$

Example 3. Consider the system of LDEs $CX = D$ and a proof of unsatisfiability R:

$$CX = D := \begin{bmatrix} 1 & -2 & 0 \\ 1 & 0 & -2 \end{bmatrix} \begin{bmatrix} x \\ y \\ z \end{bmatrix} = \begin{bmatrix} 0 \\ 1 \end{bmatrix} \qquad \begin{aligned} R &= [\tfrac{1}{2}, \tfrac{1}{2}] \\ RC &= [1, -1, -1] \\ RD &= \tfrac{1}{2} \end{aligned}$$

The above examples will be used as running examples in the paper. In section 5 we describe how a proof of unsatisfiability can be obtained in polynomial time for an unsatisfiable system of LDEs.

Definition 2. **(Implication)** *A system of LDEs $CX = D$* **implies** *a (single) LDE $AX = B$, if every integral vector X satisfying $CX = D$ also satisfies $AX = B$.*

Similarly, $CX = D$ **implies** *a (single) LME $AX \equiv_m B$, if every integral vector X satisfying $CX = D$ also satisfies $AX \equiv_m B$.*

Lemma 1. *(Linear combination) For every rational row vector U the system of LDEs $CX = D$ implies the LDE $UCX = UD$. Note that $UCX = UD$ is simply a linear combination of the equations in $CX = D$. The system $CX = D$ also implies the LME $UCX \equiv_m UD$ for any rational number m.*

Example 4. The system of LDEs $CX = D$ in Example 3 implies the LDE $[\frac{1}{2}, \frac{1}{2}]CX = [\frac{1}{2}, \frac{1}{2}]D$, which simplifies to $x - y - z = \frac{1}{2}$. The system $CX = D$ also implies the LME $x - y - z \equiv_m \frac{1}{2}$ for any rational number m.

3.1 Computing Interpolants for Systems of LDEs

Let $F \wedge G$ denote an unsatisfiable system of LDEs. The following example shows that an unsatisfiable system of LDEs does not always have a LDE as an interpolant.

Example 5. Let $F := x - 2y = 0$ and $G := x - 2z = 1$. Intuitively, F expresses the constraint that x is even and G expresses the constraint that x is odd, thus, $F \wedge G$ is unsatisfiable. We gave a proof of unsatisfiability of $F \wedge G$ in Example 3. Observe that the pair (F, G) does not have any quantifier-free interpolant that is also a LDE. The problem is that the interpolant can only refer to the variable x. We can show that there is no formula I of the form $c_1 x + c_2 = 0$, where c_1, c_2 are rational numbers, such that $F \Rightarrow I$ and $I \wedge G$ is unsatisfiable (see [10] for proof).

As shown by the above example it is possible that there exists no LDE that is an interpolant for (F, G). We show that in this case the system (F, G) always has a LME as an interpolant. In the above example an interpolant will be $x \equiv_2 0$. Intuitively, the interpolant means that x is an even integer.

We now describe the algorithm for obtaining interpolants. Let $AX = A', BX = B'$ be systems of LDEs, where $X = [x_1, \ldots, x_n]$ is a column vector of n integer variables. Suppose the combined system of LDEs $AX = A' \wedge BX = B'$ is unsatisfiable. We want to compute an interpolant for $(AX = A', BX = B')$. Let $R = [R_1, R_2]$ be a proof of unsatisfiability of $AX = A' \wedge BX = B'$ such that

$$R_1 A + R_2 B \quad \text{is integral and} \quad R_1 A' + R_2 B' \quad \text{is not an integer.}$$

Recall that a variable is said to *occur* in a system of LDEs if it occurs with a non-zero coefficient in one of the equations in the given system of LDEs. Let $V_{AB} \subseteq X$ denote the set of variables that occur in both $AX = A'$ and $BX = B'$, let $V_{A \setminus B} \subseteq X$ denote the set of variables occurring only in $AX = A'$ (and not in $BX = B'$), and let $V_{B \setminus A} \subseteq X$ denote the set of variables occurring only in $BX = B'$ (and not in $AX = A'$).

We call the LDE $R_1 AX = R_1 A'$ a **partial interpolant** for $(AX = A', BX = B')$. It is a linear combination of equations in $AX = A'$. The partial interpolant $R_1 AX = R_1 A'$ can be written in the following form

$$\sum_{x_i \in V_{A \setminus B}} a_i x_i + \sum_{x_i \in V_{AB}} b_i x_i = c \tag{1}$$

where all coefficients a_i, b_i and $c = R_1 A'$ are rational numbers. Observe that the partial interpolant does not contain any variable that occurs only in $BX = B'$ ($V_{B \setminus A}$).

Lemma 2. *The coefficient a_i of each $x_i \in V_{A \setminus B}$ in the partial interpolant $R_1 AX = R_1 A'$ (Equation 1) is an integer.*

Lemma 3. *The partial interpolant $R_1 AX = R_1 A'$ satisfies the first two conditions in the definition of an interpolant. That is,*

1. $AX = A'$ implies $R_1 AX = R_1 A'$
2. $(R_1 AX = R_1 A') \wedge BX = B'$ is unsatisfiable

If $a_i = 0$ for all $x_i \in V_{A \setminus B}$ (equation 1), then the partial interpolant only contains the variables from V_{AB}. In this case the partial interpolant is an interpolant for $(AX = A', BX = B')$. (The proof is given in [10].)

Example 6. Consider the system of LDEs $CX = D$ in Example 2. A proof of unsatisfiability for this system is $R = [\frac{1}{2}, -\frac{1}{2}, \frac{1}{2}]$. Let $AX = A'$ be the first two equations in $CX = D$, that is, $x + y = 1 \wedge x - y = 1$ (in matrix form). Let $BX = B'$ be the third equation in $CX = D$, that is, $2y + 2z = 3$. Observe that $V_{A \setminus B} := \{x\}, V_{AB} := \{y\}, V_{B \setminus A} := \{z\}$. In this case $R_1 = [\frac{1}{2}, -\frac{1}{2}]$. The partial interpolant for the pair $(AX = A', BX = B')$ is $y = 0$, which is also an interpolant because $y \in V_{AB}$.

The following example shows that a partial interpolant need not be an interpolant.

Example 7. Consider the system $CX = D$ in Example 3. A proof of unsatisfiability for this system is $R = [\frac{1}{2}, \frac{1}{2}]$. Let $AX = A'$ be the first equation in $CX = D$, that is, $x - 2y = 0$. Let $BX = B'$ be the second equation in $CX = D$, that is, $x - 2z = 1$. Observe that $V_{A \setminus B} := \{y\}, V_{AB} := \{x\}, V_{B \setminus A} := \{z\}$. In this case $R_1 = [\frac{1}{2}]$. Thus, the partial interpolant for the pair $(AX = A', BX = B')$ is $\frac{1}{2}x - y = 0$. Observe that the partial interpolant is not an interpolant as it contains the variable y, which does not occur in V_{AB}. This is not surprising since we have already seen in Example 5 that $(x - 2y = 0, x - 2z = 1)$ cannot have an interpolant that is a LDE.

We now intuitively describe how to remove variables from the partial interpolant that are not common to $AX = A'$ and $BX = B'$. In example 7 the partial interpolant is $\frac{1}{2}x - y = 0$, where $y \notin V_{AB}$. We show how to eliminate y from $\frac{1}{2}x - y = 0$ in order to obtain an interpolant. We use modular arithmetic in order to eliminate y. Informally, the equation $\frac{1}{2}x - y = 0$ implies $\frac{1}{2}x - y \equiv 0 \ (mod\ \gamma)$ for any rational number γ. Let α denote the greatest common divisor of the coefficients of variables (in $\frac{1}{2}x - y = 0$) that do not occur in V_{AB}. In this example $\alpha = 1$ (gcd of the coefficient of y). We know $\frac{1}{2}x - y = 0$ implies $\frac{1}{2}x - y \equiv 0 \ (mod\ 1)$. Since y is an integer variable $y \equiv 0 \ (mod\ 1)$. We can add $\frac{1}{2}x - y \equiv 0 \ (mod\ 1)$ and $y \equiv 0 \ (mod\ 1)$ to obtain $\frac{1}{2}x \equiv 0 \ (mod\ 1)$ (note that y is eliminated). Intuitively, the linear modular equation $\frac{1}{2}x \equiv 0 \ (mod\ 1)$ is an interpolant for $(x - 2y = 0, x - 2z = 1)$. By using basic modular arithmetic this interpolant can be written as $x \equiv 0 \ (mod\ 2)$.

We now formalize the above intuition to address the case when the partial interpolant contains variables that are not common to $AX = A'$ and $BX = B'$.

Theorem 2. *Assume that the coefficient a_i of at least one $x_i \in V_{A \setminus B}$ in the partial interpolant (Equation 1) is not zero. Let α denote the gcd of $\{a_i | x_i \in V_{A \setminus B}\}$.*
(a) α is an integer and $\alpha > 0$.

(b) Let β be any integer that divides α. Then the following linear modular equation I_β is an interpolant for $(AX = A', BX = B')$.

$$I_\beta := \sum_{x_i \in V_{AB}} b_i x_i \equiv c \ (mod\ \beta)$$

Observe that I_β contains only variables that are common to both $AX = A'$ and $BX = B'$. It is obtained from the partial interpolant by dropping all variables occurring only in $AX = A'$ ($V_{A \setminus B}$) and replacing the linear equality by a modular equality.

The complete proof can be found in [10]. Lemma 3 and Theorem 2 give us a sound and complete algorithm for computing an interpolant for unsatisfiable systems of LDEs (see [10] for algorithm pseudocode). In theorem 2, I_1 is always an interpolant for $(AX = A', BX = B')$. For $\alpha > 1$ theorem 2 allows us to obtain multiple interpolants by choosing different β. For any β that divides α, $I_\alpha \Rightarrow I_\beta$ and $I_\beta \Rightarrow I_1$.

4 System of Linear Modular Equations (LMEs)

In this section we discuss proofs of unsatisfiability and interpolation algorithm for LMEs. We first consider a system of LMEs where all equations have the same modulus l, where l is a rational number. We denote this system as $CX \equiv_l D$, where C denotes an $m \times n$ rational matrix, X denotes a column vector of n integer variables and D denotes a column vector of m rational numbers. The next theorem gives a necessary and sufficient condition for $CX \equiv_l D$ to have an integral solution.

Theorem 3. *The system $CX \equiv_l D$ has no integral solution for X if and only if there exists a rational row vector R such that RC is integral, lR is integral, and RD is not an integer.* (The proof uses reduction to LDEs and is given in [10].)

Definition 3. *We say a system of LMEs $CX \equiv_l D$ is* **unsatisfiable** *if it has no integral solution X. A* **proof of unsatisfiability** *for a system of LMEs $CX \equiv_l D$ is a rational row vector R such that RC is integral, lR is integral, and RD is not an integer.*

Example 8. Consider the system of LMEs $CX \equiv_8 D$ and a proof of unsatisfiability R:

$$CX \equiv_8 D := \begin{bmatrix} 2 & 2 \\ 2 & 1 \\ 4 & 0 \end{bmatrix} \begin{bmatrix} x \\ y \end{bmatrix} \equiv_8 \begin{bmatrix} 4 \\ 4 \\ 4 \end{bmatrix} \qquad \begin{array}{l} R \ = [\frac{1}{4}, -\frac{1}{2}, -\frac{1}{8}] \\ RC = [-1, 0] \\ lR \ = [2, -4, -1] \\ RD = -\frac{3}{2} \end{array}$$

Intuitively, $CX \equiv_8 D$ is unsatisfiable because we can take an integer linear combination of the given equations using lR to get a contradiction $0 \equiv_8 -12$.

Definition 4. (Implication) *A system of LMEs $CX \equiv_l D$* **implies** *a LME $AX \equiv_l B$, if every integral vector X satisfying $CX \equiv_l D$ also satisfies $AX \equiv_l B$.*

Lemma 4. *For every* **integral** *row vector U the system of LMEs $CX \equiv_l D$ imply $UCX \equiv_l UD$.*

4.1 Computing Interpolants for Systems of LMEs

Let $AX \equiv_l A'$ and $BX \equiv_l B'$ be two systems of LMEs such that $AX \equiv_l A' \wedge BX \equiv_l B'$ is unsatisfiable. We show that $(AX \equiv_l A', BX \equiv_l B')$ always has a LME as an interpolant. Let $R = [R_1, R_2]$ denote a proof of unsatisfiability for the system $AX \equiv_l A' \wedge BX \equiv_l B'$ such that $R_1 A + R_2 B$ is integral, $lR = [lR_1, lR_2]$ is integral, and $R_1 A' + R_2 B'$ is not an integer. The following theorem shows that we can take integer linear combinations of equations in $AX \equiv_l A'$ to obtain interpolants.

Theorem 4. *We assume $l \neq 0$. Let S_1 denote the set of non-zero coefficients of $x_i \in V_{A \setminus B}$ in $R_1 AX$. Let S_2 denote the set of non-zero elements of row vector lR_1. If $S_2 = \emptyset$, then the interpolant for $(AX \equiv_l A', BX \equiv_l B')$ is a trivial LME $0 \equiv_l 0$. Otherwise, let $S_2 \neq \emptyset$. Let α denote the gcd of numbers in $S_1 \cup S_2$. (a) α is an integer and $\alpha > 0$. (b) Let β be any integer that divides α. Let $U = \frac{l}{\beta} R_1$. Then $UAX \equiv_l UA'$ is an interpolant for $(AX \equiv_l A', BX \equiv_l B')$.* (The proof is given in [10].)

Example 9. Consider the system of LMEs $CX \equiv_l D$ in Example 8. Let $AX \equiv_l A'$ denote the first two equations in $CX \equiv_l D$ and $BX \equiv_l B'$ denote the last equation in $CX \equiv_l D$. Observe that $V_{A \setminus B} := \{y\}, V_{AB} := \{x\}, V_{B \setminus A} := \emptyset$. A proof of unsatisfiability for $CX \equiv_l D$ is $R = [\frac{1}{4}, -\frac{1}{2}, -\frac{1}{8}]$. We have $R_1 = [\frac{1}{4}, -\frac{1}{2}]$, $lR_1 = [2, -4]$, $R_1 AX$ is $-\frac{1}{2}x$, $S_1 = \emptyset$, $S_2 = \{2, -4\}$, $\alpha = 2$. We can take $\beta = 1$ or $\beta = 2$ to obtain two valid interpolants. For $\beta = 1$, $U = [2, -4]$ and the interpolant $UAX \equiv_l UA'$ is $-4x \equiv_8 -8$ (equivalently $x \equiv_2 0$). For $\beta = 2$, $U = [1, -2]$ and the interpolant $UAX \equiv_l UA'$ is $-2x \equiv_8 -4$ (equivalently $x \equiv_4 2$).

4.2 Handling LMEs with Different Moduli

Consider a system F of LMEs, where equations in F can have different moduli. In order to check the satisfiability of F, we obtain another equivalent system of equations F' such that each equation in F' has the same modulus. This is done using a standard trick. Let $m_1, \ldots, m_k$ represent the different moduli occurring in equations in F. Let m denote the least common multiple of $m_1, \ldots, m_k$. We multiply each equation $t \equiv_{m_i} c$ in F by $\frac{m}{m_i}$ to obtain another equation $\frac{m}{m_i} t \equiv_m \frac{m}{m_i} c$. Let F' represent the set of new equations. All equations in F' have same modulus m. Using basic modular arithmetic one can show that F and F' are equivalent. Suppose F is unsatisfiable. Then the interpolants for any partition of F can be computed by working with F' and using the techniques described in the previous section. For example, let F represent the following system of LMEs $x \equiv_2 1 \wedge x + y \equiv_4 2 \wedge 2x + y \equiv_8 4$. One can work with $F' := 4x \equiv_8 4 \wedge 2x + 2y \equiv_8 4 \wedge 2x + y \equiv_8 4$ instead of F.

5 Algorithms for Obtaining Proofs of Unsatisfiability

Polynomial time algorithms are known for determining if a system of LDEs $CX = D$ has an integral solution or not [20]. We review one such algorithm that is based on the computation of the *Hermite Normal Form (HNF)* of the matrix C.

Using standard Gaussian elimination it can be determined if $CX = D$ has a rational solution or not. If $CX = D$ has no rational solution, then it cannot have any integral solution. In the discussion below we assume that $CX = D$ has a rational solution. Without loss of generality we assume that the matrix C has *full row rank*, that is, all rows of C are linearly independent (linearly dependent equations can be removed).

The HNF of an $m \times n$ matrix C with full row rank is of the form $[E \;\; 0]$ where 0 represents an $m \times (n-m)$ matrix filled with zeros and E is a square $m \times m$ matrix with the following properties: 1) E is lower triangular 2) E is non-singular (invertible) 3) all entries in E are non-negative and the maximum entry in each row lies on the diagonal. The HNF of a matrix can be obtained by three elementary column operations. 1) Exchanging two columns. 2) Multiplying a column by -1. 3) Adding an integral multiple of one column to another column. Each column operation can be represented by a unimodular matrix. A *unimodular matrix* is a square matrix with integer entries and determinant +1 or -1. The product of unimodular matrices is a unimodular matrix. The inverse of a unimodular matrix is a unimodular matrix. The conversion of C to HNF can be represented as follows $CU = [E \;\; 0]$, where U is a unimodular matrix, the sizes of C, U, E are $m \times n, n \times n, m \times m$, respectively and 0 represents an $m \times (n-m)$ matrix filled with zeros ($n \geq m$ because C has full row-rank). The following result shows the use of HNF in determining the satisfiability of a system of LDEs.

Lemma 5. *(Corollary 5.3(b) in [20]) For C, X, D, E defined as above, $CX = D$ has no integral solution if and only if $E^{-1}D$ is not integral. (E^{-1} denotes the inverse of E.)*

Example 10. For the system of LDEs $CX = D$ in example 3 we have the following:

$$\underbrace{\begin{bmatrix} 1 & -2 & 0 \\ 1 & 0 & -2 \end{bmatrix}}_{C} \underbrace{\begin{bmatrix} 1 & 2 & -2 \\ 0 & 1 & -1 \\ 0 & 0 & -1 \end{bmatrix}}_{U} = \underbrace{\begin{bmatrix} 1 & 0 & 0 \\ 1 & 2 & 0 \end{bmatrix}}_{[E \;\; 0]} \qquad \underbrace{\begin{bmatrix} 1 & 0 \\ \frac{-1}{2} & \frac{1}{2} \end{bmatrix}}_{E^{-1}} \underbrace{\begin{bmatrix} 0 \\ 1 \end{bmatrix}}_{D} = \underbrace{\begin{bmatrix} 0 \\ \frac{1}{2} \end{bmatrix}}_{not\ integral}$$

5.1 Obtaining a Proof of Unsatisfiability for a System of LDEs

If a system of LDEs $CX = D$ is unsatisfiable, then we want to compute a row vector R such that RC is integral and RD is not an integer. The following corollary shows that the proof of unsatisfiability can be obtained by using the HNF of C.

Corollary 1. *Given $CX = D$ where C, D are rational matrices, and C has full row rank. Let $[E \;\; 0]$ denote the HNF of C. If $CX = D$ has no integral solution, then $E^{-1}D$ is not integral. Suppose the ith entry in $E^{-1}D$ is not an integer. Let R' denote the ith row in E^{-1}. Then (a) $R'D$ is not an integer and (b) $R'C$ is integral. Thus, R' serves as the required proof of unsatisfiability of $CX = D$.*

In example 10 the second row in $E^{-1}D$ is not an integer. Thus, the proof of unsatisfiability of $CX = D$ is the second row in E^{-1} which is $[-\frac{1}{2}, \frac{1}{2}]$.

Proofs of Unsatisfiability for LMEs: Let $CX \equiv_l D$ be a system of LMEs. Each equation $t_i \equiv_l d_i$ in $CX \equiv_l D$ can be written as an equi-satisfiable LDE, $t_i + lv_i = d_i$, where v_i is a new integer variable. In this way we can reduce $CX \equiv_l D$ to an equi-satisfiable system of LDEs $C'Z = D$. The proof of unsatisfiability of $C'Z = D$ is exactly a proof of unsatisfiability of $CX \equiv_l D$ (see the proof of theorem 3 in [10]).

If a system of LDEs or LMEs is unsatisfiable, then we can obtain a proof of unsatisfiability in polynomial time. This is because HNF computation, matrix inversion, and matrix multiplication can be done in polynomial time in the size of input [20].

6 Handling Linear Diophantine Equations and Disequations

We show how to compute interpolants in presence of linear diophantine disequations. A *linear diophantine disequation (LDD)* is of the form $c_1x_1 + \ldots + c_nx_n \neq c_0$, where $c_0, \ldots, c_n$ are rational numbers and $x_1, \ldots, x_n$ are integer variables. *A system of LDEs+LDDs* denotes conjunctions of LDEs and LDDs. For example, $x + 2y = 1 \wedge x + y \neq 1 \wedge 2y + z \neq 1$ with x, y, z as integer variables represents a system of LDEs+LDDs. We represent a conjunction of m LDDs as $\bigwedge_{i=1}^{m} C_iX \neq D_i$, where C_i is a rational row vector and D_i is a rational number. The next theorem gives a necessary and sufficient condition for a system of LDEs+LDDs to have an integral solution.

Theorem 5. *Let F denote $AX = B \wedge \bigwedge_{i=1}^{m} C_iX \neq D_i$. The following are equivalent:*
1. F has no integral solution
2. F has no rational solution or $AX = B$ has no integral solution.

The proof of $(2) \Rightarrow (1)$ in Theorem 5 is easy. The proof of $(1) \Rightarrow (2)$ is involved and relies on the following lemmas (see full proof in [10]). The first lemma shows that if a system of LDEs $AX = B$ has an integral solution, then every LDE that is implied by $AX = B$, can be obtained by a linear combination of equations in $AX = B$.

Lemma 6. *A system of LDEs $AX = B$ implies a LDE $EX = F$ if and only if $AX = B$ is unsatisfiable or there exists a rational vector R such that $E = RA$ and $F = RB$.*

We use the properties of the *cutting-plane* proof system [20,5] in order to prove lemma 6. The next lemma shows that if a system of LDEs implies a disjunction of LDEs, then it implies one of the LDEs in the disjunction (also called *convexity* [17]).

Lemma 7. *A system of LDEs $AX = B$ implies $\bigvee_{i=1}^{m} C_iX = D_i$ if and only if there exists $1 \leq k \leq m$ such that $AX = B$ implies $C_kX = D_k$.*

We use a theorem from [20] that gives a parametric description of the integral solutions to $AX = B$ in order to prove lemma 7. Let F denote $AX = B \wedge \bigwedge_{i=1}^{m} C_iX \neq D_i$. Using Theorem 5 we can determine whether F has an integral solution in polynomial time. This is because checking if $AX = B$ has an integral solution can be done in polynomial time [20,5]. Checking whether the system F has a rational solution can be done in polynomial time as well [17].

6.1 Interpolants for LDEs+LDDs

We say a system of LDEs+LDDs is **unsatisfiable** if it has no integral solution. Consider systems of LDEs+LDDs $F := F_1 \wedge F_2$ and $G := G_1 \wedge G_2$, where F_1, G_1 are

Example	Preds/Interpolants	VINT2
ex1	$y \equiv_2 1$	2.72s
ex2	$x + y \equiv_2 0$	0.83s
ex4	$x + y + z \equiv_4 0$	0.95s
ex5	$x \equiv_4 0, y \equiv_4 0$	1.1s
ex6	$4x + 2y + z \equiv_8 0$	0.93s
ex7	$4x - 2y + z \equiv_{222} 0$	0.54s
forb1	$x + y \equiv_3 0$	-

(a)

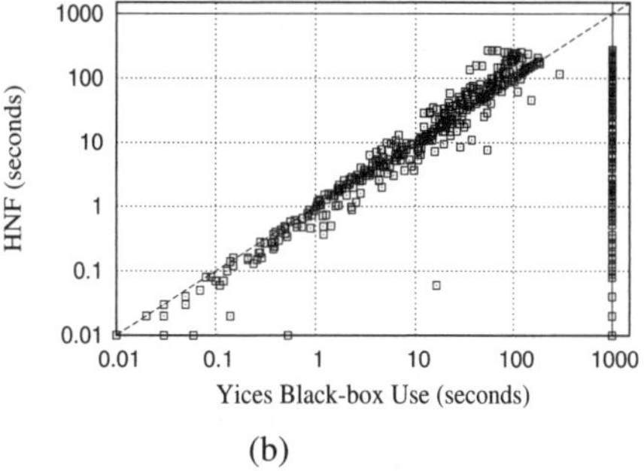

(b)

Fig. 1. (a) Table showing the predicates needed and time taken in seconds. (b) Comparing Hermite Normal Form based algorithm and black-box use of Yices for getting proofs of unsatisfiability.

systems of LDEs and F_2, G_2 are systems of LDDs. $F \wedge G$ represents another system of LDEs+LDDs. Suppose $F \wedge G$ is unsatisfiable. The interpolant for (F, G) can be computed by considering two cases (due to theorem 5):

Case 1: $F \wedge G$ is unsatisfiable because $F_1 \wedge F_2 \wedge G_1 \wedge G_2$ has no rational solution. We can compute an interpolant for (F, G) using the techniques described in [15,19,6]. The algorithms in [15,19,6] can result in interpolants containing inequalities. We describe an alternative algorithm in [10] that always produces a LDE or a LDD as an interpolant.
Case 2: $F \wedge G$ is unsatisfiable because $F_1 \wedge G_1$ has no integral solution. In this case we can compute an interpolant for the pair (F_1, G_1) using the techniques from Section 3. The computed interpolant will be an interpolant for (F, G). It can be a LDE or a LME.

7 Experimental Results

We implemented the interpolation algorithms in a tool called `INTeger INTerpolate (INT2)`. The experiments are performed on a 1.86 GHz Intel Xeon (R) machine with 4 GB of memory running Linux. `INT2` is designed for computing interpolants for formulas (LDEs, LMEs, LDEs+LDDs) that are satisfiable over rationals but unsatisfiable over integers. Currently, there are no other interpolation tools for such formulas.

Use of Interpolants in Verification: We wrote a collection of small C programs each containing a `while` loop and an ERROR label. These programs are safe (ERROR is unreachable). The existing tools based on predicate abstraction and counterexample guided abstraction refinement (CEGAR) such as `BLAST` [9], `SATABS` [1] are not able to verify these programs. This is because the inductive invariant required for the proof contains LMEs as predicates, shown in the "Preds/Interpolants" column of Figure 1(a). These predicates cannot be discovered by the interpolation engine [15,19] used in `BLAST` or by the weakest precondition based procedure used in `SATABS`. The interpolation algorithms described in this paper are able to find the right predicates by computing the interpolants for spurious program traces. Only one unwinding of the `while` loop suffices to find the right predicates in 6 out of 7 cases.

We wrote similar programs in Verilog and tried verifying them with VCEGAR [2], a CEGAR based model checker for Verilog. VCEGAR fails on these examples due to its use of weakest preconditions. Next, we externally provided the interpolants (predicates) found by INT2 to VCEGAR. With the help of these predicates VCEGAR is able to show the unreachability of ERROR labels in all examples except forb1 (ERROR is reachable in the Verilog version of forb1). The runtimes are shown in "VINT2" column.

Müller-Olm and Seidl [16] propose an abstraction technique that can infer linear invariants that are sound with respect to integer arithmetic modulo a power of 2. Their work provides an alternative way of verifying the programs listed in Figure 1(a).

Proofs of Unsatisfiability (PoU) Algorithms: We obtained 459 unsatisfiable formulas (system of LDEs) by unwinding the `while` loops for C programs mentioned above. The number of LDEs in these formulas range from 3 to 1500 with 2 to 4 variables per equation. There are two options for obtaining PoU in INT2. a) Using Hermite Normal Form (HNF) (Section 5.1). We use PARI/GP [4] to compute HNF of matrices. b) By using a state-of-the-art SMT solver Yices 1.0.11 [3] in a black-box fashion (along the lines of [19]). Given a system of LDEs $AX = B$ we encode the constraints that RA is integral and RB is not an integer by means of mixed integer linear arithmetic constraints (see [10]). The SMT solver returns concrete values to elements in R if $AX = B$ is unsatisfiable. The comparison between (a) and (b) is shown in Figure 1(b). There is a timeout of 1000 seconds per problem. The HNF based algorithm is able to solve all problems, while the black-box usage of Yices cannot solve 102 problems within the timeout. Thus, the HNF based method is superior over the black-box use of Yices.

Note that the interpolation algorithms proposed in our paper are independent of the algorithm used to generate the PoU. Any decision procedure that can produce PoU according to definitions 1, 3 can be used (we are not restricted to using HNF or Yices).

8 Conclusion

We presented polynomial time algorithms for computing proofs of unsatisfiability and interpolants for conjunctions of linear diophantine equations, linear modular equations and linear diophantine disequations. These interpolation algorithms are useful for discovering modular/divisibility predicates from spurious counterexamples in a counterexample guided abstraction refinement framework. In future, we plan to work on interpolating theorem provers for integer linear arithmetic and bit-vector arithmetic and make use of the satisfiability modulo theories framework.

Acknowledgment. We thank Axel Legay and Jeremy Avigad for their valuable comments.

References

1. SATABS 1.9 website, http://www.verify.ethz.ch/satabs/
2. VCEGAR 1.3 website, http://www.cs.cmu.edu/~modelcheck/vcegar/

3. Yices 1.0.11 website, http://yices.csl.sri.com/
4. PARI/GP, Version 2.3.2 (2006), http://pari.math.u-bordeaux.fr/
5. Bockmayr, A., Weispfenning, V.: Solving numerical constraints. In: Robinson, A., Voronkov, A. (eds.) Handbook of Automated Reasoning, pp. 751–842 (2001)
6. Cimatti, A., Griggio, A., Sebastiani, R.: Efficient interpolation in satisfiability modulo theories. In: Ramakrishnan, C.R., Rehof, J. (eds.) TACAS 2008. LNCS, vol. 4963, pp. 367–381. Springer, Heidelberg (2008)
7. Clarke, E., Grumberg, O., Jha, S., Lu, Y., Veith, H.: Counterexample-guided abstraction refinement for symbolic model checking. J. ACM 50(5) (2003)
8. Craig, W.: Linear reasoning. a new form of the herbrand-gentzen theorem. J. Symb. Log. 22(3), 250–268 (1957)
9. Henzinger, T.A., Jhala, R., Majumdar, R., McMillan, K.L.: Abstractions from proofs. In: POPL, pp. 232–244. ACM Press, New York (2004)
10. Jain, H., Clarke, E.M., Grumberg, O.: Efficient craig interpolation for linear diophantine (dis)equations and linear modular equations. Technical Report CMU-CS-08-102, Carnegie Mellon University, School of Computer Science (2008)
11. Jhala, R., McMillan, K.L.: A Practical and Complete Approach to Predicate Refinement. In: Hermanns, H., Palsberg, J. (eds.) TACAS 2006 and ETAPS 2006. LNCS, vol. 3920, pp. 459–473. Springer, Heidelberg (2006)
12. Kapur, D., Majumdar, R., Zarba, C.G.: Interpolation for data structures. In: SIGSOFT 2006/FSE-14, pp. 105–116. ACM, New York (2006)
13. Kroening, D., Weissenbacher, G.: Lifting propositional interpolants to the word-level. In: FMCAD, pp. 85–89. IEEE, Los Alamitos (2007)
14. McMillan, K.L.: Interpolation and sat-based model checking. In: Hunt Jr., W.A., Somenzi, F. (eds.) CAV 2003. LNCS, vol. 2725, pp. 1–13. Springer, Heidelberg (2003)
15. McMillan, K.L.: An Interpolating Theorem Prover. In: Jensen, K., Podelski, A. (eds.) TACAS 2004. LNCS, vol. 2988, pp. 16–30. Springer, Heidelberg (2004)
16. Müller-Olm, M., Seidl, H.: Analysis of modular arithmetic. ACM Trans. Program. Lang. Syst. 29(5), 29 (2007)
17. Nelson, G., Oppen, D.C.: Simplification by cooperating decision procedures. ACM Trans. Program. Lang. Syst. 1(2), 245–257 (1979)
18. Pudlák, P.: Lower bounds for resolution and cutting plane proofs and monotone computations. J. Symb. Log. 62(3), 981–998 (1997)
19. Rybalchenko, A., Sofronie-Stokkermans, V.: Constraint solving for interpolation. In: Cook, B., Podelski, A. (eds.) VMCAI 2007. LNCS, vol. 4349, pp. 346–362. Springer, Heidelberg (2007)
20. Schrijver, A.: Theory of linear and integer programming. John Wiley & Sons, NY (1986)
21. Yorsh, G., Musuvathi, M.: A combination method for generating interpolants. In: Nieuwenhuis, R. (ed.) CADE 2005. LNCS (LNAI), vol. 3632, pp. 353–368. Springer, Heidelberg (2005)

Linear Arithmetic with Stars

Ruzica Piskac and Viktor Kuncak

School of Computer and Communication Sciences, EPFL, Switzerland

Abstract. We consider an extension of integer linear arithmetic with a "star" operator takes closure under vector addition of the solution set of a linear arithmetic subformula. We show that the satisfiability problem for this extended language remains in NP (and therefore NP-complete). Our proof uses semilinear set characterization of solutions of integer linear arithmetic formulas, as well as a generalization of a recent result on sparse solutions of integer linear programming problems. As a consequence of our result, we present worst-case optimal decision procedures for two NP-hard problems that were previously not known to be in NP. The first is the satisfiability problem for a logic of sets, multisets (bags), and cardinality constraints, which has applications in verification, interactive theorem proving, and description logics. The second is the reachability problem for a class of transition systems whose transitions increment the state vector by solutions of integer linear arithmetic formulas.

1 Introduction

Decision procedures [1,5,7,10,15] are among key techniques that enable automated verification of infinite state systems, as, for example, in software model checkers [2,6,12]. These techniques are also increasingly used to raise the level of automation in interactive theorem provers [8,17,24]. We believe that an important step towards making such theorem provers even more effective is the development of decision procedures for new classes of formulas that go beyond the traditionally considered uninterpreted function symbols, arrays, free data structures, and linear arithmetic. In this paper we present a decision procedure for one such class, which introduces certain *unbounded* sums into linear arithmetic. Specifically, our decision procedure solves the satisfiability problem

$$F_0(\vec{u}) \wedge \exists N \geq 0. \exists \vec{x}_1, \ldots, \vec{x}_N . \vec{u} = \sum_{i=1}^{N} \vec{x}_i \wedge \bigwedge_{i=1}^{N} F(\vec{x}_i) \qquad (1)$$

where F_0 and $F(\vec{x})$ are any quantifier-free Presburger arithmetic (QFPA) formulas, all variables of F are among $\vec{x}$, and where $\vec{u}, \vec{x}_i$ are integer vectors. Because N is not known, (1) is not immediately a QFPA formula. Using notation $A^* = \{\vec{u} \mid \exists N \geq 0. \exists \vec{x}_1, \ldots, \vec{x}_N \in A. \vec{u} = \sum_{i=1}^{N} \vec{x}_i\}$ for closure of a set of vectors under addition, we denote (1) by $F_0(\vec{u}) \wedge \vec{u} \in \{\vec{x} \mid F(\vec{x})\}^*$. This paper shows that the satisfiability for the class QFPA^* of such formulas with $*$ operator (Figure 1) is in NP, generalizing the well known NP-completeness for QFPA satisfiability.

A. Gupta and S. Malik (Eds.): CAV 2008, LNCS 5123, pp. 268–280, 2008.

QFPA* formulas: $F_0 \wedge \vec{u} \in \{\vec{x} \mid F\}^*$ (free variables of F are among $\vec{x}$)
QFPA formulas:
$F ::= A \mid F_1 \wedge F_2 \mid F_1 \vee F_2 \mid \neg F_1$
$A ::= T_1 \leq T_2 \mid T_1 = T_2$
$T ::= k \mid C \mid T_1 + T_2 \mid C \cdot T_1 \mid \mathsf{ite}(F, T_1, T_2)$
terminals: k - integer variable; C - integer constant

Fig. 1. Quantifer-free Presburger arithmetic and our QFPA* extension

Previous results. We consider the satisfiability problem for QFPA* (Figure 1) where variables are interpreted over non-negative integers (the version with arbitrary integers reduces to this case by representing integers and their sums as differences of non-negative integers). The satisfiability problem for QFPA* is decidable because the set of solutions of a QFPA formula is a *semilinear set* [11], and a closure of a semilinear set under star can be expressed in QFPA [16]. However, the constructions behind these decidability results do not provide good complexity bounds because semilinear set representation can be exponentially large. First algorithm for QFPA* satisfiability that avoids explicit construction of semilinear set representation is the PSPACE algorithm in [21]. The present paper is the first to establish the exact complexity of QFPA* satisfiability, namely NP-completeness. To show this result, we will use bounds on solutions of integer linear programming problems with exponentially many variables [20], bounds on seminilinear set generators [22], and Carathéodory bounds for integer cones [9]. Our proof builds on some of the ideas previously introduced in [14, 21].

Application to reasoning about collections. Our decision procedure enables reasoning about collections of objects (sets and multisets) and their cardinalities, which was our original motivation for introducing it in [21]. Previously, Zarba [25] considered decision procedures for multiset constraints but without the cardinality operator, presenting a direct reduction to QFPA. The cardinality operator makes the reduction in [25] inapplicable. Section 2 reviews the approach [21] to reduce multiset constraints with cardinalities to QFPA*.

Application to reasoning about transition systems. In addition to reasoning about multisets, we identify another application of constraints with stars. We consider infinite-state transitions systems whose state consists of finite control and a finite number of integer counters, and whose transitions increment counters by a solution vector of a linear arithmetic formula given by the finite control. We show that the reachability problem for a class of such systems reduces to a generalization of QFPA* with multiple nested star operators. We show that our proof techniques and the NP-membership result extend to this more general case, which gives NP-completeness of the reachability problem.

Contributions. We summarize the contributions of our paper as follows:

- We present a polynomial-time algorithm (Section 3) for reducing QFPA* satisfiability to QFPA satisfiability, showing NP-completeness of QFPA*

satisfiability. This yields an algorithm for reasoning about constraints on multisets in the presence of cardinality operators (see Section 2);
- We generalize our reduction to multiple nested star operators. We show that the generalized constraints enable reasoning about reachability in a class of symbolically represented transition systems (Section 4).

2 From Multisets to Linear Arithmetic with Star

The satisfiability problem for QFPA* arises as the key step in checking the satisfiability of constraints on multisets in the presence of a cardinality operator. The reduction from multiset constraints to QFPA* is presented in [21]. We here motivate multiset constraints and use examples to illustrate the reduction.

Uses of set and multiset constraints. Sets and multisets directly arise in verification conditions for proving properties of programs in languages and paradigms such as SETL [23] and Gamma [4, Page 103]. In programming languages such as Java, data abstraction can be used to show that data structures satisfy set specifications, and then techniques based on sets become applicable for verifying data structure clients [13, 18]. Multisets and sets are also present in libraries of interactive provers Isabelle [19] and KIV [3]. Our results yield decision procedures that can increase the automation within such systems. As a simple running example, consider a verification condition for insertion of an element represented by a singleton multiset s into a container represented by multiset L. To prove that an integer field correctly maintains the size of a container, we need to prove validity of the constraint $|s| = 1 \rightarrow |L \uplus s| = |L| + 1$, that is, unsatisfiability of the corresponding negation $|s| = 1 \wedge |L \uplus s| \neq |L| + 1$.

Representation of multisets and sets. We represent sets as well as multisets (*bags*) with their characteristic functions. A multiset m is a function $E \rightarrow \mathbb{N}$, where E is the universe and $\mathbb{N}$ is the set of non-negative integers. The value $m(e)$ is multiplicity (number of occurrences) of element e in multiset m. We assume that the domain E is fixed and finite but of unknown size. We represent sets within our formulas as special multisets m for which $m(e) = 0 \vee m(e) = 1$ for all elements e.

Operations on multisets. We consider a natural class of operations and relations on multisets that are given pointwise by linear arithmetic formulas. For a relation given by QFPA formula $F(x_1, \ldots, x_k)$, we define the corresponding relation on multisets $m_1, \ldots, m_k$ by $\forall e.\, F(m_1(e), \ldots, m_k(e))$. For example, we define subset $m_1 \subseteq m_2$ by $\forall e. m_1(e) \leq m_2(e)$, multiset sum $m_1 = m_2 \uplus m_3$ by $\forall e. m_1(e) = m_2(e) + m_3(e)$, and union $m_1 = m_2 \cup m_3$ by $\forall e. m_1(e) = \max(m_2(e), m_3(e))$. To define max and other operations we use if-then-else operator $\mathsf{ite}(F, t_1, t_2)$ in QFPA, which denotes t_1 when F holds and t_2 otherwise. We define multiset difference $m_1 = m_2 \setminus m_3$ by $\forall e. m_1(e) = \mathsf{ite}(m_2(e) \leq m_3(e), 0, m_2(e) - m_3(e))$. In our example verification condition, we introduce a new multiset variable y such that $y = L \uplus s$ and we express this condition by $\forall e.\, y(e) = L(e) + s(e)$.

Cardinality operator and sums. We also permit the cardinality operator $|m|$ on multisets, given by $|m| = \sum_{e \in E} m(e)$. This operator turns a multiset expression into an integer expression, and we allow arbitrary QFPA operators on cardinalities. In our example verification condition, in addition to $\forall e.\, y(e) = L(e) + s(e)$ we have constraint $|s| = 1$, which becomes $\sum_{e \in E} s(e) = 1$, and the constraint $|y| \neq |L| + 1$, which becomes $(\sum_{e \in E} y(e)) \neq (\sum_{e \in E} L(e)) + 1$.

Without changing expressive power, we generalize the sum notation and introduce expressions of the form $(u_1, \ldots, u_d) = \sum_{e \in E}(m_1(e), \ldots, m_d(e))$ where $u_1, \ldots, u_d$ are integer variables and $m_1, \ldots, m_d$ are multiset variables. [1] By introducing non-negative integer variables u_i for the results of sums and grouping multiple sums into sum of vectors, we reduce any multiset formula to form

$$F_0 \wedge \bigwedge_{i=1}^{c} (\forall e. F_i(\vec{m}(e))) \wedge \vec{u} = \sum_{e \in E} \vec{m}(e) \tag{2}$$

where $F_0, F_1, \ldots, F_c$ are QFPA formulas, $\vec{m}(e)$ denotes $(m_1(e), \ldots, m_d(e))$, and $\vec{u} = (u_1, \ldots, u_d)$. The negation of our example verification condition becomes

$$\begin{array}{l} u_s = 1 \wedge u_y \neq u_L + 1 \wedge \\ (\forall e. y(e) = L(e) + s(e)) \ \wedge \ (u_s, u_y, u_L) = \sum_{e \in E}(s(e), y(e), L(e)) \end{array} \tag{3}$$

Reduction to QFPA*. The satisfiability of the example constraint (3) is equivalent to the satisfiability of the QFPA* constraint

$$u_s = 1 \wedge u_y \neq u_L + 1 \ \wedge \ (u_s, u_y, u_L) \in \{(s', y', L') \mid y' = L' + s'\}^* \tag{4}$$

We prove equisatisfiability of (3) and (4); see [21, Theorem 2] for the analogous proof for an arbitrary formulas on multisets. Suppose that (3) has a solution $E = \{e_1, \ldots, e_N\}$. Because functions $x, y, L : E \to \mathbb{N}$ satisfy $\forall e. y(e) = L(e) + s(e)$, the vectors $(s(e_i), y(e_i), L(e_i))$ for $1 \leq i \leq N$ represent N solutions of QFPA formula $y' = L' + s'$. Consequently, the assignment to u_s, u_y, u_L is a sum of N solutions of $y' = L' + s'$, so (4) is satisfiable with the same values of u_s, u_y, u_L. Conversely, given a solution to (4) we know that u_s, u_y, u_L is a sum of a finite number, say N, of solutions of $y' = L' + s'$, that is, there are N vectors (y'_i, L'_i, s'_i) such that $y'_i = L'_i + s'_i$. We then introduce a distinct element e_i for each of these N solutions, let $E = \{e_1, \ldots, e_N\}$ and let $s(e_i) = s'_i$, $y(e_i) = y'_i$, $L(e_i) = L'_i$. We obtain a solution of (4), as desired. Note that this proof did not depend on the structure of QFPA formulas. In general, we obtain equisatisfiability of (2) and the QFPA* formula $F_0 \wedge \vec{u} \in \{\vec{x} \mid \bigwedge_{i=1}^{c} F_i(\vec{x})\}^*$. Therefore, to check satisfiability of an expressive class of constraints on sets and multisets, we construct (in polynomial time) an equisatisfiable QFPA* formula. We next present an algorithm for checking QFPA* satisfiability.

[1] We assume that the summands are multiset variables because we introduce fresh multiset variables for subterms. However, it is easy to see that we can allow arbitrary QFPA terms as summands without changing the expressive power, see [21].

3 Linear Arithmetic with Star Operator Is in NP

We show how to reduce, in polynomial time, QFPA* satisfiability to QFPA satisfiability. This will show that QFPA* satisfiability is in NP. QFPA and therefore QFPA* subsume propositional logic, their satisfiability is therefore NP-hard, so our results establish NP-completeness of QFPA* satisfiability.

Consider satisfiability of a QFPA* formula $F_0 \wedge \vec{u} \in \{\vec{v} \mid F(\vec{v})\}^*$. Because F is a QFPA formula, its solution set $\{\vec{v} \mid F(\vec{v})\}$ is a semilinear set [11]. Therefore, there exist finitely many *generating vectors* $\vec{a}_i$, $\vec{b}_{ij}$ whose non-negative integer linear combination spans $\{\vec{v} \mid F(\vec{v})\}$. The number of generating vectors can be exponential, so we avoid explicitly constructing them. We instead apply [22] to compute an upper bound on the size of generating vectors. This gives us bounds on coefficients in an *exponentially large* QFPA formula equisatisfiable with (1). We combine the following two constructions to find a *polynomially large* equisatisfiable formula.

1. We apply a small model theorem for QFPA that follows from [20]. Because the exponential QFPA formula has only polynomially many atomic formulas, we obtain a polynomial bound on the number of bits needed for $\vec{u}$ in the smallest solution of (1).
2. We apply twice a theorem on the size of minimal generator of integer cone [9] to prove that only polynomially many vectors suffice to generate $\vec{u}$.

Finally, we show that we can group linear combinations of generating vectors into linear combination of polynomially many variables denoting solution vectors of F. Despite the multiplication of variables, we can express such linear combination as a QFPA formula because coefficients in linear combination are bounded by the bound on $\vec{u}$.

Our proof builds on several non-trivial previous results, but its algorithmic consequences are simple: we can replace (1) with a problem where N is bounded by a polynomial function of F_0 and F, and where sum over solutions of F is replaced by an integer linear combination of solutions of F, with coefficients of the linear combination polynomially bounded. We proceed to describe our construction in more detail, including concrete bounds needed to implement our algorithm.

3.1 Estimating Coefficient Bounds of Disjunctive Form

The results on which we rely are usually expressed for integer linear programming problems, so we compute dimensions and coefficient bounds for integer linear programming problems arising from QFPA formula.

Let F be a QFPA formula. We can convert F into an equivalent disjunction of integer linear programming problems $\bigvee_{i=1}^{l} A_i\vec{x} = \vec{b}_i$. Let m_i be a number of rows in A_i and let n_i be a number of columns in A_i and let a_i be a maximal absolute value of all coefficient occurring in A_i and b_i. For a given F, define $m_F = \max_{i=1}^{l} m_i$, $n_F = \max_{i=1}^{l} n_i$ and $a_F = \max_{i=1}^{l} a_i$.

Lemma 1 (Values of m_F, n_F and a_F). *Let F be a QFPA formula. If a subformula does not occur within any* ite *expression we say that it has positive polarity if it occurs under an even number of negations and say it has negative polarity if it occurs under an odd number of negations. If a subformula occurs within an* ite *expression we say that it has no polarity. Let g be the number of atomic formula occurences of the form $t_1 = t_2$ that have positive polarity in F, and let h be the number of remaining atomic formulas. Let v be the number of variables in F and a the maximum of absolute values of integer constants. Then $m_F \leq g + h$, $n_F \leq v + h$, and $a_F \leq a + 1$.*

Proof. We can transform $F[\mathsf{ite}(C, t_1, t_2)]$ into a disjunction of $C \wedge F[t_1]$ and $\neg C \wedge F[t_2]$. Repeating this transformation we eliminate all ite expressions and obtain disjuncts whose size is polynomial in the size of F. Let D be one of the disjuncts after such ite elimination. The polarity of all g atomic formulas $t_1 = t_2$ that occur positively in F remains positive in each D. Each of the remaining h atomic formulas becomes of the form $t_1 \leq t_2$, $t_1 = t_2$ or disjunction $t_1 \leq t_2 \vee t_1' \leq t_2'$. In disjunctive normal form of D, each of the h atomic formulas $t_1 \leq t_2$ may require addition of at most one fresh variable to be converted into equality $t_1 + x \leq t_2$. The resulting number of variables is therefore bounded by $v + h$ whereas the total number of atomic formulas is bounded by $g + h$. When transforming $t_1 < t_2$ into $t_1 + 1 \leq t_2$ we change the constants part of $t_2 - t_1$ by one, so $a_F \leq a + 1$. ∎

3.2 Existence and Size of Solution Set Generators

This section describes the solutions of a QFPA formula F using semilinear sets, provides bounds on the norms of vectors that represent these semilinear sets, and uses this characterization to describe the set $\{\vec{v} \mid F(\vec{v})\}^*$. Define vector set addition by $A + B = \{\vec{a} + \vec{b} \mid \vec{a} \in A, \vec{b} \in B\}$.

Definition 1. *Given a finite set $S \subseteq \mathbb{N}^n$ and $\vec{a} \in \mathbb{N}^n$, we define the* linear set *$L(a; S)$ as $\{a\} + S^*$. We call $\vec{a}$ the* base *vector, and call elements of S the* step *vectors. A* semilinear set *is a union of finitely many linear sets.*

If $Z = \cup_{i=1}^{q} L(a_i; S_i)$ is a representation of a semilinear set, we call the base vectors a_i and step vectors b_{ij} the *generators* for semilinear set. [11] showed that the set of solutions of a QFPA formula is a semilinear, so it is given by some finite set of generators. Morover, [22] shows that for formula $F(\vec{v})$ of form $A\vec{v} \leq b$ each generator $\vec{g}$ satisfies $||\vec{g}||_1 \leq (2 + ||A||_{1,\infty} + ||\vec{b}||_\infty)^m$ where A is a $m \times n$ matrix. Combining this result with Lemma 1, we obtain the following Lemma 2:

Lemma 2. *For each QFPA formula F, there exist q base vectors $\vec{a}_i$, $1 \leq i \leq q$, and for each i the corresponding q_i step vectors $\vec{b}_{ij}$ for $1 \leq j \leq q_i$, all with norms bounded by $(2+2(n_F+1)a_F)^{2m_F}$ where n_F, m_F, a_F are from Lemma 1 such that*

$$F(\vec{u}) \quad \Leftrightarrow \quad \exists \nu_{ij}. \bigvee_{i=1}^{q} (\vec{u} = \vec{a}_i + \sum_{j=1}^{q_i} \nu_{ij} \vec{b}_{ij}) \tag{5}$$

We can now express membership in the set $\{\vec{v} \mid F(\vec{v})\}^*$ using QFPA formula, using the following lemma that follows from Lemma 2 and the definition of the star operator.

Lemma 3. *Let F by a QFPA formula and $\vec{a}_i$, $\vec{b}_{ij}$ be from Lemma 2. Then $\vec{u} \in \{\vec{v} \mid F(\vec{v})\}^*$ is equivalent to*

$$\exists (\mu_i)_i, (\nu_{ij})_{ij}.\ \vec{u} = \sum_{i=1}^{q} (\mu_i \vec{a}_i + \sum_{j=1}^{q_i} \nu_{ij} \vec{b}_{ij}) \ \wedge \ \bigwedge_{i=1}^{q} (\mu_i = 0 \rightarrow \sum_{j=1}^{q_i} \nu_{ij} = 0) \qquad (6)$$

Because the number of $\vec{a}_i$ and $\vec{b}_{ij}$ vectors can be exponential, Lemma 3 shows that QFPA* satisfiability reduces to satisfiability of an exponentially larger QFPA formula. Our goal is to improve the reduction and obtain a polynomial QFPA formula.

3.3 Selecting Polynomially Many Generators

In this section we establish bounds on the number of generators needed to generate any particular solution vector $\vec{u}$: if $\vec{u}$ is a linear combination of generators, then it is also a linear combination of a polynomial subset of generators that form a smaller semilinear set. We prove this fact using a theorem about sparse solutions of integer linear programming problems. Given a set of vectors X and a vector $\vec{b} \in X^*$, the following fact determines the bound on the number of vectors sufficient for representing $\vec{b}$ as a linear combination of vectors from X.

Fact 1 (Theorem 1 (ii) in [9]). *Let $X \subseteq \mathbb{Z}^d$ be a finite set of integer vectors and let $\vec{b} \in X^*$. Then there exists a subset $\tilde{X}$ such that $\vec{b} \in \tilde{X}^*$ and $|\tilde{X}| \leq 2d \log(4dM)$, where $M = \max_{x \in X} ||x||_\infty$.*

Fact 1 has been applied in [14] in order to establishing membership in NP for constraints on sets with cardinality operators. However, in the case of multisets and QFPA* we need to generalize this idea because of dependencies between the base vectors and the corresponding step vectors.

Theorem 1. *Let F be QFPA formula and $\vec{a}_i$, $\vec{b}_{ij}$, $\vec{u}$, q, q_i be from Lemma 3. Then there exists sets $I_0, I_1 \subseteq \{1, \ldots, q\}$ and $J \subseteq \cup_{i=1}^{q} \{(i,1), \ldots, (i, q_i)\}$ such that*

$$\exists (\mu_i)_i, (\nu_{ij})_{ij}.\ \vec{u} = \sum_{i \in I_0} (\vec{a}_i + \sum_{(i,j) \in J} \nu'_{ij} \vec{b}_{ij}) + \sum_{i \in I_1} \mu'_i \vec{a}_i \qquad (7)$$

and $|I_0| \leq |J| \leq B$, and $|I_1| \leq B$, where $B = 2n_F(\log 4n_F + 2m_F \log(2 + 2(n_F + 1)a_F))$.

Proof. By assumption, $\vec{u} = \sum_{i=1}^{q} (\mu_i \vec{a}_i + \sum_{j=1}^{q_i} \nu_{ij} \vec{b}_{ij})$ and $\bigwedge_{i=1}^{q} (\mu_i = 0 \rightarrow \sum_{j=1}^{q_i} \nu_{ij} = 0)$. Removing zero indices, assume that μ_i and ν_{ij} are strictly positive. Define $\vec{a} = \sum_i \mu_i \vec{a}_i$ and $\vec{b} = \sum_{ij} \nu_{ij} \vec{b}_{ij}$, so $\vec{u} = \vec{a} + \vec{b}$. From $\vec{b} = \sum_i \nu_{ij} \vec{b}_{ij}$

and Fact 1 we conclude that there exists a set J of indices (i,j) and coefficients ν'_{ij} such that $\vec{b} = \sum_{(i,j)\in J} \nu'_{ij}\vec{b}_{ij}$ and $|J| \leq B = 2n_F \log(4n_F M)$ where M is the bound on generators. To satisfy the dependencies between $\vec{b}_{ij}$ and $\vec{a}_i$, let $I_0 = \{i \mid \exists j.(i,j) \in J\}$. Note $|I_0| \leq |J|$. Let $\vec{a}_0 = \sum_{i\in I_0} \vec{a}_i$. Then $\vec{a}_0 + \vec{b}$ is generated by vectors whose indices are I_0 and J. It remains to generate $\vec{a} - \vec{a}_0$. Note that $\vec{a} - \vec{a}_0 = \sum_{i\in I_0}(\mu_i - 1)\vec{a}_i + \sum_{i\in\{1,\ldots,q\}\setminus I_0} \mu_i\vec{a}_i$. Applying once again Fact 1 we conclude that there exists $I_1 \subseteq \{1,\ldots,q\}$ with $|I_1| \leq B$ such that $\vec{a} - \vec{a}_0 = \sum_{i\in I_1} \mu'_i a_i$. Using the bound $M = (2+2(n_F+1)a_F)^{2m_F}$ from Lemma 2 we obtain the desired value of B. ∎

3.4 Grouping Generators into Solutions

In previous two sections we have shown that if $\vec{u} \in \{\vec{v} \mid F(\vec{v})\}^*$, then $\vec{u}$ is a particular linear combination of polynomially many generating vectors $\vec{a}_i$, $\vec{b}_{ij}$ that are themselves polynomially bounded. This suggests the idea of guessing polynomially many bounded vectors, checking whether they are generators, and then checking whether $\vec{u}$ is their linear combination. We next show that we can avoid the problem of checking whether a vector is a generator and reduce the problem to checking whether a vector is a solution of F. The way we stated Theorem 1 already suggests this approach.

Lemma 4. *Let F be a QFPA formula and $\vec{u} \in \{\vec{v} \mid F(\vec{v})\}^*$. Then there exist k vectors $\vec{c}_1,\ldots,\vec{c}_k$ for $k \leq 4n_F(\log 4n_F + 2m_F \log(2 + 2(n_F + 1)a_F))$ such that $\wedge_{i=1}^k F(\vec{c}_i)\ \wedge u = \sum_{i=1}^k \lambda_i\vec{c}_i$ for some non-negative integers λ_i.*

Proof. In Theorem 1 simply note that $\vec{a}_i + \sum_{(i,j)\in J} \nu'_{ij}\vec{b}_{ij}$ are solutions of F and that their number is bounded by B. Similarly, $\vec{a}_i$ are solutions of F and their number is bounded by B. The total number of solutions is bounded by $2B$ where B is from Theorem 1. ∎

3.5 NP-Algorithm

Our NP-algorithm for checking satisfiability of QFPA* formula $F_0 \wedge \vec{u} \in \{\vec{v} \mid F(\vec{v})\}^*$ uses previously introduced bounds. First, using Lemma 1 we calculate the values of m_F, n_F and a_F. Using those values and Lemma 4 we estimate an upper bound $k = 4n_F(\log 4n_F + 2m_F \log(2 + 2(n_F + 1)a_F))$ on the number of solution vectors $\vec{x}_i$. We obtain equisatisfiable formula

$$F_0 \wedge \vec{u} = \lambda_1\vec{x}_1 + \ldots + \lambda_k\vec{x}_k \wedge \bigwedge_{i=1}^{k} F(\vec{x}_i) \tag{8}$$

Note, however, that, although it is polynomial in size, (8) is not a QFPA formula because it contains multiplication of variables $\lambda_i \cdot \vec{x}_i$. We address this problem by showing that the values of λ_i in smallest solutions have a polynomial number of bits, which allows us to express multiplication using bitwise expansion.

3.6 Multiplication by Bounded Bit Vectors

To express terms $\lambda_i \vec{c}_i$ from Lemma 4 as a QFPA term, we show that the smallest solution $\vec{u}$, if exists, is bounded [20]. Suppose that r' is a bound on $\vec{u}$ of formula $F_0 \wedge \vec{u} \in \{\vec{v} \mid F(\vec{v})\}^*$. Because λ_i in formula (8) must be a non-negative integer, $\lambda_i \leq ||\vec{u}||_\infty \leq r'$, so each λ_i is also bounded by r' and can be represented as a bit-vector of size r for $r = \lceil \log r' \rceil$. Let $\lambda_i = \overline{\lambda_{ir} \ldots \lambda_{i1} \lambda_{i0}} = \sum_{j=0}^{r} \lambda_{ij} 2^j$. Then

$$\lambda_i \vec{c}_i = (\sum_{j=0}^{r} \lambda_{ij} 2^j)\vec{c}_i = \sum_{j=0}^{r} 2^j (\lambda_{ij} \vec{c}_i) = \sum_{j=0}^{r} 2^j \mathsf{ite}(\lambda_{ij}, \vec{c}_i, 0) =$$
$$\mathsf{ite}(\lambda_{i0}, \vec{c}_i, 0) + 2(\mathsf{ite}(\lambda_{i1}, \vec{c}_i, 0) + 2(\mathsf{ite}(\lambda_{i2}, \vec{c}_i, 0) + \ldots))$$

It remains to show how to compute the estimate r'.

3.7 Estimating Solution Size Bounds

Theorem 2. *Let F_0 be a QFPA formula. Let $\vec{u} = (u_1, \ldots, u_d)$ denote a d-dimensional vector of variables ranging over non-negative integers. Let F be a QFPA formula which does not share any variable with F_0 and $\vec{u}$. If formula $F_0 \wedge \vec{u} \in \{\vec{v} \mid F(\vec{v})\}^*$ is satisfiable, then there exists a non-negative solution vector $\vec{w}$ for variables $\vec{u}$ such that $||\vec{w}||_\infty \leq r' = n(ma)^{2m+1}$ where n, m and a are defined by*

1. $m := d + m_{F_0}$
2. $n := n_{F_0} + 6d(\log(4d) + 2m_F \log(2 + (n_F + 1)a_F))$
3. $a := \max\{a_{F_0}, (2 + 2(n_F + 1)a_F)^{2m_F}\}$

Proof. We establish a bound on the size of the solution vector using two facts. First, as shown in Lemma 3, the fact that $\vec{w}$ is a solution of $\vec{u} \in \{\vec{v} \mid F(\vec{v})\}^*$ implies that $\vec{w}$ is a linear combination of generators of a semilinear set and can be expressed as

$$\vec{w} = \sum_{i=1}^{q} (\mu_i \vec{a}_i + \sum_{j=1}^{q_i} \nu_{ij} \vec{b}_{ij})$$

If we represent the above condition as form $A\vec{x} = \vec{b}$, the matrix A consists of generators of semilinear set and the negative identity matrix $-I$, while the vector $\vec{x}$ consists of $\vec{u}$ as well parameters μ_i and ν_{ij}. The dimensions of the matrix A are $d \times (n_G + d)$, where n_G is the number of generators. By Theorem 1, $n_G \leq 6d(\log(4d) + 2m_F \log(2 + 2(n_F + 1)a_F))$.

Next, observe that $\vec{w}$ is a component of the solution vector of F_0. This implies that there is a matrix B with dimensions m_{F_0} and n_{F_0} and a vector $\vec{v}$ such that $B\vec{w} = \vec{v}$.

Combining matrices A and B we obtain a new matrix C with $d + m_{F_0}$ rows and the number of columns $n_{F_0} + 6d(\log(4d) + 2m_F \log(2 + (n_F + 1)a_F))$. To establish an upper bound on the maximum of absolute values in C, we use an upper bound on the size of generating vectors in a semilinear set given by Lemma 2. We obtain the final result by applying to C the theorem on upper bounds of smallest solutions of integer linear programming problems [20]. ∎

Putting everything together, the bound k on the number of solutions of F and bounds on λ_i enables us to generate, in polynomial time, a QFPA formula equisatisfiable with the original QFPA* formula.

4 Reachability in a Class of Transition Systems

We next show another application of satisfiability checking for extensions of QFPA with star operators. We consider the reachability problem in systems whose state has finite control and an unbounded integer vector, and whose transitions increase the integer vector by a solution of QFPA formula. We show that for systems that have only one loop the problem reduces to QFPA* satisfiability and is therefore NP-complete. We show that for arbitrary graphs, the problem reduces to a generalization of QFPA* with multiple star operators, denoted QFPA^{REG}. We sketch a proof that NP-completeness for QFPA* satisfiability extends to QFPA^{REG} satisfiability.

A class of transition systems. Let $\mathcal{F}_d$ be the set of all QFPA formulas with the set of free variables $v_1, \ldots, v_d$. If $F \in \mathcal{F}_d$ is such a formula and $a_1, \ldots, a_d \in \mathbb{Z}$, we write $(a_1, \ldots, a_d) \models F$ to denote that F is true when v_i has value a_i for $1 \leq i \leq d$. We consider transition systems described by a tuple (d, Q, E, T) where 1) d is a non-negative integer, denoting the number of integer variables in the state; 2) Q is a finite set, denoting control-flow graph nodes; 3) $E \subseteq Q \times Q$, denoting control-flow graph edges; and 4) $T : E \rightarrow \mathcal{F}_d$, specifies possible increments of counters for each control-flow graph edge. Given (d, Q, E, T) we consider the set of states $S \subseteq Q \times \mathbb{Z}^d$ and define the transition relation $R \subseteq S \times S$ such that $(q, \vec{a}), (q', \vec{a}') \in R \iff (q, q') \in E \ \wedge \ (\vec{a}' - \vec{a}) \models T(q, q')$. We are interested in the question of reachability in the transition systems given by relation R. [2]

Single-loop systems. Consider first the case $Q = \{q\}$, $E = \{(q, q)\}$, $T(q, q) = F$. Our definitions then imply that $(q, \vec{a})$ reaches $(q, \vec{a}')$ precisely when the condition $(\vec{a}' - \vec{a}) \in \{\vec{v} \mid F\}^*$ holds. Therefore, the reachability problems that test QFPA relationship between initial $\vec{a}$ and final $\vec{a}'$ state in such systems reduces QFPA* satisfiability.

General case. Now consider arbitrary (d, Q, E, T) and two states $q, q' \in Q$. Let r be a regular expression over the alphabet E describing the set of all paths from q to q' in graph (Q, E), represented as a set of words over language $Q \times Q$. For example, a path q, q_1, q_2, q' is represented by word $(q, q_1)(q_1, q_2)(q_2, q')$. By

[2] Note that, unlike in Turing-complete transition systems with integer counters, the set of possible counter increments is given by formula $T(q, q')$ and does not depend on the current values of integer counters $\vec{a}$, but only on control-flow edge (q, q').

conversion algorithm from finite state machines to regular expressions we can assume that r exists and its size is polynomial in the number of elements of Q. We map r into a "commutative" regular expression with set addition acting as commutative version of concatenation and closure under vector addition acting as Kleene star. We specify this mapping using function h, defined by: $h((q_1, q_2)) = \{x \mid T(q_1, q_2)\}$, $h(r_1 r_2) = h(r_1) + h(r_2)$, $h(r_1 \cup r_2) = h(r_1) \cup h(r_2)$, $h(r*) = h(r)^*$. Due to commutativity of set addition and its consequence $A^* + B^* = (A \cup B)^*$, we can rewrite r in polynomial time to normal form stratified according to star height (the number of nested applications of $*$ operator). We call $\{v \mid T(q_1, q_2)\}$ atomic expressions and denote them a_{kij}. Then each commutative regular expression of star height $k > 0$ has the form $r_k = \cup_{i=1}^{p}(a_{ki1} + \ldots + a_{kin_i} + r_{i,k-1}^*)$ where $r_{i,k-1}$ are expressions of star height $k-1$. If $r = r_1$ i.e. r has no nested stars, then the reachability problem immediately reduces to (1) and is solvable in NP using our algorithm.

More generally, we consider formulas, denoted $\mathsf{QFPA}^{\mathsf{REG}}$, of the form $F_0 \wedge r$ where r_k is a regular expression over atomic expressions. We show that the satisfiability of $\mathsf{QFPA}^{\mathsf{REG}}$ is in NP. First, the condition $\vec{u} \in r_k$ is equivalent to

$$\exists(\vec{v}_{kij} \in a_{kij})_{kij}.\exists(\lambda_{kij})_{kij}.\vec{x} = \sum_{k,i,j} \lambda_{kij}\vec{v}_{kij} \wedge \bigwedge_{\substack{k>1 \\ i,j}} (\lambda_{kij} = 0 \Rightarrow \bigwedge_{i',j'} \lambda_{(k-1)i'j'} = 0)$$

As in Section 3 our goal is then to show that we can select a polynomial subset of vectors in this linear combination and still generate vector $\vec{u}$. The following notion of "star modulo vector dependencies" captures conditions on coefficients of linear combinations that arise from repeatedly applying star to semilinear sets. If $X = \{\vec{x}_1, \ldots, \vec{x}_N\} \subseteq \mathbb{N}^d$ is a finite set of vectors and $W \subseteq X \times X$ a dependency graph on X, define $X^{*(W)} = \{\sum_{i=1}^{N} \lambda_i \vec{x}_i \mid \forall i, j \leq N.\ \lambda_i > 0 \wedge (\vec{x}_i, \vec{x}_j) \in W \Rightarrow \lambda_j > 0\}$. The dependency graph in Theorem 1 would have an edge from each $\vec{b}_{ij}$ to $\vec{a}_i$. The generalization of Fact 1 to the class of graphs W sufficient for the more general result is the following.

Theorem 3. *Let $X \subseteq \mathbb{Z}^d$ be a finite set of integer vectors with acyclic dependency graph $W \subseteq X \times X$ such that for each node $\vec{x} \in X$ the number of nodes reachable from $\vec{x}$ in W is bounded by a constant C. If $\vec{b} \in X^{*(W)}$ then there exists $\tilde{X} \subseteq X$ such that $\vec{b} \in \tilde{X}^{*(W)}$ and $|\tilde{X}| \leq 2C^2 d \log(4dM)$, where $M = \max_{x \in X} ||x||_\infty$.*

Proof sketch. Let $B = 2d\log(4dM)$ from Fact 1. Consider a linear combination $\vec{u} = \sum_i \lambda_i \vec{v}_i$ of vectors from X that satisfies the dependencies in W. Our goal is to find a small number of vectors that generate $\vec{u}$. In the first step we consider the source nodes of W, that is, vectors $Y_0 \subseteq X$ with no incoming edges in the graph. Applying Fact 1 to $\vec{v}_0 = \sum_{\vec{v}_i \in Y_0} \lambda_i \vec{v}_i$ we obtain a subset $Z_0 \subseteq Y_0$, with $|Z_0| \leq B$, such that $\vec{v}_0 = \sum_{\vec{v}_i \in Z_0} \lambda'_i \vec{v}_i$. To enforce the constraints in the graph W, we then take closure of Z_0 under reachability in W and obtain the set Q_0 of size at most CB. Let $\vec{u}_0 = \sum_{\vec{v}_i \in Q_0} \lambda_i \vec{v}_i$.

We repeat the procedure on the vector $\vec{u} - \vec{u}_0$. Only in this step we eliminate all the sources Y_0 and vertices belonging to Q_0 from the graph and consider the vectors that are sources in the subgraph of W induced by the remaining vectors $Y_1 = X \setminus (Y_0 \cup Q_0)$. We repeat this procedure as long as there are nodes in the graph. The number of times we need to repeat it is bounded by the longest path in W, which, by assumption, is bounded by C. At each step we select CB vectors, so the total number of nodes that we need in the linear combination is bounded by C^2B. ∎

Using Fact 1 we obtain a polynomial subset of vectors that satisfy given QFPA formulas and whose linear combination is the given vector $\vec{u}$. We then use results from previous sections to show that a linear combination of solutions of a QFPA formula can be represented as a sum of a polynomial number of solutions of this QFPA formula. This allows us to generalize results of Section 3 to formulas that contain not just one star operator but any regular expression over solution sets of QFPA formulas, which in turn proves that the reachability problem for transition systems described in this section is also in NP.

Acknowledgements. We thank Nikolaj Bjørner for useful comments on a draft of this paper and CAV 2008 reviewers for their patience and useful feedback.

References

1. Ball, T., Cook, B., Lahiri, S.K., Zhang, L.: Zapato: Automatic theorem proving for predicate abstraction refinement. In: Alur, R., Peled, D.A. (eds.) CAV 2004. LNCS, vol. 3114, pp. 457–461. Springer, Heidelberg (2004)
2. Ball, T., Majumdar, R., Millstein, T., Rajamani, S.K.: Automatic predicate abstraction of C programs. In: Proc. ACM PLDI (2001)
3. Balser, M., Reif, W., Schellhorn, G., Stenzel, K., Thums, A.: Formal System Development with KIV. In: Maibaum, T.S.E. (ed.) ETAPS 2000 and FASE 2000. LNCS, vol. 1783, Springer, Heidelberg (2000)
4. Banâtre, J.-P., Métayer, D.L.: Programming by multiset transformation. Commun. ACM 36(1), 98–111 (1993)
5. Barrett, C., Berezin, S.: CVC Lite: A new implementation of the cooperating validity checker. In: Alur, R., Peled, D.A. (eds.) CAV 2004. LNCS, vol. 3114, pp. 515–518. Springer, Heidelberg (2004)
6. Basin, D., Friedrich, S.: Combining WS1S and HOL. In: FROCOS. vol. 7 of Studies in Logic and Computation (2000)
7. de Moura, L., Bjørner, N.: Efficient E-Matching for SMT Solvers. In: Pfenning, F. (ed.) CADE 2007. LNCS (LNAI), vol. 4603, pp. 183–198. Springer, Heidelberg (2007)
8. Dennis, L.A., Collins, G., Norrish, M., Boulton, R., Slind, K., Robinson, G., Gordon, M., Melham, T.: The PROSPER Toolkit. In: Schwartzbach, M.I., Graf, S. (eds.) ETAPS 2000 and TACAS 2000. LNCS, vol. 1785, Springer, Heidelberg (2000)
9. Eisenbrand, F., Shmonin, G.: Carathéodory bounds for integer cones. Operations Research Letters 34(5), 564–568 (2006), `http://dx.doi.org/10.1016/j.orl.2005.09.008`
10. Ge, Y., Barrett, C., Tinelli, C.: Solving quantified verification conditions using satisfiability modulo theories. In: Pfenning, F. (ed.) CADE 2007. LNCS (LNAI), vol. 4603, pp. 167–182. Springer, Heidelberg (2007)

11. Ginsburg, S., Spanier, E.: Semigroups, Pressburger formulas and languages. Pacific Journal of Mathematics 16(2), 285–296 (1966)
12. Henzinger, T.A., Jhala, R., Majumdar, R., Sutre, G.: Lazy abstraction. In: POPL (2002), http://doi.acm.org/10.1145/503272.503279
13. Kuncak, V.: Modular Data Structure Verification. PhD thesis, EECS Department, Massachusetts Institute of Technology (February 2007)
14. Kuncak, V., Rinard, M.: Towards efficient satisfiability checking for Boolean Algebra with Presburger Arithmetic. In: Pfenning, F. (ed.) CADE 2007. LNCS (LNAI), vol. 4603, pp. 215–230. Springer, Heidelberg (2007)
15. Lahiri, S.K., Seshia, S.A.: The UCLID decision procedure. In: Alur, R., Peled, D.A. (eds.) CAV 2004. LNCS, vol. 3114, pp. 475–478. Springer, Heidelberg (2004)
16. Lugiez, D.: Multitree automata that count. Theor. Comput. Sci. 333(1-2), 225–263 (2005)
17. McLaughlin, S., Barrett, C., Ge, Y.: Cooperating theorem provers: A case study combining HOL-Light and CVC Lite. In: PDPAR. ENTCS, vol. 144(2) (2006)
18. Nguyen, H.H., David, C., Qin, S., Chin, W.-N.: Automated verification of shape, size and bag properties via separation logic. In: Cook, B., Podelski, A. (eds.) VMCAI 2007. LNCS, vol. 4349, pp. 251–266. Springer, Heidelberg (2007)
19. Nipkow, T., Wenzel, M., Paulson, L.C., Voelker, N.: Multiset theory version 1.30 (Isabelle distribution) (2005), http://isabelle.in.tum.de/dist/library/HOL/Library/Multiset.html
20. Papadimitriou, C.H.: On the complexity of integer programming. J. ACM 28(4), 765–768 (1981)
21. Piskac, R., Kuncak, V.: Decision procedures for multisets with cardinality constraints. In: VMCAI. LNCS, vol. 4905. Springer, Heidelberg (2008)
22. Pottier, L.: Minimal solutions of linear diophantine systems: Bounds and algorithms. In: Book, R.V. (ed.) RTA 1991. LNCS. vol. 488. Springer, Heidelberg (1991)
23. Schwartz, J.T.: On programming: An interim report on the SETL project. Technical report, Courant Institute, New York (1973)
24. Shankar, N.: Using decision procedures with a higher-order logic. In: Boulton, R.J., Jackson, P.B. (eds.) TPHOLs 2001. LNCS, vol. 2152. Springer, Heidelberg (2001)
25. Zarba, C.G.: Combining multisets with integers. In: Voronkov, A. (ed.) CADE 2002. LNCS (LNAI), vol. 2392, Springer, Heidelberg (2002)

Inferring Congruence Equations Using SAT

Andy King[1] and Harald Søndergaard[2]

[1] Portcullis Computer Security Limited, Pinner, HA5 2EX, UK*
[2] The University of Melbourne, Victoria 3010, Australia.

Abstract. This paper proposes a new approach for deriving invariants that are systems of congruence equations where the modulo is a power of 2. The technique is an amalgam of SAT-solving, where a propositional formula is used to encode the semantics of a basic block, and abstraction, where the solutions to the formula are systematically combined and summarised as a system of congruence equations. The resulting technique is more precise than existing congruence analyses since a single optimal transfer function is derived for a basic block as a whole.

1 Introduction

Applications in compilation, optimisation and verification have motivated analyses that infer linear equality relationships [7,9,14] or linear congruence relationships [1,6,15] that hold between the variables of a program. For each point in a program, the former analyses discover systems of affine constraints of the form $\sum_{i=1}^{n} c_i x_i = d$ where $c_1, \ldots, c_n, d \in \mathbb{Z}$ and $x_1, \ldots, x_n$ are the program variables. The latter infer systems of congruence constraints of the form $(\sum_{i=1}^{n} c_i x_i)$ mod $m = d$ where $m \in \mathbb{Z}$ is some modulus. These analyses accurately trace relationships between variables when the assignments that arise in a program can be modeled with a linear transformation. But this precludes meaningful analysis of programs that use bitwise operators; whether written in Java, C, or assembly language. The extreme approach of treating all operands of such operators as sequences of named bits, to track all bit interrelations, does not appear attractive, owing to the large number of Boolean variables involved. However, we show that a mixture of congruence analysis and Boolean reasoning does appear to be both feasible and able to generate bit-level invariants of great precision.

We draw inspiration from the domain of congruent equations modulo 2^w [15] and the affinity between this domain and the finite-nature of the underlying computer arithmetic, to propose an extreme-precision analysis which produces tight invariants for programs with non-linear, including bitwise, operations. The idea is to express the relationship between the bits in input variables and the bits of output variables for each basic block. This technique is not new within itself [8] and programs are now routinely reduced to very large systems of propositional constraints in bounded model checking [3,20]. Our main novel contribution is in the use of a SAT solver to incrementally compute a summary (affine relaxation) of the output variables, given a summary for the input variables. This new

* Andy King is on secondment from the University of Kent, CT2 7NF, UK

A. Gupta and S. Malik (Eds.): CAV 2008, LNCS 5123, pp. 281–293, 2008.

approach is capable of discovering invariants even for programs that apply bit-twiddling; programs that have thus far thwarted automatic analysis. Summaries that are systems of congruence equations modulo 2^w naturally fit into this mix of model checking and abstract interpretation because (technically) their ascending chain property constrains the number of times the SAT-solver can be reapplied and (philosophically) the propositional encoding also makes assumptions about the finite, modulo-nature of computer arithmetic. As in conventional abstract interpretation, the summaries enable all paths through the program to be considered systematically, without enforcing a bound on depth to which loops are explored. The approach to analysis is attractive because, quite apart from providing a bridging result between SAT solving and analysis, it can compute an optimal transfer function for a whole basic block, even when the block contains non-linear assignments. This is the key to the improved precision.

The paper is structured as follows: Section 2 illustrates the key ideas of the analysis using a worked example, demonstrating how a SAT-solver can be interleaved with a relaxation technique to compute a summary. This is the primary contribution of the paper. Section 3 shows how the lattice-theoretic join of two congruence systems can be summarised merely by syntactically rearranging matrices and computing a triangular form. This is another contribution of the paper. Section 4 discusses the relationship with the wider literature and Section 5 concludes.

2 Outline of the Method

In 1960 Peter Wegner [19] reported a fast bit counting algorithm. Expressed in the language C, the method counts the number of 1-bits in a word `x`, leaving the result in a variable `c`, as follows: `y = x; for (c = 0; y; c++) y &= y - 1;` Since the method is rather devious (and the explanation pre-dates the invariant assertion principle), one may want to derive an invariant that aids understanding of the code. This is challenging because the bit-twiddling cannot be modeled with a conventional affine assignment [15], that is, `y` is not updated with a value that is a linear combination of the values of the program variables. Furthermore, modeling the update as an assignment to an arbitrary value (a so-called non-deterministic assignment [15]) is too crude to derive a useful loop invariant.

One might think that these problems are insurmountable but we show that an invariant can be derived by modeling non-linear assignments as exact operations on sequences of bits and then computing optimal congruence abstractions for a composition of bit operations. The last point warrants elaboration: numeric analyses are usually presented in terms of programs which have already been abstracted through the use of assignments that are either affine or non-deterministic. This is adequate when working at the granularity of whole numbers but the best congruent abstraction of a bit-level operation, let alone a composition of them, is somewhat less obvious. We systematise the computation of an optimal abstraction and integrate this into the analysis itself.

In the rest of this section we sketch the approach:

1. A local, bit-precise transfer function is established for each basic block.
2. These Boolean functions are then used to build a set of recursive dataflow equations, expressing the program's overall runtime behaviour.
3. In the context of a finite set of w-bit variables, a closed form of the dataflow equations can be derived using Kleene iteration. In practice, however, this iteration may need to be interleaved with steps to relax constraints, and we propose a suitable relaxation to congruence equations.

2.1 Representing Bit-Level Semantics without Abstraction

It is possible to express the semantics of the basic blocks of a program, even to the bit-level, using Boolean constraints that relate the bits of its inputs to those of its outputs. But the problem is how to do so, retain tractability, and derive *loop invariants*, that is, not just explore loops to a fixed depth [8].

Let us draw Wegner's code as a flow diagram:

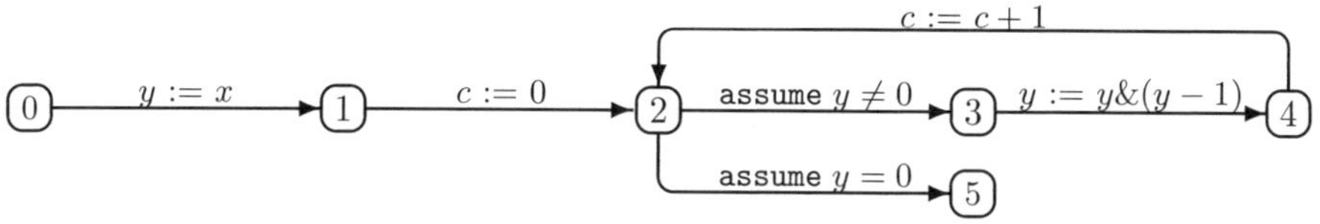

0 —($y := x$)→ 1 —($c := 0$)→ 2 —(assume $y \neq 0$)→ 3 —($y := y\&(y-1)$)→ 4 —($c := c+1$)→ 2

2 —(assume $y = 0$)→ 5

The program's basic blocks are the initial code '$y := x$; $c := 0$', the loop body '`assume` $y \neq 0$; $y := y\&(y-1)$; $c := c+1$' and the loop exit '`assume` $y = 0$'. The exact semantics of these blocks can be described relationally, as systems of propositional constraints. The idea is to represent the input and output relationships across a basic block using two systems of propositional variables $x_0, \ldots, x_{w-1}$ and $x'_0, \ldots, x'_{w-1}$ (abbreviated to $\boldsymbol{x}$ and $\boldsymbol{x}'$) that encode the input and output state of each integer variable x. We assume a twos complement integer representation and let w denote the number of bits that make up an integer. The constraints generated for the example are:

$$
\begin{aligned}
&[\![y := x; c := 0]\!] \\
&\quad = (\textstyle\bigwedge_{i=0}^{w-1} y'_i \leftrightarrow x_i) \wedge (\bigwedge_{i=0}^{w-1} \neg c'_i) \wedge (\bigwedge_{i=0}^{w-1} x'_i \leftrightarrow x_i) \\
&[\![\texttt{assume } y \neq 0; y := y\&(y-1); c := c+1]\!] \\
&\quad = (\textstyle\bigvee_{i=0}^{w-1} y_i) \wedge (\bigwedge_{i=0}^{w-1} y'_i \leftrightarrow (y_i \wedge \bigvee_{j=0}^{i-1} y_j)) \\
&\qquad \wedge (\textstyle\bigwedge_{i=0}^{w-1} c'_i \leftrightarrow (c_i \oplus \bigwedge_{j=0}^{i-1} c_j)) \wedge (\bigwedge_{i=0}^{w-1} x'_i \leftrightarrow x_i) \\
&[\![\texttt{assume } y = 0]\!] \\
&\quad = (\textstyle\bigwedge_{i=0}^{w-1} \neg y_i) \wedge (\bigwedge_{i=0}^{w-1} x'_i \leftrightarrow x_i) \wedge (\bigwedge_{i=0}^{w-1} y'_i \leftrightarrow y_i) \wedge (\bigwedge_{i=0}^{w-1} c'_i \leftrightarrow c_i)
\end{aligned}
$$

where $\oplus$ denotes exclusive-or. Elsewhere [11] we explain how these constraints can be generated automatically from the program. Suffice it to say that the constraint for an assignment `x := y + z` is derived by considering a cascade of full adders using intermediate carry bits $\boldsymbol{b}$,

$$
(\bigwedge_{i=0}^{w-1} x'_i \leftrightarrow y_i \oplus z_i \oplus b_i) \wedge \neg b_0 \wedge (\bigwedge_{i=1}^{w-1} b_i \leftrightarrow (y_{i-1} \wedge z_{i-1}) \vee (y_{i-1} \wedge b_{i-1}) \vee (z_{i-1} \wedge b_{i-1}))
$$

together with constraints to express that variables other than x do not change. The variables $\boldsymbol{b}$ are existentially quantified, and the formula can be simplified using standard Boolean quantifier elimination.

Compound expressions are handled by introducing temporary variables s and t to hold intermediate results and then applying renaming. For example,

$$[\![y := y \ \& \ (y-1)]\!] = \rho_{\boldsymbol{s}',\boldsymbol{t}}([\![s := y - 1]\!] \wedge [\![y := y \ \& \ t]\!])$$

The renaming step $\rho_{\boldsymbol{s}',\boldsymbol{t}}$ replaces the output variables $\boldsymbol{s}'$ of the first statement with the input variables $\boldsymbol{t}$ of the second. Again, the compound formula can be simplified by eliminating the remaining intermediate variables $\boldsymbol{s}, \boldsymbol{t}, \boldsymbol{t}'$.

2.2 Setting Up the Dataflow Equations

The relational semantics for the basic blocks allows us to derive the states that are reachable at program points 0, 2 and 5. They are obtained as the least (strongest) solution to the following recursive equations:

$$\begin{aligned}
f_0 &= 1 \\
f_2 &= \rho_{\boldsymbol{v}',\boldsymbol{v}}(\pi_{\boldsymbol{v}'}(f_0 \wedge [\![y := x; c := 0]\!])) \vee \\
&\quad\ \rho_{\boldsymbol{v}',\boldsymbol{v}}(\pi_{\boldsymbol{v}'}(f_2 \wedge [\![\texttt{assume}\ y \neq 0; y := y \& (y-1); c := c + 1]\!])) \\
f_5 &= \rho_{\boldsymbol{v}',\boldsymbol{v}}(\pi_{\boldsymbol{v}'}(f_2 \wedge [\![\texttt{assume}\ y = 0]\!]))
\end{aligned}$$

where $\boldsymbol{v}$ and $\boldsymbol{v}'$ are the input and output variables, that is, $\boldsymbol{v} = \boldsymbol{c} \cdot \boldsymbol{x} \cdot \boldsymbol{y}$ and $\boldsymbol{v}' = \boldsymbol{c}' \cdot \boldsymbol{x}' \cdot \boldsymbol{y}'$. The Boolean functions f_0, f_2 and f_5 represent sets of reachable states, for example, a state $\sigma = \{c_0 \mapsto 0, \ldots, c_{w-1} \mapsto 0,\ x_0 \mapsto 1, \ldots, x_{w-1} \mapsto 0,\ y_0 \mapsto 1, \ldots, y_{w-1} \mapsto 0\}$ is considered to be reachable at program point 2 iff σ satisfies f_2. The projection operation $\pi_{\boldsymbol{v}'}(f) = f'$ computes the Boolean function f' by eliminating, by existential quantification, any propositional variable y from f that does not occur in the system $\boldsymbol{v}'$. For instance, $\pi_{\langle x_1, x_2 \rangle}(f) = \exists_y(f) = x_1 \leftrightarrow x_2$ when $f = (x_1 \leftrightarrow y) \wedge (y \leftrightarrow x_2)$. Projection is used to derive a function that only expresses relations between the output variables $\boldsymbol{v}'$. The renaming operation $\rho_{\boldsymbol{v}',\boldsymbol{v}}(f) = f'$ constructs a function f' by replacing each output variable y' in f with its counterpart input variable y, for example, $\rho_{\boldsymbol{v}',\boldsymbol{v}}(c_0' \wedge (x_1' \oplus y_1')) = c_0 \wedge (x_1 \oplus y_1)$. Iteration can be used to compute f_2 from the predetermined $f_0 = 1$ and once f_2 is known, f_5 can be derived. For f_2 the iterates start:

$$\begin{aligned}
f_2^0 &= 0 \\
f_2^1 &= f_2^0 \quad \vee \quad (\textstyle\bigwedge_{i=0}^{w-1} x_i \leftrightarrow y_i) \wedge (\bigwedge_{i=0}^{w-1} \neg c_i) \\
f_2^2 &= f_2^1 \quad \vee \quad (\textstyle\bigvee_{i=0}^{w-1} x_i) \wedge (\bigwedge_{i=0}^{w-1} y_i \leftrightarrow (x_i \wedge \bigvee_{j=0}^{i-1} x_j)) \wedge c_0 \wedge (\bigwedge_{i=1}^{w-1} \neg c_i) \\
f_2^3 &= f_2^2 \quad \vee \quad \ldots
\end{aligned}$$

This sequence will eventually stabilise because a bounded number (2^{3w}) of Boolean functions are definable over $\boldsymbol{v}$. However, the $\boldsymbol{c}$ variables will enumerate all 2^w bit patterns and therefore at least 2^w iterates will be computed. This will take an impractically long time, even for $w = 16$. There is also an issue of space. A Boolean function can be represented as an ROBDD but the size of an ROBDD can be exponentially large in the number of variables (even with dynamic variable reordering [2]), and this is a pressing issue when w propositional variables are needed to represent each integer variable. Tractability can be recovered by approximating [17] or widening [10] an ROBDD when it becomes intolerably large. This would replace an ROBDD with one that could be stored more compactly and yet represented a larger set of states. The problem with this approach is that the ROBDD widenings that have been proposed thus far do not preserve sufficient information to infer useful loop invariants.

2.3 Abstracting Bit-Level Inputs and Outputs with Congruences

Considerations of tractability dictate that we look for principled ways of over-approximating solutions to systems of equations of the form

$$f = \bigvee_{m=1}^{n} \rho_{\boldsymbol{y}',\boldsymbol{y}}(\pi_{\boldsymbol{y}'}(f_m \wedge f'_m)) \tag{1}$$

without giving up too much bit-level information. We suggest that this can be achieved by restricting f, as well as each f_m, to a class of functions that can be expressed as conjunctions of congruence equations modulo m, where m is a power of 2. Each function f_m summarises the inputs to one of n basic blocks and the function f summarises all (the join of) the outputs of the n blocks. *No constraint is placed on the generality of any of the f'_m formulae.* This means that no abstraction needs to be applied to a function that describes the relational semantics of a basic block—this description is kept bit-precise.

How to solve systems of the form (1) under the restrictions just mentioned? The rest of this section explains the idea, based on the Wegner example. Let $x =_{2^w} y$ abbreviate $x = y + k2^w$ for some integer multiplier k. Observe that each of the equations $t \equiv_{2^w} t' + t''$, $t \equiv_{2^w} ny$, $t \equiv_{2^w} n$ and the disequation $t \not\equiv_{2^w} n$ can be expressed as propositional constraints when the t variables are w-bit, y is 1-bit and n is an integer. This is a consequence of the modulus being a power of 2. For instance, $t =_{2^w} ny$ and $t \neq_{2^w} ny$ can be expressed as $\bigwedge_{i=0}^{w-1} t_i \leftrightarrow (n_i \wedge y)$ and $\bigvee_{i=0}^{w-1} t_i \oplus (n_i \wedge y)$ where $n \equiv_{2^w} \sum_{i=0}^{w-1} 2^i n_i$ and $t \equiv_{2^w} \sum_{i=0}^{w-1} 2^i t_i$. Moreover, an equation $\sum_{i=1}^{k} n_i y_i \equiv_{2^w} n$ can be reduced by $\sum_{i=1}^{j} n_i y_i \equiv_{2^w} t$, $\sum_{i=j+1}^{k} n_i y_i \equiv_{2^w} t'$, $t + t' \equiv_{2^w} t''$ and $t'' \equiv_{2^w} n$ using fresh w-bit variables t, t', and t'', and hence also reduced to a propositional system. Any disequation $\sum_{i=1}^{k} m_i y_i \not\equiv_{2^w} n$ can similarly be described propositionally. Henceforth let $[\![\sum_{i=1}^{k} n_i y_i \equiv_{2^w} n]\!]$ and $[\![\sum_{i=1}^{k} n_i y_i \not\equiv_{2^w} n]\!]$ denote such propositional encodings.

To illustrate the value of these encodings, let $w = 8$ and consider computing $f_2^2 = f_2^1 \vee \rho_{\boldsymbol{y}',\boldsymbol{y}}(\pi_{\boldsymbol{y}'}(f_2^1 \wedge g))$ where $f_2^1 = \rho_{\boldsymbol{y}',\boldsymbol{y}}(\pi_{\boldsymbol{y}'}(1 \wedge [\![y := x; c := 0]\!]))$ and $g = [\![\texttt{assume}\ y \neq 0; y := y \& (y-1); c := c+1]\!]$. The encodings are used to direct the generation of truth assignments for the function $f_2^1 \wedge g$. Truth assignments are generated so as to incrementally derive the most precise congruence system describing both f_2^1 and $\rho_{\boldsymbol{y}',\boldsymbol{y}}(\pi_{\boldsymbol{y}'}(f_2^1 \wedge g))$. This system is used as the iterate, rather than the function f_2^1 itself. Since the function $f_2^0 = 0$ can be represented as a congruence system, namely $0 \equiv_{256} 1$, this construction ensures that all iterates are congruences. The number of iterates is bounded: the length of an increasing chain of congruences is at most 192, that is, the product of the power $w = 8$ and the maximum number, 24 ($3w$), of variables in each system [15].

The function f_2^1 falls into the class of formulae that can be represented congruently. This is because the satisfying assignments of f_2^1 coincide with those of the formula $(\bigwedge_{i=0}^{7} [\![c_i \equiv_{256} 0]\!]) \wedge (\bigwedge_{i=0}^{7} [\![x_i \equiv_{256} y_i]\!])$ on $\boldsymbol{c}$, $\boldsymbol{x}$ and $\boldsymbol{y}$. Hence computing $\rho_{\boldsymbol{y}',\boldsymbol{y}}(\pi_{\boldsymbol{y}'}(f_2^1 \wedge g))$ is equivalent to computing $\rho_{\boldsymbol{y}',\boldsymbol{y}}(\pi_{\boldsymbol{y}'}(g'))$ where $g' = (\bigwedge_{i=0}^{7} [\![c_i \equiv_{256} 0]\!]) \wedge (\bigwedge_{i=0}^{7} [\![x_i \equiv_{256} y_i]\!]) \wedge g$. This is convenient, because it permits the encodings to be demonstrated on a representative, non-trivial example, namely the derivation of f_2^2. To see how the congruence system for f_2^2 is incrementally constructed, observe that any model of the Boolean function

$$g' \wedge ((\bigvee_{i=0}^{7} [\![c_i' \not\equiv_{256} 0]\!]) \vee (\bigvee_{i=0}^{7} [\![x_i' \not\equiv_{256} y_i']\!])) \qquad (2)$$

defines a run of the block with an entry state that is described by f_2^1 but whose exit state is *not* characterised by f_2^1. For example, the truth assignment

$$\left\{ \begin{array}{l} c_0 \mapsto 0, c_1 \mapsto 0, \ldots, c_7 \mapsto 0, x_0 \mapsto 0, \ldots, x_6 \mapsto 0, x_7 \mapsto 1, y_0 \mapsto 0, \ldots, y_7 \mapsto 1, \\ c_0' \mapsto 1, c_1' \mapsto 0, \ldots, c_7' \mapsto 0, x_0' \mapsto 0, \ldots, x_6' \mapsto 0, x_7' \mapsto 1, y_0' \mapsto 0, \ldots, y_7' \mapsto 0 \end{array} \right\}$$

satisfies (2) and demonstrates that when c, x, and y assume values of 0, 128 and 128 on entry to the block, they can take values of 1, 128 and 0 on exit from the block (assuming an unsigned representation). By construction, the output state is not summarised by f_2^1 and therefore f_2^1 needs to be enlarged to accommodate this state. Since the output state can be represented in congruence form as

$$c_0 \equiv_{256} 1 \wedge (\bigwedge_{i=1}^{7} c_i \equiv_{256} 0) \wedge (\bigwedge_{i=0}^{6} x_i \equiv_{256} 0) \wedge x_7 \equiv_{256} 1 \wedge (\bigwedge_{i=0}^{7} y_i \equiv_{256} 0) \qquad (3)$$

this system and that for f_2^1 can be joined to obtain the summary

$$(\bigwedge_{i=1}^{7} c_i \equiv_{256} 0) \wedge (\bigwedge_{i=0}^{6} x_i \equiv_{256} y_i) \wedge x_7 \equiv_{256} c_0 + y_7 \qquad (4)$$

A model for the formula (2) can be found using standard techniques [16], translating the formula into an equi-satisfiable conjunctive normal form (CNF) representation, and presenting the CNF formula to any SAT-solver. The join can be computed by translating the two congruence systems to their sets of generators, taking the union of the two sets, then converting the union to a new congruence system [1,6,15]. Alternatively, the join can be obtained by relaxing a system of congruences constructed syntactically from the two input systems (see Section 3). Either way, whether the join describes all possible output states can be determined by solving the Boolean formula

$$g' \wedge ((\bigvee_{i=1}^{7} [\![c_i \not\equiv_{256} 0]\!]) \vee (\bigvee_{i=0}^{6} [\![x_i \not\equiv_{256} y_i]\!]) \vee [\![x_7 \not\equiv_{256} c_0 + y_7]\!]) \quad (5)$$

This formula is satisfied, for example, by a truth assignment $\{\ldots, c'_0 \mapsto 1, c'_1 \mapsto 0, \ldots c'_7 \mapsto 0, x'_0 \mapsto 0, \ldots x'_5 \mapsto 0, x'_6 \mapsto 1, x'_7 \mapsto 0, y'_0 \mapsto 0, \ldots y'_7 \mapsto 0\}$ from which the following congruence system can be derived:

$$c_0 \equiv_{256} 1 \wedge (\bigwedge_{i=1}^{7} c_i \equiv_{256} 0) \wedge (\bigwedge_{i=0, i\neq 6}^{7} x_i \equiv_{256} 0) \wedge x_6 \equiv_{256} 1 \wedge (\bigwedge_{i=0}^{7} y_i \equiv_{256} 0) \quad (6)$$

Note that the assignments to the input variables, as well as any temporary variables introduced in CNF conversion [16], are inconsequential for constructing the congruence system. Disregarding these assignments amounts to projecting onto the output variables. Notice too that the system is expressed in terms of unprimed variables, even though it encodes an output state. Constructing the congruence system thus involves renaming as well as projection, though both operations are performed on truth assignments, at which level they collapse to computationally trivial operations. Joining the previous summary (4) with the new system (6) gives the new summary

$$(\bigwedge_{i=1}^{7} c_i \equiv_{256} 0) \wedge (\bigwedge_{i=1}^{5} x_i \equiv_{256} y_i) \wedge x_6 + x_7 \equiv_{256} c_0 + y_6 + y_7 \quad (7)$$

Continuing this way, we obtain a sequence of Boolean formulae $h_0, h_1, h_2, \ldots$, the first two of which are (2) and (5), and where, more generally, h_j is

$$g' \wedge ((\bigvee_{i=1}^{7} [\![c_i \not\equiv_{256} 0]\!]) \vee (\bigvee_{i=0}^{7-j} [\![x_i \not\equiv_{256} y_i]\!]) \vee [\![\textstyle\sum_{i=7-j+1}^{7} x_i \not\equiv_{256} c_0 + \sum_{i=7-j+1}^{7} y_i]\!])$$

Of these, $h_1, \ldots, h_6$ are satisfiable, but h_7 is not, so the system

$$(\bigwedge_{i=1}^{7} c_i \equiv_{256} 0) \wedge \sum_{i=0}^{7} x_i \equiv_{256} c_0 + \sum_{i=0}^{7} y_i \quad (8)$$

summarises all reachable output states when the input states are described by f_2^1. The next iterate f_2^2 is then assigned to this system which is the most precise congruence system that describes the set of output states given an input state drawn from f_2^1.

The method is not sensitive to how the relational semantics is presented. This contrasts with previous analyses which critically depend on how the statements of a program are translated into, say, affine assignments. This is particularly pertinent when deriving invariants for assembler or obfuscated code [18].

Of course, thus far, only f_2^2 has been derived. By repeating the above process with an updated input state we obtain the sequence of iterates:

$$\begin{aligned} f_2^3 &= (\textstyle\bigwedge_{i=2}^{7} c_i \equiv_{256} 0) \wedge \sum_{i=0}^{7} x_i \equiv_{256} c_0 + 2c_1 + \sum_{i=0}^{7} y_i \\ f_2^4 &= (\textstyle\bigwedge_{i=3}^{7} c_i \equiv_{256} 0) \wedge \sum_{i=0}^{7} x_i \equiv_{256} c_0 + 2c_1 + 4c_2 + \sum_{i=0}^{7} y_i \\ f_2^6 = f_2^5 &= (\textstyle\bigwedge_{i=4}^{7} c_i \equiv_{256} 0) \wedge \sum_{i=0}^{7} x_i \equiv_{256} c_0 + 2c_1 + 4c_2 + 8c_3 + \sum_{i=0}^{7} y_i \end{aligned}$$

Interestingly, although the derivation of f_2^2 requires 8 calls to a SAT-solver and 7 join computations, the iterates f_2^3, f_2 and f_2^5 each require just two calls to a solver and one join. To check stability, that is, deduce $f_2^6 = f_2^5$, requires one call to a solver, hence 15 invocations are required in total, the largest of which involves 4507 variables and 11648 clauses (though more compact CNF conversion is possible [16]). Nevertheless, the longest time that it takes to solve any instance is 0.61 ms (wall-time) using SAT4J version 1.5 [12] on a 2.4 GHz MacBook Pro. Even without deriving $f_5 = \sum_{i=0}^{7} x_i \equiv_{256} c_0 + 2c_1 + 4c_2 + 8c_3 \wedge (\bigwedge_{i=4}^{7} c_i \equiv_{256} 0)$, it is now evident that Wegner's bit-twiddling algorithm assigns to the variable c the number of bits which are set in the variable x. As far as we aware, no other analysis is capable of deriving such an invariant. Note that the invariant includes coefficients of 4 and 8, and thus precision would be degraded if a modulo of 2 was employed rather than the word-level modulo of 2^w.

3 Joining Congruence Equations

Section 2.1 outlined the translation of basic blocks into Boolean formulae. That component preserves all information within a basic block, which is key to reasoning about non-linear operations such as bit-twiddling. We now describe the complementary component which produces the join of two congruence systems. It discards information, which is key to retaining tractability. The join ensures that the summaries reside in a finite ascending chain so they cannot be weakened forever; the maximal chain length is $w^2 n$ [15], as wn (propositional) variables are needed to represent the state of n (integer) variables of width w.

Recent work has exploited how congruence systems can be represented by sets of generators that span the solution space of the congruence system [15]. This representation is useful because it reduces the join operation to set union. However, our refined form of analysis relies on a translation mechanism from an equation $\sum_{i=1}^{k} n_i y_i \equiv_{2^w} n$ to a formula $[\![\sum_{i=1}^{k} n_i y_i \equiv_{2^w} n]\!]$ that becomes more convoluted when the generator representation is adopted. Thus it is convenient to compute the join whilst representing the input and output systems as a conjunction of equations. This can be achieved by reformulating the join of two systems as a projection operation which can, in turn, be computed by calculating a triangular form. This section explains these steps.

Basics. To state the algorithmic results of this section, it is necessary to recall some mathematical concepts and notation. The set of congruence classes modulo m is defined $\mathbb{Z}_m = \{[n] \mid n \in \mathbb{Z}\}$ where $[n] = \{n' \in \mathbb{Z} \mid n \equiv_m n'\}$ and $\equiv_m$ denotes equivalence modulo m. Henceforth, we blur the distinction between a class $[n]$ and its representative element n. The 2-fold Cartesian product $\mathbb{Z}_m^2$ is defined $\mathbb{Z}_m^2 = \mathbb{Z}_m \times \mathbb{Z}_m$ and the k-fold product $\mathbb{Z}_m^k$ is likewise defined. If $S_1, S_2 \subseteq \mathbb{Z}_m^k$ then their (Minkowski) sum is $S_1 + S_2 = \{\boldsymbol{x} \in \mathbb{Z}_m^k \mid \boldsymbol{x}_i \in S_i \wedge \boldsymbol{x} \equiv_m \boldsymbol{x}_1 + \boldsymbol{x}_2\}$. If $\lambda \in \mathbb{Z}$ and $S \subseteq \mathbb{Z}_m^k$, then $\lambda S = \{\boldsymbol{x} \in \mathbb{Z}_m^k \mid \boldsymbol{x}' \in S \wedge \boldsymbol{x} \equiv_m \lambda \boldsymbol{x}'\}$. Moreover, the linear closure of S is $\mathsf{linear}(S) = \{\sum_{i=1}^{\ell} \lambda_i \boldsymbol{x}_i \mid \lambda_i \in \mathbb{Z} \wedge \boldsymbol{x}_1, \ldots, \boldsymbol{x}_\ell \in S\}$.

```
procedure triangular(in: S, out: t)
    t := λℓ.⊥
    let {a_1, ... a_s} = S
    for i := 1 to s do
        ℓ := leading(a_i)
        while (ℓ > 0 ∧ ℓ ∈ dom(t))
            a' := t(ℓ)
            p := power(π_ℓ(a_i))
            p' := power(π_ℓ(a'))
            if p ≥ p' then
                a_i := (π_ℓ(a')/2^{p'})a_i − 2^{p−p'} a'
            else
                t := t[ℓ ↦ a_i]
                a_i := (π_ℓ(a_i)/2^p)a' − 2^{p'−p} a_i
            endif
            ℓ := leading(a_i)
        endwhile
        if ℓ > 0 then t := t[ℓ ↦ a_i]
    endfor
endprocedure
```

Fig. 1. The triangularisation method of Müller-Olm and Seidl [15]

The modules of $\mathbb{Z}_m^k$ are those subsets of $\mathbb{Z}_m^k$ that are closed under linear combination, that is, $\mathsf{Module}_m^k = \{M \subseteq \mathbb{Z}_m^k \mid \mathsf{linear}(M) = M\}$. The affine subsets of $\mathbb{Z}_m^k$ are translated modules, that is, $\mathsf{Affine}_m^k = \{\{\boldsymbol{x}\} + M \mid \boldsymbol{x} \in \mathbb{Z}_m^k \wedge M \in \mathsf{Module}_m^k\}$. The affine closure of S is the smallest affine space that encloses S and is thus defined $\mathsf{affine}(S) = \bigcap\{S' \in \mathsf{Affine}_m^k \mid S \subseteq S'\}$.

Example 1. Observe that $\emptyset$ and $\{\mathbf{0}\}$ are closed under linear combination, whence $\emptyset \in \mathsf{Module}_m^k$ and $\{\mathbf{0}\} \in \mathsf{Module}_m^k$. As $\emptyset \in \mathsf{Module}_m^k$, we have $\emptyset \in \mathsf{Affine}_m^k$. Moreover, since $\{\mathbf{0}\} \in \mathsf{Module}_m^k$, it follows that $\{\boldsymbol{x}\} \in \mathsf{Affine}_m^k$ for all $\boldsymbol{x} \in \mathbb{Z}_m^k$.

Triangularisation. While congruence equations appear as good companions for a bit-level relational semantics, Gaussian elimination cannot be immediately applied to compute a triangular form for such equations because of the need to deal with zero divisors [15]. Müller-Olm and Seidl have thus devised a triangularisation algorithm that, given an input system $A\boldsymbol{x} \equiv_{2^w} \boldsymbol{b}$, computes output system $A'\boldsymbol{x} \equiv_{2^w} \boldsymbol{b}'$ where $A' = [a_{i,j}]$ and $a_{i,j} = 0$ whenever $i > j$. Figure 1 gives the algorithm and Example 2 illustrates what the algorithm will compute for an example. (This example will, in turn, support Example 4 which demonstrates how join can be straightforwardly realised.) In the description of the algorithm, $\mathsf{leading}(\boldsymbol{a})$ returns -1 if $\boldsymbol{a} = \mathbf{0}$ and otherwise the position of the first non-zero element of the vector $\boldsymbol{a}$; $\pi_\ell(\boldsymbol{a})$ extracts the ℓ'th element from $\boldsymbol{a}$; and $\mathsf{power}(n)$ returns the largest integer p such that 2^p divides n.

Example 2. The input and output to triangularisation procedure are $A\boldsymbol{x} \equiv_2 \boldsymbol{b}$ and $A'\boldsymbol{x} \equiv_2 \boldsymbol{b}'$ respectively:

$$A = \begin{bmatrix} 1\,1\,0\,0\,0\,0\,0\,0\,0\,0\,0 \\ 0\,0\,1\,0\,0\,0\,0\,0\,0\,0\,0 \\ 1\,0\,0\,1\,0\,0\,0\,0\,0\,0\,0 \\ 0\,1\,0\,0\,0\,1\,0\,0\,0\,0\,0 \\ 0\,0\,0\,0\,0\,0\,1\,0\,0\,0\,0 \\ 0\,1\,0\,0\,0\,0\,0\,1\,0\,0\,0 \\ 0\,0\,1\,0\,0\,1\,0\,0\,1\,0\,0 \\ 0\,0\,0\,1\,0\,0\,1\,0\,0\,1\,0 \\ 0\,0\,0\,0\,1\,0\,0\,1\,0\,0\,1 \end{bmatrix} \quad \boldsymbol{b} = \begin{bmatrix} 1 \\ 0 \\ 0 \\ 0 \\ 0 \\ 0 \\ 0 \\ 0 \\ 0 \end{bmatrix} \quad A' = \begin{bmatrix} 11000000000 \\ 01010000000 \\ 00100000000 \\ 00010100000 \\ 00001001001 \\ 00000101000 \\ 00000010000 \\ 00000001100 \\ 00000000110 \end{bmatrix} \quad \boldsymbol{b}' = \begin{bmatrix} 1 \\ 1 \\ 0 \\ 1 \\ 0 \\ 0 \\ 0 \\ 0 \\ 1 \end{bmatrix}$$

By reasoning about upper triangular form, one can argue that any subset of $\mathbb{Z}_m^k$ that is closed under affine combination, can be represented congruently:

Proposition 1. $S \in \mathsf{Affine}_m^k$ iff there exists a congruence system $A\boldsymbol{x} \equiv_m \boldsymbol{b}$ such that $S = \{\boldsymbol{x} \in \mathbb{Z}_m^k \mid A\boldsymbol{x} \equiv_m \boldsymbol{b}\}$.

Projection. Quite apart from establishing this result, upper triangular form provides a way of computing arbitrary projections. Projection onto the i'th element of a k-ary vector is defined $\pi_i(\langle x_1, \ldots, x_k \rangle) = x_i$. Single element projections can be composed so that if $1 \leq i_1 < \ldots < i_j \leq k$ then the j-ary vector $\langle \pi_{i_1}(\boldsymbol{x}), \ldots, \pi_{i_j}(\boldsymbol{x}) \rangle$ is also a projection in that it also discards information pertaining to certain dimensions. Projection of an affine space is also affine:

Proposition 2. Let $S \in \mathsf{Affine}_m^k$ and $1 \leq i_1 < \ldots < i_j \leq k$. Then $T \in \mathsf{Affine}_m^j$ where $T = \{\langle \pi_{i_1}(\boldsymbol{x}), \ldots, \pi_{i_j}(\boldsymbol{x}) \rangle \mid \boldsymbol{x} \in S\}$.

If $A = [a_{i,j}]$ is in upper triangular form, the projection of $A\boldsymbol{x} \equiv_m \boldsymbol{b}$ onto a suffix $\boldsymbol{y} = \langle x_i, \ldots, x_k \rangle$ of $\boldsymbol{x}$ is found very easily. Suppose row j is the top-most row of A in which $\langle a_{j,1}, \ldots, a_{j,i-1} \rangle = \boldsymbol{0}$. Then the projection onto $\boldsymbol{y}$ is

$$\begin{bmatrix} a_{j,i} & \cdots & a_{j,k} \\ \vdots & & \vdots \\ a_{s,i} & \cdots & a_{s,k} \end{bmatrix} \boldsymbol{y} \equiv_m \begin{bmatrix} \pi_j(\boldsymbol{b}) \\ \vdots \\ \pi_s(\boldsymbol{b}) \end{bmatrix}$$

Example 3. Projecting $A\boldsymbol{x} \equiv_2 \boldsymbol{b}$ of Example 2, or equivalently the system $A(t)\boldsymbol{x} \equiv_2 \boldsymbol{b}(t)$, onto $\boldsymbol{y}_i = \langle x_i, \ldots, x_{11} \rangle$ for $i = 7, 9$ and 10 yields:

$$\begin{bmatrix} 1\,0\,0\,0\,0 \\ 0\,1\,1\,0\,0 \\ 0\,0\,1\,1\,0 \end{bmatrix} \boldsymbol{y}_7 \equiv_2 \begin{bmatrix} 0 \\ 0 \\ 1 \end{bmatrix} \qquad \begin{bmatrix} 1\,1\,0 \end{bmatrix} \boldsymbol{y}_9 \equiv_2 \begin{bmatrix} 1 \end{bmatrix} \qquad \begin{bmatrix} 0\,0 \end{bmatrix} \boldsymbol{y}_{10} \equiv_2 \begin{bmatrix} 0 \end{bmatrix} \text{ (or 1)}$$

Given a congruence system $A\boldsymbol{x} \equiv_m \boldsymbol{b}$, it is possible to project onto any subset of $\boldsymbol{x}$ merely by reordering the rows of A in synchronicity with the elements of $\boldsymbol{b}$, prior to computing the triangular form.

Join. We finally show how the join can be reduced to computing a projection (a relaxation) which, in turn, amounts to deriving an upper triangular form.

Proposition 3. Let $S_i = \{\boldsymbol{x} \in \mathbb{Z}_m^k \mid A_i\boldsymbol{x} \equiv_m \boldsymbol{b}_i\}$ and

$$A = \begin{bmatrix} 1 & 1 & 0 & 0 & 0 \\ -\boldsymbol{b}_1 & 0 & A_1 & 0 & 0 \\ 0 & -\boldsymbol{b}_2 & 0 & A_2 & 0 \\ 0 & 0 & -I & -I & I \end{bmatrix} \quad S = \left\{ \boldsymbol{x} \in \mathbb{Z}_m^k \;\middle|\; \exists \sigma_i \in \mathbb{Z}_m . \exists \boldsymbol{x}_i \in \mathbb{Z}_m^k . A \begin{bmatrix} \sigma_1 \\ \sigma_2 \\ \boldsymbol{x}_1 \\ \boldsymbol{x}_2 \\ \boldsymbol{x} \end{bmatrix} \equiv_m \begin{bmatrix} \boldsymbol{1} \\ \boldsymbol{0} \\ \boldsymbol{0} \\ \boldsymbol{0} \end{bmatrix} \right\}$$

Then $S = \mathsf{affine}(S_1 \cup S_2)$ if $S_1 \neq \emptyset$ and $S_2 \neq \emptyset$.

If a system $A_i\boldsymbol{x} \equiv_{2^w} \boldsymbol{b}_i$ has a solution set S_i, then the join of $A_1\boldsymbol{x} \equiv_{2^w} \boldsymbol{b}_1$ and $A_2\boldsymbol{x} \equiv_{2^w} \boldsymbol{b}_2$ is a system whose solutions coincide with $\mathsf{affine}(S_1 \cup S_2)$. Proposition 3 states that such a system can be obtained by rearranging A_1 and A_2 to form a new matrix A and then eliminating variables.

Example 4. Consider the join of $A_1\boldsymbol{x} \equiv_2 \boldsymbol{b}_1$ and $A_2\boldsymbol{x} \equiv_2 \boldsymbol{b}_2$ where $\boldsymbol{x} = \langle x, y, z \rangle$

$$A_1 = \begin{bmatrix} 1 & 0 & 0 \\ 0 & 1 & 0 \end{bmatrix} \quad \boldsymbol{b}_1 = \begin{bmatrix} 0 \\ 1 \end{bmatrix} \qquad A_2 = \begin{bmatrix} 1 & 0 & 0 \\ 0 & 1 & 0 \\ 0 & 0 & 1 \end{bmatrix} \quad \boldsymbol{b}_2 = \begin{bmatrix} 1 \\ 0 \\ 1 \end{bmatrix}$$

As well as minimising the size of coefficients and thereby making the presentation of large matrices manageable, a by-product of $2^w = 2$ is that $-I \equiv_2 I$ and $\boldsymbol{b}_i \equiv_2 -\boldsymbol{b}_i$. Using this, the combined system of Proposition 3 is formed—it is given below, on the left. On the right is the triangular system derived in Example 2, and we conclude that the join is $x + y \equiv_2 1$.

$$\left[\begin{array}{c|c|ccc|ccc|ccc} 1 & 1 & 0 & 0 & 0 & 0 & 0 & 0 & 0 & 0 & 0 \\ \hline 0 & 0 & 1 & 0 & 0 & 0 & 0 & 0 & 0 & 0 & 0 \\ 1 & 0 & 0 & 1 & 0 & 0 & 0 & 0 & 0 & 0 & 0 \\ \hline 0 & 1 & 0 & 0 & 0 & 1 & 0 & 0 & 0 & 0 & 0 \\ 0 & 0 & 0 & 0 & 0 & 0 & 1 & 0 & 0 & 0 & 0 \\ 0 & 1 & 0 & 0 & 0 & 0 & 0 & 1 & 0 & 0 & 0 \\ \hline 0 & 0 & 1 & 0 & 0 & 1 & 0 & 0 & 1 & 0 & 0 \\ 0 & 0 & 0 & 1 & 0 & 0 & 1 & 0 & 0 & 1 & 0 \\ 0 & 0 & 0 & 0 & 1 & 0 & 0 & 1 & 0 & 0 & 1 \end{array}\right] \begin{bmatrix} \sigma_1 \\ \sigma_2 \\ x_1 \\ y_1 \\ z_1 \\ x_2 \\ y_2 \\ z_2 \\ x \\ y \\ z \end{bmatrix} \equiv_2 \begin{bmatrix} 1 \\ 0 \\ 0 \\ 0 \\ 0 \\ 0 \\ 0 \\ 0 \\ 0 \end{bmatrix} \qquad \begin{bmatrix} 1 & 1 & 0 & 0 & 0 & 0 & 0 & 0 & 0 & 0 & 0 \\ 0 & 1 & 0 & 1 & 0 & 0 & 0 & 0 & 0 & 0 & 0 \\ 0 & 0 & 1 & 0 & 0 & 0 & 0 & 0 & 0 & 0 & 0 \\ 0 & 0 & 0 & 1 & 0 & 1 & 0 & 0 & 0 & 0 & 0 \\ 0 & 0 & 0 & 0 & 1 & 0 & 0 & 1 & 0 & 0 & 1 \\ 0 & 0 & 0 & 0 & 0 & 1 & 0 & 1 & 0 & 0 & 0 \\ 0 & 0 & 0 & 0 & 0 & 0 & 1 & 0 & 0 & 0 & 0 \\ 0 & 0 & 0 & 0 & 0 & 0 & 0 & 1 & 1 & 0 & 0 \\ 0 & 0 & 0 & 0 & 0 & 0 & 0 & 0 & 1 & 1 & 0 \end{bmatrix} \begin{bmatrix} \sigma_1 \\ \sigma_2 \\ x_1 \\ y_1 \\ z_1 \\ x_2 \\ y_2 \\ z_2 \\ x \\ y \\ z \end{bmatrix} \equiv_2 \begin{bmatrix} 1 \\ 1 \\ 0 \\ 1 \\ 0 \\ 0 \\ 0 \\ 0 \\ 1 \end{bmatrix}$$

4 Discussion

Work on deriving systems of equalities [9] and inequalities [4] between program variables dates back to the very early days of abstract interpretation. Congruence domains were pioneered by Granger [5,6] who proposed, among other things, using sets of generators for representing congruence equations and showed that congruence equations satisfied the ascending chain condition.

Recently there has been a resurgence of interest in inferring both linear [7,14] and congruence relationships [1,15], mainly from the perspective of improving efficiency, for instance, by applying randomisation [7], or fusing the domain operations with the fixed-point calculations [14], or refining the conversion between equations and generators [1], or bounding the size of the coefficients in the representation [1,15]. An interesting twist to linear equalities was given by Leroux [13] who has applied the disjunctive closure of this domain in model checking.

Our work revisits congruence analysis, not to enhance efficiency, but to improve precision. Precision is refined by capturing the semantics of a basic block accurately as a system of propositional constraints. These are combined with formulae that express congruence equations that hold upon entry to the block. The constraints that hold at the end of the block are then abstracted as a system of optimal congruence equations. This avoids the need to construct specialised transfer functions for affine assignment, nondeterministic assignment, etc. Instead all primitives, linear and non-linear, can be handled uniformly by translating them into systems of propositional constraints using transformations devised for bounded model checking [3,8,20].

An issue for future work is extending the intra-procedural analysis to inter-procedural analysis and systematic benchmarking. In its present form, the analyser consists of a Prolog and a Java component that are linked with temporary files. The Prolog component translates basic blocks, congruence equalities and congruence disequalities into propositional formulae and then applies CNF conversion [16] to construct a DIMACS file for SAT4J. The Java component implements triangularisation and join. The fixed-point is under manual control since it is both useful and pleasing to watch as the summaries converge onto a loop invariant. However, the Prolog component needs to be extended to translate other operations into formulae in order to deploy the analysis on other code and particularly programs that apply bit-level programming tricks [18].

5 Conclusion

This paper shows how congruence equations, with a modulo that is a power of two, fit elegantly with SAT solving and a relational bit-level encoding of the behaviour of the program, to derive invariants for programs that contain non-linear operations such as bit twiddling. The work calls for further research into methods in which SAT solvers are applied repeatedly to infer abstractions drawn from abstract domains that satisfy the ascending chain condition.

Acknowledgments. This work was funded by EPSRC projects EP/C015517, EP/E033105 and EP/F012896. We thank Paul Docherty for motivating discussions on reverse engineering, Neil Kettle and Axel Simon for their comments on SAT-solving and Gift Nuka for his help with Floyd-style assertions.

References

1. Bagnara, R., Dobson, K., Hill, P.M., Mundell, M., Zaffanella, E.: Grids: A Domain for Analyzing the Distribution of Numerical Values. In: Puebla, G. (ed.) LOPSTR 2006. LNCS, vol. 4407, pp. 219–235. Springer, Heidelberg (2007)
2. Bryant, R.E.: On the complexity of VLSI implementations and graph representations of Boolean functions with application to integer multiplication. IEEE Transactions on Computers 40(2), 205–213 (1991)
3. Clarke, E., Kroening, D., Lerda, F.: A tool for checking ANSI-C programs. In: Jensen, K., Podelski, A. (eds.) TACAS 2004. LNCS, vol. 2988, pp. 168–176. Springer, Heidelberg (2004)
4. Cousot, P., Halbwachs, N.: Automatic discovery of linear constraints among variables of a program. In: Symposium on Principles of Programming Languages, pp. 84–97. ACM, New York (1978)
5. Granger, P.: Static analysis of arithmetical congruences. International Journal of Computer Mathematics 30, 165–190 (1989)
6. Granger, P.: Static analyses of linear congruence equalities among variables of a program. In: Abramsky, S. (ed.) CAAP 1991 and TAPSOFT 1991. LNCS, vol. 493, pp. 167–192. Springer, Heidelberg (1991)
7. Gulwani, S., Necula, G.C.: Discovering affine equalities using random interpretation. In: Principles of Programming Languages, pp. 74–84. ACM, New York (2003)
8. Jackson, D., Vaziri, M.: Finding bugs with a constraint solver. In: International Symposium on Software Testing and Analysis, pp. 14–25. ACM, New York (2000)
9. Karr, M.: Affine relationships among variables of a program. Acta Informatica 6, 133–151 (1976)
10. Kettle, N., King, A., Strzemecki, T.: Widening ROBDDs with prime implicants. In: Hermanns, H., Palsberg, J. (eds.) TACAS 2006 and ETAPS 2006. LNCS, vol. 3920, pp. 105–119. Springer, Heidelberg (2006)
11. King, A., Søndergaard, H.: Inferring congruence equations using SAT. Technical Report 1-08, Computing Laboratory, University of Kent, CT2 7NF (2008)
12. Le Berre, D.: A satisfiability library for Java, `http://www.sat4j.org`
13. Leroux, J.: Disjunctive invariants for numerical systems. In: Wang, F. (ed.) ATVA 2004. LNCS, vol. 3299, pp. 93–107. Springer, Heidelberg (2004)
14. Müller-Olm, M., Seidl, H.: A Note on Karr's Algorithm. In: Díaz, J., Karhumäki, J., Lepistö, A., Sannella, D. (eds.) ICALP 2004. LNCS, vol. 3142, pp. 1016–1028. Springer, Heidelberg (2004)
15. Müller-Olm, M., Seidl, H.: Analysis of modular arithmetic. ACM Transactions on Programming Languages and Systems 29(5) (August 2007) (Article 29)
16. Plaisted, D.A., Greenbaum, S.: A structure-preserving clause form translation. Journal of Symbolic Compututation 2(3), 293–304 (1986)
17. Ravi, K., McMillan, K.L., Shiple, T.R., Somenzi, F.: Approximation and decomposition of binary decision diagrams. In: Design Automation Conference, pp. 445–450. IEEE Press, Los Alamitos (1998)
18. Warren Jr., H.S.: Hacker's Delight. Addison Wesley, Reading (2003)
19. Wegner, P.: A technique for counting ones in a binary computer. Communications of the ACM 3(5), 322–322 (1960)
20. Xie, Y., Aiken, A.: SATURN: A scalable framework for error detection using Boolean satisfiability. ACM Transactions on Programming Languages and Systems 29(3) (2007) (Article 16)

The Barcelogic SMT Solver*
Tool Paper

Miquel Bofill[1], Robert Nieuwenhuis[2], Albert Oliveras[2],
Enric Rodríguez-Carbonell[2], and Albert Rubio[2]

[1] Universitat de Girona
[2] Technical University of Catalonia, Barcelona

Abstract. This is the first system description of the Barcelogic SMT solver, which implements all techniques that our group has been developing over the last four years as well as state-of-the-art features developed by other research groups. We pay special attention to the theory solvers and to functionalities that are not common in SMT solvers.

1 Introduction

The importance of propositional SAT solvers in verification applications has been largely shown in the last few years. However, propositional logic is not very expressive and by encoding practical problems into SAT, sometimes important structural information is lost or substantial blow-up in the formula size is caused. A successful alternative to SAT is to consider more expressive logics that still have efficient solvers. For example, for reasoning about timed automata, it is very useful to consider *Difference Logic (DL)*, where atoms are of the form $x - y \leq k$, being x and y integer or real variables, and k a numeric constant; in hardware verification, when one wants to abstract away the concrete behavior of certain components, it is useful to consider the logic of *Equality with Uninterpreted Functions (EUF)*, where atoms are equalities between first-order terms; similarly, for software verification, one may need to reason about concrete data structures such as arrays, lists or queues. Hence, it becomes very natural to consider satisfiability modulo these concrete theories and deal with formulas that contain thousands of clauses like:

$$p \quad \vee \quad x - read(A, i) \leq y \quad \vee \quad f(write(A, j, i + 2)) = read(A, j) + 1$$

In general, the problem known as *Satisfiability Modulo Theories (SMT)* amounts to deciding the satisfiability of a typically ground formula modulo a background theory T. To achieve this goal, similarly to what is done in most state-of-the-art SMT solvers, Barcelogic combines a Boolean DPLL(X) engine, very similar in nature to a SAT solver, responsible for enumerating propositional models of the formula, with a theory solver $Solver_T$, responsible for checking that

* All authors partially supported by the by the project LogicTools-2 (TIN2007-68093-C02-01) funded by the Spanish Ministry of Science and Technology.

A. Gupta and S. Malik (Eds.): CAV 2008, LNCS 5123, pp. 294–298, 2008.

these models remain consistent with the theory T. The integration of DPLL(X) with the concrete solver $Solver_T$, produces what we call a DPLL(T) system.

In order to produce our Barcelogic SMT solver we have worked, together with some external colleagues, on developing and refining the DPLL(T) approach [NOT06, BNOT06], designing efficient solvers for distinct theories that comply with all the requirements of a $Solver_T$ [NO07, NO05] and in extending SMT solvers to give support for other uses rather than just checking the satisfiability of a formula [LNO06, NO06]. All these ideas have been incorporated in our Barcelogic system, hence producing a very efficient and robust SMT solver, as one can observe from its performance in previous editions of the SMT competition (see `http://www.smtcomp.org`).

2 System Description

In this section we discuss the main components in the Barcelogic SMT solver: the parser and preprocessor, the Boolean engine DPLL(X) and all theory solvers that allow Barcelogic to deal with EUF, DL, *Linear Arithmetic (LA)* and combinations of these theories. Finally, we sketch some additional capabilities.

2.1 Parser and Preprocessor

Given an input formula in SMT-LIB format [TR05], Barcelogic's parsing module performs two tasks. One of them is to detect which is the most efficient theory solver that is able to process all input atoms; the second task is to massage the formula so that it can be fed to the DPLL(X) engine: convert it to CNF, abstract all theory atoms taking into account T-equivalent atoms, apply Ackermann's reduction if EUF and an arithmetic theory are involved, and split the arithmetic equality constraints into conjunctions of inequalities, to name but a few.

2.2 The Boolean Engine

The DPLL(X) engine currently used is a slight modification of our Barcelogic SAT solver that took part in the 2006 SAT-Race (for detailed results, see `http://fmv.jku.at/sat-race-2006`). It is competitive with state-of-the-art SAT solvers and is indeed very similar to them, as it borrows most of the ideas present in zChaff [MMZ+01] and MiniSAT [ES04]. Additionally, it implements several adaptive heuristics to find the right frequency of calls to $Solver_T$ (both for consistency checks and theory propagations) and has been extended to accommodate the splitting-on-demand technique presented in [BNOT06] where one needs to add both new literals and clauses on the fly.

2.3 Theory Solvers

The EUF Solver. The theory solver for EUF is an extension of the congruence closure algorithm presented in [NO07]. It is an incremental algorithm that pioneered the integration of integer offsets in a congruence closure algorithm and the efficient computation of small explanations of inconsistency.

The DL Solver. The solver for DL is an implementation of the algorithm proposed by Cotton and Maler in [SM06]. It is a negative-cycle-detection algorithm that allows one to compute exhaustive theory propagation in a very efficient way. In order to improve efficiency, the infinite-precision arithmetic library GMP [GMP] is only called if C++ native arithmetic types do not suffice to ensure correctness.

The LA Solver. The solver for *Linear Real Arithmetic (LRA)* implements a primal simplex algorithm based on [RS04] that allows incremental addition and deletion of constraints. However, thanks to the work done by the preprocessor, no (dis)equalities have to be processed, and hence the costly exhaustive implicit equality propagation is no longer necessary. Further, as our algorithm can handle linear programs in general form [Mar86], bounds are dealt with in a more efficient way than in [RS04]. Finally, both the *tableau* as well as the *revised* implementations of the algorithm are available for the solver to choose depending on, e.g., the condition number or the density of the problem.

As regards *Linear Integer Arithmetic (LIA)*, a branch-and-cut algorithm has been implemented. Branching is performed by means of the cooperation of the Boolean engine and the LRA solver following the splitting-on-demand architecture [BNOT06], instead of splitting inside a stand-alone LIA solver. In combination with branch-and-bound, a cutting-planes algorithm has also been implemented along the lines of [DdM06].

Given the remarkable amount of DL literals that is typically manipulated when solving a problem in LA, the aforementioned solvers are called in a layered fashion [BBC$^+$05]: a pre-filtering DL solver is used which checks the consistency of the DL fragment of the assignment, and also propagates theory information to DPLL(X).

2.4 Further Capabilities

Model generation. When a formula is found to be satisfiable, Barcelogic outputs a model as a conjunction of atoms in the SMT-LIB format. For the theories under consideration, this amounts to (i) a truth value for each Boolean variable, (ii) a concrete integer or real for each numeric variable, and (iii) a partial mapping for each uninterpreted function symbol in the formula. Note that since the formula is ground, only a finite number of function applications appear and hence a partial mapping suffices to provide a model of the formula.

Predicate abstraction. Predicate abstraction is a technique for automatically extracting finite-state abstractions for systems with potentially infinite state space. The core operation in predicate abstraction is, given (i) a set of predicates P that express properties of the system, and (ii) a formula F that symbolically represents a transition system or a set of states, to compute the best approximation of F using the predicates P. In [LNO06] we showed that by means of a careful enumeration of satisfying assignments, state-of-the-art SMT solvers can be turned into very efficient predicate abstraction engines, obtaining important speedups wrt. previously existing techniques.

Max-SAT and Max-SMT. In several applications, one has a set of constraints which is known to be unsatisfiable in advance and wants to find an assignment that satisfies the maximum number of constraints. This is the so-called Max-SAT or Max-SMT problem, depending on whether the constraints are expressed as SAT or SMT. A further extension is the weighted version of these problems, where one assigns a weight, called the *violation cost*, to each constraint and wants to find the assignment that minimizes the sum of the costs of the unsatisfied constraints. In [NO06] we showed how SMT tools can be easily adapted to support this functionality and our Barcelogic SMT solver implements all the techniques described there.

References

[BBC+05] Bozzano, M., Bruttomesso, R., Cimatti, A., Junttila, T., van Rossum, P., Schulz, S., Sebastiani, R.: The MathSAT 3 System. In: Nieuwenhuis, R. (ed.) CADE 2005. LNCS (LNAI), vol. 3632, pp. 315–321. Springer, Heidelberg (2005)

[BNOT06] Barrett, C., Nieuwenhuis, R., Oliveras, A., Tinelli, C.: Splitting on Demand in SAT Modulo Theories. In: Hermann, M., Voronkov, A. (eds.) LPAR 2006. LNCS (LNAI), vol. 4246, pp. 512–526. Springer, Heidelberg (2006)

[DdM06] Dutertre, B., de Moura, L.: A Fast Linear-Arithmetic Solver for DPLL(T). In: Ball, T., Jones, R.B. (eds.) CAV 2006. LNCS, vol. 4144, pp. 81–94. Springer, Heidelberg (2006)

[ES04] Eén, N., Sörensson, N.: An Extensible SAT-solver. In: Giunchiglia, E., Tacchella, A. (eds.) SAT 2003. LNCS, vol. 2919, pp. 502–518. Springer, Heidelberg (2004)

[GMP] The GNU MP Bignum Library, `http://gmplib.org/`

[LNO06] Lahiri, S.K., Nieuwenhuis, R., Oliveras, A.: SMT Techniques for Fast Predicate Abstraction. In: Ball, T., Jones, R.B. (eds.) CAV 2006. LNCS, vol. 4144, pp. 424–437. Springer, Heidelberg (2006)

[Mar86] Maros, I.: Computational Techniques of the Simplex Method. Kluwer's International Series (2003)

[MMZ+01] Moskewicz, M.W., Madigan, C.F., Zhao, Y., Zhang, L., Malik, S.: Chaff: Engineering an Efficient SAT Solver. In: 38th Design Automation Conference, DAC 2001, pp. 530–535. ACM Press, New York (2001)

[NO05] Nieuwenhuis, R., Oliveras, A.: DPLL(T) with Exhaustive Theory Propagation and Its Application to Difference Logic. In: Etessami, K., Rajamani, S.K. (eds.) CAV 2005. LNCS, vol. 3576, pp. 321–334. Springer, Heidelberg (2005)

[NO06] Nieuwenhuis, R., Oliveras, A.: On SAT Modulo Theories and Optimization Problems. In: Biere, A., Gomes, C.P. (eds.) SAT 2006. LNCS, vol. 4121, pp. 156–169. Springer, Heidelberg (2006)

[NO07] Nieuwenhuis, R., Oliveras, A.: Fast Congruence Closure and Extensions. Information and Computation, IC 2005(4), 557–580 (2007)

[NOT06] Nieuwenhuis, R., Oliveras, A., Tinelli, C.: Solving SAT and SAT Modulo Theories: From an abstract Davis-Putnam-Logemann-Loveland procedure to DPLL(T). Journal of the ACM, J. ACM 53(6), 937–977 (2006)

[RS04] Rueß, H., Shankar, N.: Solving Linear Arithmetic Constraints. Technical Report CSL-SRI-04-01, SRI International (2004)

[SM06] Cotton, S., Maler, O.: Fast and Flexible Difference Constraint Propagation for DPLL(T). In: Biere, A., Gomes, C.P. (eds.) SAT 2006. LNCS, vol. 4121, pp. 170–183. Springer, Heidelberg (2006)

[TR05] Tinelli, C., Ranise, S.: SMT-LIB: The Satisfiability Modulo Theories Library (2005), http://goedel.cs.uiowa.edu/smtlib/

The MathSAT 4 SMT Solver
Tool Paper

Roberto Bruttomesso[1], Alessandro Cimatti[1], Anders Franzén[1], Alberto Griggio[2], and Roberto Sebastiani[2]

[1] FBK-IRST, Povo, Trento, Italy
{bruttomesso,cimatti,franzen}@fbk.eu
[2] DISI, Università di Trento, Italy
{griggio,rseba}@disi.unitn.it

Abstract. We present MathSAT 4, a state-of-the-art SMT solver. MathSAT 4 handles several useful theories: (combinations of) equality and uninterpreted functions, difference logic, linear arithmetic, and the theory of bit-vectors. It was explicitly designed for being used in formal verification, and thus provides functionalities which extend the applicability of SMT in this setting. In particular: model generation (for counterexample reconstruction), model enumeration (for predicate abstraction), an incremental interface (for BMC), and computation of unsatisfiable cores and Craig interpolants (for abstraction refinement).

1 Introduction

In this paper we present MathSAT 4, a modern Satisfiability Modulo Theories (SMT) solver. Despite its "traditional" name, MathSAT 4 has been completely redesigned and reimplemented from scratch, and it is thus a completely new system wrt. MathSAT 3 [3]. Unlike its predecessors, MathSAT 4 has been explicitly designed for being used in a formal verification setting: in fact, besides extending the set of "traditional" theories of interest with the theory of bit-vectors [4], it also provides novel functionalities which are particularly targeted for usage in FV. To this extent, MathSAT 4 has recently been integrated within the NuSMV model checker [7] as a workhorse engine for formal verification of word-level circuits and of timed and hybrid systems.

MathSAT 4 is available at its web page (`http://mathsat4.disi.unitn.it`), with documentation, related papers and some performance figures.

2 Architecture

MathSAT 4 is based on the lazy integration schema used in many SMT tools (see, e.g., [16]). The high-level architecture of the system is shown in Figure 1.
Interface. MathSAT 4 is written in C++. Interaction with MathSAT can be performed either via files or via a rich C API. The system supports three different input formats: a native (MSAT) one, the standard SMT-LIB one, and

A. Gupta and S. Malik (Eds.): CAV 2008, LNCS 5123, pp. 299–303, 2008.

the one used by the FOCI [13] interpolating prover. (The API supports also the possibility to use MATHSAT 4 as a Theory Context Checker (TCC) [6], see below.) MATHSAT 4 can also interface with external Boolean unsat-core extractors by exchanging purely-Boolean CNF formulas as DIMACS files.

Preprocessor. After the input formula φ is parsed (or generated through the API), a preprocessing step is performed, consisting of three parts. First, the problem is simplified by encoding equivalent theory atoms ($\mathcal{T}$-atoms) into a unique representation, and by propagating top-level information.[1] Second, the formula is converted to CNF. Finally, MATHSAT applies static learning [3], i.e. it adds to the formula small clauses representing $\mathcal{T}$-valid lemmas (e.g. transitivity constraints) which can speed up the Boolean reasoning process.

DPLL Engine. The core of the solver is the DPLL Engine. It receives as input the CNF conversion of the original problem, and drives the search by enumerating its propositional models and invoking the $\mathcal{T}$-solver(s) to check them for consistency, until either a model is found or all of them are found inconsistent. The DPLL Engine is based on the highly-efficient MINISAT 2 SAT solver.

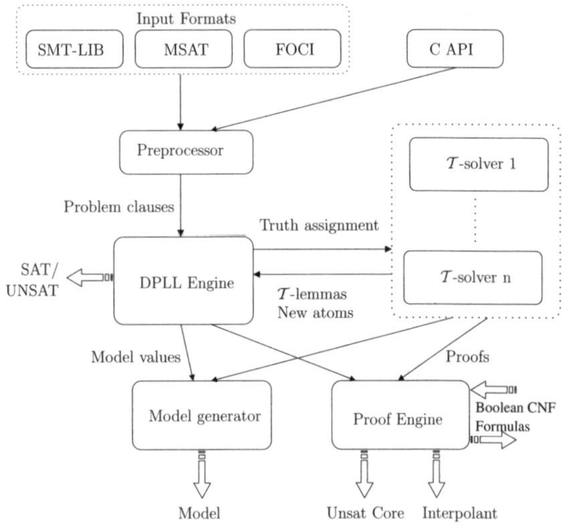

Fig. 1. MATHSAT 4 architecture

Like its predecessors, MATHSAT 4 implements most of the techniques for optimizing the interaction of DPLL and $\mathcal{T}$-solvers (see [16] for a survey). Notably, it implements also a novel adapting heuristic for controlling the interleaving between DPLL steps and $\mathcal{T}$-solver calls.

Theory solvers. In MATHSAT 4 the $\mathcal{T}$-solvers are organized as a *layered hierarchy* of solvers of increasing expressivity and complexity [3,4]: if a higher-level solver finds a conflict, then this conflict is used to prune the search at the Boolean level; if it does not, the lower level solvers are activated. These $\mathcal{T}$-solvers implement state-of-the-art procedures for the theories of equality and uninterpreted functions and predicates ($\mathcal{EUF}$) plus numeric constants and some numeric relations [14], for difference logic ($\mathcal{DL}$) [10], for the theory of linear arithmetic ($\mathcal{LA}$) over the rationals ($\mathcal{LA}(\mathbb{R})$) and over the integers ($\mathcal{LA}(\mathbb{Z})$) [12,3], and a basic procedure for a fragment of the theory of bit vectors ($\mathcal{BV}$). [2] We are currently

[1] For example, the formula $x = 5 \wedge f(x) < 3$ is rewritten into $f(5) < 3$.

[2] Currently, the $\mathcal{T}$-solver for bit-vectors can not be used in DTC with other theories.

on the way of integrating in MathSAT 4 also the $\mathcal{T}$-solver for the theory of unbounded reachability ($\mathcal{HMP}$) of [15].

A $\mathcal{T}$-solver gets in input a set of quantifier-free literals μ and checks whether μ is $\mathcal{T}$-satisfiable or not. In the first case, it also tries to perform deductions in the form $\mu' \models_{\mathcal{T}} l$, where $\mu' \subseteq \mu$ and l is a literal representing a truth assignment to a not-yet-assigned atom occurring in the input formula. In both cases, the $\mathcal{T}$-solver generates a compact explanation for the conflict or for the implication, whose negation (a $\mathcal{T}$-valid clause called $\mathcal{T}$-lemma) is then used by the DPLL engine for backjumping and learning. The $\mathcal{T}$-solvers can also generate new atoms, which allows for implementing techniques like Delayed Theory Combination (DTC) [2], Splitting On Demand [1] or Dynamic Ackermann's Expansion [11].

In order to handle problems expressed in a combination $\mathcal{T}_1 \cup \mathcal{T}_2$ of theories, MathSAT 4 applies the Delayed Theory Combination (DTC) procedure (see [2]), which is more suitable than the traditional Nelson-Oppen procedure for exploiting the synergy with the underlying DPLL engine.When $\mathcal{T}_1$ is the $\mathcal{EUF}$ theory, an alternative approach is that of reducing to a problem in $\mathcal{T}_2$ only by applying Ackermann's expansion to all the uninterpreted function symbols. In such cases, MathSAT 4 applies a simple but effective heuristic [5] to decide whether to use DTC or Ackermann's expansion.

3 Novel Functionalities

MathSAT 4 was designed primarily to be used in formal verification settings, where very often a simple "SAT/UNSAT" answer for an SMT problem is not enough, and extra information is required. Therefore, several extended functionalities are provided.

Producing models. For all theories and their combination, when φ is satisfiable, MathSAT 4 returns a satisfying interpretation $\mathcal{I}$ on domain variables with a congruent partial interpretation of uninterpreted functions and predicates. [3]

Extracting $\mathcal{T}$-unsatisfiable cores. MathSAT 4 provides two distinct techniques for extracting a $\mathcal{T}$-unsatisfiable subset of an input clause set (unsat core) described in [8]. The first ("proof-based") computes a resolution proof of $\mathcal{T}$-unsatisfiability and returns all its leaf clauses which are not $\mathcal{T}$-lemmas. The second ("lemma-lifting") invokes an external Boolean unsat-core extractor on the Boolean abstraction of the original clauses plus all $\mathcal{T}$-lemmas computed, discharging all $\mathcal{T}$-lemmas from the result. This benefits from every size-reduction techniques implemented in Boolean unsat-core extractors available off-the-shelf.

Computing Craig interpolants. MathSAT 4 allows for computing Craig interpolants of pairs of input SMT formulas [9]. This feature include an optimized interpolant generator for the full theory $\mathcal{LA}(\mathbb{R})$, an ad hoc interpolant generator for $\mathcal{DL}$, and an interpolant generator for combined theories based on DTC.

[3] E.g., in $\mathcal{EUF} \cup \mathcal{LA}$, if φ is $x = 5 \wedge f(x) < 3$, then $\mathcal{I}$ may assign x to 5 and $f(5)$ to 2.

Working incrementally. MathSAT 4 can work incrementally, that is, when invoked in sequence on similar SMT formulas, it reuses information from one run to the other so that to avoid restarting the search from scratch. This feature is very important when extending to SMT SAT-based techniques like BMC or induction-based model checking.

Enumerating all consistent assignments. MathSAT 4 implements an "All-SMT" functionality: in case of a $\mathcal{T}$-satisfiable input formula φ, it can enumerate a complete set of partial assignments satisfying φ which are consistent with the theory $\mathcal{T}$; this feature is useful for performing predicate abstraction in a SMT-based Counter-Example-Guided Abstraction-Refinement (CEGAR) context [6].

Externalizing the control of Boolean search. By means of the TCC interface, MathSAT 4 allows for an external control of the variable selection during the Boolean search in DPLL. E.g, this has been used in [6] to allow an external OBDD package for driving the enumeration of $\mathcal{T}$-consistent assignments.

References

1. Barrett, C., Nieuwenhuis, R., Oliveras, A., Tinelli, C.: Splitting on Demand in SAT Modulo Theories. In: Hermann, M., Voronkov, A. (eds.) LPAR 2006. LNCS (LNAI), vol. 4246, pp. 512–526. Springer, Heidelberg (2006)
2. Bozzano, M., Bruttomesso, R., Cimatti, A., Junttila, T., van Rossum, P., Ranise, S., Sebastiani, R.: Efficient Theory Combination via Boolean Search. Information and Computation 204(10) (2006)
3. Bozzano, M., Bruttomesso, R., Cimatti, A., Junttila, T., van Rossum, P., Schulz, S., Sebastiani, R.: MathSAT: Tight Integration of SAT and Mathematical Decision Procedures. Journal of Automated Reasoning 35(1-3) (2005)
4. Bruttomesso, R., Cimatti, A., Franzen, A., Griggio, A., Hanna, Z., Nadel, A., Palti, A., Sebastiani, R.: A Lazy and Layered SMT(BV) Solver for Hard Industrial Verification Problems. In: Damm, W., Hermanns, H. (eds.) CAV 2007. LNCS, vol. 4590, pp. 547–560. Springer, Heidelberg (2007)
5. Bruttomesso, R., Cimatti, A., Franzén, A., Griggio, A., Santuari, A., Sebastiani, R.: To Ackermann-ize or Not to Ackermann-ize? On Efficiently Handling Uninterpreted Function Symbols in SMT($\mathcal{EUF} \cup \mathcal{T}$). In: Hermann, M., Voronkov, A. (eds.) LPAR 2006. LNCS (LNAI), vol. 4246, pp. 557–571. Springer, Heidelberg (2006)
6. Cavada, R., Cimatti, A., Franzén, A., Kalyanasundaram, K., Roveri, M., Shyamasundar, R.: Computing Predicate Abstractions by Integrating BDDs and SMT Solvers. In: FMCAD, IEEE Computer Society, Los Alamitos (2007)
7. Cimatti, A., Clarke, E., Giunchiglia, E., Giunchiglia, F., Pistore, M., Roveri, M., Sebastiani, R., Tacchella, A.: NuSMV 2: An OpenSource Tool for Symbolic Model Checking. In: Brinksma, E., Larsen, K.G. (eds.) CAV 2002. LNCS, vol. 2404, Springer, Heidelberg (2002)
8. Cimatti, A., Griggio, A., Sebastiani, R.: A Simple and Flexible Way of Computing Small Unsatisfiable Cores in SAT Modulo Theories. In: Marques-Silva, J., Sakallah, K.A. (eds.) SAT 2007. LNCS, vol. 4501, pp. 334–339. Springer, Heidelberg (2007)
9. Cimatti, A., Griggio, A., Sebastiani, R.: Efficient Interpolant Generation in Satisfiability Modulo Theories. In: TACAS. LNCS, vol. 4963, Springer, Heidelberg (2008)

10. Cotton, S., Maler, O.: Fast and Flexible Difference Constraint Propagation for DPLL(T). In: Biere, A., Gomes, C.P. (eds.) SAT 2006. LNCS, vol. 4121, pp. 170–183. Springer, Heidelberg (2006)
11. de Moura, L., Bjorner, N.: Model-based theory combination. In: SMT 2007 (2007)
12. Dutertre, B., de Moura, L.: A Fast Linear-Arithmetic Solver for DPLL(T). In: Ball, T., Jones, R.B. (eds.) CAV 2006. LNCS, vol. 4144, pp. 81–94. Springer, Heidelberg (2006)
13. McMillan, K.: An interpolating theorem prover. Theor. Comp. Sci. 345(1) (2005)
14. Nieuwenhuis, R., Oliveras, A.: Proof-Producing Congruence Closure. In: Giesl, J. (ed.) RTA 2005. LNCS, vol. 3467, pp. 453–468. Springer, Heidelberg (2005)
15. Rakamarić, Z., Bruttomesso, R., Hu, A.J., Cimatti, A.: Verifying Heap Manipulating Programs in an SMT Framework. In: Namjoshi, K.S., Yoneda, T., Higashino, T., Okamura, Y. (eds.) ATVA 2007. LNCS, vol. 4762, pp. 237–252. Springer, Heidelberg (2007)
16. Sebastiani, R.: Lazy Satisfiability Modulo Theories. Journal on Satisfiability, Boolean Modeling and Computation, JSAT 3, 141–224 (2007)

CSIsat: Interpolation for LA+EUF*
Tool Paper

Dirk Beyer[1], Damien Zufferey[2], and Rupak Majumdar[3]

[1] Simon Fraser University, BC, Canada
[2] EPFL, Switzerland
[3] UCLA, CA, USA

Abstract. We present CSIsat, an interpolating decision procedure for the quantifier-free theory of rational linear arithmetic and equality with uninterpreted function symbols. Our implementation combines the efficiency of linear programming for solving the arithmetic part with the efficiency of a SAT solver to reason about the boolean structure. We evaluate the efficiency of our tool on benchmarks from software verification. Binaries and the source code of CSIsat are publicly available as free software.

1 Overview

The *Craig interpolant* for a pair (ϕ_1, ϕ_2) of formulas such that $\phi_1 \wedge \phi_2$ is not satisfiable, is a formula ψ such that ϕ_1 implies ψ, the conjunction $\psi \wedge \phi_2$ is not satisfiable, and ψ is over symbols that are common to ϕ_1 and ϕ_2 [4]. Craig interpolants have been applied successfully in formal verification and logic synthesis. For example, several software verification tools use Craig interpolants derived from infeasible counterexamples to refine their abstractions.

An *interpolating decision procedure* extends a decision procedure in the following way: it takes as input a pair (ϕ_1, ϕ_2) of formulas and has two possible outcomes: the procedure returns (1) with the answer SAT, if the conjunction $\phi_1 \wedge \phi_2$ is satisfiable, or otherwise (2) with a formula ψ that is a Craig interpolant for (ϕ_1, ϕ_2). CSIsat [1] is a new tool that implements an interpolating decision procedure for boolean combinations of linear-arithmetic expressions and equality with uninterpreted function symbols (LA+EUF).

Availability. The source code, executables, and all benchmarks for CSIsat are available online at `http://www.cs.sfu.ca/~dbeyer/CSIsat/`. The tool is free software, released under the GPLv3 license. CSIsat is the first open-source interpolating decision procedure available to verification researchers. We hope that other researchers can integrate new interpolating decision procedures into CSIsat and that developers find it easy to integrate the tool into more applications.

* Dirk Beyer and Damien Zufferey were supported in part by the Canadian NSERC grant RGPIN 341819-07; Rupak Majumdar was supported in part by the NSF grants CCF 0546170 and CCF 0720882.

[1] Available at `http://www.cs.sfu.ca/~dbeyer/CSIsat/`

A. Gupta and S. Malik (Eds.): CAV 2008, LNCS 5123, pp. 304–308, 2008.

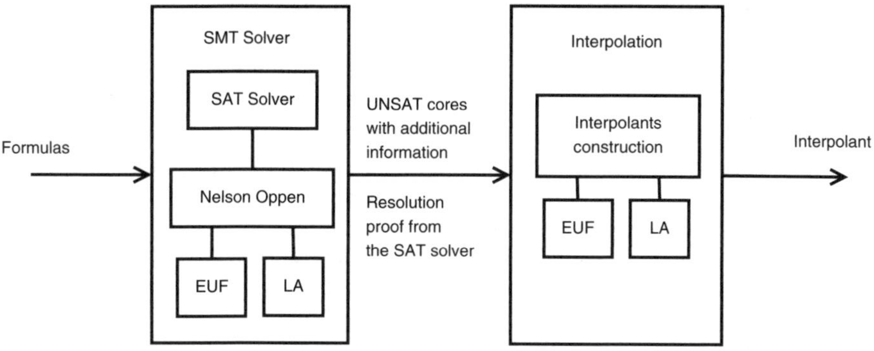

Fig. 1. Architecture of CSIsat

Related Tools. So far there are two published interpolation tools: Foci and CLPprover. McMillan's tool Foci [2] is an interpolation procedure for boolean combinations of linear-arithmetic expressions and equality with uninterpreted function symbols [6]. The tool is implemented as a proof-based theorem prover. Rybalchenko's tool CLPprover [3] is an interpolation procedure for *conjunctions* of linear-arithmetic constraints and equality with uninterpreted function symbols [9]. The tool is based on linear-constraint solving and implemented on top of the CLP($\mathbb{Q}$,$\mathbb{R}$) library [5] for SICStus Prolog.

These two existing tools have different advantages over each other: CLPprover takes advantage of linear-constraint solving and can provide an efficient solution for conjunctions of linear-arithmetic expressions, and can constrain interpolants to be only over particular variables, if possible. Foci, on the other hand, handles boolean combinations efficiently. CSIsat combines the advantages of both approaches, and uses efficient SMT algorithms to provide a fast interpolation procedure. Our experimental evaluation provides evidence of good performance. [4]

2 Architecture and Algorithm

Figure 1 illustrates the architecture of CSIsat. Our goal is to provide a tool for computing interpolants for boolean combinations of (rational) linear-arithmetic expressions (LA) and equality with uninterpreted function symbols (EUF). Interpolants for pure conjunctions of linear-arithmetic constraints can be efficiently computed using linear programming, and therefore, CSIsat uses the algorithm

[2] Available at `http://www.kenmcmil.com/foci.html`

[3] Available at `http://www.mpi-sws.mpg.de/~rybal/clp-prover/`

[4] The original motive for our work was a very practical one: Until now, we had two interpolation tools integrated in Blast: Foci and CLPprover. To verify different programs we had to use different command-line options: -foci by default, and -clp for programs that require to track linear-arithmetic expressions.

of Rybalchenko and Sofronie-Stokkermans to compute interpolants for such formulas [9].

The constraint-based algorithm cannot directly handle formulas that are not convex in their geometrical interpretation. To solve this problem, Rybalchenko and Sofronie-Stokkermans propose to convert both formulas ϕ_1, ϕ_2 to disjunctive normal form (DNF), perform multiple queries to the CLP-based algorithm and construct the final interpolant from the results of these queries [9]. Unfortunately, the DNF conversion can often blow up in practice.

As a solution to this problem, we chose a two-step approach. For the first step we use an SMT solver that integrates SAT with Nelson-Oppen style theory reasoning [7,2,8]. If the conjunction of the formulas is satisfiable, the tool stops with answer SAT and returns a satisfiable subformula that implies the conjunction of the two input formulas. If the conjunction is not satisfiable, CSIsat collects the unsatisfiable core and computes the interpolants from this. For this second step, we annotate the unsatisfiable core with additional information to avoid overhead. This information comprises the equalities deduced by theory-specific reasoning, and is used to compute partial interpolants. In addition, the resolution proof from the SAT solver is passed to the interpolation step. We use McMillan's approach to construct interpolants for the EUF part [6]; the rules were adapted to a graph-based framework. We construct interpolants for the linear-arithmetic specific part using the constraint-interpolation technique [9], and combine the interpolants using the technique of Yorsh et al. [10].

The interface to our tool is taken from Foci, i.e., CSIsat uses the same syntax for the input formulas, and the same output syntax for the interpolants, such that we can easily substitute one tool for the other. Our implementation is based on two domain-specific components. For the linear-constraint solving part, we use the GNU Linear Programming Kit (GLPK) [5]. Our SMT algorithm is based on an integrated SAT solver component. We have successfully experimented with substituting PicoSAT [6] [1] for our own SAT solver. The linear-programming component and the SAT solver component are both integrated through a wrapper interface, which can easily be adapted to other linear-programming or SAT solver components.

3 Performance Results

We show that CSIsat is competitive by comparing all three publicly available interpolation tools on some motivating examples from the software model checker Blast. Our experiments indicate that CSIsat can efficiently find interpolants.

All experiments were performed on a GNU/Linux x86_64 machine with an Intel Core 2 Duo processor and 2 GB RAM. We limited the processor speed to 1 GHz, in order to emphasize the difference. We report only the consumed User CPU Time, in order to reduce the bias from input/output operations and overhead for process setup. For all software components in our experiments,

[5] Available at `http://www.gnu.org/software/glpk/`

[6] Available at `http://fmv.jku.at/picosat/`

Table 1. Performance evaluation on Blast verification benchmarks

Program	#queries	Foci	CLPprover	CSIsat
kbfiltr	64	0.28 s	0.14 s	0.10 s
floppy	235	1.17 s	1.55 s	0.55 s
diskperf	119	0.56 s	0.61 s	0.23 s
cdaudio	130	0.60 s	0.70 s	0.26 s
ssh	6881	29 s	—	17 s
alias_swap.c	8 (908)	0.07 s	13.20 s	0.06 s

we used the latest publicly available versions, as of April 21, 2008: CSIsat 1.1; CLPprover 0.22; Foci 2003; GLPK 4.28; PicoSAT 632.

Table 1 reports the run times of the three tools on interpolation queries that occur during verification processes of different programs. The first column identifies each program that was verified by Blast. The first four programs are MS Windows device drivers, `ssh` consists of several files from the SSH software that were instrumented for verifying different properties, and `alias_swap.c` is from Blast's regression test base. During each verification run, we dumped all interpolation queries to files. Then we ran the interpolation procedures once again only on the queries, and the time in the table is the sum of the run times over all queries that were dumped for a program. The `ssh` experiment consisted of 19 verification tasks (program code and property), each resulting in about 350 interpolation queries. The row in the table reports the sum of the run times over all 19 verification tasks. The `ssh` interpolation queries contain a high number of subformulas of the form $a \neq b$. CLPprover does not support disjunctions, and transforming each such subformula to $a < b \vee a > b$, and the resulting overall formula into DNF, resulted in an intractable number of conjunctive queries. The example `alias_swap.c` requires interpolants for formulas with disjunctions, because it uses pointer aliases. In this case, we did the transformation to DNF in order to feed CLPprover. These last two examples demonstrate the importance of efficient handling of boolean combinations.

Acknowledgments. We thank Alessandro Cimatti, Alberto Griggio, and Roberto Sebastiani for interesting discussions relating to interpolation, and a pointer to their recent paper [3].

References

1. Biere, A.: PicoSAT essentials. JSAT (submitted, 2008)
2. Bozzano, M., Bruttomesso, R., Cimatti, A., Junttila, T.A., Ranise, S., Rossum, P.v., Sebastiani, R.: Efficient satisfiability modulo theories via delayed theory combination. In: Etessami, K., Rajamani, S.K. (eds.) CAV 2005. LNCS, vol. 3576, pp. 335–349. Springer, Heidelberg (2005)
3. Cimatti, A., Griggio, A., Sebastiani, R.: Efficient interpolant generation in satisfiability modulo theories. In: Ramakrishnan, C.R., Rehof, J. (eds.) TACAS 2008. LNCS, vol. 4963, pp. 397–412. Springer, Heidelberg (2008)

4. Craig, W.: Linear reasoning. A new form of the Herbrand-Gentzen theorem. J. Symb. Log. 22(3), 250–268 (1957)
5. Holzbaur, C.: OFAI clp(q,r) Manual, Edition 1.3.3. Austrian Research Institute for Artificial Intelligence, Vienna, TR-95-09 (1995)
6. McMillan, K.L.: An interpolating theorem prover. Theor. Comput. Sci. 345(1), 101–121 (2005)
7. Nelson, G., Oppen, D.C.: Fast decision procedures based on congruence closure. J. ACM 27(2), 356–364 (1980)
8. Nieuwenhuis, R., Oliveras, A., Tinelli, C.: Solving SAT and SAT modulo theories: From an abstract Davis-Putnam-Logemann-Loveland procedure to DPLL(T). J. ACM 53(6), 937–977 (2006)
9. Rybalchenko, A., Sofronie-Stokkermans, V.: Constraint Solving for Interpolation. In: Cook, B., Podelski, A. (eds.) VMCAI 2007. LNCS, vol. 4349, pp. 346–362. Springer, Heidelberg (2007)
10. Yorsh, G., Musuvathi, M.: A combination method for generating interpolants. In: Nieuwenhuis, R. (ed.) CADE 2005. LNCS (LNAI), vol. 3632, pp. 353–368. Springer, Heidelberg (2005)

Prover's Palette: A User-Centric Approach to Verification with Isabelle and QEPCAD-B
Tool Paper

Laura I. Meikle and Jacques D. Fleuriot

School of Informatics, University of Edinburgh, Appleton Tower, EH8 9LE
{lauram,jdf}@dai.ed.ac.uk

Abstract. We present Prover's Palette, a general framework for formal verification using multiple tools, centred around the user. We illustrate the framework by describing a concrete integration of the theorem prover Isabelle with the computer algebra system QEPCAD-B.

1 What Is New

The Prover's Palette is a user-centric approach to integrating theorem provers with external tools whereby the user is provided with a novel way of interacting intelligently with the different systems. Our guiding principle is that integrations should make external tools easy to use in the proof environment, but without limiting the power of the tool (by cutting out functionality) and without limiting the potential audience (by making it too difficult or intrusive). This may entail supporting both automatic and interactive usage — allowing both black-box and glass-box integrations and making it accessible to both novice and expert users. These aims are similar to those of other integration frameworks, in particular the PROSPER project [3], but recent advances in IDE systems mean these aims can now be realized to a greater extent. We build on the Eclipse Proof General Kit [1], a modular communications infrastructure for proof coupled with a rich IDE in the extensive and extensible Eclipse framework. More details on the underlying architecture and comparisons with other systems is presented in a longer paper [5].

The focus of this paper is to illustrate our concept by describing a concrete integration which can be used for non-trivial formal verification tasks involving continuous mathematics in the theorem prover Isabelle [6]. In this setting, QEPCAD-B [2], one of the most sophisticated tools for solving problems in nonlinear algebra, is used to enhance Isabelle by relieving the user from the burden of reasoning deductively about nonlinear arithmetic. Moreover, because the Prover's Palette approach allows external tools to be used in a variety of ways, our QEPCAD integration can provide proof guidance and loop invariant discovery in many situations (see §2.2). It is this versatility which distinguishes this work from other integrations with QEPCAD (*e.g.* the one with PVS [7]); this will be shown by the illustrations in this paper, however for a more comprehensive comparison see [5]. Our framework is available at `www.cognetics.org/proverspalette`.

A. Gupta and S. Malik (Eds.): CAV 2008, LNCS 5123, pp. 309–313, 2008.

2 What Is Possible

In the Prover's Palette, QEPCAD is always available in the IDE (as an Eclipse plugin, contributing a widget as shown in Fig. 1). As the prover subgoal changes, the widget updates automatically, configuring intelligently chosen defaults so that QEPCAD can be used with a single click of a button (and without any previous experience with the tool). When an Isabelle problem is not completely suitable to send to QEPCAD, *e.g.* because it is not in prenex normal form or it contains incompatible types or predicates, the widget warns the user and automatically selects the subset of assumptions and/or conclusion which can meaningfully be sent. The user also has the option to try an automatic conversion of such subgoals to ones which can be sent in their entirety to QEPCAD. Furthermore, the user can adjust the full range of QEPCAD operating parameters and select how the QEPCAD result should be used in the prover. This allows QEPCAD to be used in many different ways: as a trusted oracle, as an untrusted assistant giving insight, or even as a stand-alone client. We illustrate the first two uses through examples taken from our verification of Graham's Scan (GS) algorithm for finding convex hulls [4]. The algorithm relies heavily on the notion of "left turn", where pqr means that the point r lies to the left of the directed line from p to q, and can be defined in terms of Cartesian coordinates:

$$pqr \equiv (q_x - p_x)(r_y - p_y) - (q_y - p_y)(r_x - p_x) > 0$$

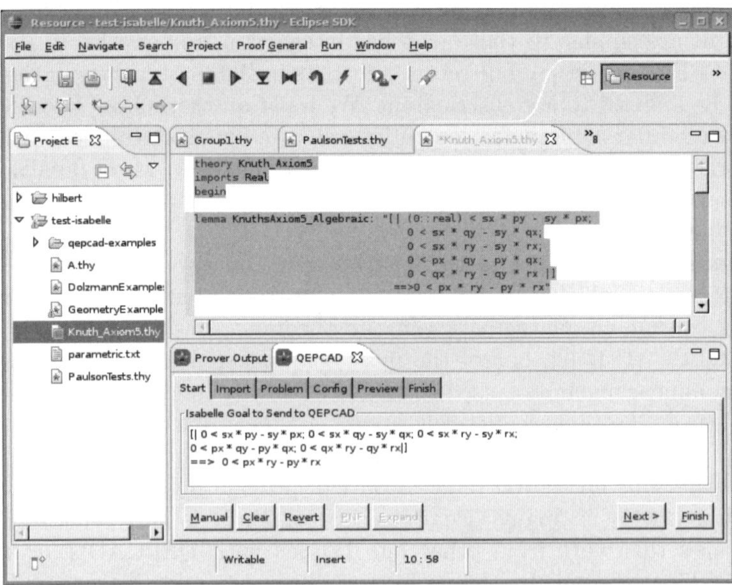

Fig. 1. Isabelle Eclipse Proof General with QEPCAD Proof Palette Widget

2.1 QEPCAD as an Automated Oracle

Using Hoare Logic to verify the GS algorithm in Isabelle entailed proving many difficult subgoals. However, using our new Isabelle/QEPCAD framework, we obtain instant help. For instance, many subgoals can quickly be checked for validity, and if the user is willing to trust QEPCAD, the integration can automatically apply any simplified result to the current proof. As an example, consider:

$$tsp \wedge tsq \wedge tsr \wedge tpq \wedge tqr \longrightarrow tpr$$

This is well known to be true, although it is not easy to prove in Isabelle alone. Taking t to be the origin (this is not necessary, but QEPCAD runs more quickly with it), we get the Isabelle lemma shown in Fig. 1 (top box). Our QEPCAD widget monitors the proof state, and whenever a subgoal looks amenable to proof using QEPCAD, the "Finish" button is enabled, indicating to the user that QEPCAD should be able to finish solving this subgoal. Clicking this button sends the translated problem to QEPCAD[1]. When a result is found — a matter of milliseconds in this case — the "Finish" tab is displayed (see Fig. 2). Selecting "Oracle" will generate the appropriate Isabelle command for the result to be trusted. In this example, the Isabelle lemma is then proved.

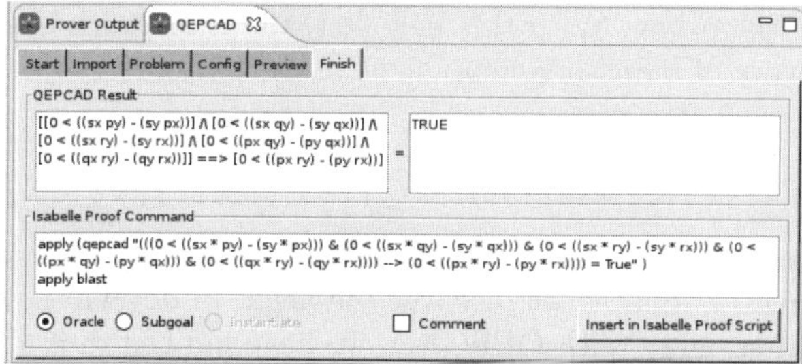

Fig. 2. QEPCAD Finish Tab

2.2 Formal Correctness: QEPCAD as Guide and Discoverer

For some applications, formal correctness requirements might disallow the use of QEPCAD as an oracle. Nevertheless, it can still be a boon to the human prover, giving insight into the problem or providing a subgoal simplification. As an example, consider the case when there are superfluous assumptions in a goal. These can obscure the relevant facts (a common difficulty of interactive proof). Our QEPCAD widget can be used to find a minimal set of assumptions which entail the conclusion of the goal. This enables many problems to be simplified without reference to an untrusted system in the formal proof.

Our framework can also guide verifications by aiding the discovery of loop invariants, a task generally accepted as non-trivial. From our experience in verifying GS, we observed that our initial loop invariant needed several refinements

[1] QEPCAD runs in the background, so a user is not blocked from using the prover.

– a process guided by failed proof attempts. Often, the root cause was a missing assumption, but identifying this was hard. One lemma we encountered was:

$$bea \wedge abd \wedge cab \wedge ade \longrightarrow ace$$

Using our new framework, the QEPCAD widget quickly tells us this is false. In this particular case, we know from the context that a subgoal similar to this is required. By using this integration interactively, we can easily identify what is missing. As QEPCAD eliminates bound variables from a problem, it can be used to yield an equivalent result in terms of the free variables only: this result may reveal useful information. By default all variables are bound when sent to QEPCAD. There is some art in selecting which variables should be free. The "Import" tab allows the user to do this interactively. In this example, a and b are used the most, so we translate a to be the origin and heuristically choose to keep b bound. With the other variables free, QEPCAD then returns:

$$d_y c_x - d_x c_y \geq 0 \vee d_x e_y - d_y e_x \leq 0 \vee e_y c_x - e_x c_y > 0$$

The second and third disjuncts are unenlightening (a negated assumption and the conclusion), but the first disjunct, however, is a hitherto missing condition to our Isabelle lemma. QEPCAD has told us that the lemma can be proven given $\neg adc$ with a at the origin. With this new fact, the framework has therefore led us to the discovery of a missing component in the loop invariant.

3 What Is Planned

We plan to continue to use the Prover's Palette to verify algorithms in computational geometry, and, as we discover the need, to develop new integrations and refine existing ones. With QEPCAD, we have noticed that although it is a powerful tool, it has one major drawback: its methods can have double exponential complexity. For some problems, performance can be improved by exploiting symmetries to reduce the number of variables, or using QEPCAD's specialised quantifiers. Currently, a user can set these quantifiers manually – which we feel is a benefit of the Prover's Palette – but it may be more desirable to automate this. We are identifying when these translations can be applied in a formally correct way. We also intend to use the software to produce witnesses where applicable: a user will then be able to use these values in a fully formal proof, even though they come from a tool which does not provide any formal proofs of its results.

Acknowledgements. We would like to thank the reviewers for their useful comments. This work was funded by the EPSRC grant EP/E005713/1.

References

1. Aspinall, D., Lüth, C., Winterstein, D.: A Framework for Interactive Proof. In: Kauers, M., Kerber, M., Miner, R., Windsteiger, W. (eds.) MKM/CALCULEMUS 2007. LNCS (LNAI), vol. 4573, pp. 161–175. Springer, Heidelberg (2007)

2. Brown, C.W.: QEPCAD B: a program for computing with semi-algebraic sets using CADs. SIGSAM Bulletin 37(4), 97–108 (2003)
3. Dennis, L.A., Collins, G., Norrish, M., Boulton, R., Slind, K., Melham, T.: The PROSPER Toolkit. Int. J Software Tools for Technology Transfer 4(2) (2003)
4. Meikle, L.I., Fleuriot, J.D.: Mechanical Theorem Proving in Computational Geometry. In: Hong, H., Wang, D. (eds.) ADG 2004. LNCS (LNAI), vol. 3763, pp. 1–18. Springer, Heidelberg (2006)
5. Meikle, L.I., Fleuriot, J.D.: Combining Isabelle and QEPCAD-B in the Prover's Palette. In: Proceedings of Calculemus (to appear, 2008)
6. Paulson, L.C. (ed.): Isabelle: A Generic Theorem Prover. LNCS, vol. 828. Springer, Heidelberg (1994)
7. Tiwari, A.: PVS-QEPCAD, www.csl.sri.com/users/tiwari/qepcad.html

Heap Assumptions on Demand

Andreas Podelski[1], Andrey Rybalchenko[2], and Thomas Wies[1]

[1] University of Freiburg
[2] MPI-SWS

Abstract. Termination of a heap-manipulating program generally depends on preconditions that express *heap assumptions* (i.e., assertions describing reachability, aliasing, separation and sharing in the heap). We present an algorithm for the inference of such preconditions. The algorithm exploits a unique interplay between counterexample-producing abstract termination checker and shape analysis. The shape analysis produces heap assumptions on demand to eliminate counterexamples, i.e., non-terminating abstract computations. The experiments with our prototype implementation indicate its practical potential.

1 Introduction

Heap-manipulating programs are prone to termination errors [2]. Manually inferring preconditions that exclude such errors is both tedious and hard, since the termination reasoning must involve the *shape* of the heap (we use the term shape in the broad sense to describe how heap locations and heap regions are aliased, inter-reachable, separated, and shared). In this paper, we present an algorithm HEAPINFER that automates this inference process. Given a heap-manipulating program, our algorithm computes a set of conditions on the shape of initial states, e.g., at the entry point of a given code fragment, that lead to terminating computations. We identify a class of *regular* programs for which the algorithm HEAPINFER is complete. An evaluation on characteristic examples practically demonstrates that the inferred preconditions are sufficiently weak.

Our algorithm iteratively applies a termination analysis to a 'shape-free' abstraction of the program. HEAPINFER avoids invocation of shape analysis until it finds a counterexample in the form of a non-terminating abstract computation, i.e., it applies shape analysis on demand. The shape analysis produces a *heap assumption*, which is an assertion describing the heap shape. This assumption refines either the abstraction or the precondition. As the result, the refinement step eliminates the counterexample. Thus, we obtain an iterative refinement scheme that applies counterexamples to guide the refinement of abstractions and preconditions.

The 'shape-free' abstraction and the demand-driven application of shape analysis rely on several specifics of termination proofs. A termination analysis synthesizes termination arguments in the form of ranking functions (whenever possible). To define a ranking function directly on heaps does not seem appropriate. The notion of a rank is intimately related to numbers. Thus, an intermediate step of our algorithm is to translate the input program over pointer variables, a *heap program* P_{H}, into a program over integer variables, which we call a *measure program* P_{M}. This translation step from heap to measure programs represents a low-cost and coarse 'shape-free' abstraction.

A. Gupta and S. Malik (Eds.): CAV 2008, LNCS 5123, pp. 314–327, 2008.

The algorithm HEAPINFER applies a termination analysis to P_M at the next step. We obtain either a termination proof for P_M and, hence, also for P_H, or a counterexample, i.e., an infinite trace of P_M. In general, the attempt to find a termination proof for P_M fails. This is not surprising as we *expect* that a termination proof must involve some amount of information that only *shape analysis* can compute. Shape analysis is notoriously expensive, however. Hence, our algorithm calls a shape analysis on demand, i.e., for a specific, isolated task: to check the validity of an invariant assertion which is crafted for the counterexample. Recent shape analysis tools can exploit this kind of specificity by adapting the degree of precision, and thus keeping the practical cost of shape analysis at a minimum [3, 25]. Furthermore, these tools can efficiently handle series of analysis requests. They reuse results obtained for previously processed queries when proving a new assertion, and thus avoid re-computation from scratch.

If the shape analysis proves the validity of the invariant assertion by checking a corresponding *assert* statement in P_H, then HEAPINFER inserts a corresponding *assume* statement into the measure program P_M. Thus, it will refine the abstraction represented by P_M. The refined version of P_M still represents a sound abstraction of P_H, but the previously discovered counterexample is no longer feasible in the program P_M. The invariant assertion, which is crafted to exclude the counterexample of P_M, is an expression over integer variables. The expression can be evaluated in P_M as well as in P_H. Thus, it is meaningful in the *assert* statement of the program P_H over pointer variables as well as in the *assume* statement in the program P_M over integer variables.

In summary, the proposed algorithm HEAPINFER exploits a unique interplay between failed abstract termination proofs and shape analysis and applies an interleaving of abstraction and precondition refinement. Thus, we obtain the (to our knowledge first) algorithm for the inference of preconditions on the heap shape that guarantee termination of heap-manipulating programs. The experiments with our prototype implementation indicate its practical potential. We applied our implementation on characteristic fragments of heap manipulating programs, see [19], including kernel code from an operating system [17]. The inferred preconditions match the intended calling environment, and were confirmed as such by the kernel developers.

Related Work. Our work fills a gap between two recent lines of research: termination proofs under given preconditions (for heap-manipulating programs), and precondition inference for correctness properties other than termination (memory safety of heap-manipulating programs and other safety properties). Our algorithm exploits the recent advances in the respected areas by utilizing the corresponding analyses as subprocedures: shape analysis for heap-manipulating programs and termination analysis of integer-manipulating programs.

The recent termination analyses for heap manipulating programs, e.g., [2, 5], do not focus on precondition inference, but rather on proving termination under given preconditions. They do not take advantage of lazy reasoning about the heap. Unlike [2], the present version of our algorithm does not account for memory safety. It can be extended to track information related to memory safety by using measures, similarly to [5, 15].

The idea of extracting ranking functions from heap-manipulating programs by translating its statements into updates of integer variables is very natural and is classical by now. The existing transformations of heap-manipulating programs into programs over integer variables in [2, 5] are sophisticated. Each transformation uses a form of shape analysis as a preliminary step, i.e., before translating to a program over integer variables. The shape analysis is used to eagerly infer strongest invariants for the whole program, and is oblivious to the actual proof obligations required for termination reasoning. The cost of the translation and the size of the resulting program over integer variables depend on the number of shapes computed by the shape analysis. In contrast, our work aims at minimizing the cost of the shape analysis by using it only for checking specially crafted assertions. The complexity of the translation step into a measure program does not depend on the number of shapes. It is cubic in the number of pointer variables and linear in the number of statements of the heap program.

The recently proposed algorithm for deriving preconditions for memory safety of list-manipulating programs [8] employs quite different technical concepts. It neither applies shape analysis lazily, nor infers to preconditions for termination.

There is a large amount of related work on shape analysis (the synthesis of invariant assertions about the heap). A partial selection of various approaches contains [4, 6, 12, 13, 22]. Our algorithm uses shape analysis as a black box. While not requiring and being dependent on any particular implementation of shape analysis, HEAPINFER can benefit from shape analyses that are property-directed, e.g. [3, 25].

To the best of our knowledge, our work is the first that applies shape analysis on demand for inferring preconditions. A graph-based heap analysis [22] can be lazily combined with predicate abstraction [14] to improve its precision in proving safety properties [3].

Our algorithm relies on a termination prover for programs over numerical domains. There exist several practical methods and tools for proving termination of such programs, e.g. [7, 9, 10, 11, 16]. All these tools can be employed by our algorithm (after adding an extension to produce counterexamples, if necessary).

2 Preconditions for Kernel Code

A major application area of termination analyses for heap manipulating programs is low-level operating systems code [1, 2]. Often the operating system kernel contains subroutines whose termination is an inevitable requirement for ensuring that the operating system remains responsive.

Figure 1 presents an example of such a subroutine. It shows a fragment of the system call handler `process_kill` found in the process scheduler of the operating system VAMOS [17]. The handler kills the process with the given process ID. The handler needs to ensure consistency of the process scheduler's data structures, e.g. *ready list*. The ready list keeps track of all processes that are ready for being scheduled. When a process with identifier `process` is killed, the handler ensures that the process is removed from the ready list (if it is contained). Furthermore, the maximal priority of the remaining ready processes is recomputed. The outer loop in the handler code traverses the ready list until either `process` is found or `NULL` is reached. If `process` is found it is removed from the list. Furthermore, if `process` has maximal priority, then the inner loop traverses the ready list once more to compute the new maximal priority of the remaining ready processes.

```
int process_kill(unsigned int pid) {
  proc_id = pid & 127u;
  process = pid2pcb(proc_id); ...
  prev_elem = NULL;
  ready_list_elem = ready_list;
  while ((ready_list_elem != NULL) && (found == false)) {
    proc_id2 = ready_list_elem->pid;
    if (proc_id == proc_id2) {
      if (prev_elem != NULL)
        prev_elem->next = ready_list_elem->next;
      else
        ready_list = ready_list_elem->next;
      ready_list_elem->next = NULL;
      if (process->priority == max_prio) {
        highest_prio = 0u;
        highest_search = ready_list;
        while (highest_search != NULL) {
          if (highest_search->priority > highest_prio)
            highest_prio = highest_search->priority;
          highest_search = highest_search->next;
        }
        max_prio = highest_prio;
      } ...
      found = true;
    }
    prev_elem = ready_list_elem;
    ready_list_elem = ready_list_elem->next;
  } ...
}
```

Fig. 1. System call handler from the process scheduler of the VAMOS kernel [17]

The execution of the handler `process_kill` may diverge if we call it from an arbitrary program state. The termination property of the code depends on the shape of the ready list. For example, if the ready list is cyclic and does not contain `process` then the outer loop does not terminate.

Our algorithm HEAPINFER automatically infers the necessary preconditions for termination: `process_kill` expects the ready list to be acyclic. At the first inference step, the algorithm automatically introduces integer variables that measure the length of paths along pointer fields in the heap. Their value may be infinity, represented by ∞, which indicates that the corresponding path does not exist in the heap. In our example, there are three measures that track the length of the paths following the `next` link from (1) `ready_list` to NULL, (2) `ready_list_elem` to NULL, and (3) `highest_search` to NULL. We refer to these measures M_1, M_2, and M_3 respectively.

Then, HEAPINFER translates the heap program into a measure program over integers. For example, the first conjunct in the loop condition of the outer loop is translated to the disequality test $M_2 \neq 0$, and the outer loop decrements the measure M_2 if its value is different from ∞. Next, the precondition inference process iteratively applies a termination analysis to the measure program and a shape analysis to the heap program. The shape analysis is used to derive new facts from the heap program that rule out spurious non-terminating computations in the measure program. Whenever such a computation cannot be ruled out, the precondition is strengthened. Both the precondition and the fact derived from the heap program are assertions over measures.

In our example, the first termination check on the measure program fails. As a counterexample, it reports an infinite computation in which the measure M_2 is initially ∞ and is never decremented in the outer loop. This is because M_1 (and thus M_2) is initially unconstrained and might have value ∞. This computation is feasible and corresponds to the infinite traversal of the ready list in case it is *cyclic*. Consequently, the inference algorithm strengthens the precondition by the assertion $\text{M}_1 < \infty$. This rules out any infinite iteration of the outer loop in the measure program, and, hence, of the heap program.

Nevertheless, the next application of the termination analysis fails and produces a counterexample that infinitely often iterates through the inner loop with the value of measure M_3 being equal to ∞. This might come as a surprise, because acyclicity of the ready list, expressed as $\text{M}_1 < \infty$, is preserved by the heap updates in the body of the outer loop. Thus, the heap program maintains $\text{M}_3 < \infty$ at entry to the inner loop. However, due to the loss of precision by the measure abstraction, this fact cannot be derived for the measure program. Now, the inference algorithm applies the shape analysis to check the validity of the assertion $\text{M}_3 < \infty$ at the entry to the inner loop. This assertion is expressible using a reachability predicate supported by the shape analysis. The shape analysis verifies that $\text{M}_3 < \infty$ holds. This fact is propagated to the measure program by assuming $\text{M}_3 < \infty$ at the inner loop entry that, in turn, makes the subsequent termination check succeed. The inference process stops and reports the precondition $\text{M}_1 < \infty$. It states that `process_kill` expects an acyclic ready list.

3 Preliminaries

We now provide necessary definitions of heap manipulating programs, their computations, and properties. To simplify presentation, we restrict ourselves to heap programs that manipulate singly-linked lists. An extension to multi-linked lists is discussed in the technical report [19].

Heap programs. We represent a *heap program* P_{H} by a tuple $(V, \mathcal{L}, \ell_0, \mathcal{T})$. Here, V is a finite set of program variables. Each variable $v \in V$ ranges over a set of memory addresses. $\mathcal{L}$ is a finite set of control locations of the program that includes the initial location ℓ_0. We assume a distinguished program variable pc that ranges over the control locations $\mathcal{L}$, and is includes in V. $\mathcal{T}$ is a finite set of program transitions. Each transition $\tau = (\ell, grd, op, \ell')$ consists of an entry and exit locations ℓ and ℓ', respectively, a guard grd, and operation op. Guards and operations are defined by the following grammar, where $v \in V \setminus \{pc\}$ and t is a data structure link name.

$$\begin{aligned} exp &::= v \mid exp.t \\ grd &::= \mathsf{true} \mid \mathsf{false} \mid exp = exp \mid grd \wedge grd \mid \neg grd \\ op &::= \mathsf{assert}(grd) \mid v := v \mid v := v.t \mid v.t := v \mid \mathsf{new}(v) \end{aligned}$$

A *state* $s = (stack, h)$ of a heap program is a valuation of the program variables $stack$ together with the heap function h. The heap function h is a *total* function from addresses to addresses. Function h models singly-linked data structures manipulated by the program. Given a variable $v \in V$, we write $s(v)$ for the valuation of v in the state s.

We write $s[v \mapsto e]$ to represent a state s' such that $s'(v) = e$ and for each $u \in V \setminus \{v\}$ we have $s'(u) = s(u)$.

Each transition $\tau = (\ell, grd, op, \ell')$ represents a transition relation ρ_τ that contains pairs of states (s, s') such that $s(pc) = \ell$, $s \models grd$ and the following conditions apply to s and s'. If op is an operation $\mathsf{assert}(grd)$, we have either $s \models grd$ and $s' = s[pc \mapsto \ell']$, or $s \not\models grd$ and $s' = s[pc \mapsto \ell_E]$. For dealing with update operations, we define an *evaluation* function $eval$ that computes the value of an expression in a given state.

$$eval(s, exp) \stackrel{\text{def}}{=} \begin{cases} s(v) & \text{if } exp = v, \\ h(eval(s, exp')) & \text{if } exp = exp'.t \end{cases}.$$

For an operation that updates a program variable $v := exp$, we have $s' = s[pc \mapsto \ell', v \mapsto eval(s, exp)]$. In case of heap update operation $v.t := exp$, we have $s' = s[pc \mapsto \ell']$ and the heap function h is modified at the address $eval(s, v.t)$ to map to the value $eval(s, exp)$. Finally, if the update operation is an allocation operation $\mathsf{new}(v)$ then $s' = s[pc \mapsto \ell', v \mapsto a]$ and h is updated to $h \cup \{a \mapsto a'\}$ where $a \notin \mathsf{dom}(h)$ is a fresh address and $a' \in \mathsf{dom}(h) \cup \{a\}$. We assume a garbage-collected heap where we always allocate a fresh address, but we put no constraint on the value of the heap function for that fresh address. For a state s and transition τ we denote by $\mathsf{post}(\tau, s)$ the set of all τ-successors of s.

A program *computation* is a (possibly infinite) sequence $\sigma = s_0 \xrightarrow{\tau_0} s_1 \xrightarrow{\tau_1} \ldots$ of states and transitions such that $s_0(pc) = \ell_0$, for each pair of consecutive states s_i and s_{i+1} we have $s_{i+1} \in \mathsf{post}(\tau_i, s_i)$. If σ is finite then for its final state, say s, and for each transitions $\tau \in \mathcal{T}$ we have $\mathsf{post}(\tau, s) = \emptyset$.

Measure Programs. A measure is a term $\mathrm{M}(e_1, e_2)$ where e_1 and e_2 are expressions. It denotes the length of the shortest (possibly empty) t-path in the heap from the address denoted by e_1 to the address denoted by e_2, and ∞ if such a path does not exist.

We extend the evaluation function $eval$ from expressions to measures as follows:

$$eval(s, \mathrm{M}(e_1, e_2)) \stackrel{\text{def}}{=} \begin{cases} \infty & \text{if for all } i \in \mathbb{N} : s \models e_1.t^i \neq e_2 \\ \min\{\, i \in \mathbb{N} \mid s \models e_1.t^i = e_2 \,\} & \text{otherwise.} \end{cases}$$

Measure assertions are defined by the following grammar:

$$\begin{aligned} rel &::= < \mid > \mid \leq \mid \geq \mid = \\ const &::= 0 \mid 1 \mid 2 \mid \cdots \mid \infty \\ mexp &::= const \mid \mathrm{M}(exp, exp) \mid mexp + mexp \mid mexp - mexp \\ atom &::= \mathsf{true} \mid \mathsf{false} \mid mexp\ rel\ mexp \\ assn &::= atom \mid \neg assn \mid assn \wedge assn \end{aligned}$$

A measure program $P_{\mathrm{M}} = (\mathcal{M}, \mathcal{L}, \ell_0, \mathcal{T})$ is a program whose program variables $\mathcal{M}$ are the set of all measures. The set of locations $\mathcal{L}$, and initial location ℓ_0 are as for heap programs. A state of a measure program is a valuation of the pc together with valuations of all measures. Transitions of measure programs are guarded by measure assertions and perform simultaneous updates of all measures. Updates of measures are expressed in terms of measure expressions $mexp$.

Memory safety. The totality of heap function h implies that in a heap program P there exists no computation that can fail because of memory manipulation error, i.e., P is memory safe. This assumption simplifies the presentation of our 'shape-free' abstraction of heap programs, and can be easily avoided in practice by using measures for proving memory safety.

4 Algorithm

We present our algorithm HEAPINFER for the automatic inference of heap assumptions for termination in Figure 2. It takes as input a heap program P_{H} and a set of measures to be tracked for proving termination of P_{H}. The output of the algorithm is a set of preconditions that guarantee termination of the input heap program.

HEAPINFER executes in two phases: the translation of the heap program into a measure program that simulates the heap program, and a counterexample-guided refinement. The refinement phase iteratively derives two kinds of new facts. First, it computes invariants of the heap program that eliminate spurious non-terminating computations in the measure program. Second, it infers preconditions that exclude feasible infinite computations in the heap program. In the following we describe the two phases of HEAPINFER in more details. Section 5 supports this description with illustrative examples.

Translation. Figure 3 presents the function Translate that is used in line 1 of HEAPINFER to translate a heap program P_{H} into a measure program under a given set of tracked measures M. The translation can be seen as a source-to-source transformation. Each transition of the heap program is translated to a set of transitions in the measure program. An update operation upd in the heap program is translated to a simultaneous update of all measures in the measure program (tracked or untracked). Tracked measures $\mathrm{M}(e_1, e_2)$ are updated according to the update function $\mathrm{M}_{upd}(e_1, e_2)$, as defined in Figure 3, while untracked measures are non-deterministically assigned a value from $\mathbb{N} \cup \{\infty\}$.

The rules in Figure 3 defining the update functions should be read in the top-down way. The rule that matches first is applied. We only provide a detailed description for the translation of heap updates $x.t := y$ and omit other cases for brevity. Such heap updates are translated into updates of measures of the form $\mathrm{M}(z.t^i, w.t^j)$. Since the heap function t occurs in the subexpressions $z.t^i$ and $w.t^j$ of the measure, the translation needs to take into account the effect of the heap update to the denotation of these subexpressions. The first two cases apply the rule recursively until x does neither occur on the path from z to $z.t^i$ nor on the path from w to $w.t^j$. Thus, eventually the third case applies. It is divided into three subcases. The first subcase handles the situation when x does not occur on the path from $z.t^i$ to $w.t^j$. Here, the measure remains unchanged. The second subcase deals with the situation when x is reachable from $z.t^i$ and the update introduces a new path from $z.t^i$ to $w.t^j$ via x and y. Finally, the third subcase accounts for the update eliminating any existing paths between $z.t^i$ and $w.t^j$. We present a soundness proof of the translation in the extended version of the paper [19].

Note that the provided updates of measures are precise with the exception of update expressions for new statements. Here, precision means that the evaluation of an update expression $\mathrm{M}_{op}(e_1, e_2)$ in a given state s determines the value of $\mathrm{M}(e_1, e_2)$ in the post

```
input
  P_H: heap program
  M: set of tracked measures
vars
  P_M : measure program
  st_i : measure statement at location ℓ_i and with guard guard_i
  PRE: measure assertion
begin
  P_M := Translate(M, P_H)
  PRE:= true
  repeat
    if P_M terminates then
      return "termination under precondition PRE"
    else
      st_1 ... st_{m-1}.(st_m ... st_n)^ω := choose infinite trace in P_M
      i := choose position in {m, ..., n}
      if under precondition PRE, P_H ∪ ℓ_i : assert(¬guard_i) is safe then
        P_M := P_M ∪ ℓ_i : assume(¬guard_i)
      else
        PRE:= PRE ∧ wlp(P_H, at_ℓ_i → ¬guard_i)
  done
end.
```

Fig. 2. Algorithm HEAPINFER for demand-driven inference of heap assumptions. The algorithm uses three oracles: 1) the termination test on a measure program, 2) the safety check on the input heap program strengthened by a measure assertion, and 3) the weakest-precondition operator on measure assertions for the input heap program.

state of s under operation op. Update expressions of new statements are not precise in this sense, because new statements translate into nondeterministic updates.

Each of the update functions $\mathrm{M}_{upd}(e_1, e_2)$ defines a set of guarded update expressions of the form $grd \Rightarrow exp$ with the following meaning. If grd is satisfied in the current state of the measure program then the next value of measure $\mathrm{M}(e_1, e_2)$ is determined by exp.

Finally, the function bifurcate transforms a single transition with guarded update expressions for each tracked measure into a set of transitions. Each of the resulting transitions corresponds to one possible choice of picking one of the guarded update expressions per tracked measure. The guard of each resulting transition is the translated guard of the original transition in the heap program conjoined with the guards of the chosen guarded update expressions.

Choosing measures to track. We determine the set of tracked measures M using a simple heuristic. Initially, we consider measures that are required for the precise translation of loop conditions. During the translation, additional measures are lazily taken into consideration if they occur in updates of existing tracked measures according to Figure 3. To ensure that the set M remains finite we only track measures of the form $\mathrm{M}(x, y)$ where x and y are program variables. Note that the precision of the inference algorithm

$$P_{\mathrm{H}} = (V, \mathcal{L}, \ell_0, \ell_E, \mathcal{T})$$

$$\mathsf{Translate}(M, P_{\mathrm{H}}) = (\mathcal{M}, \mathcal{L}, \ell_0, \ell_E, \bigcup_{\tau \in \mathcal{T}} \mathsf{trlT}(M, \tau))$$

$$\mathsf{trlT}(M, (\ell, g, op, \ell')) = \mathsf{bifurcate}(\ell, \mathsf{trlG}(g), \mathsf{trlO}(M, op), \ell')$$

$$\mathsf{trlG}(e_1 = e_2) = \mathrm{M}(e_1, e_2) = 0$$

$$\mathsf{trlG}(\mathsf{true}) = \mathsf{true}$$

$$\mathsf{trlG}(\mathsf{false}) = \mathsf{false}$$

$$\mathsf{trlG}(\neg grd) = \neg(\mathsf{trlG}(grd))$$

$$\mathsf{trlG}(grd_1 \wedge grd_2) = \mathsf{trlG}(grd_1) \wedge \mathsf{trlG}(grd_2)$$

$$\mathsf{trlO}(M, \mathsf{assert}(grd)) = \mathsf{assert}(\mathsf{trlG}(grd))$$

$$\mathsf{trlO}(M, upd) = [ms := \mathsf{trlU}(M, upd, ms) \mid ms \in \mathcal{M}]$$

$$\mathsf{trlU}(M, upd, \mathrm{M}(t_1, t_2)) = \begin{cases} \mathrm{M}_{upd}(t_1, t_2) & \text{if } \mathrm{M}(t_1, t_2) \in M \\ * & \text{otherwise} \end{cases}$$

If op is $x := y$ then

$\mathrm{M}_{op}(e_1, e_2) \stackrel{\text{def}}{=} \mathrm{M}(e_1[y/x], e_2[y/x])$

If op is $x := y.n$ then

$\mathrm{M}_{op}(x, x) \stackrel{\text{def}}{=} 0$
$\mathrm{M}_{op}(x.n^i, x.n^j) \stackrel{\text{def}}{=} \mathrm{M}_{op}(y.n^{i+1}, y.n^{j+1})$
$\mathrm{M}_{op}(e, x) \stackrel{\text{def}}{=}$
  $\mathrm{M}(e, y) = \infty \Rightarrow \mathrm{M}(e, y.n)$
  $\mathrm{M}(e, y) < \infty$
    $\mathrm{M}(y, y.n) = 1$
      $\mathrm{M}(y.n, e) \neq 0 \Rightarrow \mathrm{M}(e, y) + 1$
      $\mathrm{M}(y.n, e) = 0 \Rightarrow 0$
    $\mathrm{M}(y, y.n) = 0 \Rightarrow \mathrm{M}(e, y)$
$\mathrm{M}_{op}(x, e) \stackrel{\text{def}}{=}$
  $\mathrm{M}(y, e) = \infty \Rightarrow \infty$
  $\mathrm{M}(y, e) < \infty$
    $\mathrm{M}(y, e) > 0 \Rightarrow \mathrm{M}(y, e) - 1$
    $\mathrm{M}(y, e) = 0$
      $\mathrm{M}(y, y.n) = 1 \Rightarrow \mathrm{M}(y.n, e)$
      $\mathrm{M}(y, y.n) = 0 \Rightarrow 0$
$\mathrm{M}_{op}(x.n^i, e) \stackrel{\text{def}}{=} \mathrm{M}_{op}(y.n^{i+1}, e)$
$\mathrm{M}_{op}(e, x.n^i) \stackrel{\text{def}}{=} \mathrm{M}_{op}(e, y.n^{i+1})$
$\mathrm{M}_{op}(e_1, e_2) \stackrel{\text{def}}{=} \mathrm{M}(e_1, e_2)$

If op is $x.n := y$ then
let $e_1 = z.n^i$ and $e_2 = w.n^j$

$\mathrm{M}_{op}(e_1, e_2) \stackrel{\text{def}}{=}$
  $i > 0 \wedge \mathrm{M}(z, x) = k \wedge k < i \Rightarrow$
      $\mathrm{M}_{op}(y.n^{i-k-1}, e_2)$
  $j > 0 \wedge \mathrm{M}(w, x) = k \wedge k < j \Rightarrow$
      $\mathrm{M}_{op}(e_1, y.n^{j-k-1})$
  $(i > 0 \rightarrow \mathrm{M}(z, x) \geq \mathrm{M}(z, e_1)) \wedge$
  $(j > 0 \rightarrow \mathrm{M}(w, x) \geq \mathrm{M}(w, e_2))$
    $\mathrm{M}(e_1, e_2) \leq \mathrm{M}(e_1, x) \Rightarrow \mathrm{M}(e_1, e_2)$
    $\mathrm{M}(e_1, e_2) > \mathrm{M}(e_1, x)$
      $\mathrm{M}(y, e_2) < \infty \wedge \mathrm{M}(y, e_2) \leq \mathrm{M}(y, x) \Rightarrow$
          $\mathrm{M}(e_1, x) + 1 + \mathrm{M}(y, e_2)$
      $\mathrm{M}(y, e_2) = \infty \vee \mathrm{M}(y, e_2) > \mathrm{M}(y, x) \Rightarrow \infty$

If op is $\mathsf{new}(x)$ then

$\mathrm{M}_{op}(x, x) \stackrel{\text{def}}{=} 0$
$\mathrm{M}_{op}(e, x) \stackrel{\text{def}}{=} \infty$
$\mathrm{M}_{op}(x, e) \stackrel{\text{def}}{=} k, \quad k \in \mathbb{N}^+ \cup \{\infty\}$
$\mathrm{M}_{op}(e, x.n^i) \stackrel{\text{def}}{=} *$
$\mathrm{M}_{op}(x.n^i, e) \stackrel{\text{def}}{=} *$
$\mathrm{M}_{op}(e_1, e_2) \stackrel{\text{def}}{=} \mathrm{M}(e_1, e_2)$

Fig. 3. Translation of a heap program to a measure program. We use $*$ to denote a non-deterministically chosen element from $\mathbb{N} \cup \{\infty\}$. Here, bifurcate creates a set of transitions for each choice of measure updates, trlT, trlG, trlO, and trlU translate transitions, guards, operations and updates, respectively.

is monotonic with respect to M, *i.e.*, adding more measures to the set will result in weaker preconditions.

Refinement loop. The core of algorithm HEAPINFER is its counterexample-guided refinement loop. In each iteration of the algorithm a termination checker is applied to check whether the measure program terminates under current precondition PRE. If the termination check succeeds then HEAPINFER stops and guarantees that the heap program is guaranteed to terminate under PRE. Otherwise, there exists a non-terminating computation in the measure program. The algorithm non-deterministically chooses one of these computations: $st_1 \ldots st_{m-1}.(st_m \ldots st_n)^\omega$. Now there are two possible cases. First, the selected computation is spurious, *i.e.*, there is no corresponding computation in the heap program. Second, the computation is feasible in the heap program. To determine whether the counterexample is feasible, the algorithm chooses a guard grd_i from the loop segment $(st_m \ldots st_n)$. Then, a safety checker is called to verify whether the negation of grd_i is an invariant of the heap program at location ℓ_i under the current precondition PRE.

If this safety check succeeds then we conclude that the found counterexample is spurious. In this case, we strengthen the guards of all transitions that start at ℓ_i in the measure program using the measure assertion $\neg grd_i$, and hence eliminate the counterexample from the measure program.

If the safety check fails, then the counterexample might correspond to a feasible computation in the heap program (or some other choice of grd_i will prove its spuriousness). The algorithm invokes an oracle that computes the weakest precondition of the negated guard grd_i and adds it to the current precondition. If the same counterexample is produced in a later iteration of the refinement loop then the negation of guard grd_i is an invariant of the heap program at location ℓ_i under the new precondition. Thus, the counterexample is eliminated eventually.

If there is a counterexample in the measure program that is spurious, but all guards in its loop are reachable by some finite computation in the heap program, then the inference algorithm will produce a precondition which is too strong. In this case the safety check in line 9 will fail on all of the loop guards and the refinement will rule out the counterexample by strengthening the precondition. This incompleteness is deliberate. In such a case a ranking function based on measures simply does not exist. However, we do not expect to observe this incompleteness on program loops typically found in low-level system code.

Weakest preconditions of measure assertions. Algorithm HEAPINFER relies on an oracle wlp that computes the weakest precondition for a measure assertion and a heap program. We propose a simple solution for implementing this oracle.

Note that measure assertions are closed under weakest preconditions for loop free heap programs. In fact, we can use the update functions from Figure 3 to compute weakest preconditions for finite sequences of transitions. Assume that the current counterexample path π in the refinement loop is of the form $st_1 \ldots st_{m-1}.(st_m \ldots st_n)^\omega$. If the algorithm attempts to strengthen the precondition using a guard grd_i from a transition of the loop segment $(st_m \ldots st_n)$, then we update precondition PRE as follows:

$$\text{PRE} := \mathsf{widen}(\text{PRE} \wedge \mathsf{wlp}(st_1, \ldots, st_{i-1}, \neg grd_i)) \ .$$

The operator widen is a widening operator on measure assertions. widen(F) identifies a series of conjuncts $C(x.n^i), C(x.n^{i+1}), \ldots$ in F and replaces them by the unbounded conjunction $\forall j \geq 0 : C(x.r^{i+j})$.

If one uses update expressions of measures to compute weakest preconditions then the only source for nondeterministic updates are new statements. We use a simple quantifier elimination procedure to eliminate the resulting universal quantifiers in weakest preconditions.

The algorithm HEAPINFER has a solid theoretical foundation. We briefly sketched soundness in the discussion above. Under assumption that the oracles for the termination check, safety check, and wlp computation always terminate, there exists a backtracking strategy on the nondeterministic choices (lines 7 and 8) such that the refinement loop in algorithm HEAPINFER always terminates. Finally, we identify a class of *regular* programs for which the algorithm HEAPINFER is complete. That is, it computes the *weakest* precondition for termination of the input heap program. The details are presented in the extended version [19].

5 Example

We illustrate the algorithm HEAPINFER on a simple, yet instructive example. The left-hand side of Figure 4 shows program TRAVERSE which traverses a singly-linked list. We apply algorithm HEAPINFER to program TRAVERSE with the singleton set of tracked measures containing only $\mathrm{M}(p, q)$. Executing line 1 in the algorithm yields the measure program P_{M} shown on the right-hand side of Figure 4. For legibility, we omit the non-deterministic updates of untracked measures. Program P_{M} does not always terminate. Let us assume that the non-deterministic choice in line 7 of the algorithm HEAPINFER selects the infinite computation $\epsilon.(\ell[1])^\omega$ that repeatedly executes the loop body according to case 1. There is only one position to choose in line 8 of the algorithm, namely, the one associated with location ℓ and guard $\mathrm{M}(p, q) = \infty$. As an assertion on states of program TRAVERSE, this guard means that q is not reachable from p. Obviously, the negated guard $\mathrm{M}(p, q) < \infty$ is not an invariant of program TRAVERSE at location ℓ. Hence, the condition in line 9 does not hold. In this case, the weakest precondition of the stem $\mathsf{wlp}_\epsilon(\mathrm{M}(p, q) < \infty)$ is again the assertion $\mathrm{M}(p, q) < \infty$. Thus, line 12 assigns PRE to $\mathrm{M}(p, q) < \infty$.

One might expect that under the precondition that q is reachable from p the program TRAVERSE terminates. HEAPINFER finds that it is not sufficient. The next iteration of the algorithm produces the counterexample $\ell[2.2.1].(\ell[1])^\omega$. The loop part of this infinite trace is the same as for the previous counterexample. Thus, we again choose guard $\mathrm{M}(p, q) = \infty$. The condition in line 9 is again false. The weakest precondition of the negated guard $\mathsf{wlp}_{\ell[2.2.1]}(\mathrm{M}(p, q) < \infty)$ simplifies to the assertion

$$\mathrm{M}(p, q) > 0 \lor \mathrm{M}(p, p.n) = 0 \lor \mathrm{M}(p.n, q) < \infty \ .$$

Line 12 updates the precondition PRE to:

$$\mathrm{PRE} \equiv \mathrm{M}(p, q) < \infty \land (\mathrm{M}(p, q) > 0 \ \lor \ \mathrm{M}(p, p.n) = 0 \ \lor \ \mathrm{M}(p.n, q) < \infty) \ .$$

ℓ : **do** $\quad p := p.n;$ **while** $p \neq q$	ℓ : **do** $\quad \text{M}(p,q) :=$ $\quad\quad 1 \quad \text{M}(p,q) = \infty \quad\quad \Rightarrow \infty$ $\quad\quad 2 \quad \text{M}(p,q) < \infty$ $\quad\quad 2.1 \quad\quad \text{M}(p,q) > 0 \quad\quad \Rightarrow \text{M}(p,q) - 1$ $\quad\quad 2.2 \quad\quad \text{M}(p,q) = 0$ $\quad\quad 2.2.1 \quad\quad\quad \text{M}(p,p.n) = 1 \Rightarrow \text{M}(p.n,q)$ $\quad\quad 2.2.2 \quad\quad\quad \text{M}(p,p.n) = 0 \Rightarrow 0;$ **while** $\text{M}(p,q) > 0$

Fig. 4. Program TRAVERSE and its associated measure program P_{M}

The new precondition PRE means that q is reachable from p and either (1) p is different from q or (2) they are aliased and either (2.1) p has a self-loop or (2.2) p is on a non-trivial cycle. We expect that the program TRAVERSE terminates under the current precondition. Indeed, the termination test of the measure program P_{M} under the precondition PRE succeeds and the algorithm returns that the program terminates under the precondition PRE.

In [19], we discuss additional example programs that manipulate singly- and doubly-linked lists. These examples are inspired by code fragments found in low-level system code, such as the example in Section 2.

6 Implementation and Experiments

We developed a prototype implementation, called BOUNCER, of our algorithm for the demand-driven inference of heap assumptions. We applied BOUNCER to the example programs in [19] and a scheduling routine from the VAMOS kernel [17].

BOUNCER applies the BOHNE tool for symbolic shape analysis [24] to implement the oracle that checks assertion validity of heap programs [23,20]. For proving termination of measure programs, BOUNCER applies the ARMC tool for proving termination of transition relations in linear arithmetic [21, 18]. The oracle for wlp uses widening, as described in Section 4.

We model the value ∞ in our translation to a measure program by a negative integer constant, say c. Our translation rewrites each measure expression according to the following rules:

$$\begin{aligned} mexp = \infty &\longrightarrow mexp = c\,, \\ mexp \leq \infty &\longrightarrow mexp = c \vee mexp \geq 0\,, \\ mexp < \infty &\longrightarrow mexp \geq 0\,. \end{aligned}$$

The rewriting step allows one to apply a termination checker for programs over numerical domains as black-box.

While our implementation is preliminary, we observe that the behavior of the algorithm with respect to the number of applied measures is similar to the behavior of algorithms for predicate abstraction with respect to the number of predicates. We believe that local use of measures, similarly to localized abstraction [14], can make our tool scale to larger programs.

Our experiments with process scheduling functions from the VAMOS kernel show that BOUNCER can successfully infer preconditions for termination for interesting practical programs. In the current implementation, we had to manually abstract all non-heap operations by non-deterministic choice. The inferred preconditions are in agreement with the preconditions provided manually by the VAMOS developers.

Acknowledgments. Andrey Rybalchenko is supported in part by Microsoft Research through the European Fellowship Programme. Thomas Wies is supported by a Microsoft Research European PhD Scholarship.

References

1. Berdine, J., Calcagno, C., Cook, B., Distefano, D., O'Hearn, P.W., Wies, T., Yang, H.: Shape analysis for composite data structures. In: Damm, W., Hermanns, H. (eds.) CAV 2007. LNCS, vol. 4590, pp. 178–192. Springer, Heidelberg (2007)
2. Berdine, J., Cook, B., Distefano, D., O'Hearn, P.: Automatic Termination Proofs for Programs with Shape-Shifting Heaps. In: Ball, T., Jones, R.B. (eds.) CAV 2006. LNCS, vol. 4144, pp. 386–400. Springer, Heidelberg (2006)
3. Beyer, D., Henzinger, T.A., Théoduloz, G.: Lazy shape analysis. In: Ball, T., Jones, R.B. (eds.) CAV 2006. LNCS, vol. 4144, pp. 532–546. Springer, Heidelberg (2006)
4. Bogudlov, I., Lev-Ami, T., Reps, T.W., Sagiv, M.: Revamping TVLA: Making parametric shape analysis competitive. In: Damm, W., Hermanns, H. (eds.) CAV 2007. LNCS, vol. 4590, pp. 221–225. Springer, Heidelberg (2007)
5. Bouajjani, A., Bozga, M., Habermehl, P., Iosif, R., Moro, P., Vojnar, T.: Programs with lists are counter automata. In: Ball, T., Jones, R.B. (eds.) CAV 2006. LNCS, vol. 4144, pp. 517–531. Springer, Heidelberg (2006)
6. Bouajjani, A., Habermehl, P., Rogalewicz, A., Vojnar, T.: Abstract regular tree model checking of complex dynamic data structures. In: Yi, K. (ed.) SAS 2006. LNCS, vol. 4134, pp. 52–70. Springer, Heidelberg (2006)
7. Bradley, A., Manna, Z., Sipma, H.: The polyranking principle. In: Caires, L., Italiano, G.F., Monteiro, L., Palamidessi, C., Yung, M. (eds.) ICALP 2005. LNCS, vol. 3580, pp. 1349–1361. Springer, Heidelberg (2005)
8. Calcagno, C., Distefano, D., O'Hearn, P., Yang, H.: Footprint analysis: A shape analysis that discovers preconditions. In: Riis Nielson, H., Filé, G. (eds.) SAS 2007. LNCS, vol. 4634, pp. 402–418. Springer, Heidelberg (2007)
9. Colón, M., Sipma, H.: Practical methods for proving program termination. In: Brinksma, E., Larsen, K.G. (eds.) CAV 2002. LNCS, vol. 2404. Springer, Heidelberg (2002)
10. Cook, B., Podelski, A., Rybalchenko, A.: Termination proofs for systems code. In: PLDI (2006)
11. Cousot, P.: Proving program invariance and termination by parametric abstraction, Lagrangian relaxation and semidefinite programming. In: Cousot, R. (ed.) VMCAI 2005. LNCS, vol. 3385, pp. 1–24. Springer, Heidelberg (2005)
12. Distefano, D., O'Hearn, P., Yang, H.: A local shape analysis based on separation logic. In: Hermanns, H., Palsberg, J. (eds.) TACAS 2006 and ETAPS 2006. LNCS, vol. 3920, pp. 287–302. Springer, Heidelberg (2006)

13. Hackett, B., Rugina, R.: Region-based shape analysis with tracked locations. In: POPL (2005)
14. Henzinger, T.A., Jhala, R., Majumdar, R., McMillan, K.L.: Abstractions from proofs. In: POPL (2004)
15. Magill, S., Berdine, J., Clarke, E.M., Cook, B.: Arithmetic strengthening for shape analysis. In: Riis Nielson, H., Filé, G. (eds.) SAS 2007. LNCS, vol. 4634, pp. 419–436. Springer, Heidelberg (2007)
16. Manolios, P., Vroon, D.: Termination Analysis with Calling Context Graphs. In: Ball, T., Jones, R.B. (eds.) CAV 2006. LNCS, vol. 4144, pp. 401–414. Springer, Heidelberg (2006)
17. Maus, S.: Developing an Operating System Kernel for the VAMP Processor. Diploma thesis, Universität des Saarlandes (2005)
18. Podelski, A., Rybalchenko, A.: ARMC: The Logical Choice for Software Model Checking with Abstraction Refinement. In: Hanus, M. (ed.) PADL 2007. LNCS, vol. 4354, pp. 245–259. Springer, Heidelberg (2006)
19. Podelski, A., Rybalchenko, A., Wies, T.: Heap assumptions on demand. Technical report, University of Freiburg (2008), `http://www.informatik.uni-freiburg.de/ wies/papers/. HeapAssumptionsExtended.pdf`.
20. Podelski, A., Wies, T.: Boolean Heaps. In: Hankin, C., Siveroni, I. (eds.) SAS 2005. LNCS, vol. 3672, pp. 268–283. Springer, Heidelberg (2005)
21. Rybalchenko, A.: ARMC (2008), `http://www.mpi-sws.org/~rybal/armc`
22. Sagiv, M., Reps, T., Wilhelm, R.: Parametric shape analysis via 3-valued logic. ACM TOPLAS (2002)
23. Wies, T.: Symbolic Shape Analysis. Diploma thesis, Universität des Saarlandes, Germany (2004)
24. Wies, T.: The Bohne Tool (2008), `http://swt.informatik.uni-freiburg.de/wies/bohne`
25. Wies, T., Kuncak, V., Zee, K., Podelski, A., Rinard, M.: Verifying complex properties using symbolic shape analysis. In: HAV Workshop (2007)

Proving Conditional Termination

Byron Cook[1], Sumit Gulwani[1], Tal Lev-Ami[2,⋆], Andrey Rybalchenko[3,⋆⋆], and Mooly Sagiv[2]

[1] Microsoft Research
[2] Tel Aviv University
[3] MPI-SWS

Abstract. We describe a method for synthesizing reasonable underapproximations to weakest preconditions for termination—a long-standing open problem. The paper provides experimental evidence to demonstrate the usefulness of the new procedure.

1 Introduction

Termination analysis is critical to the process of ensuring the stability and usability of software systems, as liveness properties such "Will `Decode()` always return back to its call sites?" or "Is every call to `Acquire()` eventually followed by a call to `Release()`?" can be reduced to a question of program termination [8,22]. Automatic methods for proving such properties are now well studied in the literature, *e.g.* [1,4,6,9,16]. But what about the cases in which code only terminates for *some* inputs? What are the preconditions under which the code is still safe to call, and how can we automatically synthesize these conditions? We refer to these questions as the *conditional termination* problem.

This paper describes a method for proving conditional termination. Our method is based on the discovery of *potential ranking functions*—functions over program states that are bounded but not necessarily decreasing—and then finding preconditions that promote the potential ranking functions into valid ranking functions. We describe two procedures based on this idea: PreSynth, which finds preconditions to termination, and PreSynthPhase, which extends PreSynth with the ability to identify the phases necessary to prove the termination of phase-transition programs [3].

The challenge in this area is to find the *right precondition*: the empty precondition is correct but useless, whereas the *weakest precondition* [13] for even very simple programs can often be expressed only in complex domains not supported by today's tools (*e.g.* non-linear arithmetic). In this paper we seek a method that finds useful preconditions. Such preconditions need to be weak enough to allow interesting applications of the code in question, but also expressible in

⋆ Supported by an Adams Fellowship through the Israel Academy of Sciences and Humanities.

⋆⋆ Supported in part by Microsoft Research through the European Fellowship Programme.

A. Gupta and S. Malik (Eds.): CAV 2008, LNCS 5123, pp. 328–340, 2008.

the subset of logic supported by decision procedures, model checking tools, etc. Furthermore, they should be computed quickly (the weakest precondition expressible in the target logic may be too expensive to compute). Since we are not always computing the *weakest* precondition, in this paper we allow the reader to judge the quality of the preconditions computed by our procedure for a number of examples. Several of these examples are drawn from industrial applications.

Limitations. In this paper, we limit ourselves to the termination property and to sequential arithmetic programs. Note that, at the cost of complicating the exposition, we could use known techniques (*e.g.*, [2] and [8]) to extend our approach to programs with heap and ω-regular liveness properties. Our technique could also provide assistance when analyzing concurrent programs via [10], although we suspect that synthesizing *environment* abstractions that guarantee thread-termination is a more important problem for concurrent programs than conditional termination.

Related work. Until now, few papers have directly addressed the problem of automatically underapproximating weakest preconditions. One exception is [14], which yields constraint systems that are non-linear. The constraint-based technique in [5] could also be modified to find preconditions, but again at the cost of non-linear constraints. In contrast to methods for underapproximating weakest preconditions, techniques for weakest *liberal* preconditions are known (*e.g.*, [17, 7]). Note that weakest preconditions are so rarely considered in the literature that weakest *liberal* preconditions are often simply called weakest preconditions, *e.g.*, [17].

2 Example

In this section we informally illustrate our method by applying it to several examples. The procedures proposed by this paper are displayed in Figures 1 and 2. They will be more formally described in Section 3.

We have split our method into two procedures for presentational convenience. The first procedure illustrates our method's key ideas, but fails for the class of phase-transition programs. The second procedure extends the first with support for phase-transition programs. Note that phase-change programs and preconditions are interrelated (allowing us to solve the phase-change problem easily with our tool), as a phase-change program can be thought of as several copies of the same loop composed, but with different preconditions.

2.1 Finding Preconditions for Programs without Phase-Change

We consider the following code fragment:

```
// @requires true;
while(x>0){
    x=x+y;
}
```

We assume that the program variables x and y range over integers. The initially given `requires`-clause is not sufficient to guarantee termination. For example, if x=1 and y=0 at the loop entry then the code will not terminate. The weakest precondition for termination of this program is $\mathtt{x} \leq 0 \vee \mathtt{y} < 0$.

If we apply an existing termination prover, *e.g.*, TERMINATOR [9] or ARMC [21], on this code fragment then it will compute a counterexample to termination. The counterexample consists of 1) a stem η, which allows for manipulating the values before the loop is reached, and 2) a repeatable cycle ρ, which is a relation on program states that represents an arbitrary number of iterations of the loop.

To simplify the presentation, we represent the stem η as an initial condition θ on the variables of the loop part. (Section 4 describes this step in more detail.) In our example, the initial condition θ is *true* and the transition relation of the loop is defined by

$$\rho(\{\mathtt{x},\mathtt{y}\},\{\mathtt{x}',\mathtt{y}'\}) \ \equiv \ \mathtt{x} > 0 \wedge \mathtt{x}' = \mathtt{x} + \mathtt{y} \wedge \mathtt{y}' = \mathtt{y} \ .$$

In order to try and prove this counterexample spurious (*i.e.* to prove it well-founded, as explained in [9]), we need to find a ranking function f such that $\rho(X, X') \Rightarrow \mathcal{R}_f(X, X')$, where $\mathcal{R}_f$ is the *ranking relation* defined by f:

$$\mathcal{R}_f(X, X') \ \equiv \ f(X) \geq 0 \wedge f(X') \leq f(X) - 1 \ .$$

As the termination prover has returned the above relation ρ as a counterexample, we can assume that no linear ranking function f exists (note that there could exist a non-linear ranking function, depending on the completeness of the termination prover).

Due to the absence of a linear ranking function for ρ, we find a *potential ranking function*, *i.e.*, a function b such that one of the conjuncts defining $\mathcal{R}_b(X, X')$ holds for ρ. We compute a potential ranking function for ρ by finding an expression on the variables $\{\mathtt{x},\mathtt{y}\}$ that is bound from below. One method for finding such candiate functions ito consider only the domain (and not the range) of ρ, *i.e.*, find functions that are bounded when there is a successor. In other words, consider $\exists \mathtt{x}', \mathtt{y}'.\, \mathtt{x} > 0 \wedge \mathtt{x}' = \mathtt{x} + \mathtt{y} \wedge \mathtt{y}' = \mathtt{y}$. In practice we achieve this via a the application of a quantifier elimination procedure, *i.e.*, we have

$$\mathrm{QELIM}(\exists \mathtt{x}', \mathtt{y}'.\, \mathtt{x} > 0 \wedge \mathtt{x}' = \mathtt{x} + \mathtt{y} \wedge \mathtt{y}' = \mathtt{y}) \ \equiv \ \mathtt{x} > 0 \ .$$

We can normalize the condition $\mathtt{x} > 0$ as $\mathtt{x} - 1 \geq 0$, and thus use the function $b = \mathtt{x} - 1$. Because $\rho(\{\mathtt{x},\mathtt{y}\},\{\mathtt{x}',\mathtt{y}'\}) \Rightarrow b(\{\mathtt{x},\mathtt{y}\}) \geq 0$, which is the first conjunction required by $\mathcal{R}_b$, we can use b as our potential ranking function.[1]

Enforcing ranking with a strengthening. The function $b = \mathtt{x} - 1$ that we found only satisfies part of the requirements for proving termination with $\mathcal{R}_b$ (*i.e.*,

[1] In this simple example the result was exactly the loop condition. However, when translating the cycle returned from the termination prover to a formula, some of the conditions are not on the initial variables.

$b(X) \geq 0$ but not $b(X') \leq b(X) - 1$). We need a strengthening $s(\{\mathtt{x}, \mathtt{y}\})$ such that

$$s(\{\mathtt{x},\mathtt{y}\}) \wedge \rho(\{\mathtt{x},\mathtt{y}\}, \{\mathtt{x}',\mathtt{y}'\}) \Rightarrow \mathcal{R}_b(\{\mathtt{x},\mathtt{y}\}, \{\mathtt{x}',\mathtt{y}'\}) .$$

Since b is bounded, we find $s(\{\mathtt{x}, \mathtt{y}\})$ as follows:

$$s(\{\mathtt{x},\mathtt{y}\}) \equiv \mathrm{QELIM}(\forall \mathtt{x}',\mathtt{y}'.\, \rho(\mathtt{x},\mathtt{y},\mathtt{x}',\mathtt{y}') \Rightarrow b(\{\mathtt{x}',\mathtt{y}'\}) \leq b(\{\mathtt{x},\mathtt{y}\}) - 1) .$$

We obtain $s(\{\mathtt{x}, \mathtt{y}\}) = \mathtt{x} \leq 0 \vee \mathtt{y} < 0$. That is, if s were an invariant (and usually it is not), then ρ would be provably well-founded using b.

Synthesizing a precondition guaranteeing the strengthening. Recall that the original problem statement is to find a precondition that guarantees termination of the presented code fragment. As the strengthening s guarantees termination, we now need to find a precondition that guarantees the validity of s on every iteration of ρ. The required assertion is the weakest *liberal* precondition of s wrt. the loop statement. We use known techniques for computing underapproximations of weakest liberal preconditions to find the precondition that ensures that after any number of iterations of the loop s must hold in the next iteration. Using a tool for abstract interpretation of arithmetic programs [15], we obtain $r(\{\mathtt{x}, \mathtt{y}\}) = \mathtt{x} \leq 0 \vee \mathtt{y} < 0$. In summary, our procedure has discovered the precondition proposed above.

Note that we can alternate executions of our procedure together with successive applications of a termination prover to find a precondition that is strong enough to prove the termination of the entire program. The interaction between the tools is based on counterexamples for termination, which are discovered by the termination prover and are eliminated by the precondition synthesis procedure.

2.2 Finding Preconditions for Phase-Change Programs

Consider the following code fragment:

```
// @requires true;
while(x>0){
     x=x+y;
     y=y+z;
}
```

Again, the given `requires`-clause is not sufficient to ensure termination. For example, if x=1, y=0, and z=0 at the loop entry then the code will not terminate. However, this time, the weakest precondition is given by a non-linear assertion, which is difficult to construct automatically.

Note that the precondition $\mathtt{z} < 0$ guarantees that the loop terminates, but the termination may take place after the computation passed through two phases. The first phase is characterized by the assertion $\mathtt{y} \geq 0$. In fact, during this phase the value of x may not decrease towards zero. Nevertheless, *eventually* the value of y will decrease below zero, *i.e.*, a phase transition takes place. At this point x will start decreasing towards (and eventually reaching) zero.

In this example, the termination prover returns a stem which is the identity relation and cycle relation

$$\rho(\{\mathtt{x},\mathtt{y},\mathtt{z}\},\{\mathtt{x}',\mathtt{y}',\mathtt{z}'\}) \equiv \mathtt{x} > 0 \land \mathtt{x}' = \mathtt{x}+\mathtt{y} \land \mathtt{y}' = \mathtt{y}+\mathtt{z} \land \mathtt{z}' = \mathtt{z} .$$

The first step of the precondition inference repeats the procedure presented in the previous subsection. On this example, similarly to the previous one, the procedure computes $b = \mathtt{x} - 1$ and $s = \mathtt{x} \le 0 \lor \mathtt{y} < 0$. However, when computing a precondition that ensures that s in an invariant, we obtain a linear underapproximation[2]

$$\begin{aligned}&\mathtt{x} \le 0 \lor \mathtt{x}+\mathtt{y} \le 0 \lor (\mathtt{y} < 0 \land \mathtt{x}+2*\mathtt{y}+\mathtt{z} \le 0) \lor \\ &(\mathtt{y} < 0 \land \mathtt{x}+3*\mathtt{y}+3*\mathtt{z} \le 0) \lor (\mathtt{y} < 0 \land \mathtt{z} \le 0) .\end{aligned}$$

The first four disjuncts correspond to the cases when the loop terminates after 0, 1, 2, and 3 iterations, respectively, and in which $\mathtt{y} < 0$ holds until possibly the last iteration. The last disjunct is more interesting, as it states that if $\mathtt{y} < 0$ and $\mathtt{z} \le 0$ then the loop terminates. The expected precondition $\mathtt{z} < 0$ is not included in the disjunction, since it does not guarantee that $\mathtt{y} < 0$ from the start.

Note that each of these conditions guarantees termination when they are satisfied at any iteration of the loop, not necessarily at the first one. These conditions identify phase transition points. Once they are met, $\mathtt{x}$ will start decreasing until the loop terminates. The solution is to constrain the transition relation with the negated condition. Termination of the constrained loop ensures that these conditions are eventually met, and thus the original loop will terminate. Thus, we call the procedure recursively with the constrained transition relation, and disjoin the returned preconditions with the existing ones.

For example, constraining the loop with $\neg(\mathtt{y} < 0 \land \mathtt{z} \le 0)$ and calling the procedure recursively yields the additional preconditions $\mathtt{x}+2*\mathtt{y}+\mathtt{z} \le 0 \lor \mathtt{x}+3*\mathtt{y}+3*\mathtt{z} \le 0 \lor \mathtt{z} < 0$. The first two disjuncts are weakenings of the previous preconditions for the cases in which the loop terminates after 2 or 3 iteration, removing the condition $\mathtt{y} < 0$. The last disjunct $\mathtt{z} < 0$ is the more interesting, since it ensures eventual termination.

After simplification the precondition computed by the procedure is

$$\begin{aligned}&\mathtt{x} \le 0 \lor \mathtt{x}+\mathtt{y} \le 0 \lor \mathtt{x}+2*\mathtt{y}+\mathtt{z} \le 0 \lor \\ &\quad \mathtt{x}+3*\mathtt{y}+3*\mathtt{z} \le 0 \lor (\mathtt{y} < 0 \land \mathtt{z} \le 0) \lor \mathtt{z} < 0 .\end{aligned}$$

3 Computing Preconditions for Termination

This section formally describes the two methods for computing preconditions for termination, PreSynth and PreSynthPhase. We first define basic concepts.

[2] Any linear precondition must be an underapproximation, since the weakest liberal precondition is non-linear.

3.1 Preliminaries

We assume a program $P = (X, \theta, \rho)$ over a finite set of variables X. For simplicity, we do not single out the program counter variable that ranges over control locations of the program, and assume that it is included in X. The assertion θ represents the initial condition of the program. The transition relation ρ is represented by an assertion over the program variables X and their primed versions X'. The primed variables refer to the values after the transition is executed. For practicality, we assume that the initial condition and the transition relation are represented using a logical theory for which practical quantifier elimination procedures exist.[3]

We refer to valuations of program variables as program states. A program computation is a finite or infinite sequence of program states $s_1, s_2, \ldots$ such that s_1 is an initial state and each pair of consecutive states s_i and s_{i+1} follows the transition relation. Formally, $s_1 \models \theta$ and for each but final s_i in the computation we have $(s_i, s_{i+1}) \models \rho$. The program terminates from a state s if there is no infinite computation that starts at s. An assertion r is a precondition for termination if the program P terminates from each state s that satisfies r and the initial condition, *i.e.*, $s \models \theta \wedge r$.

Our procedure uses (an under-approximation of) the weakest liberal precondition operator WLP, which we define as usual. Given a transition relation $\rho(X, X')$ and an assertion over program variables $\varphi(X)$, this transformer yields the following conjunction when applied on the transitive closure of ρ:

$$\mathsf{WLP}(\rho^*(X, X'), \varphi(X)) \equiv \bigwedge_{n \geq 0} \forall X'. \rho^n(X, X') \Rightarrow \varphi(X') .$$

3.2 The Procedure PreSynth

Figure 1 shows the basic precondition synthesis procedure PreSynth. The input of PreSynth consists of an initial condition θ and a transition relation ρ. Note that in the most likely usage scenario, θ and ρ will represent a counterexample to termination reported by a termination prover. In this case, PreSynth is applied on a compact code fragment and not the full program, which allows one to apply precise, automated reasoning-based techniques.

We discuss the major steps of the procedure in more detail. The procedure takes a formula representing a set of initial states, together with a relation representing the transitions. Line 1 in Figure 1 strengthens the transition relation ρ with (possibly an over-approximation of) the states that are reachable from θ. This step provides necessary precision for the subsequent computations by taking into account reachability invariants. Here, we rely on an efficient abstract interpretation tool based on *e.g.*, the Octagon domain [18]. Thus, we can implement line 1 using a standard abstract reachability procedure [12] as follows (we assume that $\theta \Rightarrow \theta^\#$, $\rho \Rightarrow \rho^\#$, and lfp is the least fixpoint operator):

[3] If necessary, we can over-approximate the initial condition and the transition relation using assertions from such a theory.

```
    function PreSynth
    input
      θ(X) : initial condition
      ρ(X, X') : transition relation
    begin
0     r(X) := false
1     ρ(X, X') := ρ(X, X') ∧ QELIM(∃X₀. θ(X₀) ∧ ρ*(X₀, X))
2     B := Finite({b(X) | ∀X. (∃X'. ρ(X, X')) ⇒ b(X) ≥ 0})
3     foreach b(X) ∈ B do
4       s(X) := QELIM(∀X'.ρ(X, X') ⇒ b(X) ≥ 0 ∧ b(X') ≤ b(X) − 1)
5       r(X) := r(X) ∨ WLP(ρ*(X, X'), s(X))
6     done
7     return "precondition for termination r(X) "
    end.
```

Fig. 1. The procedure PreSynth synthesizes a precondition for the termination of a transition relation ρ from initial states θ. The procedure Finite returns a selected finite subset of its input (*i.e.* $\text{Finite}(S) \subseteq_{\text{fin}} S$). Finite may choose, for example, to return only the linear elements of S.

$$\rho(X, X') := \rho(X, X') \wedge \mathsf{lfp}(\rho^{\#}, \theta^{\#})$$

Line 2 of PreSynth computes a finite set of expressions B that are bounded by the (strengthened) transition relation ρ. In theory we could generalize the set B to include non-linear or lexicographic ranking functions, though in practice it will be limited to linear ranking functions. We will assume that each b is a function ranging over simple arithmetic types (*i.e.*, $\mathbb{N}$, $\mathbb{Z}$, $\mathbb{R}$) though in principle we could generalize the procedure to support any well-order. In practice, to compute B we apply an existential quantifier elimination procedure to eliminate the primed variables. Then, we consider the linear inequalities that appear in the result. Each bound expression $b(X) \in B$ is treated as a potential ranking function, and is used to guide the search for a strengthening $s(X)$ on the domain of the transition relation ρ that makes ρ well-founded. PreSynth only considers the well-foundedness arguments constructed using $b(X)$. Line 4 uses a quantifier elimination procedure to construct the strengthening $s(X)$ by imposing the *bounded* and *decrease* conditions

$$s(X) \wedge \rho(X, X') \Rightarrow b(X) \geq 0 \wedge b(X') \leq b(X) - 1 \ .$$

The assertion $s(X)$ guarantees that $b(X)$ is a ranking function for the given transition relation.

The strengthening imposed by $s(X)$ is effective if $s(X)$ holds for all states that are reachable from θ by applying ρ. Line 5 in Figure 1 computes the necessary precondition that guarantees the invariance of $s(X)$. All preconditions that are found using the potential ranking functions are accumulated in an assertion $r(X)$, and reported to the programmer in line 7.

Line 5 of PRESYNTH uses known abstract interpretation-based techniques for under-approximating sets and relations for the computation of $\mathsf{WLP}(\rho^*(X, X'), s(X))$. We pass the following program to the INTERPROC analyzer [15]:

```
assume θ(X');
while (*) do
   X=X';
   X'=*;
   assume ρ(X, X');
od
assume ¬s(X);
```

Using INTERPROC we first compute an abstract fixpoint using backwards analysis starting from top element in the abstract domain. Then, the abstract element at the loop entry location represents an over-approximation of the states that fail the assertion $s(X)$. The complement of this element, which we obtain by negating the corresponding assertion, provides an under-approximation of the initial set of states of the program that guarantees the invariance of $s(X)$.

Theorem 1 (PreSynth correctness). *Let $r(X)$ be an assertion computed by the procedure* PRESYNTH. *Then, r is a precondition for termination of a program with the initial condition θ and the transition relation ρ.*

Proof. Let $r(X)$ be computed by PRESYNTH. For a proof by contradiction, we assume that there is an infinite computation $s_1, s_2, \ldots$ from an initial state that satisfies $r(X)$. Let $b(X)$ be a bound expression that contributed a disjunct in $r(X)$ that holds for the state s_1, and $s(X)$ be a corresponding strengthening. From the definition of WLP, we have that $s(X)$ holds for each state s_i, where $i \geq 0$. Thus, the value of $b(X)$ decreases after each program step, while being bounded from below. We reached a contradiction to the assumption that the computation is infinite. □

3.3 The Procedure PreSynthPhase

See Figure 2. The procedure PRESYNTHPHASE extends the applicability of the basic procedure from Figure 1 to phase-transition programs (as described in Section 2). It removes the condition that the computed strengthening needs to apply immediately. Instead, we only require that the strengthening applies *eventually*. PRESYNTHPHASE implements this eventuality requirement by finding a precondition for termination of an augmented transition relation that avoids visiting states satisfying the strengthening constraint. The inferred precondition can be enforced eventually by applying PRESYNTHPHASE recursively.

We discuss the major steps of the procedure in more detail. Line 1 computes an initial precondition $r(X)$ by applying the procedure PRESYNTHPHASE. If the precondition $r(X)$ is non-empty, as checked in line 2, then PRESYNTHPHASE weakens it by a precondition that ensures the eventuality of $r(X)$. Here, we follow a standard technique for the verification of temporal liveness properties [22],

```
  function PreSynthPhase
  input
    θ(X) : initial condition
    ρ(X, X') : transition relation
  begin
1   r(X) := PreSynth(θ, ρ)
2   if ∃X. r(X) then
3     ρ(X, X') := ρ(X, X') ∧ ¬r(X)
4     r(X) := r(X) ∨ PreSynthPhase(θ, ρ)
5   endif
6   return "precondition for phase termination r(X) "
  end.
```

Fig. 2. The procedure PreSynthPhase synthesizes a precondition for the phase termination of a transition relation ρ from initial states θ. It applies the procedure PreSynth from Figure 1 when recursively identifying computation phases. PreSynthPhase can be stopped at any time, e.g., by reaching a user-provided upper bound, and it will yield a sound precondition for phase-transition termination.

and apply a translation to a termination problem. Effectively, line 3 constructs a new transition relation from ρ that avoids $r(X)$. Thus, we can apply PreSynthPhase recursively on the new transition relation, see line 4.

Theorem 2 (PreSynthPhase correctness). *Let $r(X)$ be an assertion computed by the procedure* PreSynthPhase. *Then, r is a precondition for termination of a program with the initial condition θ and the transition relation ρ.*

Proof. Let $r_1(X), \ldots, r_n(X)$ be a sequence of preconditions that are computed by applying PreSynth during the execution of PreSynthPhase. We observe that every computation that starts in an initial state that satisfies $r_i(X)$ for $i > 1$ either terminates or eventually reaches $r_{i-1}(X)$. Hence, eventually every computation either stops or reaches $r_1(X)$. From Theorem 1 follows that the program terminates on each state reachable from an initial state satisfying $r_1(X)$. □

4 Implementation and Experiments

We have built a prototype implementation of our method based on the following collection of tools: for termination proving we use ARMC [21,20], for quantifier elimination we use Cooper's procedure [11], and for abstract interpretation we use the Interproc analyzer [15].

See Figures 3 and 4, which contain example programs (both hand written and drawn from industrial examples) together with the results of our tool. We leave it to the reader to judge the usefulness of the synthesized preconditions. The running times in the figures include the times of iterating the procedure of Figure 2 and the termination prover. In the remainder of this section we highlight relevant implementation details.

Program fragment	Precondition & notes

`i = 0;`
`if (l_var >= 0) {`
`    while (l_var < 1073741824) {`
`        i++;`
`        l_var = l_var << 1;`
`    }`
`}`
	$\texttt{l\_var} > 0 \lor \texttt{l\_var} < 0 \lor \texttt{l\_var} \geq 1073741824$
	
	Example from an audio compression module: We model shift by multiplication and checked for overflow with an extra check and subtract.
	
	Time: 22 seconds.

`while (cbSrcLength >= cbHeader) {`
`  DWORD       dwHeader;`
`  UINT        cbBlockLength;`
`  cbBlockLength  = (UINT)min(cbSrcLength, nBlockAlignment);`
`  cbSrcLength   -= cbBlockLength;`
`  cbBlockLength -= cbHeader;`
`  dwHeader = *(DWORD HUGE_T *)pbSrc;`
`  pbSrc    += sizeof(DWORD);`
`  nPredSample = (int)(short)LOWORD(dwHeader);`
`  nStepIndex  = (int)(BYTE)HIWORD(dwHeader);`
`  if( !imaadpcmValidStepIndex(nStepIndex) ) return 0;`
`  *pbDst++ = (BYTE)((nPredSample >> 8) + 128);`
`  while (cbBlockLength--) {`
`    bSample = *pbSrc++;`
`    nEncSample  = (bSample & (BYTE)0x0F);`
`    nSz   = step[nStepIndex];`
`    nPredSample =`
`       imaadpcmSampleDecode(nEncSample, nPredSample, nSz);`
`    nStepIndex  =`
`       imaadpcmNextStepIndex(nEncSample, nStepIndex);`
`    *pbDst++ = (BYTE)((nPredSample >> 8) + 128);`
`    nEncSample  = (bSample >> 4);`
`    nSz   = step[nStepIndex];`
`    nPredSample =`
`        imaadpcmSampleDecode(nEncSample, nPredSample, nSz);`
`    nStepIndex  =`
`        imaadpcmNextStepIndex(nEncSample, nStepIndex);`
`    *pbDst++ = (BYTE)((nPredSample >> 8) + 128);`
`  }`
`}`
	$\texttt{cbSrcLength} < \texttt{cbHeader} \lor$
	$(\texttt{nBlockAlignment} > 0 \land \texttt{cbHeader} > 0)$
	
	Example from another audio application.
	
	Time: 106 seconds.

Fig. 3. Programs drawn from industrial examples, runtimes, and synthesized preconditions. The runtimes include the iteration of the procedure in Figure 2 together with the termination prover.

Simplification. Both the quantifier elimination and the abstract interpretation tools do not give minimal formulas. In many cases the formulas have redundant conjuncts or several disjuncts that imply each other. Simplification of these formulas was important for reducing the size of the result preconditions.

Loop termination after a bounded number of iterations. The recursive call in PreSynthPhase is problematic in the case that the precondition is the result of a loop termination in a bounded number of iteration, say 3. The problem is that the next call will find precondition that come from the loop terminating after 6 iteration and so on. To solve this problem, before the recursive call, we check whether the precondition ensures termination in a fixed number of steps, and if it does then we avoid the recursive call. That is, we unroll the relation k times (for some threshold k) and then check to see if $\rho^k = \emptyset$, which implies termination.

Stems vs. initial conditions. The termination prover returns a cycle and a stem as a counterexample. The stem is not only an initial condition, but rather a sequence of primitive statements that manipulate variables. We use existential

<table>
<tr><th>Program fragment</th><th>Precondition & notes</th></tr>
<tr><td><pre>
// @requires true;
while(x>0){
    x=x+y;
    y=y+z;
}
</pre></td><td>

$x \leq 0 \vee x + y \leq 0 \vee x + 2y + z \leq 0 \vee x + 3y + 3z \leq 0 \vee z < 0 \vee (z \leq 0 \wedge y < 0)$

This is the example from §2. The first four disjuncts cover the case where the loop terminates after 0, 1, 2, or 3 iterations respectively. The last two disjuncts ensure that the loop eventually terminates. The condition $z \leq 0 \wedge y < 0$ is the case in which the loop starts with $y < 0$ and z does not interfere.

Time: 24 seconds.
</td></tr>
<tr><td><pre>
// @requires true;
while(x<=N){
    if (*) {
        x=2*x+y;
        y=y+1;
    } else {
        x++;
    }
}
</pre></td><td>

$x > N \vee x + y \geq 0$

An example from [3]. We find preconditions that ensure that the loop is executed 0 times ($x > N$) and the precondition that ensures termination as it appears in the paper ($x + y \geq 0$).

Time: 4 seconds.
</td></tr>
<tr><td><pre>
// @requires true;
while(x>=0){
    x= -2*x + 10;
}
</pre></td><td>

$x > 5 \vee x < 0$

Example (from [19]) of a terminating linear program with no linear ranking function. Note that the program always terminates after at least 5 iterations. Our synthesized precondition provides conditions that guarantee termination after 0 or 1 iterations.

Time: 2 seconds.
</td></tr>
<tr><td><pre>
// @requires n>200 and y<9;
x = 0;
while (1) {
  if (x<n) {
    x=x+y;
    if (x>=200) break;
  }
}
</pre></td><td>

$y > 0$

Example (from [14]) after translation that allows us to disprove a safety property.

Time: 6 seconds.
</td></tr>
<tr><td><pre>
while (x!=y) {
    if (x>y) {
        x = x - y;
    } else
        y = y - x;
    }
}
</pre></td><td>

$x = y \vee (x > 0 \wedge y > 0)$

The Euclidean algorithm for GCD. Each case in the if requires a different ranking function. Demonstrates the interaction with disjunctive well foundedness.

Time: 17 seconds.
</td></tr>
</table>

Fig. 4. Handwritten programs, runtimes, and synthesized preconditions. The runtimes include the iteration of the procedure in Figure 2 together with the termination prover.

quantifier elimination to convert the stem into the initial condition for the procedure. Because there are no control-flow constructs in the stem, a straightforward application of WLP translates the precondition discovered by the procedure to a precondition applicable at the beginning of the stem.

Translating paths to formulas. The cycles returned by the termination prover often contain multiple assignments and conditions. The translation first converts the path to static single assignment form and then to a path formula (see [9] for details). The extra temporaries generated are handled in the quantifier elimination as extra variables to eliminate.

Handling disjunctions. Effective dealing with disjunctions that appear in intermediate formulas manipulated by our algorithm is crucial for its applicability. Even when applied on conjunctive inputs, the procedure PRESYNTH creates additional disjunctions. Its line 1 may split the transition relation ρ when restricting it to reachable states. Although line 5 creates disjunction by iterating over the set of candidate bound expressions B, they are tamed by the subsequent negation in PRESYNTHPHASE. On the other hand, line 5 creates a conjunction representing the weakest liberal precondition that, after its negation in PRESYNTHPHASE, may create disjuncts. Disjunctions in the preconditions are handled by calling the algorithm separately on each disjunct and accumulating the results. Disjunctions in the transition relation are handled by iterating the termination prover as explained below.

Iterating the termination prover. The constrained transition relation generated in PRESYNTHPHASE can be complicated. We found that calling the termination prover on this relation and letting the PRESYNTHPHASE only deal with the counterexamples returned improves the preconditions we compute. Thus, there is a mutual recursion between calling the termination prover and PRESYNTHPHASE.

5 Conclusion

This paper has described an automatic method for finding sound underapproximations to weakest preconditions to termination. Using illustrative examples we have shown that the method can be used to find *useful* underapproximations (*i.e.* something larger than `false`, but in cases smaller than the often complex *weakest* precondition). Beyond the direct use of proving conditional termination, we believe that our method can also be used in several areas of program verification, including the disproving of safety properties, interprocedural analysis, interprocedural termination proving, and distributed termination proving.

References

1. Balaban, I., Pnueli, A., Zuck, L.D.: Modular ranking abstraction. Int. J. Found. Comput. Sci. (2007)
2. Bouajjani, A., Bozga, M., Habermehl, P., Iosif, R., Moro, P., Vojnar, T.: Programs with Lists Are Counter Automata. In: Ball, T., Jones, R.B. (eds.) CAV 2006. LNCS, vol. 4144, pp. 517–531. Springer, Heidelberg (2006)
3. Bradley, A., Manna, Z., Sipma, H.: The polyranking principle. In: Caires, L., Italiano, G.F., Monteiro, L., Palamidessi, C., Yung, M. (eds.) ICALP 2005. LNCS, vol. 3580, pp. 1349–1361. Springer, Heidelberg (2005)

4. Bradley, A., Manna, Z., Sipma, H.: Termination of polynomial programs. In: Cousot, R. (ed.) VMCAI 2005. LNCS, vol. 3385, pp. 113–129. Springer, Heidelberg (2005)
5. Bradley, A.R., Manna, Z., Sipma, H.B.: Linear ranking with reachability. In: Etessami, K., Rajamani, S.K. (eds.) CAV 2005. LNCS, vol. 3576, pp. 491–504. Springer, Heidelberg (2005)
6. Bruynooghe, M., Codish, M., Gallagher, J.P., Genaim, S., Vanhoof, W.: Termination analysis of logic programs through combination of type-based norms. TOPLAS 29(2) (2007)
7. Calcagno, C., Distefano, D., OHearn, P., Yang, H.: Footprint Analysis: A Shape Analysis That Discovers Preconditions. In: Riis Nielson, H., Filé, G. (eds.) SAS 2007. LNCS, vol. 4634, pp. 402–418. Springer, Heidelberg (2007)
8. Cook, B., Gotsman, A., Podelski, A., Rybalchenko, A., Vardi, M.: Proving that programs eventually do something good. In: POPL (2007)
9. Cook, B., Podelski, A., Rybalchenko, A.: Termination proofs for systems code. In: PLDI (2006)
10. Cook, B., Podelski, A., Rybalchenko, A.: Proving thread termination. In: PLDI (2007)
11. Cooper, D.C.: Theorem proving in arithmetic without multiplication. Machine Intelligence (1972)
12. Cousot, P., Cousot, R.: Abstract interpretation: A unified lattice model for static analysis of programs by construction or approximation of fixpoints. In: POPL (1977)
13. Dikstra, E.W., Scholten, C.S.: Predicate Calculus and Program Semantics. Springer, Heidelberg (1989)
14. Gulwani, S., Srivastava, S., Venkatesan, R.: Program analysis as constraint solving. In: PLDI (2008)
15. Lalire, G., Argoud, M., Jeannet, B.: Interproc analyzer (2008), `http://bjeannet.gforge.inria.fr/interproc/`
16. Lee, C.S., Jones, N.D., Ben-Amram, A.M.: The size-change principle for program termination. In: POPL (2001)
17. Leino, R.: Effecient weakest preconditions. Information Processing Letters 93(6) (2005)
18. Miné, A.: The Octagon abstract domain. Higher-Order and Symbolic Computation (2006)
19. Podelski, A., Rybalchenko, A.: A complete method for the synthesis of linear ranking functions. In: Steffen, B., Levi, G. (eds.) VMCAI 2004. LNCS, vol. 2937, pp. 239–251. Springer, Heidelberg (2004)
20. Podelski, A., Rybalchenko, A.: ARMC: the logical choice for software model checking with abstraction refinement. In: Hanus, M. (ed.) PADL 2007. LNCS, vol. 4354, pp. 245–259. Springer, Heidelberg (2006)
21. Rybalchenko, A.: ARMC (2008), `http://www.mpi-sws.org/~rybal/armc/`
22. Vardi, M.Y.: Verification of concurrent programs: The automata-theoretic framework. In: LICS (1987)

Monotonic Abstraction for Programs with Dynamic Memory Heaps

Parosh Aziz Abdulla[1], Ahmed Bouajjani[2], Jonathan Cederberg[1], Frédéric Haziza[1], and Ahmed Rezine[1,2]

[1] Uppsala University, Sweden
[2] LIAFA, University of Paris 7, France

Abstract. We propose a new approach for automatic verification of programs with dynamic heap manipulation. The method is based on symbolic (backward) reachability analysis using upward-closed sets of heaps w.r.t. an appropriate preorder on graphs. These sets are represented by a finite set of minimal graph patterns corresponding to a set of bad configurations. We define an abstract semantics for the programs which is monotonic w.r.t. the preorder. Moreover, we prove that our analysis always terminates by showing that the preorder is a well-quasi ordering. Our results are presented for the case of programs with 1-next selector. We provide experimental results showing the effectiveness of our approach.

1 Introduction

Software verification needs the use of efficient algorithmic techniques for the analysis of infinite-state models. The sources of infiniteness are multiple and can be related to complex control such as (potentially recursive) procedure calls and dynamic creation of processes, or to the manipulation of (unbounded-size) dynamic data-structures and variables ranging over infinite data domains. A lot of work has been devoted in the last decade to the design of automatic verification techniques for infinite-state models, and several general approaches and formal frameworks have emerged allowing either to establish decidability results and derive verification algorithms (e.g., [20,2]), or to define generic exact/abstract analysis procedures (e.g., [7,11,22,30]).

One of the widely adopted frameworks in this context of infinite-state verification is based on the concept of *monotonic systems w.r.t. a well-quasi ordering* [2,20]. This framework provides a scheme for proving the termination of the (backward) reachability analysis, and it has been used for the design of verification algorithms for various models including Petri nets, lossy channel systems, timed Petri nets, broadcast protocols, etc. (see, e.g., [5,6,18,19]). The idea is, given a class of models, to define a preorder $\preceq$ on the configuration space such that (1) $\preceq$ is a simulation relation on the considered models, and (2) $\preceq$ is a well-quasi ordering (WQO for short). If such a preorder can be defined, then it can be proved that the reachability problem of an upward-closed set of configurations (w.r.t. $\preceq$) is decidable. Indeed, (1) monotonicity implies that for any

A. Gupta and S. Malik (Eds.): CAV 2008, LNCS 5123, pp. 341–354, 2008.

upward-closed set, the set of its predecessors is an upward-closed set, and (2) the fact that $\preceq$ is a WQO implies that every upward-closed set can be characterized by its *finite* set of minimal elements. Therefore, starting from an upward-closed set of configurations U, the iterative computation of the backward reachable configurations from U necessarily terminates since only a finite number of steps are needed to capture all minimal elements of the set of predecessors of U. Obviously, this requires that upward-closed sets can be effectively represented and manipulated (i.e., there are procedures for, e.g., computing immediate predecessors and unions, and for checking entailment). This general scheme can be applied for the verification of safety properties since this problem can be reduced to checking the reachability of a set of bad configurations which is typically an upward-closed set w.r.t. the considered preorder. (For instance, mutual exclusion is violated as soon as there are (at least) two processes in the critical section.)

Unfortunately, many systems do not fit into this framework, in the sense that there is no nontrivial (useful) WQO for which these systems are monotonic. Nevertheless, a natural approach to overcome this problem is, given a preorder $\preceq$, to define an abstract semantics of the considered systems which forces their monotonicity. Basically, the idea is to consider that a transition is possible from a configuration c_1 to c_2 if it is possible from any smaller configuration $c_1' \preceq c_1$ to c_2. This simple idea has been used recently in works concerning the verification of parametrized networks of (finite/infinite-state) processes, and surprisingly, it leads to quite efficient abstract analysis techniques which allow to handle *fully automatically* several non-trivial examples of such systems [3,4]. This encourages us to investigate its application to other classes of complex systems.

In this paper, our aim is to develop a framework based on the approach introduced above for the verification of sequential iterative programs manipulating dynamic memory heaps. The issue of verifying automatically such programs has received a lot of attention in the last few years, and many approaches and techniques have been developed including static-analysis and abstraction-based frameworks (see, e.g., [29]), logic-based frameworks(see, e.g., [25,28]), automata-based frameworks (see, e.g., [14,21]), etc. Here, we introduce a framework based on symbolic (backward) reachability analysis using upward-closed sets of heap graphs (w.r.t. some appropriate preorder). As a first step toward this framework, we present in this paper the results of our investigations concerning the case of programs manipulating heap structures with *one* next-selector, i.e., heaps of programs manipulating lists with possibility of sharing and circularity.

More precisely, we consider that heaps are represented as labeled graphs, where labels correspond to positions of program variables. We propose a preorder $\preceq$ between heap graphs which corresponds basically to the following: Given two graphs g_1 and g_2, we have $g_1 \preceq g_2$ if g_1 can be obtained from g_2 by a sequence of transformations consisting of either deleting an edge, a variable, or an isolated vertex, or of contracting segments (i.e., sequence of vertices) without sharing in the graph.

Actually, our graph representations correspond in general to sets of heaps instead of a single one. They can be seen as minimal patterns (w.r.t. $\preceq$), and they represent all the heaps that subsume (w.r.t. $\preceq$) these patterns. Therefore, our graph representations define upward-closed sets of heap graphs.

Then, we provide procedures for computing sets of predecessors w.r.t. the abstract semantics we consider (introduced above), and for checking entailment. These procedures allow to define a *simple algorithm* which computes an over-approximation of the set of backward reachable configurations starting from an upward-closed set of heap graphs (effectively given as a finite set of minimal elements). We show that this algorithm *always terminates* by proving that the preorder we have defined on heap graphs is a WQO.

Our analysis allows to check properties such as absence of null dereferencing as well as absence of garbage creation. Moreover, it allows to check shape (well-formedness) properties of the heaps (for instance the fact that the output is always a list without sharing). We show indeed that these kinds of verification problems can be reduced to the problem of reaching sets of bad configurations corresponding to the existence in the heap graph of some *minimal bad patterns*. We also provide experimental results showing the effectiveness of our approach.

Related work. As mentioned before, several approaches to the automatic analysis of programs with dynamic linked data structures have been proposed (see, e.g., [14,17,21,29]). Shape analysis as introduced in [29] is based on the computation of abstract shape graphs using the so-called instrumentation predicates. An automata-based approach using abstract regular model checking (ARMC) [15] has been proposed in [13,14]. In [10,17], an automatized analysis approach based on separation logic combined with abstraction techniques (close to widening techniques) has been proposed. With respect to these approaches, the one we present in this paper is conceptually and technically different and simpler. In particular, the ARMC-based approach needs the manipulation of quite complex encodings of the heap graphs into words or trees (in order to represent sets of heap encodings using finite-state automata), and use a sophisticated machinery for manipulating these encodings based on representing program statements as (word/tree) transducers. In contrast, the approach presented here uses a natural representation of heaps as graphs and employs direct procedures for computing operations on these graphs. This direct approach has already shown its advantages w.r.t. the approach using transducers in the context of regular model checking for parametrized networks of processes [3]. Also, the approach we present uses a built-in abstraction principle which is different from the ones used in the existing approaches, and which makes the analysis fully automatic.

The existing approaches mentioned above (shape analysis, abstract regular model checking, separation logic) can handle some classes of general heap structures (including doubly linked lists, lists of lists, trees, etc.). Although the techniques presented in this paper concern the case of heap structures with 1-next selector, our approach (based on upward-closed abstractions w.r.t. preorders on graphs) can in principle be extended to more general classes of heaps.

Concerning the particular class of programs manipulating heaps with 1-next selector, there are many other verification approaches which have been developed recently (see, e.g., [8,12,13,16,23,24]). Almost all these works use the fact that in this case (1) the heap graphs are collections of reversed trees potentially having their roots connected to a loop, and moreover (2) the number of leaves and shared vertices in these graphs is bounded linearly in terms of the number of program variables. For instance, in [24], these properties are used to define an abstraction which consists of contracting all segments without sharing. In our case, we use these properties in order to prove that the preorder we propose on graph representations is a WQO. However, our abstraction is different from the one proposed for instance in [24] since we can have graphs which are not minimal w.r.t. to contraction (e.g., we can express the fact that the length of a segment is at least some given natural number), and we can also have graphs corresponding to a partial description of the heap where only a *part* of the reachable heap from *some* of the program variables is constrained.

In [9,12], translations from programs with lists to counter automata have been defined based on the representation of heap graph as its contracted version supplied with the information about the length of each contracted sharing-free segments. These translations allow to use various existing techniques for the analysis of counter systems in order to check safety properties involving constraints relating the lengths of different lists, or to check termination. Such analysis involving quantitative reasoning cannot be done with the techniques presented in this paper. As said above, these techniques can handle some reasoning about the sizes of the lists, but only concerning constraints on minimal lengths. However, extensions of our techniques to more general constraints (e.g., gap-order constraints [27]) are possible.

Outline. In the next section, we introduce the class of programs we consider together with their graph representations. In Section 3 we describe a set of graph operations which we use in the subsequent sections. Section 4 introduces the ordering on configurations. In Section 5, we introduce a relation which we use as the basic step in the reachability algorithm. Section 6 introduces the backward reachability algorithm, and proves its partial correctness. The termination of the algorithm is shown in Section 7. Section 8 reports the results of applying a prototype, based on the method, to a number of simple programs. Finally, in Section 9 we give some conclusions.

2 Preliminaries

We consider programs that operate on data structures with one next-pointer such as traditional singly-linked lists and circular lists (possibly sharing their parts). We represent the store as a graph, where the vertices represent the list cells, and the successor of a vertex represents the cell pointed to by the current one. The graphs are of a special form in the sense that each vertex has at most one successor. A program also uses a finite set of pointers which we call *variables*. A cell is labeled by the (possibly empty) set of variables pointing to it.

For simplicity of presentation, we will treat the constant *null* as a variable, with the special property that whenever a vertex is labeled by *null*, the successor of the cell is undefined.

For a partial function f, we write $f(a) = \perp$ to denote that $f(a)$ is undefined. For a (partial) function f , we use $f[a \leftarrow b]$ to denote the function f' such that $f'(a) = b$ and $f'(x) = f(x)$ if $x \neq a$.

Formally, we assume a finite set X of variables including the element *null*. A *program* P is a pair (Q,T) where Q is a finite set of *control states* and T is a finite set of *transitions*. A transition is a triple (q_1, a, q_2) where $q_1, q_2 \in Q$ are control states and a is an *action*. An *action* is of one of the following forms $x = y$, $x \neq y$, $x := y$ where $x \neq null$, $x.next = y$ where $x \neq null$, or $x := y.next$ where $x, y \neq null$. The transition corresponds to the program changing control state from q_1 to q_2 while performing the operation described in a on the data structure. We choose to work with the above minimal set of operations. Other operations, e.g., $x = y.next$, $x \neq y.next$, etc, can be expressed using the given set.

A *graph* g is a triple $(V, succ, \lambda)$ where V is a finite set of *vertices*, $succ$ is a partial function from V to V, and λ is a partial function from X to V. Furthermore, it is always the case that $succ(\lambda(null)) = \perp$. Intuitively, the vertices correspond to the list cells. The function $succ$ defines the successors of the cells. If $succ(v) = \perp$, the cell represented by v has currently no successor. The function λ defines the cell to which a given variable points. If $\lambda(x) = \perp$, the value of variable (pointer) x is undefined.

A *configuration* c is a pair (q, g) where $q \in Q$ is a control state and g is a graph. We define a transition relation on configurations as follows. Let $t = (q_1, a, q_2)$ be a transition and let $c = (q, g)$ and $c' = (q', g')$ be configurations. We write $c \xrightarrow{t} c'$ to denote that $q = q_1$, $q' = q_2$, and $g \xrightarrow{a} g'$, where $g \xrightarrow{a} g'$ holds if one of the following conditions is satisfied:

- a is of the form $x = y$, $\lambda(x) \neq \perp$, $\lambda(y) \neq \perp$, $\lambda(x) = \lambda(y)$, and $g' = g$.
- a is of the form $x \neq y$, $\lambda(x) \neq \perp$, $\lambda(y) \neq \perp$, $\lambda(x) \neq \lambda(y)$, and $g' = g$.
- a is of the form $x := y$, $\lambda(y) \neq \perp$, $succ' = succ$, and $\lambda' = \lambda[x \leftarrow \lambda(y)]$.
- a is of the form $x := y.next$, $\lambda(y) \neq \perp$, $succ(\lambda(y)) \neq \perp$, $succ' = succ$, and $\lambda' = \lambda[x \leftarrow succ(\lambda(y))]$.
- a is of the form $x.next := y$, $\lambda(x) \neq \perp$, $\lambda(y) \neq \perp$, $\lambda(x) \neq \lambda(null)$, $\lambda' = \lambda$, and $succ' = succ[\lambda(x) \leftarrow \lambda(y)]$.

We define $\longrightarrow$ as $\bigcup_{t \in T} \xrightarrow{t}$ and use $\xrightarrow{*}$ to denote the reflexive transitive closure of $\longrightarrow$. For sets C_1 and C_2 of configurations, we use $C_1 \longrightarrow C_2$ to denote that $c_1 \longrightarrow c_2$ for some $c_1 \in C_1$ and $c_2 \in C_2$. By $c_1 \longrightarrow C_2$ we mean $\{c_1\} \longrightarrow C_2$. We define $c_1 \xrightarrow{*} C_2$, $C_1 \xrightarrow{*} C_2$, etc in a similar manner to above.

3 Operations on Graphs

In this section, we define a number of operations on graphs which we use in the subsequent sections. In the rest of the section, we assume a graph $g = (V, succ, \lambda)$.

For $v_1, v_2 \in V$, we use $(g.succ)[v_1 \leftarrow v_2]$ to denote the graph $g' = (V', succ', \lambda')$ where $V' = V$, $\lambda' = \lambda$, and $succ' = succ[v_1 \leftarrow v_2]$. Intuitively, we only modify g so that v_2 becomes the successor of v_1. We define $(g.\lambda)[x \leftarrow v]$ analogously. That is, we make x point to v.

For a vertex $v \in V$, we say that v is *simple* if $|succ^{-1}(v)| = 1$, $succ(v) \neq \perp$, and there is no $x \in X$ with $\lambda(x) = v$. In other words, v has exactly one predecessor, one successor and no label. We say that v is *isolated* in g if $succ(v) = \perp$, $succ^{-1}(v) = \emptyset$, and there is no $x \in X$ with $\lambda(x) = v$. In other words, v has no successors or predecessors and it is not labeled by any variable.

Operations on vertices. For $v \notin V$, we use $g \oplus v$ to denote the graph $g' = (V', succ', \lambda')$ such that $V' = V \cup \{v\}$, $succ' = succ$, and $\lambda' = \lambda$, i.e. we add a new vertex to g. Observe that the added vertex is then isolated.

For an *isolated* vertex $v \in V$, we use $g \ominus v$ to denote the graph $g' = (V', succ', \lambda')$ such that $V' = V - \{v\}$, $succ' = succ$, and $\lambda' = \lambda$.

Operations on variables. We define $g \oplus x$ to be the set of graphs we get from g by letting x point anywhere inside g. Formally, we define $g \oplus x$ to be the smallest set containing each graph g' such that one of the following conditions is satisfied: (i) there is a $v \notin V$ such that $g' = ((g \oplus v).\lambda)\,[x \leftarrow v]$, *i.e.* we add a vertex to g and make x point to it. (ii) there is a $v \in V$ such that $g' = (g.\lambda)\,[x \leftarrow v]$, *i.e.* we make x point to some vertex in g. (iii) there are $v_1 \in V$, $v_2 \notin V$, and graphs $g_i = (V_i, succ_i, \lambda_i)$ for $i = 1, 2, 3$, such that $succ(v_1) \neq \perp$, $g_1 = g \oplus v_2$, $g_2 = (g_1.succ)\,[v_2 \leftarrow succ_1(v_1)]$, $g_3 = (g_2.succ)\,[v_1 \leftarrow v_2]$, and $g' = (g_3.\lambda)\,[x \leftarrow v_2]$, *i.e.* we add a new vertex v_2 in between v_1 and its successor and make x point to v_2.

For variables x and y with $\lambda(x) \neq \perp$, we define $g \oplus_{=x} y$ to be the graph $g' = (g.\lambda)\,[y \leftarrow \lambda(x)]$, *i.e.* we make y point to the same vertex as x. Furthermore, we define $g \oplus_{\neq x} y$ to be the smallest set containing each graph g' such that $g' \in (g \oplus y)$ and $\lambda'(y) \neq \lambda'(x)$, *i.e.* we make y point anywhere inside g except to the vertex pointed to by x.

For variables x and y with $\lambda(x) \neq \perp$ and $succ(\lambda(x)) \neq \perp$, we use $g \oplus_{x\rightarrow} y$ to denote the graph $(g.\lambda)\,[y \leftarrow succ(\lambda(x))]$, *i.e.* we make y point to the successor of the vertex pointed to by x. For variables x and y with $\lambda(x) \neq \perp$, we define $g \oplus_{x\leftarrow} y$ to be the set of graphs we get from g by letting y point to any vertex where the successor is either undefined or pointed to by x. Formally, we define $g \oplus_{x\leftarrow} y$ to be the smallest set containing each graph g' such that one of the following conditions is satisfied: (i) there is a $v \notin V$ such that $g' = ((g \oplus v).\lambda)\,[y \leftarrow v]$. (ii) there is a $v \in V$ such that $v \neq \lambda(null)$, either $succ(v) = \perp$ or $succ(v) = \lambda(x)$, and $g' = (g.\lambda)\,[y \leftarrow v]$. That is, we place y on the vertices without a successor or the ones whose successor is pointed to by x. (iii) there are $v_1 \in V$, $v_2 \notin V$, and graphs $g_i = (V_i, succ_i, \lambda_i)$ for $i = 1, 2, 3$, such that $succ(v_1) = \lambda(x)$, $g_1 = g \oplus v_2$, $g_2 = (g_1.succ)\,[v_2 \leftarrow \lambda(x)]$, $g_3 = (g_2.succ)\,[v_1 \leftarrow v_2]$, and $g' = (g_3.\lambda_3)\,[y \leftarrow v_2]$. Intuitively, we add a new vertex v_2 in between the vertex pointed by x and its predecessors and make y point to v_2.

For a variable x, we use $g \ominus x$ to denote $(g.\lambda)\,[x \leftarrow \perp]$.

Operations on edges. If $\lambda(x) \neq \perp$, $\lambda(y) \neq \perp$ and $\lambda(x) \neq \lambda(null)$, we use $g \boxplus (x \rightarrow y)$ to denote $(g.succ)\,[\lambda(x) \leftarrow \lambda(y)]$, *i.e.* we delete the edge between the vertex $\lambda(x)$ and its successor (if any) and add an edge from $\lambda(x)$ to $\lambda(y)$.

If $\lambda(x) \neq \perp$ and $\lambda(x) \neq \lambda(null)$, we define $g \boxplus (x \rightarrow)$ to be the set of graphs we get from g by letting $x.next$ point anywhere inside g. Formally, we define $g \boxplus (x \rightarrow)$ to be the smallest set containing each graph g' such that one of the following conditions is satisfied: (i) there is a $v \notin V$ such that $g' = ((g \oplus v).succ)\,[\lambda(x) \leftarrow v]$. (ii) there is a $v \in V$ such that $g' = (g.succ)\,[\lambda(x) \leftarrow v]$. (iii) there are $v_1 \in V$, $v_2 \notin V$, and graphs $g_i = (V_i, succ_i, \lambda_i)$ for $i = 1, 2, 3$, such that $succ(v_1) \neq \perp$, $g_1 = g \oplus v_2$, $g_2 = (g_1.succ)\,[v_2 \leftarrow succ_1(v_1)]$, $g_3 = (g_2.succ)\,[v_1 \leftarrow v_2]$, and $g' = (g_3.succ)\,[\lambda_3(x) \leftarrow v_2]$.

If $\lambda(x) \neq \perp$, we denote $g \boxminus (x \rightarrow)$ as $(g.succ)\,[\lambda(x) \leftarrow \perp]$, *i.e.* we remove the edge from the vertex pointed to by x and its successor (if any).

4 Ordering

In this section, we introduce an ordering on configurations. Based on the ordering, we will define the coverability problem which we use to check safety properties, and define the abstract transition relation. The latter is an over-approximation of the concrete transition relation.

Ordering. Let $g = (V, succ, \lambda)$ and $g' = (V', succ', \lambda')$. We write $g \lhd g'$ to denote that one of the following properties is satisfied: (i) *Variable Deletion*: $g = g' \ominus x$ for some variable x, (ii) *Vertex Deletion*: $g = g' \ominus v$ for some isolated vertex $v \in V'$, (iii) *Edge Deletion*: $g = (g'.succ)\,[v \leftarrow \perp]$ for some $v \in V'$, or (iv) *Contraction*: there are vertices $v_1, v_2, v_3 \in V'$ and graphs g_1, g_2 such that v_2 is simple, $succ'(v_1) = v_2$, $succ'(v_2) = v_3$, $g_1 = (g'.succ)\,[v_2 \leftarrow \perp]$, $g_2 = (g_1.succ)\,[v_1 \leftarrow v_3]$ and $g = g_2 \ominus v_2$.

We write $g \preceq g'$ to denote that there are $g_0 \lhd g_1 \lhd g_2 \lhd \cdots \lhd g_n$ with $n \geq 0$, $g_0 = g$, and $g_n = g'$. That is, we can obtain g from g' by performing a finite sequence of variable deletion, vertex deletion, edge deletion and contraction operations. For configurations $c = (q, g)$ and $c' = (q', g')$, we write $c \preceq c'$ to denote that $q' = q$ and $g \preceq g'$.

For a configuration c, we use $c\!\uparrow$ to denote the *upward closure* of c, *i.e.* $c\!\uparrow = \{c' \,|\, c \preceq c'\}$. We use $c\!\downarrow$ to denote the *downward closure* of c, *i.e.* $c\!\downarrow = \{c' \,|\, c' \preceq c\}$ For a set C of configurations, we define $C\!\uparrow$ as $\bigcup_{c \in C} c\!\uparrow$. We define $C\!\downarrow$ analogously.

Safety Properties. In order to analyze safety properties, we study the *coverability problem* defined below.

> Coverability
> **Instance:**
> Sets C_{Init} and C_F of configurations.
> **Question:** Is it the case $C_{Init} \xrightarrow{*} C_F\!\uparrow$?

Intuitively, $C_F\!\uparrow$ represents a set of "bad" states which we do not want to reach during the execution of the program. This set is represented by a set C_F of minimal elements.

In Section 8, we describe how to encode properties such as garbage generation, dereferencing and shape violation as reachability of upward closed sets of configurations represented by finite sets of minimal elements. Therefore, checking safety with respect to these properties amounts to solving the coverability problem

Abstract Transition Relation. We write $c_1 \xrightarrow{t}_A c_2$ to denote that there is a c_3 such that $c_3 \preceq c_1$ and $c_3 \xrightarrow{t} c_2$. In other words, a step of the abstract transition relation consists of first moving to a smaller configuration (wrt $\preceq$) and then performing a step of the concrete transition relation. Notice that the abstraction corresponds to an over-approximation and therefore any safety property which holds in the abstract system will also hold in the concrete one.

5 Computing Predecessors

The main idea behind our algorithm to solve the coverability problem, is to perform backward reachability analysis. The basic step of the algorithm uses a relation $\rightsquigarrow$ defined on the set of configurations. Intuitively, $c \rightsquigarrow c'$ means that, from c', we can perform a transition and reach a configuration in the upward closure of c. First, we give the formal definition of $\rightsquigarrow$, and then describe some of its properties, and in particular how it relates to the transition relation $\longrightarrow$.

For a graph $g = (V, succ, \lambda)$, a graph g', and an action a, we write $g \overset{a}{\rightsquigarrow} g'$ to denote that one of the following conditions is satisfied:

1. a is of the form $x = y$ and one of the following conditions is satisfied:
   (a) $\lambda(x) \neq \bot$, $\lambda(y) \neq \bot$, $\lambda(x) = \lambda(y)$ and $g' = g$.
   (b) $\lambda(x) \neq \bot$, $\lambda(y) = \bot$ and $g' = g \oplus_{=x} y$.
   (c) $\lambda(x) = \bot$, $\lambda(y) \neq \bot$ and $g' = g \oplus_{=y} x$.
   (d) $\lambda(x) = \bot$, $\lambda(y) = \bot$ and $g' = g_1 \oplus_{=x} y$ for some $g_1 \in (g \oplus x)$.
   In order to be able to perform the action, the variables x and y should point to the same vertex. If one (or both) of them are missing, then we add them to the graph (with the restriction that they point to the same vertex).
2. a is of the form $x \neq y$ and one of the following conditions is satisfied:
   (a) $\lambda(x) \neq \bot$, $\lambda(y) \neq \bot$, $\lambda(x) \neq \lambda(y)$ and $g' = g$.
   (b) $\lambda(x) \neq \bot$, $\lambda(y) = \bot$ and $g' \in g \oplus_{\neq x} y$.
   (c) $\lambda(x) = \bot$, $\lambda(y) \neq \bot$ and $g' \in g \oplus_{\neq y} x$.
   (d) $\lambda(x) = \bot$, $\lambda(y) = \bot$ and $g' \in g_1 \oplus_{\neq x} y$ for some $g_1 \in (g \oplus x)$.
   We proceed as in case 1, but now under the restriction that x and y point to different vertices (rather than to the same vertex).
3. a is of the form $x := y$ and one of the following conditions is satisfied:
   (a) $\lambda(x) \neq \bot$, $\lambda(y) \neq \bot$, $\lambda(x) = \lambda(y)$ and $g' = g \ominus x$.
   (b) $\lambda(x) \neq \bot$, $\lambda(y) = \bot$ and $g' = g_1 \ominus x$ where $g_1 = g \oplus_{=x} y$.
   (c) $\lambda(x) = \bot$, $\lambda(y) \neq \bot$ and $g' = g$.
   (d) $\lambda(x) = \bot$, $\lambda(y) = \bot$ and $g' \in (g \oplus y)$.
   In difference to case 1 is that the variable x may have had any value before performing the assignment. Therefore, we remove x from the graph.
4. a is of the form $x := y.next$ and one of the following conditions is satisfied:
   (a) $\lambda(x) \neq \bot$, $\lambda(y) \neq \bot$, $succ(\lambda(y)) \neq \bot$, $succ(\lambda(y)) = \lambda(x)$ and $g' = g \ominus x$.
   (b) $\lambda(x) \neq \bot$, $\lambda(y) \neq \bot$, $\lambda(y) \neq \lambda(null)$, $succ(\lambda(y)) = \bot$ and $g' = g_1 \ominus x$, where $g_1 = g \boxplus (y \rightarrow x)$.
   (c) $\lambda(x) \neq \bot$, $\lambda(y) = \bot$ and there are graphs g_1, g_2 such that $g' = g_2 \ominus x$, $g_2 = g_1 \boxplus (y \rightarrow x)$ and $g_1 \in g \oplus_{x \leftarrow} y$.

(d) $\lambda(x) = \bot$, $\lambda(y) \neq \bot$, $succ(\lambda(y)) \neq \bot$ and $g' = g$.
(e) $\lambda(x) = \bot$, $\lambda(y) \neq \bot$, $\lambda(y) \neq \lambda(null)$, $succ(\lambda(y)) = \bot$ and $g' \in g \boxplus (y \rightarrow)$.
(f) $\lambda(x) = \bot$, $\lambda(y) = \bot$ and there are graphs g_1, g_2, g_3 such that $g_1 \in g \oplus x$, $g_2 \in g_1 \oplus_{x \leftarrow} y$, $g_3 = g_2 \boxplus (y \rightarrow x)$ and $g' = g_3 \ominus x$.

Similarly to case 3 we remove x from the graph. The successor of y should be defined and point to the same vertex as x. In case the successor is missing, we add an edge explicitly from the vertex labeled by y to the vertex labeled by x. Furthermore, if x is missing then the successor of y may point anywhere inside the graph.

5. a is of the form $x.next := y$ and one of the following conditions is satisfied:
   (a) $\lambda(x) \neq \bot$, $succ(\lambda(x)) \neq \bot$, $\lambda(y) \neq \bot$, $succ(\lambda(x)) = \lambda(y)$ and $g' = g \boxminus (x \rightarrow)$.
   (b) $\lambda(x) \neq \bot$, $succ(\lambda(x)) \neq \bot$, $\lambda(y) = \bot$ and $g' = g_1 \boxminus (x \rightarrow)$, where $g_1 = g \oplus_{x \rightarrow} y$.
   (c) $\lambda(x) \neq \bot$, $succ(\lambda(x)) = \bot$, $\lambda(y) \neq \bot$, $\lambda(x) \neq \lambda(null)$ and $g' = g$.
   (d) $\lambda(x) \neq \bot$, $succ(\lambda(x)) = \bot$, $\lambda(y) = \bot$, $\lambda(x) \neq \lambda(null)$ and $g' \in g \oplus y$.
   (e) $\lambda(x) = \bot$, $\lambda(y) \neq \bot$ and $g' = g_1 \boxminus (x \rightarrow)$, where $g_1 \in g \oplus_{y \leftarrow} x$.
   (f) $\lambda(x) = \bot$, $\lambda(y) = \bot$ and there are graphs g_1, g_2 such that $g_1 \in g \oplus y$, $g_2 \in g_1 \oplus_{y \leftarrow} x$ and $g' = g_2 \boxminus (x \rightarrow)$.

   After performing the action, the successor of the vertex labeled by x should be the same vertex as the one labeled by y. Before performing the action, the successor could have been anywhere inside the graph, and the corresponding edge is therefore removed.

Remark. In the above definition, we assume that x and y are different variables. It is straightforward to handle the case where they are the same variable.

For a transition $t = (q_1, a, q_2)$ and configurations $c = (q, g)$ and $c' = (q', g')$, we write $c \overset{t}{\rightsquigarrow} c'$ to denote that $q = q_1$, $q' = q_2$ and $g \overset{a}{\rightsquigarrow} g'$. We use $c \rightsquigarrow c'$ to denote that $c \overset{t}{\rightsquigarrow} c'$ for some $t \in T$. For a set C of configurations and a configuration c, we define $Rank(C)(c)$ to be the smallest n such that there is a sequence $c_0 \rightsquigarrow c_1 \rightsquigarrow \cdots \rightsquigarrow c_n$ where $c_0 = c$ and there is a $c' \in C$ such that $c_n \preceq c'$.

The following lemma states that small configurations simulate larger ones with respect to the backward relation.

Lemma 1. *For configurations c_1, c_2 and c_3, if $c_1 \rightsquigarrow c_2$ and $c_3 \preceq c_1$ then there is a configuration c_4 such that $c_3 \rightsquigarrow c_4$ and $c_4 \preceq c_2$.*

The following lemma relates the backward and forward transition relations.

Lemma 2. *Consider configurations c_1 and c_2. If $c_1 \rightsquigarrow c_2$ then $c_2 \longrightarrow c_1\!\uparrow$. If $c_1 \longrightarrow c_2\!\uparrow$ then $c_2 \rightsquigarrow c_1\!\downarrow$.*

6 Algorithm

We present here the reachability algorithm and show its partial correctness.

The algorithm inputs two sets C_{Init} and C_F of configurations and checks whether $C_{Init} \xrightarrow{*}_A C_F\uparrow$. The algorithm maintains two sets of configurations: a set `ToExplore`, initialized to C_F, of configurations that have not yet been analyzed; and a set `Explored`, initialized to the empty set, of configurations that contains information about the configurations that have already been analyzed. The algorithm preserves the following two invariants: (i) $C_{Init} \xrightarrow{*}_A$ $(\texttt{ToExplore} \bigcup \texttt{Explored})\uparrow$ implies $C_{Init} \xrightarrow{*}_A C_F\uparrow$; and (ii) If $C_{Init} \xrightarrow{*}_A C_F\uparrow$, then there is $c \in \texttt{ToExplore}$ such that both $Rank(C_{Init})(c) < \infty$ and $\forall c' \in \texttt{Explored}.\ Rank(C_{Init})(c) < Rank(C_{Init})(c')$.

```
Input: Two sets C_Init and C_F of configurations.

Output: C_Init -*->_A C_F↑?
________________________________________
ToExplore := C_F
Explored := ∅
while ToExplore ≠ ∅ do
  remove some c from ToExplore
  if ∃c' ∈ C_Init. c ⪯ c' then
    return true
  else if ∃c' ∈ Explored. c' ⪯ c then
    discard c
  else
    ToExplore := ToExplore ⋃ {c' | c ⇝ c'}
    Explored :=
    {c} ⋃ {c' | c' ∈ Explored ∧ (c ⋠ c')}
  end if
end while
return false
```

Due to the invariants, the following two conditions can be checked during each step of the algorithm: (i) From the second invariant, if `ToExplore` becomes empty then the algorithm terminates with a negative answer; and (ii) From the first invariant and the definition of $\longrightarrow_A$, if a configuration in $C_{Init}\downarrow$ is detected then the algorithm terminates with a positive answer.

The following theorem follows immediately from the invariants together with Lemmas 1 and 2.

Theorem 1. *The reachability algorithm is partially correct.*

7 Termination

In this section, we give an overview of the termination proof for the reachability algorithm. The full details can be found in [1]. Let $\mathbb{N}^{>0}$ denote the set of positive integers. For a set A and a preorder on A, we say that $\preceq$ is a *well quasi-ordering (WQO)* on A if the following property is satisfied: for any infinite sequence $a_0, a_1, a_2, \ldots$ of elements in A, there are i, j such that $i < j$ and $a_i \preceq a_j$. A simple example of a WQO is the standard ordering on natural numbers. We extend the ordering $\preceq$ to an ordering $\preceq^*$ on the set A^* of finite words over A as follows: $w_1 \preceq^* w_2$ if there is an order-preserving injection from w_1 to w_2 such that each element in w_1 is mapped to an element in w_2 which is larger wrt $\preceq$. It is well-known that $\preceq^*$ is a WQO (see e.g.[2]). Since multisets and vectors are special cases of words it follows, for instance, that vectors of multisets of vectors of natural numbers are WQO (this particular property will be used later in the proof).

Consider a graph $g = (V, succ, \lambda)$. A graph b is said to be a *block* of g if b is a maximal part of g which is connected. A vertex is said to be *unguarded* if it is a leaf and there is no variable $x \in X$ with $\lambda(x) = v$. For a graph g, we define the *degree* of g, denoted $deg\,(g)$, to be the number of unguarded vertices in g. A graph is said to be *compact* if it does not contain simple vertices. Intuitively, a graph is *compact* if it cannot be reduced due to contraction. An *encoding* is a tuple $e = (V, succ, \lambda, \#)$ where $g = (V, succ, \lambda)$ is a compact graph, and $\# : V \times V \to \mathbb{N}^{>0}$ is a partial mapping such that $\#(v_1, v_2) \neq \perp$ iff $v_2 = succ(v_1)$. In other words, $\#$ associates a positive integer to each edge in g.

Fix a graph $g = (V, succ, \lambda)$. We define $enc\,(g)$ to be the encoding we get from g by applying contraction as much as possible (until the resulting graph cannot be reduced any more using contraction). Furthermore, for vertices v and v', the value of $\#(v, v')$ gives the number of edges between v and v' in g. We will define an ordering $\sqsubseteq$ on graphs (the formal definition of the relation is in [1]). Consider graphs g, g' with encodings $enc\,(g) = (V, succ, \lambda, \#)$ and $enc\,(g') = (V', succ', \lambda', \#')$. Roughly speaking, $g \sqsubseteq g'$ means that for each block b in $enc\,(g)$ there is an corresponding isomorphic block b' in $enc\,(g')$ such that such that $\#(v_1, v_2) \leq \#'(v_1', v_2')$ for all vertices v_1, v_2 in b and their images v_1', v_2' in b'. The ordering $\sqsubseteq$ is a WQO on the set of compact graphs whose degrees are bounded by some $k \in \mathbb{N}^{>0}$. The reason is that in any such a graph, there number of leaves is bounded by $|X| + k$, and hence there are only finitely many types of blocks which may occur in any such graph. This means that an encoding of such a graph can be represented by vectors of multisets of vectors of natural numbers as follows. Suppose that there are ℓ types of blocks. Then, an element of the representation will be of the form $(m_1, \ldots, m_\ell)$, where each m_i is a multiset of vectors of natural numbers corresponding to one type of block: an entry of the vector corresponds to an edge in the block; the natural numbers correspond to the ones which appear on the edges. We need multisets since there is no bound on the number of blocks (of a certain type) which may appear in the graph. This means that $\sqsubseteq$ is a WQO. Also, $g \sqsubseteq g'$ implies $g \preceq g'$, since if $g \sqsubseteq g'$ then we can derive g from g' through the application of a finite sequence of variable deletion, vertex deletion, edge deletion, and contraction operations. Finally, we observe that in the definition $\rightsquigarrow$ no unguarded vertices are introduced. This means that all the configurations which are generated in the reachability algorithm are bounded by some k, and are therefore WQO. This gives the following theorem.

Theorem 2. *The reachability algorithm is guaranteed to terminate.*

8 Experimental Results

Based on our method, we have implemented a prototype in Java. We consider three classes of properties: null-dereferencing, well-formedness of output, and garbage creation. We consider a set of programs taken from the PALE website [26]. The results, obtained on a 1.1 GHz Pentium M, are summarised below.

Prog.	Prop.	Time	#$\mathbf{C}^{ons.}$	#Iter.	Prog.	Prop.	Time	#$\mathbf{C}^{ons.}$	#Iter.
`Concat`	Deref	0.4 s	7	3	`Delete`	Deref	0.4 s	8	4
`Fumble`	Deref	0.3 s	3	2	`Reverse`	Deref	0.3 s	2	1
`Walk`	Deref	0.4 s	9	3	`Zip`	Deref	1.9 s	206	12
`Fumble`	Garbage	0.7 s	38	14	`Reverse`	Garbage	0.8 s	55	24
`Reverse`	Well-form.	1.7 s	48	20					

The entry #$\mathbf{C}^{ons.}$ gives the total number of minimal configurations added to `ToExplore` in the analysis. The entry **#Iter.** is the number of iterations of the **while**-loop of the algorithm.

For each of the three properties, we give a finite set of minimal configurations violating the property. For instance, for null-dereferencing, the set contains all configurations of the form $c = (q, g)$ defined as follows. There is a transition of the form (q, a, q') where a is of one of the forms $y := x.next$ or $x.next := y$. Also, g is the graph consisting of a single vertex labeled with $null$ and x.

9 Conclusions

We have presented a new approach for automatic verification of programs with dynamic heaps. The proposed approach is based on a simple algorithmic principle, and is fully automatic. The main idea is to perform an abstract (over-approximate) reachability analysis using upward-closed sets w.r.t. a suitable preorder on heap graphs. This preorder is shown to be a well-quasi ordering, which guarantees the termination of the analysis.

The results of this paper concern the case of heap structures with 1-next selector. Our approach can however be generalized to heap structures with multiple next selectors. Several extensions of our framework can be done by refining the considered preorder (and the abstraction it induces). For instance, it could be possible (1) to take into account data values attached to objects in the heap, (2) to consider constraints on (and relating) the lengths of (contracted) paths, and (3) to consider in integer program variables whose values are related to the lengths of paths in the heap. Such extensions with arithmetical reasoning can be done in our framework by considering preorders involving for instance gap-order constraints.

References

1. Abdulla, P.A., Bouajjani, A., Cederberg, J., Haziza, F., Rezine, A.: Monotonic abstraction for programs with dynamic memory heaps. Technical Report 2008-015, Dept. of Information Technology, Uppsala University, Sweden (April 2008)
2. Abdulla, P.A., Cerans, K., Jonsson, B., Tsay, Y.-K.: Algorithmic analysis of programs with well quasi-ordered domains. Inf. Comput. 160(1-2), 109–127 (2000)

3. Abdulla, P.A., Delzanno, G., Henda, N.B., Rezine, A.: Regular model checking without transducers (on efficient verification of parameterized systems). In: Grumberg, O., Huth, M. (eds.) TACAS 2007. LNCS, vol. 4424, pp. 721–736. Springer, Heidelberg (2007)
4. Abdulla, P.A., Delzanno, G., Rezine, A.: Parameterized Verification of Infinite-State Processes with Global Conditions. In: Damm, W., Hermanns, H. (eds.) CAV 2007. LNCS, vol. 4590, pp. 145–157. Springer, Heidelberg (2007)
5. Abdulla, P.A., Jonsson, B.: Verifying programs with unreliable channels. Inf. Comput. 127(2), 91–101 (1996)
6. Abdulla, P.A., Jonsson, B.: Model checking of systems with many identical timed processes. Theor. Comput. Sci. 290(1), 241–264 (2003)
7. Abdulla, P.A., Jonsson, B., Nilsson, M., Saksena, M.: A Survey of Regular Model Checking. In: Gardner, P., Yoshida, N. (eds.) CONCUR 2004. LNCS, vol. 3170, pp. 35–48. Springer, Heidelberg (2004)
8. Balaban, I., Pnueli, A., Zuck, L.D.: Shape analysis of single-parent heaps. In: Cook, B., Podelski, A. (eds.) VMCAI 2007. LNCS, vol. 4349, pp. 91–105. Springer, Heidelberg (2007)
9. Bardin, S., Finkel, A., Lozes, É., Sangnier, A.: From pointer systems to counter systems using shape analysis. In: Proceedings of the 5th Intern. Workshop on Automated Verification of Infinite-State Systems (AVIS 2006) (2006)
10. Berdine, J., Calcagno, C., Cook, B., Distefano, D., O'Hearn, P.W., Wies, T., Yang, H.: Shape analysis for composite data structures. In: Damm, W., Hermanns, H. (eds.) CAV 2007. LNCS, vol. 4590, pp. 178–192. Springer, Heidelberg (2007)
11. Bouajjani, A.: Languages, rewriting systems, and verification of infinite-state systems. In: Orejas, F., Spirakis, P.G., van Leeuwen, J. (eds.) ICALP 2001. LNCS, vol. 2076, pp. 24–39. Springer, Heidelberg (2001)
12. Bouajjani, A., Bozga, M., Habermehl, P., Iosif, R., Moro, P., Vojnar, T.: Programs with Lists Are Counter Automata. In: Ball, T., Jones, R.B. (eds.) CAV 2006. LNCS, vol. 4144, pp. 517–531. Springer, Heidelberg (2006)
13. Bouajjani, A., Habermehl, P., Moro, P., Vojnar, T.: Verifying Programs with Dynamic 1-Selector-Linked Structures in Regular Model Checking. In: Halbwachs, N., Zuck, L.D. (eds.) TACAS 2005. LNCS, vol. 3440, pp. 13–29. Springer, Heidelberg (2005)
14. Bouajjani, A., Habermehl, P., Rogalewicz, A., Vojnar, T.: Abstract Regular Tree Model Checking of Complex Dynamic Data Structures. In: Yi, K. (ed.) SAS 2006. LNCS, vol. 4134, pp. 52–70. Springer, Heidelberg (2006)
15. Bouajjani, A., Habermehl, P., Vojnar, T.: Abstract Regular Model Checking. In: Alur, R., Peled, D.A. (eds.) CAV 2004. LNCS, vol. 3114, pp. 372–386. Springer, Heidelberg (2004)
16. Distefano, D., Berdine, J., Cook, B., O'Hearn, P.: Automatic Termination Proofs for Programs with Shape-shifting Heaps. In: Ball, T., Jones, R.B. (eds.) CAV 2006. LNCS, vol. 4144, pp. 386–400. Springer, Heidelberg (2006)
17. Distefano, D., O'Hearn, P., Yang, H.: A Local Shape Analysis Based on Separation Logic. In: Hermanns, H., Palsberg, J. (eds.) TACAS 2006 and ETAPS 2006. LNCS, vol. 3920, pp. 287–302. Springer, Heidelberg (2006)
18. Emerson, E.A., Namjoshi, K.S.: On model checking for non-deterministic infinite-state systems. In: LICS, pp. 70–80 (1998)
19. Esparza, J., Finkel, A., Mayr, R.: On the verification of broadcast protocols. In: Proceedings of LICS 1999, pp. 352–359. IEEE Computer Society, Los Alamitos (1999)

20. Finkel, A., Schnoebelen, P.: Well-structured transition systems everywhere! TCS 256(1-2), 63–92 (2001)
21. Jensen, J., Jørgensen, M., Klarlund, N., Schwartzbach, M.: Automatic Verification of Pointer Programs Using Monadic Second-order Logic. In: Proc. of PLDI 1997 (1997)
22. Kesten, Y., Maler, O., Marcus, M., Pnueli, A., Shahar, E.: Symbolic model checking with rich assertional languages. Theor. Comput. Sci., 93–112 (2001)
23. Lahiri, S.K., Qadeer, S.: Verifying properties of well-founded linked lists. In: POPL, pp. 115–126 (2006)
24. Manevich, R., Yahav, E., Ramalingam, G., Sagiv, M.: Predicate Abstraction and Canonical Abstraction for Singly-Linked Lists. In: Cousot, R. (ed.) VMCAI 2005. LNCS, vol. 3385, pp. 181–198. Springer, Heidelberg (2005)
25. O'Hearn, P.W.: Separation logic and program analysis. In: Yi, K. (ed.) SAS 2006. LNCS, vol. 4134, p. 181. Springer, Heidelberg (2006)
26. PALE - the Pointer Assertion Logic Engine, `http://www.brics.dk/PALE/`
27. Revesz, P.Z.: A closed-form evaluation for datalog queries with integer (gap)-order constraints. Theor. Comput. Sci. 116(1&2), 117–149 (1993)
28. Reynolds, J.: Separation Logic: A Logic for Shared Mutable Data Structures. In: Proc. of LICS 2002. IEEE CS Press, Los Alamitos (2002)
29. Sagiv, S., Reps, T., Wilhelm, R.: Parametric Shape Analysis via 3-valued Logic. TOPLAS 24(3) (2002)
30. Wolper, P., Boigelot, B.: Verifying systems with infinite but regular state spaces. In: Vardi, M.Y. (ed.) CAV 1998. LNCS, vol. 1427, pp. 88–97. Springer, Heidelberg (1998)

Enhancing Program Verification with Lemmas

Huu Hai Nguyen and Wei-Ngan Chin

Department of Computer Science, National University of Singapore

Abstract. One promising approach to verifying heap-manipulating programs is based on *user-defined* inductive predicates in separation logic. This approach can describe data structures with complex invariants and sound reasoning based on unfold/fold. However, an important component towards more expressive program verification is the use of *lemmas* that can soundly relate predicates beyond their original definitions. This paper outlines a new *automatic* mechanism for proving and applying *user-specified lemmas* under separation logic.

Keywords: Lemma Proving, Lemma Application, Program Verification, Separation Logic, Entailment.

1 Introduction

Inductive predicates based on separation logic [16,22] offer an important approach to the specification of data structures that make extensive use of pointers and require sophisticated invariants. The technique brings the conveniences of algebraic data structures to the imperative settings, including precise yet simple and intuitive data structure definitions. It also enables effective and automatic reasoning based on the folding and unfolding of predicate definitions, and can verify programs over a wide range of interesting data structures. However, there are some crucial limitations in existing automated verification systems that rely solely on the unfold/fold mechanism. Firstly, it constrains traversals of a data structure to links explicitly allowed by the recursively defined predicates. These are typically top-down unravelling of the data structures, in that a program first accesses the "root" of a data structure, then any of its (non-dangling) fields that can be shown pointing to other objects or data structures. Secondly, the unfold/fold reasoner cannot discover auxiliary relations between predicates that may require inductive proofs.

In this work, we propose a new mechanism that aims to address the aforementioned shortcomings. The main idea is to explicitly state any auxiliary relations between predicate definitions, so that a deductive mechanism based on unfold/fold can prove and use them. This information is presented to the system in the form of *lemmas* that can be viewed as auxiliary relations for the predicates, apart from their definitions. These auxiliary properties can capture different linkage patterns in the data structure. They can also reveal complex relations between distinct but related predicates. Currently, user effort is required in stating the lemmas. Nevertheless, once stated, each of these lemmas is automatically *proven* once and *applied* many times, without further user assistance.

As the need for lemmas in theorem proving is well-known, our contribution is not on the lemmas per se, but rather on the mechanisms to prove and apply them for automated

A. Gupta and S. Malik (Eds.): CAV 2008, LNCS 5123, pp. 355–369, 2008.

verification via separation logic. These mechanisms are non-trivial, especially for handling more complex lemmas. We shall show that our procedure is sound, terminates and is directed. Our specific contributions are:

- **Alternative Traversals.** Lemmas provide different ways to reason about inductive predicates, which allows *alternative traversals* of data structures that are not captured by the original predicate definitions.
- **Complex Subsumption.** Predicates are often related to one another in complex ways (possibly involving multiple predicates from a heap state with side conditions) that may require inductive proofs. Lemmas provide an explicit way to capture such *complex subsumptions* between heap states for use through the deductive mechanism based on unfold/fold reasoning.
- **Lemma Proving.** To prove lemmas automatically, we use the same deductive mechanism as our entailment checker, after an initial unfold on the *base* predicate in the antecedent. The lemma itself can be applied during proving, when needed, which corresponds to a cyclic proof by infinite descent [5,4]. Our proposal can be viewed as a special case of [4] since it is based on a fragment of separation logic. However, we have succeeded in providing an automated procedure for cyclic proving under this fragment which is highly suited for program verification via forward reasoning.
- **Lemma Application.** Our program verifier can also apply the lemmas describing auxiliary relationships between predicates by automatically coercing one predicate to another, as needed. Coercion provides suitable transformations on formulas that facilitate *proof search* to enhance the capability of automated verification. Our coercion mechanism is goal-directed and terminates.

2 Examples

We now illustrate the usefulness of lemmas in program verification with an example which shows the ability of lemmas to provide alternative unfoldings and foldings of predicates, thereby providing different ways to reveal points-to facts not apparent in the original definitions of predicates. Let us consider the following class and predicate definitions.

$\texttt{class node} \; \{ \; \texttt{int val; node next} \}$

$\texttt{class node2} \; \{ \; \texttt{int val; node2 prev; node2 next} \}$

$\texttt{root::ll}\langle \texttt{s} \rangle \equiv \texttt{root=null} \wedge \texttt{s=0} \; \vee \; \exists \texttt{r} \cdot \texttt{root::node}\langle \_, \texttt{r} \rangle * \texttt{r::ll}\langle \texttt{s}-1 \rangle \; \textbf{inv} \; \texttt{s} \geq 0;$

$\texttt{root::dsegN}\langle \texttt{s}, \texttt{p}, \texttt{n}, \texttt{t} \rangle \equiv \texttt{root=n} \wedge \texttt{p=t} \wedge \texttt{s=0} \; \vee$
$\quad \exists \texttt{r} \cdot \texttt{root::node2}\langle \_, \texttt{p}, \texttt{r} \rangle * \texttt{r::dsegN}\langle \texttt{s}-1, \texttt{root}, \texttt{n}, \texttt{t} \rangle \; \textbf{inv} \; \texttt{s} \geq 0;$

$\texttt{root::dcl}\langle \texttt{s} \rangle \equiv \texttt{root=null} \wedge \texttt{s=0} \; \vee$
$\quad \exists \texttt{r}_1, \texttt{r}_2 \cdot \texttt{root::node2}\langle \_, \texttt{r}_1, \texttt{r}_2 \rangle * \texttt{r}_2\texttt{::dsegN}\langle \texttt{s}-1, \texttt{root}, \texttt{root}, \texttt{r}_1 \rangle \; \textbf{inv} \; \texttt{s} \geq 0;$

Predicate `ll` defines a linear-linked list of length `s`. Predicate `dsegN`, adopted from [11], defines a doubly-linked list segment. Parameter `s` denotes its length, while `p` is the dangling `prev` field of the first element, `n` is the dangling `next` field of the last element which is also pointed to by `t`. The `dcl` predicate defines a circular list by making the dangling pointer of the `dsegN` predicate point to the same distinguished `root` node, thereby making a cycle.

Details of our specification language is given in Sec 3. Briefly, each predicate describes a data structure, which is a collection of objects reachable from a base pointer denoted by `root` in the predicate definition. `root` also serves as an implicit argument of the predicate. The expression after **inv** keyword captures a *pure*, i.e. *heap-independent*, formula that always holds for the given predicate. Formula $p{::}c\langle v^* \rangle$ denotes either a points-to fact if `c` is a class name, or an instance of predicate `c` with `p`, v^* as its arguments, where `p` is the actual argument for `root` and v^* are arguments for the explicit parameters.

The `dsegN` predicate, by its definition, favors one direction of linkage. Traversing the list in a forward manner by following the `next` field is naturally supported by the definition with unfold/fold reasoning. However, traversing the list in a backward manner using the `prev` field is not as easily done. The problem manifests itself in, for example, the following `delete` procedure for a circular doubly-linked list. The procedure deletes the element pointed by `x` and updates `x`. The precondition requires the circular list to be non-empty, and the postcondition asserts that the updated `x` points to a circular list with one fewer element.

```
1  void delete(ref node2 x)
2      requires x::dcl⟨s⟩ ∧ s ≥ 1
3      ensures x'::dcl⟨s−1⟩;
4  {
5      if (x.next == x) x = null;
6      else {
7          // x::node2⟨_, r1, r2⟩ ∗ r2::dsegN⟨s − 1, x, x, r1⟩ ∧ s ≥ 2
8          node tmp = x.prev;
9          // x::node2⟨_, r1, r2⟩ ∗ r2::dsegN⟨s − 1, x, x, r1⟩ ∧ tmp = r1 ∧ s ≥ 2
10         tmp.next = x.next;
11         // x::node2⟨_, r1, r2⟩ ∗ r2::dsegN⟨s − 2, x, r1, r3⟩ ∗ r1::node2⟨_, r3, r2⟩
12         //    ∧ tmp = r1 ∧ s ≥ 2
13         x.next.prev = x.prev;
14         //    x::node2⟨_, r1, r2⟩ ∗ r1::node2⟨_, r1, r2⟩ ∧ r2 = r1 ∧ x = r3 ∧ s = 2 ∧ tmp = r1
15         // ∨ x::node2⟨_, r1, r2⟩ ∗ r2::node2⟨_, r3, r4⟩ ∗ r4::dsegN⟨s − 3, r2, r1, r3⟩
16         //       ∗ r1::node2⟨_, r3, r2⟩ ∧ s ≥ 3
17         x = x.next ; }}
```

Fig. 1. Delete from circular list

For exposition purpose, intermediate program states are given as comments (after //) in the code, though they are automatically derived from the initial precondition. To verify the assignment to `tmp.next` at line `10`, the program verifier requires an explicit points-to fact $\mathtt{tmp}{::}\mathtt{node2}\langle \_, \_, \_ \rangle$. This is enforced by the following entailment where Φ_R is inferred.

$$\texttt{x::node2}\langle\_,\texttt{r}_1,\texttt{r}_2\rangle * \texttt{r}_2\texttt{::dsegN}\langle\texttt{s}-1,\texttt{x},\texttt{x},\texttt{r}_1\rangle \wedge \texttt{tmp}{=}\texttt{r}_1 \wedge \texttt{s}{\geq}2 \\ \vdash \texttt{tmp::node2}\langle\_,\_,\_\rangle * \Phi_R$$

This proof obligation is challenging for reasoning based on unfolding and folding of inductive definitions [16], since the dsegN predicate does not explicitly state that the parameter t points to an object when the data structure is non-empty. Fortunately, the problem can be solved by adopting the following two-way equivalence lemma.

$$\texttt{root::dsegN}\langle\texttt{s},\texttt{p},\texttt{n},\texttt{t}\rangle\wedge\texttt{s}{>}0 \leftrightarrow \exists\texttt{r}{\cdot}\texttt{root::dsegN}\langle\texttt{s}-1,\texttt{p},\texttt{t},\texttt{r}\rangle{*}\texttt{t::node}\langle\_,\texttt{r},\texttt{n}\rangle \quad (1)$$

3 Specification Language

Figure 2 shows the grammar for our specification language that has been mostly adopted from [16] except for lemma specifications. Shape predicate spred is the main specification construct that provides data structure descriptions. Formulas are compiled to an internal representation in which arguments for heap formulas are distinct and fresh. Additional existentially quantified variables are introduced if necessary to obtain this normal form.

$$\begin{array}{rrl}
\textit{Predicate} & \textsf{spred} & ::= [\texttt{root::}]c\langle v^*\rangle \equiv \Phi\ [\textbf{inv}\ \pi] \\
\textit{Formula} & \Phi & ::= \bigvee \exists v^* \cdot (\kappa \wedge \pi) \\
\textit{Pure form.} & \pi & ::= \gamma \wedge \phi \\
\textit{Pointer form.} & \gamma & ::= v_1 = v_2 \mid v = \textbf{null} \mid v_1 \neq v_2 \mid v \neq \textbf{null} \mid \gamma_1 \wedge \gamma_2 \\
\textit{Heap form.} & \kappa & ::= \textbf{emp} \mid v{::}c\langle v^*\rangle \mid \kappa_1 * \kappa_2 \\
\textit{Presburger arith.} & \phi & ::= \textsf{arith} \mid \phi_1 \wedge \phi_2 \mid \phi_1 \vee \phi_2 \mid \neg\phi \mid \exists v \cdot \phi \mid \forall v \cdot \phi \\
& \textsf{arith} & ::= \textsf{a}_1 = \textsf{a}_2 \mid \textsf{a}_1 \neq \textsf{a}_2 \mid \textsf{a}_1 < \textsf{a}_2 \mid \textsf{a}_1 \leq \textsf{a}_2 \\
& \textsf{a} & ::= k \mid v \mid k \times \textsf{a} \mid \textsf{a}_1 + \textsf{a}_2 \mid -\textsf{a} \mid \textbf{max}(\textsf{a}_1,\textsf{a}_2) \mid \textbf{min}(\textsf{a}_1,\textsf{a}_2) \\
\textit{Lemma} & L & ::= H \wedge G \bowtie B \\
\textit{Complex Lemma} & L' & ::= \forall v^* \cdot ((H * E) \wedge G \rightarrow B) \\
\textit{Head} & H & ::= [\texttt{root::}]c\langle v^*\rangle \\
\textit{ExtraHeap} & E & ::= \kappa \\
\textit{Body} & B & ::= \Phi \\
\textit{Guard} & G & ::= \pi \\
& \bowtie & ::= \rightarrow \mid \leftarrow \mid \leftrightarrow \\
& k & \in \text{Integer constants} \\
& v,c & \in \text{Identifiers}
\end{array}$$

Fig. 2. Grammar for Shape Predicates and Lemmas

Recursive shape predicate definitions need to satisfy certain syntactic restrictions, namely *well-formed* and *well-founded* conditions, to ensure soundness and termination of static reasoning. *Well-formed* conditions ensure that shape predicates and formulas do not admit garbage. They thus disallow predicates such as $\texttt{root::p}\langle\rangle \equiv \exists\texttt{x} \cdot \texttt{root::node}\langle\_,\_\rangle * \texttt{x::node}\langle\_,\_\rangle$ where $\texttt{x::node}\langle\_,\_\rangle$ is garbage as it is inaccessible from the free variables. *Well-founded* conditions disallow root to be passed as argument to a recursive predicate invocation. That means `root` either is `null`, dangles, or points to an object which ensures a decreasing heap with each recursive predicate instance.

We now describe a special class of lemmas L allowed by our new specification language. Each L lemma consists of a *head* H and a *body* B. The head H is a single predicate. The *guard* G is a pure formula whose variables are solely from H, which can be omitted if it is `true`. The body B is a formula in separation logic. The direction $\bowtie$ of a lemma constrains its applicability. The lemmas are divided into three groups, namely : (i) *weakening* lemmas using $\rightarrow$, (ii) *strengthening* lemmas using $\leftarrow$, and (iii) *equivalence* lemmas using $\leftrightarrow$. We expect lemmas to be *well-formed* and *well-founded*, but allows the `root` parameter to reference a predicate. These lemmas have a similar format as user-defined predicates and can therefore be handled by the same unfold/fold mechanism of our prover, except that it can be goal-directed.

However, we are also interested to support lemmas with more general LHS and with universally quantified variables in the guard. These more complex lemmas are captured by L' in Fig 2 as a weakening lemma. There is no need to consider a strengthening version of complex lemma since it can be converted to L'-form by swapping the two sides. To illustrate the utility of complex lemma, consider a list segment predicate below:

$$\texttt{root::lseg}\langle\texttt{p},\texttt{s}\rangle \equiv \texttt{root=p} \wedge \texttt{s=0} \vee \exists \texttt{r}\cdot\texttt{root::node}\langle\_,\texttt{r}\rangle * \texttt{r::lseg}\langle\texttt{p},\texttt{s}-1\rangle \ \mathbf{inv}\ \texttt{s}\geq 0;$$

One simple L-form lemma to support list segment breaking and joining is:

$$\texttt{root::lseg}\langle\texttt{p},\texttt{n}\rangle \leftrightarrow \exists \texttt{a},\texttt{b},\texttt{r}\cdot\texttt{root::lseg}\langle\texttt{r},\texttt{a}\rangle * \texttt{r::lseg}\langle\texttt{p},\texttt{b}\rangle \wedge \texttt{n=a+b} \wedge \texttt{a,b}\geq 0$$

However, this lemma cannot support entailment proving that requires the capture of size properties for broken segments, such as the following:

$$\texttt{x::lseg}\langle\texttt{p},\texttt{n}\rangle \wedge \texttt{n=8} \vdash \exists \texttt{r}\cdot\texttt{x::lseg}\langle\texttt{r},\texttt{a}\rangle * \texttt{r::lseg}\langle\texttt{p},\texttt{b}\rangle \wedge \texttt{a=2} \wedge \texttt{b=6} * \Phi_R$$

To support the above entailment, we require a more general L'-form lemma where some variables in the guard, such as `a` and `b`, are universally quantified, as follows:

$$\forall \texttt{a},\texttt{b}\cdot(\texttt{root::lseg}\langle\texttt{p},\texttt{n}\rangle \wedge \texttt{n=a+b} \wedge \texttt{a,b}\geq 0 \rightarrow \exists \texttt{r}\cdot\texttt{root::lseg}\langle\texttt{r},\texttt{a}\rangle * \texttt{r::lseg}\langle\texttt{p},\texttt{b}\rangle)$$

Such lemmas allow universally quantified variables to be instantiated which can crucially increase the expressive power of our entailment prover. They can be provided for the list segment with length property, but not for the list segment with bag of values property. Furthermore, there are also lemmas with multiple predicates on the LHS. An example of this was used in [1] for a decidable fragment of separation logic to safely break a class of non-touching list segments. (Our thanks to Peter O'Hearn for highlighting the importance of complex lemmas to us.)

4 Entailment

Given formulas Φ_1 and Φ_2, our entailment prover checks if Φ_1 entails Φ_2, that is if in any heap satisfying Φ_1, we can find a subheap satisfying Φ_2. Moreover, we determine a formula Φ_R for the residue heap state which captures the frame condition. Formally, our entailment relation is defined as follows:

Definition 4.1 (Entailment). *A formula Φ_1 entails a formula Φ_2 with residue Φ_R iff*

$$\forall s, h_1 \cdot s, h_1 \models \Phi_1 \Rightarrow \exists h_2, h_R \cdot h_1 = h_2 * h_R \wedge s, h_2 \models \Phi_2 \wedge s, h_R \models \Phi_R$$

The main features of our entailment prover are that, besides determining if the above relation holds, it also *infers* the *residual heap* of the entailment, that is a formula Φ_R such that $s, h_R \models \Phi_R$ and *derives* the predicate parameters. These two features are important for program verification tasks using forward analysis. The relation is formalized using judgment of the form where κ denotes the consumed heap and V is the set of existential variables encountered :

$$\Phi_1 \vdash^{\kappa}_{V} \Phi_2 * \Phi_R$$

A sound and terminating proof system for the above entailment relation is presented in [16]. That system relies on unfolding and folding of the predicate definitions to compute the subheap of Φ_1 that matches Φ_2 and the residue Φ_R. In the current paper, additional proof rules that handle user-supplied lemmas shall be presented which greatly enhances our entailment prover. We provide a re-cap on the unfold/fold mechanisms.

We apply an *unfold* operation on a predicate in the antecedent that matches with an object in the consequent. For instance, when checking:

$$\texttt{x::ll}\langle\texttt{n}\rangle \wedge \texttt{n}{>}3 \vdash (\exists \texttt{r} \cdot \texttt{x::node}\langle\_, \texttt{r}\rangle * \texttt{r::node}\langle\_, \texttt{y}\rangle \wedge \texttt{y}{\neq}\texttt{null}) * \Phi_R$$

where Φ_R is the residue, we unfold the $\texttt{x::ll}\langle\texttt{n}\rangle$ heap formula in the antecedent twice to match two objects in the consequent. This results in the following reductions towards a residual state:

$$\begin{array}{ll}
\exists \texttt{q}_1 \cdot \texttt{x::node}\langle_, \texttt{q}_1\rangle * \texttt{q}_1\texttt{::ll}\langle\texttt{n}{-}1\rangle \wedge \texttt{n}{>}3 & \vdash (\exists \texttt{r} \cdot \texttt{x::node}\langle_, \texttt{r}\rangle * \texttt{r::node}\langle_, \texttt{y}\rangle \wedge \texttt{y}{\neq}\texttt{null}) * \Phi_R \\
\texttt{q}_1\texttt{::ll}\langle\texttt{n}{-}1\rangle \wedge \texttt{n}{>}3 & \vdash (\texttt{q}_1\texttt{::node}\langle_, \texttt{y}\rangle \wedge \texttt{y}{\neq}\texttt{null}) * \Phi_R \\
\exists \texttt{q}_2 \cdot \texttt{q}_1\texttt{::node}\langle_, \texttt{q}_2\rangle * \texttt{q}_2\texttt{::ll}\langle\texttt{n}{-}2\rangle \wedge \texttt{n}{>}3 & \vdash \texttt{q}_1\texttt{::node}\langle_, \texttt{y}\rangle \wedge \texttt{y}{\neq}\texttt{null} * \Phi_R \\
\texttt{q}_2\texttt{::ll}\langle\texttt{n}{-}2\rangle \wedge \texttt{n}{>}3 \wedge \texttt{q}_2{=}\texttt{y} & \vdash \texttt{y}{\neq}\texttt{null} * \Phi_R
\end{array}$$

We apply a *fold* operation when an object in the antecedent is aliased with a predicate in the consequent. An example is:

$$\texttt{x::node}\langle 1, \texttt{q}_1\rangle * \texttt{q}_1\texttt{::node}\langle 2, \texttt{null}\rangle * \texttt{y::node}\langle 3, \texttt{null}\rangle \vdash \texttt{x::ll}\langle\texttt{n}\rangle \wedge \texttt{n}{>}1 * \Phi_R$$

The fold step may be recursively applied but is guaranteed to terminate for well-founded predicates. Furthermore, the fold operation may introduce bindings for free parameters of the folded predicate. In the above, we obtain $\texttt{n}{=}2$ which may be transferred to the antecedent since $\texttt{n}$ is free. This allows our folding step to finally derive $\texttt{y::node}\langle 3, \texttt{null}\rangle \wedge \texttt{n}{=}2 \vdash \texttt{n}{>}1 * \Phi_R$ from which we will obtain $\Phi_R = \texttt{y::node}\langle 3, \texttt{null}\rangle \wedge \texttt{n}{=}2$.

5 Lemma Application

User-supplied lemmas are *proved* and *applied* to support sound proof search by the entailment prover. Since the proof of a lemma may apply the lemma itself inductively, we first present the proof rules that apply lemmas. Depending on whether the lemma is applied to the antecedent or the consequent of the entailment, our entailment prover treats it as an unfolding or folding, respectively. $\leftarrow$ lemmas can be applied to only the consequent of an entailment, $\rightarrow$ to only the antecedent, and $\leftrightarrow$ to both.

5.1 Weakening the Antecedent by Lemma Unfolding

A lemma $H \wedge G \bowtie B$ where $\bowtie$ is $\rightarrow$ or $\leftrightarrow$ can be seen as an alternative way to unfold a predicate. Its application is formalized below which says that the lemma is applied if we can find a substitution ρ that matches H to $p_1{::}c_1\langle v_1^*\rangle$ and satisfies the guard.

$$\frac{\begin{array}{c}\boxed{\mathbf{L-LEFT}}\\ \mathit{IsPred}(c_1) \qquad p_1{::}c_1\langle v_1^*\rangle * \kappa_1 \wedge \pi_1 \vdash \rho G\\ \rho = \mathit{match}(H, p_1{::}c_1\langle v_1^*\rangle) \quad (\rho B) * \kappa_1 \wedge \pi_1 \vdash_V^{\kappa * p_1::c_1\langle v_1^*\rangle} (\kappa_2 \wedge \pi_2) * \Phi\end{array}}{p_1{::}c_1\langle v_1^*\rangle * \kappa_1 \wedge \pi_1 \vdash_V^{\kappa} (\kappa_2 \wedge \pi_2) * \Phi}$$

where $\Phi \vdash \pi$ checks if guard π holds under Φ, and *match* is defined as:

$$\mathit{match}(p_1{::}c\langle v_1^*\rangle, p_0{::}c\langle v_0^*\rangle) \stackrel{\text{def}}{=} [p_1 \mapsto p_0, v_1^* \mapsto v_0^*]$$

For a goal-directed lemma application, we shall only apply this rule when there exists a predicate $\mathtt{p_2}{::}\mathtt{c_2}\langle \mathtt{v_2^*}\rangle \in \kappa_2$ in the consequent that would (subsequently) match up via aliasing with a $\mathtt{p_2}{::}\mathtt{c_2}\langle \mathtt{v_3^*}\rangle$ in the RHS of lemma $\rho\mathtt{B}$ where $\mathtt{p_2} \in \{\mathtt{p_1}, \mathtt{v_1^*}\}$.

We now show how a lemma can help verify the `delete` procedure, in particular during an assignment to the `prev` field of the `tmp` object at line `10`. As part of the verification, the following entailment needs to be checked, where the antecedent denotes program state at that program point.

$$\begin{array}{l} \quad \mathtt{x{::}node2}\langle \_, \mathtt{r_1}, \mathtt{r_2}\rangle * \mathtt{r_1{::}dsegN}\langle \mathtt{r_3}, \_, \_, \mathtt{r_2}\rangle \wedge \mathtt{tmp} = \mathtt{r_2}\\ \quad \wedge\, \mathtt{r_3} = \mathtt{s} - 1 \wedge \mathtt{s} > 1 \vdash \mathtt{tmp{::}node2}\langle \_, \_, \_\rangle * \Phi_R\\ \rightsquigarrow (\boxed{\mathbf{L-LEFT}})\\ \quad \mathtt{x{::}node2}\langle \_, \mathtt{r_1}, \mathtt{r_2}\rangle * \mathtt{r_1{::}dsegN}\langle \mathtt{r_4}, \_, \mathtt{r_2}, \_\rangle * \mathtt{r_2{::}node2}\langle \_, \_, \_\rangle\\ \quad \wedge\, \mathtt{tmp} = \mathtt{r_2} \wedge \mathtt{r_4} = \mathtt{r_3} - 1 \wedge \mathtt{r_3} = \mathtt{s} - 1 \wedge \mathtt{s} > 1 \vdash \mathtt{tmp{::}node2}\langle \_, \_, \_\rangle * \Phi_R\\ \rightsquigarrow (\boxed{\mathbf{ENT-MATCH}})\\ \quad \mathit{Success}\end{array}$$

After the above goal-directed lemma application, we can reveal a match up between $\mathtt{r_2{::}node2}\langle \_, \_, \_\rangle$ (from the lemma) and $\mathtt{tmp{::}node2}\langle \_, \_, \_\rangle$ (from the consequent), before successfully proving the entailment.

Our proposal also handles the more complex lemma form: $\forall v^* \cdot (H * E \wedge G \rightarrow B)$. We have designed and implemented it as follows:

$$\frac{\begin{array}{c}\boxed{\mathbf{L-LEFT-COMPLEX}}\\ \mathit{IsPred}(c_1) \qquad \rho = \mathit{match}(H, p_1{::}c_1\langle v_1^*\rangle)\\ \kappa_1 \wedge \pi_1 \vdash_V^{\kappa * p_1::c_1\langle v_1^*\rangle} \rho E * \Phi_1\\ \rho B * ([(v \mapsto ?)^*] \Vdash \rho G) * \Phi_1 \wedge \pi_1 \vdash_V^{\kappa * p_1::c_1\langle v_1^*\rangle} (\kappa_2 \wedge \pi_2) * \Phi\end{array}}{p_1{::}c_1\langle v_1^*\rangle * \kappa_1 \wedge \pi_1 \vdash_V^{\kappa} (\kappa_2 \wedge \pi_2) * \Phi}$$

To support the above proof rule, we provide a new delayed guard $([(v \mapsto ?)^*] \Vdash \rho G)$ that is used to support the instantiations of v^* when the body ρB is being matched by our entailment procedure. Once v^* have been instantiated, we test the guard G before its instantiations are added to the antecedent. The use of lemmas with universal variables, where possible, allows stronger proofs to be asserted than what is possible using corresponding lemmas with existentially quantified variables. In our approach, this is realised by a novel instantiation mechanism from the delayed guard construct.

5.2 Strengthening the Consequent by Lemma Folding

A lemma $H \wedge G \bowtie B$ where $\bowtie$ is $\leftarrow$ or $\leftrightarrow$ provides an alternative way to fold a predicate. Its application is formalized as follows:

$$\frac{\begin{array}{c}\textit{IsPred}(c_2) \qquad \rho = \textit{match}(H, p_2{::}c_2\langle v_2^*\rangle) \qquad \kappa_1 \wedge \pi_1 \vdash \rho G \\ (\Phi^r, \kappa^r, \pi^r) \in \textit{foldL}^{\kappa}(\kappa_1 \wedge \pi_1, p_2{::}c_2\langle v_2^*\rangle, \rho B) \\ (\pi^a, \pi^c) = \textit{split}_V^{\{v_2^*\}}(\pi^r) \qquad \Phi^r \wedge \pi^a \vdash_V^{\kappa^r} (\kappa_2 \wedge \pi_2 \wedge \pi^c) * \Phi\end{array}}{\kappa_1 \wedge \pi_1 \vdash_V^{\kappa} (p_2{::}c_2\langle v_2^*\rangle * \kappa_2 \wedge \pi_2) * \Phi}\ \boxed{\textbf{L-RIGHT}}$$

foldL performs folding using a lemma instead of the body of a predicate.

$$\frac{W_i = V_i - \{v^*, p\} \qquad \kappa \wedge \pi \vdash_{\{p,v^*\}}^{\kappa'} \rho B * \{(\Phi_i, \kappa_i, V_i, \pi_i)\}_{i=1}^n}{\textit{foldL}^{\kappa'}(\kappa \wedge \pi, p{::}c\langle v^*\rangle, \rho B) \stackrel{\text{def}}{=} \{(\Phi_i, \kappa_i, \exists W_i \cdot \pi_i)\}_{i=1}^n}\ \boxed{\textbf{L-FOLD}}$$

Note that the folding function *foldL* uses a specialized entailment checking procedure. The checker returns a set of quadruples $(\Phi^r, \kappa^r, V, \pi^r)$, each being the result of a successful folding against a disjunct of the predicate definition or the lemma body. The meaning of each component of a triple is as following:

- Φ^r is the residue (frame) not consumed by the folded disjunct.
- κ^r is the part of the heap consumed by the folded disjunct. By definition, $\kappa^r * \Phi^r$ equals the heap in the first argument of $foldL$.
- V is the set of existential variables generated from unfoldings of the predicate definition.
- π^r is the pure constraint of the folded disjunct. It is used to obtain information, such as bindings to values, for predicate parameters. This information is especially useful for forward verification.

This use of a *set of states* can be generalized to the entire system which results in entailment proving of the form $\Phi_{\mathtt{A}} \vdash \Phi_{\mathtt{C}} * \mathtt{S}$ that has been implemented in our tool. Here, S denotes a set of residual heap states that arise from proof search for successful entailment. Failure of entailment is denoted by S=$\{\}$, while multiple answers denote alternative successful outcomes of entailment with the respective residual heaps. Proof search (with the help of lemmas) increases the expressivity of our verifier.

5.3 An Example of Entailment with Lemma Capability

An interesting application of lemma involves the list-with-tail predicate which is defined as follows:

$$\begin{array}{l}\mathtt{root{::}ll\_tail\langle tx, n\rangle} \equiv \mathtt{root{::}node\langle \_, null\rangle \wedge n = 1 \wedge tx = root} \\ \quad \vee\ \mathtt{root{::}node\langle \_, r\rangle * r{::}ll\_tail\langle tx, n-1\rangle} \quad \textbf{inv}\ \mathtt{n \geq 1}\end{array}$$

The predicate captures a list of n objects, with tx pointing to the last one. It can be coerced to a list segment, and vice versa, via the lemma:

$$\mathtt{root{::}ll\_tail\langle tx, n\rangle} \leftrightarrow \mathtt{root{::}lseg\langle tx, n-1\rangle * tx{::}node\langle \_, null\rangle} \qquad (2)$$

By applying this lemma, our verifier can easily prove the following specification for the concatenation of two lists with tail pointers:

$$
\begin{array}{l}
\{\texttt{x::ll\_tail}\langle\texttt{tx},\texttt{n}\rangle * \texttt{y::ll\_tail}\langle\texttt{ty},\texttt{m}\rangle\} \\
\texttt{tx.next} = \texttt{y}; \\
\{\texttt{x::ll\_tail}\langle\texttt{ty},\texttt{m}+\texttt{n}\rangle\}
\end{array}
$$

Separation logic semantics requires $\texttt{tx::node}\langle\_,\_\rangle$ to be present in the program state in order to safely perform the dereference operation via `tx.next`. Such an object can be exposed via an unfolding of the `ll_tail` predicate using the lemma, resulting in the following program state prior to the assignment:

$\{\texttt{x::lseg}\langle\texttt{tx},\texttt{n}-1\rangle * \texttt{tx::node}\langle\_,\texttt{null}\rangle * \texttt{y::ll\_tail}\langle\texttt{ty},\texttt{m}\rangle\}$

which is then updated by the assignment to:

$\{\texttt{x::lseg}\langle\texttt{tx},\texttt{n}-1\rangle * \texttt{tx::node}\langle\_,\texttt{y}\rangle * \texttt{y::ll\_tail}\langle\texttt{ty},\texttt{m}\rangle\}$

The weakening on the postcondition is done via an entailment, whose proof is sketched below. This proof is performed automatically by our system.

$$
\dfrac{
\dfrac{
\dfrac{
\dfrac{\left(\begin{array}{l}\text{recursive entailment}\\ \text{described below}\end{array}\right)}{\begin{array}{l}\texttt{tx::node}\langle\_,\texttt{y}\rangle * \texttt{y::ll\_tail}\langle\texttt{ty},\texttt{m}\rangle\\ \vdash \texttt{tx::lseg}\langle\texttt{ty},\texttt{m}\rangle\\ \quad *\{\texttt{ty::node}\langle\_,\texttt{null}\rangle\}\end{array}}\,(\underline{\textbf{FOLD}})
\quad
\dfrac{\left(\begin{array}{l}\text{match ty with}\\ \text{residue from fold}\end{array}\right)}{\begin{array}{l}\texttt{ty::node}\langle\_,\texttt{null}\rangle\\ \vdash \texttt{ty::node}\langle\_,\texttt{null}\rangle\\ \quad *\{\texttt{emp}\}\end{array}}
}{\begin{array}{l}\texttt{tx::node}\langle\_,\texttt{y}\rangle * \texttt{y::ll\_tail}\langle\texttt{ty},\texttt{m}\rangle\\ \vdash \texttt{tx::lseg}\langle\texttt{ty},\texttt{m}\rangle * \texttt{ty::node}\langle\_,\texttt{null}\rangle * \{\texttt{emp}\}\end{array}}\,(\underline{\textbf{FOLD}})
}{\begin{array}{l}\texttt{x::lseg}\langle\texttt{tx},\texttt{n}-1\rangle * \texttt{tx::node}\langle\_,\texttt{y}\rangle * \texttt{y::ll\_tail}\langle\texttt{ty},\texttt{m}\rangle\\ \vdash \texttt{x::lseg}\langle\texttt{ty},\texttt{m}+\texttt{n}-1\rangle * \texttt{ty::node}\langle\_,\texttt{null}\rangle * \{\texttt{emp}\}\end{array}}\,(\underline{\textbf{L-RIGHT}})
}{\begin{array}{l}\texttt{x::lseg}\langle\texttt{tx},\texttt{n}-1\rangle * \texttt{tx::node}\langle\_,\texttt{y}\rangle * \texttt{y::ll\_tail}\langle\texttt{ty},\texttt{m}\rangle\\ \vdash \texttt{x::ll\_tail}\langle\texttt{ty},\texttt{m}+\texttt{n}\rangle * \{\texttt{emp}\}\end{array}}\,(\underline{\textbf{L-RIGHT}})
$$

Our entailment prover first converts the list with tail pointer in the consequent to a list segment and a node. It then breaks the list segment into two and match the first segment with the aliased segment in the antecedent. Subsequently, it performs a fold on a $\texttt{tx::lseg}\langle\texttt{ty},\texttt{m}\rangle$ predicate which invokes a recursive entailment, as follows:

$$
\dfrac{
\dfrac{
\dfrac{
\dfrac{\text{(derive residue)}}{\texttt{ty::node}\langle\_,\texttt{null}\rangle \vdash \texttt{emp} * \{\texttt{ty::node}\langle\_,\texttt{null}\rangle\}}
}{\begin{array}{l}\texttt{y::lseg}\langle\texttt{ty},\texttt{m}-1\rangle * \texttt{ty::node}\langle\_,\texttt{null}\rangle\\ \vdash \texttt{y::lseg}\langle\texttt{ty},\texttt{m}-1\rangle * \{\texttt{ty::node}\langle\_,\texttt{null}\rangle\}\end{array}}\,(\underline{\textbf{MATCH}})
}{\texttt{y::ll\_tail}\langle\texttt{ty},\texttt{m}\rangle \vdash \texttt{y::lseg}\langle\texttt{ty},\texttt{m}-1\rangle * \{\texttt{ty::node}\langle\_,\texttt{null}\rangle\}}\,(\underline{\textbf{L-LEFT}})
}{\begin{array}{l}\texttt{tx::node}\langle\_,\texttt{y}\rangle * \texttt{y::ll\_tail}\langle\texttt{ty},\texttt{m}\rangle\\ \vdash (\exists \texttt{r}\cdot \texttt{tx::node}\langle\_,\texttt{r}\rangle * \texttt{r::lseg}\langle\texttt{ty},\texttt{m}-1\rangle) * \{\texttt{ty::node}\langle\_,\texttt{null}\rangle\}\end{array}}\,(\underline{\textbf{MATCH}})
}{\begin{array}{l}\texttt{tx::node}\langle\_,\texttt{y}\rangle * \texttt{y::ll\_tail}\langle\texttt{ty},\texttt{m}\rangle\\ \vdash \texttt{tx::lseg}\langle\texttt{ty},\texttt{m}\rangle * \{\texttt{ty::node}\langle\_,\texttt{null}\rangle\}\end{array}}\,(\underline{\textbf{FOLD}})
$$

Such applications of lemmas are critical for automatically deriving non-trivial proofs to support program verification.

5.4 Termination

To prevent non-termination during lemma applications, we assign a history to each heap constraint $p{::}c\langle v^*\rangle$ where c is a predicate name. The history is a set of predicate names which are transitively rewritten to $p{::}c\langle v^*\rangle$. Lemma application is possible only if it does not rewrite a predicate to some predicate already in the former's history. Initially the history is empty. After each predicate application, the predicate name in the head H is added to the history of each and every predicate $p{::}c\langle v^*\rangle$ in the body ρB, in addition to the history of the matching predicate instance $p{::}c\langle v^*\rangle$. Folding and unfolding predicate instances pass the predicate history on to the predicate instances in the body.

Theorem 5.1 (Termination). *Entailment proving is terminating, even in the presence of lemma applications.*

Proof Sketch. Termination is guaranteed by the fact that only a finite number of lemma applications can occur when proving an entailment. This is the case since there is a finite number of lemmas, and each predicate instance maintains a history of predicates that are rewritten by lemma applications to the current predicate instance. Therefore lemma applications cannot occur after a finite number of steps in the entailment checking process. Termination is then guaranteed by the entailment checking as in [16].

6 Lemma Proving

Correctness of lemmas is *automatically proved* by our system via the entailment prover. A weakening lemma is proved by showing that the predicate in the head of the lemma entails the body. A strengthening lemma needs an entailment in the reverse direction. An equivalence lemma needs both. During this entailment proving, the lemma being proved can be soundly used in the proof itself as an instance of cyclic proof. Formally, proving $\rightarrow$ and $\leftrightarrow$ lemmas amount to discharging the following proof obligation:

$$\mathit{unfold}(H * E \wedge G, \texttt{root}) \vdash B * \texttt{emp} \tag{3}$$

whereas $\leftarrow$ and $\leftrightarrow$ generate the following obligation:

$$\mathit{unfold}(B, \texttt{root}) \vdash (H * E \wedge G) * \texttt{emp} \tag{4}$$

At the start of lemma proving, we always unfold the head predicate in the antecedent. This ensures that *infinite descent* occurs for the resulting cyclic proof which guarantees a progress condition needed for sound induction. During lemma proving, the lemma being proved may be applied to the unfolded formulas as an instance of cyclic proving. Furthermore, we also check that the entailment derives an **empty** residual heap. This ensures that both sides of the lemma cover the same heap region.

7 Implementation

We have built a prototype system using Objective Caml. The proof obligations generated by this verification system are discharged by our entailment proving procedure with the help of Omega Calculator [21] and CVC [23]. These two arithmetic solvers have complementary strengths. In many cases, CVC Lite is faster; but Omega is more complete. We therefore run both of them and get the timing of the first returning prover; or use Omega's when CVC Lite fails.

Programs	LOC	Timing	
		with lemmas	without lemmas
List with Tail		verifies size/length	
append	1	0.18	*failed*
Circular Linked List		verifies size + circularity	
delete_first	15	0.07	0.04
count	15	0.13	*failed*
Doubly Linked Circular List		verifies size + double links + circularity	
delete	12	0.26	*failed*
Doubly Linked List		verifies size + double links	
append	26	0.16	0.12
flatten (from tree)	34	0.35	0.33
Sorted List		verifies size + min + max + sortedness	
delete	21	0.16	0.15
insertion_sort	36	0.37	0.32
selection_sort	52	0.34	0.31
bubble_sort	42	0.64	*failed*
merge_sort	105	0.61	0.56
quick_sort	85	0.67	0.65
File Manager		verifies directory structure	
search_name	18	1.71	1.49
mkdir	43	3.02	*failed*
remove	50	4.66	*failed*
copy_folder	67	7.50	*failed*
AVL Tree		verifies size + height + height-balanced	
insert	169	5.06	5.00
Red-Black Tree		verifies size + black-height + height-balanced	
insert	167	1.53	1.39

Fig. 3. Verification Times (in seconds) for Data Structures with Arithmetic Constraints

We tested our system on a suite of examples summarized in Figure 3. These examples are small but handle data structures with sophisticated shape and size properties such as sorted lists, balanced trees, etc. in a uniform way. Verification time for each function includes time to verify all functions that it calls. We compare the timings obtained with and without lemmas. Lemma proving time is not included, since they are proven once and applied many times. Preliminary results indicate that proof search with lemmas does not incur much overhead due to the directed nature of search. On the other hand, lemmas are important to verify a number of examples that would fail otherwise. For example, the bubble-sort algorithm requires sorted list to be coerced into an unsorted list expected for its precondition, whenever a swap has occurred for the bubble procedure. Also, the file manager traverses its doubly-linked lists in two directions. while circular lists are built using list segments that may require breaking and joining.

8 Related Work and Concluding Remarks

The general framework of separation logic is highly expressive but undecidable. Thus, in the search for a decidable fragment of separation logic, Berdine *et al.* [1] supports only a limited set of lemmas and predicates *without* size properties, disjunctions and existential quantifiers. This fragment forms the basis of a program verifier called Smallfoot [2]. Jia and Walker [13] also identified a decidable logic but without recursive predicates for automated reasoning of pointer programs. Preoteasa [20] showed that separation logic rules such as the frame rule are correct with respect to the predicate transformer semantics for a language with recursive procedures, local variables, value and value-result parameters via the PVS theorem prover [18]. Marti et. al. [15] verifed the heap manager of a small embedded operating system, while Feng et. al. [9] showed how the effects of interrupts and thread preemptions can be soundly modelled through ownership transfers. These approaches are based on separation logic but currently require hand-written Coq proofs. Separation logic has also been used to automatically reason about heap-manipulating programs in various contexts, e.g. locality [8], termination [3], concurrency [19]. Similar to [1], most of these works only support a limited predefined set of predicates and lemmas. Our recent work [16] allowed user-specified inductive predicates in separation logic, which are then automatically verified via a sound, terminating but incomplete verification system. Building on this prior work, the current paper proposes a new mechanism based on *user-specified lemmas* that can be *automatically proven* and *applied* by our program verifier. This feature can greatly enhance the capability of our automated program verification system, and is an important step towards building a more complete program verifier. Compared to traditional theorem provers, like Isabelle [17], our approach attained the following improvements: 1) it is based on separation logic (not classical logic), 2) it is automatically proven (via cyclic proof), 3) it is automatically applied (during entailment), and 4) it always terminates. In contrast, traditional theorem prover handles lemmas (for classical logic) using either user-specifiable tactics/heuristics or requires manual proofs, and is not guaranteed to terminate.

On the inference front, Lee et al. [14] has formalized an intraprocedural analysis for loop invariants using grammar approximation under separation logic. Their analysis can handle a wide range of shape predicates with local sharing but is restricted to predicates with two parameters and *without* size properties. Another work [10] has also formulated interprocedural shape inference but is restricted to just the list segment shape predicate. More recently, Guo et al [12] showed how fairly complex shapes can be inferred with the help of a technique based on *truncation point* which can be viewed as a lemma for cutting (or grafting) a subheap of the same predicate from (or into) a given shape.

However, the presence of numerical properties makes the truncation point technique difficult for more general user-defined predicates. The reason is that, after cutting a sub-heap and then grafting back a piece of heap of the same shape, the shape of the original heap is restored, but not necessarily its content or other quantitative properties. Another recent work by Chang and Rival [6] proposes a backward unfolding technique that requires an in-built (but generic) lemma for splitting inductive segments. This hardwired use of a lemma can be viewed as a special case to our user-defined approach. While our system does not focus on the inference aspect, we provide better support for automated verification via an expressive data structures and lemmas specification mechanism. For example, data structures with strong invariants, such as balanced heights, sortedness and graph-like pointer links, are easily captured by our specification mechanism prior to automatic verification.

To the best of the authors' knowledge, most past works in automated program verification have not made systematic provision for *user-specified lemmas* that can be automatically proven and applied, so as to widen the class of programs that can be automatically verified. However, the use of user-specified lemmas can be found in works based on dependent type systems and proof checkers. An example of this is the Applied Type System (ATS) [7] that was proposed for combining programs with proofs. In ATS, dependent types for capturing program invariants and lemmas are highly expressive, but users must supply all expected properties, associated proofs, and precisely state where they are to be applied, with ATS playing the role of a proof-checker. On the contrary, our proposed technique performs lemma proving and program verification automatically, without the need for such detailed guidance.

To summarise, we have introduced a new mechanism to support user-supplied lemmas for automated program verification via separation logic. This approach is *directed* and is guaranteed to *terminate*. It is *directed* because the lemmas are applied selectively, as guided by the need for the eventual matching up of heap predicates during entailment proving. It terminates since we use well-formed and well-founded heap formulae for both shape predicates and lemmas, together with a cycle detection technique. One strength of our approach is that users are allowed to add relevant lemmas to further enhance the capability of the automated program verification system. This puts creative control back into users' hands. Nevertheless, we provide machine support for automatically proving and then applying these given lemmas. With the appropriate use of universal quantifiers, these lemmas can be quite expressive. We believe that lemmas can greatly enhance the capability of automated program verification in general, and separation logic in particular; as they play the role of cut rules in proof systems.

Acknowledgments. Our thanks to Shengchao Qin for providing insights on how lemmas should be represented, and to Cristina David for bravely carrying the mantle for our SLEEK prover sub-system. Anonymous reviewers of CAV 2008 provided critical but fair comments that have helped us improve the presentation of this paper. This work is supported by an A*STAR-funded research project R-252-000-233-305 on "A Constructive Framework for Dependable Software".

References

1. Berdine, J., Calcagno, C., O'Hearn, P.W.: Symbolic Execution with Separation Logic. In: Yi, K. (ed.) APLAS 2005. LNCS, vol. 3780, pp. 52–68. Springer, Heidelberg (2005)
2. Berdine, J., Calcagno, C., O'Hearn, P.W.: Smallfoot: Modular automatic assertion checking with separation logic. In: FMCO. LNCS. Springer, Heidelberg (2006)
3. Berdine, J., Cook, B., Distefano, D., O'Hearn, P.: Automatic termination proofs for programs with shape-shifting heaps. In: Ball, T., Jones, R.B. (eds.) CAV 2006. LNCS, vol. 4144, pp. 386–400. Springer, Heidelberg (2006)
4. Brotherston, J.: Formalised inductive reasoning in the logic of bunched implications. In: Riis Nielson, H., Filé, G. (eds.) SAS 2007. LNCS, vol. 4634, pp. 87–103. Springer, Heidelberg (2007)
5. Brotherston, J., Simpson, A.: Complete sequent calculi for induction and infinite descent. In: LICS, pp. 51–62 (2007)
6. Chang, B.-Y.E., Rival, X.: Relational inductive shape analysis. In: POPL, pp. 247–260 (2008)
7. Chen, C., Xi, H.: Combining Programming with Theorem Proving. In: ICFP, Tallinn, Estonia (September 2005)
8. Distefano, D., O'Hearn, P.W., Yang, H.: A Local Shape Analysis based on Separation Logic. In: Hermanns, H., Palsberg, J. (eds.) TACAS 2006 and ETAPS 2006. LNCS, vol. 3920, pp. 287–302. Springer, Heidelberg (2006)
9. Feng, X., Shao, Z., Dong, Y., Guo, Y.: Certifying Low-Level Programs with Hardware Interrupts and Preemptive Threads. In: PLDI, Tucson, Arizona, June 2008. ACM Press, New York (2008)
10. Gotsman, A., Berdine, J., Cook, B.: Interprocedural Shape Analysis with Separated Heap Abstractions. In: Yi, K. (ed.) SAS 2006. LNCS, vol. 4134, pp. 240–260. Springer, Heidelberg (2006)
11. Gotsman, A., Berdine, J., Cook, B., Sagiv, M.: Thread-modular shape analysis. In: PLDI, pp. 266–277 (2007)
12. Guo, B., Vachharajani, N., August, D.I.: Shape analysis with inductive recursion synthesis. In: PLDI, pp. 256–265 (2007)
13. Jia, L., Walker, D.: ILC: A Foundation for Automated Reasoning About Pointer Programs. In: Sestoft, P. (ed.) ESOP 2006 and ETAPS 2006. LNCS, vol. 3924, pp. 131–145. Springer, Heidelberg (2006)
14. Lee, O., Yang, H., Yi, K.: Automatic verification of pointer programs using grammar-based shape analysis. In: Sagiv, M. (ed.) ESOP 2005. LNCS, vol. 3444, pp. 124–140. Springer, Heidelberg (2005)
15. Marti, N., Affeldt, R., Yonezawa, A.: Formal Verification of the Heap Manager of an Operating system using Separation Logic. In: Liu, Z., He, J. (eds.) ICFEM 2006. LNCS, vol. 4260, pp. 400–419. Springer, Heidelberg (2006)
16. Nguyen, H.H., David, C., Qin, S.C., Chin, W.N.: Automated Verification of Shape and Size Properties via Separation Logic. In: Cook, B., Podelski, A. (eds.) VMCAI 2007. LNCS, vol. 4349, pp. 251–266. Springer, Heidelberg (2007)
17. Nipkow, T., Paulson, L.C., Wenzel, M.: Isabelle/HOL — A Proof Assistant for Higher-Order Logic. LNCS, vol. 2283. Springer, Heidelberg (2002)
18. Owre, S., Rushby, J.M., Shankar, N., Stringer-Calvert, D.W.J.: PVS: An experience report. In: FM-Trends, pp. 338–345 (1998)
19. Parkinson, M., Bornat, R., O'Hearn, P.: Modular verification of a non-blocking stack. In: POPL, Nice, France (January 2007)

20. Preoteasa, V.: Mechanical verification of recursive procedures manipulating pointers using separation logic. In: Misra, J., Nipkow, T., Sekerinski, E. (eds.) FM 2006. LNCS, vol. 4085, pp. 508–523. Springer, Heidelberg (2006)
21. Pugh, W.: The Omega Test: A fast practical integer programming algorithm for dependence analysis. CACM 8, 102–114 (1992)
22. Reynolds, J.: Separation Logic: A Logic for Shared Mutable Data Structures. In: LICS, Copenhagen, Denmark (July 2002)
23. Stump, A., Barrett, C.W., Dill, D.L.: CVC: A cooperating validity checker. In: Brinksma, E., Larsen, K.G. (eds.) CAV 2002. LNCS, vol. 2404, pp. 500–504. Springer, Heidelberg (2002)

A Numerical Abstract Domain Based on *Expression Abstraction* and *Max Operator* with Application in Timing Analysis

Bhargav S. Gulavani[1] and Sumit Gulwani[2]

[1] Indian Institute of Technology, Bombay, India
bhargav@cse.iitb.ac.in
[2] Microsoft Research, Redmond, US
sumitg@microsoft.com

Abstract. This paper describes a precise numerical abstract domain for use in timing analysis. The numerical abstract domain is parameterized by a linear abstract domain and is constructed by means of two domain lifting operations. One domain lifting operation is based on the principle of *expression abstraction* (which involves defining a set of expressions and specifying their semantics using a collection of directed inference rules) and has a more general applicability. It lifts any given abstract domain to include reasoning about a given set of expressions whose semantics is abstracted using a set of axioms. The other domain lifting operation incorporates disjunctive reasoning into a given linear relational abstract domain via introduction of `max` expressions. We present experimental results demonstrating the potential of the new numerical abstract domain to discover a wide variety of timing bounds (including polynomial, disjunctive, logarithmic, exponential, etc.) for small C programs.

1 Introduction

The oldest trick for proving termination of loops has been that of finding a ranking function [25]. A ranking function for a loop is a function (over loop variables) whose value decreases in each loop iteration and is bounded below by some finite quantity. Earlier work on proving termination of loops focused on synthesizing *linear* ranking functions [11,22]. However, not all programs have linear ranking functions (e.g., Figure 1(a) and 2(a)). This led to more sophisticated proposals for proving termination like the principle of disjunctive well-foundedness of ranking functions (which can handle Figure 1(a)), and work on a richer class of ranking functions like lexicographic linear ranking functions [7] and polyranking functions [8] (which can handle Figure 2(a)).

In contrast to this recent literature on multiple methodologies for proving termination, we present a numerical abstract domain that can be used to uniformly prove the termination of a large class of programs (including the ones in Figure 1 and Figure 2), and more importantly establish precise timing bounds, a richer piece of information than simply establishing termination. Computing

A. Gupta and S. Malik (Eds.): CAV 2008, LNCS 5123, pp. 370–384, 2008.

```
Disjunction(int x0, y, z0)
  x := x0; z := z0; i := 0;
  while (x < y) do
     i := i + 1;
     if (z > x) {
        x := x + 1;
     else
        z := z + 1;
```
(a)

```
Sequential(int n, m)
  x := 0; i := 0;
  while (x < n) do
     i := i + 1;
     x := x + 1;
  while (x < m) do
     i := i + 1;
     x := x + 1;
```
(b)

```
Simple(int x0, n)
  x := x0; i := 0;
  while (x < n) do
     i := i + 1;
     x := x + 1;
```
(c)

Fig. 1. Examples that illustrate the importance of the max operator in our numerical abstract domain for both representing timing bounds as well as computing the invariants required to establish timing bounds. The examples have been instrumented with the monitor variable i. For (a) (taken from [12] which uses disjunctive well-foundedness to prove termination), our tool computes the bound $\texttt{max}(0, y - x_0) + \texttt{max}(0, y - z_0)$ on the monitor variable i after establishing the inductive loop invariant $i = (x - x_0) + (z - z_0) \wedge z \leq \texttt{max}(z_0, y) \wedge x \leq \texttt{max}(x_0, y)$. For (b), and (c) our tool computes the bounds $\texttt{max}(0, n, m)$ and $\texttt{max}(0, n - x_0)$ respectively on the monitor variable i.

timing bounds is much more useful than simply proving termination in several settings such as embedded systems and performance critical software. The basic idea of our methodology is to instrument the program with a monitor variable that increases in each loop iteration and then establish bounds on the monitor variable using abstract interpretation over a numerical abstract domain. Essentially, we have instrumented a candidate ranking function $-i$ inside the loop since the value of $-i$ always decreases and establishing an upper bound u on i will imply a lower bound on $-i$, thereby making $-i$ a ranking function and implying termination, but more importantly yielding a timing bound of u.

One key feature of our numerical abstract domain is that it incorporates (some level of) disjunctive reasoning by use of max operator (which returns the largest of its arguments). This allows our abstract domain to naturally express bounds for loops with complex control flow inside, as is the case for the program in Figure 1(a). (Observe that even proving termination of this program is non-trivial; one way to prove termination of this program is to use the principle of disjunctive well-foundedness [23], which involves splitting the termination argument into multiple ranking functions corresponding to different branches in the program.) However, it is important to note that the presence of max operator in our abstract domain not only caters to handling loops with complex control structure, but it is equally important to express bounds for programs even with simple loops (which entail linear ranking functions, a simpler termination argument), like the ones in Figure 1(b) and (c).

Another key feature of our numerical abstract domain is that it incorporates reasoning about other operators (like multiplication, exponentiation, logarithm, square-root etc) whose semantics is specified using some set of inference rules. This allows our abstract domain to naturally represent bounds for loops with inherent non-linear behavior as is the case for the program in Figure 2(a). (Observe that even proving termination of this program is non-trivial; one way to

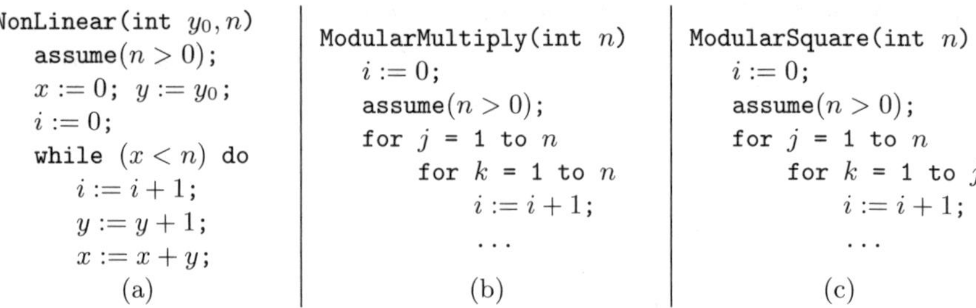

Fig. 2. Examples that illustrate the importance of using non-linear operators like multiplication and square root in our numerical abstract domain for both representing timing bounds as well as computing the invariants required to establish timing bounds. The examples have been instrumented with the monitor variable i. For (a) (taken from [6] which uses the principle of second-order differences to establish a lexicographic polyranking function for proving termination), our tool computes the bound $\sqrt{2n} + \texttt{max}(0, -2y_0) + 1$ on the monitor variable i after establishing the inductive loop invariant $i = y - y_0 \wedge y^2 \leq y_0^2 + 2x$. For (b) and (c), our tool computes the bound n^2 and $n(n+1)/2$ respectively on the monitor variable i.

prove termination is to use lexicographic polyranking functions [7,8].) However, it is important to note that the flexibility of using arbitrary expressions (from a given set) allows our abstract domain to precisely represent precise bounds of programs with simple termination arguments. For example, we can prove that mergesort has a complexity of $n \log n$ and that Fibonacci has a complexity of 2^n. It is important to note that our technique not only aims to find precise computational complexity, but also precise constant factors. For example, it can precisely establish the n^2 complexity of the doubly nested loop used for modular multiplication and $n(n+1)/2$ complexity of the doubly nested loop used for modular squaring as shown in Figure 2(b) and (c). This difference was the source of a real timing attack on implementations of RSA protocol [19].

Our numerical abstract domain is parameterized by a linear arithmetic numerical domain like intervals, difference constraints, or polyhedron domain [13]. We present two domain lifting operations that extend the base linear arithmetic domain to reason about the max operator and other operators whose semantics is specified using a set of inference rules. One of the domain lifting operation extends the linear arithmetic domain to represent linear relationships over variables as well as max-expressions (an expression of the form $\texttt{max}(e_1, \ldots, e_n)$ where e_i's are linear expressions). Another domain lifting operation lifts an abstract domain to represent constraints not only over program variables, but also over expressions from a given finite set of expressions S. The semantics of the operators used in constructing expressions in S is specified as a set of inference rules. (Our abstract domain retains efficiency by treating these expressions just like any other variable, while relying on the inference rules to achieve precision.) The rationale behind this choice is that it is easy to specify or heuristically infer the set of base expressions, but specifying linear combinations of those expressions and specifying which subsets of those linear combinations should be grouped under a max operator is a cumbersome process. Fortunately, the latter process can be automated by means of a domain lifting operation that we describe in this paper.

This paper has three main technical contributions:

- We introduce a domain constructor operation based on the notion of expression abstraction. Given a base abstract domain A and a set of expressions S whose semantics is specified using a set of rewrite rules, we show how to construct a more precise abstract domain A_S (Section 3).
- We introduce another domain constructor operation for linear arithmetic domains. Given a linear arithmetic abstract domain A, we show how to construct a new arithmetic domain $\tilde{A}$ that can represent linear relations over *max* expressions (Section 4).
- Given a linear arithmetic domain A, and a set of expressions S, we use our domain constructor operations to construct the numerical domain $\tilde{A}_S$ (Section 5), and show how it can be used to compute precise bounds for a wide variety of programs (Section 6). We discuss preliminary experimental results in Section 7.

We start with a description of the operations that need to be supported by an abstract domain for performing abstract interpretation in Section 2.

2 Preliminaries

An abstract domain A needs to be equipped with four operators (or transfer functions), $\texttt{Join}_A$, $\texttt{Widen}_A$, $\texttt{Eliminate}_A$, and $\texttt{PostPredicate}_A$ to enable (forward) abstract interpretation over the flowchart nodes of a program.

The join operator $\texttt{Join}_A$ for an abstract domain A takes two abstract elements E_1 and E_2 and computes the least upper bound of E_1 and E_2 in the abstract domain A. In other words $\texttt{Join}_A(E_1, E_2)$ denotes the most precise element E in the abstract domain A such that $E_1 \Rightarrow E$ and $E_2 \Rightarrow E$. The join operator is used to obtain the abstract element after a join node by merging the abstract elements before the join node.

The widen operator $\texttt{Widen}_A$ for an abstract domain A takes two abstract elements E_1 and E_2 such that $E_1 \Rightarrow E_2$ and computes another element E such that $E_2 \Rightarrow E$. The sequence of widen operations converges in a bounded number of steps, i.e., for any strictly increasing sequence $E_0, E_1, \ldots$ (such that $E_i \Rightarrow E_{i+1}$ for all i) , if we define $E'_0 = E_0$, $E'_1 = \texttt{Widen}_A(E'_0, E_1)$, $E'_2 = \texttt{Widen}_A(E'_1, E_2), \ldots$, then there exists $i \geq 0$ such that $E'_j = E'_i$ for all $j > i$.

The postpredicate operator $\texttt{PostPredicate}_A$ takes as input an abstract element E' and a predicate p and computes the most precise element E expressible in the domain A such that $E' \wedge p \Rightarrow E$ (meaning that $\gamma(E') \cap S_p \subseteq \gamma(E)$, where S_p is the set of states that satisfy predicate p). The postpredicate operator is used to incorporate the information provided by the predicate inside an assume statement or a conditional guard.

The existential elimination operator $\texttt{Eliminate}_A$ takes as input an abstract element E' and a variable x and computes the most precise element E expressible in the domain A such that $E' \Rightarrow E$ and E does not refer to variable x. The existential elimination operator is used to transform the abstract element E' before an assignment statement $x := e$ [1] to the element E as follows (assuming that x does not occur in e [2]): $E = \texttt{PostPredicate}_A(\texttt{Eliminate}_A(E', x), x = e)$.

In Section 3 and Section 4 below, we show how to construct the transfer functions for the richer abstract domains A_S and $\tilde{A}$ from the transfer functions of the base abstract domain A.

3 Domain Lifting Using Expression Abstraction

In this section, we introduce the notion of *expression abstraction* and use it to define a more precise abstract domain given any base abstract domain.

The process of *expression abstraction* involves defining a set of expressions S over program variables using some operators, and defining the abstract semantics of those operators using a set of directed inference rules R. We additionally assume that the set S is closed under sub-expressions. (See Section 5 for an example of S and R.) For every expression $e \in S$, we introduce a fresh variable denoted by Z_e. The elements of the abstract domain A_S represent the same kind of constraints as the abstract domain A but over an extended set of variables that includes Z_e's. The transfer functions for the abstract domain A_S (defined in Section 3.2) make use of the $\texttt{Saturate}$ operator, which we define next.

3.1 The Saturate Operator

The saturate operator $\texttt{Saturate}$ takes as input an abstract element E and a set of expressions S and returns another abstract element E' that contains constraints from E as well as includes constraints over expressions from S obtained by applying the rewrite rules that define the semantics of the expressions in S. The pseudo-code for the saturate function is shown below.

```
Saturate(E, S) =
  E_old := ⊥;
  while (E ≢ E_old) do
     E_old := E;
     foreach instantiation of an inference rule: P1 ⇒ P2
        If E ⇒ P1, then E := PostPredicate(E, P2);
  return E;
```

[1] Without loss of generality, we can assume that all assignment statements are of the form $x := e$. Memory reads and writes can be modeled using select and update expressions, without losing any precision.

[2] Without loss of generality, we can assume that x does not occur inside e since an arbitrary assignment $x := e$ can be split into two assignments $t := e; x := t$ with this property, where t is a fresh variable.

The for loop in Line 4 considers all instantiations of an inference rule $P_1 \Rightarrow P_2$ such that all terms that occur in both P_1 and P_2 are from the given set of expressions S. If the set of inference rules R has the property that the number of applications of the inference rules is bounded in any context (i.e., given any context, the number of applications of the inference rules that yield a predicate not implied by the context and other derived predicates is bounded), then the while loop in Line 2 is terminating. If not, then we simply use the heuristic of iterating a bounded number of times.

Example 1. Let A be the polyhedron abstract domain. Let R consist of the following useful inference rule for reasoning about the product operator. The rule multiplies both sides of an equality by some term u.

$$(\sum_i a_i x_i = a) \ \Rightarrow\ (\sum_i a_i Z_{x_i u} = Z_{au})$$

Let S be the set of expressions $\{y^2, y_0^2, yy_0\}$. Let E_1 be $y = y_0 \wedge x = 0$ and E_2 be $y = y_0 + 1 \wedge x = y_0 + 1$. E_1 and E_2 denote (part of) the abstract elements at the loop entry and at the loop back-edge after one loop iteration for the example in Figure 2(a). Then,

$$\texttt{Saturate}(E_1, S) = (y = y_0 \ \wedge\ x = 0 \ \wedge\ Z_{y^2} = Z_{yy_0} \ \wedge\ Z_{yy_0} = Z_{y_0^2})$$
$$\texttt{Saturate}(E_2, S) = (y = y_0 + 1 \wedge x = y_0 + 1 \wedge Z_{yy_0} = Z_{y_0^2} + y_0 \wedge Z_{y^2}{=}Z_{yy_0}{+}y)$$

3.2 Transfer Functions

In this section, we describe how to construct the transfer functions for the abstract domain A_S using the transfer functions for the base domain A. The key idea in the construction is to simply saturate the input abstract elements using the Saturate algorithm (described in Section 3.1) and then apply the corresponding transfer function from the base abstract domain.

– Join Operator

$$\texttt{Join}_{A_S}(E_1, E_2) = \texttt{Join}_A(\texttt{Saturate}(E_1, S), \texttt{Saturate}(E_2, S))$$

Example 2. Let E_1 and E_2 be the abstract elements as in Example 1. Then,

$$\texttt{Join}_{A_S}(E_1, E_2) \ = \ (y_0 \leq y \leq y_0 + 1 \ \wedge\ Z_{y^2} \leq Z_{y_0^2} + 2x)$$

Observe that the above join operation gives us one of the desired inductive invariants $y^2 \leq y_0^2 + 2x$ required for proving bounds in Figure 2(a). □

– PostPredicate Operator

$$\texttt{PostPredicate}_{A_S}(E, p) = \texttt{PostPredicate}_A(\texttt{Saturate}(E, S), p)$$

– Eliminate Operator
The eliminate operator for the new domain A_S involves saturating the input abstract element and then eliminating not only the given variable x, but also all the variables corresponding to expressions from S that involve x.

$$\texttt{Eliminate}_{A_S}(E, x) \quad = \quad \texttt{Eliminate}_A(\texttt{Saturate}(E, S), V_x)$$

where $V_x = \{x\} \cup \{Z_e \mid e \in S \text{ and change to } x \text{ results in change to } e\}$.
– Widen Operator

$$\texttt{Widen}_{A_S}(E_1, E_2) = \texttt{Widen}_A(\texttt{Saturate}(E_1, S), \texttt{Saturate}(E_2, S))$$

4 Linear Domain Lifting Using Max Operator

In this section, we define a domain lifting operation that takes a linear arithmetic abstract domain A that represents linear constraints over some set of variables V, and a subset U of V, and produces a domain $\tilde{A}$ that can represent linear constraints over $V - U$ as allowed by A, but allowing for a richer constant term - one that is constructed using linear combinations of max-linear expressions over U. A *max-linear expression* over U is of the form $\texttt{max}(e_1, .., e_n)$, where each e_i is some linear expression over U. For eg., if A is the difference constraints domain, then the domain $\tilde{A}$ can represent constraints like $v_1 - v_2 \leq \texttt{max}(u_1, 2u_2 + u_3)$, where $v_1, v_2 \in V$ and $u_1, u_2, u_3 \in U$.

The transfer functions for the abstract domain $\tilde{A}$ (defined in Section 4.2) make use of the `Witness` operator, which we define next.

4.1 Witness Coefficients

Lemma 1. *(Farkas Lemma) Let e, e_i be some linear arithmetic expressions without* the constant term. *If* $\left(\bigwedge_{i=1}^{n} (e_i \leq 0)\right) \Rightarrow e \leq 0$, *then it must be the case that there exist non-negative λ's such that* $e \equiv \sum_{i=1}^{n} \lambda_i e_i$.

For an implication $\left(\bigwedge_{i=1}^{n} (e_i \leq 0)\right) \Rightarrow e \leq 0$, we define $\texttt{Witness}(\bigwedge_{i=1}^{n} (e_i \leq 0), e \leq 0)$ to be $(\lambda_1, \ldots, \lambda_n)$. Note that there may exist multiple witnesses but any single witness is sufficient for soundness of the transfer functions described below.

Example 3

$$\texttt{Witness}(i \leq 0 \ \wedge \ -x \leq 0, i \leq 0) = (1, 0)$$
$$\texttt{Witness}(i - x \leq 0 \ \wedge \ x \leq 0, i \leq 0) = (1, 1)$$

4.2 Transfer Functions

In this section, we describe how to construct the transfer functions for the abstract domain $\tilde{A}$ using the transfer functions for the base domain A. The key idea in the construction is to remove the constant and the part corresponding to variables in U from each inequality in the input(s), and then apply the corresponding transfer function from the base domain, and then add back an appropriate symbolic max expression to each inequality in the result.

– Join Operator.

$\texttt{Join}_{\tilde{A}}(E_1, E_2)$ =
Let E_1 be $\bigwedge_i (e_i \leq f_i)$ and let E_2 be $\bigwedge_i (e'_i \leq f'_i)$;
(where e_i, e'_i are linear over V-U and f_i, f'_i are max-linear over U)
$E := \top$; $E' := \texttt{Join}_A(\bigwedge_i (e_i \leq 0), \bigwedge_i (e'_i \leq 0))$;
Foreach inequality $e \leq 0 \in E'$,
    $(\lambda_i) := \texttt{Witness}(\bigwedge_i (e_i \leq 0), e \leq 0)$; $(\lambda'_i) := \texttt{Witness}(\bigwedge_i (e'_i \leq 0), e \leq 0)$;
    $f := \sum_i \lambda_i f_i$; $f' := \sum_i \lambda'_i f'_i$;
    $f'' := \texttt{ComputeMax}(f, f')$;
    if $f'' \neq \top$, $E := E \wedge (e \leq f'')$;
return E;

The function $\texttt{ComputeMax}(f, f')$ either returns $\top$ denoting that there are too many arguments to the max function, or returns a possibly max function.

Example 4. Let $V = \{i, x, x_0, n\}$ and $U = \{x_0, n\}$. Let E_1 be $i \leq 0 \;\wedge\; -x \leq -x_0$ and E_2 be $i - x \leq -x_0 \;\wedge\; x \leq n$. E_1 and E_2 denote (part of) the abstract elements at the loop entry and at the loop back-edge after one loop iteration for the example in Figure 1(c.) Then, using the result of the witness functions from Example 3, we obtain

$$\texttt{Join}_{\tilde{A}}(E_1, E_2) \quad = \quad i \leq \texttt{max}(0, n - x_0)$$

Observe that the above join operation provides us with the invariant $i \leq \texttt{max}(0, n - x_0)$ required for computing bounds in Figure 1(a). □

– PostPredicate Operator.

$$\texttt{PostPredicate}_{\tilde{A}}(E, p) = \texttt{PostPredicate}_A(E, p)$$

– Eliminate Operator.

$\texttt{Eliminate}_{\tilde{A}}(E, x)$ =
Let E be $\bigwedge_i (e_i \leq f_i)$;
(where e_i are linear over $V - U$ and f_i are max-linear over U)
$E_1 := \texttt{Eliminate}_A(\bigwedge_i (e_i \leq 0))$; $E_2 := \top$;
Foreach inequality $e \leq 0 \in E_1$:
    $(\lambda_i) := \texttt{Witness}(\bigwedge_i (e_i \leq 0), e \leq 0)$;
    $f := \sum_i \lambda_i f_i$;
    $E_2 := E_2 \wedge (e \leq f)$;
return E_2;

Note that we require the variables to be eliminated be from the set $V - U$. This puts the restriction that the variables in set U are never modified in the program.

Example 5. Let $V = \{i, x, x_0, n\}$ and $U = \{x_0, n\}$. Let E be $i - x \leq -x_0 \wedge x \leq n$. Consider computing $\texttt{Eliminate}_{\tilde{A}}(E, x)$. Line 2 of the above algorithm computes E_1 as $i \leq 0$. Using the witness $(1, 1)$, we obtain the result $i \leq -x_0 + n$.

– Implication Operator.

$\texttt{Implies}_{\tilde{A}}(E_1, E_2)$ =
  Let E_1 be $\bigwedge_i e_i \leq f_i$ and let E_2 be $\bigwedge_j e'_j \leq f'_j$;
  (where e_i, e'_j are linear over V-U and f_i, f'_j are max-linear over U)
  Let result $:=$ true;
  Foreach inequality $e'_j \leq f'_j \in E_2$:
    $(\lambda_i) :=$ Witness$(\bigwedge_i e_i \leq 0, e'_j \leq 0)$; $f := \sum_i \lambda_i f_i$;
    if not LessEq(f, f'_j), result $:=$ false;
  return result;

Let $f = \texttt{max}(e_1, \ldots, e_n)$ and $f' = \texttt{max}(e'_1, \ldots, e'_n)$ be max-linear expressions. $\texttt{LessEq}(f, f')$ returns true iff for each e_i there exists e'_j such that $e_i \leq e'_j$ is valid. For example $\texttt{LessEq}(\texttt{max}(x, n), \texttt{max}(x + 1, n))$ returns true, whereas $\texttt{LessEq}(\texttt{max}(x, n), \texttt{max}(x + 1, n - 1))$ returns false.

– Widen Operator.
The widen operator $\texttt{Widen}_{\tilde{A}}(E_1, E_2)$ simply returns the conjunction of those constraints from E_1 that are also implied by E_2.

5 A New Numerical Abstract Domain

In this section, we discuss the design choices made while applying the domain lifting operators for obtaining the numerical abstract domain that we use for timing analysis. We pick any linear relational abstract domain A and lift it using the domain lifting operation based on expression abstraction (as described in Section 3). We use the operators multiplication ($x \times y$), logarithm ($\lceil \log x \rceil$), square-root ($\lceil \sqrt{x} \rceil$), and exponentiation (2^x) to construct the set S of expressions. The set S of expressions involving these operators can either be provided by the programmer [3] (since they may have a better idea of what kinds of bounds the program may entail) or it can be constructed automatically using some initial heuristic such as we can apply the unary operators (logarithm, square-root,

[3] Note that we are not requiring the programmer to provide the exact bound. We are simply requiring the programmer to provide the base expressions and the bounds would be automatically computed by taking linear combinations of these expressions and additionally the expressions obtained by applying max operator. Furthermore, computation of bounds requires establishing the inductive invariants, which are usually much harder than the bound itself.

exponentiation) to all program variables once and then apply the only binary operator (multiplication) to all pairs of resulting expressions. This will allow our abstract domain to represent linear relationships over expressions like $n \times m$, $n\lceil \log m \rceil$, etc. We use the following inference rules R to reason about these operators. (These specific rules were chosen because they appear to capture the reasoning required to compute bounds over the chosen set of non-linear operators for a large class of programs.) For simplicity, we overload the notation Z_x to denote x, if x is a program variable. (Recall that the notation Z_x normally denotes the special variable associated with an expression $x \in S$).

1. $(Z_x \leq c) \Rightarrow (Z_{2^x} \leq 2^c)$
2. $(Z_x \leq Z_y + c) \Rightarrow (Z_{2^x} \leq 2^c \times Z_{2^y})$
3. $(Z_x \leq c) \wedge (Z_x > 0) \Rightarrow (Z_{\lceil \log x \rceil} \leq \lceil \log c \rceil)$
4. $(Z_x \leq c \times Z_y) \wedge (Z_x > 0) \wedge (Z_y > 0) \Rightarrow (Z_{\lceil \log x \rceil} \leq \lceil \log c \rceil + Z_{\lceil \log y \rceil})$
5. $(\sum_i a_i Z_{x_i} \geq a) \wedge (Z_y \geq 0) \Rightarrow (\sum_i a_i Z_{x_i y} \geq ay)$
6. $(\sum_i a_i Z_{x_i} = a) \Rightarrow (\sum_i a_i Z_{x_i y} = aZ_y)$
7. $\texttt{true} \Rightarrow Z_{\sqrt{x^2}} = \texttt{max}(Z_x, -Z_x)$
8. $\texttt{true} \Rightarrow Z_{x^2} \geq 0$
9. $(Z_{x^2} \leq \sum_i a_i Z_{x_i}) \wedge \bigwedge_i a_i \geq 0 \wedge \bigwedge_i Z_{x_i} \geq 0 \Rightarrow (Z_x \leq \sum_i \sqrt{a_i} Z_{\sqrt{x_i}})$
   This rule is useful to compute an upper bound on a variable if an upper bound has been computed on its square.

The application of the above rules requires querying the abstract domain for constraints between a specific set of variables. These queries can be performed by existential elimination of all variables other than the specific set of variables.

We then apply the domain construction discussed in Section 4 on the domain A_S obtained above to obtain the domain $\widetilde{A_S}$. For this purpose, we choose U to be the set of the input variables since we are ultimately interested in finding bounds with possibly max expressions over only the input variables. (Recall that U was the set of variables over which the abstract domain computes max expressions.)

6 Timing Analysis

In this section, we consider the problem of computing upper bounds on the time complexity of a program expressed as a function of the program inputs. We assume that each atomic statement is annotated with the units of time that it takes to execute. (To reduce cluttering in our examples, we assume that a recursive procedure call instruction and a backward jump instruction takes unit amount of time, while all other instructions take zero time. In other words, we estimate a bound on the total number of loop iterations and total number of recursive procedure call invocations).

Given a program P, we instrument the program with a monitor variable i that keeps track of the time consumed by the program. The monitor variable i is initialized to 0 at the beginning of the program, and is incremented by t units after execution of any instruction that takes t units of time to execute.

Table 1. Experimental Results: Column 4 describes the upper bounds computed by our tool on the number of loop iterations and recursive procedure call invocations. Column 3 gives the expression set used for computing the bounds shown in Column 4.

Program	Time (s)	S	Upper Bound
Disjunction (Fig1)	0.030	-	$\max(0, y - x_0) + \max(0, y - z_0)$
Sequential (Fig1)	0.010	-	$\max(0, n, m)$
NonLinear (Fig2)	0.018	y^2, y_0^2, $\sqrt{n}$	$\sqrt{2n} + \max(0, -2y_0) + 1$
ModularMultiply (Fig2)	0.105	jn, n^2	n^2
ModularSquare (Fig2)	0.098	j^2, n^2	$n(n+1)/2$
Log	0.084	$\lceil \log n \rceil$, $\lceil \log x \rceil$	$\lceil \log n \rceil$
Fibonacci	0.138	2^n	2^n
MergeSort	0.065	$n\lceil \log n \rceil$	$n\lceil \log n \rceil$
p1	0.141	-	41
p2	0.022	-	$\max(0, z_0)$
p3	0.215	$\lceil \log i \rceil$, $\lceil \log max \rceil$, $\lceil \log size \rceil$	$\lceil \log size \rceil$
p4	0.163	s^2, bs, ts	s^2
p5	0.025	-	$m + 1$
p6	0.031	-	N

Claim. Let $u_1, .., u_n$ be the upper bounds on the instrumented monitor variable i at different locations where i is incremented, expressed as a function of program inputs. Then, $\max(0, u_1, .., u_n)$ denotes an upper bound on the timing complexity of the program. [4]

Note that we simply cannot compute an upper bound on the monitor variable i at the end of the program to obtain an upper bound on the timing complexity of the program. For example, consider a program with a non-terminating loop. 0 is a valid upper bound on i at the end of the loop (since at an unreachable program location, any fact holds), but does not describe an upper bound on the timing complexity of the program.

We compute upper bounds on the monitor variable i at different program locations by performing abstract interpretation of the program over the numerical domain $\widetilde{A_S}$ described in Section 5, which provides us invariants at different program locations. An appropriate upper bound on i at a program location π is then obtained by considering the invariant I at π and existentially quantifying out all variables except i and the input variables from I. We can use a similar strategy for computing lower bounds on i. (An advantage of computing lower bounds is that they can be used as a measure of precision of our analysis for computing upper bounds.)

7 Experiments

We have implemented a prototype of our numerical abstract domain on top of the APRON [1] numerical abstract domains library. We have used this abstract domain in an abstract interpreter to bound the total number of loop iterations and the total number of recursive procedure calls invocations in several

[4] This assumes that the input variables are not modified in the program. (If they are, then we can create their copies and modify them instead.)

C programs [5]. Our abstract interpreter is implemented in `ocaml` and uses the CIL infrastructure to parse input C programs. We summarize the results of running our tool on a set of benchmarks in Table 1.

The programs p* are taken from some benchmarks (originally from Octagon library distribution) presented in a paper that describes and compares some state-of-the-art techniques for proving program termination[10]. Most of the remaining programs are presented in Section 1. `Fibonacci` and `MergeSort` are recursive programs. `Log` uses a multiplicative counter for loop iteration. The programs were analyzed using (cartesian-product) combination of polyhedra and octagon abstract domains lifted with *expression abstraction* and interval domain lifted with *max operator*.

For most of the programs shown in Table 1, the computed upper bounds are precise (i.e., they match the lower bounds computed by our tool on the monitor variable). These benchmarks include programs whose termination cannot be established by simple linear ranking functions [22], but requires more sophisticated techniques as in [23,7,8,10]. This shows that not only our abstract domain is precise enough to express exact upper bounds but also that the abstract operations are precise enough to compute these bounds. The number of required input expressions is relatively small and simple heuristics (like the one described in Section 5) can be used to infer these automatically.

Our analyzer takes for each program a set of interesting expressions to track during the analysis. Although currently we provide these expressions manually, we intend to use some heuristics in the future to automatically infer these expressions.

8 Related Work

Inference-rule based reasoning has often been used for building efficient (but incomplete) decision procedures for otherwise intractable logics [18,4]. The idea is to partially axiomatize the semantics of the underlying operators such that it leads to efficient reasoning as well as is precise enough to capture the reasoning required in the common case. In this paper, we show how to apply inference-rule based reasoning in the context of abstract interpreters as opposed to simply decision procedures. Secondly, we focus on a different domain, one that involves numerical operators.

There has been work on extending linear arithmetic abstract domains to also represent linear constraints over expressions constructed using uninterpreted functions [9,14]. In contrast, our domain constructor allows extension to expressions with arbitrary operators (as opposed to only uninterpreted functions), but represents linear constraints only over the given set of expressions.

There has been work on discovering restricted form of quadratic inequalities [5], polynomial inequality invariants [2], and equality invariants [24,21] of bounded degree. In contrast, our domain lifting operation allows for discovering arbitrary polynomial inequalities as well as non-polynomial inequalities in a uniform setting, but over a given set of expressions.

---

[5] Available at http://www.cfdvs.iitb.ac.in/~bhargav/timing.html

[20] describes a technique based on solving recurrences for computing a non-negative constant that represents the number of loop iterations required for reaching a particular error state. In contrast, we produce symbolic bounds.

There is a large body of work on estimating worst case execution time (WCET) in the embedded and real-time systems community [3,15,16,17,26]. The WCET research is more orthogonally focused on distinguishing between the complexity of different codepaths and low-level modeling of architectural features such as caches, branch prediction, instruction pipelines. For establishing loop bounds, the WCET techniques either require user annotation, or use simple techniques based on pattern matching or a simple interval analysis. In contrast, we present a path-insensitive analysis, but one that automatically estimates precise (non-linear and disjunctive) bounds on loop iterations.

For example, the AiT-WCET tool uses combination of interval-based abstract interpretation and pattern matching. Loop bound analysis of BoundT-WCET tool is based on Presburger arithmetics. Both these are much less precise than our abstract domain which is not only relational but can also represent non-linear bounds. [15] describes an interval analysis based approach (as opposed to our more precise relational linear analysis) for automatic computation of loop bounds. However, it analyzes single-path executions of programs (i.e., using input data corresponding to one execution). Hence, their bounds are in real seconds, while our bounds are symbolic and functions of inputs. The analysis described in [16] is aimed at synchronous programs and linear hybrid systems. The only similarity is that they model delays in such programs using simple counters. We also use counter instrumentation; however our abstract domain construction allows us to compute disjunctive and non-linear bounds using a base linear abstract domain. Bagnara and Zaccagnini describe how to solve a class of recursive equations that are used to express complexity measures in several systems [3]. However, the class of equations that they handle are far apart from cost relations generated from real programs.

9 Conclusion

We have presented two domain lifting operations to make linear numerical abstract domains more precise. Domain lifting via expression abstraction enables computation of non-linear invariants. Domain lifting by max expressions provides a compact representation for disjunctive bounds. The importance of these domain lifting operations is reflected by the fact that we have been able to automatically compute precise timing bounds for several benchmark programs that have recently been used by the state-of-the-art techniques for proving termination of programs.

Acknowledgments. The first author was supported by Microsoft Corporation and Microsoft Research India under the Microsoft Research India PhD Fellowship Award.

References

1. APRON. Numerical abstract domain library (2007), `http://apron.cri.ensmp.fr/library`
2. Bagnara, R., Rodríguez-Carbonell, E., Zaffanella, E.: Generation of Basic Semi-algebraic Invariants Using Convex Polyhedra. In: Hankin, C., Siveroni, I. (eds.) SAS 2005. LNCS, vol. 3672, pp. 19–34. Springer, Heidelberg (2005)
3. Bagnara, R., Zaccagnini, A.: Checking and bounding the solutions of some recurrence relations. Quaderno 344, Università di Parma, Italy (2004)
4. Bingham, J.D., Rakamaric, Z.: A logic and decision procedure for predicate abstraction of heap-manipulating programs. In: Emerson, E.A., Namjoshi, K.S. (eds.) VMCAI 2006. LNCS, vol. 3855, pp. 207–221. Springer, Heidelberg (2005)
5. Blanchet, B., Cousot, P., Cousot, R., Feret, J., Mauborgne, L., Miné, A., Monniaux, D., Rival, X.: Design and implementation of a special-purpose static program analyzer for safety-critical real-time embedded software. In: Mogensen, T.Æ., Schmidt, D.A., Sudborough, I.H. (eds.) The Essence of Computation. LNCS, vol. 2566, pp. 85–108. Springer, Heidelberg (2002)
6. Bradley, A., Manna, Z., Sipma, H.: Termination of polynomial programs. In: Cousot, R. (ed.) VMCAI 2005. LNCS, vol. 3385, pp. 113–129. Springer, Heidelberg (2005)
7. Bradley, A.R., Manna, Z., Sipma, H.B.: Linear ranking with reachability. In: Etessami, K., Rajamani, S.K. (eds.) CAV 2005. LNCS, vol. 3576, pp. 491–504. Springer, Heidelberg (2005)
8. Bradley, A.R., Manna, Z., Sipma, H.B.: The polyranking principle. In: Caires, L., Italiano, G.F., Monteiro, L., Palamidessi, C., Yung, M. (eds.) ICALP 2005. LNCS, vol. 3580, pp. 1349–1361. Springer, Heidelberg (2005)
9. Chang, B.-Y.E., Leino, K.R.M.: Abstract interpretation with alien expressions and heap structures. In: Cousot, R. (ed.) VMCAI 2005. LNCS, vol. 3385, pp. 147–163. Springer, Heidelberg (2005)
10. Chawdhary, A., Cook, B., Gulwani, S., Sagiv, M., Yang, H.: Ranking abstractions. In: Drossopoulou, S. (ed.) ESOP 2008. LNCS, vol. 4960, pp. 148–162. Springer, Heidelberg (2008)
11. Colón, M., Sipma, H.: Practical methods for proving program termination. In: Brinksma, E., Larsen, K.G. (eds.) CAV 2002. LNCS, vol. 2404, pp. 442–454. Springer, Heidelberg (2002)
12. Cook, B., Podelski, A., Rybalchenko, A.: Termination proofs for systems code. In: PLDI, pp. 415–426 (2006)
13. Cousot, P., Halbwachs, N.: Automatic discovery of linear restraints among variables of a program. In: POPL, pp. 84–97 (1978)
14. Gulwani, S., Tiwari, A.: Combining abstract interpreters. In: PLDI, pp. 376–386 (2006)
15. Gustafsson, J., Ermedahl, A., Sandberg, C., Lisper, B.: Automatic derivation of loop bounds and infeasible paths for wcet analysis using abstract execution. In: RTSS, pp. 57–66 (2006)
16. Halbwachs, N., Proy, Y.-E., Roumanoff, P.: Verification of real-time systems using linear relation analysis. Form. Methods Syst. Des. 11(2), 157–185 (1997)
17. Healy, C.A., Sjodin, M., Rustagi, V., Whalley, D.B., van Engelen, R.: Supporting timing analysis by automatic bounding of loop iterations. Real-Time Systems 18(2/3), 129–156 (2000)

18. Jhala, R., McMillan, K.: Array abstractions from proofs. In: Damm, W., Hermanns, H. (eds.) CAV 2007. LNCS, vol. 4590, pp. 193–206. Springer, Heidelberg (2007)
19. Kocher, P.C.: Timing attacks on implementations of diffie-hellman, rsa, dss, and other systems. In: Koblitz, N. (ed.) CRYPTO 1996. LNCS, vol. 1109, pp. 104–113. Springer, Heidelberg (1996)
20. Kroening, D., Weissenbacher, G.: Counterexamples with loops for predicate abstraction. In: Ball, T., Jones, R.B. (eds.) CAV 2006. LNCS, vol. 4144, pp. 152–165. Springer, Heidelberg (2006)
21. Müller-Olm, M., Seidl, H.: Precise interprocedural analysis through linear algebra. In: POPL, pp. 330–341 (2004)
22. Podelski, A., Rybalchenko, A.: A complete method for the synthesis of linear ranking functions. In: Steffen, B., Levi, G. (eds.) VMCAI 2004. LNCS, vol. 2937, pp. 239–251. Springer, Heidelberg (2004)
23. Podelski, A., Rybalchenko, A.: Transition invariants. In: LICS, pp. 32–41. IEEE, Los Alamitos (2004)
24. Sankaranarayanan, S., Sipma, H., Manna, Z.: Non-linear loop invariant generation using gröbner bases. In: POPL, pp. 318–329 (2004)
25. Turing, A.: Checking a large routine, pp. 70–72 (1989)
26. Wilhelm, R., Engblom, J., Ermedahl, A., Holsti, N., Thesing, S., Whalley, D., Bernat, G., Ferdinand, C., Heckmann, R., Mueller, F., Puaut, I., Puschner, P., Staschulat, J., Stenström, P.: The Determination of Worst-Case Execution Times—Overview of the Methods and Survey of Tools. ACM Transactions on Embedded Computing Systems (TECS) (2007)

Scalable Shape Analysis for Systems Code

Hongseok Yang[1], Oukseh Lee[2], Josh Berdine[3], Cristiano Calcagno[4], Byron Cook[3], Dino Distefano[1], and Peter O'Hearn[1]

[1] Queen Mary, Univ. of London
[2] Hanyang University, Korea
[3] Microsoft Research
[4] Imperial College

Abstract. Pointer safety faults in device drivers are one of the leading causes of crashes in operating systems code. In principle, shape analysis tools can be used to prove the absence of this type of error. In practice, however, shape analysis is not used due to the unacceptable mixture of scalability and precision provided by existing tools. In this paper we report on a new join operation $\sqcup\!\!\!\!\!\mid$ for the separation domain which aggressively abstracts information for scalability yet does not lead to false error reports. $\sqcup\!\!\!\!\!\mid$ is a critical piece of a new shape analysis tool that provides an acceptable mixture of scalability and precision for industrial application. Experiments on whole Windows and Linux device drivers (firewire, pci-driver, cdrom, md, etc.) represent the first working application of shape analysis to verification of whole industrial programs.

1 Introduction

Pointer safety faults in device drivers are one of the leading causes of operating system crashes. The reasons for this are as follows:

- The average Windows or Linux computer has numerous (*i.e.* >15) device drivers installed,
- Most device drivers manage relatively complex combinations of shared singly- and doubly-linked lists,
- Device drivers are required to respect many byzantine invariants while manipulating data structures (*e.g.* pieces of data structures that have been paged out can only be referenced at low thread-priority). This results in complex and nonuniform calling conventions, unlike typical benchmark code.

By pointer safety we mean that a program does not dereference null or a dangling pointer, or produce a memory leak. In principle a shape analysis tool can be used to prove the absence of pointer safety violations: shape analysis is a heap-aware program analysis with accurate handling of deep update. Furthermore, device drivers are small (*e.g.* <15k LOC) and usually do not use trees or DAGs—thus making device drivers the perfect application for shape analysis.

So, why aren't shape analysis tools regularly applied to device drivers? The reason is that today's shape analysis tools are either scalable, or precise, but

A. Gupta and S. Malik (Eds.): CAV 2008, LNCS 5123, pp. 385–398, 2008.

Table 1. Results with the $\sqcup$ extension of SpaceInvader on Windows and Linux device drivers. Experiments were performed on an Intel Core Duo 2.0GHz with 2GB. Each error found was confirmed manually. Errors in the Windows device driver (`t1394Diag.c`) were confirmed by the Windows kernel team. The time and space columns contain the numbers for the analysis of fixed versions of the drivers (and so report time to find proofs of pointer safety).

Program	LOC	Sec	MB	Memory leaks	Dereference errors	False error rate
`scull.c`	1010	0.36	0.25	1	0	0%
`class.c`	1983	8.21	7.62	2	1	0%
`pci-driver.c`	2532	0.97	1.72	0	0	0%
`ll_rw_blk.c`	5469	887.94	485.87	3	1	0%
`cdrom.c`	6218	103.26	71.52	0	2	0%
`md.c`	6635	1585.69	847.63	6	5	0%
`t1394Diag.c`	10240	135.05	68.81	33	10	0%

not both. Numerous papers have reported on the application of accurate shape analysis to small examples drawn from real systems code; other papers have reported on very imprecise analysis on large code bases. The verification of whole industrial programs, however, requires both.

Towards the elusive goal of finding a scalable *and* precise analysis, in this paper we describe a new join operation, $\sqcup$, for shape analysis tools based on the separation domain [4,10,17]. $\sqcup$ provides a mixture of scalability and precision sufficient for the problem of proving pointer safety of whole industrial device drivers. A join operation (in the terminology of abstract interpretation [9]) takes a disjunction of two abstract states, each of which describes (in our setting) a set of concrete heaps that may arise during program execution. $\sqcup$ attempts to construct a common generalization of the states. In case the attempt succeeds ($\sqcup$ is a *partial* join operator) the generalization subsequently replaces the original disjunction, leading to fewer cases to consider during the shape analysis.

In order to demonstrate the scalability and accuracy of $\sqcup$, we have implemented it in our shape analysis tool SpaceInvader, together with an abstract model of the operating system environment that we have developed. Then, we have applied the resulting tool to numerous Windows and Linux device drivers.

2 Experiments

Before describing the technical details of $\sqcup$, we first present the results of an experimental evaluation that demonstrates its scalability and precision. Table 1 displays the results of experiments with the $\sqcup$ extension of SpaceInvader on seven device drivers. Each of the drivers manipulates multiple, sometimes nested, sometimes circular, linked lists. One driver, `t1394Diag.c`, is the IEEE 1394 (firewire) driver for the Windows operating system. The drivers `pci-driver.c`, `ll_rw_blk.c`, `cdrom.c` and `md.c` are from an industrial version of Embedded Linux, given to us by ETRI. The driver `class.c` is from a standard Linux distribution, and `scull.c` is a Linux char driver used in the experiments in [7][1] Each of these drivers is analyzed in the context of environment code which nondeterministically generates input data structures, and calls the driver's dispatch routines repeatedly. In essence, each driver is supplied with a particular precondition (expressed as C code, as in [7]) but the model of system calls can be reused from driver to driver.

During our experiments SPACEINVADER was used in a stop-first configuration, where the analyzer halts if it cannot prove that a dereferencing operator is safe or if it cannot prove that a cell is reachable. When we encountered bugs we would fix them, and then run our tool again. The time and space columns in Table 1 report the numbers for the analysis of our bug-fixed versions of the drivers. Note that, during our experiments, *no false errors were found.* Also, note that for the fixed drivers SPACEINVADER *proved* pointer safety. No known tool with scalability reported to programs up to 10k LOC can match that precision.

Caveats. Device drivers often use circular doubly-linked lists. The first caveat is that, in several cases, we modified the examples in order to operate over singly-linked lists, in order to aid our analysis. Pointer safety can often be proved using singly-linked semantics even though the code is designed to operate over doubly-linked lists (it is rare for code to actually make use of the back pointers). Second, there is a significant caveat regarding arrays. SPACEINVADER currently presumes memory safety of arrays, by returning a nondeterministic value for any array dereferencing. The treatment of pointer safety can still be sound under such an assumption, and in the (slightly modified) Linux drivers our analyzer encountered no false alarms. However, the 1394 device driver contains arrays of pointers, which are beyond what our method can handle: we modified the code such that those arrays have size 1 and can be treated as pointer variables. This, of course, is just one instance of the fact that the problems of analyzing arrays and pointers are not independent. We regard this issue as an avenue for interesting future work. Finally, note that SPACEINVADER currently only implements shape analysis for sequential programs, whereas device drivers of course are multi-threaded. As reported in [12], a sequential shape analysis tool such as SPACEINVADER can be used to find and then verify resource invariants for device drivers, thereby proving pointer safety for the concurrent program. However, we emphasize that developing a scalable, precise shape analysis for concurrent programs is an open problem; only very recently, some interesting ideas such as [5,12,18] have been proposed, which give promising new lines of attack, but on which further, especially experimental, work is needed.

3 Abstract States and Setting

In this section we describe the abstract states that SPACEINVADER analysis operates over. In the next section we will describe the details of $\sqcup\!\!\!\!+$. Due to space

[1] This is a modified version of the Linux scull driver, where arrays are assumed to be of size 1.

constraints we will assume that the reader is somewhat familiar with the basics of program analysis and shape analysis.

SPACEINVADER operates over abstract states expressed as separation logic formulae. Following [4,10,17], we call these abstract states *symbolic heaps*. The symbolic heaps q, are defined by the following grammar:

$$\begin{array}{ll} e ::= x \mid x' \mid 0 & \mathcal{P} ::= \cdots \\ \Pi ::= \Pi \wedge \Pi \mid e{=}e \mid e{\neq}e \mid \mathsf{true} & \Sigma ::= \Sigma * \Sigma \mid \mathsf{emp} \mid \mathcal{P} \mid \mathsf{true} \\ q ::= \mathsf{err} \mid \Pi \wedge \Sigma & \end{array}$$

A symbolic heap q can be err, denoting the error state, or it has the form $\Pi \wedge \Sigma$, where Π and Σ describe properties of variables and the heap, respectively. The separating conjunction $\Sigma_0 * \Sigma_1$ holds for a heap if and only if the heap can be split into two disjoint parts, one making Σ_0 true and the other making Σ_1 true. emp means the empty heap, and true holds for all heaps. The primed variables x' in a symbolic heap are assumed to be (implicitly) existentially quantified.

$\mathcal{P}$ is a collection of basic predicates. One instantiation is

$$k \ ::= \ \mathsf{PE} \mid \mathsf{NE} \qquad\qquad \mathcal{P} \ ::= \ (e \mapsto e) \mid \mathsf{ls}\, k\, e\, e$$

Here, $e \mapsto f$ means a heap with only one cell e that stores f. The list segment predicate $\mathsf{ls}\, k\, e_0\, e_1$ denotes heaps containing one list segment from e_0 to e_1 only. This list segment starts at cell e_0 and its last cell stores e_1. The list is possibly empty if $k = \mathsf{PE}$; otherwise (*i.e.*, $k = \mathsf{NE}$), the list is not empty. The meanings of the segment predicates can be understood in terms of the definitions

$$\begin{array}{l} \mathsf{ls}\,\mathsf{PE}\, e\, f \iff (e = f \wedge \mathsf{emp}) \vee (\mathsf{ls}\,\mathsf{NE}\, e\, f), \\ \mathsf{ls}\,\mathsf{NE}\, e\, f \iff (e \mapsto f) \vee (\exists y'.\, e \mapsto y' * \mathsf{ls}\,\mathsf{NE}\, y'\, f). \end{array}$$

These definitions are not *within* the shape domain (*e.g.*, the domain does not have $\vee$), but are mathematical definitions in the metalanguage, used to verify soundness of operations on the predicates. Note that there is no problem with the recursion in $\mathsf{ls}\,\mathsf{NE}$: the recursive instance is in a positive position, and the definition satisfies monotonicity properties sufficient to ensure a solution.

A different instantiation of $\mathcal{P}$ gives us a variation on [3].[2]

$$k \ ::= \ \mathsf{PE} \mid \mathsf{NE} \qquad\qquad \mathcal{P} \ ::= \ (e \mapsto \vec{\mathtt{f}} \colon \vec{e}) \mid \mathsf{ls}\, k\, \phi\, e\, e$$

Here, the points-to predicate $(e \mapsto \vec{\mathtt{f}} \colon \vec{e})$ is for records with fields $\vec{\mathtt{f}}$, and ϕ is a binary predicate that describes the shape of each node in a list. The definition of the nonempty list segment here is

$$\mathsf{ls}\,\mathsf{NE}\, \phi\, e\, f \iff \phi(e, f) \vee (\exists y'.\, \phi(e, y') * \mathsf{ls}\,\mathsf{NE}\, y'\, f)$$

and the ϕ predicate gives us a way to describe composite structures.

[2] This instantiation assumes the change of the language where we have heap cells with multiple fields, instead of unary cells.

For example, if $\mathtt{f}$ is a field, let $\phi_{\mathtt{f}}$ be the predicate where $\phi_{\mathtt{f}}(x,y)$ is $x \mapsto \mathtt{f}\colon y$. Then using $\phi_{\mathtt{f}}$ as ϕ, the formula $\mathsf{ls}\,\mathsf{NE}\,\phi\,e\,f$ describes lists linked by the $\mathtt{f}$ field. The formula

$$(x \mapsto \mathtt{f}\colon y', \mathtt{g}\colon z') \;*\; \mathsf{ls}\,\mathsf{PE}\,\phi_{\mathtt{f}}\,y'\,x \;*\; \mathsf{ls}\,\mathsf{PE}\,\phi_{\mathtt{g}}\,z'\,x$$

describes two circular linked lists sharing a common header, where one list uses $\mathtt{f}$ for linking and the other uses $\mathtt{g}$. Finally, if ϕ itself describes lists, as when $\phi(x,y)$ is the predicate $\exists x'.\,(x \mapsto \mathtt{g}\colon x', \mathtt{f}\colon y) * \mathsf{ls}\,\mathsf{PE}\,\phi_{\mathtt{g}}\,x'\,0$, then $\mathsf{ls}\,\mathsf{NE}\,\phi\,e\,f$ describes a nonempty linked list where each node points to a possibly empty sublist, and where the sublists are disjoint. Combinations of these kinds of structures, nested lists and multiple lists with a common header node, are common in device drivers.

The experiments in this paper are done using this second instantiation of $\mathcal{P}$. It is similar to the domain from [3], but uses predicates for both possibly empty and necessarily nonempty list segments. The reader might have noticed that having $\mathsf{ls}\,\mathsf{PE}$ does not give us any extra expressive power: its meaning can be represented using two abstract states, one a emp and the other a $\mathsf{ls}\,\mathsf{NE}$. However, having $\mathsf{ls}\,\mathsf{PE}$ impacts performance, as it represents disjunctive information, succinctly.

SpaceInvader implements a context sensitive, flow sensitive analysis, using a variant of the RHS interprocedural dataflow analysis algorithm [22,11]. It employs join to make procedure summaries smaller. Following [21,23], SpaceInvader also passes only the reachable portion of the heap to a procedure and aggressively discards intermediate states. The mixture of these optimizations—join, locality, discarding states—is key; turning off any one of the optimizations results in the analysis using more than the 2GB RAM on at least one of the examples, causing disk thrashing, and then leading to timeout (which we set at 90min). Thus we do not claim that $\uplus$ alone is the root cause for the performance found in Table 1, but it is a critical ingredient (c.f., §4.3).

4 A Join for Symbolic Heaps

We now discuss $\uplus$. We begin with an intuitive explanation. Later, in §4.1, we provide a formal definition.

In the framework of abstract interpretation [9], a join operator takes two symbolic states in a program analysis and attempts to find a common generalization. To see the issue, consider the program

```
x=0;  while (NONDET) { d=malloc(sizeof(Node)); d->next=x; x=d; }
```

which nondeterministically generates acyclic linked lists. When we run our basic analysis algorithm, without $\uplus$, it returns three symbolic heaps at the end: $(\mathsf{ls}\,\mathsf{NE}\,x\,0) \vee (x \mapsto 0) \vee (x = 0 \wedge \mathsf{emp})$. (Here, for simplicity in the presentation, we have elided the ϕ parameter of the ls predicates.)

Now, if you look at the first two disjuncts there is evident redundancy: If you know that either x points to 0 or a nonempty linked list, then that is the same as knowing you have a nonempty linked list. So, $\uplus$ replaces the first two

Table 2. Creation routines. Reports the number of states in the postcondition with join turned on or off, and the base list predicates chosen to be either nonempty ls only (NE), or both nonempty and possibly empty ls (PE).

	NO JOIN		JOIN	
Program	NE	PE	NE	PE
onelist_create.c	3	3	2	1
twolist_create.c	9	9	4	1
firewire_create.c	3969	3087	32	1

disjuncts with just the list segment formula, giving us $(\mathsf{ls}\,\mathsf{NE}\, x\, 0) \vee (x = 0 \wedge \mathsf{emp})$. It is possible to take yet a further step, using the notion of a possibly empty list segment. If you know that either you have a nonempty list, or that $x = 0 \wedge \mathsf{emp}$, then that is the same as having $\mathsf{ls}\,\mathsf{PE}\, x\, 0$, and ⧺ produces this formula from the previous two. Thus, using ⧺ we have gone from a position where we have three disjuncts in our postcondition, to where we have only one. The saving that this possibly gives us is substantial, especially for more complicated programs or more complicated data structures.

Table 2 gives an indication. `onelist_create.c` in the table is the C program above that nondeterministically creates a list and `twolist_create.c` is a similar C program that creates two disjoint linked lists. `firewire_create.c` is the environment code we use in the analysis of the 1394 firewire driver: it creates five cyclic linked lists, which share a common header node, with head pointers in some of the lists, and with nested sublists.

There are two points to note. The first is just the great saving, in number of states (*e.g.*, from 3087 down to 1). This is particularly important with environment code, like `firewire_create.c`, which is run as a harness to generate heaps on which driver routines will subsequently be run. The second is the distinction between NE and PE. In the table we keep track of two versions of our analysis, one where ls NE is the only list predicate used by the analysis, and another where we use both ls NE and ls PE.

This illustration shows some of the aspects of ⧺, but not all. In the illustration ⧺ worked perfectly, never losing any information, but this is not always the case. Part of the intuition is that you generalize points-to facts by list segments when you can. So, considering $y \mapsto 0 * (\mathsf{ls}\,\mathsf{NE}\, x\, 0) \vee (\mathsf{ls}\,\mathsf{NE}\, y\, 0) * x \mapsto 0$, ⧺ will produce $(\mathsf{ls}\,\mathsf{NE}\, y\, 0) * (\mathsf{ls}\,\mathsf{NE}\, x\, 0)$. This formula is less precise than the disjunction, in that it loses the information that one or the other of the lists pointed to by x and y has length precisely 1. Fortunately, it is unusual for programs to rely on this sort of disjunctive information.

We have tried to keep the intuitive description simple, but the truth is that ⧺ must deal with disequalities, equalities, and generalization of "nothing" by ls PE in ways that are nontrivial. It also must deal with the existential (primed) variables specially. In the end, for instance, when ⧺ is given

$$q_0 \equiv x \neq y \wedge (\mathsf{ls}\,\mathsf{NE}\, x\, 0 * y \mapsto 0) \qquad \text{and}$$
$$q_1 \equiv x \neq y \wedge x' \neq y \wedge (x \mapsto x' * \mathsf{ls}\,\mathsf{NE}\, y\, x' * \mathsf{ls}\,\mathsf{NE}\, x'\, 0),$$

it will produce $x{\neq}y \wedge \mathsf{ls}\,\mathsf{NE}\,x\,v' * \mathsf{ls}\,\mathsf{NE}\,y\,v' * \mathsf{ls}\,\mathsf{PE}\,v'\,0$. Now we turn to the formal definition.

4.1 Formal Definition

In this section, we define the (partial) binary operator $\boxplus$ on symbolic heaps, considering only the simple linked lists (the first instantiation of $\mathcal{P}$). The $\boxplus$ for nested lists will be described in the next section.

$\boxplus$ works in two stages. Suppose that it is given symbolic heaps $(\Pi_0 \wedge \Sigma_0)$ and $(\Pi_1 \wedge \Sigma_1)$ that do not share any primed variables. In the first stage, $\boxplus$ constructs Σ and a ternary relation ϵ' on expressions such that

$$\forall i \in \{0,1\}.\ \ \left(\bigwedge\{e_i{=}x' \mid (e_0,e_1,x') \in \epsilon'\}\right) \wedge \Sigma_i \implies \Sigma. \tag{1}$$

Intuitively, this condition means that Σ overapproximates both Σ_0 and Σ_1, and that ϵ' provides witnesses of existential (primed) variables of Σ for this overapproximation. For instance, if $\Sigma_0 \equiv (\mathsf{ls}\,\mathsf{NE}\,x\,0 * y \mapsto 0)$ and $\Sigma_1 \equiv (x \mapsto x' * \mathsf{ls}\,\mathsf{NE}\,y\,x' * \mathsf{ls}\,\mathsf{NE}\,x'\,0)$, then $\boxplus$ returns

$$\Sigma \equiv \mathsf{ls}\,\mathsf{NE}\,x\,v' * \mathsf{ls}\,\mathsf{NE}\,y\,v' * \mathsf{ls}\,\mathsf{PE}\,v'\,0, \qquad\qquad \epsilon' \equiv \{(0,x',v')\}. \tag{2}$$

In this case, the condition (1) is

$$\begin{aligned} 0{=}v' \wedge (\mathsf{ls}\,\mathsf{NE}\,x\,0 * y \mapsto 0) &\implies (\mathsf{ls}\,\mathsf{NE}\,x\,v' * \mathsf{ls}\,\mathsf{NE}\,y\,v' * \mathsf{ls}\,\mathsf{PE}\,v'\,0) \\ x'{=}v' \wedge (x \mapsto x' * \mathsf{ls}\,\mathsf{NE}\,y\,x' * \mathsf{ls}\,\mathsf{NE}\,x'\,0) &\implies (\mathsf{ls}\,\mathsf{NE}\,x\,v' * \mathsf{ls}\,\mathsf{NE}\,y\,v' * \mathsf{ls}\,\mathsf{PE}\,v'\,0). \end{aligned}$$

This means that both Σ_0 and Σ_1 imply Σ when 0 and x' are used as witnesses for the (implicitly) existentially quantified variable v' of Σ.

After constructing Σ and ϵ', the $\boxplus$ operator does one syntactic check on ϵ', in order to decide whether it has lost crucial sharing information of input symbolic heaps. Only when the check succeeds does $\boxplus$ move on to the second stage. (We will describe the details of the first stage, including the check on ϵ', later.)

In the second stage, the $\boxplus$ operator computes an overapproximation Π of Π_0 and Π_1:

$$\Pi \ \stackrel{def}{=}\ \bigwedge \left(\begin{array}{l} \{e{=}f \mid e{=}f \text{ has no primed vars, it occurs in } \Pi_0 \text{ and } \Pi_1\} \\ \cup\, \{e{\neq}f \mid e{\neq}f \text{ has no primed vars, it occurs in } \Pi_0 \text{ and } \Pi_1\} \\ \cup\, \{x'{\neq}0 \mid (e_0,e_1,x') \in \epsilon' \text{ and } e_i{\neq}0 \text{ occurs in } \Pi_i\} \end{array}\right).$$

This definition says that $\boxplus$ keeps an equality or disequality in Π if it appears in both Π_0 and Π_1 and does not contain any primed variables, or if it is of the form $x'{\neq}0$ and its witness e_i for the i-th symbolic heap is guaranteed to be different from 0. Both cases are considered here in order to deal with programming patterns found in device drivers. For instance, $x'{\neq}0$ in the second case should be included, because some drivers store 0 or 1 to a cell, say, x, depending on whether a linked list y is empty, and subsequently, they use the contents of cell x to decide the emptiness of the list y. The computed Π and the result Σ of the first stage become the output of $\boxplus$.

Computation of Σ, ϵ': We now describe the details of the first stage of $\sqcup\!\!\!\!\sqcap$. For this, we need a judgment

$$\Sigma_0, \Sigma_1, \epsilon \rightsquigarrow \Sigma, \epsilon', \delta_0, \delta_1$$

where δ_i is a binary relation on expressions in Σ_i. This judgment signifies that Σ_0 and Σ_1 can be joined to give Σ and a ternary relation ϵ' for witnesses. Furthermore, the judgment ensures that ϵ' extends the given ϵ, and that δ_i records (e_i, f_i) of all $\mathsf{ls}\, k\, e_i\, f_i$ in Σ_i that have been generalized to a possibly empty list during the join; these δ_i components are used later to decide whether this join to Σ has lost too much information and should, therefore, be discarded. For instance, we have

$$\begin{aligned} &(\mathsf{ls}\,\mathsf{NE}\,x\,0 * y \mapsto 0),\ \ (x \mapsto x' * \mathsf{ls}\,\mathsf{NE}\,y\,x' * \mathsf{ls}\,\mathsf{NE}\,x'\,0),\ \ \emptyset \\ &\quad \rightsquigarrow\ (\mathsf{ls}\,\mathsf{NE}\,x\,v' * \mathsf{ls}\,\mathsf{NE}\,y\,v' * \mathsf{ls}\,\mathsf{PE}\,v'\,0),\ \{(0, x', v')\},\ \emptyset,\ \{(x', 0)\}. \end{aligned}$$

which means that $\Sigma_0 \equiv (\mathsf{ls}\,\mathsf{NE}\,x\,0 * y \mapsto 0)$ and $\Sigma_1 \equiv (x \mapsto x' * \mathsf{ls}\,\mathsf{NE}\,y\,x' * \mathsf{ls}\,\mathsf{NE}\,x'\,0)$ are joined to $\Sigma \equiv (\mathsf{ls}\,\mathsf{NE}\,x\,v' * \mathsf{ls}\,\mathsf{NE}\,y\,v' * \mathsf{ls}\,\mathsf{PE}\,v'\,0)$. The judgment also says that v' in Σ corresponds to 0 in Σ_0 and x' in Σ_1. Note that the δ_1 component of the judgment is $\{(x', 0)\}$, and it reflects the fact that $\mathsf{ls}\,\mathsf{NE}\,x'\,0$ in Σ_1 is generalized to a possibly empty list and results in $\mathsf{ls}\,\mathsf{PE}\,v'\,0$ in Σ.

The derivation rules of the $\rightsquigarrow$ predicate are given in Figure 1. The first two rules deal with the cases when emp or true appear in both Σ_0 and Σ_1. The third rule has to do with generalizing two lists or abstracting a points-to to a list, and the last two rules are about generalizing (or synthesizing) possibly empty lists. Note that when possibly empty lists are introduced by the last two rules, the appropriate δ_i component is extended with the information about the ls predicate of Σ_i that supports this generalization.

The first stage of $\sqcup\!\!\!\!\sqcap$ works as follows:

1. $\sqcup\!\!\!\!\sqcap$ searches for $\Sigma, \epsilon', \delta_0, \delta_1$ for which $\Sigma_0, \Sigma_1, \emptyset \rightsquigarrow \Sigma, \epsilon', \delta_0, \delta_1$ can be derived using the rules in Figure 1. This proof search proceeds by viewing rules backward from conclusion to premise. It searches for a rule whose conclusion has the left hand side matching with $\Sigma_0, \Sigma_1, \epsilon$ and whose side condition is satisfied with this matching. Once such a rule is found, the search modifies $\Sigma_0, \Sigma_1, \epsilon$ such that they fit the left hand side of the $\rightsquigarrow$ judgment in the premise. The search continues with this modified $\Sigma_0, \Sigma_1, \epsilon$, until it hits the base case (*i.e.*, the first rule in Figure 1). Figure 2 shows an example proof search. If the search fails, the join fails.
2. $\sqcup\!\!\!\!\sqcap$ checks whether for all $(e_0, e_1, e), (f_0, f_1, f) \in \epsilon' \cup \{(e, e, e) \mid e \text{ not primed var}\}$ and all $i \in \{0, 1\}$,

$$\big(e_i = f_i \wedge e_i \neq 0 \implies (e_{1-i}, f_{1-i}) \in \mathsf{eq}(\delta_{1-i})\big),$$

where $\mathsf{eq}(\delta_i)$ is the least equivalence relation containing δ_i. Intuitively this condition amounts to the following: consider Σ_0 and Σ_1 viewed as graphs with edges for $\mapsto$ and ls, and then identify vertices according to the returned δ's, then they should be isomorphic via $\epsilon' \cup \{(e, e, e) \mid e \text{ not primed var}\}$.

$$A(e,f) ::= (e \mapsto f) \mid \mathsf{ls}\, k\, e\, f \qquad \mathsf{EQ} = \{(e,e,e) \mid e \text{ is not a primed var}\}$$

$$\mathsf{PE} \sqcup \mathsf{NE} = \mathsf{NE} \sqcup \mathsf{PE} = \mathsf{PE} \sqcup \mathsf{PE} = \mathsf{PE} \qquad \mathsf{NE} \sqcup \mathsf{NE} = \mathsf{NE}$$

$$A(e,f) \sqcup A(e,f) = A(e,f) \quad (\mathsf{ls}\, k_0\, e\, f) \sqcup (\mathsf{ls}\, k_1\, e\, f) = (\mathsf{ls}\, (k_0 \sqcup k_1)\, e\, f)$$
$$(e \mapsto f) \sqcup (\mathsf{ls}\, k\, e\, f) = (\mathsf{ls}\, k\, e\, f) \sqcup (e \mapsto f) = \mathsf{ls}\, k\, e\, f$$

$$\frac{}{\mathsf{emp},\ \mathsf{emp},\ \epsilon \rightsquigarrow \mathsf{emp},\ \epsilon,\ \emptyset,\ \emptyset}\,\mathsf{emp} \qquad \frac{\Sigma_0\ ,\ \Sigma_1\ ,\ \epsilon \rightsquigarrow \Sigma\ ,\ \epsilon'\ ,\ \delta_0\ ,\ \delta_1}{\mathsf{true} * \Sigma_0,\ \mathsf{true} * \Sigma_1,\ \epsilon \rightsquigarrow \mathsf{true} * \Sigma,\ \epsilon',\ \delta_0,\ \delta_1}\,\mathsf{true}$$

$$\frac{\Sigma_0\ ,\ \Sigma_1\ ,\ \mathsf{ext}(\epsilon, f_0, f_1, f) \rightsquigarrow \Sigma\ ,\ \epsilon'\ ,\ \delta_0\ ,\ \delta_1}{A_0(e_0, f_0) * \Sigma_0\ ,\ A_1(e_1, f_1) * \Sigma_1\ ,\ \epsilon \rightsquigarrow (A_0(e,f) \sqcup A_1(e,f)) * \Sigma\ ,\ \epsilon'\ ,\ \delta_0\ ,\ \delta_1}\,\mathsf{match}$$
$$(\text{when } (e_0, e_1, e) \in (\epsilon \cup \mathsf{EQ}) \ \wedge\ \mathsf{comb}_\epsilon(f_0, f_1) = f)$$

$$\frac{\Sigma_0\ ,\ \Sigma_1\ ,\ \mathsf{ext}(\epsilon, f_0, e_1, f) \rightsquigarrow \Sigma\ ,\ \epsilon'\ ,\ \delta_0\ ,\ \delta_1}{(\mathsf{ls}\, k\, e_0\, f_0) * \Sigma_0\ ,\ \Sigma_1\ ,\ \epsilon \rightsquigarrow (\mathsf{ls}\, \mathsf{PE}\, e\, f) * \Sigma\ ,\ \epsilon'\ ,\ \delta_0 \cup (e_0, f_0)\ ,\ \delta_1}\,\mathsf{PE\text{-}left}$$
$$(\text{when } (e_0, e_1, e) \in (\epsilon \cup \mathsf{EQ}) \ \wedge\ e_1 \notin \mathsf{MayAlloc}(\Sigma_1) \ \wedge\ \mathsf{comb}_\epsilon(f_0, e_1) = f)$$

$$\frac{\Sigma_0\ ,\ \Sigma_1\ ,\ \mathsf{ext}(\epsilon, e_0, f_1, f) \rightsquigarrow \Sigma\ ,\ \epsilon'\ ,\ \delta_0\ ,\ \delta_1}{\Sigma_0\ ,\ (\mathsf{ls}\, k\, e_1\, f_1) * \Sigma_1\ ,\ \epsilon \rightsquigarrow (\mathsf{ls}\, \mathsf{PE}\, e\, f) * \Sigma\ ,\ \epsilon'\ ,\ \delta_0\ ,\ \delta_1 \cup (e_1, f_1)}\,\mathsf{PE\text{-}right}$$
$$(\text{when } (e_0, e_1, e) \in (\epsilon \cup \mathsf{EQ}) \ \wedge\ e_0 \notin \mathsf{MayAlloc}(\Sigma_0) \ \wedge\ \mathsf{comb}_\epsilon(e_0, f_1) = f)$$

Here (a) we write $X \cup x$ instead of $X \cup \{x\}$; (b) $\mathsf{ext}(\epsilon, e_0, e_1, e)$ is $(\epsilon \cup (e_0, e_1, e)) - \mathsf{EQ}$; (c) $\mathsf{MayAlloc}(\Sigma)$ is the set of expressions that appear on the left hand side of a points-to predicate or as a first expression argument of ls in Σ; (d) comb_ϵ is a function defined as:

$$\mathsf{comb}_\epsilon(e_0, e_1) = \begin{cases} e & \text{if } (e_0, e_1, e) \in \epsilon \text{ for some } e \\ e_0 & \text{if } e_0 = e_1 \text{ and } e_0 \text{ is not a primed var} \\ x' \text{ for some } x' \notin \mathsf{FV}(\epsilon, e_0, e_1) & \text{otherwise} \end{cases}$$

Fig. 1. Rules for $\rightsquigarrow$

Only when the check succeeds does the first stage of $\sqcup\!\!\!\!\sqcap$ return Σ, ϵ'. For instance, given $\Sigma_0 \equiv (x \mapsto y) * \mathsf{ls}\, \mathsf{NE}\, y\, 0$ and $\Sigma_1 \equiv \mathsf{ls}\, \mathsf{NE}\, x\, 0 * (y \mapsto 0)$, the proof search in the previous step succeeds with

$$\Sigma \equiv \mathsf{ls}\, \mathsf{NE}\, x\, y' * \mathsf{ls}\, \mathsf{NE}\, y\, 0, \qquad \epsilon' \equiv \{(y, 0, y')\}, \qquad \delta_0 \equiv \delta_1 \equiv \emptyset.$$

However, the final check on ϵ' fails, since y in the Σ_0 symbolic heap is related to both 0 (by ϵ') and y (by default) in Σ_1. Thus, the join fails. Note that the failure is desired in this case since Σ_0 and Σ_1 describe heaps with different shapes.

4.2 Composite Structures

In order to handle composite structures, such as nested lists, we adjust the definition of $\sqcup\!\!\!\!\sqcap$ in the previous section. Specifically, we change the rules for the $\rightsquigarrow$ relation in Figure 1. Firstly, we modify the third rule, which is used to generalize

$$\dfrac{\dfrac{\dfrac{\dfrac{}{\mathsf{emp}\ ,\ \mathsf{emp}\ ,\ \epsilon'\ \leadsto\ \mathsf{emp}\ ,\ \epsilon'\ ,\ \emptyset\ ,\ \emptyset}\ \mathsf{emp}}{\mathsf{emp}\ ,\ (\mathsf{ls\,NE}\,x'\,0)\ ,\ \epsilon'\ \leadsto\ \mathsf{ls\,PE}\,v'\,0\ ,\ \epsilon'\ ,\ \emptyset\ ,\ \{(x',0)\}}\ \mathsf{PE\text{-}right}}{(y\mapsto 0)\ ,\ (\mathsf{ls\,NE}\,y\,x' * \mathsf{ls\,NE}\,x'\,0)\ ,\ \epsilon'\ \leadsto\ \mathsf{ls\,NE}\,y\,v' * \mathsf{ls\,PE}\,v'\,0\ ,\ \epsilon'\ ,\ \emptyset\ ,\ \{(x',0)\}}\ \mathsf{match}}{\begin{array}{l}(\mathsf{ls\,NE}\,x\,0 * y\mapsto 0)\ ,\ (x\mapsto x' * \mathsf{ls\,NE}\,y\,x' * \mathsf{ls\,NE}\,x'\,0)\ ,\ \emptyset \\ \qquad\qquad \leadsto\ \mathsf{ls\,NE}\,x\,v' * \mathsf{ls\,NE}\,y\,v' * \mathsf{ls\,PE}\,v'\,0\ ,\ \epsilon'\ ,\ \emptyset\ ,\ \{(x',0)\}\end{array}}\ \mathsf{match}$$

Fig. 2. Example proof search, where $\epsilon' = \{(0, x', v')\}$

two ls or points-to predicates, such that it can deal with points-to predicates with multiple fields $\vec{\mathtt{f}}$ and a parameterized list-segment predicate. Each of the new rules, shown in Figure 3, corresponds to one of the four cases of $A_0 \sqcup A_1$ in the third rule of Figure 1. The first rule combines two points-to predicates with multiple fields, by extending ϵ with the targets of all the fields. The other rules generalize two list-segment predicates (the second rule) or a list segment and its length-one instance (the third and fourth rules), by looking inside the two available descriptions of list nodes (denoted ϕ_0 and ϕ_1), and chooses the more general one (denoted $\phi_0 \sqcup \phi_1$). In the third rule of Figure 3, the first input symbolic heap is decomposed into $\phi_0(e_0, f_0)[\vec{e}/\vec{x'}] * \Sigma_0$ using a frame inference algorithm [4] to subtract a symbolic heap $\phi_0(e_0, f_0)[\vec{e}/\vec{x'}]$ such that ϕ_0 can be $\sqcup$-joined with ϕ_1, leaving Σ_0 as a remainder. And similarly in the fourth rule. Secondly, we change the remaining rules in Figure 1 such that they work with parameterized list-segment predicates. We simply replace all unparameterized list-segment predicates $\mathsf{ls}\,k\,e\,e'$ in the rules by parameterized ones $\mathsf{ls}\,k\,\phi\,e\,e'$.

After these changes, $\boxplus$ works for composite structures. For instance, let $\phi_{\mathsf{d}}(x, y) \equiv (x \mapsto \mathsf{d}{:}y)$, $\phi(x, y) \equiv \exists x'.\,(x \mapsto \mathsf{d}{:}x', \mathtt{f}{:}y) * (\mathsf{ls\,PE}\,\phi_{\mathsf{d}}\,x'\,0)$, and $\psi(x, y) \equiv (x \mapsto \mathsf{d}{:}0, \mathtt{f}{:}y)$. Given two symbolic heaps

$$(\mathsf{ls\,NE}\,\phi\,x\,y) * (y \mapsto \mathsf{d}{:}y', \mathtt{f}{:}0) * (y' \mapsto \mathsf{d}{:}0) \quad \vee \quad (\mathsf{ls\,PE}\,\psi\,x\,y) * (\mathsf{ls\,PE}\,\phi\,y\,0),$$

the $\boxplus$ generalizes the list segments from x to y to a possibly empty ϕ-shaped list since $\psi(x, y) \vdash \phi(x, y)$. Then, it views the two points-to facts on y and y' as an instantiation $\phi'(x, y)[y'/x']$ of $\phi'(x, y) \equiv \exists x'.\,(x \mapsto \mathsf{d}{:}x', \mathtt{f}{:}y) * (x' \mapsto \mathsf{d}{:}0)$, combines these facts with the list y since $\phi'(x, y) \vdash \phi(x, y)$, and produces

$$\mathsf{ls\,PE}\,\phi\,x\,y * \mathsf{ls\,PE}\,\phi\,y\,0.$$

4.3 Incorporating $\boxplus$ into the Analysis

SpaceInvader incorporates $\boxplus$ together with RHS [22], a now-standard interprocedural analysis algorithm. RHS associates a set of symbolic heaps with each program point, which represents the disjunction of those heaps. $\boxplus$ is applied to reduce the number of disjuncts in those sets.

Given a set of symbolic heaps at a program point, the analysis takes two symbolic heaps in the set and applies $\boxplus$ to them. If the application succeeds, the

$$\phi_0 \sqcup \phi_1 = \begin{cases} \phi_0 & \text{if } \phi_1(x,y) \vdash \phi_0(x,y) \\ \phi_1 & \text{if } \phi_0(x,y) \vdash \phi_1(x,y) \\ \text{undefined} & \text{otherwise} \end{cases} \quad \begin{array}{l}\text{where } \phi_0(x,y) \vdash \phi_1(x,y) \text{ denotes} \\ \text{a call to a sound theorem prover} \\ \text{for fresh } x, y\end{array}$$

$$\frac{\Sigma_0, \Sigma_1, \mathsf{ext}(\mathsf{ext}(\epsilon, f_0, f_1, f), g_0, g_1, g) \leadsto \Sigma, \epsilon', \delta_0, \delta_1}{(e_0 \mapsto \mathtt{f}{:}f_0, \mathtt{g}{:}g_0) * \Sigma_0,\ (e_1 \mapsto \mathtt{f}{:}f_1, \mathtt{g}{:}g_1) * \Sigma_1,\ \epsilon\ \leadsto\ (e \mapsto \mathtt{f}{:}f, \mathtt{g}{:}g) * \Sigma,\ \epsilon',\ \delta_0,\ \delta_1}\ \mathsf{match1}$$
$$(\text{when } (e_0, e_1, e) \in (\epsilon \cup \mathsf{EQ})\ \wedge\ \mathsf{comb}_\epsilon(f_0, f_1) = f\ \wedge\ \mathsf{comb}_\epsilon(g_0, g_1) = g)$$

$$\frac{\Sigma_0,\ \Sigma_1,\ \mathsf{ext}(\epsilon, f_0, f_1, f)\ \leadsto\ \Sigma,\ \epsilon',\ \delta_0,\ \delta_1}{\mathsf{ls}\,k_0\,\phi_0\,e_0\,f_0 * \Sigma_0,\ \mathsf{ls}\,k_1\,\phi_1\,e_1\,f_1 * \Sigma_1,\ \epsilon\ \leadsto\ \mathsf{ls}\,(k_0 \sqcup k_1)\,(\phi_0 \sqcup \phi_1)\,e\,f * \Sigma,\ \epsilon',\ \delta_0,\ \delta_1}\ \mathsf{match2}$$
$$(\text{when } (e_0, e_1, e) \in (\epsilon \cup \mathsf{EQ})\ \wedge\ \phi_0 \sqcup \phi_1 \text{ is defined}\ \wedge\ \mathsf{comb}_\epsilon(f_0, f_1) = f)$$

$$\frac{\Sigma_0,\ \Sigma_1,\ \mathsf{ext}(\epsilon, f_0, f_1, f)\ \leadsto\ \Sigma,\ \epsilon',\ \delta_0,\ \delta_1}{\phi_0(e_0, f_0)[\vec{e}/\vec{x'}] * \Sigma_0,\ (\mathsf{ls}\,k\,\phi_1\,e_1\,f_1) * \Sigma_1,\ \epsilon\ \leadsto\ (\mathsf{ls}\,k\,(\phi_0 \sqcup \phi_1)\,e\,f) * \Sigma,\ \epsilon',\ \delta_0,\ \delta_1}\ \mathsf{match3}$$
$$(\text{when } (e_0, e_1, e) \in (\epsilon \cup \mathsf{EQ})\ \wedge\ \phi_0 \sqcup \phi_1 \text{ is defined}\ \wedge\ \mathsf{comb}_\epsilon(f_0, f_1) = f)$$

$$\frac{\Sigma_0,\ \Sigma_1,\ \mathsf{ext}(\epsilon, f_0, f_1, f)\ \leadsto\ \Sigma,\ \epsilon',\ \delta_0,\ \delta_1}{(\mathsf{ls}\,k\,\phi_0\,e_0\,f_0) * \Sigma_0,\ \phi_1(e_1, f_1)[\vec{e}/\vec{x'}] * \Sigma_1,\ \epsilon\ \leadsto\ (\mathsf{ls}\,k\,(\phi_0 \sqcup \phi_1)\,e\,f) * \Sigma,\ \epsilon',\ \delta_0,\ \delta_1}\ \mathsf{match4}$$
$$(\text{when } (e_0, e_1, e) \in (\epsilon \cup \mathsf{EQ})\ \wedge\ \phi_0 \sqcup \phi_1 \text{ is defined}\ \wedge\ \mathsf{comb}_\epsilon(f_0, f_1) = f)$$

Here $-[\vec{e}/\vec{x'}]$ in $\phi(e, f)[\vec{e}/\vec{x'}]$ is the substitution of all the existentially quantified primed variables $\vec{x'}$ in $\phi(e, f)$ by $\vec{e}$.

Fig. 3. Sample rules for $\leadsto$. Composite structure case.

result of the join replaces those heaps. Otherwise, those two symbolic heaps are returned to the set.

In order to maintain precision in the analysis, we restrict the application of $\boxplus$ to only those program points where controlling the number of disjuncts is crucial. They are (a) the beginning of loops, (b) the end of conditional statements when those statements are not inside loops, (c) the call sites of procedures, and (d) the exit points of procedures. The first case accelerates the analysis of the usual fixed-point computation for loops, and the second prevents the combinatorial explosion caused by a sequence of conditional statements; for instance, the procedure `register_cdrom` in `cdrom.c` uses 25 conditional statements to adjust values of a structure for cdrom, which makes the analysis without join suffer from a serious performance problem. The other two cases aim for computing small procedure summaries; the third reduces the number of input symbolic heaps to consider for each procedure, and the last reduces the analysis results of a procedure with respect to each symbolic heap.

We have measured the effects of $\boxplus$ on the performance of SpaceInvader, using our seven driver examples. Table 3 reports the results of our measurements. The third and fourth columns of the table record the time of analyzing the drivers without using $\boxplus$: without $\boxplus$, we cannot analyze our example drivers except the simplest one, `scull.c`. The next two columns concern a pivotal design decision for $\boxplus$, looking at variations on the ls predicate; the fifth column considers the necessarily non-empty ls predicate only, and the sixth column considers both

Table 3. Experimental results on the effects of ⫵. Timeout (X) set at 90min. Experiments run on Intel Core Duo 2.0GHz with 2GB RAM. The "Opt. except ⫵" column records the results of the analysis runs without ⫵ nor possibly empty ls predicates, but with two optimizations: discarding the intermediate analysis results and passing only the reachable portion of the heap to a procedure. The next column contains the analysis time with these two optimizations and ⫵, but without possibly empty ls predicates. The last column contains the analysis time with all the optimizations.

Program	LOC	No Opt. (sec)	Opt. except ⫵ (sec)	Opt. including ⫵, with NE only (sec)	Opt. including ⫵, with NE and PE (sec)
`scull.c`	1010	1.41	1.15	0.59	0.36
`class.c`	1983	X	X	48.24	8.21
`pci-driver.c`	2532	X	X	2.69	0.97
`ll_rw_blk.c`	5469	X	X	X	887.94
`cdrom.c`	6218	X	X	193.01	103.26
`md.c`	6653	X	X	X	1585.69
`t1394Diag.c`	10240	X	X	3415.76	135.05

the necessarily non-empty and possibly empty ls predicates. These experimental results confirm the benefit of using the $\mathsf{ls}\,\mathsf{PE}$ predicate in ⫵.

5 Related Work

Device driver verification has attracted considerable interest due to the realization that most OS failures arise from bugs in device drivers [8,24,2]. Tools like SLAM [2] and BLAST [15] have been effectively applied in verification of properties of real device drivers, especially properties describing the calling conventions of OS kernel APIs. Unfortunately these tools use coarse models of the heap; SLAM, for example, *assumes* memory safety. Other tools are known to prove memory safety, but with the restriction that the input programs do not perform dynamic memory allocation (*e.g.* ASTRÉE [6]). Proving full memory safety (which includes array bounds errors as well as what we have termed pointer safety) of entire systems programs is thus a more difficult problem than that considered in this paper, or in work that concentrates on array bounds errors.

Several papers report on the results of applying shape analysis to the source code of substantial, real-world systems programs. The analysis in [14] has been applied to non-trivial code, but the abstract domain there is purposely much less precise than here, and it could not be used to verify pointer safety of the device drivers that we consider. [7] includes an analysis of a restricted and modified version of the Linux scull driver. Our analysis terminates on the modified scull code (which they kindly supplied to us) in 0.36sec, where [7] terminated in 9.71sec when using user-supplied assertions (which we did not use) to help the analysis along. It is also worth mentioning [13], which uses slicing to remove heap-irrelevant statements. An earlier version of SPACEINVADER [3] analyzed several procedures from the 1394 driver used in Table 1. It timed out on an 1800 LOC subset of the driver, and this drove us to consider ⫵.

The very idea of a join operator is of course not novel, and many other joins have been successfully applied in their application domain. The problem is always one of balancing precision and speed. The claim that $\sqcup$ does not lose too much precision is backed up with experimental results. $\sqcup$ is not unrelated to other join operators that have been proposed in shape analysis [19,1,7]. For instance, Chang *et al.* define a partial join operator for separation logic formulas, and Arnold [1] develops a notion of "loose embedding" in TVLA [16] which is in an intuitive sense related to our use of predicates for possibly-empty, rather than only nonempty, lists. However, our $\sqcup$ is different in its detailed formulation; unlike Chang *et al.*, we simplify symbolic heaps before applying $\sqcup$, and unlike Arnold and Manevich [19,1], our $\sqcup$ keeps the structure of composite data structures precisely. The latter difference, in particular, is crucial to verifying the drivers.

Marron *et al.* reports on shape analyses of several Java programs of up to 3705 LOC [20]. They use an aggressive join operator which *always* merges several abstract states into one. Such a join operator would lead to many false alarms when applied to our device drivers (for example, when dealing with exceptional conditions), and so is too imprecise for our goal of *proving* pointer safety.

6 Conclusions

This paper has presented the first application of shape analysis to a real-world industrial verification problem: proving pointer safety of entire Windows and Linux device drivers. We have achieved this milestone by enhancing our separation domain based shape analysis tool with a sophisticated new join operation, $\sqcup$. This paper has made two contributions: $\sqcup$, and a demonstration that shape analysis can be scaled to real-world industrial verification problems. The second contribution is, in a sense, the most important one. We hope, now that we know that whole device drivers can be accurately handled by today's shape analysis tools, that future research papers on the subject will use device drivers and other substantial systems programs as a part of their experimental evaluations.

Acknowledgments. We would like to thank Viktor Vafeiadis for helpful discussions on the OCaml garbage collector. The London authors acknowledge the support of the EPSRC. Lee was supported by Brain Korea 21. Distefano was supported by a Royal Academy of Engineering research fellowship. O'Hearn was supported by a Royal Society Wolfson Research Merit Award.

References

1. Arnold, G.: Specialized 3-valued logic shape analysis using structure-based refinement and loose embedding. In: Yi, K. (ed.) SAS 2006. LNCS, vol. 4134, pp. 204–220. Springer, Heidelberg (2006)
2. Ball, T., Bounimova, E., Cook, B., Levin, V., Lichtenberg, J., McGarvey, C., Ondrusek, B., Rajamani, S.K., Ustuner, A.: Thorough static analysis of device drivers. In: EuroSys (2006)

3. Berdine, J., Calcagno, C., Cook, B., Distefano, D., O'Hearn, P., Wies, T., Yang, H.: Shape analysis of composite data structures. In: Damm, W., Hermanns, H. (eds.) CAV 2007. LNCS, vol. 4590, pp. 178–192. Springer, Heidelberg (2007)
4. Berdine, J., Calcagno, C., O'Hearn, P.: Symbolic execution with separation logic. In: Yi, K. (ed.) APLAS 2005. LNCS, vol. 3780, pp. 52–68. Springer, Heidelberg (2005)
5. Berdine, J., Lev-Ami, T., Manevich, R., Ramalingam, G., Sagiv, M.: Thread quantification for concurrent shape analysis. In: CAV (2008)
6. Blanchet, B., Cousot, P., Cousot, R., Feret, J., Mauborgne, L., Miné, A., Monniaux, D., Rival, X.: A static analyzer for large safety-critical software. In: PLDI (2003)
7. Chang, B., Rival, X., Necula, G.: Shape analysis with structural invariant checkers. In: Riis Nielson, H., Filé, G. (eds.) SAS 2007. LNCS, vol. 4634, pp. 384–401. Springer, Heidelberg (2007)
8. Chou, A., Yang, J., Chelf, B., Hallem, S., Engler, D.R.: An empirical study of operating system errors. In: SOSP (2001)
9. Cousot, P., Cousot, R.: Abstract interpretation: A unified lattice model for static analysis of programs by construction or approximation of fixpoints. In: POPL (1977)
10. Distefano, D., O'Hearn, P., Yang, H.: A local shape analysis based on separation logic. In: Hermanns, H., Palsberg, J. (eds.) TACAS 2006 and ETAPS 2006. LNCS, vol. 3920, pp. 287–302. Springer, Heidelberg (2006)
11. Gotsman, A., Berdine, J., Cook, B.: Interprocedural shape analysis with separated heap abstractions. In: Yi, K. (ed.) SAS 2006. LNCS, vol. 4134, pp. 240–260. Springer, Heidelberg (2006)
12. Gotsman, A., Berdine, J., Cook, B., Sagiv, M.: Thread-modular shape analysis. In: PLDI (2007)
13. Guo, B., Vachharajani, N., August, D.: Shape analysis with inductive recursion synthesis. In: PLDI (2007)
14. Hackett, B., Rugina, R.: Region-based shape analysis with tracked locations. In: POPL (2005)
15. Henzinger, T., Jhala, R., Majumdar, R., McMillan, K.: Abstractions from proofs. In: POPL (2004)
16. Lev-Ami, T., Sagiv, M.: TVLA: A system for implementing static analyses. In: Palsberg, J. (ed.) SAS 2000. LNCS, vol. 1824, pp. 280–302. Springer, Heidelberg (2000)
17. Magill, S., Nanevski, A., Clarke, E., Lee, P.: Inferring invariants in Separation Logic for imperative list-processing programs. In: SPACE (2006)
18. Manevich, R., Lev-Ami, T., Ramalingam, G., Sagiv, M., Berdine, J.: Heap decomposition for concurrent shape analysis. In: SAS (2008)
19. Manevich, R., Sagiv, M., Ramalingam, G., Field, J.: Partially disjunctive heap abstraction. In: Giacobazzi, R. (ed.) SAS 2004. LNCS, vol. 3148, pp. 265–279. Springer, Heidelberg (2004)
20. Marron, M., Hermenegildo, M., Kapur, D., Stefanovic, D.: Efficient context-sensitive shape analysis with graph based heap models. In: CC (2008)
21. O'Hearn, P., Reynolds, J., Yang, H.: Local reasoning about programs that alter data structures. In: Fribourg, L. (ed.) CSL 2001 and EACSL 2001. LNCS, vol. 2142. Springer, Heidelberg (2001)
22. Reps, T., Horwitz, S., Sagiv, M.: Precise interprocedural dataflow analysis via graph reachability. In: POPL (1995)
23. Rinetzky, N., Bauer, J., Reps, T., Sagiv, M., Wilhelm, R.: A semantics for procedure local heaps and its abstractions. In: POPL (2005)
24. Swift, M.M., Bershad, B.N., Levy, H.M.: Improving the reliability of commodity operating systems. In: SOSP (2003)

Thread Quantification for Concurrent Shape Analysis

J. Berdine[1], T. Lev-Ami[2,⋆], R. Manevich[2,⋆⋆], G. Ramalingam[3], and M. Sagiv[2]

[1] Microsoft Research Cambridge
jjb@microsoft.com
[2] Tel Aviv University
{tla,rumster,msagiv}@post.tau.ac.il
[3] Microsoft Research India
grama@microsoft.com

Abstract. In this paper we address the problem of shape analysis for concurrent programs. We present new algorithms, based on abstract interpretation, for automatically verifying properties of programs with an unbounded number of threads manipulating an unbounded shared heap.

Our algorithms are based on a new abstract domain whose elements represent *thread-quantified* invariants: i.e., invariants satisfied by all threads. We exploit existing abstractions to represent the invariants. Thus, our technique lifts existing abstractions by wrapping universal quantification around elements of the base abstract domain. Such abstractions are effective because they are thread modular: e.g., they can capture correlations between the local variables of the same thread as well as correlations between the local variables of a thread and global variables, but forget correlations between the states of distinct threads. (The exact nature of the abstraction, of course, depends on the base abstraction lifted in this style.)

We present techniques for computing sound transformers for the new abstraction by using transformers of the base abstract domain. We illustrate our technique in this paper by instantiating it to the Boolean Heap abstraction, producing a Quantified Boolean Heap abstraction. We have implemented an instantiation of our technique with Canonical Abstraction as the base abstraction and used it to successfully verify linearizability of data-structures in the presence of an unbounded number of threads.

1 Introduction

This paper is concerned with verifying (basic safety and other functional correctness) properties of dynamically-allocated data structures in programs with an unbounded number of threads. For example, the techniques in this paper enable, for the first time, automatic verification of linearizability of various implementations of concurrent data structures *even when an unbounded number of client threads manipulate these data structures concurrently*.

Our approach is based on abstract interpretation, which requires us to address the standard two principal challenges:

⋆ Supported by an Adams Fellowship through the Israel Academy of Sciences and Humanities.
⋆⋆ This research was partially supported by the Clore Fellowship Programme.

A. Gupta and S. Malik (Eds.): CAV 2008, LNCS 5123, pp. 399–413, 2008.

- to define a finite representation of infinite sets of program states that can concisely and precisely express the properties of interest, and
- to compute sound transformers, which over-approximate a program's semantics using this representation.

Quantification-Based Abstract Domain. The basis of our approach is the use of an abstract domain whose elements represent quantified invariants of the form $\forall t.\varphi(t)$, where the quantification is over threads. The formulas $\varphi(t)$ correspond to an abstraction of the program state from the perspective of a thread t. A second aspect of our approach is that we exploit existing abstractions to capture the component $\varphi(t)$ inside the quantifier. Informally, assume that we have an underlying abstraction where the abstract domain corresponds to a set of formulas A_{Voc} over a vocabulary *Voc*. (Usually, the vocabulary captures the dependence of the abstract domain on the program being analyzed.) We refine the abstraction and work with the set of formulas $L_{Voc} = \{\forall t.\varphi(t) \mid \varphi(t) \in A_{Voc \cup \{t\}}\}$. Thus, our technique may be seen as a domain constructor. The thread-quantified domain we construct is bounded to the degree that the underlying domain is: e.g., a finite-height base domain yields a finite-height quantified domain.

Transformers. We show how we can compute sound transformers for our new domain using sound transformers for the base domain. We present a simple technique for computing a *basic* sound transformer. The basic transformer works well when a thread's action does not (potentially) affect the invariants (or state) observed by other threads. We also present a more sophisticated technique for computing a *refined* transformer, which is useful for thread actions that affect other threads.

The basic ideas underlying the construction of such a quantified abstract domain have appeared in various forms in recent work, see Sec. 5. One of the novel contributions of our work is the use of such quantification for concurrent *shape analysis* by using suitable shape analysis abstractions such as *Canonical Abstraction* [24] and *Boolean Heaps* [23] as the base domain. We have implemented our technique on top of the TVLA [16] system,[1] which is based on Canonical Abstraction, and used it to verify *linearizability* of fine-grained concurrency algorithms [1]. However, we illustrate our ideas in this paper using Boolean Heaps as the base domain, as its simplicity allows us to focus on the essence of our approach.

The thread-quantified abstract domain is a natural domain to use for reasoning about programs with an unbounded number of threads. It permits expressing properties that correlate a thread's local variables with each other and with shared global state, but not ad-hoc properties that correlate distinct, threads' local variables. (By "ad-hoc", we mean properties that cannot be captured using quantification.) Note that when the underlying base domain is disjunctive, as is the case with Canonical Abstraction and Boolean Heaps, the new domain permits disjunctions inside the quantifier, which is quite useful.

2 Overview

In this section, we present an informal overview of our method.

[1] We actually implemented our technique in HeDec [18], which generalizes Canonical Abstraction by allowing coarser and more scalable abstractions.

```
Object g = null; // global variable
threadProc() {
      Object x = null, y = null;
 [1]  x = new Object();
 [2]  y = x;
 [3]  assert(x == y);
      g = x;
 [4]
assert(g != null);
}
```

Fig. 1. A simple multithreaded program. The program consists of an unbounded number of threads executing `threadProc`.

A Motivating Example. Fig. 1 shows a toy concurrent program used to illustrate the ideas in this paper. Sec. 4 reports on applying these ideas to more realistic programs. The program satisfies a couple of very simple invariants (expressed as assertions) that we would like to automatically infer. The first invariant is that when a thread is at statement `[3]`, the values of it's `x` and `y` variables are equal. This is an example of a *thread-local* invariant (which cannot be affected by the execution of other threads). The second invariant is that when a thread is at statement `[4]`, the global variable `g` is non-null. This is an example of a *non-local thread* invariant, and can be affected by the execution of other threads. In general, a non-local thread invariant could involve global as well as thread-local variables. As an example, consider an assertion that when a thread is at statement `[4]`, the value of `g` and it's `x` are equal. This is an assertion that fails to hold for the given program (because of interaction with other threads).

Background: The Boolean Heap Abstraction. As explained in Sec. 1, our approach is to lift an existing abstraction to produce a more precise abstraction that is suitable for programs with an unbounded number of threads. We will illustrate our idea using Boolean Heaps [23] as the underlying base abstraction in this paper. Boolean Heaps are abstractions targeted at shape analysis, and describe sets of states consisting of an unbounded number of heap objects using formulas of the form

$$\bigvee_{i \in I} \{\forall v : \textit{object}.\ \varphi_i'(v)\}$$

where v ranges over heap objects and $\varphi'(v)$ is a quantifier-free formula, in which v possibly occurs free, over a set of unary predicates, kept in DNF.

The Quantified Boolean Heap Abstraction. As explained earlier, the basis of our approach is to use quantified invariants of the form $\forall t.\ \varphi(t)$,where the set of formulas $\varphi(t)$ allowed inside the invariant are determined by a base domain. Using Boolean Heaps as

Table 1. Predicates used for (Quantified) Boolean Heap Abstraction

Predicate	Intended meaning
$x(t,v)$	local variable `x` of thread t points to object v
$y(t,v)$	local variable `y` of thread t points to object v
$g(v)$	global variable `g` points to object v
$null(v)$	v is a special null object
$at[l](t)$	thread t is at program label l

Table 2. Part of the computed quantified invariant for the running example

$$
\begin{array}{ll}
\forall t. & \ldots \vee at[4](t) \wedge \\
& \forall v \{ \ x(t) \wedge \ y(t) \wedge \ g \wedge \neg null\}(v) \vee \{\neg x(t) \wedge \neg y(t) \wedge \neg g \wedge null\}(v) \\
& \vee \forall v \{ \ x(t) \wedge \ y(t) \wedge \neg g \wedge \neg null\}(v) \vee \{\neg x(t) \wedge \neg y(t) \wedge \neg g \wedge null\}(v) \vee \\
& \quad \{\neg x(t) \wedge \neg y(t) \wedge \ g \wedge \neg null\}(v) \\
& \vee \forall v \{ \ x(t) \wedge \ y(t) \wedge \ g \wedge \neg null\}(v) \vee \{\neg x(t) \wedge \neg y(t) \wedge \neg g \wedge null\}(v)
\end{array}
$$

our base domain leads to the following definition of *Quantified Boolean Heaps*. Quantified Boolean Heaps approximate sets of states by formulas of the form

$$\forall t\!:\!thread.\ \bigvee_{i\in I}\{\forall v\!:\!object.\ \varphi_i(t,v)\}$$

where t ranges over threads, v ranges over non-thread objects, $\varphi(t,v)$ is a quantifier-free formula, in which t and v possibly occur free, over a set of unary and binary predicates, kept in DNF.[2] For the running example, we will use the predicates shown in Tab. 1. We assume that a *null* value is represented by a special heap object.

Notice that a quantified boolean heap is a universally quantified disjunction of standard Boolean heaps, but where some previously unary predicates have been indexed by the universal variable. This increases the expressive power of the abstract elements.

For brevity we use the following, not disjunctive normal, form

$$\forall t\!:\!thread.\ \bigvee_{l\in \mathsf{Labels}} at[l](t) \wedge \left(\bigvee_{i\in I}\{\forall v\!:\!object.\ \varphi_i(t,v)\}\right)$$

where Labels is the set of program labels and the predicates of the form $at[l](t)$ are implicitly mutually exclusive.[3] We use the following notational conventions: Logical variables $\{sc, t, t_1, t_2, \ldots\}$ range over thread objects and other logical variables range over non-thread objects. We use $\{p \wedge q\}(v)$ as shorthand for $p(v) \wedge q(v)$ and curry binary predicates, i.e., $\{p(t) \wedge q\}(v)$ is a shorthand for $p(t,v) \wedge q(v)$.

Tab. 2 shows the part of the Quantified Boolean Heap describing the invariant of our running example for the threads at program location `[4]`, as computed by our analysis.

Abstract Transformers. Computing abstract transformers is very challenging, especially in the presence of concurrency, as the execution of one thread may affect the state observable by other threads. In Sec. 3 we present effective techniques for computing sound transformers for our lifted abstractions, utilizing transformers of the base abstraction. The main idea is to "instantiate" two symbolic threads, one for the scheduled thread, and one representing another arbitrary thread, and to utilize the transformer of the underlying base domain to compute the change in the abstract state as observed by each of these threads.

Discussion. For comparison, consider the Quantified Boolean Heaps abstraction just described and the abstractions used by the original Boolean Heaps and 3VMC [27], which naturally models unbounded objects and threads in a uniform fashion using

---

[2] This is similar to Indexed Predicate Abstraction [15], except that the number of index variables is limited to 2, and that we allow a disjunction between the quantifiers.

[3] The location predicates are written outside the internal universal quantifier because they are independent of v.

Canonical Abstraction. For *the set of predicates described in this section*, our new analysis is capable of inferring the invariants mentioned in the program: namely, that for any thread t at program location 3, we have $x(t) = y(t)$, and that $\mathtt{g}$ is not null at the end (for any thread's execution). On the other hand, without adding different predicates, neither the Boolean Heaps analysis nor 3VMC can infer the above invariants. Indeed, these abstractions cannot even express these invariants. If a richer set of predicates is used, especially instrumentation predicates, these abstractions can be made more expressive and be used to prove the above invariants. An advantage of the new abstraction is that it can reduce the need for nonstandard or program-specific predicates, or the number of predicates, in a very natural way.

3 The Thread Quantification Domain Constructor

In this section, we describe how thread quantification can be used as a domain construction operator to generate a more precise abstract domain from an existing abstract domain. We illustrate this by applying it to the Boolean Heap domain to obtain the Quantified Boolean Heap domain.

3.1 The Concrete Semantics

We start by defining operational concrete semantics useful for describing concurrent programs without procedures. For simplicity of presentation, we concentrate on reference variables and fields. Let $\mathsf{Threads}$ and $\mathsf{Locations}$ (containing a distinguished *null* value) be countable sets representing threads and heap locations, respectively. Let LVars, GVars, and Fields be finite sets of local variables, global variables, and heap fields, respectively. Finally, let Labels be a finite set of program labels. Let Σ be the set of possible states. A state $\sigma \in \Sigma$ maps the following: for each global variable $\mathtt{g}$, $\sigma(\mathtt{g}) \in \mathsf{Locations}$; for each local variable $\mathtt{x}$, $\sigma(\mathtt{x}) : \mathsf{Threads} \to \mathsf{Locations}$; for each field $\mathtt{f}$, $\sigma(\mathtt{f}) : \mathsf{Locations} \to \mathsf{Locations}$; and for $\mathtt{pc}$, $\sigma(\mathtt{pc}) : \mathsf{Threads} \to \mathsf{Labels}$.

Being interested in invariance properties, we start with a concrete powerset domain $\mathcal{P}(\Sigma)$, for which we assume a concrete semantics of programs $\mathsf{spost}_{(\cdot)} : \mathsf{Threads} \to \mathcal{P}(\Sigma) \to \mathcal{P}(\Sigma)$ that maps individual threads to their semantics. This induces the semantics of the overall concurrent program $\mathsf{cpost} : \mathcal{P}(\Sigma) \to \mathcal{P}(\Sigma)$ by

$$\mathsf{cpost}(S) = \bigcup_{\mathsf{sc} \in \mathsf{Threads}} \mathsf{spost}_{\mathsf{sc}}(S) \ .$$

3.2 The Base Abstraction

We present the lifted abstract interpreter as well as the base abstract interpreter as operating on formulas in a normal form. This is done for simplicity of presentation. For example, Boolean Heaps are already presented using these terms. Details on how Canonical Abstraction can be presented in such terms can be found in [29].

In Sec. 3.6 we will specify the assumptions on the abstract domain input to the thread quantification construction, but it is useful to present in several steps.

Base Domain. Consider a base abstract domain with elements drawn from a set A of formulas, where $(\mathcal{P}(\Sigma), \alpha_A, \gamma_A, A)$ is a Galois Connection, with meet $\sqcap_A$ and join $\sqcup_A$, and sound sequential transformer $\mathsf{spost}^\sharp_{(\cdot)} : \mathsf{Threads} \to A \to A$. As in the concrete semantics, this induces the abstract concurrent semantics $\mathsf{cpost}^\sharp : A \to A$, which over-approximates cpost, by

$$\mathsf{cpost}^\sharp(a) = \bigsqcup_{A\ \mathsf{sc} \in \mathsf{Threads}} \mathsf{spost}^\sharp_{\mathsf{sc}}(a) .$$

Open Formulas. Abstract elements often correspond to formulas without free variables. E.g., the formula $\forall v.g(v) \Leftrightarrow \mathit{null}(v)$ represents states where g is *null*. The first step toward thread-quantified formulas is to permit formulas with free variables (e.g., $\forall v.x(t, v) \Leftrightarrow \mathit{null}(v)$) as abstract domain elements.

For a set V of variables, let $A[V]$ denote the set of formulas in normal form with free variables contained in V. Let $\mathsf{Assign}_V = V \to \mathsf{Threads}$ be the set of assignments of (thread) variables in V to threads. A state $\sigma \in \Sigma$ and an assignment $\theta \in \mathsf{Assign}_V$ satisfy $\varphi(V) \in A[V]$, denoted $\sigma, \theta \models \varphi(V)$, when assigning the parameters according to θ and interpreting the predicates according to σ yields true. Define Σ_V to be $\Sigma \times \mathsf{Assign}_V$.

Example. The open formula $\forall v.x(t, v) \Leftrightarrow \mathit{null}(v)$ represents the set of all pairs $\langle \sigma, \theta \rangle$ such that the local variable x of thread $\theta(t)$ is null in σ, i.e., $\sigma(x)(\theta(t)) = \mathit{null}$.

By defining $\gamma_{A[V]} : A[V] \to \mathcal{P}(\Sigma_V)$ by $\gamma_{A[V]}(\varphi(V)) = \{\langle \sigma, \theta \rangle \mid \sigma, \theta \models \varphi(V)\}$, the satisfaction relation determines a Galois Connection $(\mathcal{P}(\Sigma_V), \alpha_{A[V]}, \gamma_{A[V]}, A[V])$.

Transformers for Open Formulas. Since the states Σ_V for open formulas are related to program states Σ simply by the first projection, the concrete semantics can be lifted to $\mathsf{spost}_{V,(\cdot)} : \mathsf{Threads} \to \mathcal{P}(\Sigma_V) \to \mathcal{P}(\Sigma_V)$ by defining

$$\mathsf{spost}_{V,\mathsf{t}}(S) = \bigcup_{\langle \sigma, \theta \rangle \in S} (\mathsf{spost}_{\mathsf{t}}(\{\sigma\}) \times \{\theta\}) .$$

The concurrent semantics $\mathsf{cpost}_V : \mathcal{P}(\Sigma_V) \to \mathcal{P}(\Sigma_V)$ is induced by $\mathsf{spost}_{V,(\cdot)}$ in the same way as cpost is induced by $\mathsf{spost}_{(\cdot)}$:

$$\mathsf{cpost}_V(S) = \bigcup_{\mathsf{sc} \in \mathsf{Threads}} \mathsf{spost}_{V,\mathsf{sc}}(S) .$$

The thread quantification domain construction requires transformers for open formulas $\mathsf{cpost}^\sharp_V : A[V] \to A[V]$ that over-approximate cpost_V. While the definition of $\mathsf{cpost}^\sharp_V$ from $\mathsf{cpost}^\sharp$ varies from one domain to another, note that $\Pi_1(\mathsf{cpost}_V(S)) = \mathsf{cpost}(\Pi_1(S))$ (where Π_1 is the first projection of a pair, lifted pointwise to sets of pairs), and so an abstract transformer is sound with respect to cpost_V if and only if it is sound with respect to cpost. Also note that since cpost_V always leaves the thread assignment unchanged, sound over-approximations must also. Hence the free thread variables can be treated as constant symbols, and binary predicates such as $x(t, v)$ can be curried and then interpreted as unary predicates $(x(t))(v)$, which many base domains A directly support. In particular, assuming the base domain A can handle constant symbols, a domain $A[V]$ can be produced systematically.

We will specifically be interested in the case of formulas with a single free variable t: i.e., the case where $V = \{t\}$. The method can be generalized to multiple free variables and thus multiple universal quantifiers. This is outside the scope of the paper. Note that the union over all threads sc in the concrete transformer cpost_V captures the effect of a *single transition performed by an arbitrary thread* sc. A sound abstract transformer $\mathsf{cpost}^\sharp_{\{t\}}$ must handle two cases: where thread variable t is the same as the scheduled thread sc, and where t is different from sc.

Example. We now illustrate the application of a sound transformer for the transition corresponding to the single statement y=x on the open formula $\varphi(t) = \forall v.x(t,v) \Leftrightarrow null(v)$. This formula represents states σ and assignments $[t : \mathtt{t}]$ where the local variable x of thread t is null in σ. If thread t executes the statement y=x, the resulting state can be described by the formula $\varphi_1(t) = \forall v.y(t,v) \Leftrightarrow x(t,v) \Leftrightarrow null(v)$. If some thread other than t is scheduled, then the local variables of t are not affected, and the resulting state can be described by $\varphi(t)$ itself. We account for these two cases by taking the disjunction $\varphi_1(t) \vee \varphi(t)$, which simplifies to $\varphi(t)$, yielding the result of the transformer.

3.3 The Lifted Abstraction (with Basic Transformers)

We define the lifted domain $L = \{\forall t.\, \varphi(t) \mid \varphi(t) \in A[\{t\}]\}$, i.e., with the base domain instantiated with $V = \{t\}$. The lifted domain inherits meet and (an over-approximation of) join operations from $A[\{t\}]$: e.g., $(\forall t.\, \varphi_1) \sqcup_L (\forall t.\, \varphi_2) = \forall t.\, (\varphi_1 \sqcup_{A[\{t\}]} \varphi_2)$. Defining $\gamma_L : L \to \mathcal{P}(\Sigma)$ by

$$\gamma_L(\forall t.\, \varphi(t)) = \left\{\sigma \mid \sigma, \theta \models \varphi(t) \text{ for every } \theta \in \mathrm{Assign}_{\{t\}}\right\}$$

produces a Galois Connection from $\mathcal{P}(\Sigma)$ to L. We obtain a sound transformer $\mathsf{cpost}^\sharp_L : L \to L$ from the sound abstract transformer $\mathsf{cpost}^\sharp_{\{t\}}$ for formulas with a free variable t discussed earlier as follows:

$$\mathsf{cpost}^\sharp_L(\forall t.\varphi(t)) = \forall t.\mathsf{cpost}^\sharp_{\{t\}}(\varphi(t)) \,.$$

Example. Consider the statement y=x from the example program in Fig. 1 and the abstract state $S1$:

$$\begin{array}{l} S1 = \forall t.S1_a(t) \vee S1_b(t) \\ S1_a(t) \;=\; at[1](t) \wedge \\ \qquad \forall v.\ \{\ x(t) \wedge\ y(t) \wedge\ g \wedge\ null\}(v) \vee \{\neg x(t) \wedge \neg y(t) \wedge \neg g \wedge \neg null\}(v) \\ S1_b(t) \;=\; at[2](t) \wedge \\ \qquad \forall v.\ \{\neg x(t) \wedge\ y(t) \wedge\ g \wedge\ null\}(v) \vee \{\ x(t) \wedge \neg y(t) \wedge \neg g \wedge \neg null\}(v) \vee \\ \qquad\quad \{\neg x(t) \wedge \neg y(t) \wedge \neg g \wedge \neg null\}(v) \end{array}$$

Applying the Boolean Heap transformer for y=x to $S1_a(t)$ leaves $S1_a(t)$ unchanged no matter whether t was the scheduled thread or not. Applying the Boolean Heap transformer for y=x to $S1_b(t)$ yields the heaps $S1'_{b_1}(t)$ for the case where t is the scheduled thread, and leaves $S1_b(t)$ unchanged for the complementary case. The final result is obtained by universally quantifying over t, resulting in $S1'$:

$$\begin{array}{l} S1' = \forall t.S1_a(t) \vee S1_b(t) \vee S1'_{b_1}(t) \\ S1'_{b_1}(t) \;=\; at[3](t) \wedge \\ \qquad \forall v.\ \{\neg x(t) \wedge \neg y(t) \wedge\ g \wedge\ null\}(v) \vee \{\ x(t) \wedge\ y(t) \wedge \neg g \wedge \neg null\}(v) \vee \\ \qquad\quad \{\neg x(t) \wedge \neg y(t) \wedge \neg g \wedge \neg null\}(v) \end{array}$$

Let $\phi_{\texttt{x==y}} = \forall t.at[3](t) \Rightarrow \forall v.x(t,v) \Leftrightarrow y(t,v)$ be the assertion at line [3]. Now, $S1' \models \phi_{\texttt{x==y}}$ (the only disjunct where *at*[3] holds is $S1'_{b_1}(t)$, in which x and y point to the same node). The statement y=x changes only information local to one thread and therefore this kind of reasoning is sufficiently precise.

Let $\phi_{\texttt{g!=null}} = \forall t.at[4](t) \Rightarrow \forall v.\neg(g(v) \wedge null(v))$ be the assertion at line [4]. Now, however, consider the statement g=x and the abstract state $S2$:

$$\begin{aligned} S2 &= \forall t.S2_a(t) \\ S2_a(t) &= at[3](t) \wedge \\ &\quad \forall v.\ \{\neg x(t) \wedge \neg y(t) \wedge \ g \wedge \ null\}(v) \vee \{\ x(t) \wedge \ y(t) \wedge \neg g \wedge \neg null\}(v) \vee \\ &\qquad \{\neg x(t) \wedge \neg y(t) \wedge \neg g \wedge \neg null\}(v) \end{aligned}$$

When t is not the scheduled thread, applying the Boolean Heap transformer to the Boolean Heap $S2_a(t)$ yields many Boolean Heaps. This is because of the lack of information about the status of other threads, which we get by dropping the universal quantification over t. The scheduled thread may be different from t. Thus, $S2_a(t)$ has no information about it. In particular, $S2_a(t)$ represents a state where x and y are null for the scheduled thread. As a result, the assignment g=x also creates the Boolean Heap $S_{bad} = at[4](t) \wedge \forall v.\{x(t) \wedge y(t) \wedge g \wedge null\}(v)$. Obviously, $S_{bad} \not\models \phi_{\texttt{g!=null}}$ and thus the transformer is not precise enough for the purpose of our analysis.

The reason is that g=x changes a global variable. This change is visible by other threads and thus the thread-local reasoning used above does not model the effect of the other threads using the information captured by the Quantified Boolean Heap.

3.4 The Semantics of Non-deterministic Scheduling

In order to obtain a more precise sound transformer for our lifted abstract domain, we exploit the internal structure of the concrete semantics and the base abstract transformer imposed by the semantics of non-deterministic scheduling.

Recall that the concurrent semantics of a program cpost is defined in terms of spost_t, which gives the sequential semantics of the individual threads. This function indicates the transitions a given thread t can take. The semantics of non-deterministic scheduling of threads is captured by the union over all threads in the definition of cpost.

While the basic transformer $\mathsf{cpost}^\sharp_L$ was defined in terms of $\mathsf{cpost}^\sharp_V$, for the refined transformer we will not use the naive definition of the concurrent semantics in terms of the sequential semantics but will instead define the refined transformer directly in terms of the sequential abstract transformer.

In particular, we assume an abstract transformer $\mathsf{spost}^\sharp_{V,sc} : A[V \cup \{sc\}] \to A[V \cup \{sc\}]$ for $sc \notin V$ that over-approximates $\mathsf{spost}_{V,sc} : \mathcal{P}(\Sigma_{V \cup \{sc\}}) \to \mathcal{P}(\Sigma_{V \cup \{sc\}})$ given by

$$\mathsf{spost}_{V,sc}(S) = \textstyle\bigcup_{\langle \sigma,\theta \rangle \in S} (\mathsf{spost}_{\theta(sc)}(\{\sigma\}) \times \{\theta\}) \,.$$

The difference between this semantics and $\mathsf{spost}_{V,(\cdot)}$ above is that $\mathsf{spost}_{V,sc}$ looks up sc in the assignment in the input state to determine which thread to execute. In essence, we are assuming that the scheduled thread is specified as an extra parameter for the transformer of open formulas in the base domain. Lifting the transformers of the base domain to support the scheduled thread as an extra parameter is usually straightforward. Inducing the concrete semantics from $\mathsf{spost}_{V,sc}$ by

$$\mathsf{cpost}_V(S) = \bigcup_{\mathsf{sc} \in \mathsf{Threads}} \{\langle \sigma', \theta'|_V \rangle \mid \langle \sigma', \theta' \rangle \in \mathsf{spost}_{V,sc}\, \{\langle \sigma, [\theta | sc{:}\mathsf{sc}] \rangle \mid \langle \sigma, \theta \rangle \in S\}\}$$

(where $\theta'|_V$ is θ' restricted to domain V) yields the same definition of cpost_V as above.

We also assume that the base abstract domain has an operation $\mathsf{project}(sc, (\cdot)) : A[V \cup \{sc\}] \rightarrow A[V]$ for projecting away a thread parameter sc. This is equivalent to over-approximating existential elimination. For example, in Boolean Heaps, we can simply throw away all literals (positive and negative) that contain sc.

Using these operations, the transformer for the overall concurrent program $\mathsf{cpost}^\sharp_V : A[V] \rightarrow A[V]$ is defined, for $sc \notin V$, by

$$\mathsf{cpost}^\sharp_V(\varphi(V)) = \mathsf{project}(sc, \mathsf{spost}^\sharp_{V,sc}(\varphi(V)))\ .$$

Note how this definition allows an arbitrary thread to execute since sc does not occur in $\varphi(V)$, hence $\varphi(V)$ does not constrain the thread assigned to sc, and hence the set of states that satisfy $\varphi(V)$ will include assignments that map sc to any element of $\mathsf{Threads}$.

3.5 A More Precise Transformer for the Lifted Domain

We will now present a more precise sound transformer for the lifted domain. The basic transformer presented in Sec. 3.3 transformed a quantified formula $\forall t.\varphi(t)$ by applying the base domain's (open formula) transformer to $\varphi(t)$. This leads to a loss of precision because the base domain transformer knows only that t satisfies $\varphi(t)$. *It does not know and cannot use the fact that both the scheduled thread sc and another arbitrary thread t satisfy the invariant.* We now show how we can incorporate this extra piece of information, *while still reusing the base domain's transformer*, producing a more precise transformer for the lifted domain.

We define the refined transformer $\mathsf{cpost}'^\sharp_L : L \rightarrow L$ by

$$\mathsf{cpost}'^\sharp_L(\forall t.\varphi(t)) = \forall t.\mathsf{project}(sc, \mathsf{spost}^\sharp_{\{t\},sc}(\varphi(t) \sqcap_{A[\{t,sc\}]} \varphi(sc)))\ .$$

Specifically, we apply the base domain's transformer to $\varphi(t) \sqcap_{A[\{t,sc\}]} \varphi(sc)$, exploiting the base domain's meet operation to "inform" the base domain's transformer that both $\varphi(t)$ and $\varphi(sc)$ are true in the input state.

Example. We demonstrate the refined transformer by computing $\mathsf{cpost}'^\sharp_L(S2_a(t))$. The first step of the transformer is to compute the meet of $S2_a(t)$ and $S2_a(sc)$ (where for brevity, we have not converted the formula to DNF):

$$\begin{aligned}
\varphi(sc, t) = S2_a(t) \sqcap S2_a(sc) = &\ at[3](t) \wedge at[3](sc) \wedge \\
&\forall v.\ \{\neg x(t) \wedge \neg y(t) \wedge g \wedge null\}(v) \vee \{x(t) \wedge y(t) \wedge \neg g \wedge \neg null\}(v) \vee \\
&\quad \{\neg x(t) \wedge \neg y(t) \wedge \neg g \wedge \neg null\}(v) \\
\wedge &\ \forall v.\ \{\neg x(sc) \wedge \neg y(sc) \wedge g \wedge null\}(v) \vee \{x(sc) \wedge y(sc) \wedge \neg g \wedge \neg null\}(v) \vee \\
&\quad \{\neg x(sc) \wedge \neg y(sc) \wedge \neg g \wedge \neg null\}(v)
\end{aligned}$$

Next, we apply the Boolean Heaps transformer $\mathsf{spost}^\sharp_{\{t\},sc}$ to $\varphi(sc, t)$. As explained earlier, this is a sound transformer of a single transition taken by thread sc. As before, we obtain the result as a disjunction of two heaps, $\varphi'_a(sc, t)$ for the case in which $sc = t$ and $\varphi'_b(sc, t)$ for the case in which $sc \neq t$.

$$\begin{aligned}
&\varphi'(sc,t) = \varphi'_a(sc,t) \vee \varphi'_b(sc,t)\\
&\varphi'_a(sc,t) = at[4](t) \wedge at[4](sc) \wedge\\
&\quad \forall v.\ \{\neg x(t) \wedge \neg y(t) \wedge \neg g \wedge null\}(v) \vee \{x(t) \wedge y(t) \wedge g \wedge \neg null\}(v) \vee\\
&\qquad \{\neg x(t) \wedge \neg y(t) \wedge \neg g \wedge \neg null\}(v)\\
&\quad \wedge \forall v.\ \{\neg x(sc) \wedge \neg y(sc) \wedge \neg g \wedge null\}(v) \vee \{x(sc) \wedge y(sc) \wedge g \wedge \neg null\}(v) \vee\\
&\qquad \{\neg x(sc) \wedge \neg y(sc) \wedge \neg g \wedge \neg null\}(v)\\
&\varphi'_b(sc,t) = at[3](t) \wedge at[4](sc) \wedge\\
&\quad \forall v.\ \{\neg x(t) \wedge \neg y(t) \wedge \neg g \wedge null\}(v) \vee \{x(t) \wedge y(t) \wedge \neg null\}(v) \vee\\
&\qquad \{\neg x(t) \wedge \neg y(t) \wedge \neg null\}(v)\\
&\quad \wedge \forall v.\ \{\neg x(sc) \wedge \neg y(sc) \wedge \neg g \wedge null\}(v) \vee \{x(sc) \wedge y(sc) \wedge g \wedge \neg null\}(v) \vee\\
&\qquad \{\neg x(sc) \wedge \neg y(sc) \wedge \neg g \wedge \neg null\}(v)
\end{aligned}$$

We project away sc, by removing all literals containing it, which yields:

$$\begin{aligned}
&\varphi''(t) = \varphi''_a(t) \vee \varphi''_b(t)\\
&\varphi''_a(t) = at[4](t) \wedge\\
&\quad \forall v.\ \{\neg x(t) \wedge \neg y(t) \wedge \neg g \wedge null\}(v) \vee \{x(t) \wedge y(t) \wedge g \wedge \neg null\}(v) \vee\\
&\qquad \{\neg x(t) \wedge \neg y(t) \wedge \neg g \wedge \neg null\}(v)\\
&\varphi''_b(t) = at[3](t) \wedge\\
&\quad \forall v.\ \{\neg x(t) \wedge \neg y(t) \wedge \neg g \wedge null\}(v) \vee \{x(t) \wedge y(t) \wedge \neg null\}(v) \vee\\
&\qquad \{g \wedge \neg null\}(v) \vee \{\neg x(t) \wedge \neg y(t) \wedge \neg null\}(v)
\end{aligned}$$

The interesting observation here is that `g` and `null` are not aliased in both conjuncts of $\varphi'_b(sc,t)$. Thus, after the projection we retain this information.

Finally, the result is universally quantified, i.e., $S_2{}' = \forall t.\varphi''(t)$. As expected, $S_2{}' \models \phi_{\texttt{g!=null}}$.

3.6 Summary of Construction

In summary, the thread quantification domain construction requires implementations of: an abstract domain $A[V]$ of open formulas that is in a Galois Connection with $\mathcal{P}(\Sigma_V)$ induced by the satisfaction relation; meet and join operations on $A[V]$; sequential transformers $\mathsf{spost}^\sharp_{V,sc}$ of open formulas, parameterized by the scheduled thread, which over-approximate $\mathsf{spost}_{V,sc}$ as in Sec. 3.4; and an over-approximation of existential elimination $\mathsf{project}(sc, (\cdot))$ as in Sec. 3.4. From this the construction produces an implementation of an abstract domain L of quantified formulas, which is in a Galois Connection with $\mathcal{P}(\Sigma)$, with basic transformers $\mathsf{cpost}^\sharp_L$ and refined transformers $\mathsf{cpost}'^\sharp_L$ for concurrent programs that over-approximate the concrete semantics cpost.

4 Case Study: Proving Linearizability

As a case study for the approach, we have verified linearizability of three well-known concurrent data structure implementations that use fine-grained concurrency.

4.1 Implementation

We have implemented the approach on top of TVLA [16]. (Actually, we implemented our technique in HeDec [18], which generalizes Canonical Abstraction by allowing coarser and more scalable abstractions.) The thread parameters were implemented as unary predicates. Support for treating a binary formula of the form $x(t, v)$ as a unary predicate was done by adding an appropriate instrumentation predicate (i.e., predicate defined using a formula from other predicate and automatically updated by the system).

The meet and join operations required from the base domain are already implemented in TVLA. Thread projection is done by forgetting all information about the unary predicate representing the thread and all instrumentation predicates based on it.

In TVLA, it is easier to implement separate transformers for each statement and let the engine deal with constructing the full post operator. As a result, we are able to use the basic transformer for some statements and the more expensive refined transformer only for statements that require the extra precision. We use the basic transformer for statements that modify only the local state of the scheduled thread and leave the global state intact. In these cases the abstract state of any thread that is not the scheduled thread is unchanged by the operation, thus the precision of the basic transformer is enough.

4.2 Proving Linearizability

Linearizability [12] is one of the main correctness criteria for implementations of concurrent data structures. Informally, a concurrent data structure is linearizable if the concurrent execution of a set of operations on it is equivalent to some sequential execution of the same operations, in which the global order between non-overlapping operations is preserved. The equivalence is based on comparing the arguments and results of operations (responses). The permitted behavior of the concurrent object is defined in terms of a specification of the desired behavior of the object in a sequential setting.

Verifying linearizability is challenging because it requires correlating any concurrent execution with a corresponding permitted sequential execution. Verifying linearizability for concurrent dynamically allocated linked data structures is particularly challenging, because it requires correlating executions that may manipulate memory states of unbounded size.

Intuitively, we verify linearizability by representing in the concrete state both the state of the concurrent program and the state of the reference sequential program. Each element entered into the data structure is correlated at linearization points with the matching object from the sequential execution. The details are described in [1].

We have taken the instantiation of Canonical Abstraction presented in [1] as the base abstraction for the analysis. That analysis was performed for a bounded number of threads, by using specialized predicates treating each local variable of each thread as a distinct predicate. We removed these extra predicates, instead treating the thread local variables as binary predicates. The analysis has predicates for local and global variables, heap fields and program labels. Finally, we use as is two extra predicates that capture the correlation between the concurrent and sequential executions (see [1]).

4.3 Experimental Results

Tab. 3 summarizes the experimental results of running our linearizability analysis on the algorithms. These benchmarks were run a 2.4GHz E6600 Core 2 Duo processor with 2 GB of memory running Linux. We used two abstractions to analyze these examples. The first is vanilla canonical abstraction as described in Sec. 4.2. The second abstraction is an extension of canonical abstraction with decomposition of the heap as described in [18]. With this abstraction, the state space is significantly reduced, yielding fewer states and better times. The adaptation of the transformer for the decomposing abstraction was no harder than that for vanilla canonical abstraction.

Treiber's stack algorithm [25] is lock-free, and uses a Compare And Swap (CAS) operation for synchronization. The two-lock queue algorithm [19] has Head and Tail pointers, each protected with its own lock. It allows benign data-races when the queue is empty, i.e., the Head and Tail pointers are aliased. The non-blocking queue algorithm [6] is lock-free and uses CAS for synchronization. It is more complicated than the other two algorithms and has a much larger state space with our abstraction. Canonical Abstraction without decomposition, on this example, resulted in state space explosion.

5 Related Work

The abstract interpretation presented in this paper inherits from, and combines, two lines of prior work: (1) Prior work on abstract domains of quantified formulas, especially in the context of verification of parametrized concurrent systems, and (2) Prior work on shape analysis.

Process-Centric Abstraction. The general approach we use of reasoning about concurrent programs in terms of an *abstraction of the program state relative to a thread* is classic in work on program logic: assertions within the code of a thread refer to the state from that thread's perspective, and the thread's concurrent environment is over-approximated by, for instance, invariants [13,21] or relations [14] on the shared state. This idea has also been used early on for automatic compositional verification [4]. More recently, this approach has led to the notion of thread-modular verification for model checking systems with finitely-many threads [8], and has also been applied more closely to our present domain of heap-manipulating programs with coarse-grained concurrency [9], and less automatically to fine-grained concurrency [2]. This general principle has also previously been used in the context of verification of sequential programs in the form of abstractions of program state relative to one or more non-deterministically chosen objects (e.g., in the heap or an array) [7,28,23,26].

Abstract Interpretation with Open Formulas and Quantified Invariants. In this paper, we realize such a reference-object-centric perspective within the framework of abstract interpretation, using abstract domains consisting of formulas with free variables as a stepping stone toward abstract domains consisting of quantified formulas. This approach has

Table 3. Experimental results of proving linearizability for an unbounded number of threads

	Canonical Abstraction		with decomposition	
Algorithms	States	secs.	States	secs.
Stack [25]	4,097	53	764	7
Two-lock queue [19]	4,897	47	3,415	17
Non-blocking queue [6]	MemOut	MemOut	10,333	252

been previously formalized in the work on Indexed Predicate Abstraction [15] and also appears in the work on Environment Abstraction [5,3]. Indices, or free variables, in the indexed predicate abstraction work can range over anything, depending on the application. Our use of a single variable for a reference process is similar to the approach in Environment Abstraction. A similar quantified invariants approach has also been used in the analysis of heap properties [23] and properties of collections [10] in sequential programs.

Transformers for Quantified Formulas. The chief difficulty, particularly for domain *constructions*, is defining the transformers: semantics of program statements on elements of the abstract domain. In their work on Indexed Predicate Abstraction, Lahiri et al., outline the idea of using *quantifier instantiation* to compute abstract transformers of quantified formulas. They use a tool to generate candidate instantiations (based on the subexpressions that appear in the predicate and next-state expressions) for this purpose. We use a very specific and fixed quantifier instantiation strategy: namely, we instantiate it for the reference process and for the executing process.

Concurrent Shape Analysis. One aspect of our work that distinguishes it from the prior work referenced above is that we apply these ideas to the problem of concurrent shape analysis. In particular, to address the heap, we use abstractions that can more readily make distinctions that are not directly expressible in terms of the program (for instance, the distinction between heap cells to which there are and there are not multiple incoming pointers). Also, the abstraction we use expresses correlations between a single thread's local state and the global shared state, but does not directly express relations between the state of multiple threads. Relations between multiple threads are captured only by the transformers, unlike in Environment Abstraction, which can additionally use predicates that have been chosen to explicitly relate threads. In the way that our abstractions (partially) correlate locals to globals, but not locals to locals, they exhibit a thread-modular character, except that threads need not be entirely uncorrelated.

The most closely related prior work on concurrent shape analysis is that of Yahav [27], which uses Canonical Abstraction for this purpose. The Quantified Canonical Abstraction domain we use is more precise than Canonical Abstraction, and it allows us to automatically verify, for the first time, linearizability of concurrent data structures in the presence of an unbounded number of threads.

Other Related Work. Counter Abstraction [17] (which has been applied to programs in e.g. [11]) provides a reduction from systems with unboundedly-many processes to finite state, though does not offer much help with the abstract transformers for that finite-state system. Invisible Invariants [22] is another technique that employs thread variables, and works by considering systems with a small number of processes and then attempting to generalize the results to unboundedly-many processes. Work on Split Invariants [20] extends Invisible Invariants using a connection with compositional techniques (such as [21]), yielding an analysis with a process-centric abstraction that computes universally quantified invariants using transformers that resemble ours. In particular, if the assertion logic has a small model property with bound k, then an invariant for unboundedly-many threads can be computed using k instantiations of the invariant. In contrast, we define transformers that are sound (but incomplete) for unboundedly-many threads without a small model property, and using many fewer instantiations.

6 Conclusion

In this paper, we have developed a new shape analysis for fine-grained concurrent programs with an unbounded number of threads and demonstrated that it is precise enough to prove linearizability of useful data structure implementations. This is done by a universal lifting domain construction applied to existing shape analysis domains.

References

1. Amit, D., Rinetzky, N., Reps, T., Sagiv, M., Yahav, E.: Comparison under abstraction for verifying linearizability. In: Damm, W., Hermanns, H. (eds.) CAV 2007. LNCS, vol. 4590, pp. 477–490. Springer, Heidelberg (2007)
2. Calcagno, C., Parkinson, M.J., Vafeiadis, V.: Modular safety checking for fine-grained concurrency. In: Riis Nielson, H., Filé, G. (eds.) SAS 2007. LNCS, vol. 4634, pp. 233–248. Springer, Heidelberg (2007)
3. Clarke, E., Talupur, M., Veith, H.: Proving Ptolemy right: The environment abstraction framework for model checking concurrent systems. In: Ramakrishnan, C.R., Rehof, J. (eds.) TACAS 2008. LNCS, vol. 4963, pp. 33–47. Springer, Heidelberg (2008)
4. Clarke, E.M.: Synthesis of resource invariants for concurrent programs. TOPLAS 2(3) (1980)
5. Clarke, E.M., Talupur, M., Veith, H.: Environment abstraction for parameterized verification. In: Emerson, E.A., Namjoshi, K.S. (eds.) VMCAI 2006. LNCS, vol. 3855, pp. 126–141. Springer, Heidelberg (2005)
6. Doherty, S., Groves, L., Luchangco, V., Moir, M.: Formal verification of a practical lock-free queue algorithm. In: de Frutos-Escrig, D., Núñez, M. (eds.) FORTE 2004. LNCS, vol. 3235, pp. 97–114. Springer, Heidelberg (2004)
7. Flanagan, C., Qadeer, S.: Predicate abstraction for software verification. In: POPL (2002)
8. Flanagan, C., Qadeer, S.: Thread-modular model checking. In: Ball, T., Rajamani, S.K. (eds.) SPIN 2003. LNCS, vol. 2648, pp. 213–224. Springer, Heidelberg (2003)
9. Gotsman, A., Berdine, J., Cook, B., Sagiv, M.: Thread-modular shape analysis. In: PLDI (2007)
10. Gulwani, S., McCloskey, B., Tiwari, A.: Lifting abstract interpreters to quantified logical domains. In: POPL (2008)
11. Henzinger, T.A., Jhala, R., Majumdar, R.: Race checking by context inference. In: PLDI (2004)
12. Herlihy, M.P., Wing, J.M.: Linearizability: a correctness condition for concurrent objects. TOPLAS 12(3) (1990)
13. Hoare, C.A.R.: Towards a theory of parallel programming. Operating System Techniques (1972)
14. Jones, C.B.: Specification and design of (parallel) programs. In: IFIP Congress (1983)
15. Lahiri, S.K., Bryant, R.E.: Predicate abstraction with indexed predicates. TOCL 9(1) (2007)
16. Lev-Ami, T., Sagiv, M.: TVLA: A framework for implementing static analyses. In: Palsberg, J. (ed.) SAS 2000. LNCS, vol. 1824, pp. 280–302. Springer, Heidelberg (2000), http://www.cs.tau.ac.il/~tvla/
17. Lubachevsky, B.D.: An approach to automating the verification of compact parallel coordination programs I. Acta Inf. 21 (1984)

18. Manevich, R., Lev-Ami, T., Sagiv, M., Ramalingam, G., Berdine, J.: Heap Decomposition for Concurrent Shape Analysis. In: Alpuente, M., Vidal, G. (eds.) SAS 2008. LNCS, vol. 5079, pp. 363–377. Springer, Heidelberg (2008)
19. Michael, M.M., Scott, M.L.: Simple, fast, and practical non-blocking and blocking concurrent queue algorithms. In: PODC (1996)
20. Namjoshi, K.S.: Symmetry and completeness in the analysis of parameterized systems. In: Cook, B., Podelski, A. (eds.) VMCAI 2007. LNCS, vol. 4349, pp. 299–313. Springer, Heidelberg (2007)
21. Owicki, S.S., Gries, D.: Verifying properties of parallel programs: An axiomatic approach. CACM 19(5) (1976)
22. Pnueli, A., Ruah, S., Zuck, L.D.: Automatic deductive verification with invisible invariants. In: Margaria, T., Yi, W. (eds.) ETAPS 2001 and TACAS 2001. LNCS, vol. 2031. Springer, Heidelberg (2001)
23. Podelski, A., Wies, T.: Boolean Heaps. In: Hankin, C., Siveroni, I. (eds.) SAS 2005. LNCS, vol. 3672, pp. 268–283. Springer, Heidelberg (2005)
24. Sagiv, M., Reps, T., Wilhelm, R.: Parametric shape analysis via 3-valued logic. TOPLAS 24(3) (2002)
25. Treiber, R.K.: Systems programming: Coping with parallelism. Technical Report RJ 5118, IBM Almaden Research Center (1986)
26. Wachter, B., Westphal, B.: The spotlight principle. In: Cook, B., Podelski, A. (eds.) VMCAI 2007. LNCS, vol. 4349, pp. 182–198. Springer, Heidelberg (2007)
27. Yahav, E.: Verifying safety properties of concurrent Java programs using 3-valued logic. ACM SIGPLAN Notices 36(3) (2001)
28. Yahav, E., Ramalingam, G.: Verifying safety properties using separation and heterogeneous abstractions. In: PLDI (2004)
29. Yorsh, G., Reps, T., Sagiv, M.: Symbolically computing most-precise abstract operations for shape analysis. In: Jensen, K., Podelski, A. (eds.) TACAS 2004. LNCS, vol. 2988, pp. 530–545. Springer, Heidelberg (2004)

The Scyther Tool: Verification, Falsification, and Analysis of Security Protocols*

Tool Paper

Cas J.F. Cremers

Department of Computer Science, ETH Zurich, 8092 Zurich, Switzerland
cas.cremers@inf.ethz.ch

1 Introduction

With the rise of the Internet and other open networks, a large number of security protocols have been developed and deployed in order to provide secure communication. The analysis of such security protocols has turned out to be extremely difficult for humans, as witnessed by the fact that many protocols were found to be flawed after deployment. This has driven the research in formal analysis of security protocols. Unfortunately, there are no effective approaches yet for constructing correct and efficient protocols, and work on concise formal logics that might allow one to easily prove that a protocol is correct in a formal model, is still ongoing. The most effective approach so far has been automated falsification or verification of such protocols with state-of-the-art tools such as ProVerif [1] or the Avispa tools [2]. These tools have shown to be effective at finding attacks on protocols (Avispa) or establishing correctness of protocols (ProVerif).

In this paper we present a push-button tool, called *Scyther*, for the verification, the falsification, and the analysis of security protocols. Scyther can be freely downloaded, and provides a number of novel features not offered by other tools, as well as state-of-the-art performance. Novel features include the possibility of unbounded verification with guaranteed termination, analysis of infinite sets of traces in terms of patterns, and support for multi-protocol analysis.

Scyther is based on a pattern refinement algorithm, providing concise representations of (infinite) sets of traces. This allows the tool to assist in the analysis of classes of attacks and possible protocol behaviours, or to prove correctness for an unbounded number of protocol sessions. The tool has been successfully applied in both research and teaching.

2 The Scyther Tool

The tool provides a graphical user interface (Fig. 1), that complements the command-line and Python scripting interfaces. The GUI is aimed at users interested in verifying or understanding a protocol. The command-line and scripting interfaces facilitate the use of Scyther for large-scale protocol verification tests.

* This work was partially supported by the Hasler Foundation, ManCom project 2071.

A. Gupta and S. Malik (Eds.): CAV 2008, LNCS 5123, pp. 414–418, 2008.

Scyther combines a number of novel features with state-of-the art performance. First, Scyther is guaranteed to terminate whilst allowing to prove correctness of protocols for an unbounded number of sessions, and can optionally output the proof tree (by using the backend). In contrast to other unbounded verification tools, the tool provides useful results even in the case that no attack is found but also no unbounded correctness can be established. In such cases, the results from Scyther have a similar interpretation as bounded verification tools, stating that no attack exists within a certain bound.

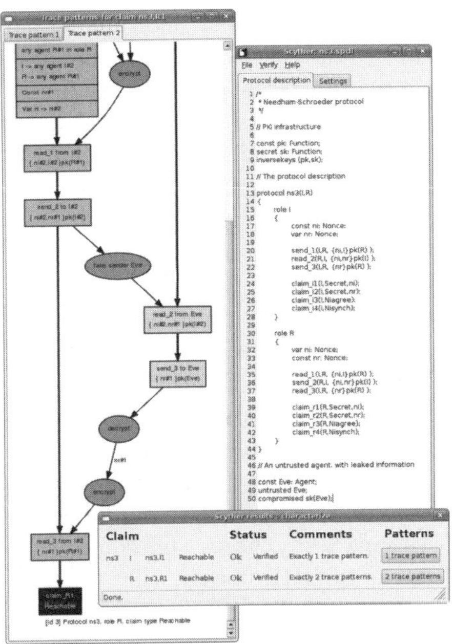

Fig. 1. The graphical user interface

Second, Scyther assists in protocol analysis by providing classes of protocol behaviour (or classes of attacks), as opposed to just single attack traces provided by other tools.

Third, Scyther facilitates so-called multi-protocol analysis. In such an analysis, the parallel composition of two (sub)protocols is analyzed, as in [3]. Traditionally, such an analysis has been infeasible for protocol tools because of state-space explosion. With the performance provided by the Scyther tool, multi-protocol analysis has become feasible, and can be performed by simply verifying the concatenation of multiple protocol description files.

Given the description of a protocol in the *spdl* language, which is based on the operational semantics found in [4], the tool can be used in three ways: to verify whether the security claims in the protocol description hold or not; to automatically generate appropriate security claims for a protocol and verify these; to analyze the protocol by performing *complete characterization*. We describe each of these modes below.

Verification of claims. The input language of Scyther allows for specification of security properties in terms of claim events, i.e., in a role specification one can claim that a certain value is confidential (secrecy) or certain properties should hold for the communication partners (authentication). Scyther can be used to verify these properties or falsify them.

Automatic claims. If the protocol specification contains no security claims, Scyther can automatically generate claims. At the end of each role, authentication claims are added, claiming that the supposed communication partners must have performed the protocol as expected. Similarly, secrecy claims are added for all locally generated values (nonces) and variables. This augmented protocol description is then analyzed by Scyther as in the previous case. This enables users to quickly assess the properties of a protocol.

Characterization. For protocol analysis, each protocol role can be "characterized", which means that Scyther analyzes the protocol, and provides a finite representation of all traces that contain an execution of the protocol role. In most cases, this representation consists of a small number (1-5) of possible execution patterns. By manually inspecting these patterns, one can quickly gain insight in the potential problems with the protocol and modify it if necessary. For example, given the Needham-Schroeder protocol, Scyther determines that there are only two patterns for the responder role: one is the correct behaviour of the protocol, and the other is the well-known man-in-the-middle attack. Hence, there are no other possible ways of executing the responder role.

The algorithm developed for the Scyther tool extends on ideas described in [5], and the idea of analyzing protocols in terms of trace classes was published first in [6]. Scyther addresses the undecidability of the security problem by (1) significantly improving and extending the class pruning theorems from [5] and (2) introducing a parameter that limits the pattern size, ensuring termination. Even though the pattern depth size is limited, Scyther can perform unbounded verification of the majority of protocols, as each pattern represents an infinite class of traces. In practice, with protocols from libraries such as SPORE [7], Scyther is known to provide in about 80 percent of cases either unbounded verification or falsification, and in the other 20 percent provides bounded verification. Further details about the underlying methods are given in [8]. The Scyther tool is freely available for Windows, Linux, and Max OS X platforms. It can be downloaded at `http://people.inf.ethz.ch/cremersc/scyther/`, and comes with a library of example protocols modeled after the SPORE library [7].

3 Performance and Applications

We have extensively investigated the performance of Scyther compared to other state-of-the-art protocol verification tools, which is reported in [9]. In these tests, Scyther outperformed the state-of-the-art Avispa tools [2]. Even though no abstraction techniques are used by Scyther, it offered a level of performance similar to the abstraction-based ProVerif tool [1]. In practice this means small (e.g. Needham-Schroeder, Yahalom, Otway-Rees) to medium-sized (e.g. TLS, Kerberos) protocols are usually verified in less than a second. To the best of our knowledge, Scyther is currently the fastest protocol verification tool that does not use approximation methods.

Scyther has been successfully used for the analysis and design of protocols, and has also been used for theoretical research and teaching. Exploiting the state-of-the-art performance of Scyther, we have discovered a number of previously unreported attacks, e.g. as in [10, 3]. Scyther has also been used to verify theoretical results regarding protocol composition in [11], and was used for finding the counterexample that led to the main theorem of [12]. The tool is currently being used for teaching purposes at several universities, including the Eindhoven University of Technology, ETH Zurich, University of Luxembourg, University of Twente, and the University of Grenoble. For teaching, the clear relation between the protocol specification and the protocol semantics has proven useful in explaining the fundamentals of protocol design and analysis. The concise protocol descriptions help in focussing on the protocol as opposed to tool details, in contrast to other protocol tools, which require the specification of error-prone scenarios for the verification of properties. For teaching purposes, a set of example exercises for students is available at `http://people.inf.ethz.ch/cremersc/scyther/scyther-exercises.html`.

In future work we aim to turn the informal Scyther proof output into a proof object that can be verified by mechanical theorem provers. The underlying protocol model has already been modeled in Isabelle/HOL, as described in [13].

References

1. Blanchet, B.: An efficient cryptographic protocol verifier based on Prolog rules. In: Proc. 14th IEEE Computer Security Foundations Workshop (CSFW), pp. 82–96. IEEE, Los Alamitos (2001)
2. Armando, A., Basin, D., Boichut, Y., Chevalier, Y., Compagna, L., Cuellar, L., Drielsma, P., Heám, P., Kouchnarenko, O., Mantovani, J., Mödersheim, S., von Oheimb, D., Rusinowitch, M., Santiago, J., Turuani, M., Viganò, L., Vigneron, L.: The AVISPA Tool for the Automated Validation of Internet Security Protocols and Applications. In: Etessami, K., Rajamani, S.K. (eds.) CAV 2005. LNCS, vol. 3576, pp. 281–285. Springer, Heidelberg (2005)
3. Cremers, C.: Feasibility of multi-protocol attacks. In: Proc. of The 1st Int. Conf. on Availability, Reliability and Security (ARES), pp. 287–294. IEEE, Los Alamitos (2006)
4. Cremers, C., Mauw, S.: Operational semantics of security protocols. In: Leue, S., Systä, T.J. (eds.) Scenarios: Models, Transformations and Tools. LNCS, vol. 3466, pp. 66–89. Springer, Heidelberg (2005)
5. Song, D.: An Automatic Approach for Building Secure Systems. PhD thesis, UC Berkeley (December 2003)
6. Doghmi, S., Guttman, J.D., Thayer, F.: Skeletons, homomorphisms, and shapes: Characterizing protocol executions. In: Proc. of the 23rd Conf. on the Mathematical Foundations of Programming Semantics (MFPS XXIII). ENTCS, vol. 173, pp. 85–102. Elsevier ScienceDirect, Amsterdam (2007)
7. Security Protocols Open Repository, `http://www.lsv.ens-cachan.fr/spore`
8. Cremers, C.: Scyther - Semantics and Verification of Security Protocols. Ph.D. dissertation, Eindhoven University of Technology (2006)
9. Cremers, C., Lafourcade, P.: Comparing state spaces in automatic protocol verification. In: Proc. of the 7th Int. Workshop on Automated Verification of Critical Systems (AVoCS 2007). ENTCS (September 2007) (to appear)
10. Cremers, C., Mauw, S.: Generalizing Needham-Schroeder-Lowe for multi-party authentication, CSR 06-04, Eindhoven University of Technology (2006)

11. Andova, S., Cremers, C., Gjøsteen, K., Mauw, S., Mjølsnes, S., Radomirović, S.: A framework for compositional verification of security protocols. Information and Computation 206, 425–459 (2008)
12. Cremers, C.: On the protocol composition logic PCL. In: Abe, M., Gligor, V. (eds.) Proc. of the ACM Symposium on Information, Computer & Communication Security (ASIACCS 2008), Tokyo, pp. 66–76. ACM Press, New York (2008)
13. Meier, S.: A formalization of an operational semantics of security protocols. Diploma thesis, ETH Zurich (August 2007), http://people.inf.ethz.ch/meiersi/fossp/index.html

The CASPA Tool: Causality-Based Abstraction for Security Protocol Analysis
Tool Paper

Michael Backes[1,2], Stefan Lorenz[1], Matteo Maffei[1], and Kim Pecina[1]

[1] Saarland University, Saarbrücken, Germany
[2] MPI-SWS

Abstract. CASPA constitutes a push-button tool for automatically proving secrecy and authenticity properties of cryptographic protocols. The tool is grounded on a novel technique for causality-based abstraction of protocol executions that allows establishing proofs of security for an unbounded number of concurrent protocol executions in an automated manner. We demonstrate the expressiveness and efficiency of the tool by drawing a comparison with T4ASP, the static analyzer for secrecy properties offered by the AVISPA tool. CASPA is capable of coping with a substantially larger set of protocols, and excels in performance.

1 Introduction

Proofs of security protocols are known to be error-prone and, owing to the distributed-system aspects of multiple interleaved protocol runs, awkward to do. In fact, vulnerabilities have accompanied the design of such protocols ever since early authentication protocols like the Needham-Schroeder protocol, to carefully designed de-facto standards like SSL and PKCS, up to current widely deployed products like Microsoft Passport. Formal methods have proved to be salient tools for dealing with such flaws, by helping both to securely design and to analyze security protocols, and even to formally establish their security properties.

A central intricacy that these tools have to tackle is to concisely treat the potentially very large number of concurrent protocol executions. We have developed CASPA (Causality-based Abstraction for Security Protocol Analysis), a tool for establishing formal security proofs of cryptographic protocols for an unbounded number of concurrent protocol executions in a mechanized manner. The tool is grounded on a recently proposed abstract interpretation of cryptographic protocols [3] based on causal graphs. Causal graphs are finite dependency graphs in which nodes represent process events and edges express the causality among events. These graphs constitute a sound abstraction of an unbounded number of protocol executions and, interestingly, they serve as a graphical illustration of the actual protocol behavior. A quick inspection of these graphs often suffices to identify unintended, and possibly harmful, interactions among parties. This facilitates protocol design and error detection even on the human level.

A. Gupta and S. Malik (Eds.): CAV 2008, LNCS 5123, pp. 419–422, 2008.

Related Work. We demonstrate the expressiveness and efficiency of our tool by drawing a comparison with T4ASP [13,9], the static analyzer for secrecy properties offered by AVISPA [2], the well-known tool suite for security protocol analysis. CASPA is capable of coping with a substantially larger set of protocols than TA4SP, and it furthermore excels in terms of performance; moreover, CASPA is capable of verifying both secrecy and authenticity properties in contrast to only secrecy properties in the case of TA4SP. Note that TA4SP so far constitutes the only tool of the AVISPA tool suite that is capable of producing proofs of security; the remaining tools such as OFMC [5] rely on state-space exploration techniques and are constraint to either taking into account only a limited number of concurrent protocol executions, typically two or three, or giving up guaranteed termination in general. The same holds for other techniques based on state-space exploration such as Athena [16], CPSA [12], and Scyther [11]. ProVerif [6] is a tool based on Horn-clause resolution that verifies trace-based security properties, such as secrecy and authenticity, as well as observational equivalence of process behavior, is very efficient, but provides guaranteed termination only for tagged protocols [7]. The Lysa tool [8] implements a control-flow analysis that offers termination and security proofs for an unbounded number of protocol executions, but does not support security properties based on correspondence assertions [17].

2 The CASPA Tool

CASPA is written in Objective CAML and equipped with a graphical user interface, assisting the user in the specification and in the analysis of cryptographic protocols, see Figure 1. CASPA provides an *editor* for protocol specifications, offering a quick loading procedure for the protocols specified in underlying protocol libraries, and a convenient parsing procedure for user-defined protocol specifications. In addition, the tool features a *graph management system* that automatically generates and displays causal graphs. Finally, CASPA offers a fully mechanized *analyzer* that verifies secrecy and authenticity properties on a given causal graph and displays the results. More precisely, CASPA allows for analyzing the security properties *secrecy*, *weak authenticity*, and *strong authenticity* [14]: as in AVISPA, authenticity properties are specified in terms of correspondence assertions [17]. Since causal graphs are of finite size, the analysis is assured to terminate. The analysis entails security proofs that establish the safety of the protocol for a potentially unbounded

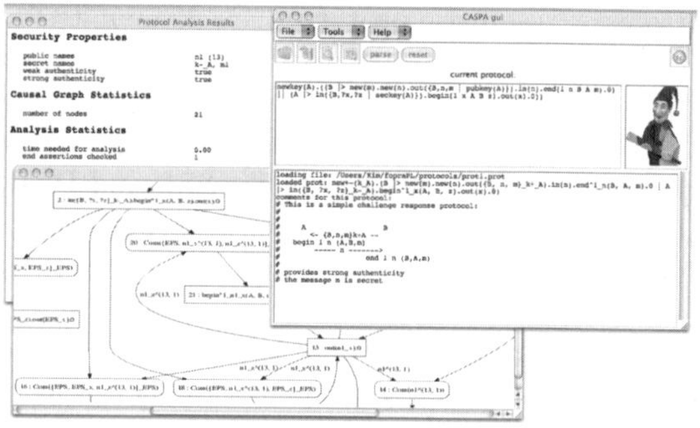

Fig. 1. The CASPA Tool

Table 1. Protocol results, conducted on a Pentium-IV 3GHz 1GB under linux

Protocol	CASPA	TA4SP	OFMC	Protocol	CASPA	TA4SP	OFMC
CHAPv2	0,93s	10,59s	0,32s	NSPK	0,13s	7,56s	0,01s
CRAM-MD5	0,09s	-	0,71s	NSPK-KS	28m	-	1,1s
EKE	0,81s	7,56s	0,19s	NSPK-fix	0,08s	0,98s	0,18s
IKEv2-CHILD	0,31s	-	1,19s	NSPK-KS-fix	7m	-	24,86s
ISO1	0,05s	×	0,02s	SHARE	0,4s	14,38s	0,08s
ISO3	1,08s	×	0,04s	UMTS-AKA	0,04s	0,51s	0,02s
LPD-MSR	0,05s	-	0,02s	APOP	0,44s	×	2,94s
LPD-IMSR	0,37s	-	0,08s	DHCP-DA	1,03s	-	0,06s

number of protocol executions. As usual for static analysis techniques and due to the undecidability of the security problem, false positives may occur caused by an insufficient precision of the analysis, hence potentially classifying secure protocols as insecure. The CASPA tool is freely available at [4].

3 Performance Evaluation of CASPA

We evaluate the performance of our tool by running it on a subset of the AVISPA library. The facilitate our experiments, we developed a translator from the Intermediate Format protocol language [2] that the AVISPA suite is based upon into the dialect of the spi-calculus [1] used in our tool. The translation is only partially automated in that it requires some manual steps, such as specifying the owners of the keys and defining suitable correspondence assertions. These steps were straightforward in all protocols considered so far, but they admittedly require basic familiarity with our language and understanding of the protocol.

The results are reported in Table 1. The tool succeeded in the analysis of safe protocols (i.e., we did not get any false positives), and it failed to establish security proofs of flawed protocols as expected. For each protocol, the table reports the running time for CASPA and TA4SP. The performance evaluation shows that CASPA is capable of dealing with a substantially larger set of protocols than TA4SP: the symbol - means that the protocol is not supported by the tool, while the symbol × means that the protocol guarantees only authenticity properties, which can be verified by CASPA and not by TA4SP. In addition, the evaluation shows that even for the protocols that are in scope of both TA4SP and CASPA, the CASPA tool improves in terms of performance. Note that the subset of protocols we consider comprises all protocols of the AVISPA library for which an analysis with TA4SP succeeds. Some protocols in the AVISPA library employ non-standard equational theories that are not supported by our tool (e.g., NSPKxor that is based on xor) and for some other protocols the analysis did not succeed due to memory exhaustion (e.g., SET and TSL).

For comparison, we additionally depict the running times of OFMC [5], the most advanced model checker in the AVISPA tool, when the analysis is constrained to three executions. In this restricted setting, OFMC shows significantly better performances. Scaling this approach to more executions rapidly becomes infeasible, hence leaving potential attacks undetected (e.g., cf. [10,15]).

References

1. Abadi, M., Gordon, A.D.: A calculus for cryptographic protocols: The spi calculus. Information and Computation 148(1), 1–70 (1999)
2. Armando, A., Basin, D., Boichut, Y., Chevalier, Y., Compagna, L., Cuellar, L., Drielsma, P., Heám, P., Kouchnarenko, O., Mödersheim, J.M.S., von Oheimb, D., Rusinowitch, M., Santiago, J., Turuani, M., Viganò, L., Vigneron, L.: The avispa tool for the automated validation of internet security protocols and applications. In: Etessami, K., Rajamani, S.K. (eds.) CAV 2005. LNCS, vol. 3576, pp. 281–285. Springer, Heidelberg (2005)
3. Backes, M., Cortesi, A., Maffei, M.: Causality-based abstraction of multiplicity in cryptographic protocols. In: Proc. 20th IEEE Symposium on Computer Security Foundations (CSF), pp. 355–369. IEEE, Los Alamitos (2007)
4. Backes, M., Lorenz, S., Maffei, M., Pecina, K.: The CASPA tool, `www.infsec.cs.uni-sb.de/caspa`
5. Basin, D.A., Mödersheim, S., Viganò, L.: Ofmc: A symbolic model checker for security protocols. IJIS 4(3), 181–208 (2005)
6. Blanchet, B.: An efficient cryptographic protocol verifier based on Prolog rules. In: Proc. 14th IEEE Computer Security Foundations Workshop (CSFW), pp. 82–96. IEEE, Los Alamitos (2001)
7. Blanchet, B., Podelski, A.: Verification of cryptographic protocols: Tagging enforces termination. In: Gordon, A.D. (ed.) ETAPS 2003 and FOSSACS 2003. LNCS, vol. 2620, pp. 136–152. Springer, Heidelberg (2003)
8. Bodei, C., Buchholtz, M., Degano, P., Nielson, F., Nielson, H.R.: Static validation of security protocols. Journal of Computer Security 13(3), 347–390 (2005)
9. Boichut, Y., Genet, T.: Feasible trace reconstruction for rewriting approximations. In: Pfenning, F. (ed.) RTA 2006. LNCS, vol. 4098, pp. 123–135. Springer, Heidelberg (2006)
10. Chadha, R., Kremer, S., Scedrov, A.: Formal analysis of multi-party contract signing. In: Proc. 17th IEEE Computer Security Foundations Workshop (CSFW), pp. 266–279. IEEE, Los Alamitos (2004)
11. Cremers, C.: Scyther - Semantics and Verification of Security Protocols. Ph.D. dissertation, Eindhoven University of Technology (2006)
12. Doghmi, S., Guttman, J., Thayer, F.: Searching for shapes in cryptographic protocols. In: Grumberg, O., Huth, M. (eds.) TACAS 2007. LNCS, vol. 4424, pp. 523–537. Springer, Heidelberg (2007)
13. Genet, T., Tong, V.: Reachability Analysis of Term Rewriting Systems with Timbuk. In: Nieuwenhuis, R., Voronkov, A. (eds.) LPAR 2001. LNCS (LNAI), vol. 2250, pp. 695–706. Springer, Heidelberg (2001)
14. Lowe, G.: A Hierarchy of Authentication Specification. In: Proc. 10th IEEE Computer Security Foundations Workshop (CSFW), pp. 31–44. IEEE, Los Alamitos (1997)
15. Millen, J.: A necessarily parallel attack. In: Proc. of Workshop on Formal Methods and Security Protocols (1999)
16. Song, D.X.: Athena: a new efficient automatic checker for security protocol analysis. In: Proc. 12th IEEE Computer Security Foundations Workshop (CSFW), pp. 192–202. IEEE, Los Alamitos (1999)
17. Woo, T.Y.C., Lam, S.S.: A lesson on authentication protocol design. Operation Systems Review 28(3), 24–37 (1994)

Jakstab: A Static Analysis Platform for Binaries*

Tool Paper

Johannes Kinder and Helmut Veith

Technische Universität Darmstadt, 64289 Darmstadt, Germany

Abstract. For processing compiled code, model checkers require accurate model extraction from binaries. We present our fully configurable binary analysis platform JAKSTAB, which resolves indirect branches by multiple rounds of disassembly interleaved with dataflow analysis. We demonstrate that this iterative disassembling strategy achieves better results than the state-of-the-art tool IDA Pro.

Introduction. While most of today's model checkers operate on source code, there are various settings where we need to verify binary code. First, when source code is not available, e.g., when a software manufacturer wants to verify the conformance of third party modules, such as drivers or plugins, to the API specification. Second, to be able to detect errors introduced in the compiling process [1], which is of particular importance in the field of embedded systems, where compilers can be unreliable. Third, binary level analysis results can supplement execution traces collected by testing and vice versa, as demonstrated by the SYNERGY algorithm [2]. And finally, our original motivation for this research stems from using model checking to detect malicious code inside executables [3].

Extracting a control flow graph (CFG) from an executable is not simply a matter of implementing a language front-end for assembly. Compiled code lacks many comfortable properties of structured high level languages and poses several challenges for analysis tools. Function pointers are only seldom handled by source-level verification tools, but on assembly level, calls and jumps to pointers are too abundant to be ignored. The treatment of function pointers requires dataflow analysis on an incomplete CFG. Thus, the traditional sequence, in which an analyzer builds the CFG first and only then performs dataflow analysis, has to be replaced by an iterative process. Another challenge is the loss of structure in compiled code. For accurate analysis results, procedures, along with their calling conventions, need to be explicitly detected. Compiler optimizations and, worse, obfuscation techniques can further mangle the control flow structure of an executable and impede correct disassembly and control flow extraction [4].

Existing disassemblers can be divided into two categories [4]: Linear sweep disassemblers, such as GNU objdump, simply sequentially translate machine code into assembly instructions. Recursive traversal disassemblers, such as IDA Pro, follow direct branches and decode the program by depth first search. We extend this classification by defining an *iterative disassembler* as one that interleaves multiple disassembly rounds with dataflow analysis to achieve accurate and complete CFG extraction.

* Supported by DFG grant FORTAS – Formal Timing Analysis Suite for Real Time Programs (VE 455/1-1) and the European Commission under Contract IST-2002-507932 ECRYPT.

A. Gupta and S. Malik (Eds.): CAV 2008, LNCS 5123, pp. 423–427, 2008.

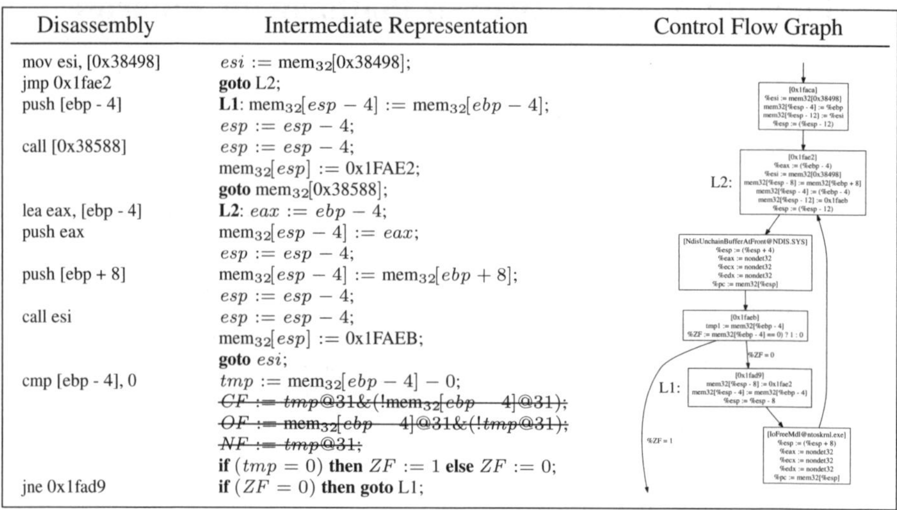

Fig. 1. Part of procedure 0x1FACA in fwdrv.sys. The second call is not resolved by IDA Pro.

Our tool JAKSTAB [1] (**Ja**va tool**k**it for **st**atic **a**nalysis of **b**inaries) serves as a flexible front end to make executables accessible to static analysis and model checking. To this end, JAKSTAB contains an iterative disassembler and a library of semantic descriptions that translates assembly instructions to an RTL-style intermediate representation. Disassembler and semantic descriptions are fully configurable to support multiple target platforms. Using the intermediate representation, JAKSTAB iteratively creates the CFG, calculating and resolving indirect branch targets using results from dataflow analysis. JAKSTAB is implemented in Java and can be either used as a library or via its command line interface, which outputs plain disassembly or the intermediate representation as a CFG in graphviz-format. The intermediate representation, consisting of assignments, if, and goto statements, is independent of the target hardware and provides a natural interface to model checkers and program analysis tools.

Today's de facto industry standard for disassembly is IDA Pro. Its heuristic matches common prologue bytes to identify procedures and assumes that every call returns to its original site, regardless of the call target, which can lead to erroneous fall-through edges. Furthermore, the CFG is usually incomplete, since IDA Pro has only a very basic ability to resolve indirect branch instructions (function pointers): It propagates constants just within a basic block, and decorates calls to such constants with comments containing the actual target. While this is enough to aid human engineers, it is insufficient for automated analysis. Figure 1 shows an exemplary piece of assembly code from a Windows driver executable (fwdrv.sys from *Sunbelt Personal Firewall*), where IDA Pro (v4.7) fails to identify an indirect call to an imported function, whose address is stored at a memory location pointed to by the register *esi*. Finally, even though IDA Pro offers an (unsupported) SDK for plugin development, it is closed source software and thus cannot be easily integrated with an analysis tool.

[1] Project page online at `http://www.jakstab.org`

To the best of our knowledge, the most successful approach to static analysis of executables currently is the CodeSurfer/x86 project [5]. CodeSurfer/x86 uses IDA Pro to access binaries, and combines two program analysis algorithms, value set analysis (VSA) and aggregate structure identification (ASI). In recent work, they combined VSA with a property automaton that encodes certain usage rules for the Windows driver API [6]. Generally, they assume a standard compilation model for binaries, which guarantees correct disassembly by IDA Pro. They acknowledge that IDA Pro's output can be incomplete and do connect missing edges from indirect calls, yet they lack a complete loop to disassemble previously unprocessed branch targets.

Closely related to executable analysis is the idea of building a *decompiler*, which transforms an executable back to source code [7,8]. Chang et al. describe an architecture of communicating decompilers at different language levels [9]. Their implementation propagates static analysis facts through all language levels one instruction at a time, instead of strictly separating decompilation stages by language level. The prototype targets assembly source files generated by a set of compilers, and thus requires access to source code. We believe that JAKSTAB would fit nicely into this tool-chain as a provider of well-formed CFGs from generic executables.

Control Flow Reconstruction. In most assembly languages, instructions can affect multiple registers and status flags. The x86 architecture, which we first focused on, features an especially rich instruction set where instructions often represent non-trivial operation sequences. To fully capture instruction semantics and enable easy extensibility, JAKSTAB is designed to read Semantic Specification Language (SSL) files supplied with the Boomerang decompiler, which are available for several architectures including x86, PowerPC, 68K, and SPARC [10,8]. Figure 1 shows the intermediate representation JAKSTAB produces from the assembly snippet using SSL definitions for the x86 architecture. Mapping every assembly instruction to its semantic specification creates a program representation with obvious pieces of dead code. In particular, most of the status flags are not used but simply overwritten by later instructions. To reduce the program size, our tool executes a live variable analysis and afterward removes any dead code. In our experiments, usually about 30% of the statements are identified as dead code and removed from the control flow graph. In the example in Figure 1, three flag updates are removed (crossed out text), and only one relevant update remains.

JAKSTAB recreates the control flow graph in an iterative process. Starting from the entry point of the executable, it propagates and folds constants through registers and memory cells to resolve indirect branch targets. JAKSTAB supports indirect memory access, which is common for local variables stored on the stack. Whenever Jakstab cannot resolve the address of an indirect write, it currently assumes that every memory cell can become undefined. Calls to shared libraries, which, in the Windows PE-format, appear as indirect calls to memory locations, are handled by creating stub procedures in the control flow graph. Constant propagation and folding is performed on all parts of the CFG already known, which allows JAKSTAB, in contrast to IDA Pro, to successfully recover the CFG of the example in Figure 1. Note that the results of constant propagation can theoretically be incorrect if incoming edges to *existing* nodes are discovered in later iterations. In such cases, the CFG reconstruction process has to be restarted.

	cmd.exe		dnsrslvr.dll		faultrep.dll		ftp.exe		nmnt.sys		rcp.exe		svchost.exe	
IDA Pro	74%	9.4s	81%	36.2s	73%	5.4s	88%	2.4s	74%	3.1s	42%	1.4s	56%	1.5s
JAKSTAB	91%	32.4s	92%	3.2s	98%	9.0s	94%	2.7s	96%	4.5s	100%	1.1s	88%	1.0s

Fig. 2. Success rates and processing times for resolving indirect branches in executables

Any target location that has been successfully resolved in one iteration is scheduled for disassembly in the next one. Newly detected procedures are inlined to ensure correct interprocedural results in the next round of constant propagation. Figure 1 shows the CFG extracted from the example code, including stubs for imported library functions. The stubs non-deterministically assign those registers which might be overwritten by library functions (*eax, ecx, edx* according to the Intel application binary interface).

We compared JAKSTAB's and IDA Pro's capabilities in resolving indirect branches on Microsoft Windows system binaries. The results we present in Fig. 2 clearly show that JAKSTAB is able to provide significantly more accurate CFGs than IDA Pro at similar, and in some cases even faster, execution speeds.

Applications and Future Work. Our goal is to use JAKSTAB as a versatile platform for different verification tasks on binary level. Currently, we are building a bounded model checker on top of the existing framework to allow better resolution of indirect jumps and the extraction of all targets from jump tables. Besides the internal use of the bounded model checker for improving the CFG, we will investigate what kind of specifications can be verified on binary level, with particular focus on API usage specifications.

JAKSTAB, unlike IDA Pro, does not assume a standard compilation model. Therefore it is well suited to process code protected against disassembly, in particular malicious code. Anti-disassembly patterns that obscure the control flow of a program will thwart traditional recursive traversal disassemblers [4]. For example, return instructions are commonly misused as generic jumps by pushing the desired target address on the stack immediately beforehand. Since JAKSTAB supports local constant propagation through the stack, it can retarget disassembly correctly in these cases and is able to recover the real control flow. A CFG extracted from such a potentially malicious program can then be used as input to a semantic malware detector [3].

References

1. Balakrishnan, G., Reps, T., Melski, D., Teitelbaum, T.: WYSINWYX: What You See Is Not What You eXecute. In: VSTTE, Zurich, Switzerland (2005)
2. Gulavani, B., Henzinger, T., Kannan, Y., Nori, A., Rajamani, S.: SYNERGY: a new algorithm for property checking. In: SIGSOFT FSE 2006, pp. 117–127. ACM, New York (2006)
3. Kinder, J., Katzenbeisser, S., Schallhart, C., Veith, H.: Detecting malicious code by model checking. In: Julisch, K., Krügel, C. (eds.) DIMVA 2005. LNCS, vol. 3548, pp. 174–187. Springer, Heidelberg (2005)
4. Linn, C., Debray, S.: Obfuscation of executable code to improve resistance to static disassembly. In: CCS 2003, pp. 290–299. ACM, New York (2003)
5. Balakrishnan, G., Reps, T.: Analyzing memory accesses in x86 executables. In: Duesterwald, E. (ed.) CC 2004. LNCS, vol. 2985, pp. 5–23. Springer, Heidelberg (2004)

6. Balakrishnan, G., Reps, T.: Analyzing stripped device-driver executables. In: TACAS 2008. LNCS, pp. 124–140. Springer, Heidelberg (2008)
7. Cifuentes, C.: Reverse Compilation Techniques. PhD thesis, Queensland University of Technology (1994)
8. van Emmerik, M., Waddington, T.: Using a decompiler for real-world source recovery. In: WCRE 2004, pp. 27–36. IEEE Computer Society, Los Alamitos (2004)
9. Chang, B., Harren, M., Necula, G.: Analysis of low-level code using cooperating decompilers. In: Yi, K. (ed.) SAS 2006. LNCS, vol. 4134, pp. 318–335. Springer, Heidelberg (2006)
10. Cifuentes, C., Sendall, S.: Specifying the semantics of machine instructions. In: International Workshop on Program Comprehension (IWPC 1998), pp. 126–133. IEEE Computer Society, Los Alamitos (1998)

THOR: A Tool for Reasoning about Shape and Arithmetic*

Tool Paper

Stephen Magill[1], Ming-Hsien Tsai[2], Peter Lee[1], and Yih-Kuen Tsay[2]

[1] Carnegie Mellon University
[2] National Taiwan University

Abstract. We describe THOR (**T**ool for **H**eap-**O**riented **R**easoning), a tool based on separation logic that is capable of reasoning automatically about heap-manipulating programs. There are several such systems in development now. However, THOR is unique in that it provides not only shape analysis, but also arithmetic reasoning via a novel combination procedure. Also, considerable effort has been put into making the output clear and easy to understand. THOR uses Javascript and HTML to produce an interactive representation of the analysis results.

1 Introduction

There has been a surge of interest in the use of separation logic to automatically prove memory safety. This has resulted in a number of program analysis tools that use separation logic to describe program states. One such tool is under development at Berkeley [5]. Another is the Space Invader tool [6], developed at Queen Mary's, which is able to scale to programs with over 10,000 lines of code [12]. SLAYER [2], developed at Microsoft Research, also has scalability as its primary goal and is focused on proving memory safety of large programs.

We have taken a different approach with our tool, THOR. Instead of trying to see how far we can scale a memory safety analysis, we are interested in seeing how precise an analysis we can develop while maintaining full automation. In particular, we have focused on the combination of list reasoning and arithmetic. THOR implements a shape analysis based on separation logic that is capable of reasoning about doubly-linked lists. It then adds support for arithmetic reasoning involving stack-based integers, integers in the heap, and the lengths of lists. This arithmetic support is provided by utilizing an "off-the-shelf" arithmetic analysis tool as described in [9]. This is interesting for two reasons. First, it provides very robust arithmetic reasoning, since the precision of the combination improves with the precision of the arithmetic tool being used. Secondly, the integer programs produced by the shape analysis phase provide a new source of test programs for arithmetic analysis tools. And our experiments indicate that the arithmetic programs produced by our method pose a challenge for some existing tools.

* This work was partially supported by the iCAST project sponsored by the National Science Council, Taiwan, under the Grant No. NSC96-3114-P-001-002-Y.

A. Gupta and S. Malik (Eds.): CAV 2008, LNCS 5123, pp. 428–432, 2008.

```
int i = malloc(sizeof(int));
List *curr = NULL;   *i = 0;
while(*i < n) {
    t = (List*) malloc(sizeof(List));
    t->next = curr;
    curr = t;
    *i = *i + 1;
}
free(i);  int j = 0;
while(j < n) {
   t = curr->next;
   free(curr);
   curr = t;
   j++;
}
```

Fig. 1. Example showing motivation for combined shape and arithmetic reasoning

```
int a = 0;
int k = 0;
while(a < n) {
  a = a + 1;
  k = k + 1;
}
int j = 0;
while(j < n) {
  if(k = 0)
    goto ERROR;
  else
    k = k - 1;
  j++;
}
```

Fig. 2. Arithmetic counterexample program produced by the shape analysis

Figure 1 contains an example of the sort of programs that THOR is targeting. In this program, a list of length `n` is constructed and then deallocated. The deallocation code relies on the fact that the list being freed is of length `n`. Proving that this loop is memory safe involves showing that at the beginning of each loop iteration, `n - j` is the length of the still-allocated portion of the list.

Ours is not the first approach to a static analysis combining shape and arithmetic reasoning. MUTANT [3] is similar in spirit to our approach but does not output arithmetic programs and relies on weaker arithmetic invariants. Also, THOR's treatment of list lengths can be viewed as a realization of the connection between list programs and counter automata that Bouajjani et al. described in [4]. However, we believe that leveraging a separation logic-based shape analysis to provide the connection gives a more generally applicable method of going from a pointer program to an arithmetic program.

2 Interacting with THOR

THOR is a command-line program written in OCaml. It takes as input a C program and a function name. It then runs a shape analysis that proceeds by exploring the state-space of the program, symbolically executing all paths through the code. For loops, a join operator similar to that in [10] is used to ensure that the symbolic description of the program state converges to an overapproximation of the reachable states. If the program can be proved memory safe with only shape reasoning, then no further processing is required. However, if safety of the program depends on arithmetic information, as in Figure 1, then the second phase of THOR's combined analysis will be invoked.

This second phase translates the results of the shape analysis into a purely arithmetic program. This translation is such that if the arithmetic program can be shown to be safe (where safety means non-reachability of a designated "error" location), then the original program is guaranteed to be memory safe. Viewed another way, the shape analysis implemented in THOR translates memory safety of heap-manipulating code to assertion safety of purely stack-based code. Details are given in [9], but we mention here two of the most interesting cases.

The first involves variables present only in the analysis, such as variables representing the lengths of lists. If the source program contains a loop that modifies a list during each iteration, the translation will insert a variable k into the generated program along with initialization and update statements for k that ensure this variable always tracks the length of the list. This allows the arithmetic analysis tool to discover relationships involving these length variables. We can see this occuring in the first loop in Figure 2 with the variable k.

The second interesting case involves branches on length variables. Sometimes a command is memory safe only if length variable k is positive, indicating that a certain list is non-empty. In such cases, a branch is inserted in the analysis and in the generated program that tests whether $k = 0$. In the *true* branch, we go to the error state, as this case corresponds to a memory fault. In the *false* branch, we continue exploring the state space, as this case is memory safe. This is the role of the `if` statement in the second loop in Figure 2.

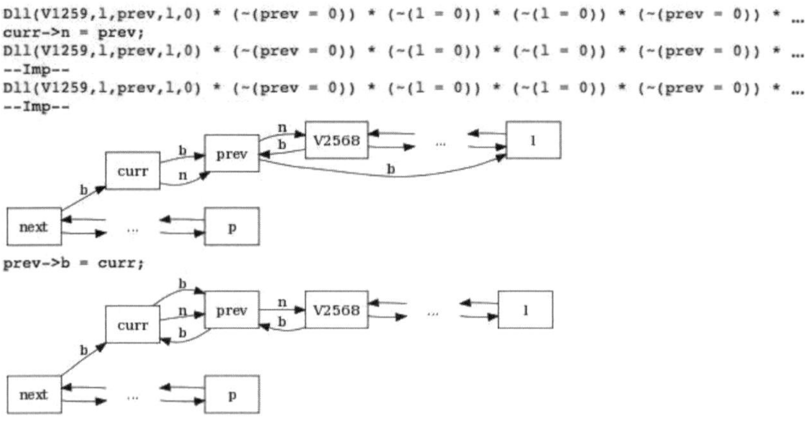

Fig. 3. Fragment of THOR's HTML output

By default, the arithmetic program is output as C code, so any analysis tool capable of handling C can be used to check the arithmetic program and thus prove memory safety of the original code. We have tried both BLAST [8] and ARMC [11] on the C code generated by THOR. We have also implemented an option to output the program in the FAST file format for use with that tool [1].

In addition to the arithmetic program, the tool also outputs a representation of the execution tree that was constructed while proving the program. The tree is represented as C-style source code. Branches in the analysis are represented as *if* statements and inductive invariants are represented by *gotos*.[1] Annotations are inserted between every command, which is helpful both when analyzing counter-example paths and when debugging the tool itself.

The tool can also generate graphical depictions of the program state. This is accomplished by using the Graphviz library [7] to render "box and pointer" diagrams that provide an intuitive view of the contents of memory. These graphical renderings are linked to the textual descriptions via Javascript such that by clicking on a memory state, the user can toggle between the text-based separation logic description and the Graphviz rendering. Figure 3 shows an example with the first three states described using a separation logic formula and the last two rendered as diagrams.

3 Conclusion

We have described THOR, our **T**ool for **H**eap-**O**riented **R**easoning. THOR is an implementation of the combination procedure given in [9]. It provides a means of analyzing programs that require both shape and arithmetic reasoning and also serves as a source of interesting test programs for arithmetic tools. THOR is available for download at `http://www.cs.cmu.edu/~smagill/thor`. The distribution also includes a number of example programs.

References

1. Bardin, S., Finkel, A., Leroux, J., Petrucci, L.: Fast: Fast acceleration of symbolic transition systems. In: Hunt Jr., W.A., Somenzi, F. (eds.) CAV 2003. LNCS, vol. 2725, pp. 118–121. Springer, Heidelberg (2003)
2. Berdine, J., Calcagno, C., Cook, B., Distefano, D., O'Hearn, P., Wies, T., Yang, H.: Shape analysis for composite data structures. In: Damm, W., Hermanns, H. (eds.) CAV 2007. LNCS, vol. 4590, pp. 178–192. Springer, Heidelberg (2007)
3. Berdine, J., Cook, B., Distefano, D., O'Hearn, P.: Automatic termination proofs for programs with shape-shifting heaps. In: Ball, T., Jones, R.B. (eds.) CAV 2006. LNCS, vol. 4144, pp. 386–400. Springer, Heidelberg (2006)
4. Bouajjani, A., Bozga, M., Habermehl, P., Iosif, R., Moro, P., Vojnar, T.: Programs with lists are counter automata. In: Ball, T., Jones, R.B. (eds.) CAV 2006. LNCS, vol. 4144, pp. 517–531. Springer, Heidelberg (2006)
5. Chang, B.-Y.E., Rival, X., Necula, G.C.: Shape analysis with structural invariant checkers. In: Riis Nielson, H., Filé, G. (eds.) SAS 2007. LNCS, vol. 4634, pp. 384–401. Springer, Heidelberg (2007)
6. Distefano, D., O'Hearn, P.W., Yang, H.: A local shape analysis based on separation logic. In: Hermanns, H., Palsberg, J. (eds.) TACAS 2006 and ETAPS 2006. LNCS, vol. 3920, pp. 287–302. Springer, Heidelberg (2006)
7. Gansner, E., North, S.: An open graph visualization system and its applications to software engineering. Software — Practice and Experience 30(11) (2000)

[1] If I is an inductive invariant for the code c, then there will be a path in the output of the form `label: {I} c; goto label;`

8. Henzinger, T., Jhala, R., Majumdar, R., Sutre, G.: Lazy abstraction. In: POPL 2002: Principles of Programming Languages, pp. 58–70. ACM Press, New York (2002)
9. Magill, S., Berdine, J., Clarke, E., Cook, B.: Arithmetic strengthening for shape analysis. In: Riis Nielson, H., Filé, G. (eds.) SAS 2007. LNCS, vol. 4634, pp. 419–436. Springer, Heidelberg (2007)
10. Magill, S., Nanevski, A., Clarke, E., Lee, P.: Inferring invariants in separation logic for imperative list-processing programs. In: SPACE (2006)
11. Podelski, A., Rybalchenko, A.: ARMC: the logical choice for software model checking with abstraction refinement. In: Hanus, M. (ed.) PADL 2007. LNCS, vol. 4354, pp. 245–259. Springer, Heidelberg (2006)
12. Yang, H., Lee, O., Berdine, J., Calcagno, C., Cook, B., Distefano, D., O'Hearn, P.: Scalable shape analysis for systems code. In: Gupta, A., Malik, S. (eds.) CAV 2008. LNCS, vol. 5123, pp. 385–398. Springer, Heidelberg (2008)

Functional Verification of Power Gated Designs by Compositional Reasoning

Cindy Eisner, Amir Nahir, and Karen Yorav

IBM Haifa Research Laboratory

Abstract. Power gating is a technique for low power design in which whole sections of the chip are powered off when they are not needed, and powered back on when they are. Functional correctness of power gating is usually checked as part of system-level verification, where the most widely used technique is simulation using pseudo-random stimuli. We propose instead to perform a sequential equivalence check between the power gated design and a version of itself in which power gating is disabled. We take a compositional approach that looks for partial equivalence of each unit under a suitable set of assumptions, guaranteed by the neighboring units. We make use of so-called circular reasoning rules to compose the partial equivalences proved on the individual units back into total equivalence on the whole chip.

1 Introduction

Power consumption is an important consideration in modern chip design. A portable device should use as little energy as possible, in order to extend battery life, and a non-portable device should use as little energy as possible in order to save electricity costs – the expense of powering a server farm can easily outstrip the cost of the servers themselves. Thus design teams go to great lengths to design systems that use as little power as possible.

Many power saving techniques are purely electrical, such as changing the type of transistor used. Other techniques, while electrical, can be understood at the logical level. Among them are multiple power domains, in which different areas of the chip get different voltages; dynamic frequency/voltage scaling, in which the frequency and/or voltage is changed while the chip is working; clock gating, in which the clock input of a memory element is prevented from "ticking" if a tick would be redundant; and power gating, in which whole sections of the chip are powered off when not needed, and powered back on when they are. In addition to the physical design challenges, i.e., getting the electronics right, each of these techniques adds a new dimension to the functional verification problem.

In this paper we focus on power gating [2]. Power gating is one of the most powerful techniques for saving power, as it reduces both dynamic power (the power used when the design transitions from state to state) as well as static, or leakage, power (the power used by the design when it is in a steady state). Functional correctness of power gating is usually checked as part of system-level

A. Gupta and S. Malik (Eds.): CAV 2008, LNCS 5123, pp. 433–445, 2008.

verification, where the most widely used technique is simulation with pseudo-random stimuli. However, because power gating is such a complicated technique, this approach can be a real barrier to its use – either because there is not enough confidence that the current methods can verify correctness, or because the effort required to do so is not economically feasible. Therefore a practical methodology to formally verify the correctness of power gating fulfills a real need.

We present a methodology that addresses functional verification of a design in the presence of power gating, in which we break the verification task into two parts. In the first, correct functionality of the design is checked when power gating is disabled, using the usual techniques (formal and/or dynamic, as the case may be). The second part is a sequential equivalence check between a version of the design with power gating enabled and one with it disabled.

We focus on the second part, which due to the size problem is not trivial. At the level of a single unit, we do not necessarily expect complete sequential equivalence. Rather, we expect that the cases in which the two versions of the unit differ are don't cares for the neighboring units. Thus we require only partial equivalence of each unit, and we make use of so-called circular reasoning rules to allow us to compose the units and the partial equivalences proved on them into total equivalence on the whole chip.

At a very high level, our method can be understood as follows: identify the conditions under which the interface between a power gated unit and its neighbors is "active", and require that the power gated unit preserve functionality only then. Separately, prove that the neighbors are not affected by a difference in behavior when the interface is not active. We note that implicit in our method is the assumption that power gating does not change the external behavior of the chip, but this assumption does not hold in some cases – for example, pipelines that stall when the needed unit is not available. Overcoming this limitation is future work.

In general, compositional reasoning is difficult because of the manual effort involved in coming up with appropriate assumptions, and this has severely limited its use in industrial settings. In our case, the specification is not an arbitrary formula but rather sequential equivalence between outputs of the (composed) design. This simplifies the task of figuring out what the assumptions should be. We require only that the user supply us with an *observer*, a piece of code that observes the chip but does not influence its behavior. The observations that it needs to make are of a specific and limited nature, thus manual coding of an observer is not an unduly difficult task to expect of the user.

Related Work. In this paper we do not invent a new compositional reasoning rule, but rather make use of the one presented in [4]. Our contribution to the compositional reasoning community is the identification of a class of real-life problems for which compositional reasoning is relatively easy, thus barriers to industrial adoption of it are minimal.

As defined in [5], sequential equivalence solves the problem of deciding whether two arbitrary designs are equivalent without knowledge of their initial states or a reset sequence that will bring each of them to a single known (initial) state.

Other works (e.g., [7]) use the term sequential equivalence to describe a more restricted problem, in which some knowledge about the initial state exists. We use *alignability equivalence* to describe the more general problem examined in [5], and *sequential equivalence* for the more restricted problem as described in [7].

The problem of *compositional* alignability equivalence is examined in [3]. In contrast, our concern is with sequential equivalence. Like us, [3] look for a partial equivalence, but their partial equivalence includes assumptions about the environment of the full chip, which they do not discharge. We also allow such assumptions, if necessary, but the main source of partiality in our work is knowledge about how a particular power gated design is intended to work. Furthermore, we discharge all assumptions that are derived from such knowledge.

Sequential equivalence checking is widely used to verify the correctness of local changes introduced by techniques such as retiming and clock gating [1]. It is used to verify that the change preserved the functionality of the original design. We propose something similar for power gating, however our equivalence check is complicated by the fact that the change causes non-local side effects.

To the best of our knowledge, there is no prior literature on functional verification in the presence of power gating. Technical articles available online discuss various limited aspects of the problem, for instance, that the power-up and power-down sequences work as expected, that the correct parts of the state are retained by the power retention logic, or that the power management unit itself functions correctly. However, no prior work has considered the problem of functional correctness of the chip as a whole in the presence of power gating.

2 Power Gating

Recall that in power gating, a unit or a large part of it is powered off when it is not needed, and powered back on when it is. For instance, the power to a floating point unit may be shut off when there are no floating point instructions to be processed, and turned back on when there is a need. The memory elements of a power gated unit lose their memory when powered off. Thus, power gating requires either that the power gated unit be *memoryless*, i.e., such that when turned back on it is able to function with no memory of the past, or else that there is some circuitry that takes care of *state retention* – remembering enough of the state to allow the unit to function correctly when turned back on.

When a logic gate is powered off, its output should not be read by a logic gate that is powered on, for physical design reasons that are beyond the scope of this paper. This results in two complications. First, the inputs of a powered on unit must be insulated from the outputs of a powered off unit. This is achieved by inserting an isolation device, which we will term a *fence*, between the outputs of a powered-off unit and the inputs of a powered-on unit. The physical implementation of a fence is beyond the scope of this paper. Logically, when the fence is "up" it drives a constant value and when it is down, it simply passes on the value of the output that it is fencing. A fence must be put up before a power gated unit is powered off, and must be left up until the power gated unit has been powered on and sufficient time (some number of clock cycles – the exact number is design dependent) has passed to ensure that all outputs are reliable.

The second complication is that state retention is not simply a matter of powering off some memory elements while leaving others powered on, because the powered on memory elements must be fenced off from the powered off elements. Therefore, state retention is normally accomplished by copying part of the state to dedicated memory elements before powering off a unit, and then copying them back after the unit has been repowered (and before it is used).

The process of raising the fences and copying a partial state to the dedicated state retention before powering off is part of the *power down sequence*, while the process of powering up the design, copying the retained state back into the functional memory elements, and then lowering the fences is part of the *power up sequence*. Triggering the power down and power up sequences is the responsibility of the *power management unit*.

Conceptually, there are two parts to verifying the correct implementation of power gating. One part is syntactic, or structural: checking that fences exist at the right places, and checking that if the power is off, the fence is up and stays up for a sufficient number of clock cycles to ensure the (electrical) reliability of the gate. This is easily checked, and is orthogonal to the matter we examine in this paper. The other part is semantic, or functional: checking whether the implementation of power gating preserves the behavior of the chip. The latter is the problem that we are concerned with in this paper.

3 Methodology

We first show the setup of our methodology and the inputs required from the user, then show how we use these inputs to prove sequential equivalence between a power gated design and a version of itself in which power gating is disabled.

Preliminaries. Given a set Σ of signals, a *path* is a function $\pi : \Sigma \times \mathbb{N} \rightarrow \{0,1\}$ that assigns each signal a Boolean value at each time point. We can constrain the behavior of signals on a path by associating a predicate with each signal. Such a predicate can be either an *input predicate*, which imposes no constraint at all, a *gate predicate* such that the value of signal $\sigma \in \Sigma$ at time t is a function of the values of other signals at time t (including the constant functions 0 and 1), or a *latch predicate* such that the value of signal $\sigma \in \Sigma$ at time t is a function of the values of other signals at time $t-1$. Formally,

Definition 1 (Input, gate and latch predicates). *Let σ_i for $1 \leq i \leq n$ be signals, and denote the value of signal σ_i at time t by $\sigma_i(t)$. Let f be an n-ary function and let init_values be a subset of $\{0,1\}$. Then:*

- $p(t) \in \{0,1\}$ *is an* input predicate.
- $p(t) \in f(\sigma_1(t), \ldots, \sigma_n(t))$ *is a* gate predicate
- $p(t) \in \begin{cases} \mathtt{init\_values} & : t = 0 \\ f(\sigma_1(t-1), \ldots, \sigma_n(t-1)) & : t > 0 \end{cases}$ *is a* latch predicate

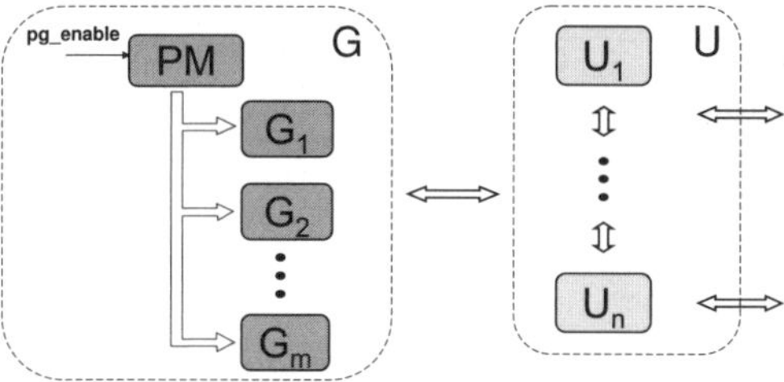

Fig. 1. Partitioning of the design for power gating verification

Let Σ be the set of signals in a non-power gated design D, and describe each signal $\sigma \in \Sigma$ as an input, gate, or latch predicate p_σ in the obvious way. The set D of predicates $\{p_\sigma | \sigma \in \Sigma\}$ describes the design D. Note that a multiply clocked design can be described as easily as a singly clocked design (as is usual when building a model of a design for model checking or equivalence checking).

In a power gated design, each gate and latch has an additional input pin that represents the state of the power supply – either on or off. Thus to describe a power gated design, we modify the "obvious way" of the preceding paragraph as follows. An input or gate predicate is not affected by whether the power is on or off, whereas a latch predicate has its "normal" value when the power is on, and a non-deterministic value otherwise. This gives us that when the power comes back on, the entire unit is in some arbitrary state.

Although this does not completely reflect reality (to do so, we would have to modify gate predicates as well), it is enough for our purposes assuming that each output of the power gated domain is fenced and that the fence is guaranteed to be up when the gate or latch driving the output is powered off. As stated in Section 2, this is orthogonal to the issue we examine in this paper and thus we can assume that it will be checked by other means.

Setup of Our Methodology. We partition the design as shown in Figure 1. Here and in the rest of this paper we use the word "unit" to refer to a piece of the design of any size that has a well-defined interface, and may or may not be composed of other units or blocks of code. G consists of one or more power gated units G_i plus the power management unit PM, which controls the powering on and off of each G_i (when power is turned off to a G_i, not all other G_j's are necessarily powered off). We require the existence of one specific signal in G: a signal *pg_enable* that is read only by PM and whose role is to enable the power gating. For simplicity we assume that *pg_enable* is a constant within G.

U consists of a number of U_i's. These are every non-power gated unit in the design that directly interfaces with G. Any part of the design that interfaces with U but is not contained in G (call it the environment of U) exists intuitively to the right of U in the figure. Usually it will be completely abstract – that is, we will make no assumptions on it at all, although we may relax this slightly if necessary. Finally, the outputs of U to the left (in the direction of G) and those to the right (in the direction of U's environment) are not necessarily disjoint.

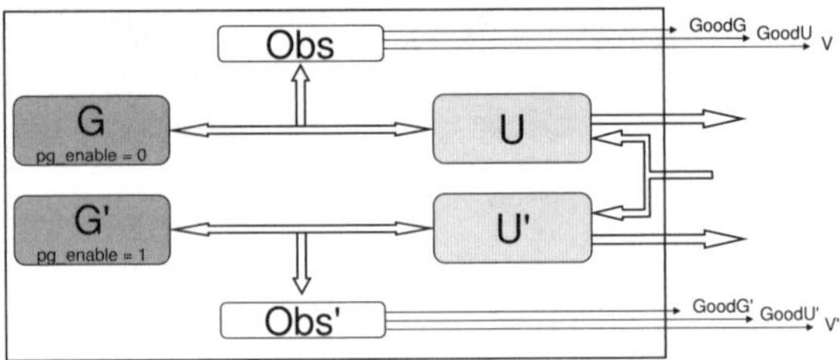

Fig. 2. Setup for equivalence checking

Note that G has no interface other than with U. That is, if G receives inputs directly from the chip interface or drives outputs directly to it, we assume for simplicity that they are buffered (with possibly zero delay) through U.

Now, if we can show that the design $G||U$ is equivalent to the design $G'||U'$, where the only difference between the primed and unprimed versions is that $pg\_enable = 1$ in $G||U$ whereas $pg\_enable = 0$ in $G'||U'$, then we will have shown that power gating does not affect the functionality of the design as a whole. Note that this strategy as stated is not complete – it is possible to create a correct design in which the effects of the power gating penetrate to the interface of U with its environment, in which case our method will not be able to prove that the design is correct. However, in such a case we can move part of U's environment into U. In the worst case, $G||U$ will include the entire chip (recall that we assume that power gating does not change the external behavior of the chip).

Recall that our goal is to show that $G||U$ is equivalent to $G'||U'$, and due to size problems we would like to do it compositionally – that is, compare each G_i with G'_i and each U_i with U'_i. For simplicity of the explication, we first show how to break the problem into comparing G with G' and U with U', and only afterwards how to break the problem down further.

Obviously we do not have complete equivalence – when G is powered off, its outputs are not necessarily equivalent to those of G'. And although U and U' will surely behave the same if they receive the same inputs (because there is no difference between them), in our scheme U will get its inputs from G and U' from G', thus showing equivalence between them is not trivial. Furthermore, when comparing G with G' we have to be careful: if the inputs of the power management unit PM "misbehave", it might shut off some G_i at an inappropriate time – for instance, when it is in the middle of processing a transaction. Thus we may need some minimal assumption over the inputs that influence PM, and we need to be able to guarantee this assumption.

We therefore ask the user to supply a simple *observer*, a piece of code that monitors the interface between G and U and outputs flags that indicate properties of the interface. Each flag is used as an assumption by one of $G||G'$ or $U||U'$ and is guaranteed by the other, and the apparent circularity is broken by induction over time. Thus the setup of our methodology is as shown in Figure 2, where the flags are signals partitioned into sets $GoodU$, $GoodG$ and V, as follows:

- **GoodU.** Each flag in this set has the value 1 as long as some assumption about the behavior of U is preserved. These assumptions do not specify the exact correct behavior of U on this interface, only the minimal needed restrictions. As soon as a violation of these assumptions is detected the flag goes to 0 and stays so forever.
- **GoodG.** This set is similar to *goodU*, but over G.
- **V.** Conceptually, this set contains a single flag v, which is a "valid" signal that indicates whether the interface between G and U is active. When $v = 1$ we expect the outputs of G and G' to be equivalent, and when $v = 0$ we do not. For example, v could be $ready \wedge transmitting$, where $ready$ is an output of U signifying that U is ready to receive data and $transmitting$ is an output of G signifying that G has data ready on the bus. In fact, V is not a single flag but a set of flags, because each U_i may have its own interface with each G_j, and even across a single interface not all signals necessarily follow the same protocol.

Note that the observer that generates the flags is coded manually. At first glance this may seem to be a very complicated task, one that perhaps should be automated. However, an observer is a direct result of the interface protocol between G and U, thus should be very easy for a human to code, and indeed all the designers we have spoken with understand intuitively what the observer for their design should look like. On the other hand, because it requires an understanding of the design intent, coding an observer would be extremely difficult to automate. We note that if the user makes incorrect assumptions that are either too strong or too weak, then some proof obligation (that will be presented below) will fail with a counterexample. Thus, wrong assumptions in the coding of *GoodU*, *GoodG* or V will never lead to an incorrect claim of equivalence.

Regarding the sets *GoodU* and *GoodG*, we expect the typical user to start off with empty sets and gradually add constraints refining them as needed. Examples of such sets are presented in Section 4. In the general case of assume-guarantee reasoning for functional correctness, this refinement process is difficult because it requires some kind of semantic understanding of how the design is intended to work. In our simplified setting, these conditions will typically be simple translations from the English specification of the interface, such as "there are no requests during reset". Moreover, we need much weaker assumptions than those necessary to check functional correctness, because we don't care if the designs misbehave as long as the two copies (mis)behave in exactly the same way.

Note that it is possible to code a correct design in which the interface between G and U is always active (despite the fact that G can be powered down), and that this does not break our methodology. In such a case the fences and the state retention logic of G will be such that the valid signal has the constant value 1, and the equivalence between U and U' will be trivial.

Proving Sequential Equivalence. We base our approach on the compositional reasoning rule presented by McMillan in [4], and borrow our notation from there. Following [4], we abuse notation by using Q to denote the conjunction of all predicates in the set Q.

Let P be a set of predicates describing the design and let S be a set of predicates defining the specification. For each predicate $s \in S$ let $\mathcal{E}_s \subseteq P \cup S$ be the environment of s. Intuitively, this is the set of predicates needed in order to show that s holds. We assume a well-founded order $\prec$ on S that defines for each predicate s which other predicates will be assumed up to time i when proving s at time i (this is Z_s), and which will be assumed only up to time $i-1$ (this is $\bar{Z}_s$, the complement of Z_s). Then by [4] we can use Theorem 1 below.

Theorem 1 (McMillan [4]). *Let P and S be sets of predicates, for each $s \in S$ let $\mathcal{E}_s \subseteq P \cup S$, and let $\prec$ be a well-founded order on S. Let $Z_s = P \cup \{s' \in S : s' \prec s\}$, and for a predicate p let $p \uparrow^\tau$ stand for $\bigwedge_{t \leq \tau} p(t)$. Then, if for all $s \in S$,*

$$(\mathcal{E}_s \cap Z_s) \uparrow^\tau \wedge (\mathcal{E}_s \cap \bar{Z}_s) \uparrow^{\tau-1} \Rightarrow s(\tau)$$

is valid, then $(\forall t.P(t)) \Rightarrow \forall t.S(t)$ is valid.

Our goal is to use Theorem 1 to prove sequential equivalence between $G \| U$ and $G' \| U'$. Since we have assumed that all outputs of $G \| U$ are outputs of U it is enough to show that the predicate

- $EqU(t) \stackrel{\text{def}}{=} \{o(t) \leftrightarrow o'(t) : o \text{ is an output of } U\}$

holds at all times t. We will need the following auxiliary sets of predicates:

- $P_{GoodU}(t) \stackrel{\text{def}}{=} \{s(t) = 1 | s \in GoodU\}$
- $P_{GoodG}(t) \stackrel{\text{def}}{=} \{s(t) = 1 | s \in GoodG\}$
- $P_V(t) \stackrel{\text{def}}{=} \{v(t) \leftrightarrow v'(t) | v \in V\}$
- $EqG(t) \stackrel{\text{def}}{=} \{v_o(t) \rightarrow (o(t) \leftrightarrow o'(t)) : o \text{ is an output of } G \text{ and } v_o \in V \text{ is its associated valid bit}\}$

Let G, G', U, U', Ob and Ob' be the sets of predicates describing the respective designs of Figure 2. Let $\hat{G} = G \cup G' \cup Ob \cup Ob'$ and $\hat{U} = U \cup U' \cup Ob \cup Ob'$. Let $P = \hat{G} \cup \hat{U}$, and $S = P_V \cup P_{GoodU} \cup P_{GoodG} \cup EqU \cup EqG$.

To begin with, let's assume that the relation $\prec$ is empty, thus for every element s of S, we have $Z_s = P$ and $\bar{Z}_s = S$. Therefore proving the following

$$\hat{G} \uparrow^\tau \wedge (EqU \cup P_{GoodU}) \uparrow^{\tau-1} \quad \Rightarrow \quad (EqG \cup P_{GoodG} \cup P_V)(\tau) \qquad (1)$$

$$\hat{U} \uparrow^\tau \wedge (EqG \cup P_{GoodG} \cup P_V) \uparrow^{\tau-1} \quad \Rightarrow \quad (EqU \cup P_{GoodU})(\tau) \qquad (2)$$

will allow us to conclude that $(\forall t.P(t)) \Rightarrow \forall t.S(t)$, and in particular that $(\forall t.P(t)) \Rightarrow \forall t.EqU(t)$, which is our goal.

In practice there will usually be some combinational paths from inputs to outputs in one or more of G, U and Ob, in which case we will need stronger assumptions for some of the proof obligations. That is, we will need $s \uparrow^\tau$ as opposed to $s \uparrow^{\tau-1}$ for some element $s \in S$ used on the left-hand side of Obligation (1) or (2). Thus we will need to set an order, easily determined from the topology of the design, between the elements of S. As noted by McMillan in [4], such an order is guaranteed to exist when there are no combinatorial loops in the design. Since a combinatorial loop is a basic structural design error, we are guaranteed the existence of a well-founded order. Using the well-founded order, each of the Obligations (1) and (2) will be split into a number of proof obligations, one for each predicate in the conjunction on the right hand side. For example, let one such predicate be $s(t) = (v_o \rightarrow (o(t) \leftrightarrow o'(t))) \in EqG$, and let $A = \{s'(t) | s' \prec s \text{ and } s' \in EqU \cup P_{GoodU}\}$ and $B = (EqU \cup P_{GoodU}) \setminus A$. The corresponding proof obligation for s is then:

$$(\hat{G} \cup A) \uparrow^{\tau} \wedge B \uparrow^{\tau-1} \Rightarrow (v_o \rightarrow (o(\tau) \leftrightarrow o'(\tau))) \tag{3}$$

Conceptually, it has been convenient up till now to view G and U as monolithic units. However, in reality each will typically consist of a number of smaller units, as shown in Figure 1. Thus we would like to decompose the verification problem further by considering each G_i and U_i separately. For an output o of some U_i, we would like to use only U_i rather than all of U on the left hand side of its proof obligation. To do so, we must add the following predicates to S:

- $EqIntU(t) \stackrel{\text{def}}{=} \{s(t) \leftrightarrow s'(t)$: s is an interface signal between U_i and U_j for some $i \neq j\}$

The situation for a single G_i is slightly more complicated: we must include the power management unit PM together with each G_i, and the predicates that we add for the outputs of G_i will be conditional, thus we might need to add some new valid signals. Denote the new valid signals by V_{new}. Then we add the following additional predicates to S:

- $P_{V_{new}}(t) \stackrel{\text{def}}{=} \{v(t) \leftrightarrow v'(t) | v \in V_{new}\}$
- $EqIntG(t) \stackrel{\text{def}}{=} \{v_s(t) \rightarrow (s(t) \leftrightarrow s'(t))$: s is an interface signal between G_i and G_j for some $i \neq j$ and $v_s \in \{V \cup V_{new}\}$ is its associated valid bit$\}$

The order $\prec$ is easily extended to the new predicates by a topological analysis of the design. For each output o of some G_i or U_i, we verify its proof obligation using $\hat{G}_i$ or $\hat{U}_i$ in place of $\hat{G}$ or $\hat{U}$, where $\hat{G}_i = PM \| G_i \| G_i'$ and $\hat{U}_i = U_i \| U_i'$.

Note that the theory supports multiply clocked designs as well as singly clocked ones. In the case of a singly clocked design, each time t is simply a tick of the clock. In the case of a multiply clocked design, each time t is a tick of the smallest granularity of time as seen by the verification tool (this is exactly the same as in model checking or equivalence checking of multiply clocked designs).

4 Case Study on an Execution Unit

We applied our methodology to Calc3 [8], an RT-level implementation of a simple execution unit. Calc3 has 4 user ports through which commands are entered. It can process up to 4 commands from each port in parallel, supporting out-of-order execution of commands. The unit has 16 internal 32-bit registers and supports 4

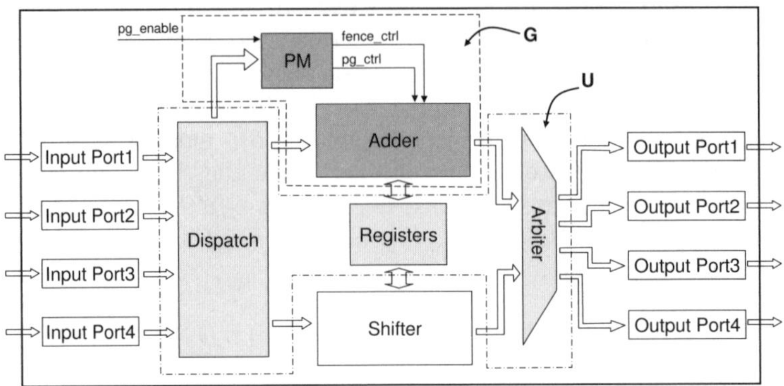

Fig. 3. The architecture of Calc3 enhanced with power gating

types of commands: *load-store*, *branch*, *shift*, and *add-subtract* commands. The branch commands contain a condition on a register; if this condition is met then the next command on that port is skipped.

Although Calc3 is relatively large for formal verification, it is relatively small compared to a typical execution unit. Still, the implementation is far from trivial. For example, some parts of the design work on the rising edge of the clock and others on the falling edge. There is also some amount of speculative calculation in which operations are performed before it is known whether they will be needed, and then results are ignored if they are not. These optimizations add complexity to the design, and of course to the verification. We chose it because it is simple enough to serve as a first application of a new methodology but still contains enough complications and design optimizations to make it interesting.

Figure 3 gives an overview of the internal structure of Calc3. Commands are injected into the unit through the four input ports, and are held in the dispatch queue until they are sent to either the adder or the shifter, depending on their type. The results pass to an arbiter, which distributes them to the four output ports. The adder is responsible for all add/subtract and branch commands, while the shifter executes shift and load/store commands.

For our case study we added power gating to the adder. The new power management unit *PM* keeps track of the types of commands that are in the dispatch queue by monitoring the inputs and outputs of the dispatch. Whenever there are no pending commands for the adder, *PM* initiates the power down sequence for it, which involves fencing the outputs and one clock cycle later powering down the adder. Whenever *PM* detects an incoming command destined for the adder, it initiates the power up sequence for it, which consists of restoring power and one cycle later lowering the fences. When powered down, most of the adder unit is inactive, except for one block which must remain powered at all times (this is the part that remembers that a branch command was taken and outputs the appropriate indication towards the arbiter). The power gated version of Calc3 has 2200 state variables, 200 input bits and 144 output bits.

We divided Calc3 into the G side, containing the adder and PM, and the U side, containing the dispatch unit, the registers, and the arbiter, as shown in Figure 3. Note that the shifter does not directly interface with G and therefore is not part of U, but belongs to U's environment, as described in Section 3. We also built an observer as described in Section 3, that outputs the required flags.

GoodU. This set contains several simple properties, as follows: interface signals must change value only on the appropriate clock edge and no commands enter during reset. In addition, there is a flag indicating when the outputs of the dispatch that go to PM and those that go to the adder are inconsistent (i.e. the adder is given a command to execute and the PM was not informed that this command ever entered the dispatch).

GoodG. The equivalence on the U side is much easier, and does not require any assumptions on the behavior of G. This set contains only a flag indicating that there is no activity during reset.

V. We have two valid signals. The first is a Boolean combination of the outputs of the adder; it is 1 iff the adder is outputting a command towards the arbiter. The second is associated with outputs that signal whether or not the next command should be skipped, and it is a constant '1'.

$\hat{G}$, $\hat{U}$ and S are built as per the methodology described in the previous section. In order to make the sequential equivalence work we needed a few assumptions on the external environment: input signals must change on the correct clock edge, no activity during reset, and no commands are injected into the unit when there are 16 active commands inside. The last requirement was due to the fact that illegally injecting too many commands caused internal counters in the power management machine to overflow. We note that these properties were enough to prove equivalence, but are not sufficient to prove even the simplest functional properties of the design. Furthermore we note that such assumptions are common in the world of practical equivalence checking. The equalities generated 697 proof goals. GoodU was broken down to 4 proof goals, and GoodG had one proof goal.

The well-founded order $\prec$ was defined solely by examining the topology of the design. The full list of signals and dependencies is too large to be described here. However, we will note that there were combinational paths from inputs to outputs in each of the units G, U and Ob. This, of course, did not break the well-foundedness of the order since there are no combinational loops in the design. Overall there were 37 signal dependencies.

The required proof obligations are easily translated into model checking of temporal logic specifications. However, as we have described, many of our predicates actually describe a form of conditional sequential equivalence, thus for reasons of capacity we would prefer to use a dedicated equivalence checking algorithm. We therefore translated those proof obligations into sequential equivalence problems. We explain the translation for the $\hat{G}$ side; the method for the $\hat{U}$ side is similar.

Our translation is a modification of a standard equivalence check between G and G'. The first adjustment is that we leave the inputs of G and G' assumed to be equivalent up to time $\tau - 1$ free, rather than tying them together as if we wanted to check standard equivalence. Those assumed equivalent up to τ are tied together as usual. Notice that a practical implication of this is that it is useful to add additional dependencies to the order $\prec$ (while maintaining well-foundedness) in addition to those resulting from the topology of the design, because this results in more inputs tied together, which makes the equivalence checking problem easier. We added dependencies on almost all of G's inputs, and many of U's inputs; without this, the $\hat{G}$ side would be intractable.

The second adjustment is to factor in the assumptions (including that untied inputs should be equivalent up to time $\tau - 1$). We define a signal *mask_outputs* that is asserted when an assumption is violated, and stays that way forever. We then modify each output of G and G' to have (the same) constant value when *mask_outputs* is asserted. Proving the equivalence of an output o between the two modified versions implies that there does not exist a computation that violates the proof obligation for o (because it implies there can not be a reachable state such that all assumptions hold and $o \neq o'$).

We verified Calc3 using IBM's internal tools. Equivalences (including conditional equivalences) were proven using SixthSense [1], IBM's sequential equivalence checking tool. The proof of each pair of outputs was done separately, since this lets SixthSense utilize its optimizations to their full power. The other properties (*GoodU* and *GoodG*) were proven using RuleBasePE [6], IBM's model checking tool.

The proof obligations on the U side went through relatively easily. On this side there were 493 proof obligations, and each took between 60 to 8500 seconds to prove. Overall the proof of the U side ran 124 hours on a 2.4GHz Opteron dual core machine with 8GB, running Red Hat Enterprise Linux 4.[1]

On the G side there were only 209 proof obligations, but it was more difficult to prove. The overall runtime was 91 hours, with a single goal taking up to 21000 seconds (1500 seconds on average). There were several measures we took in order to reduce the problem size. For certain internal bits we were able to prove that the fence being down implies equality between the two versions. We then wired the two versions to use the same copy of the signal when the fence is down, and each their own copy when the fence is up. This resulted in a significant performance boost. Also, the tuning of SixthSense was extremely important. The first runs ran out of memory even for the simplest goals. It took some fiddling with parameters to be able to prove all goals.

During the verification of $\hat{G}$ we discovered three bugs in the implementation of PM. Each resulted in traces in which the power gated version of the adder failed to output a command when the non-power gated version did. In each trace it was obvious that the reason was because PM failed to turn on the adder on time, and from there it was easy to diagnose the error in PM.

[1] This is the cumulative time of the whole verification effort on a non-trivial industrial example, and is negligible compared to years of CPU time used for simulation.

5 Conclusions and Future Work

We have presented a methodology for the verification of power gating, based on comparing a power gated design to a version of itself in which power gating is disabled. In order to be able to deal with real world designs, we take a compositional approach in which we check each unit for partial equivalence across a suitable set of assumptions. In contrast to general assume-guarantee reasoning, which can be difficult to employ because of the manual effort involved in coming up with appropriate assumptions, the goal of our verification effort is one specific and simple formula – sequential equivalence – and this greatly simplifies the task of figuring out what the assumptions should be. We have shown the feasibility of our method by applying it to a non trivial execution unit.

Future work is to extend our methodology to designs in which power gating changes the external behavior of the chip (for instance, pipelines that stall when the needed unit is not available), to apply it to real-life designs under development at IBM, and to develop a tool that automates much of the work of coming up with the assumptions (for example, those that deal with the clock edges and activity during reset can be derived automatically) and deciding on the well-founded order.

Acknowledgments. Thank you to Jason Baumgartner and Hari Mony of the SixthSense team for their help in getting the equivalence check on G to run.

References

1. Baumgartner, J., Mony, H., Paruthi, V., Kanzelman, R., Janssen, G.: Scalable sequential equivalence checking across arbitrary design transformations. In: ICCD 2006 (October 2006)
2. Keating, M., Flynn, D., Aitken, R., Gibbons, A., Shi, K.: Low Power Methodology Manual. Springer, US (2007)
3. Khasidashvili, Z., Skaba, M., Kaiss, D., Hanna, Z.: Theoretical framework for compositional sequential hardware equivalence verification in presence of design constraints. In: ICCAD 2004, pp. 58–65 (2004)
4. McMillan, K.L.: Verification of an implementation of Tomasulo's algorithm by compositional model checking. In: Vardi, M.Y. (ed.) CAV 1998. LNCS, vol. 1427, pp. 110–121. Springer, Heidelberg (1998)
5. Pixley, C.: A theory and implementation of sequential hardware equivalence. IEEE Trans. on CAD of Integrated Circuits and Systems 11(12), 1469–1478 (1992)
6. RuleBasePE, `http://www.haifa.ibm.com/projects/verification/RB_Homepage/`
7. van Eijk, C.A.J.: Sequential equivalence checking based on structural similarities. IEEE Trans. on CAD of Integrated Circuits and Systems 19(7), 814–819 (2000)
8. Wile, B., Goss, J.C., Roesner, W.: Comprehensive Functional Verification - The Complete Industry Cycle. Elsevier, Amsterdam (2005)

A Practical Approach to Word Level Model Checking of Industrial Netlists

Per Bjesse

Advanced Technology Group
Synopsys Inc.
bjesse@synopsys.com

Abstract. In this paper we present a word-level model checking method that attempts to speed up safety property checking of industrial netlists. Our aim is to construct an algorithm that allows us to check both bounded and unbounded properties using standard bit-level model checking methods as back-end decision procedures, while incurring minimum runtime penalties for designs that are unsuited to our analysis. We do this by combining modifications of several previously known techniques into a static abstraction algorithm which is guaranteed to produce bit-level netlists that are as small or smaller than the original bitblasted designs. We evaluate our algorithm on several challenging hardware components.

1 Introduction

Word-level methods, which leverage design information captured at a higher level than that of individual wires and primitive gates, are the next frontier in hardware verification. At the word level, data-path elements and data packets are viewed as entities in their own right as opposed to a group of bit-level signals without any special semantics.

There has been a lot of activity lately around word-level formula decision procedures such as SMT solvers [10] and reduction-based procedures like `UCLID` [2] and `BAT` [7]. However, as promising as this direction of research is, the use of these procedures for model checking is inherently restricted in that they analyze formulas rather than sequential systems. This has two consequences: First of all, sequential properties can only be checked by these procedures by relying on methods such as induction and interpolation that employ bounded checks to infer unbounded correctness. Second, these procedures do not fit into a *transformation-based* approach to sequential system verification [1], where sequential verification problems are iteratively simplified and processed by any of a large set of back-end model checkers.

In this paper, we introduce a method for practical word-level model checking of both bounded and unbounded properties for hardware designs. Our aim is to (1) not require any additional input from the user, (2) never perform worse than a straight bit-level sequential analysis of a given netlist, and (3) to provide the possibility of speedups when there are significant parts of the design that can be treated on the word level. Our solution is engineered as a reduction method

A. Gupta and S. Malik (Eds.): CAV 2008, LNCS 5123, pp. 446–458, 2008.

where the word-level netlist is abstracted to an equivalent but smaller gate-level netlist. In the worst case, our analysis generates results that are no larger than the gate-level version of the original design. We tune the runtime of our solution so that it runs very quickly even on industrial size problems, and so that we incur little penalty in the case of netlists that are not amenable to our analysis.

In order to demonstrate the utility of our approach, we analyze an academic high-performance router and two industrial designs (a FIFO and a content addressable memory). We demonstrate significantly reduced netlists, while spending less than a second in the reduction part of our analysis.

2 Preliminaries

We assume a standard frontend flow that compiles problems into netlists by processing a hardware design with properties and constraints into combinational logic over a set of unconstrained inputs I, state variables S, and constants. The top of the resulting forest of combinational logic contain next-state variables S' and single bit outputs O. We assume that (1) the properties we are interested in are all safety properties whose failure is signaled by some output assuming the value `false`, and that (2) each state variable has a known initial state.

In Section 4 we will use an alternate way of describing netlists in terms of three formulas $\mathit{Init}(S)$, $\mathit{Next}(I, S, S')$, and $\mathit{Prop}(S)$. The correspondence between these formulas and the netlist is simple: The $\mathit{Init}(S)$ formula describes the initial states, $\mathit{Next}(I, S, S')$ describes the next-state functions, and $\mathit{Prop}(S)$ captures a particular safety property output function.

3 Netlist Reduction

We use a word-level frontend to compile a model of the given netlist problem. The resulting data is represented as a Directed Acyclic Graph (DAG) of operators over a set of inputs I, state variables S, and binary constant vectors. Every node ϕ in the graph has an associated signal width k; we sometimes annotate nodes with superscripts that denote the size of the bit vector they represent.

The top level nodes of the DAG are partitioned into circuit outputs O and next-state variables S'. Internal nodes in the graph have the following form:

- $\phi^k = \mathtt{not}(\alpha^k)$.
- $\phi^k = \mathtt{and}(\alpha^k, \beta^k)$.
- $\phi^k = \mathtt{arithOp}(\alpha^k, \beta^k)$ for $\mathtt{arithOp} \in \{+, -, \ldots\}$.
- $\phi^1 = \mathtt{compOp}(\alpha^k, \beta^k)$ for $\mathtt{compOp} \in \{<, \leq, =, \neq, \geq, >\}$.
- $\phi^k = \mathtt{mux}(\alpha^1, \beta^k, \gamma^k)$.
- $\phi^k = \mathtt{extract}(l, \alpha^m)$.
- $\phi^k = \mathtt{concat}(\alpha_1^i, \alpha_2^j, \ldots)$.

The `not` and `and` operators are *bitwise* operators in the sense that bit i of the result is generated by applying the boolean operator to bit i of the input nodes.

The `mux` node returns β_k if α_1 is true, and γ_k otherwise. The `extract` node constructs a smaller bit vector by projecting out k bits from position l to $l{+}k{-}1$ of its operand. Finally, the `concat` node forms a larger signal by concatenating its operands to form a larger bit vector. Earlier operands in the argument list to `concat` become higher order bits, so $\texttt{concat}(01, 00) = 0100$.

The select signal of `mux` and the output of comparison operator nodes are restricted to have bit width one. Such signals are said to be *bit-level signals.* We refer to signals that are not bit level as *word-level signals*, and use the term *segment* to denote a group of contiguous bits.

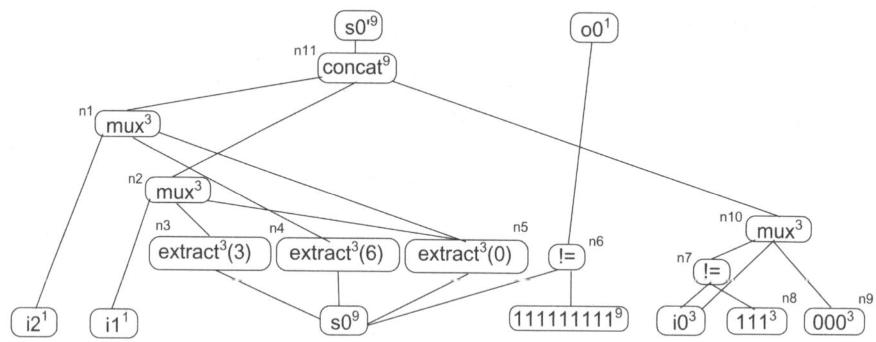

Fig. 1. An example word level netlist

Example 1. Consider the word-level netlist in Figure 1. The initial state for $\texttt{s0}^9$ is 000000000^9, and the property of interest is that $\texttt{o0}^1$ is always true. The circuit generates the next state for `s0` by concatenating three parts. The lowest part is a fresh input value, but only if it is not equal to 111 (otherwise we switch in 000). The other two segments are the result of either keeping the current segment of `s0`, or swapping in the low order segment 0..2 of `s0` depending on the value of two externally controlled inputs. Hence, the system is safe in that the output can never become false. □

Our analysis will be performed in three steps. First, we will rewrite the netlist into a design where the datapath is completely separated from all boolean control logic, and where word-level registers and inputs have been broken up into smaller parts that do not intermingle control and data. Second, we will analyze the datapath portions of the circuit, and find reduced safe sizes for all the word-level entities. Third, we generate a smaller final netlist which can be analyzed by standard gate-level reductions and model checking algorithms.

3.1 Selective Bitblasting

It is easy to see that every word-level netlist can be *bitblasted* into an equivalent bit-level netlist by splitting all variables into single bit segments, and implementing the internal nodes in terms of boolean logic. The result is a netlist where all signals have width one, and the only internal nodes are the two boolean operators. In the first step of our analysis we will perform *selective bitblasting*, with the aim of ending up with a graph over constants, variable nodes, boolean operators of width one, comparison operators from the set $\{=, \neq\}$, and `mux` operators. All other operators will be removed by a translation into bit-level constructs.

In order to perform this analysis we will annotate each node in the graph with information on which of its segments are treated as word-level *packages*—units of data that are treated uniformly. To do this, we make use of a modification of a data flow algorithm introduced in [6]. We perform the analysis by maintaining a partition of each node in the network into bit segments. For each bit segment of every node, we maintain an equivalence class of other segments of nodes that either depends on it, or that it depends on.

We make use of some primitive operations on equivalence classes and intervals: $registerNode(n)$, $split(n, j)$, $mkCompatible(n_1, n_2, \ldots)$, $bitblast(n)$ and $unionNodes(n_1, n_2, \ldots)$:

- The creation operator, $registerNode(n)$ takes a node, and constructs a singleton equivalence class containing the segment $(0..k-1)$ if the node n has k bits.
- The refining operators *split*, *mkCompatible*, and *bitblast* perform the following functions:
  1. $split(n, j)$ finds the segment equivalence class for node n that contains the bit j. If the bit j falls internally to the segment interval $i \ldots k$ so that $i < j < k$, then the equivalence class is split into two new classes; the first containing the $j - i$ first bits of each segment, and the other containing the remaining bits of each segment.
  2. $mkCompatible(n_1, n_2, \ldots)$ applies the *split* operator to its operands until their segmentations match.
  3. $bitblast(n)$ applies *split* to a node until it is segmented up into one bit slices.
- The merge operator $unionNodes(n_1, n_2, \ldots)$ takes a number of nodes whose segmentation match. If the nodes all have k segments, then the merge operator generates k new equivalence classes by merging the equivalence classes for all first segments of its operands, then merging the equivalence classes for all second segments of its operands, and so on until all k new classes have been formed.

We will also use the following terminology: We say that the segmentation of a signal ϕ^k is *consistent* with the segmentation of another signal ψ^k if all cuts in ψ^k exists in ϕ^k. We *transfer* segments from ϕ^k to ψ^k by using *split* to introduce cuts in ψ^k at all positions where there are cuts in ϕ^k.

Our dataflow analysis is performed in a depth first recursive manor. Each node ϕ_k encountered is registered using *registerNode*, and then processed as follows:

- ϕ_k is a constant: Use *split* to partition the node into maximal segments of consecutive bits of the form $00\ldots0$ and $11\ldots1$. The constant 000100 generates the segments $(0,1)$,$(2,2)$, and $(3,5)$.

- ϕ^k is a variable: Do nothing.
- $\phi^k = \texttt{not}(\alpha^k)$: Bitblast ϕ^k and α^k, and perform $unionNodes(\phi^k, \alpha^k)$.
- $\phi^k = \texttt{and}(\alpha^k, \beta^k)$: Bitblast ϕ^k, α^k, and β^k. Perform $unionNodes(\phi^k, \alpha^k, \beta^k)$.
- $\phi^k = \texttt{arithOp}(\alpha^k, \beta^k)$: Bitblast ϕ^k, α^k, and β^k. Perform $unionNodes(\phi^k, \alpha^k, \beta^k)$.
- $\phi^1 = \texttt{compOp}(\alpha^k, \beta^k)$ and
  - compOp $\in \{=, \neq\}$: Perform $mkCompatible(\alpha^k, \beta^k)$, and $unionNodes(\alpha^k, \beta^k)$.
  - compOp $\in \{<, \leq, >, \geq\}$: Bitblast α^k and β^k. $unionNodes(\alpha^k, \beta^k)$.
- $\phi^k = \texttt{mux}(\alpha^1, \beta^k, \gamma^k)$: Perform $mkCompatible(\phi^k, \beta^k, \gamma^k)$ and $unionNodes(\phi^k, \beta^k, \gamma^k)$.
- $\phi^k = \texttt{extract}(l, \alpha^m)$: Use *split* to introduce cuts at bit l and $l+k$ for α^m. Next, transfer all segment cuts in the region between bit l and $l+k-1$ from α^m to the corresponding positions in ϕ^k, and union the corresponding segments of ϕ^k and α^m.
- $\phi^k = \texttt{concat}(\alpha_1^i, \alpha_2^j, \ldots)$: Segment ϕ^k to match the operand borders. Then transfer the internal segment cuts from the operands to the corresponding points in ϕ^k, and union the corresponding segments.

When all nodes have been processed, we traverse all present-state and next-state pairs (ϕ^k, ϕ'^k) and perform $mkCompatible(\phi^k, \phi'^k)$ and $unionNodes(\phi^k, \phi'^k)$. We also use *split* to ensure that the segmentation of each current-state node is consistent with the segmentation of its initial state.

Example 2. Consider the verification problem from Example 1. Assume we traverse the netlist by first visiting s0. This creates the partition information s0 : (0..8) in a singleton equivalence class. After visiting nodes n3, n4, n5, i1, n2, i2 and n1 we have the new segmentation s0 : (0..2), (3..5), (6..8). The equivalence class of s0 : (0..2), now contain the other elements n1 : (0..2), n2 : (0..2), n3 : (0..2), n4 : (0..2), n5 : (0..2), s0 : (3..5) and s0 : (6..8). □

After we have performed the data flow analysis, we will have segment information for each node, and we are assured that (1) the segmentation of current and next-state variables is consistent, (2) the segmentation of current-state variables and initial-state variables is consistent, and (3) the segment sources of our netlist DAG of size greater than one will only be propagated through multiplexor networks or be compared using the operators $\{=, \neq\}$.

We can now create a modified word-level netlist from the bottom up, by converting each node in the original netlist into a list of new nodes (one per segment). We do this as follows. If a variable or constant node has n segments, we generate a list of n fresh nodes of appropriate type and size. Nodes of type not, and, arithOp or comparison operators from the set $\{<, \leq, >, \geq\}$ are guaranteed to have been bitblasted, so we just return the list of signals corresponding to the bit-level implementation of the operator in terms of its inputs. Any remaining node ϕ^k is handled as follows:

- $\phi^1 = \texttt{compOp}(\alpha^k, \beta^k)$ and compOp $\in \{=, \neq\}$: implement ϕ^1 as a boolean network of equalities over the respective segments.

- $\phi^k = \texttt{mux}(\alpha^1, \beta^k, \gamma^k)$: generate a list of multiplexors $\texttt{mux}(\alpha^1, x, y)$. Each multiplexor takes x and y to be the resulting segment nodes generated for the corresponding position in β^k and γ^k.
- $\phi^k = \texttt{extract}(l, \alpha^m)$: Project out the list of new nodes generated for the desired interval segments of α^m.
- $\phi^k = \texttt{concat}(\alpha_1^i, \alpha_2^j, \ldots)$: Return the concatenation of the lists of new nodes generated for the operands.

Example 3. Assume that a node $\texttt{mux}^{32}(\texttt{n1}^1, \texttt{n2}^{32}, \texttt{n3}^{32})$ has been segmented as $(0..7), (8..31)$, and that the result of reimplementing $\texttt{n1}^1$ was $[\texttt{m1}^1]$, and that the result of reimplementing $\texttt{n2}^{32}$ and $\texttt{n3}^{32}$ was $[\texttt{m2}^{24}, \texttt{m3}^8]$ and $[\texttt{m4}^{24}, \texttt{m5}^8]$, respectively. Then we return

$$[\texttt{mux}^{24}(\texttt{m1}^1, \texttt{m2}^{24}, \texttt{m4}^{24}), \texttt{mux}^8(\texttt{m1}^1, \texttt{m3}^8, \texttt{m5}^8)]$$

□

When we reach the top of the new DAG, we create one new next-state variable or output per segment from the list of implementations of the feeders.

The signals in the resulting netlist graph come in two different flavors. The first type of signal has bitwidth one, and are thus bit-level signals. These signals get processed using standard boolean logic. The second type has bitwidth greater than one. These word-level signals are moved around using multiplexor networks, and generate bit-level signals using comparison operators. Note that an original input or state variable in the design may very well be split up into several parts, some of which are bit level, and some which are word level.

Theorem 1. *The selective bitblasted netlist is equivalent to the original netlist.*

Proof. Our segment analysis is the formula analysis in [6], modified to bitblast more operators, and force consistent segmentation of current-state variables, next-state variables, and initial-state values. Correctness follows from the correctness of [6], together with an induction over time. Due to limited space we omit the detailed proof.

3.2 Abstraction of Word-Level Variables

The selectively bitblasted netlist now has two components: (1) A word-level component that reads packages from the inputs and word-level registers, moves them around using multiplexors, and performs package comparisons. (2) A bit-level component that reads bit-level signals from the inputs, controls the multiplexors (possibly based on the outputs from comparison operators), and computes bit-level outputs.

As the word-level variables are only compared for equality and inequality and moved around, it seems like we should be able to abstract them somehow. This is indeed true. In a 1995 paper, Hojati and Brayton introduce a reduction for designs they refer to as *Data Comparison Controllers* (DCCs) [4]. These designs are partitioned into a boolean part and a datapath part that manipulates infinite packets modeled as integers by moving them around and comparing them, just like our selectively bitblasted designs. It is shown in Hojati and Brayton's paper that for every DCC, there always exists a finite smallest package size that preserve the status of the properties of the design. In fact, if the system has N infinite integer variables and M integer constant nodes, the integers can safely be modeled using length $S_{min} = \lceil \log_2(N + M) \rceil$ bit vectors.

Unfortunately we can not apply this result directly, for two different reasons:

1. Our packages do not have infinite initial size.
2. We have more than one package size.

Let us first deal with issue 1, and momentarily assume that we have a system where all word-level variables have a single bitwidth S.

Lemma 1. *The DCC sizing theorem from [4] can be restated as follows: Any package size $S \geq S_{min}$ bits gives the same status to all design properties.*

Proof. Reducing the size of one or more variable domains can only reduce the number of design behaviors. The number of provable properties will hence grow monotonically. As the DCC sizing theorem implies that exactly the same properties are provable in the infinite case as for package size S_{min}, finite package sizes larger than S_{min} proves exactly the same properties. □

As a result of Lemma 1, as long as the initial package size S is larger than or equal to S_{min}, size S_{min} packages will preserve the status of all properties. However, if the original package size is *smaller* than S_{min}, we have no guarantees about what will happen if we size them up. In order to circumvent this problem, we choose $S_{new} = \min(S, S_{min})$, which will guarantee safety regardless of initial package size. Also note that in our case, M (the number of constants of a particular length) always equals two, due to our partitioning of constants.

In order to solve the second issue, one potential solution could be to choose our S_{min} based on the size of the largest package in the system. This would clearly be conservative. However, this is not necessary: After our selective bitblasting, the resulting netlist has no facility for converting a size N word-level segment into some other size segment. Segments of a different width can hence not be compared, or registered in the same word-level register slices. Our converted designs are therefore *generalized DCCs*, with one bit-level component, and finite number of separate word-level components that only communicate with each other using bit-level signals. By iterating the argument in [4] it is easy to show that we can abstract each of these word-level components individually. By combining this fact with Lemma 1 we arrive at our master theorem:

Theorem 2. *For each S, assume that there exists N_S state variable and input segments of size S. The status of all properties of our selectively bitblasted design D are then preserved if we resize size S segments to have* $\min(S, \lceil \log_2(N_S + 2) \rceil)$ *bits.*

Example 4. Again consider the verification problem from Example 1. Our selective bit blasting generated the final segmentation `s0` : (0..2), (3..5), (6..9). Together with the input segment `i0` : (0..2), there are a total of four segments of size three. A safe reduced bit size for these segments is hence $\min(3, \lceil \log_2(4+2) \rceil) = 3$ bits. In this case we can not perform any reduction. However, note that any larger size system with the same structure could safely be scaled down to represent `s0` with nine bits. □

The next and final step of our analysis is to traverse all state variables of the selectively bitblasted netlist and compute a tally of segment size populations. For each segment size, we compute a new reduced size using Theorem 2. When we have sized all word-level state variables and constants, we compute our abstracted netlist by rewriting the word-level component of the selectively bitblasted design to use variables and constants of the new correct size, and adjusting the width of the internal operators.

4 Impossibility Results

The reduction we presented in Section 3 forces the word-level parts of the design to only contain multiplexors and negated and unnegated equality comparisons. Could this be extended to allow inequalities, bit-parallel boolean operators, or arithmetic? The answer, as we will see, is no.

Let us first investigate the extension of DCCs to allow inequality comparisons between signals.

Theorem 3. *If we allow inequality comparisons between word-level variables, there exists a system whose smallest reduction is the system itself for any bitwidth N. We can hence not find a static safe package size based only on the number of word-level state variables.*

Proof. Take the system to be constructed from the width N state variable v and input i, and the output o. Let $Init(v) = 11\ldots1$, $Next(v) \equiv \mathtt{mux}(i < v, i, 00\ldots0)$, and $Prop(v) \equiv v \neq 00\ldots0$. For a given bitwidth N this system has a trace where it takes $2^N - 1$ steps before the property output becomes zero. No smaller bitwidth will preserve this trace. □

However, it *is* safe to allow inequality comparisons where one operator is a constant. To see this, realize that the only constant segments that exist after our analysis have the form $00\ldots0$ and $11\ldots1$. The comparison can hence be implemented as a boolean network whose leaves only contain inequality comparisons between variable segments and $00\ldots0$ and $11\ldots1$. These leaves, in turn can be rewritten in terms of equality operators (for example $x_k < 00\ldots0_k$ is equivalent to `false` and $x_k < 11\ldots1_k$ is equivalent to $x_k \neq 11\ldots1_k$. The resulting transformed system is a DCC, so our main theorem applies.

It is also unsafe to allow the use of bit-parallel operators in the word level partition:

Theorem 4. *If we allow bitwise boolean operators in our word-level component, there exists a system whose smallest reduction is the system itself for any bitwidth N. We can hence not find a static safe package size based only on the number of word-level state variables.*

Proof. Take the system to be constructed from the width N state variable v and input i, and the output o. Let $Init(v) \equiv 11\ldots1$, $Next(v) \equiv \mathtt{mux}(v \neq \mathtt{and}(i,v), \mathtt{and}(i,v), 00\ldots0)$, and $Prop(v) \equiv v \neq 00\ldots0$. For a given bitwidth N this system has a trace where it takes $N-1$ steps before the property output becomes zero. No smaller bitwidth will preserve this trace. □

Finally, we could imagine allowing the use of arithmetic nodes in our word-level machinery. Again, this would not be a good idea:

Theorem 5. *If we allow arithmetic operators in our word-level partition, there exists a system whose smallest reduction is the system itself for any bitwidth N. We can hence not find a static safe package size based only on the number of word-level state variables.*

Proof. Take the system to be the system containing a single state variable v, a single input i and the output o. Let $Init(v) = 00\ldots0$, $Next(v) \equiv v+1$, and $Prop(v) \equiv v \neq 11\ldots1$. For a given bitwidth N this system has a trace where it takes $2^N - 1$ steps before the property output becomes zero. No smaller bitwidth will preserve this trace. □

It is easy to modify the arguments in our impossibility theorems to work for a system with some single other inequality between variables, a single `or` or `xor` operator, or a single other arithmetic operator. In essence, our analysis hence allows the richest word-level components possible, while still allowing a static analysis.

5 Implementing the Reduction

Our chief concern in implementing our formula reduction is to make the generation of the reduced system as fast as possible. The worst case for the reduction is that the design becomes completely bitblasted, which in our flow would be a necessary precondition of further processing anyway. As long as we tune our reduction to take a very small amount of time, we can hence always apply it safely.

In order to make the reduction as fast as possible, we must choose effective data structures for the signal segment information and the equivalence class information.

We maintain the segmentation information using *skip lists* [9], which is a probabilistic data structure that allow $\log(N)$ average case insertion and deletions of a set of ordered elements, with a better constant factor than many balanced tree implementations. Each segment in the skip list is indexed by the start of the segment. In each segment we store equivalence set information using standard union-find algorithms [3], which allows equivalence class operations in close to constant amortized time.

In our implementation, each splitting of a segment equivalence class takes a linear number of equivalence class and skip list operations in the size of the class. There are systems with N nodes with word-level variables of width bounded by M where every variable gets partitioned into one bit segments, and the equivalence classes contain every signal in the system, so that $O(N * M)$ equivalence set and skip list operations are necessary. However, our experimental experience is that with the efficient equivalence class and segment maintenance algorithms, processing of industrial netlists is in practice not noticeably slower than a linear node traversal.

There are several important practical implementation details that affect the reduction strength of our analysis. The most important detail in our implementation has to do with constant sharing: Any reasonable word-level DAG representation for the netlists use hashing to share nodes maximally. This is a good thing, but complicates things for our analysis in the case of constants: Constants with multiple fanout force the merging of some equivalence classes that otherwise would have been kept separate (if two logic cones only share a constant, segmentation propagation from one to the other is unnecessary). In our implementation, we work around this issue by introducing fresh variable nodes on the fly for each reference to a constant. At the end of the analysis, these variables gets transformed back into constant nodes.

Another improvement we have found empirically important is to preprocess the representation we get from the HDL frontend to provide an optimum starting point for our analysis. The reason for this is that in certain designs, some or all of the logic that moves packets is implemented by instantiating bit-level modules for each packet bit. This means that words coming in get broken up into individual bits, moved around in a uniform way, and then recombined into words. Such designs will fragment segments into bit-level signals and weaken the results of our reduction. In our experimental results, the high-performance router [8] is such a design. Without this preprocessing no netlist reduction is possible. We avoid this problem by sweeping the initial netlists and detecting subgraphs where words are split up, routed, and recombined. Each such subgraph is automatically reimplemented on the word level.

In order to cope with industrial designs, we have also implemented some extensions to the techniques presented in Section 3. Notably, we handle symbolic memories with abstract read and write nodes.

6 Related Work

As can be seen in Section 3, our reduction leverages a dataflow analysis for extracting a DCC from a general word-level netlist. The basic approach used in this analysis was introduced in [6], where it was used to perform a formula reduction. In order to apply similar techniques on the netlist level, we have had to change the analysis to include partitioning from initial states, and to make sure that the current and next-state variable segmentation corresponds. Moreover, due to the impossibility results presented in Section 4, we have to bitblast the bit-parallel boolean operators and inequality operators which are kept in [6].

Rather than automatically compute the segmentation of signals into word level and bit-level parts, we could imagine that we could let the user annotate the design with this information. As we can compute the necessary segmentation quickly and accurately, we believe that this is a bad proposition; in our experience, any piece of extra data that a user has to provide decrease the utility of a given algorithm in industrial tools.

Several different word-level formula solvers such as `UCLID` [2] and `BAT` [7] have been used for hardware verification. These procedures can in theory be used as a foundation for induction or interpolation-based unbounded property verification, but they do not provide a general way to leverage an arbitrary bit-level model checker as a back-end decision oracle. Moreover, performing the word-level reduction on netlists rather than on formulas has several benefits even if we are focused on bounded checks only: First of all, there is no need to re-abstract a newly unfolded formula every time the bound is increased. Second, by performing the reduction once and for all, we can apply standard performance optimizations in the back-end SAT solver such as incremental solving and the reuse of conflict clauses without having to resort to special tricks.

In terms of netlist reductions, the most closely related work we are aware of is the DCC reduction work [4] we make use of in our analysis. The chief problem with this work as it stands is that it requires the design to be partitioned up into a legal DCC to begin with, and this is unlikely to be true in the case of many systems where control data is integrated into the same packets as the word-level payload (the router we analyze in Section 7 is one such design). We solve this problem using our selective bitblasting approach. We also modify the DCC analysis to work with finite packets, and extend it so that it can deal with several different package sizes in the same design.

There are approaches to data type reduction that have some similarities to the DCC analysis, such as symmetry reduction based on the use of *scalarsets* [5]. For these type of reductions, it is typically up to the user to manually introduce reducible data types when modeling a given design, which would be very time consuming for industrial size netlists that intermingle control and data information. Moreover, these reductions are generally not straight netlist transformations, so they would be hard to use in a transformation-based verification environment.

7 Experimental Results

Our aim with this work is to provide a word-level reduction method for a transformation-based verification system. We are hence focusing on simplifying the netlist problem as much as possible, and providing output that can be processed with an interleaving of model checking methods and simplification procedures like retiming, rewriting and reparametrization. We thus strive for maximal reduction at minimal runtime cost.

We present results on three designs, which represent classes of problems where our customers have found that traditional model checking provides little or no traction today. In all of our experiments, we use standard SAT-based bounded model checking and BDD-based reachability analysis as our back-end engines.

The first benchmark is an industrial FIFO memory. We check that if a slot has been written and not overwritten, the correct data is read out when the entry is popped. We check two version of this design. The first version has 16 slots, each 16 bits wide. Before this reduction, this design can not be proved correct in one hour of compute time using BDD-based model checking. After the reduction, this design is proved correct in less than one minute. The second version has 75 slots, each 32 bits wide. Before the reduction the design has 2594 registers, which is reduced to 646 registers after analysis. The reduction takes less than .5 seconds of compute time. Bounded model checking of the unreduced design for 18 cycles takes 45 minutes of CPU time, whereas it takes 180 seconds after the reduction.

The second benchmark is an industrial Content Addressable Memory (CAM) with three data ports and 48 slots each containing 20 bits of data. We are checking that if a piece of data has been written to some particular slot in the CAM, and has not been overwritten, the data is reported as existing in the CAM. The original design has 1111 registers, which is reduced to 383 registers post reduction. The transformation takes less than .5 seconds of compute time. A bounded check of depth eight takes approximately 60000 seconds before the reduction, compared to 1780 seconds after the reduction.

Our final benchmark is a pipelined high-performance router created at Stanford [8]. The router has four port connections to adjacent routers plus an inject node and an eject node. The router's main crossbar is implemented on the bit level. It has two virtual channels per port, and moves packets broken up into units called *flits* that it reads from the environment ports each cycle. Each flit contains destination data, type field control information, and data payload packed into 32 bits. We check the partial correctness property that a packet that is injected on a port when the router is in an neutral state appears on the correct output port within a predetermined time. The word-level netlist contains 7516 registers before the abstraction, and is reduced to 4816 registers after the abstraction. The total reduction analysis time is less than one second.

8 Conclusions

In this paper, we have introduced a word-level model checking approach aimed at unbounded property checking for industrial netlists. Our approach is based on a two-step method, where a quick analysis rewrites the netlist into a design where the word-level datapath that manipulates packages is completely separated from the boolean control logic. We then resize all packages using statically computed safe lower bounds that guarantee that we preserve the properties being checked. The resulting system can be analyzed using any standard bit-level model checking technique, or further processed using transformational verification simplifications.

Our contributions include the combination of a modified word-level extraction algorithm previously only used on formulas [6] with some modified lower bound computation theory [4]. We have also showed that our analysis is tight in the sense that the obvious extensions of operators allowed in the word-level part of the circuit all preclude a static analysis. Finally, we provided key insights into how to implement the algorithms efficiently, and demonstrated the utility of our reduction on a number of industrial designs.

Acknowledgments. Many thanks to Tamir Heyman, who participated in discussions and helped with the work necessary to integrate our analysis into the frontend flow.

References

1. Baumgartner, J., Gloekler, T., Shanmugam, D., Seigler, R., Huben, G.V., Mony, H., Roessler, P., Ramanandray, B.: Enabling large-scale pervasive logic verification through multi-algorithmic formal reasoning. In: Proc. of the Formal Methods in CAD Conf. (2006)
2. Bryant, R.E., Lahiri, S.K., Seshia, S.A.: Modeling and verifying systems using a logic of counter arithmetic with lambda expressions and uninterpreted functions. In: Brinksma, E., Larsen, K.G. (eds.) CAV 2002. LNCS, vol. 2404, pp. 78–92. Springer, Heidelberg (2002)
3. Galler, B., Fischer, M.: An improved equivalence algorithm. Communications of the ACM (May 1964)
4. Hojati, R., Brayton, R.: Automatic datapath abstraction in hardware systems. In: Wolper, P. (ed.) CAV 1995. LNCS, vol. 939, pp. 98–113. Springer, Heidelberg (1995)
5. Ip, C.N., Dill, D.L.: Better verification through symmetry. Formal Methods in System Design (August 1996)
6. Johannesen, P.: Speeding up hardware verification by automated data path scaling. PhD thesis, Christian-Albrechts-Universität zu Kiel (2002)
7. Manolios, P., Srinivasan, S.K., Vroon, D.: BAT: The Bit-Level Analysis Tool. In: Damm, W., Hermanns, H. (eds.) CAV 2007. LNCS, vol. 4590, pp. 303–306. Springer, Heidelberg (2007)
8. Peh, L.-S., Dally, W.: A delay model and speculative architecture for pipelined routers. In: Proc. Intl. Symposium on High-Performance Computer Architecture (2001)
9. Pugh, W.: Skip lists: a probabilistic alternative to balanced trees. Communications of the ACM (June 1990)
10. Ranise, S., Tinelli, C.: Satisfiability modulo theories. Trends and Controversies - IEEE Intelligent Systems Magazine (December 2006)

Validating High-Level Synthesis*

Sudipta Kundu, Sorin Lerner, and Rajesh Gupta

University of California, San Diego, La Jolla, CA 92093-0404
{skundu,lerner,rgupta}@cs.ucsd.edu

Abstract. The growing design-productivity gap has made designers shift toward using high-level languages like C, C++ and Java to do system-level design. High-Level Synthesis (HLS) is the process of generating Register Transfer Level (RTL) design from these initial high-level programs. Unfortunately, this translation process itself can be buggy, which can create a mismatch between what a designer intends and what is actually implemented in the circuit. In this paper, we present an approach to validate the result of HLS against the initial high-level program using insights from translation validation, automated theorem proving and relational approaches to reasoning about programs. We have implemented our validating technique and have applied it to a highly parallelizing HLS framework called SPARK. We present the details of our algorithm and experimental results.

1 Introduction

While hardware designer productivity has grown at an impressive rate over the past few decades, the rate of improvement has not kept pace with chip capacity growth. High-Level Synthesis (HLS) [10,17,22] is often seen as a solution to bridge the design-productivity gap. HLS is the process of generating Register Transfer Level (RTL) design consisting of a data path and a control unit from a high-level behavioral description of a digital system, expressed in languages like C, C++ and Java. The synthesis process consists of several inter dependent sub-tasks such as: specification, compilation, scheduling, allocation, binding and control generation. HLS is an area that has been widely explored and relatively mature implementations of various HLS algorithm have started to emerge [10,17,22]. HLS tools are large and complex software systems, often with hundreds of thousands of lines of code, and as with any software of this scale, they are prone to logical and implementation errors. Errors in these tools lead to the synthesis of RTL designs with bugs in them, which often have expensive ramifications if they go undetected until after fabrication or large-scale production. Hence, correctness of these HLS tools has always been an important concern.

Despite significant amount of work in the area of verification we are still far from being able to prove automatically that a given optimizing HLS tool always produces target programs that are semantically equivalent to their source versions. However, even if one cannot prove an HLS tool correct once and for all, one can try

* This research was supported in part by NSF CAREER Grant 0644306.

A. Gupta and S. Malik (Eds.): CAV 2008, LNCS 5123, pp. 459–472, 2008.

to show, for each translation that the HLS tool performs, that the output program produced by the tool has the same behavior as the original program. Although this approach does not guarantee that the HLS tool is bug free, it does guarantee that any errors in translation will be caught when the tool runs, preventing such errors from propagating any further in the hardware fabrication process. This approach to verification, called *translation validation*, has previously been applied with success in the context of optimizing compilers [8,19,20,21,23], and for automatically proving refinements of CSP programs [14].

The main contribution of this paper is to show how translation validation can effectively be implemented in a previously unexplored setting: an HLS tool. In particular, we present an algorithm for validating all the phases (except for parsing, binding and code generation) of the SPARK HLS tool [10] against the initial behavioral description. With over 4,000 downloads, and over 100 active members in the user community, SPARK is a widely used tool. Although commercial HLS tools exists, these tools are not available for academic experimentation – SPARK represents the state of the art in the academic community.

Our algorithm uses a bisimulation relation approach to prove equivalence. In particular, we automatically establish a bisimulation relation that states what points in the specification program are related to what points in the implementation program. This bisimulation relation guarantees that for each execution sequence in the specification, a related and equivalent execution sequence exists in the implementation and vice versa. To deal with the parallelism introduced by the scheduling step of SPARK, we exploit the structure of the transformations that SPARK applies during scheduling. These transformations convert a sequential program to a program that contains instruction-level parallelism. Our algorithm deals with this parallelism using standard techniques for computing weakest preconditions and strongest postconditions of parallel programs [4]. Furthermore, the algorithm we present also draws insights and techniques from various areas, including translation validation [19,20], theorem proving [5], and relational approaches to reasoning about programs [11,15].

We implemented our algorithm in a tool that validates SPARK's HLS process. We used our tool to verify the translation of a variety of benchmarks. Because our verification approach works on one procedure at a time, it is modular. Furthermore, our validation tool took on average 6 seconds to run per procedure, showing that translation validation of HLS transformations can be fast enough to be practical. Finally, in running our tool, two failed validation runs have lead us to discover two previously unknown bugs in the SPARK tool. These bugs cause SPARK to generate *incorrect* RTL for a given high-level program. This demonstrates that translation validation of the HLS process can catch bugs that even testing and long-term use may not uncover.

2 Overview

We start by presenting a simple example that illustrates our approach (Figure 1). The specification is a sequential function shown in Figure 1(a) using a transition

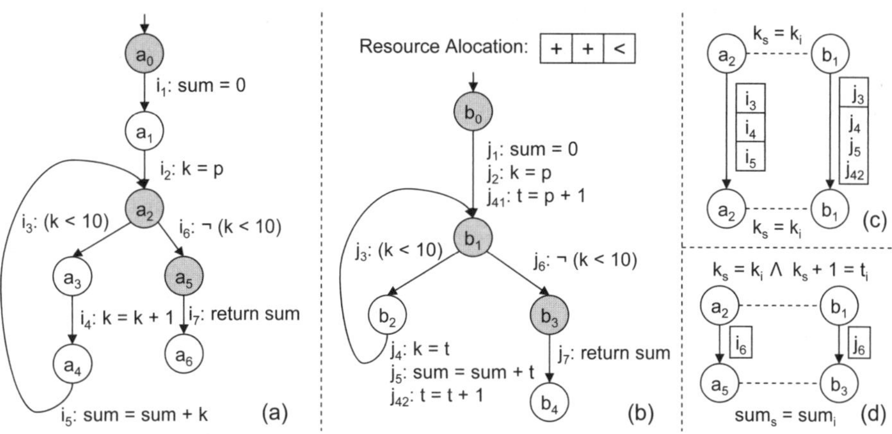

Fig. 1. Our running example (a) Specification (b) Implementation (c) and (d) Parts of 2nd iteration

Table 1. Iterations for computing the bisimulation relation

(l_1, l_2)	1st iteration	2nd iteration	3rd iteration (ϕ)
1. (a_0, b_0)	$\mathsf{p_s} = \mathsf{p_i}$	$\mathsf{p_s} = \mathsf{p_i}$	$\mathsf{p_s} = \mathsf{p_i}$
2. (a_2, b_1)	$\mathsf{k_s} = \mathsf{k_i}$	$(\mathsf{k_s} = \mathsf{k_i}) \wedge$ $(\mathsf{sum_s} = \mathsf{sum_i}) \wedge$ $((\mathsf{k_s} + 1) = \mathsf{t_i})$	$\mathsf{k_s} = \mathsf{k_i} \wedge$ $\mathsf{sum_s} = \mathsf{sum_i} \wedge$ $(\mathsf{k_s} + 1) = \mathsf{t_i}$
3. (a_5, b_3)	$\mathsf{sum_s} = \mathsf{sum_i}$	$\mathsf{sum_s} = \mathsf{sum_i}$	$\mathsf{sum_s} = \mathsf{sum_i}$

diagram. This function takes p as input and computes the sum from $(\mathsf{p} + 1)$ to 10 using a loop and returns it. A parallelizing HLS tool will apply various kinds of transformations to this sequential function, with the goal of scheduling each operation based on some resource constraints.

Figure 1(b) shows the result of running SPARK's HLS algorithm given resource constraints of 2 adders and a comparator. Instructions on the same transition edge are executed in *parallel.* For this example SPARK has performed several transformations. First, it applied a loop-shifting transformation that moves the operation $\mathsf{i_4}$ from the beginning of the loop body to the end of the loop body ($\mathsf{j_{42}}$), while also placing a copy of the operation in the loop header ($\mathsf{j_{41}}$) using the temporary variable t. The effect of this loop-shifting transformation is a form of software pipelining [16]. Notice that without this pipelining transformation it would not have been possible to schedule the operation $\mathsf{i_4}$ and $\mathsf{i_5}$ together due to data dependence between them. In addition to loop-shifting, SPARK also performed copy propagation of instruction $\mathsf{j_2}$ to $\mathsf{j_{41}}$ and instruction $\mathsf{j_4}$ to $\mathsf{j_{42}}$.

Bisimulation Relation. In order to show that the implementation is equivalent to the specification, our approach computes a bisimulation relation between the two programs. The goal of the bisimulation relation is to guarantee that the specification and the implementation perform the same set of visible instructions. In our case, we consider visible instructions to be function calls and return statements. Our technique thus guarantees that the specification and the implementation perform the same sequence of function calls (with the same arguments) and returns (with the same returned values).

The bisimulation relation (defined formally in Section 4) consists of a set of entries of the form (l_1, l_2, ϕ), where l_1 and l_2 are locations in the specification and implementation respectively, and ϕ is a predicate over variables of the specification and implementation. The pair (l_1, l_2) captures how the control state of the specification is related to the control state of the implementation, whereas ϕ captures how the data is related. For instance, Table 1 shows the bisimulation relation for our running example. The control component of entries in the bisimulation relation are shown in the first column and the data component in the last column of the table.

The first entry in the bisimulation relation relates the start location of the specification and the implementation. For this entry, the relevant data invariant is $\mathsf{p_s} = \mathsf{p_i}$, which states that the value of the input argument p in the specification is equal to the value of the input argument p in the implementation. We use subscript s to denote variables in the specification and subscript i for variables in the implementation. The second entry in the bisimulation relation relates the loop head (a_2) in the specification with the loop head (b_1) of the implementation. This entry represent two loops that run in synchrony, one loop being in the specification and the other being in the implementation. The invariant can be seen as a loop invariant across the specification and the implementation, which guarantee that the two loops produce the same effect on the visible instructions. The control part of this entry guarantee that the two loops are in fact synchronized. The last entry in the bisimulation relation relates the location a_5 in the specification with the location b_3 of the implementation. The relevant invariant for this entry is $\mathsf{sum_s} = \mathsf{sum_i}$, since the value returned by both the program should be same (our equivalence criterion).

The entries in the bisimulation relation must satisfy some simple local requirements (which are made precise in Section 4). Intuitively, for any entry (l_1, l_2, ϕ) in the bisimulation relation, if the specification and implementation start executing in parallel at control locations l_1 and l_2 in states where ϕ holds, and in doing so reach another bisimulation entry (l_1', l_2', ϕ'), then ϕ' must hold in the resulting states.

Our Approach. Our technique for equivalence checking starts by finding pairs of locations in the implementation and the specification that need to be related in the bisimulation. This amounts to computing the first column of Table 1. In the given example, our algorithm first adds (a_0, b_0) as a pair of interest, which is the entry location of both programs. Then it moves forward simultaneously in the implementation and the specification until it reaches a branch, a function call or a return instruction. In the example from Figure 1, our algorithm finds that there is a branch and a return instruction that must be matched (the specification locations a_2 and a_5 should match, respectively, with the implementation location b_1 and b_3). While finding these pairs of locations, our algorithm correlates the branch in the specification and the implementation (the details of how we establish branch correlations is explained in Section 5).

Once the related pairs of locations have been collected we define, for each pair of locations (l_1, l_2), a constraint variable $\psi_{(l_1,l_2)}$ to represent the state-relating formula that will be computed in the bisimulation relation for that pair. We then define a set of constraints over these variables that must be satisfied in order for the would-be bisimulation relation to in fact be a bisimulation.

There are two kinds of constraints. First, for each pair of locations (l_1, l_2) that are related, we want $\psi_{(l_1,l_2)}$ to imply that any visible instructions about to execute at (l_1, l_2) behave the same way. For example, $\psi_{(a_5,b_3)}$ should imply $\mathsf{sum_s} = \mathsf{sum_i}$, so that the returned values are the same. Such constraints guarantee that the computed bisimulation relation is strong enough to show that the visible instructions behave the same way in the specification and the implementation. A second kind of constraint is used to state the relationship between one pair of related locations and other pairs of related locations. For example, if starting at (l_1, l_2) in states satisfying $\psi_{(l_1,l_2)}$, the specification and implementation can execute in parallel to reach another related pair of locations (l'_1, l'_2), then $\psi_{(l'_1,l'_2)}$ must hold in the resulting states. As shown in Section 5, such constraints can be stated over the variables $\psi_{(l_1,l_2)}$ and $\psi_{(l'_1,l'_2)}$ using the weakest precondition operator (wp). This second kind of constraint guarantees that the computed bisimulation relation is in fact a bisimulation.

Once the constraints are generated, we solve them using an iterative algorithm that starts with all constraint variables set to *true* and then iteratively strengthens the constraint variables until a theorem prover is able to show that all constraints are satisfied. Although in general this constraint-solving algorithm is not guaranteed to terminate, in practice it can quickly find the required bisimulation relation.

The constraint solving for our example is shown in Table 1. Our algorithm first initializes the constraint variables with the conditions that are required for the visible instructions to be equivalent. Then it chooses any entry from the table, say (a_2, b_1) and finds the entries that can reach it (i.e. (a_2, b_1) and (a_0, b_0)). Consider the synchronized loop from (a_2, b_1) to (a_2, b_1) shown in Figure 1(c). Our algorithm computes the weakest precondition of the formula at the bottom $(\mathsf{k_s} = \mathsf{k_i})$ over the instructions in the implementation and in the specification, which happens to be $\delta = [(\mathsf{k_s} < 10) \Rightarrow (\mathsf{k_i} < 10) \Rightarrow (\mathsf{k_s} + 1) = \mathsf{t_i}]$. Next, it asks a theorem prover if the condition at the top i.e. $\mathsf{k_s} = \mathsf{k_i}$ implies δ. Since it does not, our algorithm strengthens the condition at the top with $(\mathsf{k_s} + 1) = \mathsf{t_i}$ which is a stronger condition than δ. A similar pass through Figure 1(d) strengthens the condition at (a_2, b_1) with $(\mathsf{sum_s} = \mathsf{sum_i})$. Our constraint solving continues in this manner until a fixpoint is reached.

3 Definition of Equivalence

Having illustrated our approach using a simple example, we now present a formal description. Our approach verifies each procedure from the specification against the corresponding procedure from the implementation. We represent each procedure in the specification and the implementation using a *transition diagram* that describes the control structure of the procedure in terms of *program locations* and *program transitions*. A program location represents a point of control in the procedure, and a transition describes how the program state changes from one program location to another. We represent these transitions by *instructions*.

More formally, we define a program state to be a function $VAR \rightarrow VAL$ assigning values to variables, where VAR denotes the set of variables and VAL denotes the domain of values. We denote by Σ the a set of all program states. We define an instruction to be a pair (c, f) where $c : \Sigma \rightarrow \mathcal{B}$ is a predicate and $f : \Sigma \rightarrow \Sigma$ is a state transformation function. The predicate c is the condition under which the state transformation function f can happen. For instance, in Figure 1(a) the instruction i_3 has $c = (\mathsf{k} < 10)$ and $f(\sigma) = \sigma$, whereas the instruction i_2 has $c = \mathit{true}$ and $f(\sigma) = \sigma[\mathsf{k} \mapsto \sigma(\mathsf{p})]$.

Finally a transition diagram is defined as follows.

Definition 1 (Transition Diagram). *A transition diagram π is a tuple $(\mathcal{L}, \mathcal{I}, \rightarrow, \iota)$, where $\mathcal{L}$ is a finite set of locations, $\mathcal{I}$ is a finite set of instructions, $\rightarrow \subseteq \mathcal{L} \times \mathcal{I} \times \mathcal{L}$ is a finite set of triples (l, a, l') called transitions, and $\iota \in \mathcal{L}$ is the entry location. We write $l \xrightarrow{i} l'$ to denote $(l, i, l') \in \rightarrow$.*

Definition 2 (Semantic Step). *Given a transition diagram $\pi = (\mathcal{L}, \mathcal{I}, \rightarrow, \iota)$, we define a configuration to be a pair $\langle l, \sigma \rangle$, where $l \in \mathcal{L}$ and $\sigma \in \Sigma$. Given two configurations $\langle l, \sigma \rangle$ and $\langle l', \sigma' \rangle$, and an instruction $i \in \mathcal{I}$, the semantic step relation is defined as follows:*

$$\langle l, \sigma \rangle \overset{i}{\rightsquigarrow} \langle l', \sigma' \rangle \quad \mathit{iff} \quad l \xrightarrow{i} l' \text{ and } i = (c, f) \text{ and } c(\sigma) = \mathit{true} \text{ and } \sigma' = f(\sigma)$$

Definition 3 (Execution Sequence). *For a given transition diagram $\pi = (\mathcal{L}, \mathcal{I}, \rightarrow, \iota)$, an execution sequence η starting in $\sigma_0 \in \Sigma$ is a sequence of configurations such that:*

$$\langle l_0, \sigma_0 \rangle \overset{i_1}{\rightsquigarrow} \langle l_1, \sigma_1 \rangle \overset{i_2}{\rightsquigarrow} \cdots \overset{i_n}{\rightsquigarrow} \langle l_n, \sigma_n \rangle$$

We denote by $\mathcal{N}$ the set of all execution sequences.

We define ϑ to be the set of *visible instructions*. These are the instructions whose semantics we would like preserved between the specification and implementation. In our system we consider visible instructions to be function calls and returns. For $v_1, v_2 \in \vartheta$, we write $\langle v_1, \sigma_1 \rangle \equiv \langle v_2, \sigma_2 \rangle$ to represent that v_1 in program state σ_1 is equivalent to v_2 in program states σ_2. For two visible instructions to be equivalent, they must both be returns, or both calls. Furthermore, returns are equivalent if the returned value and the state of the memory are the same. Two function calls are equivalent if the state of globals, the arguments and the address of the called function are the same. This concept of equivalence for visible instruction can be extended to execution sequences as follows.

Definition 4 (Equivalence of Execution Sequences). *Two execution sequences η_1 and η_2 are said to be equivalent, written $\eta_1 \equiv \eta_2$, if the two sequences contain visible instructions that are pairwise equivalent.*

Definition 5 (Equivalence of Transition Diagrams). *For given initial states $\sigma_1 \in \Sigma_1$ and $\sigma_2 \in \Sigma_2$, two transition diagrams $\pi_1 = (\mathcal{L}_1, \mathcal{I}_1, \rightarrow_1, \iota_1)$ and $\pi_2 = (\mathcal{L}_2, \mathcal{I}_2, \rightarrow_2, \iota_2)$ are said to be equivalent if for every execution sequence of π_1 starting in configuration $\langle \iota_1, \sigma_1 \rangle$ there is an equivalent execution sequence of π_2 starting in configuration $\langle \iota_2, \sigma_2 \rangle$ and vice-versa.*

4 Bisimulation Relation

A *verification relation* between two transition diagrams π_1 and π_2 is a set of triples (l_1, l_2, ϕ), where $l_1 \in \mathcal{L}_1$, $l_2 \in \mathcal{L}_2$ and ψ is a predicate over the variables live at locations l_1 and l_2. Let the set of such predicates be denoted by $\Phi \stackrel{def}{=} \Sigma \times \Sigma \rightarrow \mathcal{B}$. We write $\phi(\sigma_1, \sigma_2) = true$ to indicate that ϕ is satisfied in $(\sigma_1, \sigma_2) \in \Sigma \times \Sigma$.

Simulation relations and *bisimulation relations* are verification relations with a few additional properties. To define these properties, we make use of a *cumulative semantic step* relation $\rightsquigarrow^*$, which works like $\rightsquigarrow$, except that it can take multiple steps at once, and it accumulates the steps taken into an execution sequence.

Definition 6 (Cumulative Semantic Step). *Given configurations $\langle l_0, \sigma_0 \rangle$ and $\langle l_n, \sigma_n \rangle$, and an execution sequence η that contains at least one transition, we define $\rightsquigarrow^*$ as follows:*

$$\langle l_0, \sigma_0 \rangle \overset{\eta}{\rightsquigarrow^*} \langle l_n, \sigma_n \rangle \quad \textit{iff} \quad \eta = \langle l_0, \sigma_0 \rangle \overset{i_1}{\rightsquigarrow} \cdots \overset{i_n}{\rightsquigarrow} \langle l_n, \sigma_n \rangle$$

Definition 7 (Simulation Relation). *A simulation relation R for two transition diagrams $\pi_1 = (\mathcal{L}_1, \mathcal{I}_1, \rightarrow_1, \iota_1)$ and $\pi_2 = (\mathcal{L}_2, \mathcal{I}_2, \rightarrow_2, \iota_2)$ is a verification relation such that:*

$$R(\iota_1, \iota_2, true)$$

$$\begin{aligned} &\forall (l_1, l_2, l_1', \sigma_1, \sigma_2, \sigma_1', \phi, \eta_2) \in \mathcal{L}_1 \times \mathcal{L}_2 \times \mathcal{L}_1 \times \Sigma \times \Sigma \times \Sigma \times \Phi \times \mathcal{N} \,. \\ &\quad \left[\langle l_1, \sigma_1 \rangle \overset{\eta_1}{\rightsquigarrow_1^*} \langle l_1', \sigma_1' \rangle \wedge R(l_1, l_2, \phi) \wedge \phi(\sigma_1, \sigma_2) = true \right] \Rightarrow \\ &\quad\quad \exists (l_2', \sigma_2', \phi', \eta_2) \in \mathcal{L}_2 \times \Sigma \times \Phi \times \mathcal{N} \,. \\ &\quad\quad\quad \left[\langle l_2, \sigma_2 \rangle \overset{\eta_2}{\rightsquigarrow_2^*} \langle l_2', \sigma_2' \rangle \wedge R(l_1', l_2', \phi') \wedge \phi'(\sigma_1', \sigma_2') = true \wedge \eta_1 \equiv \eta_2 \right] \end{aligned}$$

Intuitively, these conditions respectively state that (1) the entry location of π_1 must be related to the entry location of π_2; and (2) if π_1 and π_2 are in a pair of related configurations, and π_1 can proceed one or more steps producing an execution sequence η_1, then π_2 must also be able to proceed one or more steps, producing a sequence η_2 that is equivalent to η_1, and the two resulting configurations must be related.

Even though in the above definition, the state-relating predicate for the entry locations is *true*, dummy assignments to a procedure's arguments allow us to prove that the arguments in the specification are equal to those in the implementation, at the beginning of each procedure.

Definition 8 (Bisimulation Relation). *A verification relation R is a bisimulation relation for π_1, π_2 iff R is a simulation relation for π_1, π_2 and $R^{-1} = \{(l_2, l_1, \phi) \mid R(l_1, l_2, \phi)\}$ is a simulation relation for π_2, π_1.*

The following lemma and theorem connect the definition of bisimulation with our definition of equivalence for transition diagrams (Definition 5).

Lemma 1. *If R is a bisimulation relation for π_1, π_2, then for each element $(l_1, l_2, \psi) \in R$, all pairs of executions of π_1 started at l_1 and of π_2 started at l_2, in states that satisfy the predicate ψ, are equivalent.*

Theorem 1. *If there exists a bisimulation relation for π_1, π_2, then π_1 and π_2 are equivalent.*

The conditions from Definition 7 are the base case and the inductive case of a proof by induction showing that π_2 is equivalent to π_1. Thus, a bisimulation relation is a witness that two transition diagrams are equivalent.

5 Translation Validation Algorithm

Our translation validation algorithm works by inferring a bisimulation relation. Given a transition diagram π, we define $\mathcal{P}_\pi$ to be the set of locations for which our approach will try to infer bisimulation entries. These include all locations before visible events and also all locations before branch statements. To focus our attention on only those locations for which our approach infers bisimulation entries, we define the *skipping transition* relation $\hookrightarrow$, which is a version of $\longrightarrow$ that skips over all locations not in $\mathcal{P}_\pi$.

Definition 9 (Skipping Transition). *Let $\pi = (\mathcal{L}, \mathcal{I}, \rightarrow, \iota)$ be a transition diagram, $l, l' \in \mathcal{P}_\pi$, and $w \in \mathcal{I}^*$, where $w = i_0 \cdots i_n$. We define $\hookrightarrow$ as follows:*

$$l \overset{w}{\hookrightarrow} l' \text{ iff there exists } l_1, \cdots, l_n \in (\mathcal{L} - \mathcal{P}_\pi) \text{ such that } l \overset{i_0}{\longrightarrow} l_1 \cdots l_n \overset{i_n}{\longrightarrow} l'$$

Throughout the rest of this section, we assume that $\pi_1 = (\mathcal{L}_1, \mathcal{I}_1, \rightarrow_1, \iota_1)$ represents the procedure in the specification whose translation we want to verify, and $\pi_2 = (\mathcal{L}_2, \mathcal{I}_2, \rightarrow_2, \iota_2)$ represents the corresponding procedure in the implementation. We let $\mathcal{P}_1 = \mathcal{P}_{\pi_1}$ and $\mathcal{P}_2 = \mathcal{P}_{\pi_2}$. We also let $\hookrightarrow_1$ and $\hookrightarrow_2$ be the skipping transitions for π_1 and π_2 respectively.

We now define a parallel transition relation $\hookrightarrow\!\!\!\rightarrow$ that essentially traverses the two transition diagrams (specification and implementation) in synchrony.

Definition 10 (Parallel Transition). *Given $(l_1, l_2) \in \mathcal{P}_1 \times \mathcal{P}_2$, $(l_1', l_2') \in \mathcal{P}_1 \times \mathcal{P}_2$, $w_1 \in \mathcal{I}_1^*$ and $w_2 \in \mathcal{I}_2^*$, we define $\hookrightarrow\!\!\!\rightarrow$ as follows:*

$$(l_1, l_2) \overset{(w_1, w_2)}{\hookrightarrow\!\!\!\rightarrow} (l_1', l_2') \quad \textit{iff} \quad l_1 \overset{w_1}{\hookrightarrow}_1 l_1' \textit{ and } l_2 \overset{w_2}{\hookrightarrow}_2 l_2' \textit{ and } \mathsf{Rel}(w_1, w_2, l_1, l_2)$$

The predicate $\mathsf{Rel} : \mathcal{I}^* \times \mathcal{I}^* \times \mathcal{P}_1 \times \mathcal{P}_2 \rightarrow \mathcal{B}$ used in the above definition is a heuristic that tries to estimate when a path in the specification is related to a path in the implementation. Consider for example the branch in the specification of Figure 1 and the corresponding branch in the implementation. For any two such branches, the Rel function uses heuristics to guess a correlation between them: either they always go in the same direction, or they always go in opposite direction. Using these correlations, $\mathsf{Rel}(w_1, w_2, l_1, l_2)$ returns true only if the paths w_1 and w_2 follow branches in a correlated way. Although Rel makes guesses about the correlation of branches, the later constraint solving phase of our approach makes sure that these guesses are correct.

Our implementation of Rel correlates branches in two ways. First, using the results of a strongest postcondition pre-pass over the specification and the implementation, Rel tries to use a theorem prover to prove that certain branches are correlated. If the theorem prover is not able to determine a correlation, Rel uses the structure of the branch predicate and the structure of the instructions on each side of the branch to guess a correlation. For instance, in the example of Figure 1, since the strongest postcondition involves the input parameter p, the theorem prover is unable to reason about it. However, because SPARK does not change the structure of the branch predicate, Rel can conclude that the two branches go in the same direction.

We now define the relation $\mathcal{R} \subseteq \mathcal{P}_1 \times \mathcal{P}_2$ of location pairs that will form the entries of our bisimulation relation.

Definition 11 (Pairs of Interest). *The relation $\mathcal{R} \subseteq \mathcal{P}_1 \times \mathcal{P}_2$ is defined to be the minimal relation that satisfies the following two properties:*

$$\mathcal{R}(\iota_1, \iota_2)$$

$$\left[\mathcal{R}(l_1, l_2) \ \wedge \ (l_1, l_2) \overset{(w_1, w_2)}{\hookrightarrow\!\!\!\rightarrow} (l_1', l_2')\right] \Longrightarrow \mathcal{R}(l_1', l_2')$$

The set $\mathcal{R}$ defined above can easily be computed by starting with the empty set, and applying the above two rules exhaustively.

For our approach to successfully validate a translation, the computed set $\mathcal{R}$ must relate locations where the instructions to be executed are similar. This is made precise by the following definition of *well-matching* of $\mathcal{R}$. If the computed set $\mathcal{R}$ is *not* well-matched, then our validation approach immediately rejects the translation from specification to implementation.

Definition 12 (Well-matching). *For each $(l_1, l_2) \in \mathcal{R}$, if we let i_1 and i_2 be the instructions to be executed after l_1 and l_2, respectively, then for $\mathcal{R}$ to be well-matched, the following must hold: i_1 is a branch iff i_2 is a branch; i_1 is a function call iff i_2 is a function call; and i_1 is a return iff i_2 is a return.*

We describe our translation validation approach in terms of constraint solving. In particular, for each $(l_1, l_2) \in \mathcal{R}$ we define a constraint variable $\psi_{(l_1,l_2)}$ representing the predicate that we want to compute for the bisimulation entry (l_1, l_2). We denote by Ψ the set of all such constraint variables. Using these constraint variables, the final bisimulation relation will have the form $\{(l_1, l_2, \psi_{(l_1,l_2)}) \mid \mathcal{R}(l_1, l_2)\}$.

1. **function** SolveConstraints($\mathcal{C}$)
2.   **for each** $(l_1, l_2) \in \mathcal{R}$ **do**
3. $\psi_{(l_1,l_2)} := \mathit{true}$
4.   **let** $\mathit{worklist} := \mathcal{C}$
5.   **while** *worklist* not empty **do**
6.   **let** $[\psi_{(l_1,l_2)} \Rightarrow f(\psi_{(l_1',l_2')})] := \mathit{worklist}.\mathsf{Remove}$
7.   **if** $\mathsf{ATP}(\psi_{(l_1,l_2)} \Rightarrow f(\psi_{(l_1',l_2')})) \neq \mathit{Valid}$ **then**
8.   **if** $(l_1, l_2) = (\iota_1, \iota_2)$ **then**
9.   Error("Start Condition not strong enough")
10. $\psi_{(l_1,l_2)} := \psi_{(l_1,l_2)} \wedge f(\psi_{(l_1',l_2')})$
11. $\mathit{worklist} := \mathit{worklist} \cup \{c \in \mathcal{C} \mid \exists \psi, g \ . \ c = [\psi \Rightarrow g(\psi_{(l_1,l_2)})]\}$

Fig. 2. Algorithm for solving constraints

To compute the predicates that the constraint variables $\psi_{(l_1,l_2)}$ stand for, we define a set of constraints on these variables, and then solve the constraints. The constraints are defined as follows.

Definition 13 (Constraint). *A constraint is a formula of the form* $\psi_1 \Rightarrow f(\psi_2)$*, where* $\psi_1, \psi_2 \in \Psi$*, and* f *is a boolean function.*

Definition 14 (Set of Constraints). *The set* $\mathcal{C}$ *of constraints is defined by:*

$$\text{For each } (l_1, l_2) \text{ in } \mathcal{R}\text{:} \quad \left[\psi_{(l_1,l_2)} \Rightarrow \mathsf{CreateSeed}(l_1, l_2)\right] \in \mathcal{C}$$

$$\text{For each } (l_1, l_2) \overset{(w_1,w_2)}{\hookrightarrow\!\!\!\rightarrow} (l_1', l_2')\text{:} \quad \left[\psi_{(l_1,l_2)} \Rightarrow \mathsf{wp}(w_1, \mathsf{wp}(w_2, \psi_{(l_1',l_2')}))\right] \in \mathcal{C}$$

The CreateSeed function above creates for each pair of locations (l_1, l_2) a formula (which does not refer to any constraint variables) that captures the condition under which the instructions about to execute at l_1 and l_2 are equivalent. Because $\mathcal{R}$ is well-matched (see Definition 12), there are three cases: if the instructions about to execute at l_1 and l_2 are calls, then the formula returned by CreateSeed states that the parameters of the calls are equal; if the two instructions are returns, then the formula states that the returned values are equal; if the two instructions are branches, then the formula states the two branches are correlated (either they both go in the same direction, or in opposite directions).

The other function wp used above computes the weakest precondition with respect to w_2 and then with respect to w_1. When computing wp with respect to one sequence, we treat all variables from the other sequence as constants. As a result, the order in which we process the two sequences does not matter.

Having created a set of constraints $\mathcal{C}$, our validation approach now solves these constraints using the algorithm in Figure 2. The algorithm starts by setting each constraint variable to *true* (line 3) and initializing a *worklist* with the set of all constraints (line 4). Next, while the *worklist* is not empty, it removes a constraint from the worklist (line 6), and checks using a theorem prover if it is *Valid* (line 7). If not, then it appropriately strengthens the left-hand-side variable of the constraint (line 10) and adds to the worklist all the constraints that have this variable in the right-hand side (line 11).

6 Evaluation

We implemented our validation algorithm on the intermediate representation (IR) of the SPARK HLS framework [10]. SPARK is a C-to-VHDL parallelizing high-level synthesis framework that employs a set of compiler, parallelizing compiler, and synthesis transformations to improve the quality of high-level synthesis results. SPARK starts with a behavioral description in ANSI-C as input – currently with the restrictions of no pointers, no recursion, and no irregular control-flow jumps. It converts the input program into its own IR, and then applies a set of code transformations, including loop unrolling, loop fusion, common subexpression elimination, copy propagation, dead code elimination, loop-invariant code motion, induction variable analysis, and operation strength reduction. Following these transformations, SPARK performs a scheduling phase using resource allocation information provided by the user. This scheduling phase also performs a variety of transformations, including speculative code motion, dynamic renaming of variables, dynamic branch balancing, chaining of operations across conditional blocks, and scheduling on multi-cycle operations. The scheduling phase is followed by a resource binding phase and finally by a back-end code generation pass that produces RTL VHDL.

We implemented our translation validation algorithm using the Simplify theorem prover [5] in a tool that validates SPARK's HLS process. Our tool takes as input the IR program that is produced by the parser, and the IR program right before resource binding, and verifies that the two are equivalent. Our tool therefore validates the entire HLS process of SPARK, except for parsing, resource binding and code generation. Our tool is around 7,500 lines of C++ code, whereas SPARK's implementation excluding the parser consists of over 125,000 lines of C++ code. Thus, with around 15 times less effort compared to SPARK's implementation we can build a framework that validates its synthesis process.

We tested our tool on 12 benchmarks obtained from SPARK's test suite. Of these benchmarks, 10 passed and 2 failed. For the ones that passed, our tool was able to quickly find the simulation relation, taking on average around 6 seconds per procedure, and a maximum of 27 seconds for the largest procedure (80 lines of code). Furthermore, the computed bisimulation relations were small, ranging in size from 6 to 29 entries, with an average of about 14. To infer these bisimulation relations, our approach made an average of 189 calls to the theorem prover per procedure (with a minimum of 9 and a maximum of 797). Our approach is compositional since it works on one procedure at a time, and the above results show that our approach can handle realistically size procedures.

As mentioned previously, two benchmarks failed our validation test. Upon further analysis each of them lead us to discover previously unknown bugs in SPARK. One bug occurs in a particular corner case of copy propagation for array elements. The other bug is in the implementation of the code motion algorithm in the scheduler. The fact that our translation validation approach found two previously unknown bugs in a widely-used HLS framework emphasizes the usefulness and bug-isolating capabilities of our tool.

In general, our tool will perform well when the transformations that are performed preserve most of the program's control flow structure. Such transformations are called *structure-preserving transformations* [23]. The only non structure-preserving transformation that SPARK performs is loop unrolling, but in our examples this transformation did not trigger.

7 Related Work

Our work is related to translation validation [8,14,19,20,21,23], HLS verification [1,6,13,18], and relational approaches to reasoning about programs [7,3,15,2,11]. Despite long lines of work in each one of these areas, our work distinguishes itself in the following way: it the first to show that translation validation can be effective in the context of a realistic HLS tool. We now discuss each area in more detail.

Translation Validation. The technique described in this paper is similar to our previous translation-validation algorithm for CSP programs [14]. The algorithm presented here, however, runs in the context of a realistic HLS tool, as opposed to the more theoretical results from our previous work. Furthermore, our current algorithm handles a more restricted form of concurrency than found in CSP, which allows it to run more efficiently. Our work also bears similarities to Necula's translation-validation algorithm for inferring simulation relations that prove equivalence of sequential programs [19]. Unlike Necula's approach, our algorithm must take into account statements running parallel, since one of the main tasks that HLS tools perform is to schedule statements for parallel execution. Furthermore our algorithm is expressed in terms of calls to a general theorem prover, rather than using specialized solvers and simplifiers. In this sense our algorithm is more modular, since the theorem proving part of the algorithm has been modularized into a component with a very simple interface (it takes a formula and returns *Valid* or *Invalid*).

HLS Verification. Techniques like correctness-preserving transformations [6], formal assertions [18] and relational approaches for functional equivalence of FSMDs [12,13] have been used to validate the scheduling step of HLS. However, all these techniques assume that the scheduler does not move code across basic blocks and variable names do not change, which would prevent them from validating SPARK's HLS process. In work that is complementary to ours, model checking was used to validate the binding step of HLS [1], which is the only internal step of SPARK that our tool does not validate.

Relational Approaches. Relational approaches have been used for a variety of verification tasks, including model checking [3,7], translation validation [19,20], and reasoning about optimizations once and for all [2,15]. In this context, our work can be seen as automating Joseph's relational approach for proving refinement of concurrent systems [11].

8 Conclusion and Future Work

We have presented an algorithm for translation validation of the HLS process, and have implemented it within the context of a HLS tool called SPARK. The innovation in our work lies in showing that translation validation approaches work well in the application domain of high-level synthesis. In particular, with only a fraction of the development cost of SPARK, our algorithm can validate the translations performed by SPARK, and it also uncovered bugs that eluded long-term use. Moving forward, we intend to implement translation validation in SPARK for the remaining phases: parsing, binding and code generation. We also intend to adapt our translation validation techniques to SystemC [9] programs.

References

1. Ashar, P., Bhattacharya, S., Raghunathan, A., Mukaiyama, A.: Verification of RTL generated from scheduled behavior in a high-level synthesis flow. In: ICCAD, pp. 517–524 (1998)
2. Benton, N.: Simple relational correctness proofs for static analyses and program transformations. In: POPL 2004 (January 2004)
3. Bustan, D., Grumberg, O.: Simulation based minimization. In: McAllester, D.A. (ed.) CADE 2000. LNCS, vol. 1831. Springer, Heidelberg (2000)
4. Chandy, K.M.: Parallel program design: a foundation. Addison-Wesley Longman Publishing Co. Inc., Boston, MA, USA (1988)
5. Detlefs, D., Nelson, G., Saxe, J.B.: Simplify: A theorem prover for program checking. Journal of Association Computing Machinery 52(3), 365–473 (2005)
6. Eveking, H., Hinrichsen, H., Ritter, G.: Automatic verification of scheduling results in high-level synthesis. In: DATE 1999, NY, USA (1999)
7. Fisler, K., Vardi, M.Y.: Bisimulation and model checking. Correct Hardware Design and Verification Methods (September 1999)
8. Goldberg, B., Zuck, L., Barrett, C.: Into the loops: Practical issues in translation validation for optimizing compilers. Electronic Notes in Theoretical Computer Science 132(1), 53–71 (2005)
9. Grötker, T.: System Design with SystemC. Kluwer Academic Publishers, Dordrecht (2002)
10. Gupta, S., Dutt, N., Gupta, R., Nicolau, A.: Spark: A high-level synthesis framework for applying parallelizing compiler transformations. In: VLSI Design 2003 (2003)
11. Josephs, M.B.: A state-based approach to communicating processes. Distributed Computing 3(1), 9–18 (1988)
12. Karfa, C., Mandal, C., Sarkar, D., Pentakota, S.R., Reade, C.: A formal verification method of scheduling in high-level synthesis. In: ISQED, pp. 71–78 (2006)
13. Kim, Y., Kopuri, S., Mansouri, N.: Automated formal verification of scheduling process using finite state machines with datapath (fsmd). In: ISQED 2004 (2004)
14. Kundu, S., Lerner, S., Gupta, R.: Automated refinement checking of concurrent systems. In: ICCAD 2007 (2007)
15. Lacey, D., Jones, N.D., Wyk, E.V., Frederiksen, C.C.: Proving correctness of compiler optimizations by temporal logic. In: POPL 2002 (January 2002)
16. Lam, M.: Software pipelining: an effective scheduling technique for VLIW machines. In: PLDI 1988 (June 1988)

17. Lin, Y.-L.: Recent developments in high-level synthesis. ACM Transactions on Design Automation of Electronic Systems 2(1), 2–21 (1997)
18. Narasimhan, N., Teica, E., Radhakrishnan, R., Govindarajan, S., Vemuri, R.: Theorem proving guided development of formal assertions in a resource-constrained scheduler for high-level synthesis. Form. Methods Syst. Des. 19(3), 237–273 (2001)
19. Necula, G.C.: Translation validation for an optimizing compiler. In: PLDI 2000 (June 2000)
20. Pnueli, A., Siegel, M., Singerman, E.: Translation Validation. In: Steffen, B. (ed.) ETAPS 1998 and TACAS 1998. LNCS, vol. 1384, pp. 151–166. Springer, Heidelberg (1998)
21. Rinard, M., Marinov, D.: Credible compilation. In: Proceedings of the FLoC Workshop Run-Time Result Verification (July 1999)
22. Walker, R., Camposano, R.: A Survey of High-Level Synthesis Systems. Kluwer Academic Publishers, Boston, MA, USA (1991)
23. Zuck, L., Pnueli, A., Fang, Y., Goldberg, B.: VOC: A methodology for the translation validation of optimizing compilers. Journal of Universal Computer Science 9(3), 223–247 (2003)

An Algebraic Approach for Proving Data Correctness in Arithmetic Data Paths

Oliver Wienand, Markus Wedler, Dominik Stoffel, Wolfgang Kunz, and Gert-Martin Greuel

University of Kaiserslautern/Germany
wedler@eit.uni-kl.de

Abstract. This paper proposes a new approach for proving arithmetic correctness of data paths in System-on-Chip modules. It complements existing techniques which are, for reasons of complexity, restricted to verifying only the control behavior. The circuit is modeled at the arithmetic bit level (ABL) so that our approach is well adapted to current industrial design styles for high performance data paths. Normalization at the ABL is combined with the techniques of computer algebra. We compute normal forms with respect to Gröbner bases over rings $\mathbb{Z}/\langle 2^n \rangle$. Our approach proves tractable for industrial data path designs where standard property checking techniques fail.

1 Introduction

Property checking has become well-established in modern design flows for Systems-on-Chip (SoCs). Its main application domain is ensuring the correctness of the individual SoC blocks. This does not only lead to high quality IP (intellectual property) modules but also reduces the costs for system integration and chip-level simulation. Given IP modules of provably high quality, chip-level simulation may concentrate on true system-level aspects and is relieved from hunting bugs in local modules. Therefore, in recent years, a lot of effort has been made to develop sophisticated methodologies and tools for formal module verification based on property checking. Today, formal property checking can handle almost all types of modules that can be found in today's SoCs. Nonetheless, a few pathological cases remain that sometimes limit the application of property checking in industrial practice. In particular, data paths are often a challenge for formal techniques, especially, if not only the correctness of the control flow but also correctness of the data is to be proved.

For complex arithmetic data paths simulation is, therefore, still prevailing in industrial verification environments. This is due to the inability of standard proving procedures based on satisfiability solving (SAT) or binary decision diagrams (BDDs) to handle arithmetic functions. Especially multiplication — as it is part of nearly all data paths for signal processing applications — has remained a severe problem for standard tools. This deficiency has motivated the research community to investigate alternative proof methods with focus on arithmetic.

A. Gupta and S. Malik (Eds.): CAV 2008, LNCS 5123, pp. 473–486, 2008.

In case the validity of a property can be proven without consideration of the exact functionality of the data path, abstraction and refinement techniques have shown superiority over pure Boolean SAT techniques. A survey on these techniques can be found in [1]. However, for properties that depend on the exact functionality of the datapath a suitable abstraction is not likely to be found.

Another direction of research investigates SAT-modulo-theory (SMT) solvers. These solvers combine a SAT solver with specialized solvers for certain well-selected theories. An example for such a theory is the theory of equality with uninterpreted functions used in UCLID [2]. In case the problem at hand really depends on the exact functionality of a datapath, as is typically the case, most SMT solvers resort to bit blasting [1] for the corresponding problem parts. In this case SMT solvers show the same performance limitations as pure SAT solvers as soon as these datapaths include multiplication operations. The decision problems in RTL-property checking could be expressed as SAT problems for formulas of the quantifier free logic (QF-BV) and in principle be solved using solvers such as Yices [3], MathSat[4], Z3 [5] or Spear [6]. For sophisticated datapath implementations involving multiplication, however, our experience is that the problems are still beyond the capacity of such solvers.

Recently, techniques from symbolic computer algebra have entered the verification arena. The authors of [7] present a procedure to determine whether a multivariate polynomial with fixed word length operands is vanishing. By this means a comparison of polynomial representations for bit vector functions is feasible. This procedure is extended towards multiple word length operands in [8,9]. However, both approaches require a word-level representation of the datapaths under comparison. This limits their applicability in RTL property checking. Due to performance and area requirements RTL designers typically design specialized arithmetic components. These components are often designed using bit level arithmetic circuitry to build addition trees and partial products. The smallest entities in an addition tree can be described using half and full adders in general. An approach for verification of such bit level implementations using Gröbner basis theory over fields is reported in [10]. This approach requires polynomial specifications for every building block in the hierarchy of the arithmetic circuit design. After proving that a block, e.g., a CSA adder, fulfills its local specification, the polynomial representation is used to verify the block in the next level of the hierarchy. However, as the correctness proof includes a range check the intermediate results at the block boundary are required to have sufficient bit width to represent every possible result. For designs implementing integer arithmetic with fixed bit width this is often not the case.

A heuristic approach to exploit the availability of arithmetic bit level (ABL) information in RTL designs has been reported in [11]. In this work a data structure called ABL description for representation of addition networks and bitwise multiplication is transformed into a reduced normal form. By canceling out common addends from addition networks in the fanin of a comparator the normalization approach relieves the SAT solver from reasoning in structurally different implementations for the same arithmetic function.

In order to overcome the limitations of [10] we use computer algebra algorithms for rings $\mathbb{Z}/\left\langle 2^N \right\rangle$ to solve decision problems at the arithmetic bit level. This extends the normalization approach of [11] with a clean and well-understood mathematical foundation. We show that an ABL description [11] can directly be transformed into a set of equivalent *variety subset problems*. We exploit the observation that under certain monomial orderings the set G of polynomials generated from the ABL components forms a Gröbner basis of the ideal $I = \langle G \rangle$ generated by these polynomials with special properties. This allows to solve the variety subset problem and hence decide problems at the arithmetic bit level.

The remainder of the paper is organized as follows: Section 2 briefly reviews the notion of an ABL description and describes how such a description can be generated given a design under verification and a property. Section 3 details the mathematical modeling for decision problems at the ABL. The proposed techniques are evaluated by experiments summarized in Section 4. Finally, Section 5 concludes the paper.

2 ABL Description

Arithmetic bit level (ABL) descriptions as introduced in [11] have proven to be useful for modeling the arithmetic parts of a property checking instance. In this section we briefly review this notion as far as it is required for this paper. We use the following notations:

- For $a \in \mathbb{Z}$, $b > 0$ the remainder, $a \mod b$, of the integer division a/b denotes the smallest $k \geq 0$ with $k = a - mb$ for some $m \in \mathbb{Z}$.
- For $n > 0$ and $a \in \mathbb{Z}$ the uniquely determined bit vector $(a_{n-1}, \ldots, a_0)$ with $a \mod 2^n = \sum_{i=0}^{n-1} 2^i a_i$ is denoted as $\langle a, n \rangle = (a_{n-1}, \ldots, a_0)$, i.e., $\langle a, n \rangle$ is the n-bit binary unsigned integer representation of a.
- $\mathbb{B} = \{0, 1\} \subset \mathbb{Z}$ denotes the Boolean space.

The combinatorial transition function of an RTL circuit design is usually modeled by a directed acyclic graph where the vertices are labeled with bit vector functions. It is common practice to translate verification problems for RTL circuits into such bit vector netlists with a single output indicating whether, e.g., a certain property holds for a design. For the arithmetic problem parts we extract an *ABL description* from this netlist. This description again is a directed acyclic graph where the vertices can be of type "partial product generator", "addition network" or "comparator". These vertex types are defined as follows:

Definition 1. *Let $n, m \in \mathbb{N}$, $w : \{0, \ldots, m\} \to \mathbb{Z}$ and $c \in \mathbb{Z}$. The bit vector function $r : \mathbb{B}^m \to \mathbb{B}^n$ with*

$$r(x_1, \ldots, x_m) = \langle (c + \sum_{i=0}^{m-1} w(i) \cdot x_i), n \rangle$$

is called addition network with addend set $A = \{x_1, \ldots, x_m\}$. n *is called* result width, c *is called* constant offset *of the network and* $w(i)$ *is called* weight of the addend x_i.

The bit vector function $pp : \mathbb{B}^n \times \mathbb{B}^m \to \mathbb{B}^{nm}$ *with*

$$pp(x_1, \ldots, x_n, y_1, \ldots, y_m) = (x_i \cdot y_j | i = 1, \ldots, n \text{ and } j = 1, \ldots, m)$$

is called partial product generator.

Every bit vector function $cmp : \mathbb{B}^n \times \mathbb{B}^n \to \mathbb{B}$ *with*

$$cmp(< x + k, n >, < y + k, n >) = cmp(< x, n >, < y, n >)$$

for all $k \in \mathbb{Z}$ *is called* comparator.

Partial product generators model bit-wise multiplication and comparators model comparison of bit vectors. Bit level addition units like *half adders (HA)* or *full adders (FA)* are modeled as addition networks. By construction, addition networks can be used to model any addition circuit ranging from *HAs* and *FAs* up to the entire addition scheme of a multiplier or a multiply-accumulate unit. This is true for both signed and unsigned arithmetic.

Example 1. An signed 2×2-bit multiplier can be modeled with the partial product generator

$$pp(x_0, x_1, y_0, y_1) = (x_0 y_0, x_1 y_0, x_0 y_1, x_1 y_1)$$

and the addition network

$$r(p_{0,0}, p_{1,0}, p_{0,1}, p_{1,1}) = \langle p_{0,0} - 2p_{1,0} - 2p_{0,1} + 4p_{1,1}, 4 \rangle$$

A simple bit-level implementation of this multiplier may implement the addition network using a fulladder and two halfadders. They can be modeled by the addition networks $fa(a, b, c) =< a + b + c, 2 >$ and $ha(a, b) =< a + b, 2 >$, respectively.

For reasons of space we omit the formal definition of *ABL descriptions* as a DAG. The interested reader is referred to [11]. Basically, the nodes of the graph are labelled with their vertex type and the edges describe the interconnections between them. Here, we explain this concept by continuing Example 1.

Example 2. The ABL description for the comparison of the bit level multiplier implementation discussed in Example 1 against its word level specification is depicted in Figure 1.

The vertices of this graph are labeled with the bit vector function defined in the previous example. The edges (v, v') are labeled with bit vectors that propagate the result of v to the inputs of v'. In other words, the variables are defined by the following equations:

- $(p_0, p_1, p_2, p_3) = pp(x_0, x_1, y_0, y_1) = (x_0 y_0, x_1 y_0, x_0 y_1, x_1 y_1)$
- $(z_0, z_1, z_2, z_3) = r(p_0, p_1, p_2, p_4) = \langle p_0 - 2p_1 - 2p_2 + 4p_3, 4 \rangle$

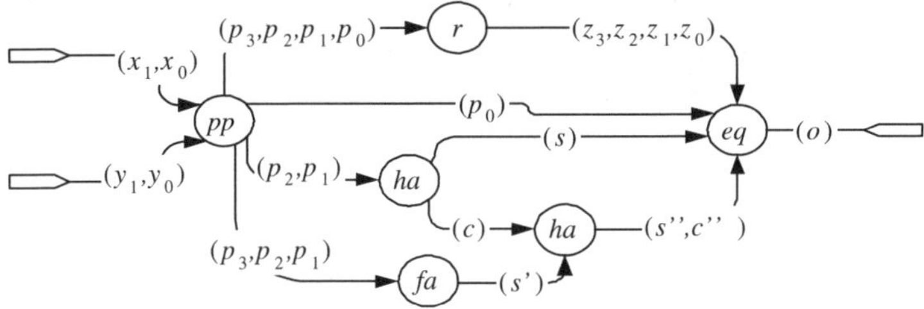

Fig. 1. ABL description for Example 1

- $(s, c) = ha(p_1, p_2) = \langle p_1 + p_2, 2\rangle$
- $(s', c') = fa(p_1, p_2, p_3) = \langle p_1 + p_2 + p_3, 2\rangle$
- $(s'', c'') = ha(c, s') = \langle c + s', 2\rangle$
- $(o) = eq((z_0, z_1, z_2, z_3), (p_0, s, s'', c'')) = ((z_0, z_1, z_2, z_3) == (p_0, s, s'', c''))$

This example illustrates that ABL descriptions may contain structurally dissimilar representations for one and the same arithmetic function. To simplify the comparison of such representations a heuristic ad-hoc algorithm called *ABL normalization* was proposed in [11]. This algorithm performs a series of local equivalence transformations on the ABL description that are based on the commutative and distributive laws.

However, in the next section we will describe how to obtain a variety subset problem that is equivalent to the decision problem resulting from the comparison of such ABL representations. This paves the way for the application of generic computer algebra algorithms for which efficient implementations are available.

3 Mathematical Background

Application of computer algebra techniques to ABL verification problems requires ABL components to be modeled by polynomials over a unique ring. Due to the operation mod used to specify ABL components, the ring $\mathbb{Z}/2^n$ seems to be the natural choice. However, the mapping of ABL descriptions on sets of polynomials $G \subset \mathbb{Z}/2^n[X]$ over such a ring is not trivial and will be detailed in this section. The key observation is that the constructed set G is a Gröbner basis of the generated ideal $I = \langle G\rangle$. This makes the proposed approach computational feasible.

We start with a set of equations $G_j, j = 1, \ldots, m$ given by polynomials $f_j \in \mathbb{Z}[X]$, X a finite set of variables, which are of the form

$$G_j : \sum_{i=0}^{n_j-1} 2^i r_i^{(j)} = f_j\left(a_1^{(j)}, a_2^{(j)}, \ldots, a_{m_j}^{(j)}\right) \mod 2^{n_j}.$$

For the variables $r_i^{(j)}, a_k^{(l)} \in X$ in this equation we assume $r_i^{(j)} \neq a_k^{(l)}$ for $1 \leq l \leq j$ and all i, k. We call the variables $a_i^{(j)}$ *inputs* and $r_i^{(j)}$ *outputs* of G_j.

Note that the equations G_j can be easily generated from the vertices of an ABL description and that the condition $r_i^{(j)} \neq a_k^{(l)}$ is fulfilled as the ABL description is acyclic by definition. For illustration we give a few examples.

Example 3. The partial products of a non-Booth-encoded $n \times m$ multiplier can be modeled by the polynomial equations

$$G_{i,k} : p_{i,k} = a_i b_k \bmod 2, (k = 0, \ldots, n-1, i = 0, \ldots, m-1)$$

Example 4. A full adder with inputs a_0, a_1, a_2 and outputs s and c for sum and carry is modeled by the equation

$$G_{\text{FA}} : 2c + s = a_0 + a_1 + a_2 \bmod 4$$

Example 5. A k-bit adder with inputs $a = (a_0, \ldots, a_{k-1})$ and $b = (b_i)$ and result $r = (r_i)$ is modeled by

$$G_{\text{adder}} : \sum_{i=0}^{k-1} 2^i r_i = \sum_{i=0}^{k-1} 2^i (a_i + b_i) \bmod 2^k$$

For every proof goal, we obtain an additional polynomial g depending on a subset of variables $\{a_1, \ldots, a_t\} \subset X$ and need to check whether

$$g(a_1, \ldots, a_t) = 0 \mod 2^n$$

for all solutions of the set of equations $\{G_j\}$.

Example 6. A k-bit comparator of operands a and b is modeled by the polynomial

$$g = \sum_{i=0}^{k-1} 2^i (a_i - b_i)$$

Denote the set of all solutions to $\{G_j\}$ as $V(\{G_j\})$. Analogously let $V(g)$ be the set of all roots of g. Usually the equations G_j and the polynomial g are given $\bmod 2^k$ for different k. We apply a number of transformations to create an equivalent *variety subset problem* $V(\{h_i\}) \subset V(g)$ where h_i and g are polynomials over a single ring $\mathbb{Z}/2^N$ with appropriate N, which is necessary in order to apply computer algebra. To solve the problem we construct a Gröbner basis and then use normal form computations with respect to this basis.

For the reader's convenience we recall some basic facts about Gröbner basis theory (cf. [12,13]). We need a *monomial ordering* $<$, i.e., a well ordering on the set of monomials s.t. multiplication with a monomial respects the ordering. Here a *monomial* is a power product of variables and a term is the product of a monomial with a coefficient, i.e., an element of the ring $\mathbb{Z}/2^N$. Any polynomial $f \neq 0$ can be written as a finite sum of terms, $f = c_1 m_1 + \cdots + c_r m_r$ with c_i coefficients $\neq 0$ and m_i monomials s.t. $m_1 > m_2 > \cdots > m_r$. The largest term plays a special role and we call $\text{LM}(f) := m_1$ resp. $\text{LC}(f) := c_1$ resp. $\text{LT}(f) := c_1 m_1$ the *leading monomial* resp. the *leading coefficient* resp. the *leading term* of f.

Let $G \subset \mathbb{Z}/2^N[X]$ be a finite set of polynomials and $f \in \mathbb{Z}/2^N[X]$. If cm is any (non-zero) term of f and if cm is divisible by the leading term of an element $h \in G$ we say that f is reducible to $f' := f - (cm/LT(h)) \cdot h$ and write $f \underset{h}{\rightarrow} f'$. The transitive and reflexive closure of the relation $\underset{h}{\rightarrow}$ is denoted by $\overset{*}{\underset{G}{\rightarrow}}$. If $f \overset{*}{\underset{G}{\rightarrow}} g$ and if g is not reducible by any h of G we call g a *normal form* of f w.r.t. G. This notion is, however, only useful if G is a Gröbner basis. In order to define a Gröbner basis we need the ideal $I = \langle G \rangle := \{\sum_{h\in G} f_h h | f_h \in \mathbb{Z}/2^N[X]\}$ generated by an arbitrary set G of polynomials. Note that for the set of solutions we have $V(I) = V(G)$ for any set of generators G. A set of generators G is called a (strong) Gröbner basis (of I) if $f \overset{*}{\underset{G}{\rightarrow}} 0$ for all $f \in I$. If G is a Gröbner basis then the normal form of any element $g \in \mathbb{Z}/2^N[X]$ is essentially unique and equal to 0 if and only if $f \in \langle G \rangle$.

3.1 Problem Formulation over a Single Ring

Instead of directly converting the equations G_j into a set of polynomials over a single ring, we generate some additional equations. These equations are redundant in the sense that they can be derived from the original equations G_j. However, they will play an important role for the efficiency of the solution techniques described in Section 3.2. More precisely, these equations ensure that the polynomial system generated from them is a Gröbner basis of the corresponding ideal. This will be discussed later.

For every G_j we generate n_j equations

$$G_j^{(t)} : \sum_{i=0}^{t-1} 2^i r_i^{(j)} = f_j^{(t)} \left(a_1^{(j)}, a_2^{(j)}, \ldots, a_{m_j}^{(j)} \right) \mod 2^t$$

with $t = 1, \ldots, n_j$ and with $f_j^{(t)} = f_j \mod 2^t$ being the minimal polynomial [14] representing the same polynomial function $(\mathbb{Z}/2^t)^{m_j} \rightarrow \mathbb{Z}/2^t$ as f_j.

Obviously, every solution of the G_j is also a solution of the system $\{G_j^{(t)} \mid t = 1, \ldots, n_j\}$ and vice versa.

Let S be the set of variables (signals) occurring in g saturated with respect to the property that if $r_{t-1}^{(j)} \in S$ then all variables of $G_j^{(t)}$ are also in S. For the further course of action only the equations $G_j^{(t)}$ with $r_{t-1}^{(j)} \in S$ are relevant. The solution set for the variables in S does not change when omitting the other equations. Note that this corresponds to a cone-of-influence reduction on the netlist of a circuit.

Example 7. Suppose the n bit final adder of a multiply/accumulate unit is reused for computation of an m-bit addition ($m < n$). In a property checking instance for this addition only the lowermost m bits of the adder take influence on the arithmetic result. By the above construction we only instantiate the equations

$$G^{(t)}_{\text{adder}} : \sum_{i=0}^{t-1} 2^i r_i = \sum_{i=0}^{t-1} 2^i (a_i + b_i) \bmod 2^t$$

for $t < m$.

So far the equations $G^{(t)}_j$ use the operation $\bmod 2^t$ for different t, that is, we work over different rings $\mathbb{Z}/2^t$ and none is contained in the other (we have only surjections of rings $\mathbb{Z} \twoheadrightarrow \mathbb{Z}/2^{t'} \twoheadrightarrow \mathbb{Z}/2^t$ if $t' \geq t$). In order to apply Gröbner basis techniques to our problem we need to generate a set of polynomials over a single ring.

Let $N := n + \max\{n_j \mid j = 1, \ldots, m\}$ with n_k, n, m as above. We want to transform every equation into an element of the polynomial ring over $\mathbb{Z}/2^N$. To achieve this, we introduce new variables $s^{(j)}_t$ (called *slack variables*) and consider the polynomials

$$\widetilde{G}^{(t)}_j := \sum_{i=0}^{t-1} 2^i r^{(j)}_i - f^{(t)}_j \left(a^{(j)}_1, a^{(j)}_2, \ldots, a^{(j)}_{m_j}\right) - 2^t s^{(j)}_t.$$

The set of common roots for the $\widetilde{G}^{(t)}_j$ projected on the variables in S corresponds to $V(\{G_j\})$. We can omit some of the extra variables $s^{(j)}_t$ if we know that $0 \leq f^{(t)}_j \leq 2^t - 1$ holds over $\mathbb{Z}$. If this condition cannot be guaranteed and we need to know the exact value of $s^{(j)}_t$ during the computation we can replace $s^{(j)}_t$ by a polynomial in the variables $a^{(j)}_1, a^{(j)}_2, \ldots, a^{(j)}_{m_j}$, i.e., a subset of the inputs of G_j. For example, the polynomial modeling a half adder $r_0 - a_0 - a_1 + 2s$ results in the polynomial $s = a_0 a_1$ for the slack variable. However, often it is better to introduce the slack variables because, in general, the polynomials for the slack variables will be very large even for small polynomials $f^{(t)}_j$.

Let $G = \{\widetilde{G}^{(t)}_j \mid j = 1, \ldots, m \text{ and } t = 1, \ldots, n_j\}$ and $I = \langle G \rangle$ be the ideal generated by this set. Using the language of computer algebra our decision problem can be formulated by the following question:

Is $V(I) \subset V(2^{N-n} g)$, where $V(I)$ and $V(f)$ denote the set of all common roots (in $(\mathbb{Z}/2^N)^k$, where k is the number of variables) of the polynomials in I and the set of roots of the polynomial $2^{N-n} g$, respectively?

In the next section we will detail how to efficiently solve this problem.

3.2 Solving Decision Problems at the ABL

The following proposition turns out to be the key for an effective solution of the presented problem.

Proposition 2. *The set $G = \{\widetilde{G}^{(t)}_j\}$ is a Gröbner basis with respect to any monomial order refining the following partial order*

$$r^{(j)}_i > \textit{every monomial in the variables } a^{(j)}_k, s^{(j)}_t, r^{(j)}_l$$

for all i, k, t, j and $l < i$.

Proof. Let $<$ be a monomial order as required in the statement. We need to show that it is not possible to generate a polynomial from the polynomials in G with a leading term that is not divisible by any leading term of the polynomials in G. It is sufficient to show (cf. [15], Theorem 30)

(1) For any two polynomials $f, g \in G$ the normal form of

$$\frac{\operatorname{lcm}(\operatorname{LT}(f), \operatorname{LT}(g))}{\operatorname{LT}(f)} f - \frac{\operatorname{lcm}(\operatorname{LT}(f), \operatorname{LT}(g))}{\operatorname{LT}(g)} g$$

with respect to G is zero.

(2) For any $f \in G$ the normal form of $\frac{2^N}{\operatorname{LC}(f)} f$ is zero.

A slight generalization of the product criterion (cf. [15], Lemma 35) states that (1) is fulfilled, as our polynomials have different variables in their leading terms and these variables do not occur in any other term of the corresponding polynomials. Now let $f = G_j^{(t)}$. We obtain

$$\begin{aligned}
\operatorname{LT}\left(\frac{2^N}{\operatorname{LC}(f)} f\right) &= \operatorname{LT}\Big(\underbrace{2^{N-t+1} G_j^{(t)}}_{\|} \Big) \\
&= \operatorname{LT}\Big(\overbrace{2^{N-t+1} G_j^{(t-1)}} \Big) \\
&= 2^{N-t+1} \operatorname{LT}\left(G_j^{(t-1)}\right)
\end{aligned}$$

as $2^{N-t+1} \cdot 2^{t-1} r_{t-1}^{(j)} = 0 = 2^{N-t+1} \cdot 2^{t-1} s_{t-1}^{(j)}$ and $2^{N-t+1} \cdot 2^{t-2} r_{t-2}^{(j)} \neq 0$, and since the polynomials $f_j^{(t)}$ appearing in $\widetilde{G}_j^{(t)}$ are chosen minimal. In the first step of the normal form algorithm we will select $G_j^{(t-1)}$ and this reduces $2^{N-t+1} G_j^{(t)}$ to zero. This shows (2).

By Lemma 3 we prove that normal form computation can be used as an effective solution procedure for our problem at hand.

Lemma 3. *Let G be a Gröbner basis of an ideal $I \subset \mathbb{Z}/2^N[\mathbf{x}]$, $\mathbf{x} = (\mathbf{x}', \mathbf{x}'')$, and g a polynomial such that h, the normal form of g w.r.t. G, is in $\mathbb{Z}/2^N[\mathbf{x}']$. Assume that for all $\mathbf{x}'$ there exist $\mathbf{x}''$ with $f(\mathbf{x}', \mathbf{x}'') = 0$ for all $f \in G$. Then h defines the zero function if and only if $V(G) \subset V(g)$.*

Proof. If h defines a constant zero function the set $V(h) = V(g)$ contains all points and therefore $V(G) \subset V(g)$ is trivial. Assume that for the variables $\mathbf{x}'$ of h a valuation exists such that h is not zero. By assumption we can extend this valuation to a valuation on all variables such that $g(\mathbf{x}) = 0, g \in G$. It follows $V(G) \not\subset V(f)$.

Let $g \in \mathbb{Z}/2^n[\mathbf{x}]$ and h be the normal form of $2^{N-n} g$ with respect to G, which can be computed [15] by Algorithm 1. Since we are only interested in the function of h on $V(I)$ we can always replace portions of h by equivalent polynomials with respect to $V(I)$. In particular, we can replace every slack variable in the normal form by a polynomial expression in the inputs of the corresponding equation G_j. Therefore we may assume that h does not contain any slack variables. Furthermore, the output variables of the equations G_j do not occur in h as otherwise h would be reducible by some of the generated sub-identities $G_j^{(t)}$, hence h satisfies the assumptions of Lemma 3.

This guarantees that the variables present in h are inputs to the ABL description. Every valuation of these variables can be extended to a consistent valuation for the signals of the ABL. Further we can effectively decide whether h defines the zero function for all rings $\mathbb{Z}/m$ (cf. [14]) and therefore decide the ABL problem by Lemma 3.

As already noted in Section 3.1 it is not always efficient to replace all remaining slack variables by polynomial expressions in terms of the input variables of the corresponding equations. Therefore we use special procedures for the practical computations, which we do not detail here.

Require: *f a polynomial, G a finite set of polynomials,*
$>$ a monomial ordering
Ensure: *A normal form of f*
while $f \neq 0$ *and* $\emptyset \neq G' = \{g \in G \ : \ \mathrm{LT}(g) \mid \mathrm{LT}(f)\}$
do
Select $g \in G'$
Let $\mathrm{LT}(f) = m \cdot \mathrm{LT}(g)$ *with* $m \cdot \mathrm{LC}(g) \neq 0$
$f := f - m \cdot g_i$
end while
return f

Algorithm 1. Normal form algorithm

4 Experimental Results

In order to evaluate the techniques presented in the previous sections we conducted a series of experiments. Except for one experiment explicitly indicated in the sequel, all experiments were carried on a machine running Suse Linux 10.3 on a Intel Core 2 Duo E6400 with 8 GB RAM.

The algorithms presented in Section 3 have been implemented within the framework of the general purpose computer algebra system Singular [16]. We used the industrial formal property checker Onespin 360 MV [17] to generate bit vector netlists for the considered verification problems. From these bit-vector netlists we extracted an arithmetic bit level description for the arithmetic parts of the decision problem and dumped out the resulting ABL description. The resulting problem file is used to generate the variety subset problem that is handed over to Singular in order to find a solution.

As a first step of the evaluation we used a number of parameterized benchmarks to evaluate the scalability of the proposed approach with respect to the bit-width of the datapath under verification. The benchmark suite consists of two instances (distrib and commute) for word-level implementations of the functions $ab + ac$ and $(ab)c$ where commutative and distributive laws have been applied to the word level operands, a bit level implementation of an unsigned multiplier with Booth-encoded partial products (mult_ub) and a sequential implementation for the multiplication of four values with a single multiplier (shared).

We compare the performance in terms of run-time of our solution based on Singular against the normalization approach of [11], a SAT-based decision procedure based on bit blasting, and the SMT solver Spear v.2.0 for the theory of fixed-size bit-vector functions (QF-BV). Note that an earlier version of Spear showed the best performance in this category on the 2007 SMT competition.

Table 1. CPU-times(s) of scalability experiments

Instance	Bit-Width	Normalizer	SAT	SMT	Singular
distrib	4	0,01	0,28	0,40	0,69
distrib	8	0,03	> 3600s	> 3600s	0,66
distrib	16	0,10	> 3600s	> 3600s	0,97
distrib	32	0,81	> 3600s	> 3600s	2,19
distrib	64	14,33	> 3600s	> 3600s	11,30
commute	4	0,02	0,55	1,01	0,69
commute	8	0,08	> 3600s	> 3600s	0,67
commute	16	1,40	> 3600s	> 3600s	1,09
commute	32	57,17	> 3600s	> 3600s	3,56
commute	64	2794,67	> 3600s	> 3600s	26,03
mult_ub	4	0,02	0,02	0,15	0,66
mult_ub	8	0,13	41,53	> 3600s	0,96
mult_ub	16	2,21	> 3600s	> 3600s	3,87
mult_ub	32	53,55	> 3600s	> 3600s	79,04
mult_ub	64	1136,14	> 3600s	> 3600s	> 8 GB
shared	4	0,04	2,83	16,78	0,97
shared	8	0,46	> 3600s	> 3600s	0,64
shared	16	39,79	> 3600s	> 3600s	1,00
shared	32	2707,72	> 3600s	> 3600s	20,25

Table 1 summarizes the results of these experiments. The table is organized as follows. Columns one and two contain instance and operand bit-width of the datapath. The remaining columns show the CPU times required by the particular tool to prove the instance. In case the memory limit or timeout limit was reached this is indicated by "> 8 GB" and "> 3600", respectively.

In order to evaluate the performance of Singular with respect to other computer algebra systems we also report results for solving the generated variety subset problems with the industrial computer algebra tool Magma [18]. However, due to license restrictions, these results were obtained using another machine, namely an AMD Dual Opteron 2.2 GHz with 16 GB RAM running Linux. We re-ran the Singular problems on this machine in order to allow for comparison of

Table 2. CPU-times(s) of scalability experiments

Instance	Bit-Width	Singular	Magma
distrib	16	1,08	2,33
distrib	32	2,70	15,61
distrib	64	14,53	> 16 GB
commute	16	1,35	5,53
commute	32	4,71	46,07
commute	64	38,96	> 16 GB
mult_ub	4	0,56	> 3600s
mult_ub	16	4,07	> 3600s
mult_ub	32	85,77	> 3600s
shared	4	0,46	1,08
shared	8	0,66	1,35
shared	16	1,37	3,09
shared	32	32,27	108,01

the run times. For the comparison we also increased the memory limit to 16GB. Table 2 summarizes the results for this comparison.

The presented results of the scalability experiments indicate that the proposed modeling and the proposed algorithms are adequate to solve verification problems with industrial impact. To demonstrate this we investigated a property suite originating from the verification of Infineon's Tricore 2 processor. The processor has advanced DSP features including a sophisticated integer pipeline that provides a large variety of multiply and multiply/accumulate instructions. The properties in the investigated property suite verify that every variant of these instructions causes the integer pipeline of the processor to deliver the expected arithmetic result according to the architectural manual. In order to obtain a high degree of resource sharing large portions of the datapath have been implemented at the arithmetic bit level and sophisticated control logic is used for configuration according to the executed instructions.

We used the techniques of [11] to generate the decision problems at the arithmetic bit level. All the resulting decision problems could be solved with Singular when modeled by polynomials as presented in this paper. Table 3 shows the results for a representative subset of the problem instances derived from the Tricore 2 property suite. It is organized as follows. The first column shows the commitment of the property specifying the arithmetic result of the integer instruction under verification. Columns two and three show the run-time of the normalization approach and the corresponding Singular run-time. Unless explicitly indicated all operations are considered as signed operations on the specified bit-vectors.

In essence, all our experiments show that the presented approach outperforms the ad-hoc normalization approach in terms of CPU time. Moreover, algorithms and modeling rely on a well-understood mathematical foundation which opens ample opportunities for further extensions of this framework.

However, the use of a generic computer algebra system as Singular for solving the normalization problems is paid with a price in terms of memory consumption.

Table 3. Results for selected Tricore 2 properties

Datapath result	Normalizer	Singular
res[31:0]=op3[31:0]+(op1[31:0]*op2[31:0]);	49,94	4,42
res[31:0]=op3[31:0]+(op1[15:0]*op2[31:16]<<1); res[63:32]=op3[63:32]+(op1[31:16]*op2[15:0]<<1);	39,72	2,28
res[31:0]=op3[31:0]+(op1[31:16]*op2[31:16]<<1);	18,39	2,47
res[15:0]=rnd16(op3[31:0]+(7FFFFFFF)); res[31:16]=rnd16(op3[63:32] +(op1[31:16]*op2[15:0]<<1));	19,90	2,46
res[63:0]=op3[63:0]+(op1[31:16]*op2[15:0]<<16) -(op1[31:16]*op2[31:16]<<16)	31,04	8,77
res[63:0]=op3[31:0]-(op1[31:0]*op2[31:0]);	55,19	20,01
res[63:0]=op3[63:0]-(op1[31:16]*op2[15:0]<<16) -(op1[15:0]*op2[15:0]<<16)	27,18	9,64
res[63:0]=op1[31:0]*op2[31:0]; (unsigned)	57,33	14,73
res[31:16]=rnd16(op1[31:16]*op2[31:16]);	17,41	2,21

Except for some of the problems where the ABL description is generated from word-level problems Singular typically requires 3–8 GB of memory. This is caused by the data structures used inside Singular to represent polynomials.

These data structures are not optimized with respect to the characteristics of the problems considered here. Compared to problems typically considered in computer algebra, we consider a large number of variables, and many polynomials. On the other hand the individual polynomials have low degree and only use a small fraction of the variables. With application-specific implementations of the employed algorithms such as the normal form computation a great improvement of the memory efficiency can be obtained easily.

5 Conclusion and Future Work

Decision problems at the arithmetic bit level have been modeled using polynomials over rings $\mathbb{Z}/\langle 2^n \rangle$. It has been proven that the generated sets of polynomials form a Gröbner basis with respect to certain monomial orderings that can easily be determined using the topological ordering of design signals. This allows for utilization of the normal form algorithm to efficiently solve a variety subset problem that is equivalent to the original decision problem.

By this means we provide a solid mathematical foundation to the ad-hoc technique of arithmetic bit level normalization. The developed techniques have proven to be applicable to verification problems of industrial size.

As the datastructures for polynomial representation of in-the-box computer algebra systems do not exploit the typical characteristics of the generated polynomial sets, we are working on a specialized implementation of the employed algorithms that will dramatically reduce the memory consumption.

References

1. Kroening, D., Seshia, S.A.: Formal verification at higher levels of abstraction. In: Proc. International Conference on Computer-Aided Design (ICCAD) (2007)
2. Seshia, S.A., Lahiri, S.K., Bryant, R.E.: A hybrid SAT-based decision procedure for separation logic with uninterpreted functions. In: Proc. International Design Automation Conference (2003)
3. Dutertre, B., de Moura, L.: A Fast Linear-Arithmetic Solver for DPLL(T). In: Ball, T., Jones, R.B. (eds.) CAV 2006. LNCS, vol. 4144, pp. 81–94. Springer, Heidelberg (2006)
4. Bruttomesso, R., Cimatti, A., Franzén, A., Griggio, A., Hanna, Z., Nadel, A., Palti, A., Sebastiani, R.: A lazy and layered smt($\{BV\}$) solver for hard industrial verification problems. In: Damm, W., Hermanns, H. (eds.) CAV 2007. LNCS, vol. 4590, pp. 547–560. Springer, Heidelberg (2007)
5. de Moura, L.M., Bjoerner, N.: Efficient e-matching for smt solvers. In: Pfenning, F. (ed.) CADE 2007. LNCS (LNAI), vol. 4603, pp. 183–198. Springer, Heidelberg (2007)
6. Spear, `http://www.cs.ubc.ca/~babic/index.htm`
7. Shekhar, N., Kalla, P., Enescu, F., Gopalakrishnan, S.: Equivalence verification of polynomial datapaths with fixed-size bit-vectors using finite ring algebra. In: Proc. International Conference on Computer-Aided Design (ICCAD) (2005)
8. Shekhar, N., Kalla, P., Enescu, F.: Equivalence verification of arithmetic datapath with multiple word-length operands. In: Proc. International Conference on Design, Automation and Test in Europe (DATE) (2006)
9. Shekhar, N., Kalla, P., Enescu, F.: Equivalence verification of polynomial datapaths using ideal membership testing. IEEE Transactions on Computer-Aided Design 26(7), 1320–1330 (2007)
10. Watanabe, Y., Homma, N., Aoki, T., Higuchi, T.: Application of symbolic computer algebra to arithmetic circuit verification. In: Proc. International Conference on Computer Design (ICCD), pp. 25–32 (October 2007)
11. Wedler, M., Stoffel, D., Brinkmann, R., Kunz, W.: A normalization method for arithmetic data-path verification. IEEE Transactions on Computer-Aided Design 26(11), 1909–1922 (2007)
12. Adams, W., Loustaunau, P.: An introduction to Gröbner bases. (Graduate studies in mathematics) AMS (2003)
13. Greuel, G.M., Pfister, G.: A SINGULAR Introduction to Commutative Algebra, 2nd edn., 705 pages. Springer, Heidelberg (2007)
14. Wienand, O.: The Gröbner basis of the ideal of vanishing polynomials (2007) arXiv: arXiv:0709.2978v1 [math.AC]
15. Brickenstein, M., Dreyer, A., Greuel, G.M., Wedler, M., Wienand, O.: New developments in the theory of Gröbner bases and applications to formal verification. Journal of Pure and Applied Algebra (accepted for publication)
16. Greuel, G.M., Pfister, G., Schönemann, H.: Singular 3.0.4. - A Computer Algebra System for Polynomial Computations. Centre for Computer Algebra, University of Kaiserslautern (2007), `http://www.singular.uni-kl.de`
17. Onespin Solutions GmbH Munich, Germany, `www.onespin.com`
18. Bosma, W., Cannon, J., Playoust, C.: The MAGMA algebra system I: the user language. J. Symb. Comput. 24(3-4), 235–265 (1997)

Application of Formal Word-Level Analysis to Constrained Random Simulation

Tool Paper

Hyondeuk Kim[1,2], Hoonsang Jin[2], Kavita Ravi[2], Petr Spacek[2], John Pierce[2], Bob Kurshan[2], and Fabio Somenzi[1]

[1] University of Colorado at Boulder
[2] Cadence Design Systems

Abstract. Constrained random simulation is supported by constraint solvers integrated within simulators. These constraint solvers need to be fast and memory efficient to maintain simulation performance. Binary Decision Diagrams (BDDs) have been successfully applied to represent constraints in this context. However, BDDs are vulnerable to size explosions depending on the constraints they are representing and the number of Boolean variables appearing in them. In this paper, we present a word-level analysis tool *DomRed* to reduce the number of Boolean variables required to represent constraints by reducing the domain of constraint variables. *DomRed* employs static analysis techniques to obtain these reductions. We present experimental results to illustrate the impact of this tool.

1 Introduction

Constrained random simulation is in increasing demand with hardware designers and verification engineers. As the name indicates, it is the simulation of a design under specified constraints. The user is required to capture the behavior of the environment of the design as constraints and the simulation tools simulate the design under these constraints with the aid of constraint solvers embedded in them. Commercial tools, such as Specman, have been popular for providing this capability. To address the need for constrained random simulation, modern hardware description languages (HDL), such as System Verilog, have incorporated constraint specification as part of their syntax.

The overwhelming benefit of constrained random simulation over the traditional writing of testbenches is the automation. Once the constraints are specified, the constraint solver in the simulator enumerates the valid scenarios instead of a human. Further, by specifying weights on the search space, the user can indicate whether the constrained space should be sampled uniformly or specific areas should be focused on.

Given that constraint solving comprises the bulk of constrained random simulation time, the efficiency and performance of constraint solvers is critical. Traditional constraint solving techniques, such as integer linear programming and constraint programming, far lag the performance of simulators. Boolean engines, e.g., BDDs, have been applied quite successfully to this problem[3] by taking advantage of the finite state nature of HDL constraints. More recently, Kitchen and Keuhlmann[2] have provided a

A. Gupta and S. Malik (Eds.): CAV 2008, LNCS 5123, pp. 487–490, 2008.

word-level technique based on Markov-chain Monte Carlo methods. The scalability of this technique to industrial strength designs is yet to be proven.

In our constraint solver, *ValueGen*, we have incorporated both BDD and SAT-based Boolean engines. BDDs provide the advantage of fast generation of uniformly distributed solutions. However, some constraints have very large BDDs that cause memory explosion during simulation. SAT solvers are less vulnerable to size explosion. On the other hand, each solution generation could be exponentially slower than BDDs.

In this paper, we present a word-level pre-processor, *DomRed*, that *ValueGen* applies to the constraints to reduce the size of their representation in the Boolean engines. The pre-processing is a static analysis technique that uses an SMT-like framework. *DomRed* combines a SAT solver and a linear arithmetic solver that handles primarily integer difference logic, with a minor extension to positive and negative coefficient inequalities. The input to the tool is a Boolean combination of linear arithmetic constraints and bit-vector constraints. The output is a set of variables and their reduced domains. The constraints with reduced-domain variables are then passed on to the Boolean engines, resulting in smaller Boolean representations for constraint solving. We present experimental results of applying *DomRed* within *ValueGen* on our simulation testcases.

2 Constraint Solving in Simulation

Constraints are Boolean combinations of linear arithmetic and bit-vector expressions on design variables. The expressiveness of the specified constraints is limited by the HDL being used. For example, a System Verilog constraint is

```
constraint c1 {src_addr >= 0 && src_addr < 65536 &&
               payload_len >= 0 && payload_len < 4096 &&
               dest_addr - src_addr >= 4096 && dest_addr < 65536}
```

Constraint solving is the task of generating values for the design variables that satisfy the constraints. In the above example, `src_addr` $= 512$, `payload_len` $= 1024$, `dest_addr` $= 4608$ is a set of legal values. Our constraint solver, *ValueGen*, is invoked dynamically during simulation i.e., every time the simulator encounters a user call to generate new values for variables appearing in constraints, the simulator calls the constraint solver. Tight integration is required between the two to maintain efficiency.

Constraints are typically written on the inputs of the design and may depend on some internal design signals (*state* variables). During constraint solving, the solver is required to generate values that satisfy both the constraints as well as the states values.

Each set of related HDL constraints, when encountered, is parsed by the simulator, and sent to *ValueGen* through a word-level API along with the state values. Internally, *ValueGen* maintains a applies several optimizations at the word-level, including partitioning based on non-overlapping variable support and constant propagation. Finally, it bit-blasts the word-level constraints and calls the Boolean engines (BDD or SAT) on the Boolean representation.

The optimizations in *ValueGen* result from syntactic and very minor semantic analysis of the constraints. They do not include the ability to deduce that the tightest ranges of dest_addr and src_addr in the above example. *DomRed* addresses exactly this deficiency. It extracts a subset of invariants from semantic analysis of the constraints. If an invariant yields variable bound reductions, then the reduced number of bits are applied to encode the respective variables, the default number of bits are used otherwise.

3 *DomRed*: Technical Details

ValueGen provides *DomRed* with a quantifier-free first order logic formula with linear arithmetic constraints. An *LA* constraint is of the form $a_1x_1 + \ldots + a_nx_n \bowtie c$, where $\bowtie \in \{=, \leq, <, >, \geq, \neq\}$. A *difference* constraint is a special case of an LA constraint whose form is $x_i - x_j \bowtie c$. A positive-(negative-)inequality is another special case of an LA constraint where $\forall i.a_i \geq 0, x_i \geq 0, c \geq 0$ ($\forall i.a_i \leq 0, x_i \leq 0, c \leq 0$). We are working on the extension to bit-vector constriants.

As in the SMT-framework, the first order logic formula is abstracted conservatively into a propositional formula and given to the SAT solver. The SAT solver extracts a set of level-zero assignments, which corresponds to a set of LA constraints. From this set, we gather *difference* constraints, analyze them with the Bellman-Ford algorithm described in [1] and derive reduced bounds for the variable domains if possible. Among the LA constraints left over, positive- and negative-coeffient inequalites may also yield reduced upper (lower) bounds of x_i equal to c/a_i. The remaining LA constraints are conservatively marked as not yielding any domain reduction.

Example: Users commonly declare design inputs as int, meaning a 32-bit finite integer, causing the Boolean representation of the example in Section 2 to contain 96 bits. In applying *DomRed*, the equality constraint is translated into two inequalities in the usual manner. Inequalities are encoded with one bit each in the SAT solver. All these bits appear in the set of level-zero assignments. Since they all correspond to difference constraints, the Bellman-Ford algorithm yields the intervals $[0, 61439]$ for src_addr, $[0, 4095]$ for payload_len and $[4096, 65535]$ for dest_addr. The Boolean encoding will then require 16, 12 and 16 bits respectively, totalling 44 bits in the resulting Boolean expression (more than 2X reduction).

DomRed may also indicate to *ValueGen* that the constraints are infeasible (overconstrained situation) if the SAT solver or the LA solver detects it. This is of great value to *ValueGen* since it can avoid building the Boolean representations altogether.

4 Experimental Results

We integrated our tool *DomRed* into *ValueGen*, which, in turn, is integrated with our simulator. Our benchmark set includes both System-C and System Verilog examples. The System-C examples are smaller in size; 40 out of 68 showed improvements, the rest showed no degradation. The detailed table of results is not presented here for lack of space. The System Verilog examples consist of industrial-strength customer benchmarks. Of the 34 System Verilog examples that we experimented with, 11 showed improvement and are presented in Table1, the remaining 23 showed no degradation.

We use three parameters to measure the performance impact of applying *DomRed*—number of bits, CPU times and memory used. *ValueGen* switches between the BDD and

Table 1. Comparison Table of without and with Bound Reduction

Design	Sim. cycles	# of bits without	with	%	CPU (sec) without	with	%	MEM (Mbytes) without	with	%
design1	5000000	112	101	10	683.0	549.4	**20**	40.8	34.2	**16**
design2	1000000	335	321	4	325.5	319.2	**2**	70.6	53.9	**24**
design3	50000	491	301	39	412.3	333.4	**19**	103.1	93.5	**9**
design4	1000000	54	40	26	180.9	174.1	**4**	37.2	37.8	-2
design5	1000000	64	60	6	86.1	44.0	**49**	33.2	33.6	-1
design6	1000000	64	60	6	75.9	48.1	**37**	33.2	33.7	-1
design7	1000000	16	14	12	340.2	344.6	-1	37.0	33.8	**9**
design8	44000	7	5	29	967.2	966.7	0	115.0	116.4	-1
design9	400000	8484	8428	1	607.1	559.6	**8**	62.3	62.0	0
design10	40	160	97	39	648.5	603.3	**7**	809.1	756.2	**7**
design11	2500	374	335	10	234.6	186.3	**21**	370.7	282.1	**24**

SAT solver based on the Boolean representation size to maximize the size constraints that can be solved and optimize the speed of constraint solving (better with BDDs). Our experimental results show the improvement over the default optimized algorithm. However, this makes comparing the Boolean representation sizes harder since different solvers may be used when *DomRed* is applied. We are working on addressing this problem to obtain a tighter comparison.

Column 1 of Table1 specifies the design, Column 2 shows the number of simulation cycles, Columns 3–5 show the reduction of the number of bits in the constraints. Note that the number of bits is measured for the constraints only and the design may have several thousand more bits. Columns 6–8 show the CPU times and Columns 9–11 the memory reduction. The time taken by *DomRed* is negligibly small and hence, not presented here. The CPU time includes simulation time only in 2/11 cases, hence the CPU time improvement for most examples is for constraint solving alone.

The table shows that the reduction in the number of bits is sometimes substantial, upto 39%. Smaller constraints yield better CPU times and memory reductions. Given that *DomRed* takes negligible time, 11/34 examples show improvement on applying *DomRed* and the remaining 23 examples are no worse off, we conclude that *DomRed* is a cheap preprocessing technique and that it is always beneficial to apply it. These results are encouraging and as part of future work, we hope to apply more powerful static analysis to reduce the size of the Boolean representation even further.

References

[1] Kim, H., Somenzi, F.: Finite instantiations for integer difference logic. In: Formal Methods in Computer Aided Design (FMCAD 2006), San Jose, CA, pp. 31–38 (November 2006)

[2] Kitchen, N., Kuelhmann, A.: Stimulus generation for constrainted random simulation. In: ICCAD (2007)

[3] Yuan, J., Shultz, K., Pixley, C., Miller, H., Aziz, A.: Modeling design constraints and biasing in simulation using bdds. In: ICCAD, pp. 584–590 (1999)

Producing Short Counterexamples Using "Crucial Events"

Sujatha Kashyap[1] and Vijay K. Garg[2,⋆]

[1] ECE Department, University of Texas at Austin, Austin TX 78712, USA
kashyap@ece.utexas.edu
[2] IBM India Research Lab., New Delhi, India
vijgarg1@in.ibm.com

Abstract. Ideally, a model checking tool should successfully tackle state space explosion for complete system validation, while providing short counterexamples when an error exists. Techniques such as partial order (p.o.) reduction [1,2] are very effective at tackling state space explosion, but do not produce short counterexamples. On the other hand, directed model checking [3,4] techniques find short counterexamples, but are prone to state space explosion in the absence of errors. To the best of our knowledge, there is currently no single technique that meets both requirements. We present such a technique in this paper.

For a subset of CTL, which we call CETL (Crucial Event Temporal Logic), we show that there exists a unique **minimum** set of events in each program trace whose execution is both **necessary and sufficient** to lead to an error state. These events are called "crucial events". We show how crucial events can be used to produce short counterexamples, while also providing state space reduction.

We have implemented the techniques presented here as an extension to the model checker SPIN, called SPICED (Simple PROMELA Interpreter with Crucial Event Detection). Experimental results are presented.

1 Introduction

Partial order reduction techniques [1,2] successfully tackle state space explosion, but tend to produce lengthy error trails [3,5]. Short error trails greatly reduce debugging effort. Also, the ability to find errors at shorter depths can make it possible to verify larger models, by finding the error before the model checker runs out of computational resources. Recently, there has been much interest in the use of heuristic search techniques to produce short error trails [3,4]. Heuristic search techniques calculate a cost function for each outgoing transition from a state, then explore these transitions in the order of increasing cost. Lower cost transitions are considered to be "closer" to the error state. However, in the absence of errors, these techniques do not reduce state space explosion because they only change the order in which nodes are expanded without reducing the number of nodes to be expanded. While there has been some effort to combine heuristic search with state space reduction techniques, the combination can interfere with the efficiency of the individual techniques[5]. To the best of our knowledge, there is currently

⋆ Part of the work reported here was done while this author was at the Univ. of Texas at Austin.

A. Gupta and S. Malik (Eds.): CAV 2008, LNCS 5123, pp. 491–503, 2008.

no single technique that achieves both objectives - state space reduction for complete validation, while narrowing down on error states quickly to produce short error trails. We present such a technique in this paper.

The set of reachable states in a (Mazurkiewicz) trace [6] of a program forms a lattice [7]. A lattice is a partial order in which every pair of elements has a unique meet (infimum) and join (supremum). A property is said to be *meet-closed* [8] in a trace if, whenever it holds at any two states in the trace, it also holds at the state given by their lattice meet. It was shown in [8] that, given a trace and a meet-closed property, there exists a unique minimum set of events in the trace whose execution is both *necessary and sufficient* to reach a state satisfying the property. We call these events *crucial events*. Executing crucial events in any order consistent with the dependency relation results in the same state [2]. Thus, for a single trace, it is sufficient to explore any one interleaving comprising entirely of crucial events. We call such an interleaving a *crucial path*. If an error state exists, any crucial path will lead to it through the fewest possible transitions. We show how crucial paths can be used to improve the efficiency of explicit-state model checking, and show how crucial events can be identified.

We identify a subset of CTL, called Crucial Event Temporal Logic (CETL), which contains only meet-closed formulae. CETL includes the existential until and release operators of CTL, and allows conjunction. Atomic propositions are limited to process-local variables. CETL does not allow negation, except for atomic propositions, nor does it allow disjunction. Despite these limitations, CETL can express many reachability, safety, liveness and response properties. In fact, of the 131 properties in the BEEM database [9], which is a large repository of benchmarks for explicit-state model checkers, 101 (77%) can be expressed in CETL.

We have implemented a CETL model checker called SPICED (Simple PROMELA Interpreter with Crucial Event Detection), using the techniques presented here. SPICED is based on the popular model checker SPIN [10]. We provide experimental results from a wide range of examples from the BEEM database [9], and from the SPIN distribution [10]. We ran experiments on 75 different variations (with differences in problem sizes and the location of errors) of 15 different models from the BEEM database. SPICED achieved trail reduction greater than 1x in 93% of the cases, greater than 10x in 55% of the cases, and greater than 100x in 19% of the cases. We completed verification faster than SPIN (with p.o. reduction) in 44% of the cases, with a 10x reduction in time in 9% of the cases. For 3 of the 15 models, we were able to verify problem sizes for which SPIN ran out of resources. We also provide results to show that we achieve state space reduction comparable to p.o. techniques even in the absence of errors.

Lattice theory has previously been applied to the verification of *finite* program traces. A survey of these applications was presented in [11]. In [12], a logic called RCTL was defined, which included the CTL operators EG, AG and EF. RCTL formulae were shown to be meet-closed, and an efficient verification algorithm for RCTL formulae was presented. However, these applications were limited to a single *finite* trace of a program, and required a partial order (implicit) representation of the state space. To the best of our knowledge, this paper is the first time these lattice theoretic concepts have been applied to explicit-state model checking of an entire program.

This paper is organized as follows. Section 2 introduces relevant concepts and notation. In Section 3, we introduce some CTL operators that preserve meet-closure, define the logic CETL, and introduce the notion of crucial events. In Section 4, we show how crucial events can be used for model checking CETL formulae within a single trace of a program, then extend this to model checking the complete program in Section 5. In Section 6, we show how crucial events are identified. Experimental results are presented in Section 7, followed by concluding remarks in Section 8.

2 Preliminaries

A *finite-state program* P is a triple (S, T, s_0) where S is a finite set of states, T is a finite set of operations, and $s_0 \in S$ is the initial state. The set of transitions that are executable from a given state $s \in S$ is denoted by $enabled(s)$. A transition $\alpha \in enabled(s)$ transforms the state s into a *unique* state s', denoted by $s' = \alpha(s)$. A state s is *reachable* in a program P iff it can be reached from s_0 by executing only enabled transitions at each state. The *full state space graph* of P is a directed graph whose vertices are the reachable states of P. An edge exists from vertex s to t iff $\exists \alpha \in enabled(s)$ such that $t = \alpha(s)$. A path through the full state space graph consists of a (finite or infinite) sequence of states. Each path has a corresponding *transition sequence*, consisting of the transitions executed along the path.

An independence relation [6,1] $I \subseteq T \times T$ is a symmetric, irreflexive relation such that $(\alpha, \beta) \in I$ iff for every state $s \in S$, (a) if $\alpha \in enabled(s)$, then $\beta \in enabled(s)$ if and only if $\beta \in enabled(\alpha(s))$, and (b) if $\alpha, \beta \in enabled(s)$, then $(\alpha(\beta(s)) = \beta(\alpha(s)))$. Simply put, the execution of α does not affect the enabledness of β, and executing α and β in either order results in the same state. We say that α, β are independent iff $(\alpha, \beta) \in I$. The dependency relation, D, is the reflexive, symmetric relation given by $D = (T \times T) \setminus I$. The independence relation I partitions the set of all transition sequences (correspondingly, paths) of a program into equivalence classes called *traces* [6]. Given two finite transition sequences u and v, we say that u and v are *trace equivalent*, denoted $u \equiv v$, iff they have the same starting state, and v can be derived from u by repeatedly commuting adjacent independent transitions.

Trace equivalence for infinite transition sequences is defined with the help of the relation $\preceq$. Given two (finite or infinite) transition sequences u and v, $u \preceq v$ iff for each finite prefix u' of u, there exists a prefix v' of v, and some w such that $v' \equiv w$, and u' is a prefix of w. For infinite sequences u, v, we have $u \equiv v$ iff $u \preceq v$ and $v \preceq u$.

Each occurrence of a transition in a transition sequence is called an *event*. For example, the transition sequence $\alpha\beta\alpha\beta$ consists of four events. We say that two events are dependent (correspondingly, independent) iff their corresponding transitions are dependent (independent). Every path of a trace starts from the same state, and consists of the same set of events. We will use the notation $\sigma = [s, v]$ to denote a trace with starting state s, and representative transition sequence v. All paths of a trace have the same length, and the same final state [6,1].

The concatenation of a finite trace $\sigma_1 = [s, v]$ with a finite or infinite trace $\sigma_2 = [t, w]$ is defined when t is also the final state of σ_1, and is given by $\sigma_1.\sigma_2 = [s, vw]$. We say that $\sigma_2 = [s, v]$ *subsumes* $\sigma_1 = [s, u]$, denoted $\sigma_1 \sqsubseteq \sigma_2$, iff $u \preceq v$. If σ_1 is finite, then $\sigma_1 \sqsubseteq \sigma_3$ iff there exists σ_2 such that $\sigma_3 = \sigma_1.\sigma_2$. If $\sigma \sqsubseteq \sigma'$, then the reachable states of σ is a subset of the reachable states of σ'. We say that a trace of a program P is *maximal* iff no other trace of P subsumes it. Clearly, the set of maximal traces of a program contains all its reachable states.

2.1 Traces, Posets and Lattices

A 1-1 correspondence exists between traces and partially ordered sets (posets)[7,6]. Let $\sigma = [s, v]$ be a trace, and E be the set of events in v. We can define a poset $(E, \rightarrow)$, where $\forall e, f \in E : e \rightarrow f$ iff $(e, f) \in D$ and e occurs before f in v. The relation $\rightarrow$ expresses causal dependence. Every transition sequence of σ is a linearization of $(E, \rightarrow)$, and conversely every linearization of this poset is a valid transition sequence of σ. We will use the notation $\sigma = (E, \rightarrow)$ for the poset corresponding to a trace σ.

The same state can be visited multiple times during the execution of a transition sequence, for example, in the case of a cycle in the state space graph. However, each *occurrence* of the state corresponds to a unique prefix of the transition sequence. If an event e is executed as part of a transition sequence, then the events that causally precede e must have been executed before e. A subset $G \subseteq E$ of a poset $(E, \rightarrow)$ is called a *down-set* if, whenever $f \in G$, $e \in E$ and $e \rightarrow f$, we have $e \in G$. In a trace $\sigma = (E, \rightarrow)$, there exists a correspondence between occurrences of states and down-sets. That is, an occurrence of a state in σ corresponds to executing the set of events in some down-set of $(E, \rightarrow)$. Conversely, every state in σ can be reached by executing the events in some down-set of $(E, \rightarrow)$. For simplicity of presentation, in this paper we overload the term "down-set" to mean both a set of events, and an occurrence of a state.

Progress in a computation is measured by the execution of additional events from the current state. For down-sets G and H of a trace $(E, \rightarrow)$, $G \subseteq H$ iff H is reachable from G in the full state space graph. The set of all down-sets of $(E, \rightarrow)$ forms a lattice under the $\subseteq$ relation, with the meet and join operations given by set intersection and union, respectively [13,7]. That is, if G and H are down-sets of $(E, \rightarrow)$, so are $(G \cap H)$ and $(G \cup H)$. We will use $\mathcal{L}(\sigma)$ to denote the lattice of down-sets of a trace σ. Note that, while a vertex in the full state space graph corresponds to a program state, a vertex in $\mathcal{L}(\sigma)$ corresponds to an *occurrence* of a state. Figure 1 illustrates these concepts.

We say that a formula (property) is meet-closed (correspondingly, join-closed) if, whenever any two states of a trace σ satisfy the formula, the state corresponding to their meet (correspondingly, join) in $L(\sigma)$ also satisfies it. For a down-set G and formula ϕ, the notation "$G \models \phi$" means that the state corresponding to G satisfies ϕ.

Definition 1. Meet-closed *[8]: A formula ϕ is meet-closed iff, for every trace σ of a program P: $\forall G, H \in \mathcal{L}(\sigma) : [(G \models \phi) \wedge (H \models \phi) \Rightarrow (G \cap H) \models \phi]$.*

Definition 2. Join-closed*: A formula ϕ is join-closed iff, for every trace σ of a program P: $\forall G, H \in \mathcal{L}(\sigma) : [(G \models \phi) \wedge (H \models \phi) \Rightarrow (G \cup H) \models \phi]$.*

Definition 3. Regular *[14]: A formula ϕ is regular iff it is meet- and join- closed.*

In the next section, we present some CTL operators that preserve meet- and join-closure.

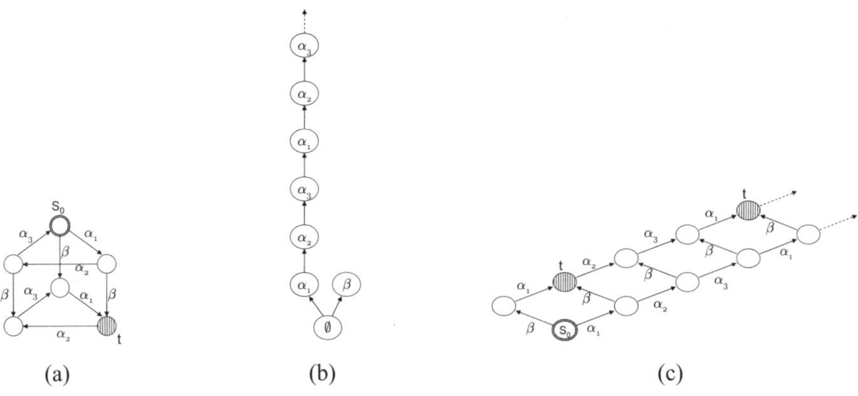

Fig. 1. (a) The full state space graph of a program P. (b) The poset corresponding to a maximal trace $\sigma = [s_0, \beta(\alpha_1\alpha_2\alpha_3)^\omega]$. (c) The lattice $\mathcal{L}(\sigma)$, showing two occurrences of a state t.

3 Meet- and Join-Closure of CTL Operators

We consider concurrent systems, where the system is modeled as a set of processes. Each process P_i has a set of transitions T_i, and a set of *local* variables V_i that can only be changed by transitions in T_i. All the transitions in T_i are pairwise dependent, that is, if $\alpha, \beta \in T_i$, then $(\alpha, \beta) \in D$. A transition in T_i can also change the values of shared (global) variables. A formula ϕ is called a *process-local state formula* iff its truth value is purely determined by the current values of the local variables V_i of some process P_i. Theorems 1 and 3 in this section are proved in [15].

Theorem 1. *Process-local state formulae are regular.*

The following theorem was proved in [14], using set union and intersection.

Theorem 2. *If ϕ_1 and ϕ_2 are regular, then $(\phi_1 \wedge \phi_2)$ is regular.*

On the other hand, disjunction does not preserve meet-closure [14].

Let π^i denote the ith state on the path π. We consider the following CTL operators:

- $s \models EG(\phi)$ iff there exists a path π starting from s such that $\forall i : i \geq 0 : \pi^i \models \phi$.
- $s \models E[\phi_1 U \phi_2]$ iff there exists a path π starting from s such that $\exists j : j \geq 0 : \pi^j \models \phi_2$, and $\forall i : i < j : \pi^i \models \phi_1$.
- $EF(\phi) = E[true\ U\ \phi]$
- $E[\phi_2 R \phi_1] = E[\phi_1 U(\phi_1 \wedge \phi_2)] \vee EG(\phi_1)$

It can be shown that the existential until operator, $E[\phi_1 U \phi_2]$, does not preserve meet-closure [15]. However, a specific flavor of this operator does, as shown in the following theorem. In most cases, the system specification makes it equally valid to check for $E[\phi_1 U(\phi_1 \wedge \phi_2)]$ instead of $E[\phi_1 U \phi_2]$.

Theorem 3. *If ϕ_1 and ϕ_2 are regular, then so are $E[\phi_2 R \phi_1]$ and $E[\phi_1 U(\phi_1 \wedge \phi_2)]$.*

As $EF(\phi_1) = E[true\ U(true \wedge \phi_1)]$, and $EG(\phi_1) = E[false\ R\ \phi_1]$, and $true$ and $false$ are trivially regular, we have:

Corollary 1. *If ϕ_1 is a regular formula, so are $EF(\phi_1)$ and $EG(\phi_1)$.*

We define a logic in which every formula is regular, as are all its subformulae.

Definition 4 Crucial Event Temporal Logic (CETL). *A CETL formula is one that can be generated from the following rules:*

1. *The trivial propositions true and false are CETL formulae.*
2. *Every process-local state formula is a CETL formula.*
3. *If ϕ_1 and ϕ_2 are CETL formulae, so are $(\phi_1 \wedge \phi_2)$, $E[\phi_2 R \phi_1]$, and $E[\phi_1 U(\phi_1 \wedge \phi_2)]$.*

We now explore the relation between meet-closure and *crucial events*.

3.1 Crucial Events

Let G be any down-set of a trace $\sigma = (E, \rightarrow)$. Let ϕ be some meet-closed formula, and $G \not\models \phi$. Let $\mathcal{G}$ be the set of all ϕ-satisfying states that are reachable from G in σ. That is, $\mathcal{G} = \{H \in \mathcal{L}(\sigma) | G \subseteq H \wedge H \models \phi\}$. Now, $\mathcal{G}$ can be an infinite set. Let $\mathcal{H}$ be the set of elements of $\mathcal{G}$ that are minimal under $\subseteq$:

$$\mathcal{H} = \{H \in \mathcal{G} | \forall H' : H \subset H' \Rightarrow H' \notin \mathcal{H}\} \tag{1}$$

$\mathcal{H}$ is necessarily finite for finite-state programs. We now define $K = \bigcap_{H \in \mathcal{H}} H$. By the meet-closure of ϕ, $K \models \phi$. Also, $G \subseteq K$. That is, K is the unique and well-defined ϕ-satisfying state that is reachable from G by executing the *fewest* events. In particular, $K \setminus G$ is the minimum set of events that *must* be executed along any path starting from G, to reach a ϕ-satisfying state in σ. The events in $K \setminus G$ are called **crucial events** [8].

Definition 5. Crucial event: *In a trace σ, an event e is said to be crucial from a state G with respect to a meet-closed formula ϕ, denoted $e \in crucial(G, \phi, \sigma)$ iff:*

$$\forall H \in \mathcal{L}(\sigma) : (G \subseteq H) \wedge (G \not\models \phi) \wedge (H \models \phi) \Rightarrow (e \in H \setminus G)$$

A transition sequence starting from G and comprising exactly of the events in $crucial$ (G, ϕ, σ) gives us a path of shortest length from G to a ϕ-satisfying state in σ. Such a path is called a **crucial path**. A special case arises when $\mathcal{H} = \emptyset$. In this case, we define $K = E$ (the set of all events), and any maximal path starting from G in $\mathcal{L}(\sigma)$ constitutes a crucial path. The proof for the following theorem is straightforward.

Theorem 4. *Let $\mathcal{H}$ be as defined in (1). If $\mathcal{H} \neq \emptyset$, then a crucial path for ϕ starting from G cannot contain a cycle.*

Recall that a down-set is an occurrence of a state. Suppose the down-set G is an occurrence of the state s. Executing the events in $crucial(G, \phi, \sigma)$ from s will lead to a ϕ-satisfying state in the full state space graph. The state s can have multiple occurrences in σ (for example, in Figure 1(c), the state t occurs multiple times in σ_2). Let G' be another down-set of σ that is also an occurrence of s. It is easy to see that $crucial(G, \phi, \sigma) = crucial(G', \phi, \sigma)$. Thus, every occurrence of s in σ has the same set of crucial events w.r.t. ϕ. Based on this observation, we define $crucial(s, \phi, \sigma) \equiv crucial(G, \phi, \sigma)$, where G is any down-set of σ that is an occurrence of s.

The complexity of identifying the exact set of events that constitutes $crucial(s, \phi, \sigma)$ for a given CETL formula ϕ is an open problem. However, we can identify a *subset* of $crucial(s, \phi, \sigma)$ in most cases, as we shall see in Section 6.

If G is a down-set of $\mathcal{L}(\sigma)$, and H is an immediate successor of G in $\mathcal{L}(\sigma)$, we denote this by $G \triangleright H$. Formally, if $G, H \in \mathcal{L}(\sigma)$, and $\exists e \notin G$, and $H = G \cup \{e\}$, then $G \triangleright H$. The notation $G \trianglerighteq H$ means $(G \triangleright H) \vee (G = H)$. The following lemmas are used in the proofs presented in Sections 4.1 and 4.2, and are proved in [15].

Lemma 1. *For a trace σ and $C, D, F \in \mathcal{L}(\sigma)$, if $C \triangleright F$ and $D \subseteq F$, then $(C \cap D) \trianglerighteq D$.*

Lemma 2. *For a trace σ and $C, D, F \in \mathcal{L}(\sigma)$, if $F \triangleright C$ and $F \subseteq D$, then $D \trianglerighteq (C \cup D)$.*

We now show how the concepts presented so far can be used to prune the state space while model checking CETL formulae. In particular, we show that it is sufficient to explore only crucial paths in order to verify a CETL formula. For better presentation, we start with the problem of verifying a CETL formula on a single trace of a program. We will consider CETL model checking for the complete program in Section 5.

4 Model Checking CETL in a Program Trace

The approach we present here can be used to enhance any local, recursive CTL model checking algorithm, such as ALMC [16]. A *local* model checking algorithm starts from an initial program state and performs a state space exploration using either depth-first or breadth-first search. *Recursive* means that the truth value of subformulae are determined at a state before the top formula is evaluated. In this section, we show that rather than exploring all enabled events from a state, it is sufficient to explore only a subset of these events in order to verify a CETL formula in a given trace. This explored subset is called an "ample set" [1]. The ample set chosen at a state s while verifying a CETL formula ϕ is denoted by $ample(s, \phi)$. In the non-reduced (baseline) case, $ample(s, \phi) = enabled(s)$.

4.1 $E[\phi_1 U(\phi_1 \wedge \phi_2)]$

Let G_0 be some down-set of σ that satisfies $E[\phi_1 U(\phi_1 \wedge \phi_2)]$. Let π be the corresponding witness path with $\pi^l = H$ as its final state. Then, $\forall j : 0 \leq j \leq l : \pi^j \models \phi_1$, and $H \models (\phi_1 \wedge \phi_2)$. Let $\mathcal{J}$ be the set of all down-sets of σ that are reachable from G_0, are minimal under $\subseteq$ (this ensures that $\mathcal{J}$ is finite), and satisfy $(\phi_1 \wedge \phi_2)$. Define:

$$G = \bigcap_{J \in \mathcal{J}} J \tag{2}$$

Since $(\phi_1 \wedge \phi_2)$ is regular, $G \models (\phi_1 \wedge \phi_2)$.

Theorem 5. *There exists a path from G_0 to G such that every state along the path satisfies ϕ_1.*

Proof. We will construct a path λ from G_0 to G, consisting entirely of ϕ_1-satisfying states. We construct this path backwards, starting from $\lambda^k = G$, towards $\lambda^0 = G_0$.

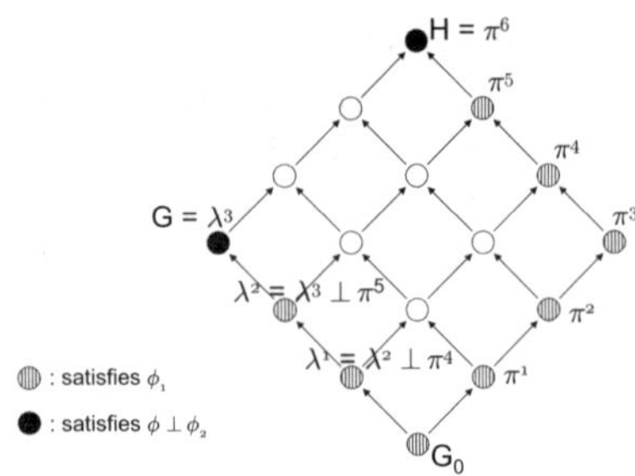

Fig. 2. Example illustrating the construction of Theorem 5

We show that, if $\lambda^i \models \phi_1$ for any $1 \leq i \leq k$, there exists a $G' \triangleright \lambda^i$ such that $G' \models \phi_1$. We can then extend λ with $\lambda^{i-1} = G'$, and proceed with our construction. For the base case, we have $\lambda^k = G$, and $G \models \phi_1$.

Let $1 \leq j \leq l$ be the *least* j such that $\lambda^i \subseteq \pi^j$. First, we show that such a j must exist. Recall that $\pi^l = H$, and $\lambda^i \subseteq G \subseteq H$. Therefore, for some $j \leq l$, $\lambda^i \subseteq \pi^j$. Also, $\pi^0 = \lambda^0 = G_0$, so $\forall i : i \geq 1 : \lambda^i \not\subseteq \pi^0$. Therefore, $j \geq 1$. Since j is the least such, we have $\lambda^i \not\subseteq \pi^{j-1}$. So, we have $\pi^{j-1} \triangleright \pi^j$, and $\lambda^i \subseteq \pi^j$. From Lemma 1, this implies $(\lambda^i \cap \pi^{j-1}) \trianglerighteq \lambda^i$. We cannot have $(\lambda^i \cap \pi^{j-1}) = \lambda^i$, because this would imply $\lambda^i \subseteq \pi^{j-1}$, which is a contradiction. Therefore, $(\lambda^i \cap \pi^{j-1}) \triangleright \lambda^i$. Set $G' = (\lambda^i \cap \pi^{j-1})$. Since $\lambda^i \models \phi_1$, and $\pi^{j-1} \models \phi_1$, by the meet-closure of ϕ_1, $G' \models \phi_1$. □

Theorem 5 tells us that if $G_0 \models E[\phi_1 U(\phi_1 \wedge \phi_2)]$, then a crucial path for $(\phi_1 \wedge \phi_2)$ can act as a witness. Since $G_0 \models \phi_1$, and every state along the witness path satisfies ϕ_1, it is easy to see that $crucial(G_0, (\phi_1 \wedge \phi_2), \sigma) = crucial(G_0, \phi_2, \sigma)$. The following theorem shows how we can construct this path "forward", that is, starting from G_0.

Theorem 6. *To construct the path of Theorem 5, starting from G_0, at each state H we execute a single enabled event α such that $\alpha \in crucial(H, \phi_2, \sigma)$, and $H \cup \{\alpha\} \models \phi_1$.*

Proof. Let G be as in (2). From Theorem 5, there exists *some* path λ such that $\lambda^0 = G_0$, $\lambda^k = G$, and $\forall j : 0 \leq j \leq k : \lambda^j \models \phi_1$. We need to show that we can construct such a path by choosing, at each state, any crucial event that leads to a ϕ_1-satisfying successor.

Clearly, if every event along our path is crucial for ϕ_2, then our path will lead to G. We need to show that at any state H along our constructed path, there exists a successor J such that $J \models \phi_1$. To begin with, $H = G_0$. Of course, our construction ends when $H = G$, so any H for which a successor needs to be found must be a strict subset of G.

Let $0 \leq i < k$ be the *greatest* i such that $\lambda^i \subseteq H$. We first show that such an i exists. Note that $\lambda^0 = G_0 \subseteq H$. Thus, for some $i \geq 0 : \lambda^i \subseteq H$. Also, $\lambda^k = G$, and $H \subset G$. Therefore, $\lambda^k \not\subseteq H$, so $i < k$. Since i is the greatest such, we have $\lambda^{i+1} \not\subseteq H$. Now, $\lambda^i \triangleright \lambda^{i+1}$, and $\lambda^i \subseteq H$. By Lemma 2, $H \trianglerighteq (\lambda^{i+1} \cup H)$. If $H = (\lambda^{i+1} \cup H)$, then $\lambda^{i+1} \subseteq H$, which is a contradiction. Therefore, $H \triangleright (\lambda^{i+1} \cup H)$. Also, $H \models \phi_1$, and $\lambda^{i+1} \models \phi_1$, so by the join-closure of ϕ_1, $\lambda^{i+1} \cup H \models \phi_1$. Hence, $J = \lambda^{i+1} \cup H$ is the required successor for H. □

4.2 $E[\phi_2 R\phi_1]$

Recall that $E[\phi_2 R\phi_1] \stackrel{def}{\equiv} E[\phi_1 U(\phi_1 \wedge \phi_2)] \vee EG(\phi_1)$. Theorem 6 showed how to construct a witness for $G_0 \models E[\phi_1 U(\phi_1 \wedge \phi_2)]$. The following theorem shows how to construct a witness for $G_0 \models EG(\phi_1)$.

Theorem 7. *Let $G_0 \in \mathcal{L}(\sigma)$ such that $G_0 \models EG(\phi_1)$ in σ. We can construct a witness path as follows. Starting from G_0, at each state H, we execute a single enabled event α such that $H \cup \{\alpha\} \models \phi_1$.*

Proof. We simply need to show that, for every state H on the constructed path, there exists a ϕ_1-satisfying successor state. The proof is similar to that of Theorem 6. □

Theorems 6 and 7 show that, given a formula ϕ of the form $E[\phi_1 U(\phi_1 \wedge \phi_2)]$ or $E[\phi_2 R\phi_1]$, and a trace $\sigma = [s, v]$, we can decide if $s \models \phi$ by exploring only ample sets satisfying the following condition:
(C0) If $ample(s, \phi) \neq enabled(s)$, then $ample(s, \phi) = \{\alpha\}$, where $\alpha \in crucial(s, \phi_2, \sigma)$, and $\alpha(s) \models \phi_1$.

5 Model Checking CETL in a Program

Let ϕ be a CETL formula of the form $E[\phi_1 U(\phi_1 \wedge \phi_2)]$ or $E[\phi_2 R\phi_1]$. We now consider the problem of deciding whether a given state s satisfies ϕ in a program. In this case, we need to explore a crucial path for every maximal program trace starting from s. That is, for every maximal trace σ starting from s, $ample(s, \phi)$ must contain some event from $crucial(s, \phi_2, \sigma)$. In [1], it was shown that if $ample(s, \phi)$ satisfies the following condition (C1), then it contains a successor for each maximal trace starting from s.

(C1) Along every path starting from s in the full state space graph, a transition that is dependent on a transition from $ample(s, \phi)$ cannot be executed without a transition from $ample(s, \phi)$ occurring first.

Theorem 8. *[1] If $ample(s, \phi)$ satisfies condition (C1), then for every maximal trace σ starting from s, there exists some $\alpha \in ample(s, \phi)$ such that $[s, \alpha] \sqsubseteq \sigma$.*

By Theorem 8, an ample set satisfying condition (C1) will generate some successor for each maximal trace starting from s. We now need to ensure that each event in the ample set is crucial in *every* trace to which it belongs.

Definition 6. Universally crucial event: *An event α is said to be universally crucial from a state s for a meet-closed formula ϕ_2, denoted $\alpha \in ucrucial(s, \phi_2)$, iff for every trace σ such that $[s, \alpha] \sqsubseteq \sigma$, $\alpha \in crucial(s, \phi_2, \sigma)$.*

The following is a straightforward extension of condition (C0) from Section 4.

(C2) If $ample(s, \phi) \neq enabled(s)$, then for each $\alpha \in ample(s, \phi)$, $\alpha \in ucrucial(s, \phi_2)$ and $\alpha(s) \models \phi_1$.

The following theorem is proved in [15].

Theorem 9. *To determine whether $s \models \phi$, it is sufficient to explore ample sets satisfying (C1) and (C2).*

To construct an ample set satisfying (C1) and (C2), we need to identify a subset of $enabled(s)$ that satisfies these conditions. Condition (C1) is used in p.o. reduction [1], and we use the techniques from [17] to construct a subset of $enabled(s)$ that satisfies (C1). Condition (C2) requires us to identify a subset of $enabled(s)$ that consists of universally crucial events. We now show how universally crucial events can be identified.

6 Identifying Universally Crucial Events

In the previous section, we derived the conditions for $ample(s, \phi)$, where ϕ is a CETL formula of the form $E[\phi_1 U(\phi_1 \wedge \phi_2)]$ or $E[\phi_2 R \phi_1]$. Note that our construction of a witness path for $s \models \phi$ ends with success when we encounter a state which satisfies ϕ_2. Therefore, we are only interested in constructing ample sets for states at which ϕ_2 does not hold. That is, we are interested in $ucrucial(s, \phi_2)$ when $s \not\models \phi_2$. The problem of identifying $ucrucial(s, \phi_2)$ for a general CETL formula ϕ_2 remains open. In this section, we identify some cases for which we can determine universally crucial events. Recall that T_i is the set of transitions of process P_i.

Theorem 10. *If ϕ_2 is a process-local state formula in process P_i, then $T_i \cap enabled(s) \subseteq ucrucial(s, \phi_2)$.*

Proof. Only transitions in T_i can change the truth value of ϕ_2. Therefore, we must execute some transition in T_i from s in order to reach a state in which ϕ_2 holds. Recall that transitions from the same process are pairwise-dependent. Therefore, each transition in $T_i \cap enabled(s)$ gives rise to a different trace. Therefore, each event $\alpha \in T_i \cap enabled(s)$ is crucial in every trace that subsumes $[s, \alpha]$. Thus, $T_i \cap enabled(s) \subseteq ucrucial(s, \phi)$. □

Recall, from Section 4, that our approach applies to *recursive* model checking algorithms. Thus, in the following two theorems, ψ_1 and ψ_2 have already been evaluated at s before ϕ_2 is evaluated. The proof of Theorem 11 is straightforward.

Theorem 11. *Let $\phi_2 = \psi_1 \wedge \psi_2$. If $s \not\models \psi_1$ then $ucrucial(s, \psi_1) \subseteq ucrucial(s, \phi_2)$, else $ucrucial(s, \psi_2) \subseteq ucrucial(s, \phi_2)$.*

Theorem 12. *Let ϕ_2 be of the form $E[\psi_1 U(\psi_1 \wedge \psi_2)]$ or $E[\psi_2 R \psi_1]$. If $s \not\models \psi_1$, then $ucrucial(s, \psi_1) \subseteq ucrucial(s, \phi_2)$. Else, if $s \models \psi_1$ and $\neg\psi_1$ is meet-closed, then $ucrucial(s, \neg\psi_1) \subseteq ucrucial(s, \phi_2)$.*

Proof.
- **Case 1:** $s \not\models \psi_1$. Any state that satisfies ϕ_2 must also satisfy ψ_1. Therefore, we first need to execute the minimum set of events that will lead to a state satisfying ψ_1. Hence, $ucrucial(s, \psi_1) \subseteq ucrucial(s, \phi_2)$.
- **Case 2:** $s \models \psi_1$. Recall that $s \not\models \phi_2$, and we are interested in reaching a state that satisfies ϕ_2. Assume there exists a state t, reachable from s, that satisfies ϕ_2. Let w be a witness for $t \models \phi_2$. Then, along every path v from s to t, there must exist some state s' such that $s' \not\models \psi_1$. If not, then $v.w$ would have been a witness for $s \models \phi_2$. Thus, in order to reach a state that satisfies ϕ_2, it is necessary to go through some state that satisfies $\neg\psi_1$. If $\neg\psi_1$ is meet-closed, then the execution of the events in $ucrucial(s, \neg\psi_1)$ is necessary to reach a state satisfying $\neg\psi_1$. Therefore, $ucrucial(s, \neg\psi_1) \subseteq ucrucial(s, \phi_2)$.

7 Implementation and Experimental Results

We have implemented our approach as an extension to the SPIN model checker [10], called SPICED (Simple PROMELA Interpreter with Crucial Event Detection). Our implementation incorporates the ample set selection techniques presented in this paper into ALMC [16], a local CTL model checking algorithm based on depth-first search. The complete algorithm can be found in [16], as well as in the technical report version of this paper [15]. Our implementation of SPICED, along with detailed experimental results, is available at: `http://maple.ece.utexas.edu/spiced`.

We ran SPICED against a large set of examples from the BEEM database [9], which contains PROMELA (the input language for SPIN) models with errors injected into them, and lists the properties to be verified on these models. All experiments were performed on a 1-cpu 2.8 GHz Intel Pentium 4 machine with 512 MB RAM, running Red Hat Enterprise Linux WS Rel 4.

Table 1. Trail reduction with SPICED, compared to SPIN with p.o. reduction

Model	Tool	Time (sec)	States	Memory (MB)	Formula	Trail length
phils.7	SPICED	0.01	15	3.15	$EF(P_0.req \wedge EG(!P_0.grant))$	6
	SPIN	**Could not complete**			$\neg\Box(req0 \Rightarrow \Diamond grant0)$	-
szymanski.9	SPICED	0.02	256	3.15	$EF(P_0.wait \wedge EG(!P_0.cs))$	43
	SPIN	**Could not complete**			$\neg\Box(wait0 \Rightarrow \Diamond cs0)$	-
fischer.18	SPICED	0.02	28	3.15	$EF(P_0.wait \wedge EG(!P_0.cs))$	19
	SPIN	**Could not complete**			$\neg\Box(wait0 \Rightarrow \Diamond cs0)$	-
mcs.5	SPICED	0.09	30227	4.89	$EF(P_0.wait \wedge EG(!P_0.cs))$	14
	SPIN	0.03	2821	2.72	$\neg\Box(wait0 \Rightarrow \Diamond cs0)$	5646
anderson.7	SPICED	0.03	65387	7.03	$EF(P_0.wait \wedge EG(!P_0.cs))$	82
	SPIN	0.13	15692	6.63	$\neg\Box(wait0 \Rightarrow \Diamond cs0)$	31389
peterson.7	SPICED	0.09	29080	4.89	$EF(P_0.wait \wedge EG(!P_0.cs))$	159
	SPIN	0.1	9992	9.93	$\neg\Box(wait0 \Rightarrow \Diamond cs0)$	19984
lamport.7	SPICED	0.06	6850	3.45	$EF(P_0.wait \wedge EG(!P_0.cs))$	30
	SPIN	0.02	665	2.62	$\neg\Box(wait0 \Rightarrow \Diamond cs0)$	1330
at.7	SPICED	0.02	19	3.15	$EF(P_0.wait \wedge EG(!P_0.cs))$	11
	SPIN	0.01	182	2.62	$\neg\Box(wait0 \Rightarrow \Diamond cs0)$	370
bakery.6	SPICED	0.01	69	3.15	$EF(P_0.wait \wedge EG(!P_0.cs))$	46
	SPIN	0.02	896	2.62	$\neg\Box(wait0 \Rightarrow \Diamond cs0)$	856
gear.2	SPICED	0.03	4185	3.13	$EF(Clutch.err\_open)$	5056
	SPIN	0.13	22386	5.5	local assert()	19396
needham.4	SPICED	0.01	27	2.72	$EF(init0.fin \wedge resp0.fin)$	15
	SPIN	0.04	4003	3.03	$\neg\Diamond(init\_fin \wedge resp\_fin)$	52
msmie.2	SPICED	0.02	83	2.72	$EF(P_0.wait \wedge EG(!P_0.cs))$	63
	SPIN	0.01	370	2.62	$\neg\Box(wait0 \Rightarrow \Diamond cs0)$	214
loyd.2	SPICED	0.19	50931	9.24	$EF(Check.done)$	52597
	SPIN	0.63	166133	17.61	local assert()	84418
driving_phils.4	SPICED	0.01	212	3.15	$EF(P_0.req \wedge EG(!P_0.grant))$	123
	SPIN	0.01	85	2.62	$\neg\Box(req0 \Rightarrow \Diamond grant0)$	170

Table 2. State space reduction with SPICED

Model	Tool	Time (sec)	States	Memory (MB)	Formula
sort	SPIN, no reduction	1.19	107713	20.64	-
	SPIN, p.o. reduction	0.1	135	2.62	-
	SPICED	0.1	148	2.72	$EG(!left.tstvar)$
leader	SPIN, no reduction	0.17	15779	3.35	-
	SPIN, p.o. reduction	0.01	97	2.62	-
	SPICED	0.05	104	2.72	$EG(!node[4].tstvar)$
eratosthenes	SPIN, no reduction	0.52	49790	9.07	-
	SPIN, p.o. reduction	0.02	3437	3.03	-
	SPICED	0.02	2986	3.13	$EG(!sieve[0].tstvar)$
snoopy	SPIN, no reduction	0.53	81013	11.34	-
	SPIN, p.o. reduction	0.06	14169	4.06	-
	SPICED	0.4	58081	9.69	$EF(cpu0.tstvar)$

Table 1 shows the results for the largest problem sizes, for each of the verified models. A comprehensive list of results is available from our website: `http://maple.ece.utexas.edu/spiced`. In our experiments, SPICED produced error trails that were shorter than SPIN's in 93% of the cases, with no deterioration in trail length for the remaining 7% of cases. Our error trails were >10x shorter in 55% of the cases and >100x shorter in 19% of the cases. For 44% of the cases, SPICED completed verification faster than SPIN, with >10x reduction in time in 9% of the cases. Although CETL is a branching-time logic, in these examples, the properties were in LTL ∩ CETL, so the error trails were non-branching. The error trails were produced in the same format as those of SPIN's, and can be examined using SPIN's guided simulation feature. For SPIN, never claims were used for the verification of LTL properties, and simple assert() statements were used for reachability detection. For SPICED, the CETL formulae were specified a separate file, and fed directly as input to our model checking algorithm.

Table 2 shows the state space reduction achieved by SPICED, compared to SPIN with p.o. reduction, in the absence of errors. The examples in Table 2 are from the SPIN distribution [10], and have previously been used to showcase the effectiveness of p.o. reduction [18]. For SPIN, no LTL properties were specified during verification, which is optimal for maximizing the effectiveness of p.o. reduction. Since our algorithm is based on choosing crucial events, it requires the specification of a property. For each example, we chose a property that is never satisfied in the program, and forces exhaustive validation. Our results show state space reduction comparable to p.o. techniques.

8 Conclusions and Future Work

We have presented a model checking technique that produces short error trails, while achieving state space reduction. Experimental results confirm that our approach can significantly outperform SPIN in the presence of errors, while providing state space reduction comparable to partial order techniques. The effectiveness of our approach depends on the ability to identify crucial events. We have shown how crucial events can be identified in some cases. The problem of finding crucial events for a general CETL formula is a direction for future research.

References

1. Peled, D.: Combining partial order reductions with on-the-fly model-checking. In: Dill, D.L. (ed.) CAV 1994. LNCS, vol. 818, pp. 377–390. Springer, Heidelberg (1994)
2. Godefroid, P.: Partial-Order Methods for the Verification of Concurrent Systems. LNCS, vol. 1032, p. 142. Springer, New York (1996)
3. Edelkamp, S., Leue, S., Lluch-Lafuente, A.: Directed explicit-state model checking in the validation of communication protocols. International Journal on Software Tools for Technology Transfer 6(4) (2004)
4. Tan, J., Avrunin, G.S., Clarke, L.A., Zilberstein, S., Leue, S.: Heuristic-guided counterexample search in FLAVERS. SIGSOFT Softw. Eng. Notes 29(6), 201–210 (2004)
5. Lluch-Lafuente, A., Edelkamp, S., Leue, S.: Partial order reduction in directed model checking. In: Proceedings of the 9th International SPIN Workshop on Model Checking of Software, London, UK, pp. 112–127. Springer, Heidelberg (2002)
6. Mazurkiewicz, A.W.: Basic notions of trace theory. In: Linear Time, Branching Time and Partial Order in Logics and Models for Concurrency, School/Workshop, London, UK, pp. 285–363. Springer, Heidelberg (1989)
7. Winskel, G.: Event structures. In: Advances in Petri nets 1986, part II on Petri nets: applications and relationships to other models of concurrency, New York, NY, USA, pp. 325–392. Springer-Verlag New York, Inc, Heidelberg (1987)
8. Chase, C.M., Garg, V.K.: Efficient detection of restricted classes of global predicates. In: Helary, J.-M., Raynal, M. (eds.) WDAG 1995. LNCS, vol. 972, pp. 303–317. Springer, Heidelberg (1995)
9. Pelanek, R.: BEEM: BEnchmarks for Explicit Model checkers (2007), `http://anna.fi.muni.cz/models/index.html`
10. Holzmann, G.: On-the-fly LTL model checking with SPIN (2007), `http://spinroot.com/spin/`
11. Garg, V.K., Mittal, N., Sen, A.: Applications of lattice theory to distributed computing. ACM SIGACT Notes 34(3), 40–61 (2003)
12. Sen, A., Garg, V.K.: Detecting temporal logic predicates in distributed programs using computation slicing. In: Papatriantafilou, M., Hunel, P. (eds.) OPODIS 2003. LNCS, vol. 3144, pp. 171–183. Springer, Heidelberg (2004)
13. Davey, B., Priestley, H.: Introduction to Lattices and Order. Cambridge University Press, Cambridge (1990)
14. Garg, V.K., Mittal, N.: On slicing a distributed computation. In: ICDCS 2001: Proceedings of the The 21st International Conference on Distributed Computing Systems, p. 322. IEEE Computer Society, Washington (2001)
15. Kashyap, S., Garg, V.K.: Producing short counterexamples using "crucial events". Technical Report TR-PDS-2008-002, ECE Dept, University of Texas at Austin (2008), `http://maple.ece.utexas.edu/TechReports/2008/TR-PDS-2008-002.pdf`
16. Vergauwen, B., Lewi, J.: A linear local model checking algorithm for CTL. In: Best, E. (ed.) CONCUR 1993. LNCS, vol. 715, pp. 447–461. Springer, Heidelberg (1993)
17. Clarke, E.M., Grumberg, O., Peled, D.A.: Model Checking. MIT Press, Cambridge (1999)
18. Clarke, E.M., Grumberg, O., Minea, M., Peled, D.: State space reduction using partial order techniques. International Journal on Software Tools for Technology Transfer (STTT) 2(3), 279–287 (1999)

Discriminative Model Checking

Peter Niebert[1], Doron Peled[2], and Amir Pnueli[3]

[1] Laboratoire d'Informatique Fondamentale de Marseille
CMI, 39, rue Joliot Curie, 13453 Marseille Cedex 13, France
[2] Department of Computer Science, Bar Ilan University
Ramat Gan 52900, Israel
Computing Science Department, Courant Institute of Mathematical Sciences,
New York University, 251 Mercer Street,
New York, NY 10012

Abstract. Model checking typically compares a system description with a formal specification, and returns either a counterexample or an affirmation of compatibility between the two descriptions. Counterexamples provide evidence to the existence of an error, but it can still be very difficult to understand what is the cause for that error. We propose a model checking methodology which uses two levels of specification. Under this methodology, we group executions as *good* and *bad* with respect to satisfying a *base* LTL specification. We use an *analysis specification*, in CTL* style, quantifying over the good and bad executions. This specification allows checking not only *whether* the base specification holds or fails to hold in a system, but also *how* it does so. We propose a model checking algorithm in the style of the standard CTL* decision procedure. This framework can be used for comparing between good and bad executions in a system and outside it, providing assistance in locating the design or programming errors.

1 Introduction

A 20 years old debate in the model checking community exists between the use of a branching or state-centric specification (e.g., CTL [3,2,13]) or using linear or path-centric specification (e.g., LTL [11]). One can also combine the approaches (using, e.g., CTL* [4]). We promote here a view where the specification of a system is given using linear formalism, as its *primary* or *base* specification. However, unlike the traditional linear approach, we are not satisfied with an affirmation or a counter example. Instead, we embed the base specification in a branching *analysis specification* that can express *how* it fails. The use of an analysis specification helps us obtain more refined information than the yes/no and counterexample situation of traditional model checking.

The base specification φ is given in LTL. This is the property that we are interested that the system (hardware or software) will satisfy. The formalism we use for the analysis specification is termed EmCTL*, for *embedded* CTL*. This is a CTL*-based specification, where the path quantifiers can range over the executions that satisfy or do not satisfy some base property. These will also be

A. Gupta and S. Malik (Eds.): CAV 2008, LNCS 5123, pp. 504–516, 2008.

referred to as *good* or *bad* executions, respectively. The analysis specification thus has some past flavor in it; one needs to observe some information of the prefix of the current execution so far in order to decide whether it can be continued into a good or a bad execution, rather than just looking at the possible future(s) from the current state.

We show the relation of EmCTL* to existing formalisms, in particular to its closest logic mCTL*, from which we can also obtain relevant expressiveness and lower bound complexity results. In fact, one can use a linear translation into mCTL*, and from there using the transformation into alternating hesitant tree automata with satellites word automata for model checking, presented in [8]. It turns out that CTL*, mCTL* and EmCTL* have the same expressive power, although expressing the same property in the different logics may vary in form and size.

We provide an alternative model checking algorithm in the style of standard CTL* decision procedure [4]. One advantage of using our algorithm is that it can be easily programmed as a variant of the CTL* algorithm. Moreover, its complexity analysis takes into account the different components of the specification, namely the base LTL part and the EmCTL* analysis specification. Specifically, we show that the complexity of the model checking is in PSPACE complete w.r.t. the analysis specification, and in EXSPACE complete w.r.t. the base property.

Our decision procedure suggests a generic method of exploiting an ability of finitary (regular-based) encoding of the past that can be combined with a nondeterministic choice for the future part, in order to add past-related quantification to branching temporal logics. We show how to use this capability in order to extend the quantification so that one can reason not only about the executions of a system, but also on executions that are not generated by the system, and may or may not satisfy the given specification.

In the analysis specification we use $\forall^{\varphi}$, $\exists^{\varphi}$ to range over the good executions, $\forall^{\neg\varphi}$, $\exists^{\neg\varphi}$ to range over the bad executions, and $\forall$, $\exists$ to range over all the executions. We use $\Diamond$, $\Box$, U, W for *eventually*, *always*, *until* and *unless* (weak until), respectively. Some examples follow:

Every prefix of a bad execution is also a prefix of some good execution: $(\forall^{\neg\varphi}\Box\exists^{\varphi}\mathit{true})$.

A prefix of a good execution always has a point where it cannot turn bad: $(\forall^{\varphi}\Diamond\neg\exists^{\neg\varphi}\mathit{true})$.

There are always a possibility of a good and a bad execution: $(\forall\Box(\exists^{\neg\varphi}\mathit{true}\wedge\exists^{\varphi}\mathit{true}))$.

Before executing α, there are good and bad executions. Once α is executed for the first time, there are only bad executions. In order to assert about the execution of particular transition, we assume that the predicate Exec_{α} holds in a state if α is the last transition executed[1]: $\forall((\neg\mathit{Exec}_{\alpha}\wedge\exists^{\varphi}\mathit{true})W(\mathit{Exec}_{\alpha}\wedge\forall^{\varphi}\mathit{false}))$.

[1] This is either a transition predicate, or a predicate on states, where the state information includes the last executed transition, hence may separate formerly identical states.

The paper is structured as follows: In Section 2, the syntax and semantics of EmCTL* is introduced. In Section 3, we describe the model checking procedure. More detailed applications are given in Section 4, in particular, the formalism is extended to deal with execution sequences beyond those allowed by executing the system. In Section 5, we give details on the relation between EmCTL* and mCTL*. Conclusions appear in Section 6.

2 Embedding LTL Properties in Branching Time Specification

Syntax

An EmCTL* formula is expressed with respect to an LTL formula φ called the base specification. The syntax of EmCTL* is similar to CTL*, including state formulas (ψ) and path formulas (μ). The only difference is that the path quantifiers $\forall$ and $\exists$ may be superscripted by φ, referring to the base LTL formula. (In calculating the size of an EmCTL* formula we do not count the size of the superscripting formula φ, which is provided separately.)

$$\psi ::= p|\psi \vee \psi|\neg\psi|\exists\mu|\exists^{\varphi}\mu|\exists^{\neg\varphi}\mu \tag{1}$$

where p is a propositional letter over some set of propositions $\mathcal{P}$.

$$\mu ::= \psi|\mu \vee \mu|\neg\mu|X\mu|\mu U\mu \tag{2}$$

LTL syntax is restricted to a path formula without state subformulas.

Semantics

Let M be a finite structure $(S, \{s_0\}, E, \mathcal{P}, L)$ with states S, an initial state[2] $s_0 \in S$, edges $E \subseteq S \times S$, set of propositions $\mathcal{P}$, and labeling function $L : S \mapsto 2^{\mathcal{P}}$. (We may also assume a mapping on the edges from a fixed domain of *transitions* T, e.g., in order to define predicates such as $Exec_\alpha$, used in the introduction.) For simplicity, we assume that each state in S has a successor. This can be forced by adding a self loop, marked with a special symbol ϵ that is not in T, for each node without successors. A *path* in S is a finite or infinite sequence $\langle g_0 g_1 g_2 \ldots\rangle$, where $g_0 \in S$ and for each $i \geq 0$, $g_i E g_{i+1}$. An *execution* is an infinite path, starting with $g_0 = s_0$.

We denote the ith state of a path π by π_i, the suffix of π from the ith state by π^i and the prefix of π up to the ith state by $\hat{\pi}^i$. The concatenation of two paths ρ and π, where the last state of ρ is the same as first state of π, is denoted by $\rho \frown \pi$. Note that $\langle\pi_0\rangle \frown \pi = \pi$, and that $\rho \frown \langle s\rangle = \rho$, where s is necessarily the last state of ρ.

[2] The initial state is denoted as a singleton *set* for future compatibility when taking the product with other structures.

We first recall the LTL semantics, for the base property φ. The semantics is defined for a suffix of an execution π of M as follows:
$\pi \models p$ if $p \in L(\pi_0)$.
$\pi \models \mu_1 \vee \mu_2$ if either $\pi \models \mu_1$ or $\pi \models \mu_2$.
$\pi \models \neg\mu$ if it is not the case that $\pi \models \mu$.
$\pi \models X\mu$ if $\pi^1 \models \mu$.
$\pi \models \mu U \eta$ if there exists some i such that $\pi^i \models \eta$ and for each $0 \leq j < i$, $\pi^j \models \mu$.

The semantics given for an *EmCTL** state formula is of the form $M, \rho, s \models \psi$, where ρ is a finite prefix of an execution in M, leading to (i.e., ending with) the state s. The semantics given for a path formula μ is of the form $M, \rho, \pi \models \mu$, where ρ is again finite prefix of an execution of M, leading up to a state from which an infinite path π of M starts (thus, $\rho \frown \pi$ is an infinite execution of M). A path subformula μ is then evaluated in $M, \rho, \pi \models \mu$ according to the path π.

Intuitively, in the semantic definition of in $M, \rho, s \models \psi$ and, similarly, in the definition of $M, \rho, \pi \models \mu$, we keep the path so far ρ, so that we can assert whether the base property holds from the beginning of the execution or not. As we progress with the temporal operators in time over some finite fragment of a path, this fragment is appended to ρ and removed from π. We can either talk about the existence of an execution path, with $\exists$, one that satisfies φ, with $\exists^\varphi$, or one that does not satisfy φ, with $\exists^{\neg\varphi}$. The formal semantics is given as follows.
$M, \rho, s \models p$ if $p \in L(s)$.
$M, \rho, s \models \psi_1 \vee \psi_2$ if either $M, \rho, s \models \psi_1$ or $M, \rho, s \models \psi_2$.
$M, \rho, s \models \neg\psi$ if it is not the case that $M, \rho, s \models \psi$.
$M, \rho, s \models \exists\mu$ if there exists a path π of M such that $\pi_0 = s$ and $M, \rho, \pi \models \mu$.
$M, \rho, s \models \exists^\varphi\mu$ if there exists a path π of M such that $\pi_0 = s$ and $M, \rho, \pi \models \mu$, and furthermore, $\rho \frown \pi \models \varphi$ in LTL.
$M, \rho, s \models \exists^{\neg\varphi}\mu$ if there exists a path π of M such that $\pi_0 = s$ and $M, \rho, \pi \models \mu$, and furthermore, $\rho \frown \pi \models \neg\varphi$ in LTL.
$M, \rho, \pi \models \psi$ for a state formula ψ if $M, \rho, \pi_0 \models \psi$.
$M, \rho, \pi \models \mu_1 \vee \mu_2$ if either $M, \rho, \pi \models \mu_1$ or $M, \rho, \pi \models \mu_2$.
$M, \rho, \pi \models \neg\mu$ if it is not the case that $M, \rho, \pi \models \mu$.
$M, \rho, \pi \models X\mu$ if $M, \rho \frown \langle \pi_0 \pi_1 \rangle, \pi^1 \models \mu$.
$M, \rho, \pi \models \mu U \eta$ if there exists some i such that $M, \rho \frown \hat{\pi}^i, \pi^i \models \eta$ and for each $0 \leq j < i$, $M, \rho \frown \hat{\pi}^j, \pi^j \models \mu$.
We write $M \models \psi$ when $M, \langle s_0 \rangle, s_0 \models \psi$.

Other operators can be obtained using equivalences, e.g., $true = p \vee \neg p$, $false = \neg true$, $\psi_1 \wedge \psi_2 = \neg((\neg\psi_1) \vee (\neg\psi_2))$, $\mu \rightarrow \eta = (\neg\mu) \vee \eta$, $\forall^\varphi\mu = \neg\exists^\varphi\neg\mu$, $\Diamond\mu = true U \mu$, $\Box\mu = \neg\Diamond\neg\mu$, $\mu W \eta = (\mu U \eta) \vee \Box\mu$, etc.

3 Model Checking

Büchi and Generalized Büchi Automata

A *Büchi automaton* is similar to a finite structure, with an additional component $F \subseteq S$ of *accepting states*. An *accepted execution* (or *accepted run*) must satisfy in addition to the conditions described for an execution of a state space above, that at least one accepting state from F appears in the execution infinitely many times.

A *generalized Büchi automaton* can have *multiple* accepting states F_1, F_2, ..., $F_m \subseteq S$. An accepted execution satisfies that at least one state from *each* accepting set appears on it infinitely often. Hence, this generalizes the singleton accepting set of a Büchi automaton. The generalization does not increase the expressive power, but makes the following presentation simpler. Translation from generalized Büchi automata to (simple) Büchi automata is standard. We identify the case of having no accepting set (i.e., when $m = 0$) with the case of having one accepting set consisting of all the states.

The product of two generalized Büchi automata is obtained in a standard way, where the state space consists of pairs of states that agree on their labeling. The acceptance is defined by taking together the accepting sets of each automata. Define $Q \bowtie R$ when $Q \subseteq S^1$, $R \subseteq S^2$ as $\{(q,r) | q \in S^1 \wedge r \in S^2 \wedge L^1|_{\mathcal{P}^1 \cap \mathcal{P}^2}(q) = L^2|_{\mathcal{P}^1 \cap \mathcal{P}^2}(r)\}$ (where $L|_{\mathcal{P}}(s) = L(s) \cap \mathcal{P}$). That is, the set of pairs that agree on the labeling of mutual propositional values. Let $M^1 = (S^1, S^1_0, E^1, \mathcal{P}^1, L^1, F^1_1, \ldots F^1_m)$, $M^2 = (S^2, S^2_0, E^2, \mathcal{P}^2, L^2, F^2_1, \ldots F^2_n)$. Then $M^1 \times M^2 = (S^1 \bowtie S^2, S^1_0 \bowtie S^2_0, \hat{E}, \mathcal{P}^1 \cup \mathcal{P}^2, \hat{L}, F^1_1 \bowtie S^2, \ldots, F^1_m \bowtie S^2, S^1 \bowtie F^2_1, \ldots, S^1 \bowtie F^2_n)$. The relation $\hat{E}$ is defined so that $(s,q)\hat{E}(s',q')$ iff $(s,q),(s',q') \in S^1 \bowtie S^2$, sE^1s' and qE^2q' and $\hat{L}$ is defined such that $\hat{L}(s,q) = L^1(s) \cup L^2(q)$.

A translation from an LTL formula φ into a Büchi automaton that accepts exactly the sequences satisfying φ is quite standard (see, e.g., [10]).

Model Checking Algorithm

We use the following components:

- Let A_φ be a Büchi automaton that accepts the sequences satisfying φ.
- Let $A_{\neg\varphi}$ be a Büchi automaton that accepts the sequences satisfying $\neg\varphi$.
- Let $Det(A_\varphi)$, $Det(A_{\neg\varphi})$ be the determinized *subset construction* for A_φ, $A_{\neg\varphi}$, respectively. A state of $Det(A_\varphi)$ represents a maximal set of possible states where control can be in A_φ after some finite run. Note that all the states of A_φ that participate in one state of $Det(A_\varphi)$ agree on their labeling. Moreover, $Det(A_\varphi)$ has no accepting component (equivalently, all its states are accepting).
- Let A_δ be a Büchi automaton for a formula δ, where $\exists\delta$, $\exists^\varphi\delta$ or $\exists^{\neg\varphi}\delta$ is a subformula of the analysis specification formula and where the maximal state subformulas of δ are treated as new propositional variables. Note that δ an LTL formula.

Suppose that we want to perform model checking for an analysis specification ψ over some base LTL property φ. We perform model checking on several types of automata products:

- $P = M \times Det(A_\varphi) \times Det(A_{\neg\varphi})$. Each state of P is annotated with state subformulas of ψ in a recursive way to be described below. Thus, subformulas can be marked based on earlier marking of their subformulas. Note that there is no acceptance component in this product (thus, all states are considered accepting). If the analysis specification does not use positive quantification ($\exists^\varphi$) or negative quantification ($\exists^{\neg\varphi}$) then we may omit the product component $Det(A_\varphi)$ or $Det(A_{\neg\varphi})$, respectively[3].

- $P_{\varphi,\delta} = P \times A_\varphi \times A_\delta$. This component is designed to mark the $\exists^\varphi\delta$ subformulas that hold in states of P (note that a state subformula of EmCTL* depends not only on the current state of the checked structure M, but also on information depending on the prefix of the execution, as summarized in P by the $Det(A_\varphi)$ and $Det(A_{\neg\varphi})$ components). This product requires that in each of its states, the component from A_φ will be a member of the subset of states in the $Det(A_\varphi)$ component. Since both A_φ and A_δ contribute acceptance conditions, an accepted execution will have to visit infinitely often accepting states from both Büchi automata (this corresponds to the generalized Büchi automata acceptance condition).
- $P_{\varphi,\neg\delta} = P \times A_{\neg\varphi} \times A_\delta$. This product is similar to the previous one, but the product is with a Büchi automaton for $\neg\varphi$, and is designed to mark the $\exists^{\neg\varphi}\delta$ subformulas that hold in states of P.
- $P_\delta = P \times A_\delta$. This product is design to mark the $\exists\delta$ subformulas in P.

The subset construction components $Det(A_\varphi)$ and $Det(A_{\neg\varphi})$ are used to represent the information that is needed in order to check whether a path that starts from some given state, together with the prefix that is leading to it from the start of the execution, satisfy or does not satisfy the base property φ. This is required due to the use of the modalities $\exists^\varphi$ and $\exists^{\neg\varphi}$. Intuitively, for $\exists^\varphi$ it is sufficient to find a state of A_φ that is inside the subset construction, from which one can continue the execution according to the automaton A_φ. Since we do not know which such state can be used to complete the execution into one that satisfies φ, the subset construction keeps all the possibilities. In this way, there is no need to keep the entire prefix executed so far in order to see if there is a continuation that satisfies φ (or $\neg\varphi$, respectively).

A preparatory step before model checking, is to translate the formula to eliminate universal quantification $\forall^\varphi$, $\forall^{\neg\varphi}$ or $\forall$. This can be done based on the equivalence $\forall\delta = \neg\exists\neg\delta$, which carries over also to the other two universal quantifiers. The model checking is performed now recursively on the structure of the CTL* subformula ξ of the formula ψ. In each stage, we recursively mark states of P with either ξ or, otherwise, with $\neg\xi$ (hence in the list of cases below, there is no separate case for a subformula starting with the negation symbol). We identify $\neg\neg\xi$ with ξ.

ξ is a propositional variable. Then mark each state in P with ξ if ξ is in the labeling of the M component (we will, henceforce, omit mentioning repeatedly the fact that if we deal with ξ and a state is not marked with it, we will mark it with $\neg\xi$).

[3] For example, it would be beneficial to rewrite the first example in the introduction in an equivalent form as $(\forall\Box\exists^\varphi \mathit{true})$.

- ξ is a state formula of the form $\psi_1 \vee \psi_2$. Then recursively mark P according to ψ_1 and according to ψ_2, and then mark a state in P with $\psi_1 \vee \psi_2$ if it is marked with either ψ_1 or with ψ_2.
- $\xi = \exists^\varphi \delta$ where δ is a path formula, when replacing each maximal state subformula of δ by a new propositional variable (the names of these variables can be the same as the subformulas they represent). For example consider $\xi = \exists^\varphi (X\kappa \wedge \mu U \eta)$ with state subformulas κ, μ and η. After the replacement, δ becomes an LTL formula, with Büchi automaton A_δ. First, we recursively call the marking procedure with the maximal state subformulas. We then construct $P_{\varphi,\delta} = P \times A_\varphi \times A_\delta$, as described above. A state of P is marked with $\exists^\varphi \delta$ if there exists a state in $P_{\varphi,\delta}$ with first component as in P (and recall that the second component is in accordance with the subset construction of $Det(A_\varphi)$ in P, as described above), and third component an initial state of A_δ. In addition, from this state we need to be able to reach a strongly connected component of $P_{\varphi,\delta}$ where there is an accepting state for A_φ and an accepting state for A_δ (as defined in the product construction).
- $\xi = \exists^{\neg\varphi} \delta$ where δ is a path formula. The procedure is similar to the previous case, replacing $P_{\varphi,\delta}$ with $P_{\neg\varphi,\delta}$.
- $\xi = \exists \delta$. In this case, we use the product P_δ and mark a state in P with δ if there exists a state in the product with the same P component, and an initial state of A_δ, from which a strongly connected component with an accepting state of A_δ is reached.

At the end of the marking, we check whether the initial states of the product P are marked with ψ.

The last three possibilities need an additional explanation. In order to check that a subformula, say $\exists^\varphi \delta$, holds in a state, we first mark all the states of P recursively with the maximal state subformulas in δ, or their negation. We treat the maximal state subformulas of δ as propositional variables, which are already marked in P by our recursive procedure. The satisfaction $\exists^\varphi \delta$ depends not only on the current state, but also on the path so far. As explained before, that information is encapsulated in components of $Det(A_\varphi)$, which are also embedded in the states of P. In a similar way, $\exists^{\neg\varphi} \delta$ holds in a state depending on the information embedded using $Det(A_{\neg\varphi})$.

We then construct the structure $P_{\varphi,\delta}$ and look for strongly connected components that satisfy both acceptance conditions of A_φ and A_δ. We now seek the states of $P_{\varphi,\delta}$ from which such a strongly connected component is reachable with its A_δ component being at its initial state and with P being at state s. We can then conclude that from the corresponding P state s, there is a path that satisfies δ. Moreover, recall that our product construction of $P_{\varphi,\delta}$ guarantees that the A_φ component agrees with one of the possible choices of $Det(A_\varphi)$. Thus, the fact that the acceptance condition of A_φ holds on the reachable strongly connected component guarantees that the formula φ holds when prefixing this path with the finite execution that reached s, taking care of the $\exists^\varphi$ quantification.

Complexity

In a naive implementation, our decision procedure is exponential in the size of the EmCTL* formula, and doubly exponential in the size of the embedded LTL property φ in both time and space. The latter is because of the need to determinize the Büchi automaton constructed for φ in order to keep track of all the possibilities where the control can reside under the current prefix. Note that some LTL specifications result in deterministic Büchi automata, in which case we will not incur the additional exponential cost.

The proof of EXSPACE lower bound of mCTL* can be used here. In particular, the formula used in the reduction (from *exponential tiling*) in [8] can be expressed in EmCTL* as $\forall\Box\exists^{\varphi}\mathit{true}$. Encouragingly, the additional exponent of the model checking is only in the size of the LTL base property φ, and not the checked structure M or the EmCTL* specification analysis formula.

Achieving the EXSPACE complexity with our construction can be done without constructing the automaton P explicitly, nor the components $P_{\varphi,\delta}$, $P_{\neg\varphi,\delta}$, P_{δ}. Instead, we perform a binary search recursively on the structure of the formula (according to the above cases) through the state space of these components. For a path subformula, this involves searching for a lasso shape of a bounded size. One needs to represent in this way states of the construction, where the size of a state of any one of the above components being exponential in the size of the base specification (and polynomial in the size of the analysis specification). Using the standard Savitch construction [14], one needs not represent the entire state space.

Using Multiple Base Properties

A simple modification to the decision procedure that was described earlier in this paper allows us to generalize our framework and quantify over different sets of path, satisfying separate linear properties. In general, we can allow multiple path quantifiers in an *EmCTL** formula to be indexed by different LTL formulas, e.g., $\exists^{\varphi_1}\ldots\exists^{\varphi_m}$. In this case, we need in the translation to take the product of the state space with a determinized version per each indexing property φ_i that appears on a path.

We may also want to use multi-indexed quantifiers, e.g., $\forall^{\varphi_1,\varphi_2}$, which will allow for selecting sequences satisfying several LTL conditions.[4] Note that this is not the same as $\forall^{\varphi_1}\forall^{\varphi_2}$, as, according to the semantic definition, the 1st quantifier is amalgamated into the 2nd one, since the 2nd quantifier provides a state formula, rather than a linear formula for the 1st quantifier.

In this case, we need to generalize the automata products in our construction. That is, we need to take components of the subset construction $Det(A_{\varphi_1})$ and $Det(A_{\varphi_2})$ in P, and take care of the acceptance conditions of both φ_1 and φ_2

[4] Semantically, this is the same as $\forall^{\varphi_1\wedge\varphi_2}$, but it might be beneficial to separate the subformulas, e.g., when two quantifiers share some but not all the subformulas, e.g., in $\forall^{\varphi_1,\varphi_2}\psi_1\wedge\exists^{\varphi_1,\varphi_3}\psi_2$.

(as well as δ) in $P_{\varphi_1,\varphi_2,\delta}$. The complexity will be then doubly exponential in the sum of the lengths of the formulas that appear together in such a quantifier.

We later show how to combine this extension with the ability to assert about both executions inside and outside the checked system.

4 Applications

4.1 Level of Failure

In [7], an application of model checking for the generation of correct mutual exclusion algorithms through genetic programming was described. In order to select versions (mutations) of the code that are better than others, for creating new mutations, one assigns a level of failure, based on model checking analysis. Practice shows that it is not sufficient to do simple model checking with yes/no result. In [7], a deeper analysis, based on converting the LTL property into Streett automata, was performed. We can express such levels of failures and many others using EmCTL* specification, and use the model checking algorithm described here, instead of providing ad-hoc verification procedures. Some examples for formulating such levels of failure are as follows:

All the executions satisfy the base LTL property φ. This is simply expressed as $\forall\varphi$ (or, just φ in LTL). Note that we avoid writing this as $\forall^{\neg\varphi}\mathit{false}$, since this would incur an exponential blowout in model checking.

There exists a finite prefix of an execution from which all the executions are bad. $\exists\Diamond\forall^{\varphi}\mathit{false}$.

For all bad executions, after any prefix, one can always make a decision that would complete the behavior into a good execution. $\forall^{\neg\varphi}\Box\exists^{\varphi}\mathit{true}$. (Note again that this can be written equivalently as $\forall\Box\exists^{\varphi}\mathit{true}$.) Intuitively, this means that in order to have a bad execution, one needs to schedule infinitely many bad choices.

Note that, as in the model checking procedure presented here, the decision procedure in [7] also incurs an additional exponential explosion in the size of the checked property over LTL model checking. This is due to the use of deterministic Street automata in the analysis.

4.2 Reasoning Outside the Checked System

One of the directions for trying to locate errors in a system is based on comparing related executions, e.g., similar pairs of executions such that one of them satisfies the specification and the other does not [5,16]. This is usually done in the context of the system's executions. One of the examples in the introduction already demonstrated how we can use our framework to find, e.g., the point when the execution of a transition terminates the possibility of having further good executions. A related approach [6] searches for the moves that forces the system towards the error in the sense of game theory. We propose that such analysis for locating errors can benefit from comparing not only the executions of the system, but also executions potentially not possible in the system.

One can use multiple indexed quantification to reason within the same formula about executions of the system, as well as executions not belonging to the system.

The quantifiers $\forall, \exists, \forall^{\varphi}, \exists^{\varphi}$ now range over both, executions within the system, and executions leaving the system. To recover the possibility of reasoning within the system, let new quantifiers $\forall^{S}, \exists^{S}, \forall^{S,\varphi}, \exists^{S,\varphi}$ refer to sequences that stay in the system.

The first step for performing model checking is to expand the state space M to represent also executions that do not result from the execution of the checked system, although a prefix might be possible in the system. Intuitively, once the actual system is left, "anything is possible". An efficient coding can be done as follows:

- We assume a labeling of edges of the state space, according to the executed transition.
- The state labeling function is extended to allow a conjunction of propositional variables, either negated or non-negated, which does not necessarily includes all the propositional variables. This allows representing states in a more compact way, i.e., one state can represent all the states that satisfy such a given formula (e.g., as done in [10]).
- A new propositional variable r will mark nodes as *reached.* Its negation $\neg r$ will denote nodes that were not reached through an execution of the checked system.
- We add a new *sink* state τ, labeled with the formula $\neg r$ (with no reference to the other propositional variables). The rest of the nodes of M retain their original label, with the added conjunct r. The state τ is *not* initial if we want to perform model checking with respect to the given initial state s_0.
- From each state s of the structure M we add an edge from s to τ marked with ϵ. There is also a self loop marked with ϵ from τ to itself.

It is straightforward to adapt the constructions of Section 3 to deal with these modified quantifiers. When quantifying over executions of the system and checking an LTL subformula, we must still have a reachable strongly connected component satisfying all the Büchi conditions as in the algorithm before. But now, in addition, we need such a strongly connected component also to have a state marked with r. This is sufficient due to the monotonicity of the $\neg r$ marking.

Safety and Liveness

As an example of properties that can be expressed and checked with quantifiers going beyond the system specification, let us consider safety and liveness.

Halpern and Schneider formally defined the classification of properties in [1] based on Lamport's informal characterization as follows:

Safety. A property φ is a safety property if $\rho \models \psi$ iff for any decomposition $\rho = \rho_1 \frown \rho_2$ there exists a σ such that $\rho_1 \frown \sigma \models \varphi$.

Liveness. A property φ a liveness property iff for any ρ there exists a σ such that $\rho \frown \sigma \models \varphi$.

For properties given as Büchi automata, Alpern and Schneider moreover give constructions for splitting a temporal property φ into two parts, $safe(\varphi)$, which is the safety part related to φ, and $live(\varphi)$¡ which is the liveness part. We can express the liveness and safety parts in our setting as follows: $safe(\varphi)$ states that any finite prefix of an execution of the system can be extended (within the system or not) to satisfy φ:

$$safe(\varphi) = \forall^S \Box \exists^\varphi true$$

$live(\varphi)$ states that any system execution that does not satisfy φ passes through a prefix that has no extension (in the system or not) that satisfies φ:

$$live(\varphi) = \forall^S (\varphi \vee \Diamond \forall^\varphi false)$$

Indeed, the conjunction of $\Box\exists^\varphi true$ and $\Diamond\forall^\varphi false$ is contradictory so that the conjunction of $safe(\varphi)$ and $live(\varphi)$ is equivalent to $\forall^S\varphi$.

$safe(\varphi)$ and $live(\varphi)$ can then be directly checked using our decision procedure. Thus, when a property φ fails to hold in a system, we can check whether this is already due to its safety or liveness part failing to hold. Note, that our decision procedure (the naive version) is doubly exponential in the size of φ. The constructions of [1] is based on modifications of the property automaton, but in order to do model checking, the automaton needs to be complemented – at exponential cost. If model checking is the aim, the worst case complexity of either the original construction of [1] or the use of our algorithm is the same.

5 Related Work and Expressiveness

The closest logic to ours, as far as we know, is mCTL*, presented in [8]. In that logic, quantifying is relativized to the current prefix of execution. But as opposed to EmCTL*, path quantification is done always with respect to the beginning of the path. A special symbol, *present*, keeps track of the current state in a prefix of an execution, from which quantification forces the subformula to be interpreted from the initial state again. One can then refer to this symbol inside the subformula, in order to have the ability to assert about the continuation of that finite prefix (care should be taken, as multiple occurrences of the *present* symbol may refer to different states in the same formula).

It is shown that one can translate every CTL* property into an mCTL* property with linear blowup, where each subformula of the form $\exists\varphi$ (where universal quantification is first eliminated) is translated into $\exists\Diamond(present \wedge \varphi)$. A reverse translation is also available, but may explode the formula (inherently) in a nonelementary way [8].

We can translate each EmCTL* formula into mCTL* in the following way: each subformula of the form $\exists^\varphi\psi$ will be translated into $\exists(\varphi \wedge \Diamond(present \wedge \psi))$

and $\exists^{\neg\varphi}\psi$ will be translated into $\exists((\neg\varphi) \wedge \Diamond(present \wedge \psi))$. This allows one to use the decision procedure from [8]. The translation goes via models called *alternating hesitant tree automata with satellites word automata.* However, that construction is doubly exponential in the size of the overall property.

We suggested in this paper a generalized Büchi automata based construction, which incurs a small modification on the traditional CTL* model checking algorithm [4], and separated the complexity analysis; the additional exponent is only due to the LTL embedded part, and not related to the CTL* part (including path subformulas appearing in the embedding analysis specification formula. The main gain is in the cases where the embedded LTL part is rather small, or translates efficiently into a deterministic Büchi automaton. Of course, in general, there are cases, even for simple safety properties (take e.g., the property that p holds some fixed number of steps before q holds), where the deterministic automaton for the property is exponentially bigger than the nondeterministic version [12,9].

The collection of the following three facts:

- every CTL* formula is in particular an EmCTL* formula (when not using indexed path quantification),
- each EmCTL* formula is translatable to mCTL* (as we showed), and
- each mCTL* formula can be translated into CTL* (proved in [8]),

proves that the expressive power of the three logics is the same. Nevertheless, this does not mean that they are interchangeable for all purposes, nor that the naive approach of picking one formalism, translating the specification into it and performing the model checking, is a sensible choice. One should select the appropriate formalism according to the application (e.g., depending on ease of expressiveness, complexity of model checking with respect to the typically used formulas, etc).

6 Conclusions

We have outlined a specification and model checking methodology that allows to reason in CTL*-style about how an LTL property is satisfied (or not) by a finite state transition system. This is achieved using a formalism we named EmCTL* with path quantifiers $\forall^{\varphi}$, $\exists^{\varphi}$ limiting the quantification to system executions that meet an LTL property φ. Model checking is then in EXPSPACE-complete with respect to the indexing base formula φ.

We provided several examples of analysis specifications that can be used to gain further information about the checked system. This can be interesting in particular after model checking of the base property was performed and an error trace was found, and thence one would like to gain some additional information about the failure. Such an analysis requires an additional exponent in the size of the base specification on top of ordinary LTL model checking. However, it still in PSPACE-complete with respect to the checked system (and also in size of the analysis specification).

We propose that even a restricted subset of our logic and decision procedure can be a useful extension for model checking. In particular, we observe that our choice of examples point out to such a useful subset of formulas: the nesting of indexed quantification is at most two, and the deepest level of subformulas is followed by the trivial state subformulas *true* or *false*. This restricted form contains however the formula $\forall\Box\exists^{\varphi} true$, whose model checking is in EXSPACE complete in the size of the base property φ.

References

1. Alpern, B., Schneider, F.B.: Recognizing Safety and Liveness. Distributed Computing 2, 117–126 (1987)
2. Clarke, E.M., Emerson, E.A.: Design and Synthesis of Synchronization Skeletons using Branching Time Temporal Logic. In: Kozen, D. (ed.) Logic of Programs 1981. LNCS, vol. 131, pp. 52–71. Springer, Heidelberg (1982)
3. Emerson, E.A., Clarke, E.M.: Characterizing Correctness Properties of Parallel Programs using Fixpoints. In: de Bakker, J.W., van Leeuwen, J. (eds.) ICALP 1980. LNCS, vol. 85, pp. 169–181. Springer, Heidelberg (1980)
4. Emerson, E.A., Lei, C.L.: Modalities for Model Checking. Science of Computer Programming 8, 275–306 (1987)
5. Groce, A., Chaki, S., Kroening, D., Strichman, O.: Error Explanation with Distance Metrics. STTT 8, 229–247 (2006)
6. Jin, H., Ravi, K., Somenzi, F.: Fate and Free Will in Error Traces. In: Katoen, J.-P., Stevens, P. (eds.) ETAPS 2002 and TACAS 2002. LNCS, vol. 2280, pp. 445–459. Springer, Heidelberg (2002)
7. Katz, G., Peled, D.: Model Checking Based Genetic Programming with an Application to Mutual Exclusion. In: Ramakrishnan, C.R., Rehof, J. (eds.) TACAS 2008. LNCS, vol. 4963, pp. 141–156. Springer, Heidelberg (2008)
8. Kupferman, O., Vardi, M.Y.: Memoryful Branching Time Logic. In: LICS 2006, Seatle, USA, pp. 265–274 (2006)
9. Kupferman, O., Vardi, M.Y.: Model Checking Safety Properties. In: Halbwachs, N., Peled, D.A. (eds.) CAV 1999. LNCS, vol. 1633, pp. 172–183. Springer, Heidelberg (1999)
10. Gerth, R., Peled, D., Vardi, M.Y., Wolper, P.: Simple on-the-fly automatic verification of linear temporal logic. In: PSTV 1995, pp. 3–18 (1995)
11. Pnueli, A.: The Temporal Logic of Programs. In: 18th IEEE Symposium on Foundations of Computer Science, pp. 46–57 (1977)
12. Pnueli, A., Rosner, R.: On the Synthesis of Reactive Systems. In: POPL 1989, Austin, Texas, pp. 179–190 (1989)
13. Quielle, J.P., Sifakis, J.: Specification and Verification of Concurrent Systems in CESAR. In: 5th International Symposium on Programming, pp. 337–350 (1981)
14. Savitch, W.J.: Relationships between Nondeterministic and Deterministic Tape Complexities. Journal of Computer and System Science 4, 177–192 (1970)
15. Shahar, E.: The TLV System and its Applications, M.Sc. Thesis, The Weizmann Institute of Science
16. Sharygina, N., Peled, D.: A Combined testing and Verification Approach for Software Reliability. In: Oliveira, J.N., Zave, P. (eds.) FME 2001. LNCS, vol. 2021, pp. 611–628. Springer, Heidelberg (2001)

Correcting a Space-Efficient Simulation Algorithm*

Rob van Glabbeek[1,2] and Bas Ploeger[3,**]

[1] National ICT Australia, Locked Bag 6016, Sydney, NSW1466, Australia
[2] School of Computer Science and Engineering, The University of New South Wales, Sydney, NSW 2052, Australia
[3] Eindhoven University of Technology, Design and Analysis of Systems Group, P.O. Box 513, 5600 MB Eindhoven, The Netherlands

Abstract. Although there are many efficient algorithms for calculating the simulation preorder on finite Kripke structures, only two have been proposed of which the space complexity is of the same order as the size of the output of the algorithm. Of these, the one with the best time complexity exploits the representation of the simulation problem as a generalised coarsest partition problem. It is based on a fixed-point operator for obtaining a generalised coarsest partition as the limit of a sequence of partition pairs. We show that this fixed-point theory is flawed, and that the algorithm is incorrect. Although we do not see how the fixed-point operator can be repaired, we correct the algorithm without affecting its space and time complexity.

1 Introduction

The *simulation preorder* [17] is a behavioural refinement relation on concurrent systems, represented as Kripke structures or labelled transition systems, that plays a crucial rôle in compositional verification and model checking. It preserves the existential and universal fragments of temporal and modal logics. For CTL* [6] this is shown in [5], and for the modal μ-calculus [14] it is shown in [16]. This makes it possible to combat the state explosion problem in model checking by minimising the state space of a given system modulo simulation equivalence before checking the validity of relevant properties within those fragments. Given that the simulation preorder is a precongruence for parallel composition [11], components in parallel compositions can even be minimised individually.

Simulation equivalence is also used directly in equivalence checking [15] of finite-state processes. Often deciding the simulation preorder between processes is the most appropriate method of showing that two systems are related by another preorder, that may be appropriate for the task at hand. In applications where deadlock behaviour plays a crucial rôle, the *ready simulation preorder* [1] is widely regarded to be an appropriate behavioural refinement relation for matching an implementation with a specification. Via a straightforward reduction (the computation of the initial partition ER_1 in [2]),

* This is an extended abstract; all proofs are omitted. They can be found in the full version [10].

** This author is partially supported by the Netherlands Organisation for Scientific Research (NWO) under VoLTS grant number 612.065.410.

A. Gupta and S. Malik (Eds.): CAV 2008, LNCS 5123, pp. 517–529, 2008.

finding a ready simulation between two processes is as hard as finding a plain simulation. In applications where deadlock behaviour plays no rôle, *trace inclusion* is often proposed as an appropriate refinement relation. However, deciding trace inclusion on finite-state processes is PSPACE-hard [19], and as the simulation preorder is the coarsest preorder included in trace inclusion that is known to be decidable in polynomial time [2,3,8,12,18,20], establishing a simulation between two processes is a favourite way of showing that they are related by trace inclusion.

In many crucial applications, space rather than time becomes the bottleneck as the input graph grows [4,7,8,13]. Hence, simulation algorithms with minimal space complexity are of particular interest. These are the ones by Bustan and Grumberg [3] and by Gentilini, Piazza and Policriti [8]. For an input graph with N states, T transitions and S simulation equivalence classes, the space complexity of both algorithms is $\mathcal{O}(S^2 + N \log S)$. This can be considered minimal: $\mathcal{O}(S^2)$ space is needed for storing the simulation preorder as a partial order on simulation equivalence classes and $\mathcal{O}(N \log S)$ space is needed to store for every state, the equivalence class to which it belongs. Of these algorithms, the one by Gentilini *et al.* has a better time complexity: $\mathcal{O}(S^2 T)$. A more time-efficient algorithm is the one by Ranzato and Tapparo [18], but it is less space efficient.

The approach of Gentilini *et al.* represents the simulation problem as a generalised coarsest partition problem (GCPP). According to the authors, this problem can be solved by approximating the greatest fixed point of a decreasing operator on partition pairs that they define in their paper. They give a partitioning algorithm to compute this fixed point for any legal input. We recite this definition and a part of the algorithm in Sect. 3. In Sect. 4 we show that the operator is flawed because it is not uniquely defined for all partition pairs. We give an instance of the GCPP for which repeated application of the operator does not lead to a unique fixed point. We also show that on this example the partitioning algorithm irrevocably allocates two simulation-equivalent states to different simulation-equivalence classes, and subsequently deadlocks.

In Sect. 5 we define a simple, yet inefficient fixed-point operator for which we prove correctness. This operator is not meant to be an improvement over the original one, but merely serves as an expedient for establishing correctness of the algorithm that we present in Sect. 6. This algorithm is obtained from that of Gentilini *et al.* by means of a few simple corrections; consequently, it benefits from the key ideas behind the original partitioning algorithm and has the same time and space complexities. Yet its correctness proof requires entirely new techniques and is surprisingly non-trivial. We also show that no fixed-point operator can be defined that captures the behaviour of this algorithm.

2 Preliminaries

Partitions and relations. For any set S, a *partition* over S is a set $\Sigma \subseteq \mathcal{P}(S)$ such that $\bigcup \Sigma = S$ and $\forall \alpha \in \Sigma \,.\, \alpha \neq \emptyset \wedge \forall \beta \in \Sigma \,.\, \alpha \neq \beta \Rightarrow \alpha \cap \beta = \emptyset$. For any $s \in S$ we denote by $[s]_\Sigma$ the block $\alpha \in \Sigma$ such that $s \in \alpha$. Given two partitions Σ and Π we say Π *is finer than* Σ iff for every $\alpha \in \Pi$ there exists an $\alpha' \in \Sigma$ such that $\alpha \subseteq \alpha'$. For any set S, we denote by $\mathcal{I}(S)$ the *identity relation* over S, *i.e.* $\mathcal{I}(S) = \{(s,s) \mid s \in S\}$. For any relation P, we denote by P^+ the *transitive closure* of P.

Graphs. A *(directed) graph* is a tuple $(N, \rightarrow)$ where N is a finite set of nodes and $\rightarrow \subseteq N \times N$ is a set of directed transitions between those nodes. A *labelled graph* is a tuple $(N, \rightarrow, \Sigma)$ where $(N, \rightarrow)$ is a graph and Σ is a partition over N. For a graph $(N, \rightarrow)$, $a \in N$ and $\beta \subseteq N$, we write $a \rightarrow \beta$ if $\exists b \in \beta \,.\, a \rightarrow b$. Moreover, we define the relations $\rightarrow_\exists$ and $\rightarrow_\forall$ over $\mathcal{P}(N)$ as follows, for any $\alpha, \beta \subseteq N$:

$$\alpha \rightarrow_\exists \beta \Leftrightarrow \exists a \in \alpha \,.\, a \rightarrow \beta \qquad\qquad \alpha \rightarrow_\forall \beta \Leftrightarrow \forall a \in \alpha \,.\, a \rightarrow \beta.$$

Simulations. For any labelled graph $(N, \rightarrow, \Sigma)$ a relation $R \subseteq N \times N$ is a *simulation* iff for any $a, b \in N$, $(a, b) \in R$ implies:

- $[a]_\Sigma = [b]_\Sigma$ and
- $\forall c \in N \,.\, a \rightarrow c \Rightarrow \exists d \in N \,.\, b \rightarrow d \wedge (c, d) \in R$.

We say that a is *simulated by* b, denoted $a \subsetneq_{\rightarrow} b$, iff there exists a simulation R such that $(a, b) \in R$. It is well known and easy to check that $\subsetneq_{\rightarrow}$ is a preorder, *i.e.* a reflexive and transitive relation, on N, and moreover the largest simulation. We say that a and b are *simulation equivalent*, denoted $a \rightleftarrows b$, iff $a \subsetneq_{\rightarrow} b$ and $b \subsetneq_{\rightarrow} a$.

The simulation problem. Given a labelled graph $G = (N, \rightarrow, \Sigma)$, the *simulation problem* over G consists in finding the simulation preorder $\subsetneq_{\rightarrow}$ on G.

A variant of the simulation problem asks, given a labelled graph $(N, \rightarrow, \Sigma)$ and two nodes $a, b \in N$, whether $a \subsetneq_{\rightarrow} b$. In general, no methods to solve this problem are known that are more efficient than computing the entire relation $\subsetneq_{\rightarrow} \subseteq N \times N$ and looking up whether $(a, b) \in \subsetneq_{\rightarrow}$. Another variant of the simulation problem merely asks to find the simulation equivalence relation $\rightleftarrows$ rather than the preorder $\subsetneq_{\rightarrow}$. Again, no methods to solve that problem are known that do not amount to finding $\subsetneq_{\rightarrow}$ as well.

Typically, the simulation problem arises in the context of *Kripke structures* or *labelled transition systems*. It is trivial to encode a Kripke structure as a labelled graph in such a way that the simulation preorder on the Kripke structure agrees with the one on its labelled graph representation. Likewise, it is not hard to reduce the simulation problem for labelled transition systems to that for labelled graphs. Alternatively one can enrich the theory in a straightforward way to deal with transition labels as well, so that it is applicable to labelled transition systems directly.

The generalised coarsest partition problem. Given a graph $G = (N, \rightarrow)$, a *partition pair* over G is a pair $\langle \Sigma, P \rangle$ where Σ is a partition over N and $P \subseteq \Sigma \times \Sigma$ is a reflexive, acyclic relation over Σ. A partition pair $\langle \Sigma, P \rangle$ is called *transitive* if P is transitive, and hence a partial order. Given a partition Σ, a partition Π finer than Σ, and a relation P over Σ, we denote by $P(\Pi)$ the *induced relation* of P on Π:

$$P(\Pi) = \{(\alpha, \beta) \in \Pi \times \Pi \mid \exists (\alpha', \beta') \in P \,.\, \alpha \subseteq \alpha' \wedge \beta \subseteq \beta'\}.$$

We define a *partial order* $\leq$ on partition pairs by writing, for any partition pairs $\langle \Sigma, P \rangle$ and $\langle \Pi, Q \rangle$: $\langle \Pi, Q \rangle \leq \langle \Sigma, P \rangle$ iff Π is finer than Σ and $Q \subseteq P(\Pi)$. Given a graph $G = (N, \rightarrow)$, we say a partition pair $\langle \Sigma, P \rangle$ over G is *stable with respect to* $\rightarrow$ [8] iff:

$$\forall \alpha, \beta, \gamma \in \Sigma \,.\, ((\alpha, \beta) \in P \wedge \alpha \rightarrow_\exists \gamma) \Rightarrow \exists \delta \in \Sigma \,.\, (\gamma, \delta) \in P \wedge \beta \rightarrow_\forall \delta.$$

Given a graph $G = (N, \rightarrow)$ and a partition pair $\langle \Sigma, P \rangle$ over G, the *generalised coarsest partition problem* (GCPP) [8] consists in finding a $\leq$-maximal partition pair $\langle \Xi, \preceq \rangle$ such that $\langle \Xi, \preceq \rangle \leq \langle \Sigma, P^+ \rangle$ and $\langle \Xi, \preceq \rangle$ is stable with respect to $\rightarrow$.

The simulation problem as a GCPP. Let $G = (N, \rightarrow, \Sigma)$ be a labelled graph. Any preorder $\sqsubseteq$ on N can be represented as a partition pair $\mathsf{PP}(\sqsubseteq) := \langle \Pi, \preceq \rangle$, as follows: Π is the set of equivalence classes of N w.r.t. the equivalence relation $\equiv := \sqsubseteq \cap \sqsubseteq^{-1}$ induced by $\sqsubseteq$, and $\preceq$ is given by $[a]_\Pi \preceq [b]_\Pi$ iff $a \sqsubseteq b$. Note that $\preceq$ is a partial order. Moreover, if $\sqsubseteq$ is a simulation then $\mathsf{PP}(\sqsubseteq)$ is stable w.r.t. $\rightarrow$ and $\mathsf{PP}(\sqsubseteq) \leq \langle \Sigma, \mathcal{I}(\Sigma) \rangle$.

Any partition pair $\langle \Pi, Q \rangle$ over the graph $(N, \rightarrow)$ can be represented as a relation $R_{\langle \Pi, Q \rangle} \subseteq N \times N$ as follows: $(a, b) \in R_{\langle \Pi, Q \rangle}$ iff $\exists (\alpha, \beta) \in Q \,.\, a \in \alpha \wedge b \in \beta$. Note that if $\langle \Pi, Q \rangle$ is stable w.r.t. $\rightarrow$ and $\langle \Pi, Q \rangle \leq \langle \Sigma, \mathcal{I}(\Sigma) \rangle$ then $R_{\langle \Pi, Q \rangle}$ is a simulation. Moreover, $\langle \Pi, Q \rangle \leq \langle \Pi', Q' \rangle$ iff $R_{\langle \Pi, Q \rangle} \subseteq R_{\langle \Pi', Q' \rangle}$. Also note that $R_{\mathsf{PP}(\sqsubseteq)} = \sqsubseteq$.

Hence $\mathsf{PP}(\subsetneq)$ is the solution of the GCPP on $(N, \rightarrow)$ and $\langle \Sigma, \mathcal{I}(\Sigma) \rangle$. In particular, the GCPP, when applied to partition pairs of the form $\langle \Sigma, \mathcal{I}(\Sigma) \rangle$ (plain partitions), always has a unique solution $\langle \Xi, \preceq \rangle$, in which moreover $\preceq$ is always a partial order.[1]

3 The GCPP Solution of Gentilini, Piazza and Policriti

To solve the GCPP, Gentilini, Piazza and Policriti [8] introduce the following operator:

Definition 4.11 in [8] (Operator σ). Let $G = (N, \rightarrow)$ and $\langle \Sigma, P \rangle$ be a partition pair over G. The partition pair $\langle \Pi, Q \rangle = \sigma(\langle \Sigma, P \rangle)$ is defined as follows:

(1σ) Π is the coarsest partition finer than Σ such that

(a) $\forall \alpha \in \Pi \; \forall \gamma \in \Sigma (\alpha \rightarrow_\exists \gamma \Rightarrow \exists \delta \in \Sigma ((\gamma, \delta) \in P \wedge \alpha \rightarrow_\forall \delta))$;

(2σ) Q is maximal such that $Q \subseteq P(\Pi)$ and if $(\alpha, \beta) \in Q$, then

(b) $\forall \gamma \in \Sigma (\alpha \rightarrow_\forall \gamma \Rightarrow \exists \gamma' \in \Sigma ((\gamma, \gamma') \in P \wedge \beta \rightarrow_\exists \gamma'))$ and

(c) $\forall \gamma \in \Pi (\alpha \rightarrow_\forall \gamma \Rightarrow \exists \gamma' \in \Pi ((\gamma, \gamma') \in Q \wedge \beta \rightarrow_\exists \gamma'))$.

They argue that applying σ iteratively on an initial partition pair $\langle \Sigma_0, P_0 \rangle$ yields a sequence of partition pairs $\langle \Sigma_i, P_i \rangle_{i \geq 0}$ with $\langle \Sigma_{i+1}, P_{i+1} \rangle = \sigma(\langle \Sigma_i, P_i \rangle)$. By construction, this sequence is decreasing, in the sense that $\langle \Sigma_{i+1}, P_{i+1} \rangle \leq \langle \Sigma_i, P_i \rangle$. Hence it will reach a fixed point $\langle \Sigma_k, P_k \rangle = \sigma(\langle \Sigma_k, P_k \rangle)$. This is the solution to the GCPP.

Applying this, they give a partitioning algorithm to solve the GCPP. We have included it here as Algorithm 1 and call it PA_{GPP}. It takes as input a graph $(N, \rightarrow)$ and a transitive partition pair $\langle \Sigma, P \rangle$ and repeatedly calls the following functions to compute σ until a fixed point is reached: $\text{REFINE}_{\text{GPP}}$ which computes the partition Π of (1σ) and $\text{UPDATE}_{\text{GPP}}$ which computes the relation Q of (2σ). The boolean variable *change* is set to $\top$ by $\text{REFINE}_{\text{GPP}}$ iff its output partition differs from its input partition. We have

[1] The same reasoning extends to the GCPP applied to any partition pairs, but this requires considering simulations on structures of the form $(N, \rightarrow, \Sigma, \preceq)$ with $(N, \rightarrow, \Sigma)$ a labelled graph, and $\preceq$ a partial order on Σ; the first clause in the definition of simulation then becomes $[a]_\Sigma \preceq [b]_\Sigma$.

Algorithm 1. The partitioning algorithm of [8]: $\text{PA}_{\text{GPP}}((N, \rightarrow), \langle \Sigma, P \rangle)$

$\mathit{change} := \top; i := 0; \Sigma_0 := \Sigma; P_0 := P;$
while change **do**
  $\mathit{change} := \bot;$
  $\Sigma_{i+1} := \text{REFINE}_{\text{GPP}}(\Sigma_i, P_i, \mathit{change});$
  $P_{i+1} := \text{UPDATE}_{\text{GPP}}(\Sigma_i, P_i, \Sigma_{i+1});$
  $i := i + 1;$
end while

Algorithm 2. The refine function of [8]: $\text{REFINE}_{\text{GPP}}(\Sigma_i, P_i, \mathit{change})$

1: $\Sigma_{i+1} := \Sigma_i;$
2: **for all** $\alpha \in \Sigma_{i+1}$ **do** $\mathit{Stable}(\alpha) := \emptyset$; **end for**
3: **for all** $\gamma \in \Sigma_i$ **do** $\mathit{Row}(\gamma) := \{\gamma' \mid (\gamma, \gamma') \in P_i\}$; **end for**
4: Let Sort be a reverse topological sorting of Σ_i w.r.t. P_i;
5: **while** $\mathit{Sort} \neq \emptyset$ **do**
6:   $\gamma := \mathit{dequeue}(\mathit{Sort});$
7:   $A := \emptyset;$
8:   **for all** $\alpha \in \Sigma_{i+1}, \alpha \rightarrow_{\exists} \gamma, \mathit{Stable}(\alpha) \cap \mathit{Row}(\gamma) = \emptyset$ **do**
9:     $\alpha_1 := \alpha \cap \rightarrow^{-1}(\gamma);$
10:     $\alpha_2 := \alpha \setminus \alpha_1;$
11:     **if** $\alpha_2 \neq \emptyset$ **then** $\mathit{change} := \top$; **end if**
12:     $\Sigma_{i+1} := \Sigma_{i+1} \setminus \{\alpha\};$
13:     $A := A \cup \{\alpha_1, \alpha_2\};$
14:     $\mathit{Stable}(\alpha_1) := \mathit{Stable}(\alpha) \cup \{\gamma\};$
15:     $\mathit{Stable}(\alpha_2) := \mathit{Stable}(\alpha);$
16:   **end for**
17:   $\Sigma_{i+1} := \Sigma_{i+1} \cup A;$
18:   $\mathit{Sort} := \mathit{Sort} \setminus \{\gamma\};$
19: **end while**
20: **return** $\Sigma_{i+1};$

included the $\text{REFINE}_{\text{GPP}}$ function as Algorithm 2. In line 4 of this algorithm, a "reverse topological sorting of Σ_i w.r.t. P_i" indicates an ordered listing of the elements of Σ_i such that if $(\gamma, \delta) \in P_i$ then δ occurs prior to γ.

4 Incorrectness of the Fixed-Point Operator

Following the definition of σ, the authors claim that for any partition pair $\langle \Sigma, P \rangle$, if $\langle \Pi, Q \rangle = \sigma(\langle \Sigma, P \rangle)$ then Q is acyclic. We give an example that counters this claim.

Counterexample 1. Consider the graph in Fig. 1(a) and the partition pair $\langle \Sigma, P \rangle$ with $\Sigma = \{\alpha, \beta, \gamma, \delta\}$ as depicted and $P = \mathcal{I}(\Sigma) \cup \{(\beta,\delta), (\delta,\gamma)\}$. Let $\langle \Pi, Q \rangle = \sigma(\langle \Sigma, P \rangle)$, then

$$\Pi = \{\alpha_1, \alpha_2, \beta, \gamma, \delta\} \qquad Q = \mathcal{I}(\Pi) \cup \{(\alpha_1, \alpha_2), (\alpha_2, \alpha_1), (\beta, \delta), (\delta, \gamma)\}$$

where $\alpha_1 = \{a_1\}$ and $\alpha_2 = \{a_2\}$. Q is not acyclic, which counters the claim. □

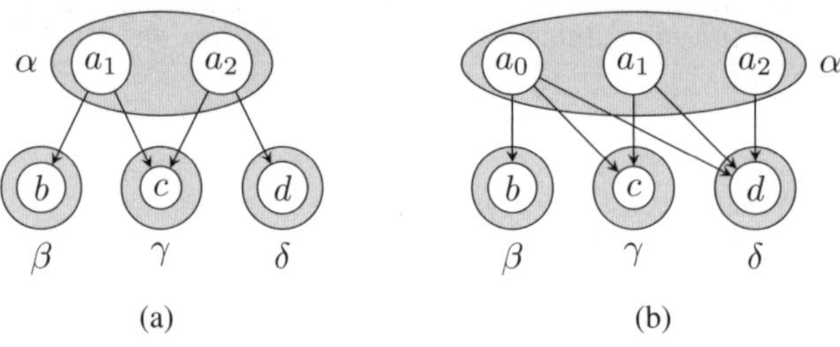

Fig. 1. Counterexamples for (a) acyclicity of Q and (b) well-definedness of σ

This counterexample shows that applying σ to a given partition pair does not necessarily yield another partition pair. After all, for that the resulting relation has to be acyclic.

However, a more fundamental theorem that the authors claim to have proven, turns out not to hold. Theorem 4.13 states that for every partition pair $\langle \Sigma, P\rangle$ there exists a unique $\leq$-maximal partition pair $\langle \Pi, Q\rangle \leq \langle \Sigma, P\rangle$ satisfying conditions (a), (b) and (c) of Definition 4.11, *i.e.* the σ operator is well-defined, and a function. This theorem is countered by the following example.

Counterexample 2. Consider the graph in Fig. 1(b) and the partition pair $\langle \Sigma, P\rangle$ with $\Sigma = \{\alpha, \beta, \gamma, \delta\}$ as depicted and $P = \mathcal{I}(\Sigma) \cup \{(\beta, \gamma), (\gamma, \delta)\}$. Let $\langle \Pi, Q\rangle$ and $\langle \Pi', Q'\rangle$ be partition pairs such that:

$$\Pi = \{\alpha_0, \alpha_1, \beta, \gamma, \delta\} \qquad Q = \mathcal{I}(\Pi) \cup \{(\alpha_0, \alpha_1), (\alpha_1, \alpha_0), (\beta, \gamma), (\gamma, \delta)\}$$
$$\Pi' = \{\alpha_0', \alpha_1', \beta, \gamma, \delta\} \qquad Q' = \mathcal{I}(\Pi') \cup \{(\alpha_0', \alpha_1'), (\alpha_1', \alpha_0'), (\beta, \gamma), (\gamma, \delta)\}$$

where $\alpha_0 = \{a_0, a_1\}$, $\alpha_1 = \{a_2\}$, $\alpha_0' = \{a_0\}$ and $\alpha_1' = \{a_1, a_2\}$. Both $\langle \Pi, Q\rangle$ and $\langle \Pi', Q'\rangle$ satisfy conditions (a), (b) and (c) of Definition 4.11, but neither is the $\leq$-largest. The only partition pair greater than both $\langle \Pi, Q\rangle$ and $\langle \Pi', Q'\rangle$ and at most as large as $\langle \Sigma, P\rangle$, is $\langle \Sigma, P\rangle$ itself, but $\langle \Sigma, P\rangle$ does not satisfy (a). Hence, this example counters Theorem 4.13 of [8] and shows that σ is not well-defined. □

Following Theorem 4.13, the authors present their main fixed-point theorem which states that the solution of the GCPP over a graph G and partition pair $\langle \Sigma, P\rangle$ can be computed by applying σ to $\langle \Sigma, P\rangle$ finitely many times until a fixed point is reached (Theorem 4.14). In this theorem, the authors demand that P be transitive. One might be inclined to think that Counterexample 2 does not affect this theorem, as we used a non-transitive P. We now show that this is not the case: the main theorem indeed loses its meaning due to our counterexample for Theorem 4.13. To do so, we first give an example in which the application of σ to a transitive partition pair produces a non-transitive partition pair.

Example 3. Consider the graph in Fig. 2(a) and the partition pair $\langle \Sigma, P\rangle$ with $\Sigma = \{\alpha, \beta, \gamma\}$ as depicted and $P = \mathcal{I}(\Sigma)$. Let $\langle \Pi, Q\rangle = \sigma(\langle \Sigma, P\rangle)$, then:

$$\Pi = \{\alpha_1, \alpha_2, \alpha_3, \beta, \gamma\} \qquad Q = \mathcal{I}(\Pi) \cup \{(\alpha_3, \alpha_1), (\alpha_1, \alpha_2)\}$$

where $\alpha_1 = \{a_0, a_1\}$, $\alpha_2 = \{a_2\}$ and $\alpha_3 = \{a_3\}$. □

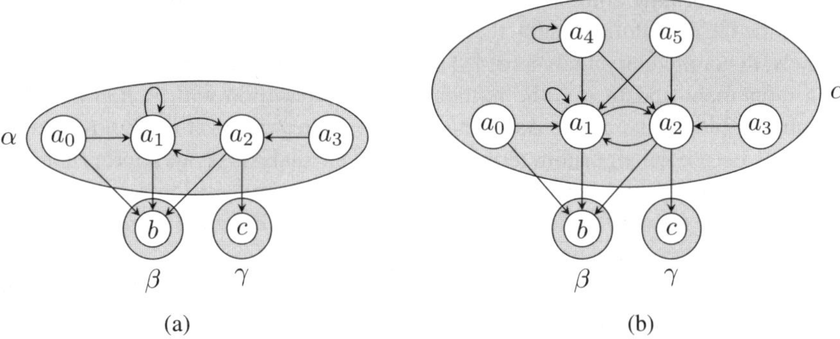

Fig. 2. (a) Example for which σ produces a non-transitive relation Q and (b) counterexample for correctness of σ

Our final counterexample shows that σ is not suitable for computing the solution of the GCPP, and is constructed by embedding Counterexample 2 in Example 3, such that the first application of σ produces a non-transitive partition pair on which σ is not well-defined.

Counterexample 4. Consider the graph in Fig. 2(b) and the partition pair $\langle \Sigma, P \rangle$ with $\Sigma = \{\alpha, \beta, \gamma\}$ as depicted and $P = \mathcal{I}(\Sigma)$. Let $\langle \Pi, Q \rangle = \sigma(\langle \Sigma, P \rangle)$, then:

$$\Pi = \{\alpha_1, \alpha_2, \alpha_3, \beta, \gamma\} \qquad\qquad Q = \mathcal{I}(\Pi) \cup \{(\alpha_3, \alpha_1), (\alpha_1, \alpha_2)\}$$

where $\alpha_1 = \{a_0, a_1\}$, $\alpha_2 = \{a_2\}$ and $\alpha_3 = \{a_3, a_4, a_5\}$. Now, in $\langle \Pi, Q \rangle$ the block α_3 has to be split, because $\alpha_3 \rightarrow_\exists \alpha_3$ but $\neg\exists \delta \in \Pi \,.\, ((\alpha_3, \delta) \in Q \wedge \alpha_3 \rightarrow_\forall \delta))$. There are two candidate partition pairs for $\sigma(\langle \Pi, Q \rangle)$: α_3 can be split into either $\alpha_{3,0} = \{a_4\}$ and $\alpha_{3,1} = \{a_3, a_5\}$ or $\alpha'_{3,0} = \{a_4, a_5\}$ and $\alpha'_{3,1} = \{a_3\}$. However, neither of these is greater than the other, so a unique $\leq$-maximal partition pair does not exist. □

When splitting α_3 in Counterexample 4, the REFINE$_{\text{GPP}}$ function of algorithm PA$_{\text{GPP}}$ splits the block into $\alpha_{3,0}$ and $\alpha_{3,1}$. Observe that this is wrong: a_4 and a_5 should not end up in different equivalence classes because $a_4 \rightleftarrows a_5$. This split also results in UPDATE$_{\text{GPP}}$'s returning a cyclic relation. In the subsequent iteration of PA$_{\text{GPP}}$, the execution of REFINE$_{\text{GPP}}$ then fails because there is no reverse topological sorting of the partition w.r.t. the cyclic relation (line 4).

5 An Auxiliary Fixed-Point Operator

In this section we introduce a fixed-point operator ρ to solve the GCPP and prove its correctness. The definition of ρ is straightforward: it is based directly on the stability condition of Sect. 2.

We emphasise that ρ is not intended to be an improvement over the σ operator of Sect. 3 in any way: it is a less advanced operator than σ aimed to be. The purpose of σ was to compute the solution to the GCPP efficiently, while ρ gives rise to an algorithm that has an inferior time complexity of $\mathcal{O}(S^3T)$ where S is the number of equivalence classes of the GCPP solution and T the number of transitions of the input graph.

Namely, the complexity analysis of [8] uses that, as long as no fixed point is reached, in each refinement-update step the refinement of the partition will be non-trivial, *i.e.* the number of blocks increases. As a consequence, there will be at most S refinement-update steps before the algorithm terminates. Such an analysis is not appropriate for ρ: applying ρ repeatedly could involve many steps in which the partition does not change. Consequently, the number of iterations of the algorithm is bounded merely by the size of a relation on the eventual partition, *i.e.* by S^2.

The sole purpose of ρ is to serve as an auxiliary operator for establishing the correctness of the algorithm that we present in Sect. 6. That algorithm has the same time complexity as PA_{GPP} and does not correspond to any fixed-point operator, as we show in the same section.

Definition 1 (Operator ρ). *Let $\langle \Sigma, P \rangle$ be a transitive partition pair over a graph $(N, \rightarrow)$. Then $\rho(\langle \Sigma, P \rangle)$ is the $\leq$-largest partition pair $\langle \Pi, Q \rangle \leq \langle \Sigma, P \rangle$ satisfying*

$$(1)\ \forall \alpha, \beta \in \Pi \,.\, \forall \gamma \in \Sigma \,.\, ((\alpha, \beta) \in Q \wedge \alpha \rightarrow_{\exists} \gamma \Rightarrow \exists \delta \in \Sigma \,.\, ((\gamma, \delta) \in P \wedge \beta \rightarrow_{\forall} \delta)).$$

Alternatively, ρ could be defined just like σ of Definition 4.11, but insisting that its input partition pair is transitive, and omitting clause (c). It is not hard to check that this definition is equivalent to the one above. The correctness of Definition 1 is ensured by the following.

Proposition 1. *Let $\langle \Sigma, P \rangle$ be a transitive partition pair over a graph $(N, \rightarrow)$. Then there exists a $\leq$-largest partition pair $\langle \Pi, Q \rangle \leq \langle \Sigma, P \rangle$ that satisfies* (1). *Moreover, Q is transitive.*

Proposition 2. *The operator ρ is monotone with respect to $\leq$: if $\langle \Sigma, P \rangle$ and $\langle \Sigma', P' \rangle$ are transitive partition pairs with $\langle \Sigma, P \rangle \leq \langle \Sigma', P' \rangle$, then $\rho(\langle \Sigma, P \rangle) \leq \rho(\langle \Sigma', P' \rangle)$.*

Since $\rho(\langle \Sigma, P \rangle) \leq \langle \Sigma, P \rangle$ and $\leq$ is a partial order on a finite set, we obtain:

Proposition 3. *Let $\langle \Sigma, P \rangle$ be a transitive partition pair over a graph. Then for some $n \geq 0$, $\rho^{n+1}(\langle \Sigma, P \rangle) = \rho^n(\langle \Sigma, P \rangle)$, i.e. repeated application of ρ leads to a fixed point.*

The solution to the GCPP over an input graph G and an initial partition pair $\langle \Sigma, P \rangle$ over G can be obtained by repeatedly applying ρ to $\langle \Sigma, P^+ \rangle$. The following lemmata say that as soon as a fixed point is reached, the resulting partition pair is stable. Moreover, each of the intermediate partition pairs is larger than or equal to the solution of the GCPP. It then follows that the obtained fixed point is in fact the solution to the GCPP.

Lemma 1. *Let $\langle \Sigma, P \rangle$ be a transitive partition pair over a graph $(N, \rightarrow)$. Then $\rho(\langle \Sigma, P \rangle) = \langle \Sigma, P \rangle$ if and only if $\langle \Sigma, P \rangle$ is stable with respect to $\rightarrow$.*

Lemma 2. *Let $\langle \Sigma, P \rangle$ and $\langle \Pi, Q \rangle$ be partition pairs over a graph G, with Q transitive, and let $\langle \Xi, \preceq \rangle$ be the solution of the GCPP over G and $\langle \Sigma, P \rangle$. If $\langle \Xi, \preceq \rangle \leq \langle \Pi, Q \rangle$ then $\langle \Xi, \preceq \rangle \leq \rho(\langle \Pi, Q \rangle)$.*

Theorem 1. *Let $\langle \Sigma, P \rangle$ be a partition pair over a graph $G = (N, \rightarrow)$ and $\langle \Xi, \preceq \rangle$ be the solution of the GCPP over G and $\langle \Sigma, P \rangle$. Let $n \geq 0$ be such that $\rho^{n+1}(\langle \Sigma, P^+ \rangle) = \rho^n(\langle \Sigma, P^+ \rangle)$. Then $\rho^n(\langle \Sigma, P^+ \rangle) = \langle \Xi, \preceq \rangle$.*

6 A Correct and Efficient Algorithm

Our repaired partitioning algorithm is called PA, see Algorithm 3. The variable *change* and the input graph $(N, \rightarrow)$ have global scope: they can be accessed from any function. Note however, that UPDATE$_{\text{GPP}}$ does not access *change*.

Algorithm 3. The repaired partitioning algorithm: PA$((N, \rightarrow), \langle \Sigma, P \rangle)$

```
1: Σ_1 := REFINE(Σ, P);
2: P_1 := UPDATE_GPP(Σ, P, Σ_1);
3: change := ⊤; i := 1;
4: while change do
5:     change := ⊥;
6:     Σ_{i+1} := REFINE(Σ_i, P_i);
7:     P_{i+1} := UPDATE_GPP(Σ_i, P_i, Σ_{i+1});
8:     i := i + 1;
9: end while
```

Our corrections of the algorithm are two. Firstly, it is ensured that at least two refinement-update steps are taken before the algorithm terminates (lines 1 and 2). The necessity of this correction is explained in Sect. 6.1. Secondly, the most important error — the one resulting from the incorrect σ operator — is repaired by the new REFINE function, Algorithm 4. It contains a few minor improvements over REFINE$_{\text{GPP}}$: using list notations for variable *Sort* and preventing empty blocks from being added to Π. However, the actual correction is in line 21: if for some $\gamma \in \Sigma$ and $\alpha \in \Pi$ with $\alpha \rightarrow_{\exists} \gamma$ we have $\mathit{Stable}(\alpha) \cap \mathit{Row}(\gamma) \neq \emptyset$ then we add γ to $\mathit{Stable}(\alpha)$.

We use the ρ operator of Sect. 5 to prove correctness of PA in Sect. 6.2. Its space and time complexities are the same as for PA$_{\text{GPP}}$: no additional space is needed and the corrections do not increase the time complexity. Finally, in Sect. 6.3 we show that there is no fixed-point operator that captures the refinement performed by our REFINE function.

6.1 The Correction of Another Mistake

Apart from the error in PA$_{\text{GPP}}$ that results from the incorrect σ operator, we found another mistake in the algorithm. We describe it in this section and propose a solution. The mistake is shown by the following example.

Example 5. Consider the graph $G = (N, \rightarrow)$ on the right and the partition pair $\langle \Sigma, P \rangle$ with $\Sigma = \{\alpha, \beta\}$ as depicted and $P = \mathcal{I}(\Sigma) \cup \{(\alpha, \beta)\}$. Observe that the solution to the GCPP over G and $\langle \Sigma, P \rangle$ is $\langle \Xi, \preceq \rangle$ with $\Xi = \{\alpha_0, \alpha_1, \beta\}$ and $\preceq = \mathcal{I}(\Xi) \cup \{(\alpha_1, \alpha_0)\}$ where $\alpha_i = \{a_i\}$. After the first iteration of PA$_{\text{GPP}}(G, \langle \Sigma, P \rangle)$, we have $\Sigma_1 = \Sigma_0 = \Sigma$ and $P_1 = \mathcal{I}(\Sigma)$. The algorithm then terminates because $\mathit{change} = \bot$, and $\langle \Sigma_1, P_1 \rangle$ is its answer to the GCPP over G and $\langle \Sigma, P \rangle$. Obviously $\langle \Sigma_1, P_1 \rangle \neq \langle \Xi, \preceq \rangle$, so this answer is wrong. □

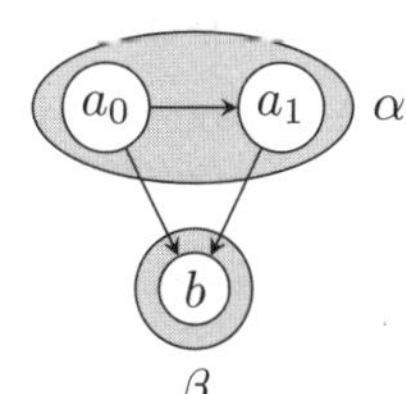

Algorithm 4. The repaired refine function: REFINE(Σ, P)

```
Π := Σ;
for all α ∈ Π do Stable(α) := ∅; end for
for all γ ∈ Σ do Row(γ) := {γ' | (γ, γ') ∈ P}; end for
Let Sort be a reverse topological sorting of Σ w.r.t. P;
while Sort ≠ [] do
    γ := head(Sort);
    A := ∅;
    for all α ∈ Π, α →∃ γ do
        if Stable(α) ∩ Row(γ) = ∅ then
            α1 := α ∩ →⁻¹(γ);
            α2 := α \ α1;
            Π := Π \ {α};
            A := A ∪ {α1};
            Stable(α1) := Stable(α) ∪ {γ};
            if α2 ≠ ∅ then
                change := ⊤;
                A := A ∪ {α2};
                Stable(α2) := Stable(α);
            end if
        else
            Stable(α) := Stable(α) ∪ {γ};
        end if
    end for
    Π := Π ∪ A;
    Sort := tail(Sort);
end while
return Π;
```

The correctness of PA$_{\text{GPP}}$ hinges on the theory that whenever REFINE$_{\text{GPP}}(\Pi, Q, change)$ returns its input partition Π, and thus fails to split any block in Π, then also the relation Q will be unaffected by UPDATE$_{\text{GPP}}$, *i.e.* UPDATE$_{\text{GPP}}(\Pi, Q, \Pi)$ returns Q. This theory is the upshot of Theorem 4.15 in [8] and is essential in the complexity analysis of the algorithm. However, the above example shows that it does not hold in general.

In the next section we show that this theory does hold under the condition that Q itself is obtained as output of UPDATE$_{\text{GPP}}$ (Proposition 5). Therefore, this error in PA$_{\text{GPP}}$ can be fixed, without violating the complexity analysis, by insisting that at least two refinement-update steps are performed prior to termination.

6.2 Correctness of PA

From here on we will use the correctness of the function UPDATE$_{\text{GPP}}$, as established by Gentilini *et al.* [9]. This correctness can be summarised as follows:

Proposition 4. *Let $\langle \Sigma, P \rangle$ be a partition pair over a graph $(N, \rightarrow)$, and Π be a partition over N that is finer than Σ. Then there exists a unique relation $Q \subseteq P(\Pi)$ satisfying condition (2σ) of Definition 4.11. Moreover, this relation is returned by* UPDATE$_{\text{GPP}}(\Sigma, P, \Pi)$.

Using this, we obtain the result promised in Sect. 6.1: the following proposition implies that if a call to REFINE in the while-loop of PA does not split any blocks, then the subsequent call to $\text{UPDATE}_{\text{GPP}}$ will return its input relation. The requirement that this relation has been computed by a previous call to $\text{UPDATE}_{\text{GPP}}$ is guaranteed by line 2.

Proposition 5. *Let $\langle \Sigma, P\rangle$ and $\langle \Pi, Q\rangle$ be partition pairs over a graph such that Π is finer than Σ and $\text{UPDATE}_{\text{GPP}}(\Sigma, P, \Pi)$ returns Q. Then $\text{UPDATE}_{\text{GPP}}(\Pi, Q, \Pi)$ also returns Q.*

Let $\langle \Sigma_i, P_i\rangle_{1\leq i\leq k}$ be the sequence of partition pairs produced by PA. The following proposition says that every P_i is acyclic and that the sequence is decreasing. The former implies that PA will never deadlock due to the inability to find a reverse topological sorting (see line 4 of REFINE). The latter implies that the algorithm terminates.

Proposition 6. *Let $\langle \Sigma, P\rangle$ be a partition pair over a graph $(N, \rightarrow)$, $\text{REFINE}(\Sigma, P)$ return Π and $\text{UPDATE}_{\text{GPP}}(\Sigma, P, \Pi)$ return Q. Then $\langle \Pi, Q\rangle$ is a partition pair with $\langle \Pi, Q\rangle \leq \langle \Sigma, P\rangle$.*

Corollary 1. *For any graph G and any partition pair $\langle \Sigma, P\rangle$ over G, the algorithm $\text{PA}(G, \langle \Sigma, P\rangle)$ terminates.* □

The following lemmata state that REFINE and $\text{UPDATE}_{\text{GPP}}$ converge towards a fixed point at least as fast as ρ without ever diverging from the path towards the GCPP solution. In combination with the monotony of ρ (Proposition 2) this implies the correctness of our algorithm.

Lemma 3. *Let $\langle \Sigma, P\rangle$ be a partition pair over a graph $(N, \rightarrow)$, $\text{REFINE}(\Sigma, P)$ return Π, and $\text{UPDATE}_{\text{GPP}}(\Sigma, P, \Pi)$ return Q. Then $\langle \Pi, Q^+\rangle \leq \rho(\langle \Sigma, P^+\rangle)$.*

Lemma 4. *Let $\langle \Sigma, P\rangle$ and $\langle \Pi, Q\rangle$ be partition pairs over a graph $G = (N, \rightarrow)$, $\langle \Xi, \preceq\rangle$ be the solution of the GCPP over G and $\langle \Sigma, P\rangle$, and $\langle \Xi, \preceq\rangle \leq \langle \Pi, Q\rangle$. Let $\text{REFINE}(\Pi,Q)$ return Π' and $\text{UPDATE}_{\text{GPP}}(\Pi,Q,\Pi')$ return Q'. Then $\langle \Xi,\preceq\rangle \leq \langle \Pi',Q'\rangle$.*

Theorem 2. *Let $\langle \Sigma, P\rangle$ be a partition pair over a graph $G = (N, \rightarrow)$. Let k be the value of variable i upon termination of $\text{PA}(G, \langle \Sigma, P^+\rangle)$. Then $\langle \Sigma_k, P_k\rangle$ is the solution of the GCPP over G and $\langle \Sigma, P\rangle$.*

6.3 No Fixed-Point Operator

We now show that there is no (functional) fixed-point operator that captures the partition refinement performed by REFINE, *i.e.* a function π such that for any partition pairs $\langle \Sigma, P\rangle$ and $\langle \Pi, Q\rangle$ with $\langle \Pi, Q\rangle = \pi(\langle \Sigma, P\rangle)$, $\text{REFINE}(\Sigma, P)$ returns Π. More specifically, we show that the partition returned by REFINE is not uniquely defined, but depends on the particular reverse topological sorting that is chosen in line 4.

Example 6. Consider the graph $G = (N, \rightarrow)$ of Fig. 3 and the partition pair $\langle \Sigma, P\rangle$ with $\Sigma = \{\alpha, \beta, \gamma, \delta, \varepsilon\}$ as depicted and $P = \mathcal{I}(\Sigma) \cup \{(\beta, \delta), (\delta, \gamma)\}$. Then $S = [\varepsilon, \gamma, \delta, \beta, \alpha]$ and $S' = [\gamma, \delta, \beta, \varepsilon, \alpha]$ are reverse topological sortings of Σ with respect to P. Let Π and Π' be the partitions returned by $\text{REFINE}(\Sigma, P)$ on sortings S and S' respectively. Then $\Pi = \{\{a_0\}, \{a_1\}, \{a_2\}\}$ and $\Pi' = \{\{a_0, a_1\}, \{a_2\}\}$. □

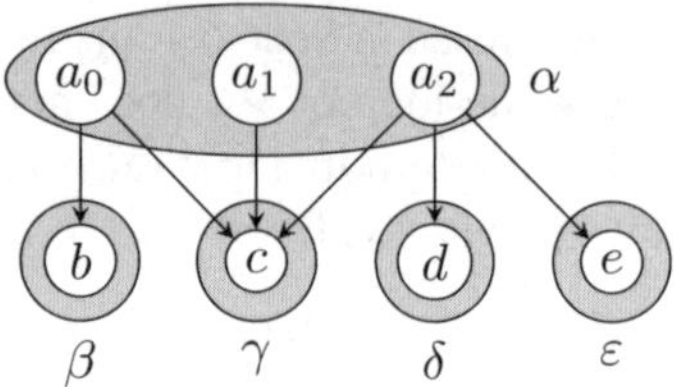

Fig. 3. Example on which REFINE does not return a uniquely defined partition

Similar to the construction of Counterexample 4, this example can be embedded in Example 3 to obtain an example with a transitive relation for which the partition after the second refinement depends on the chosen reverse topological sorting.

7 Conclusions

The correspondence between the simulation problem for finite, labelled graphs and the generalised coarsest partition problem (GCPP) for unlabelled graphs can be easily established. We have shown that the σ operator defined by Gentilini *et al.* [8] to solve the GCPP is flawed. In particular, when applied to a partition pair, the result is not necessarily another partition pair or even well-defined. Moreover, when applied repeatedly to a transitive partition pair, convergence towards a unique fixed point is not guaranteed. Thereby we have shown that σ is not suitable for solving the GCPP. On the counterexample for the latter property, the algorithm of [8] that computes σ, produces a wrong result in which two simulation-equivalent states are put in different equivalence classes.

We have repaired this algorithm such that it correctly computes the solution of the GCPP. Apart from correcting the error that results from the flaws in the σ operator, we also corrected a mistake that caused premature termination of the algorithm on certain input. Our algorithm benefits from the key ideas behind the original partitioning algorithm and has the same space and time complexities. We have proven its correctness using an auxiliary operator ρ of which we have shown that it solves the GCPP, though inefficiently. Finally, we have shown that no operator can be defined that captures the partition refinement performed in every iteration of our algorithm.

Another way to repair the algorithm of [8] may be to use the relation P^+ instead of P in REFINE$_{\text{GPP}}$. The so obtained algorithm would converge to a fixed point slightly slower than ours. More importantly, due to the cost of computing the transitive closure in each iteration, the time complexity would not match that of the original algorithm.

Acknowledgements. We would like to thank Raffaella Gentilini and Carla Piazza for answering some of our questions about their paper and providing us with their implementation of the algorithm.

References

1. Bloom, B., Istrail, S., Meyer, A.R.: Bisimulation can't be traced. Journal of the ACM 42(1), 232–268 (1995)
2. Bloom, B., Paige, R.: Transformational design and implementation of a new efficient solution to the ready simulation problem. Science of Computer Programming 24(3), 189–220 (1995)

3. Bustan, D., Grumberg, O.: Simulation-based minimization. ACM Transactions on Computational Logic 4(2), 181–206 (2003)
4. Courcoubetis, C., Vardi, M.Y., Wolper, P., Yannakakis, M.: Memory efficient algorithms for the verification of temporal properties. In: Clarke, E., Kurshan, R.P. (eds.) CAV 1990. LNCS, vol. 531, pp. 233–242. Springer, Heidelberg (1991)
5. Dams, D., Grumberg, O., Gerth, R.: Generation of reduced models for checking fragments of CTL. In: Courcoubetis, C. (ed.) CAV 1993. LNCS, vol. 697, pp. 479–490. Springer, Heidelberg (1993)
6. Emerson, E.A., Halpern, J.Y.: "Sometimes" and "Not Never" revisited: On branching versus linear time temporal logic. Journal of the ACM 33(1), 151–178 (1986)
7. Evangelista, S., Pradat-Peyre, J.-F.: Memory efficient state space storage in explicit software model checking. In: Godefroid, P. (ed.) SPIN 2005. LNCS, vol. 3639, pp. 43–57. Springer, Heidelberg (2005)
8. Gentilini, R., Piazza, C., Policriti, A.: From bisimulation to simulation: Coarsest partition problems. Journal of Automated Reasoning 31(1), 73–103 (2003)
9. Gentilini, R., Piazza, C., Policriti, A.: From bisimulation to simulation: Coarsest partition problems. RR 12-2003, Dep. of Computer Science, University of Udine, Italy (2003)
10. van Glabbeek, R.J., Ploeger, B.: Correcting a space-efficient simulation algorithm. CS-Report 08-06, Eindhoven University of Technology (2008)
11. Groote, J.F., Vaandrager, F.W.: Structured operational semantics and bisimulation as a congruence. Information and Computation 100(2), 202–260 (1992)
12. Henzinger, M.R., Henzinger, T.A., Kopke, P.W.: Computing simulations on finite and infinite graphs. In: 36th Annual Symposium on Foundations of Computer Science (FOCS 1995), pp. 453–462. IEEE Computer Society Press, Los Alamitos (1995)
13. Holzmann, G.J.: An improved protocol reachability analysis technique. Software Practice and Experience 18(2), 137–161 (1988)
14. Kozen, D.: Results on the propositional μ-calculus. Theoretical Computer Science 27, 333–354 (1983)
15. Kucera, A., Jancar, P.: Equivalence-checking on infinite-state systems: Techniques and results. Theory and Practice of Logic Programming 6(3), 227–264 (2006)
16. Loiseaux, C., Graf, S., Sifakis, J., Bouajjani, A., Bensalem, S.: Property preserving abstractions for the verification of concurrent systems. Formal Methods in System Design 6(1), 11–44 (1995)
17. Park, D.M.R.: Concurrency and automata on infinite sequences. In: Deussen, P. (ed.) GI-TCS 1981. LNCS, vol. 104, pp. 167–183. Springer, Heidelberg (1981)
18. Ranzato, F., Tapparo, F.: A new efficient simulation equivalence algorithm. In: Proc. 22nd Annual IEEE Symposium on Logic in Computer Science (LICS 2007), pp. 171–180. IEEE Computer Society Press, Los Alamitos (2007)
19. Stockmeyer, L.J., Meyer, A.R.: Word problems requiring exponential time. In: Proc. 5th Annual ACM Symposium on Theory of Computing (STOC 1973), pp. 1–9. ACM, New York (1973)
20. Tan, L., Cleaveland, R.: Simulation revisited. In: Margaria, T., Yi, W. (eds.) ETAPS 2001 and TACAS 2001. LNCS, vol. 2031, pp. 480–495. Springer, Heidelberg (2001)

Semi-external LTL Model Checking*

Stefan Edelkamp[1], Peter Sanders[2], and Pavel Šimeček[3]

[1] Faculty of Informatics, Dortmund University of Technology, Germany
[2] Faculty of Informatics, University of Karlsruhe, Germany
[3] Faculty of Informatics, Masaryk University, Brno, Czech Republic

Abstract. In this paper we establish c-bit semi-external graph algorithms, – i.e., algorithms which need only a constant number c of bits per vertex in the internal memory. In this setting, we obtain new trade-offs between time and space for I/O efficient LTL model checking. First, we design a c-bit semi-external algorithm for depth-first search. To achieve a low internal memory consumption, we construct a RAM-efficient perfect hash function from the vertex set stored on disk. We give a similar algorithm for double depth-first search, which checks for presence of accepting cycles and thus solves the LTL model checking problem. The I/O complexity of the search itself is proportional to the time for scanning the search space. For on-the-fly model checking we apply iterative-deepening strategy known from bounded model checking.

1 Introduction

Graph search algorithms such as breadth-first search (BFS), depth-first search (DFS), A*, and their variants, play an important role in model checking, as well as in other branches of computer science. All use duplicate detection in order to recognize when the same vertex is reached via alternative paths in a graph. This traditionally involves storing already explored vertices in random access memory (RAM) and checking newly generated vertices against the stored vertices. However, the available amount of RAM severely limits the range of problems that can be solved with this approach. Although many clever memory saving techniques, such as state space reduction, abstraction, and compression, have been developed, all are eventually limited in terms of scalability, and many practical graph search problems are too large to be solved using any of these techniques. Relying on the virtual memory slows down the exploration due to an excessive number of page faults. Over the past few years, several researchers have shown that the scalability of graph search algorithms can be dramatically improved by using external memory, such as disk, to store generated vertices for use in duplicate detection. However, this requires different search strategies to eliminate the impact of the several orders of magnitude difference in random access speed between RAM and disk.

External memory algorithms [20] carefully organize the access to disk. The efficiency of algorithms is then measured in number of block I/O operations performed. The frequently used I/O pattern is external file scanning, processing a stream of records

* This work has been partially supported by the Grant Agency of Czech Republic grant No. 201/06/1338, the Academy of Sciences grant No. 1ET408050503, DFG grant SA 933/3-1 and DFG grant ED 74/4-1.

A. Gupta and S. Malik (Eds.): CAV 2008, LNCS 5123, pp. 530–542, 2008.

stored consecutively on disk. If the block size is B, the number of block accesses (I/Os) for scanning N nodes is $O(scan(N)) = O(N/B)$. Another important operation is external sorting. Given that the RAM can contain M nodes it has a complexity of $O(sort(N)) = O(\frac{N}{B} \log_{M/B} \frac{N}{B})$ I/Os.

Enumerative model checking is a search in an implicitly given state space graph (*implicit graph*) $G = (V, E)$ induced by an initial vertex s and a successor generation function *succ*. Vertices correspond to states and transitions to edges. The maximal size of the vertex, v_{max}, depends on the encoding and denotes the length of a state vector. *Reachability* is one of the simplest model checking problems. If G is undirected, the complexity of the external memory variant of BFS for solving the reachability problem is bounded by $O(sort(|E|) + scan(|V|))$ I/Os [21]. For directed graphs, the complexity raises to $O(sort(|E|) + l \cdot scan(|V|))$ [26], where $l := \max\{\delta(s,u) - \delta(s,v) + 1 \mid (u,v) \in E\}$ is the length of the longest back-edge in the BFS graph or its *locality* [27]. The locality is bounded by the length of the shortest counter-example $\delta^* := \min\{\delta(s,f) \mid f \in F\}$ (in case of an error) and by the eccentricity of s (max. BFS level) $\epsilon_s := \max\{\delta(s,v) \mid v \in V\}$ (in case of no error), where $F \subseteq V$ is a set of accepting vertices.

Semi-external graph algorithms [1] are algorithms, which allocate $O(|V|)$ *machine words* in the internal memory. Thus, they can store $O(v_{max}) \geq O(\log |V|)$ bits of data per vertex. Since the internal memory is the limiting factor for such algorithms, it makes sense to further reduce the memory requirement to a small constant number of *bits* per vertex. Therefore, we define *c-bit semi-external search algorithms*, which take at most c bits per vertex in the internal memory. Considering small c, with Gigabytes of RAM, and given that state vectors in model checking are large, such algorithms allow us to handle spaces that are orders of magnitudes larger than the main memory.

We present a semi-external solution to the LTL model checking problem, which amounts to finding an accepting cycle in the state space graph of a Büchi automaton [12]. The algorithm we present has an I/O complexity equal to the complexity of I/O-efficient reachability. The approach relies on the I/O-efficient external construction of a *minimal perfect hash function* (MPHF) [7,8], i.e., a one-to-one correspondence between V and $\{0, \ldots, |V| - 1\}$. It allows compressing V to $c|V|$ bits for small constants c. Thereby, we solve a problem considered in a series of preceding papers [15,4,5] that have I/O complexities higher than the reachability analysis.

The paper is organized as follows. First, we define c-bit semi-external algorithms as needed for implicit graph search and explain the generation of space-efficient minimal perfect hash functions. Subsequently, we illustrate our semi-external solution to the LTL model checking problem using the constructed perfect hash function, and analyze its I/O complexity. Afterwards, we address the problem to find the counter-example on-the-fly. For the purpose of comparison, we give a brief overview of related work and put our complexity results into context. And finally, we provide an experimental comparison of existing algorithms for I/O efficient LTL model checking.

2 *c*-Bit Semi-external Graph Algorithms

Semi-external graph algorithms [1] allow to store $O(|V|)$ vertices in the internal memory, thus restricting graph algorithms to store internally only information about vertices, but not about all edges. Although this definition of semi-external graph algorithms appears practical for explicit graphs, it is too general for algorithms on implicit graphs. E.g., almost every internal memory graph algorithm on implicit graphs would be considered as semi-external, since graph edges are given implicitly and there is no need to store them. However, due to large state vector sizes it is apparent that $O(|V|)$ items may easily exceed the amount of available internal memory. For this reason, we study memory consumption in more detail. To derive exact bounds on internal memory consumption, we give a definition of a subclass of semi-external graph algorithms.

Definition 1. *(c-bit Semi-External Algorithm) The graph algorithm $\mathcal{A}$ is called* c-bit semi-external *for $c \in \mathbb{R}^+$, if for each implicit graph $G = (V, E)$ the internal memory requirements of $\mathcal{A}$ are at most $O(v_{max}) + c \cdot |V|$ bits.*

$O(v_{max})$ stands for the internal memory consumed by a program code, auxiliary fixed sized variables, and storage of a constant amount of vertices. The value $c \cdot |V|$ stands for the internal memory consumed by information about vertices. Including v_{max} in the complexity is necessary, since this value differs for different graphs[1] – otherwise, for a bound of $O(1) + c \cdot |V|$ bits, we could always find a graph requiring v_{max} that exceeds the constant in $O(1)$, which would prohibit storing even a constant number of vertices. Including the state vector size in the definition of semi-external algorithm also takes a lower bound of $\log \log u + (\log_2 e)|V| + O(\log |V|)$ bits [16] on the space of an MPHF into account, where u denotes the number of all possible states (including unreachable ones) and e is Euler's number.

In the remainder of this section, we refer to results on memory-efficient construction of MPHFs and give sufficient details on the space and the I/O complexity of used algorithms. We introduce a c-bit semi-external depth-first search with use of MPHF. Its I/O complexity is $O(scan(|V|))$ plus the complexity of the hash function construction. There is no known external algorithm computing the DFS order with such a low I/O complexity on general graphs. Third, we extend this DFS implementation to find accepting cycles in order to solve the LTL model checking problem with the same I/O complexity.

2.1 Memory Efficient Minimal Perfect Hash Function

Perfect hashing is a space efficient way of associating unique identifiers with the elements of a static set $V \subseteq U$. A perfect hash function maps $V \subseteq U$ to unique values in the range $\{0, \ldots, N-1\}$, for some appropriate value of N. A minimal perfect hash function is a perfect hash function with $N = |V|$. Consequently, a minimal perfect hash function is a one-to-one correspondence between V and $\{0, \ldots, |V|-1\}$.

Surprisingly, after 23 years of research, an asymptotically space optimal, practical algorithm for generating MPHFs was recently discovered [7]. The external memory variant, referred to as EPH algorithm, was given in [8]. Although the I/O complexity of this EPH algorithm is not analyzed in [8], it is clear that it is dominated by the need to

[1] The binary vertex representation $v_{\max}$ takes obviously at least $\log |V|$ bits. In the model checking case, it is usually much more.

```
Procedure Perform-DFS(s, succ)
  V := Enumerate-BFS(s, succ)
  h := Construct-MPHF(V)
  Depth-First-Search(s, succ, h)

Procedure Depth-First-Search(s, succ, h)
  Vars: visited : Internal Bit Array[1..n] = (0,..,0)
        dfsStack : External Stack of Vertices
  visited[h(s)] := 1
  dfsStack.push(s)
  while (not dfsStack.empty())
    u := dfsStack.top()
    if ∃v ∈ succ(u). visited[h(v)] = 0 then
      dfsStack.push(v)
      visited[h(v)] := 1
    else
      dfsStack.pop()
```

Fig. 1. c-bit Semi-External Depth-First Search

sort all items by their hash signature in a partitioning step. MPHFs constructed by EPH can be stored in less than 4 bits per item.[2]

The Heuristic EPH algorithm, published also in [8], differs from EPH in the choice of the hash function. It results in a substantial speed-up in both construction and search times, but incurs additional memory overhead per bucket. In our implementation, it needs 1 additional bit per vertex, i.e., our implementation of the Heuristic EPH requires 5 bits per vertex.

2.2 Depth-First Search

The main observation for graph search is that given a perfect hash function h, algorithms like plain DFS, BFS, and A* need only one bit per vertex storing whether it has already been visited. The general approach applying a bit-array for tracking reached vertices in DFS is illustrated in Fig 1 as procedure *Depth-First-Search*.

Our algorithm first enumerates all reachable vertices using external BFS, which performs $O(l \cdot scan(|V|) + sort(|E|))$ operations (*Enumerate-BFS*) [18,26]. Then the EPH algorithm constructs the MPHF with I/O complexity $O(sort(|V|))$ (*Construct-MPHF*) – this complexity follows from [8], although it is not explicitly stated there.

The stack in procedure *Depth-First-Search* can be stored on disk. The procedure performs exactly $|V|$ *push* operations and $|V|$ *pop* operations. It is easy to implement the stack in the way that the I/O-complexity of the procedure is $O(scan(|V|))$. Thus the overall I/O complexity of the algorithm (procedure *Perform-DFS*), including graph generation and hash function construction, is $O(l \cdot scan(|V|) + sort(|E|) + sort(|V|) + scan(|V|)) = O(l \cdot scan(|V|) + sort(|E|) + sort(|V|))$. In implicit graphs, we have

[2] Although [8, Table 2] shows a value higher than 4 bits per item for $n = 10^8$, it is caused by a poor choice of the *bucket* size; i.e., $b = 20$ causes buckets to contain less than 96 items per bucket on average. In our implementation of EPH, we always choose a better bucket count, which guarantees that each bucket contains at least 128 items on average. This amount assures that the fixed cost per bucket is divided into sufficiently many items to keep the overall costs below 4 bits per item.

```
Procedure Perform-DDFS(s, succ)
  Vars: V : Vertex Set;
        h: Perfect Hash Function;
  V := Enumerate-BFS(s, succ)
  h := Construct-MPHF(V)
  Double-Depth-First-Search(s, succ, h)

Procedure Double-Depth-First-Search(s, succ, h)
  Vars: visited : Internal Bit Array[1..n] := (0,..,0);
        F : List of Accepting Vertices;
  F := Depth-First-Search-1(s, succ, h)
  visited := (0, ..., 0)
  for each i in F do
    if visited[h(i)] = 0 then
      if (Depth-First-Search-2(i, succ, h))
        return 'cycle found'
  return 'no cycle'
```

Fig. 2. c-bit Semi-External Double Depth-First Search

$|V| < |E|$, because all vertices contained in V induced by s and *succ* are reachable. Therefore the I/O complexity simplifies to $O(l \cdot scan(|V|) + sort(|E|))$.

With EPH minimum perfect hashing the algorithm is 5-bit semi-external, since less than 4 bits per vertex are needed for storing h, and 1 bit per vertex is needed for *visited*.

2.3 Double Depth-First Search

The LTL model checking problem amounts to detecting accepting cycles in the global state space graph. It is possible to find an accepting cycle with the *double depth-first search* algorithm [13, Algorithm A]. The algorithm performs the first DFS to find a list F of all accepting vertices sorted in DFS postorder. The second DFS explores the graph gradually from all vertices in F. The pseudo code of this algorithm is shown in Fig. 2.

Depth-First-Search-1 is a modified version of *Depth-First-Search*, which appends an accepting vertex to F, while it is removed from *dfsStack*. *Depth-First-Search-2* is a modified version of *Depth-First-Search*, which finishes with return value *true*, if it wants to add its initial vertex to *dfsStack* in the main loop (and so it finds a path from the initial vertex to itself). For simplicity and memory efficiency, we assume that array *visited* is shared by both procedures. The correctness of the algorithm is proven in [13].

These two modifications of *Depth-First-Search* have clearly the same I/O complexity as the original procedure. Therefore, the overall I/O complexity of *Perform-DDFS* remains at $O(l \cdot scan(|V|) + sort(|E|))$. Moreover, they share the same hash function and memory space for the *visited* field. As in the DFS case, with EPH the algorithm is 5-bit semi-external, since less than 4 bits per vertex are needed for storing h, and 1 bit per vertex is needed for *visited*.

2.4 General Graph Search

We have shown a way to solve the LTL model checking problem using double depth-first search with a visited vertex set represented in form of a minimum perfect hash function. We have chosen double depth-first search, because it is the most time and memory efficient algorithm for searching accepting cycles in the internal memory and it sustains the efficiency in semi-external setting. For example, nested depth-first search [13] has higher memory demands and its on-the-fly nature is not a big advantage, since for MPHF construction, the algorithm would have to enumerate the entire vertex set anyway.

Procedure General-Search
Vars: V : Vertex Set;
h: Perfect Hash Function;
$V :=$ *External-BFS*$(s, succ)$
$h :=$ *Construct-MPHF*(V)
Search$(s, succ, h)$

Fig. 3. General c-bit Semi-External Graph Search

Besides combination of MPHF and LTL model checking, we can also consider other applications in model checking. E.g., for global CTL model checking [12], graph decomposition to strongly connected components (SCCs) is needed, which is easy using Kosaraju-Sharir's algorithm (SCC decomposition using forward and backward DFS [25]). If the implicit definition also contains backward successor generation, the algorithm for SCC decomposition is straightforward. This way, employing MPHF gives us a handle to many space efficient semi-external model checking algorithms for the prize of single state space generation needed for MPHF construction. In fact, all such general search algorithms on implicit graphs can follow the semi-external exploration procedure as outlined in Fig.3.

3 On-the-Fly LTL Model Checking

The idea of an increasing depth bound to obtain short lasso shaped counter-examples as witnesses for a falsified LTL property refers to pioneering work of [6], which searches for a counter-example in the state space graph unrolled to a fixed depth k.

A similar iterative-deepening strategy can be easily applied to our case. Since we use external breadth-first search to generate the state space, we can search for a counter-example every time a new level is generated. This approach has two main advantages

- The counter-example can be found before the entire graph is generated – it is found *on-the-fly*. Since graph generation is the main source of I/Os, performance can be significantly improved on inputs with existing counter-examples.
- It can produce a shorter counter-example, since the depth for its search is limited. However, the counter-example is not necessarily the shortest.

The algorithm is derived from the one in Section 2.3 by unwinding procedure *Enumerate-BFS* and moving MPHF construction and DDFS inside BFS levels generation as shown in Fig. 4.

Every search for a counter-example in an incomplete graph applies $O(sort(|V|))$ I/Os, determined by the I/O complexities of *Construct-MPHF* and *Double-Depth-First-Search*. The search for a counter-example is performed after the generation of each BFS level (one BFS iteration). With $\epsilon_s := \max\{\delta(s, v) \mid v \in V\}$ at most ϵ_s BFS iterations are invoked. Moreover, the generation of every BFS level requires $O(scan(|V|))$ I/Os. Therefore, the overall I/O complexity of the algorithm is

$$O(\epsilon_s \cdot sort(|V|) + l \cdot scan(|V|) + sort(|E|)) = O(\epsilon_s \cdot sort(|V|) + sort(|E|)).$$

Procedure Perform-IDDFS$(s, succ)$
Vars: V : Vertex Set;
h: Perfect Hash Function;
nextLevel: Set of Vertices;
nextLevel $:= \{s\}$
while *nextLevel* $\neq \emptyset$ **do**
$V := V \cup$ *nextLevel*
$h :=$ *Construct-MPHF*(V)
Double-Depth-First-Search$(s, succ, h)$
nextLevel $:=$ *succ*(*nextLevel*) $\setminus V$

Fig. 4. On-the-fly Semi-External Double Depth-First Search

Although the I/O complexity of double DFS is only $scan(|V|)$, in practice, due to the efforts for generating the successors of a vertex, the run time of DFS search often substantially exceeds the run time of hash function construction (even though its I/O complexity is $sort(|V|)$). Therefore, the internal memory search for a counter-example is too expensive to be invoked after the generation of each BFS level. For this reason, we implemented the algorithm in such a way that it measures run times of checks for a counter-example (including hash function generation) and tries to predict the run time on the next level. This refined threshold determination algorithm invokes a check for counter-examples, only if the predicted time sufficiently amortizes the time for graph generation.

4 Related Work

Since model checking amounts to graph search, our algorithm is strongly related to external memory graph algorithms [20,11]. Most results consider the graph to be explicitly given in the external memory. Graph algorithms on explicit graphs, however, suffer from the storage of edges in the external memory, which causes one I/O operation each time they need to access successors of a given vertex. It brings at least $|V|$ additional I/Os. The situation has been slightly improved for undirected graphs, where the I/O complexity improves to $O(\sqrt{|V| \cdot scan(|V| + |E|)} + sort(|V| + |E|))$ [19], but in general, state spaces in model checking are directed.

Fortunately, practical model checking is performed on state spaces given by a system model – i.e., implicit graph definition. Thus, it avoids the expensive fetching of edges. However, in contrast to the explicit case, the implicit representation of edges does not allow to store the information about explored edges, which is essential to avoid re-exploration. For this reason, it is not trivial to make algorithms on explicit graphs work also on implicit graphs efficiently. For example, there is no known efficient implementation of depth-first search for implicit graphs.

Therefore, algorithms refer to a breadth-first traversal through the graph and employ the *delayed duplicate detection* technique [18,21,26]. The search procedure has to maintain a set of visited vertices to prevent their re-exploration. Since the graphs are large, the visited set cannot be kept completely in main memory. Most of it is stored on an external memory device. When a new vertex is generated, it is checked against the visited set to avoid its re-exploration. The idea of the delayed duplicate detection technique is to postpone the individual checks and perform them together in a group, for the price of a single scan operation. The group of vertices waiting for checking against the visited set is called *candidate set*. There are two basic kinds of duplicate detection: The one making an internal memory a buffer for candidate set and the one storing candidate set in the external memory. The first has an advantage that no sorting is needed during duplicates removal. The latter is better, when candidate sets are too large to fit in the internal memory and thus, using the first approach they would have to be divided into several pieces and checked separately. Complexities of both approaches are different, and incomparable in general.

4.1 External LTL Model Checking

The first I/O-efficient solution for the LTL model checking problem by Edelkamp and Jabbar [15] builds on the reduction of liveness to the safety approach by Schuppan and Biere [24] designed for symbolic exploration with BDDs. It operates on-the-fly and applies guidance for checking liveness properties [15] with a set of heuristic functions.

Barnat et al. proposed another I/O efficient algorithm [4] for accepting cycle detection. It applies the OWCTY (One Way Catch Them Young) algorithm [23,10] – an accepting cycle detection algorithm based on topological sort. The algorithm itself is an off-line algorithm. It generates the whole state space and then iteratively prunes the parts of the state space that do not lead to any accepting cycle. The underlying exploration strategy is breadth-first based. Later, they also proposed an on-the-fly algorithm [5] based on the MAP (Maximal Accepting Predecessors) algorithm [9].

All three approaches were theoretically compared, experimentally evaluated and each of them has shown its practical applicability to a certain class of problems.

4.2 Complexity Comparison

In this section, we compare the new semi-external approach to the existing external LTL model checking algorithms, in terms of internal memory consumption and I/O complexity (see Table 1).

Regarding the I/O complexity, the new algorithms contributed in this paper compete much better than previous work. The off-line version (DDFS) has the same I/O complexity as the external breadth-first search, which defeats existing algorithms substantially. The on-the-fly variant (*IDDFS*) is worse than off-line, but it is still reasonable compared to the rest of the algorithms. Table 1 shows I/O complexities of all algorithms and also gives I/O complexities of their versions with candidate set stored in RAM. Note that in this case we consider a variant of the EPH algorithm with I/O complexity $O(n/M \cdot scan(n))$ (different bound for external sort) rather than $O(sort(n))$, because it simplifies resulting complexities. The existing external memory algorithms are named after their internal memory variants: L2S [15], OWCTY [4] and MAP [5].

The disadvantage of our semi-external algorithm is that it needs $\Omega(|V|)$ bits in the internal memory – thus we can always find a graph, on which it exceeds an available internal memory. For example, for 5-bit semi-external search on a computer with 2 GB RAM, the algorithm cannot handle graphs with more than $2 \cdot 2^{30} \cdot 8/5 \approx 3 \cdot 10^9$ vertices

Table 1. I/O complexities for LTL model checking

	Candidate Set on Disk	Candidate Set in RAM
L2S	$O(l \cdot scan(f \cdot n) + sort(f \cdot m))$	$O((l + f \cdot m/M) \cdot scan(f \cdot n))$
OWCTY	$O(\tau \cdot ((\epsilon_s + \psi) \cdot scan(n) + sort(m)))$	$O(\tau \cdot (\epsilon_s + \psi + m/M) \cdot scan(n))$
MAP	$O(f \cdot ((d+f) \cdot scan(n) + sort(f \cdot m)))$	$O(f \cdot ((d+m/M+f) \cdot scan(n) + sort(n)))$
DDFS	$O(l \cdot scan(n) + sort(m))$	$O((l + m/M) \cdot scan(n))$
IDDFS	$O(\epsilon_s \cdot sort(n) + sort(m))$	$O((\epsilon_s + m/M) \cdot n/M \cdot scan(n))$

$m = |E|$ ... number of edges, $n = |V|$ ... number of vertices,
$f = |F|$ ... number of accepting vertices, τ ... length of the longest path in the SCC graph,
ϵ_s ... eccentricity of the initial vertex, d ... diameter of the graph,
l ... locality, i.e., length of the longest back edge in breadth-first search graph
ψ ... length of the longest path in the graph going through trivial strongly connected components (without self-loops).

(since it stores 5 bits per vertex internally). In contrast, purely external algorithms are limited only by the capacity of the external memory. Nevertheless, considering that one vertex is v_{max} bytes long, we get that, with 2 Gigabytes of RAM our algorithm can handle state spaces which need approximately $3 \cdot v_{max}$ Gigabytes to be stored externally. For practical values of v_{max} (20-1000 from our experience on models from the Benchmark for Explicit Model Checkers [22]) the state space would be hundreds or thousands of Gigabytes large. When manipulating such a large piece of data, our algorithm takes advantage of lower I/O complexity and as a result, it can be much faster than previous algorithms, which makes a price of 5 bits of the internal memory per vertex quite reasonable.

5 Experimental Results

In order to obtain experimental evidence about the behavior of our algorithm in practice, we implemented three existing external memory LTL model checking algorithms (as introduced in Section 4) and compared their run times and allocated disk space to both versions (DDFS and IDDFS) of the new semi-external algorithm.

All algorithms have been implemented on top of the DiVinE library [3], providing the state space generator, and the STXXL library [14], providing the I/O primitives. Experiments were run on a Linux workstation with 2 GHz Intel Xeon processor, the main memory was limited to 2 GB, the disk space to 60 GB and wall clock time limit was set to 120 hours. For compilation of sources we used GNU C++ compiler with optimization level 2. Algorithm L2S was implemented as a procedure that performs the graph transformation as suggested in [15] and then employs I/O efficient breadth-first search to check for a counter-example. Note, that our implementation of L2S does not include the A* search heuristics and, hence, can be less efficient when searching for an existing counter-example. Algorithms DDFS and IDDFS were implemented using Heuristic EPH [8] (for the sake of speed), thus one additional bit per vertex is allocated in comparison to EPH.

Table 2. Experimental results for different I/O-efficient algorithms

	L2S		OWCTY		MAP		DDFS		IDDFS	
Experiment	Time	Disk	Time	Disk	Time	Disk	Time	Disk	Time	Disk
Valid Properties										
Elev.2(16),P4	(OOS)		09:54	9.2 GB	07:45	16 GB	08:01	10 GB	08:03	10 GB
Lamport(5),P4	(OOS)		02:37	5.5 GB	03:16	5.7 GB	02:15	3.3 GB	02:16	3.3 GB
MCS(5),P4	(OOS)		03:27	9.8 GB	04:59	10 GB	03:42	6.2 GB	03:59	6.2 GB
Peterson(5),P4	(OOS)		18:20	26 GB	25:09	26 GB	14:19	16 GB	18:37	16 GB
Phils(16,1),P3	(OOS)		01:49	6.2 GB	02:31	7.8 GB	02:26	6.7 GB	02:54	6.7 GB
Ret.(16,8,4),P2	53:06	12 GB	07:22	3.2 GB	12:31	6.3 GB	06:26	3.4 GB	07:52	3.4 GB
Szyman.(5),P4	(OOS)		45:52	38 GB	59:35	38 GB	30:36	24 GB	34:21	24 GB
Invalid Properties										
Bakery(5,5),P3	00:25	5.4 GB	68:23	38 GB	<1m	16 MB	36:48	29 GB	00:01	71 MB
Szyman.(4),P2	00:00	203 MB	00:20	253 MB	<1m	2 MB	00:10	237 MB	<1m	8 MB
Elev.2(7),P5	00:01	130 MB	<1m	6 MB	<1m	2 MB	<1m	4 MB	<1m	6 MB
Lifts(7),P4	00:01	59 MB	00:28	475 MB	<1m	4.6 MB	00:32	561 MB	00:07	239 MB

Times are given in hh:mm format, "OOS" = "out of space", "<1m" = "below 1 minute".

Table 3. Size of used models and internal memory used for storage of MPHF

Model	Number of Vertices	v_{max}	ϵ_s	MPHF Size (bits/vertex)
Elev.2(16),P4	173,916,122	30 bytes	94	4.941
Lamport(5),P4	74,413,141	24 bytes	99	4.941
MCS(5),P4	119,663,657	28 bytes	91	4.941
Peterson(5),P4	284,942,015	32 bytes	177	4.941
Phils(16,1),P3	61,230,206	50 bytes	47	4.941
Ret.(16,8,4),P2	31,087,573	91 bytes	553	4.941
Szyman.(5),P4	419,183,762	32 bytes	223	4.941

The experimental results are listed in Tab. 2. Names of algorithms correspond to names in Section 4. We note that just before the unsuccessful termination of L2S due to exhausting the disk space, the BFS level size still tended to grow. This suggests that the computation would last substantially longer if sufficient disk space would have been available. For the same input graphs, algorithms OWCTY, MAP, DDFS and IDDFS managed to perform the verification using a few Gigabytes of disk space only. All the models and their LTL properties are taken from the BEEM project [22].

Measurements on models with valid properties demonstrate that DDFS is able to successfully prove their correctness, while L2S fails. Additionally, DDFS is faster than OWCTY on most of inputs and outperforms MAP on all inputs except for model $Elev.2(16), P4$. We observe that DDFS is especially better than OWCTY on inputs where the eccentricity of the initial vertex (Tab. 3) is high, since the first enumeration phase costs almost the same time as the initial iteration of OWCTY, but the double depth-first search is not influenced by the eccentricity, while the other iterations of OWCTY are.

A notable weakness of DDFS is its bad performance on models with invalid properties. It does not work on-the-fly, and hence is outperformed by L2S and MAP on some inputs. For this reason, the iterative-deepening variant (IDDFS) has been proposed in

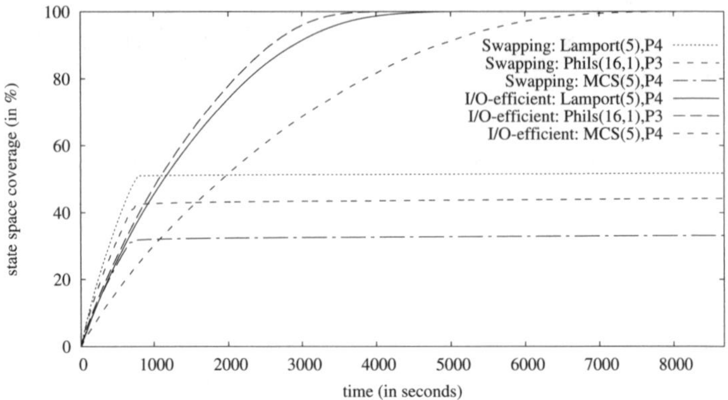

Fig. 5. Comparison of BFS with virtual memory swapping and I/O-efficient BFS

Section 3. It is a little bit slower than DDFS on inputs with valid properties (since all intermediate checks are useless in that case), but our measurements confirm that it is able to find a counter-example much sooner. Actually, its run times are close to run times of MAP, which appears to be the best choice for models with invalid properties. In our implementation, IDDFS is designed to keep costs for intermediate checks below 20% of the run time.

We also measured the internal memory taken for MPHFs representing state spaces (Tab. 3). Measured amounts of bits per vertex confirm theoretically achieved estimations for Heuristic EPH.

Finally, to support a need for I/O-efficient algorithms, we demonstrate in Fig. 5 that after exceeding the main memory, BFS with use of virtual memory swapping almost stops the exploration due to excessive amount of page faults. In contrast, I/O-efficient BFS is able to finish in reasonable time. A similar observation has been made in [2].

6 Conclusion and Discussion

With this paper we contribute c-bit semi-external DFS search for validating safety and c-bit semi-external double DFS for validating liveness properties. For bug-hunting, we implemented an iterative deepening variant of double DFS using the same amount of RAM. With minimum perfect hashing with EPH, we obtained a c-value of about 5, with Heuristic EPH used in the experiments we validated that c is less than 6.

With double DFS (DDFS) we are in many cases faster than all previous algorithms for LTL model checking L2S [15], OWCTY [4], MAP [5] (in theory and practice). Moreover, we saw that solving the LTL model checking problem off-line is not more involved than state space enumeration.

As a drawback, semi-external DDFS is neither optimal, nor on-the-fly. We discussed improvements for transforming DDFS into an on-the-fly algorithm using iterative deepening, but currently we lack an algorithm that is linear wrt. generating the search space. The algorithm by Edelkamp and Jabbar [15] operates on-the-fly, can be directed towards the error, and produces optimal counter-examples, but traverses the cross-product graph, which can be too large in many cases. MAP has a considerable I/O complexity, and is on-the-fly only for a restricted number of properties. Another open question is the design of optimal counter-examples providing LTL model checking algorithm that is linear in the size of the search space.

A notion of c-bit semi-external algorithms makes a space consumption estimates much closer to the theoretical lower bound [16]. An interesting question is how small we can get the c for c-bit semi-external DFS. Both in theory and practice, we can do somewhat better by using perfect hash functions (PHFs) with range $\{0, \ldots, m-1\}$ for $m = \Theta(n)$ rather than minimal PHFs (with $m = n$). We need to allocate m bits for storing visited-bits, but we need less space for representing the hash function. Botelho et al. [7] cites a theoretical bound of $(1+(m/n-1)\ln(1-n/m))n \log e$ bits for storing the PHF.[3] Adding m and optimizing, we obtain an optimal value of $m \approx 1.302n$ yielding a total space consumption of about $2.108n$ bits. Botelho et al. give a practical scheme based on 3-uniform hypergraphs with $m \approx 1.23n$ that uses about $1.95n$ bits for the PHF so that we need about $3.18n$ bits in total which is only about n bits off the theoretical bound. Even for the faster and simpler construction using 2-uniform hypergraphs, we get a slight improvement over the 5 bit solution than we obtain with MPHFs: using $m \approx 2n$, we need about $2n$ bits for storing the PHF yielding total space about $4n$. Note that this solution is even more computationally efficient than MPHF based schemes, since it saves a compression step needed to construct an MPHF from a PHF.

Because of external duplicate detection, vertex enumeration is time consuming. There are different possible approaches to tackle the problem. Given a sufficient number of file pointers, external sorting can be reduced to at most two scans over the search space. Moreover, pipelining [14] helps a lot in reducing the number of I/Os in BFS. Furthermore, by faster random access time, flash media might additionally reduce the run time [17].

Due to all of these techniques, we believe that external memory model checking is practical and can be made even more time and memory efficient.

References

1. Abello, J., Buchsbaum, A.L., Westbrook, J.: A functional approach to external graph algorithms. In: Bilardi, G., Pietracaprina, A., Italiano, G.F., Pucci, G. (eds.) ESA 1998. LNCS, vol. 1461, pp. 332–343. Springer, Heidelberg (1998)
2. Barnat, J.: Distributed Memory LTL Model Checking. PhD thesis, Masaryk University Brno, Faculty of Informatics (2004)
3. Barnat, J., Brim, L., Černá, I., Moravec, P., Ročkai, P., Šimeček, P.: DiVinE – A Tool for Distributed Verification (Tool Paper). In: Ball, T., Jones, R.B. (eds.) CAV 2006. LNCS, vol. 4144, pp. 278–281. Springer, Heidelberg (2006)
4. Barnat, J., Brim, L., Šimeček, P.: I/O Efficient Accepting Cycle Detection. In: Damm, W., Hermanns, H. (eds.) CAV 2007. LNCS, vol. 4590, pp. 281–293. Springer, Heidelberg (2007)
5. Barnat, J., Brim, L., Šimeček, P., Weber, M.: Revisiting Resistance Speeds Up I/O-Efficient LTL Model Checking. In: Proc. of TACAS. LNCS, vol. 4963, pp. 48–62. Springer, Heidelberg (2008)

[3] For simplicity, we drop all lower order terms and arbitrarily small constants appearing in the actual bounds.

6. Biere, A., Cimatti, A., Clarke, E.M., Zhu, Y.: Symbolic model checking without BDDs. In: Cleaveland, W.R. (ed.) ETAPS 1999 and TACAS 1999. LNCS, vol. 1579, pp. 193–207. Springer, Heidelberg (1999)
7. Botelho, F.C., Pagh, R., Ziviani, N.: Simple and space-efficient minimal perfect hash functions. In: Dehne, F.K.H.A., Sack, J.-R., Zeh, N. (eds.) WADS 2007. LNCS, vol. 4619, pp. 139–150. Springer, Heidelberg (2007)
8. Botelho, F.C., Ziviani, N.: External perfect hashing for very large key sets. In: Conference on Information and Knowledge Management (CIKM), pp. 653–662. ACM, New York (2007)
9. Brim, L., Černá, I., Moravec, P., Šimša, J.: Accepting Predecessors are Better than Back Edges in Distributed LTL Model-Checking. In: Hu, A.J., Martin, A.K. (eds.) FMCAD 2004. LNCS, vol. 3312, pp. 352–366. Springer, Heidelberg (2004)
10. Černá, I., Pelánek, R.: Distributed explicit fair cycle detection. In: Ball, T., Rajamani, S.K. (eds.) SPIN 2003. LNCS, vol. 2648, pp. 49–73. Springer, Heidelberg (2003)
11. Chiang, Y.-J., Goodrich, M.T., Grove, E.F., Tamassia, R., Vengroff, D.E., Vitter, J.S.: External-memory graph algorithms. In: Symposium on Discrete Algorithms (SODA), pp. 139–149. Society for Industrial and Applied Mathematics (1995)
12. Clarke, E.M., Grumberg, O., Peled, D.A.: Model checking. MIT Press, Cambridge (1999)
13. Courcoubetis, C., Vardi, M., Wolper, P., Yannakakis, M.: Memory-efficient algorithms for the verification of temporal properties. Form. Methods Syst. Des. 1(2-3), 275–288 (1992)
14. Dementiev, R., Kettner, L., Sanders, P.: STXXL: Standard template library for XXL data sets. In: Proc. of ESA. LNCS, vol. 3669, pp. 640–651. Springer, Heidelberg (2005)
15. Edelkamp, S., Jabbar, S.: Large-scale directed model checking LTL. In: Valmari, A. (ed.) SPIN 2006. LNCS, vol. 3925, pp. 1–18. Springer, Heidelberg (2006)
16. Fredman, M.L., Komlós, J.: On the size of separating systems and families of perfect hash functions. SIAM Journal on Algebraic and Discrete Methods 5(1), 61–68 (1984)
17. Gal, E., Toledo, S.: Algorithms and data structures for flash memories. ACM Computing Surveys 37(2), 138–163 (2005)
18. Korf, R.E.: Delayed duplicate detection: Extended abstract. In: International Joint Conference on Artificial Intelligence (IJCAI), pp. 1539–1541 (2003)
19. Mehlhorn, K., Meyer, U.: External-memory breadth-first search with sublinear I/O. In: Möhring, R.H., Raman, R. (eds.) ESA 2002. LNCS, vol. 2461, pp. 723–735. Springer, Heidelberg (2002)
20. Meyer, U.: Algorithms for Memory Hierarchies. Springer, Heidelberg (2003)
21. Munagala, K., Ranade, A.: I/O-complexity of graph algorithms. In: Symposium on Discrete Algorithms (SODA), pp. 687–694. Society for Industrial and Applied Mathematics (1999)
22. Pelánek, R.: BEEM: Benchmarks for explicit model checkers. In: Bošnački, D., Edelkamp, S. (eds.) SPIN 2007. LNCS, vol. 4595, pp. 263–267. Springer, Heidelberg (2007)
23. Ravi, K., Bloem, R., Somenzi, F.: A Comparative Study of Symbolic Algorithms for the Computation of Fair Cycles. In: Johnson, S.D., Hunt Jr., W.A. (eds.) FMCAD 2000. LNCS, vol. 1954, pp. 143–160. Springer, Heidelberg (2000)
24. Schuppan, V., Biere, A.: Efficient reduction of finite state model checking to reachability analysis. Int. Journal on Software Tools for Technology Transfer 5(2–3), 185–204 (2004)
25. Sharir, M.: A strong-connectivity algorithm and its applications in data flow analysis. Computers and Mathematics with Applications 7(1), 67–72 (1981)
26. Stern, U., Dill, D.L.: Using magnetic disk instead of main memory in the Murϕ verifier. In: Probst, D.K., von Bochmann, G. (eds.) CAV 1992. LNCS, vol. 663, pp. 172–183. Springer, Heidelberg (1993)
27. Zhou, R., Hansen, E.: Breadth-first heuristic search. In: Int. Conf. on Automated Planning and Scheduling (ICAPS) pp. 92–100. AAAI Press / The MIT Press (2004)

QMC: A Model Checker for Quantum Systems
Tool Paper

Simon J. Gay[1,⋆], Rajagopal Nagarajan[2,⋆⋆], and Nikolaos Papanikolaou[2,⋆⋆]

[1] Department of Computing Science, University of Glasgow
simon@dcs.gla.ac.uk
[2] Department of Computer Science, University of Warwick
{biju,nikos}@dcs.warwick.ac.uk

1 Introduction

The novel field of quantum computation and quantum information has been growing at a rapid rate; the study of quantum information in particular has led to the emergence of communication and cryptographic protocols with no classical analogues. Quantum information protocols have interesting properties which are not exhibited by their classical counterparts, but they are most distinguished for their applications in cryptography. Notable results include the unconditional security proof [1] of quantum key distribution. This result, in particular, is one of the reasons for the widespread interest in this field. Furthermore, the implementation of quantum cryptography has been demonstrated in non-laboratory settings and is already an important practical technology. Implementations of quantum cryptography have already been commercially launched and tested by a number of companies including MagiQ, Id Quantique, Toshiba, and NEC. The unconditional security of quantum key distribution protocols does not automatically imply the same degree of security for actual systems, of course; this justifies the need for systems modelling and verification in this setting.

The benefits of automated verification techniques are well known for classical communication protocols, especially in the cryptographic setting. Our research programme is to apply these techniques to *quantum* protocols with the expectation of gaining corresponding benefits. Our earlier work involved applying probabilistic model–checking techniques to such protocols (see e.g. [2]; it became clear that existing techniques are not satisfactory for analysing quantum systems. Today, while simulation tools for quantum systems abound, to our knowledge no other authors have developed a tool directly aimed at verification. In this paper we describe just such a tool, named QMC (Quantum Model Checker); it allows for automated verification of properties of quantum protocols.

⋆ Partially supported by the EPSRC grant EP/E00623X/1 *(Semantics of Quantum Computation)* and EP/F004184/1 *(Quantum Computation: Foundations, Security, Cryptography and Group Theory)*.

⋆⋆ Partially supported by the EU Sixth Framework Programme (Project SecoQC: *Development of a Global Network for Secure Communication based on Quantum Cryptography*) and the EPSRC grant EP/E00623X/1.

A. Gupta and S. Malik (Eds.): CAV 2008, LNCS 5123, pp. 543–547, 2008.

Properties to be verified by QMC are expressed using the logic QCTL [3].QMC enables the modelling of systems which can be expressed within the quantum *stabilizer formalism*; such systems are known to be simulable in polynomial time (viz. Gottesman–Knill theorem [4]). The systems expressible in this formalism are restricted, in the sense that the set of operations which they can perform is not universal for quantum computation. Nevertheless, stabilizers are sufficient to describe a number of systems of practical interest. We assume that the reader has an understanding of the basics of quantum computation [5].

2 Tool Description

QMC is a command–line tool, implemented in Java 5, which simulates all possible executions of a quantum protocol model, and then checks that the model satisfies any number of specification formulae supplied by the user. We will now discuss the modelling language of QMC, the simulation of models, the logic used for specifying properties, and the algorithms used for model–checking. We will briefly discuss a concrete example, namely a model of quantum coin-flipping [6], the QMC source for which is available at`http://www.dcs.warwick.ac.uk/ nikos/downloads/case-study.qmc`.

Modelling Language. Protocols are modelled using a high-level, concurrent programming language developed especially for QMC. It is an imperative language with lightweight concurrency, typed variables including channels, and commands for allocating and manipulating the qubits in a global quantum state. Qubit variables are pointers to the individual quantum bits (qubits) in the global state. The language has guarded commands, which are used to express looping and nondeterministic choices.

QMC has a statically typed language, and variables of different types must be declared at the beginning of a protocol model (shared variables), or at the beginning of the process in which they are used (local variables). The classical data types include integers (type **integer**), bits (type **bool**), and floating–point numbers (type **real**). There is a channel type corresponding to each of these base types (e.g. **channel of integer**). Qubit variables are references to qubits in QMC's internal quantum state. After declaration, these variables must be initialized with the newqubit expression. A process can apply any one of the *Clifford group operators* (see [5]): Hadamard (**had**), Controlled–not (**cnot**), Phase (**ph**) to a given qubit, as well as perform quantum measurement with respect to the computational basis. These are the operations allowed in the quantum stabilizer formalism.

Simulating QMC *programs.* As in PROMELA, any command in QMC is either *executable* or not from the current state. QMC's process scheduler interleaves executable statements from all process declarations in the input, resolving all non–deterministic choices. Quantum measurement is also treated as a source of non-determinism; a qubit in a general stabilizer state always produces one of two possible outcomes with equal probability, and we treat these equiprobable outcomes like a non–deterministic choice. Consequently, we do not analyse the probabilities of different execution paths; this is an area for future work. The scheduler ensures finiteness of models.We are currently developing an operational semantics for the QMC language. Execution of a QMC program produces a tree of execution paths, in which the branching arises from non–determinism.

Specification Logic. The properties of a given protocol model are expressed using the logic QCTL [3], which is a temporal logic especially designed for reasoning about quantum systems based on CTL [7]. This logic allows us to reason about the evolution of the global quantum state as a given protocol model is executed. It also enables reasoning about the states of classical variables and, hence, measurement outcomes. The reader is referred to [3] for full details of the syntax of the logic, which includes classical formulae ($\alpha = \bot \mid \mathsf{qb} \mid \alpha \Rightarrow \alpha$), terms ($t = x \mid (t+t) \mid (tt) \mid \mathrm{Re}(|\top\rangle_A) \mid \mathrm{Im}(|\top\rangle_A) \mid \int \phi$), quantum formulae ($\gamma = (t \leq t) \mid \bot \mid (\alpha \sqsupset \alpha)$) and temporal formulae ($\theta = \gamma \mid \theta \sqsupset \theta \mid (\mathsf{EX}\theta) \mid (\mathsf{E}[\theta \mathsf{U} \theta]) \mid (\mathsf{AF}\theta)$).

Interesting state formulas are those of the form $\int \phi \leq a$ and $[\mathsf{qb}_i, \mathsf{qb}_j]$, where ϕ is a base formula (e.g. $\neg \mathsf{qb}_0$) and $\mathsf{qb}_i, \mathsf{qb}_j, \mathsf{qb}_0$ are qubit variables. The first formula states that, the probability of formula ϕ being satisfied in the current state is less than or equal to a. The formula $[\mathsf{qb}_i, \mathsf{qb}_j]$ states that, in the current state, the qubits corresponding to the variables qb_i and qb_j are *not entangled* with the rest of the quantum system. In order to evaluate such a formula, QMC analyses the entanglement of the current quantum state.

Verification Algorithms. QMC implements algorithms for evaluating EQPL formulas over stabilizer states, which are represented internally using a matrix representation (see [8]). In order to check the truth of a particular formula, its truth needs to be determined for all possible valuations; the tool automatically extracts all valuations from the internal representation. More interestingly, the tool has been designed to explore all possible executions of a particular protocol arising from different measurement outcomes and non–deterministic choices. Entanglement formulae may be checked without converting the internal representation to the set of all valuations. In fact, it is possible to determine whether a list of qubits $q_i, q_{i+1}, \ldots$ constitutes a partition of the global quantum state at any point during execution using the *entanglement normal forms* developed by Audenaert and Plenio [9]. Temporal formulae are checked by traversing the tree produced by the execution of a protocol model.

An Example. We have built a QMC model for the quantum coin–flipping protocol due to Bennett and Brassard [6] which is available online at the URL mentioned at the beginning of Section 2. Quantum coin–flipping enables two users, Alice and Bob, to establish a common random bit x through the transmission of a single qubit q and its measurement. The protocol relies on the principle that, if Alice and Bob use compatible bases for preparation and for measurement of this qubit, their bit values will be guaranteed to match by the laws of quantum mechanics. Incompatible bases will produce a matching bit value only with probability $\frac{1}{2}$, although in this case the protocol is repeated and the bit discarded. There are various possible attacks that may be performed by an enemy, which would enable him or her to compromise the final bit value. It is such attacks that we would like to investigate using the QMC tool on this and related quantum protocols. The basic property which needs to be checked at the end of the protocol is that the bit values of Alice and Bob do indeed match, and this is only true if the measurement basis $\hat{b}$ chosen by Bob matches the preparation basis b of Alice. Also, if Alice's and Bob's bits and bases are in agreement, the protocol should not abort; this is expressed as a temporal property.

3 Conclusion and Future Work

We have described QMC, a model-checking tool for quantum protocols. As far as we know, it is the first dedicated verification tool (as opposed to simulation systems) for quantum protocols. QMC allows the modelling and verification of properties of protocols expressible in the quantum stabilizer formalism. The logic for expressing properties is QCTL. We have considered a simple example illustrating the input language of the tool. It is significant to note that the restriction to quantum stabilizer states allows QMC to simulate protocols efficiently, although it limits the expressive power of the tool. There are protocols which involve quantum states that fall outside the scope of the stabilizer formalism, and we are currently investigating ways of approximating such states using techniques due to Bravyi and Kitaev [10]. Using such techniques it will be possible to obtain an implementation which is powerful enough for the analysis of general quantum protocols and the detection of potential flaws in their design.

References

[1] Mayers, D.: Unconditional security in quantum cryptography. Journal of the ACM 48(3), 351–406 (2001)

[2] Gay, S.J., Nagarajan, R., Papanikolaou, N.: Probabilistic model–checking of quantum protocols. In: DCM 2006: Proceedings of the 2nd International Workshop on Developments in Computational Models (2006) arXiv:quant-ph/0504007

[3] Baltazar, P., Chadha, R., Mateus, P., Sernadas, A.: Towards model-checking quantum security protocols. In: Dini, P., et al. (eds.) Proceedings of the First Workshop on Quantum Security: QSec 2007, IEEE Press, Los Alamitos (2007)

[4] Gottesman, D.: The Heisenberg representation of quantum computers. In: Corney, S., Delbourgo, R., Jarvis, P. (eds.) Group22: Proceedings of the XXII International Colloquium on Group Theoretical Methods in Physics. International Press (1999)

[5] Nielsen, M.A., Chuang, I.L.: Quantum Computation and Quantum Information. Cambridge University Press, Cambridge (2000)

[6] Bennett, C.H., Brassard, G.: Quantum Cryptography: Public key distribution and coin tossing. In: Proceedings of International Conference on Computers, Systems and Signal Processing (1984)

[7] Clarke, E.M., Emerson, E.A., Sistla, A.P.: Automatic verification of finite state concurrent system using temporal logic specifications: a practical approach. In: POPL 1983: Proceedings of the 10th ACM SIGACT-SIGPLAN symposium on Principles of programming languages, pp. 117–126. ACM, New York (1983)
[8] Aaronson, S., Gottesman, D.: Improved simulation of stabilizer circuits. Physical Review A 70(52328) (2004)
[9] Audenaert, K., Plenio, M.: Entanglement on mixed stabiliser states: Normal forms and reduction procedures. New Journal of Physics 7(170) (2005)
[10] Bravyi, S., Kitaev, A.: Universal quantum computation with ideal Clifford gates and noisy ancillas. Physical Review A 71, 1–14 (2005)

T(O)RMC: A Tool for (ω)-Regular Model Checking
Tool Paper

Axel Legay

Carnegie Mellon University, Computer Science Department, Pittsbugh, PA
alegay@cs.cmu.edu

1 Introduction and Motivations

Within the context of the verification of infinite-state systems, "(ω)-Regular model checking" is the name of a family of techniques in which states are represented by words, sets of states by finite automata on these objects, and transitions by finite automata operating on pairs of state encodings, i.e. finite-state transducers. If the states are encoded by finite words, then sets of (pairs of) states can be represented by finite-word automata. This setting can be used to represent various classes of infinite-state systems [17], including parametric systems, FIFO-queue systems, and systems manipulating integer variables (those defined in Presburger arithmetic). When the states are encoded by infinite words, sets of (pairs of) states are represented by deterministic weak Büchi automata[1]. This setting is mainly used to represent systems involving both integer and real variables [4, 6], such as linear hybrid systems with a constant derivative.

It is known [7, 8] that computing the set of reachable states and verifying linear temporal properties in this automata-based framework reduces to solving the *(ω-)regular reachability problems.* Given a finite-word (resp. deterministic weak) automaton A representing the initial states and a finite-word (resp. deterministic weak) transducer T representing the transition relation, those problems amount to computing the iterative closures T^+ and $T^+(A)$. This can be done either by *specific techniques* that exploit the specific properties and representations of the domain being considered, or by *generic techniques* that consider automata-based representations and provide algorithms that operate directly on these representations.

In [5, 6], we have proposed a generic technique for computing $T^+(A)$ and T^+ for the finite and infinite word cases. Our approach consists in computing the limit of a possibly infinite sequence of automata. This is an undecidable problem to which the computation of T^+ and $T^+(A)$ as well as several others can be reduced. The technique has been evaluated and improved with the help of prototypes, which have been applied on various classes of problems [5,6,8,9,12].

[1] A weak Büchi automaton is a Büchi automaton whose strongly connected components contain either only accepting or only nonaccepting states. Deterministic weak Büchi automata are easily complementable and admit a unique minimal form.

A. Gupta and S. Malik (Eds.): CAV 2008, LNCS 5123, pp. 548–551, 2008.

Those prototypes predate the T(O)RMC toolset which uniforms them, and is the subject of this paper[2].

2 The Underlying Technique from [5, 6, 12]

Given a sequence $S = A_0, A_1, A_2, \ldots$ of minimal finite-word (resp. weak) automata, the techniques computes a finite-word (resp. weak Büchi) automaton for $\bigcup_{i=0}^{\infty} A^i$. The general idea is to extrapolate one of the finite *sampling sequence* of S, *i.e.* selected automata from one of its finite prefixes. The extrapolation step is done by comparing successive automata in the sampling sequence, trying to identify the difference between these in the form of an *increment* (*i.e.* a difference in their graph structures), and *extrapolating* the repetition of this increment by adding loops to the last automaton of the sequence. After the extrapolation has been built, one has to check whether it corresponds to the limit of the sequence[3]. If this is the case, then the computation terminates. Otherwise, another sampling sequence has to be chosen.

Testing whether two (or more) automata differ by the addition of increments is decided by a combination of forward and backward language equivalences. Choosing the sampling sequence is a rather tricky issue and there is no guarantee that this can be done in a way that ensures that the extrapolation step can be applied. However, there is a number of heuristics that are very effective for obtaining a sample sequence that can be extrapolated (see Chapter 5 of [12] to know how to choose a "good" sampling strategy for your case study). Finally, checking safety of the extrapolation (does it include the limit?) is simple, but checking preciseness (is it exactly the limit?) is a much more involved problem for which only partial solutions has been developed (see Chapter 7 of [12]). In most of cases, working with an over approximation is practically sufficient.

3 The Tool

The T(O)RMC toolset (available at [15]), builds upon the LASH toolset. LASH [11] takes the form of a set of C functions for building and manipulating (union, intersection, complementation, minimization, ...) both finite-word and weak Büchi automata. The tool also provides some domain specific techniques for solving the regular reachability problems. In addition, several compilers can be used to make easier the construction of specific classes of automata (e.g. those that represent solution of Presburger formulas). T(O)RMC improves LASH with three new packages, that are (see [12] and [15] for details):

1. *The transducer package* that provides data structures and algorithms to manipulate transducers. The package also provides several heuristics to improve

[2] States for Tool for (ω)-Regular Model Checking.

[3] In practice, the user may also be satisfied with an extrapolation whose correctness is unknown.

the efficiency of the composition between transducers (sometimes from days to seconds!).

2. *The extrapolation package* for detecting increments in a sequence of automata, and extrapolating a finite sampling sequence. The tool allows the user to precise (1) which sampling strategy has to be used, and (2) how to build the successive elements in the infinite sequence.
3. *The counter-word automata package* that provides data structures and algorithms to check the correctness of the extrapolation for several classes of problems.

4 Summary of the Experiments

The T(O)RMC toolset has been evaluated over more than 100 case studies. Due to space limitations, this section only briefly recaps the classes of problems for which T(O)RMC has been used so far. Details about the experiments (including performances in terms of time and memory, which vary from examples to examples) can be found in Chapters 7 and 13 of [12] and in [8].

We first used T(O)RMC to compute an automata-based representation of the set of reachable states of several infinite-states systems, including parametric systems, FIFO-queue systems, and systems manipulating integer variables. Others experiments concerned the computation of the transitive closure of several arithmetic relations. It is worth mentioning that the disjunctive nature of some relations sometimes prevents the direct use of specific domain-based techniques [10,3]. We also applied T(O)RMC to the challenging problem of analyzing linear hybrid systems. One of the case studies consisted of computing a precise representation of the set of reachable states of several versions of the *leaking gas burner*. To the best of our knowledge, only the technique in [3] was able to handle the cases we considered. Among the other experiments, we should also mention the computation of the set of reachable states of an augmented version of the IEEE Root Contention Protocol [12], which has been point out to be a hard problem [14]. The ability of T(O)RMC to compute the limit of an infinite sequence of automata has other applications. As an example, the tool has been used in a semi-algorithm to compute the convex hull of a set of integer vectors [9]. T(O)RMC was also used to compute a symbolic simulation over the state-space of an infinite-state system, with the aim of verifying temporal properties [8].

5 Conclusion

We presented T(O)RMC, a tool for computing the limit of an infinite sequence of finite-word or deterministic weak Büchi automata.

T(O)RMC implements a very general automata sequence extrapolation technique. As such it is slower than tools that are specific to the arithmetic domain (FAST [1], LIRA [2], LASH), but is perfectly competitive when handling other regular model checking cases (parametric systems, FIFO systems, ...) [13, 16].

The main goal of the tool is not performance improvement, but to allow experimentation with automata sequence extrapolation in a variety of context that goes beyond (omega-)Regular Model checking problems.

References

1. Bardin, S., Leroux, J., Point, G.: Fast extended release. In: Ball, T., Jones, R.B. (eds.) CAV 2006. LNCS, vol. 4144, pp. 63–66. Springer, Heidelberg (2006)
2. Becker, B., Dax, C., Eisinger, J., Klaedtke, F.: LIRA: Handling constraints of linear arithmetics over the integers and the reals. In: Damm, W., Hermanns, H. (eds.) CAV 2007. LNCS, vol. 4590, pp. 307–310. Springer, Heidelberg (2007)
3. Boigelot, B., Herbreteau, F.: The power of hybrid acceleration. In: Ball, T., Jones, R.B. (eds.) CAV 2006. LNCS, vol. 4144, pp. 438–451. Springer, Heidelberg (2006)
4. Boigelot, B., Jodogne, S., Wolper, P.: An effective decision procedure for linear arithmetic over the integers and reals. ACM Transactions on Computational Logic 6(3), 614–633 (2005)
5. Boigelot, B., Legay, A., Wolper, P.: Iterating transducers in the large (extended abstract). In: Hunt Jr., W.A., Somenzi, F. (eds.) CAV 2003. LNCS, vol. 2725, pp. 223–235. Springer, Heidelberg (2003)
6. Boigelot, B., Legay, A., Wolper, P.: Omega-regular model checking. In: Jensen, K., Podelski, A. (eds.) TACAS 2004. LNCS, vol. 2988, pp. 561–575. Springer, Heidelberg (2004)
7. Bouajjani, A., Jonsson, B., Nilsson, M., Touili, T.: Regular model checking. In: Emerson, E.A., Sistla, A.P. (eds.) CAV 2000. LNCS, vol. 1855, pp. 403–418. Springer, Heidelberg (2000)
8. Bouajjani, A., Legay, A., Wolper, P.: Handling liveness properties in (omega-)regular model checking. In: Proc. 6th Int. INFINITY. ENTCS, vol. 138(3), Elsevier, Amsterdam (2004)
9. Cantin, F., Legay, A., Wolper, P.: Computing convex hulls by automata iteration. In: Proc. 1th Int. AUTOMATHA (2007)
10. Finkel, A., Leroux, J.: How to compose presburger-accelerations: Applications to broadcast protocols. In: Agrawal, M., Seth, A.K. (eds.) FSTTCS 2002. LNCS, vol. 2556, pp. 145–156. Springer, Heidelberg (2002)
11. The Liège Automata-based Symbolic Handler (LASH), `http://www.montefiore.ulg.ac.be/~boigelot/research/lash/`.
12. Legay, A.: Generic Techniques for the Verification of Infinite-state Systems. Collection des publications de la Faculté des Sciences Appliquées de l'Université de Liège, Liège, Belgium (2007), `http://www.montefiore.ulg.ac.be/~legay/papers/index`
13. The regular model checking tool (RMC), `http://www.it.uu.se/research/docs/fm/apv/rmc`
14. Simons, D.P.L., Stoelinga, M.: Mechanical verification of the ieee 1394a root contention protocol using uppaal2k. International Journal STTT 3(4), 469–485 (2001)
15. The T(O)RMC toolset, `http://www.montefiore.ulg.ac.be/~legay/TORMC/index-tormc.html`
16. Vardhan, A., Viswanathan, M.: Lever: A tool for learning based verification. In: Ball, T., Jones, R.B. (eds.) CAV 2006. LNCS, vol. 4144, pp. 471–474. Springer, Heidelberg (2006)
17. Wolper, P., Boigelot, B.: Verifying systems with infinite but regular state spaces. In: Y. Vardi, M. (ed.) CAV 1998. LNCS, vol. 1427, pp. 88–97. Springer, Heidelberg (1998)

Faster Than UPPAAL ?
Tool Paper

Sebastian Kupferschmid, Martin Wehrle, Bernhard Nebel, and Andreas Podelski

Albert-Ludwigs-Universität Freiburg
Institut für Informatik
Freiburg, Germany
{kupfersc,mwehrle,nebel,podelski}@informatik.uni-freiburg.de

Abstract. It is probably very hard to develop a new model checker that is faster than UPPAAL for verifying *(correct)* timed automata. In fact, our tool MCTA does not even try to compete with UPPAAL in this (i. e., UPPAAL's) arena. Instead, MCTA is geared towards analyzing *incorrect* specifications of timed automata. It returns (shorter) error traces faster.

1 Our Tool: MCTA

We present MCTA, a model checking tool for real-time specifications modeled as timed automata. Although the tool can be used for verification, MCTA is rather optimized for falsification, i. e., detecting violations against safety properties fast and returning short error traces. Several types of traces can be generated, including an option to find a (guaranteed) shortest error trace. There is also the possibility to examine MCTA's traces with UPPAAL's graphical user interface.

MCTA accepts input models in the form of the UPPAAL input language (cf. [1]). So far only a fraction thereof is supported, e. g. there is no support for urgent channels, arrays, etc. yet. Internally, MCTA uses UPPAAL's timed automata parser library. For the representation of zones, MCTA uses UPPAAL's difference bound matrices library. Both libraries are released under the terms of the LGPL or GPL, respectively, and are freely available at `http://www.uppaal.com/`. All other data structures and all algorithms (and their implementation) used are genuine to MCTA.

MCTA is free software and also released under the terms of the GPL. Pre-compiled Linux executables and a snapshot of the source code of our tool are also freely available at `http://www.informatik.uni-freiburg.de/~kupfersc/mcta/`.

2 The Ingredients of MCTA

MCTA accelerates the detection of error states by using the well-known directed model checking approach [5,4]. In this approach, an abstract distance value is computed for each state encountered during the state space traversal. The abstract distance values determine the *order* in which the states are explored. Among possible successor states, the ones with a lower value are preferred. There are many different strategies to explore the state space. MCTA allows the user to choose between two strategies based on two

A. Gupta and S. Malik (Eds.): CAV 2008, LNCS 5123, pp. 552–555, 2008.

wide-spread search methods called A* and *greedy search*. The first explores states s with lowest value of $c(s) + h(s)$ first, where $c(s)$ is the length of the path from the initial state through which s was reached. Under certain conditions on the abstract distance values, one is guaranteed a shortest error path. In the second strategy, states are explored by increasing value of $h(s)$. Doing so, the length of the detected error path is not guaranteed to be as short as possible, but tends to explore less states in practice.

MCTA generates the abstract distance values *fully automatically* for each input given by a timed automaton and a safety property. This is done by efficiently computing a rather coarse abstraction (the user can choose among several kinds of abstraction, see below) and taking the distance in the abstract state space. MCTA in addition offers the possibility to automatically recognize which transitions should be penalized during the state space traversal; this is a new technique presented in [8].

Monotonicity Abstraction. Currently, MCTA comes with several kinds of abstractions as the basis for computing the abstract distance values. Here, we will only explain the distance function which is based on the *monotonicity abstraction* [7]. This abstraction is mainly an adaption of a technique from AI Planning namely *ignoring delete lists* [2]. The idea of the corresponding abstraction is to have *every state variable, once it has obtained a value, keeps that value forever*. I. e., the value of a variable is no longer an element, but a *subset* of its domain. This subset grows monotonically over transition applications, hence the name of this abstraction.

MCTA assigns to each state encountered during the state space traversal an abstract distance value by applying the monotonicity abstraction to the part of the state space that is rooted in the current state, and traversing the abstract state space. The length of the abstract error trace is the state's abstract distance value. If there is no abstract error path, then there is no concrete one either.

3 Results

We compare the performance of MCTA and UPPAAL for detecting error traces in incorrect specifications of timed automata. For both tools we chose the most powerful options. We used the current version of UPPAAL (4.0.6) with the option *randomized depth first search* (rDF). The results for rDF in Table 1 are averaged over 10 runs. For MCTA the specific options are: the strategy for the state space traversal being based on greedy search, the abstraction for the abstract distance values being the monotonicity abstraction, and the recognition of "useless transitions" and the state space traversal penalizing recognized transitions [8].

The examples $C_1, \ldots, C_9$ stem from an industrial case study called "Single-tracked Line Segment" [6]. It models a distributed real-time controller for a segment of tracks where trams share a piece of track. The examples $M_1, \ldots, M_4$ and $N_1, \ldots, N_4$ come from a case study namely "Mutual Exclusion" [3]. It models a real-time protocol to ensure mutual exclusion of states in a distributed system via asynchronous communication. Both case studies are part of the AVACS project benchmark suite.

The results in Table 1 (visualized in Fig. 1) clearly demonstrate that the algorithms employed by our tool are useful for analyzing incorrect timed automata. In comparison

Table 1. Experimental results of UPPAAL's and MCTA's most powerful options. The results are computed on an Intel Xeon with 2.66 Ghz. Dashes indicate out of memory (more than 4 GB).

Exp	explored states UPPAAL	explored states MCTA	runtime in s UPPAAL	runtime in s MCTA	memory in MB UPPAAL	memory in MB MCTA	trace length UPPAAL	trace length MCTA
M_1	8343	4256	0.37	0.07	38	56	829	97
M_2	27156	8186	1.53	0.10	40	57	3245	146
M_3	24368	10650	1.39	0.12	40	58	2991	91
M_4	70906	22412	4.93	0.24	45	64	11728	136
N_1	11115	5689	0.93	0.10	38	59	607	108
N_2	45998	15377	4.99	0.25	41	62	3788	152
N_3	31725	16332	3.31	0.26	41	65	3302	91
N_4	220262	44199	25.31	0.71	51	84	14003	118
C_1	15407	1658	0.23	0.12	38	56	945	91
C_2	31308	1333	0.32	0.16	39	56	820	91
C_3	45443	1153	0.36	0.13	39	56	541	91
C_4	366056	1001	2.90	0.19	49	57	1690	121
C_5	2629269	833	23.54	0.22	120	57	2345	114
C_6	21940802	833	230.08	0.29	761	57	3237	114
C_7	–	829	–	0.35	–	57	–	114
C_8	–	816	–	0.28	–	57	–	95
C_9	–	13423	–	3.24	–	71	–	90

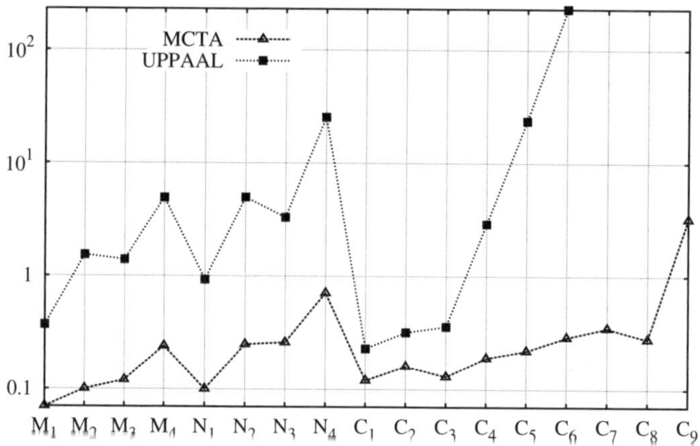

Fig. 1. Comparing UPPAAL and MCTA on incorrect specifications of timed automata ("MCTA's arena"): runtime (in s) for detecting error traces. MCTA is orders of magnitude faster.

with UPPAAL which does not employ such algorithms (based on automatically generated abstract error distances), our tool finds the error paths faster. It explores less states and uses less memory and thus scales to larger benchmarks. At the same time, it returns shorter error paths.

4 Outlook

In the future, MCTA will evolve by supporting more and more language constructs for defining (extensions of) timed automata, and by providing more and more kinds of abstractions for computing abstract distance values. Eventually, we hope, the results and the practical experience with MCTA for analyzing incorrect specifications will flow into tools that were originally geared towards analyzing *correct* timed automata.

Acknowledgements

This work was partly supported by the German Research Council (DFG) as part of the Transregional Collaborative Research Center "Automatic Verification and Analysis of Complex Systems" (SFB/TR 14 AVACS). See `http://www.avacs.org/` for more information.

We also thank the UPPAAL team for making their DBM and their parser library freely available.

References

1. Behrmann, G., David, A., Larsen, K.G.: A tutorial on Uppaal. In: Bernardo, M., Corradini, F. (eds.) SFM-RT 2004. LNCS, vol. 3185, pp. 200–236. Springer, Heidelberg (2004)
2. Bonet, B., Geffner, H.: Planning as heuristic search. Artificial Intelligence 129(1–2), 5–33 (2001)
3. Dierks, H.: Comparing model-checking and logical reasoning for real-time systems. Formal Aspects of Computing 16(2), 104–120 (2004)
4. Edelkamp, S., Leue, S., Lluch-Lafuente, A.: Directed explicit-state model checking in the validation of communication protocols. International Journal on Software Tools for Technology Transfer 5(2–3), 247–267 (2004)
5. Edelkamp, S., Lluch-Lafuente, A., Leue, S.: Directed explicit model checking with HSF-Spin. In: Dwyer, M.B. (ed.) SPIN 2001. LNCS, vol. 2057, pp. 57–79. Springer, Heidelberg (2001)
6. Krieg-Brückner, B., Peleska, J., Olderog, E.-R., Baer, A.: The UniForM workbench, a universal development environment for formal methods. In: Woodcock, J.C.P., Davies, J., Wing, J.M. (eds.) FM 1999. LNCS, vol. 1709, pp. 1186–1205. Springer, Heidelberg (1999)
7. Kupferschmid, S., Hoffmann, J., Dierks, H., Behrmann, G.: Adapting an AI planning heuristic for directed model checking. In: Valmari, A. (ed.) SPIN 2006. LNCS, vol. 3925, pp. 35–52. Springer, Heidelberg (2006)
8. Wehrle, M., Kupferschmid, S., Podelski, A.: Useless transitions are useful. Reports of SFB/TR 14 AVACS 39, SFB/TR 14 AVACS (2008)

Author Index

Printing: Mercedes-Druck, Berlin
Binding: Stein+Lehmann, Berlin

Lecture Notes in Computer Science

Sublibrary 1: Theoretical Computer Science and General Issues

For information about Vols. 1–4782
please contact your bookseller or Springer

Vol. 5130: J.v.z. Gathen, J.L. Imaña, Ç.K. Koç (Eds.), Arithmetic of Finite Fields. X, 205 pages. 2008.

Vol. 5124: J. Gudmundsson (Ed.), Algorithm Theory – SWAT 2008. XIII, 438 pages. 2008.

Vol. 5123: A. Gupta, S. Malik (Eds.), Computer Aided Verification. XVII, 558 pages. 2008.

Vol. 5104: F. Bello, E. Edwards (Eds.), Biomedical Simulation. XI, 228 pages. 2008.

Vol. 5103: M. Bubak, G.D. van Albada, J. Dongarra, P.M.A. Sloot (Eds.), Computational Science – ICCS 2008, Part III. XXVIII, 758 pages. 2008.

Vol. 5102: M. Bubak, G.D. van Albada, J. Dongarra, P.M.A. Sloot (Eds.), Computational Science – ICCS 2008, Part II. XXVIII, 752 pages. 2008.

Vol. 5101: M. Bubak, G.D. van Albada, J. Dongarra, P.M.A. Sloot (Eds.), Computational Science – ICCS 2008, Part I. XLVI, 1058 pages. 2008.

Vol. 5092: X. Hu, J. Wang (Eds.), Computing and Combinatorics. XIV, 680 pages. 2008.

Vol. 5090: R.T. Mittermeir, M.M. Syslo (Eds.), Informatics Education - Supporting Computational Thinking. XV, 357 pages. 2008.

Vol. 5073: O. Gervasi, B. Murgante, A. Laganà, D. Taniar, Y. Mun, M.L. Gavrilova (Eds.), Computational Science and Its Applications – ICCSA 2008, Part II. XXIX, 1280 pages. 2008.

Vol. 5072: O. Gervasi, B. Murgante, A. Laganà, D. Taniar, Y. Mun, M.L. Gavrilova (Eds.), Computational Science and Its Applications – ICCSA 2008, Part I. XXIX, 1266 pages. 2008.

Vol. 5065: P. Degano, R. De Nicola, J. Meseguer (Eds.), Concurrency, Graphs and Models. XV, 810 pages. 2008.

Vol. 5062: K.M. van Hee, R. Valk (Eds.), Applications and Theory of Petri Nets. XIII, 429 pages. 2008.

Vol. 5059: F.P. Preparata, X. Wu, J. Yin (Eds.), Frontiers in Algorithmics. XI, 350 pages. 2008.

Vol. 5058: A.A. Shvartsman, P. Felber (Eds.), Structural Information and Communication Complexity. X, 307 pages. 2008.

Vol. 5050: J.M. Zurada, G.G. Yen, J. Wang (Eds.), Computational Intelligence: Research Frontiers. XVI, 389 pages. 2008.

Vol. 5038: C.C. McGeoch (Ed.), Experimental Algorithms. X, 363 pages. 2008.

Vol. 5036: S. Wu, L.T. Yang, T.L. Xu (Eds.), Advances in Grid and Pervasive Computing. XV, 518 pages. 2008.

Vol. 5035: A. Lodi, A. Panconesi, G. Rinaldi (Eds.), Integer Programming and Combinatorial Optimization. XI, 477 pages. 2008.

Vol. 5029: P. Ferragina, G.M. Landau (Eds.), Combinatorial Pattern Matching. XIII, 317 pages. 2008.

Vol. 5028: A. Beckmann, C. Dimitracopoulos, B. Löwe (Eds.), Logic and Theory of Algorithms. XIX, 596 pages. 2008.

Vol. 5022: A.G. Bourgeois, S.Q. Zheng (Eds.), Algorithms and Architectures for Parallel Processing. XIII, 336 pages. 2008.

Vol. 5018: M. Grohe, R. Niedermeier (Eds.), Parameterized and Exact Computation. X, 227 pages. 2008.

Vol. 5015: L. Perron, M.A. Trick (Eds.), Integration of AI and OR Techniques in Constraint Programming for Combinatorial Optimization Problems. XII, 394 pages. 2008.

Vol. 5011: A.J. van der Poorten, A. Stein (Eds.), Algorithmic Number Theory. IX, 455 pages. 2008.

Vol. 5010: E.A. Hirsch, A.A. Razborov, A. Semenov, A. Slissenko (Eds.), Computer Science – Theory and Applications. XIII, 411 pages. 2008.

Vol. 5008: A. Gasteratos, M. Vincze, J.K. Tsotsos (Eds.), Computer Vision Systems. XV, 560 pages. 2008.

Vol. 5004: R. Eigenmann, B.R. de Supinski (Eds.), OpenMP in a New Era of Parallelism. X, 191 pages. 2008.

Vol. 5000: O. Grumberg, H. Veith (Eds.), 25 Years of Model Checking. VII, 231 pages. 2008.

Vol. 4996: H. Kleine Büning, X. Zhao (Eds.), Theory and Applications of Satisfiability Testing – SAT 2008. X, 305 pages. 2008.

Vol. 4988: R. Berghammer, B. Möller, G. Struth (Eds.), Relations and Kleene Algebra in Computer Science. X, 397 pages. 2008.

Vol. 4985: M. Ishikawa, K. Doya, H. Miyamoto, T. Yamakawa (Eds.), Neural Information Processing, Part II. XXX, 1091 pages. 2008.

Vol. 4984: M. Ishikawa, K. Doya, H. Miyamoto, T. Yamakawa (Eds.), Neural Information Processing, Part I. XXX, 1147 pages. 2008.

Vol. 4981: M. Egerstedt, B. Mishra (Eds.), Hybrid Systems: Computation and Control. XV, 680 pages. 2008.

Vol. 4978: M. Agrawal, D. Du, Z. Duan, A. Li (Eds.), Theory and Applications of Models of Computation. XII, 598 pages. 2008.

Vol. 4975: F. Chen, B. Jüttler (Eds.), Advances in Geometric Modeling and Processing. XV, 606 pages. 2008.

Vol. 4974: M. Giacobini, A. Brabazon, S. Cagnoni, G.A. Di Caro, R. Drechsler, A. Ekárt, A.I. Esparcia-Alcázar, M. Farooq, A. Fink, J. McCormack, M. O'Neill, J. Romero, F. Rothlauf, G. Squillero, A.Ş. Uyar, S. Yang (Eds.), Applications of Evolutionary Computing. XXV, 701 pages. 2008.

Vol. 4973: E. Marchiori, J.H. Moore (Eds.), Evolutionary Computation, Machine Learning and Data Mining in Bioinformatics. X, 213 pages. 2008.

Vol. 4972: J. van Hemert, C. Cotta (Eds.), Evolutionary Computation in Combinatorial Optimization. XII, 289 pages. 2008.

Vol. 4971: M. O'Neill, L. Vanneschi, S. Gustafson, A.I. Esparcia Alcázar, I. De Falco, A. Della Cioppa, E. Tarantino (Eds.), Genetic Programming. XI, 375 pages. 2008.

Vol. 4967: R. Wyrzykowski, J. Dongarra, K. Karczewski, J. Wasniewski (Eds.), Parallel Processing and Applied Mathematics. XXIII, 1414 pages. 2008.

Vol. 4963: C.R. Ramakrishnan, J. Rehof (Eds.), Tools and Algorithms for the Construction and Analysis of Systems. XVI, 518 pages. 2008.

Vol. 4962: R. Amadio (Ed.), Foundations of Software Science and Computational Structures. XV, 505 pages. 2008.

Vol. 4961: J.L. Fiadeiro, P. Inverardi (Eds.), Fundamental Approaches to Software Engineering. XIII, 430 pages. 2008.

Vol. 4960: S. Drossopoulou (Ed.), Programming Languages and Systems. XIII, 399 pages. 2008.

Vol. 4959: L. Hendren (Ed.), Compiler Construction. XII, 307 pages. 2008.

Vol. 4957: E.S. Laber, C. Bornstein, L.T. Nogueira, L. Faria (Eds.), LATIN 2008: Theoretical Informatics. XVII, 794 pages. 2008.

Vol. 4943: R. Woods, K. Compton, C. Bouganis, P.C. Diniz (Eds.), Reconfigurable Computing: Architectures, Tools and Applications. XIV, 344 pages. 2008.

Vol. 4942: E. Frachtenberg, U. Schwiegelshohn (Eds.), Job Scheduling Strategies for Parallel Processing. VII, 189 pages. 2008.

Vol. 4941: M. Miculan, I. Scagnetto, F. Honsell (Eds.), Types for Proofs and Programs. VII, 203 pages. 2008.

Vol. 4935: B. Chapman, W. Zheng, G.R. Gao, M. Sato, E. Ayguadé, D. Wang (Eds.), A Practical Programming Model for the Multi-Core Era. VI, 208 pages. 2008.

Vol. 4934: U. Brinkschulte, T. Ungerer, C. Hochberger, R.G. Spallek (Eds.), Architecture of Computing Systems – ARCS 2008. XI, 287 pages. 2008.

Vol. 4927: C. Kaklamanis, M. Skutella (Eds.), Approximation and Online Algorithms. X, 289 pages. 2008.

Vol. 4926: N. Monmarché, E.-G. Talbi, P. Collet, M. Schoenauer, E. Lutton (Eds.), Artificial Evolution. XIII, 327 pages. 2008.

Vol. 4921: S.-i. Nakano, M.. S. Rahman (Eds.), WALCOM: Algorithms and Computation. XII, 241 pages. 2008.

Vol. 4919: A. Gelbukh (Ed.), Computational Linguistics and Intelligent Text Processing. XVIII, 666 pages. 2008.

Vol. 4917: P. Stenström, M. Dubois, M. Katevenis, R. Gupta, T. Ungerer (Eds.), High Performance Embedded Architectures and Compilers. XIII, 400 pages. 2008.

Vol. 4915: A. King (Ed.), Logic-Based Program Synthesis and Transformation. X, 219 pages. 2008.

Vol. 4912: G. Barthe, C. Fournet (Eds.), Trustworthy Global Computing. XI, 401 pages. 2008.

Vol. 4910: V. Geffert, J. Karhumäki, A. Bertoni, B. Preneel, P. Návrat, M. Bieliková (Eds.), SOFSEM 2008: Theory and Practice of Computer Science. XV, 792 pages. 2008.

Vol. 4905: F. Logozzo, D.A. Peled, L.D. Zuck (Eds.), Verification, Model Checking, and Abstract Interpretation. X, 325 pages. 2008.

Vol. 4904: S. Rao, M. Chatterjee, P. Jayanti, C.S.R. Murthy, S.K. Saha (Eds.), Distributed Computing and Networking. XVIII, 588 pages. 2007.

Vol. 4878: E. Tovar, P. Tsigas, H. Fouchal (Eds.), Principles of Distributed Systems. XIII, 457 pages. 2007.

Vol. 4875: S.-H. Hong, T. Nishizeki, W. Quan (Eds.), Graph Drawing. XIII, 402 pages. 2008.

Vol. 4873: S. Aluru, M. Parashar, R. Badrinath, V.K. Prasanna (Eds.), High Performance Computing – HiPC 2007. XXIV, 663 pages. 2007.

Vol. 4863: A. Bonato, F.R.K. Chung (Eds.), Algorithms and Models for the Web-Graph. X, 217 pages. 2007.

Vol. 4860: G. Eleftherakis, P. Kefalas, G. Păun, G. Rozenberg, A. Salomaa (Eds.), Membrane Computing. IX, 453 pages. 2007.

Vol. 4855: V. Arvind, S. Prasad (Eds.), FSTTCS 2007: Foundations of Software Technology and Theoretical Computer Science. XIV, 558 pages. 2007.

Vol. 4854: L. Bougé, M. Forsell, J.L. Träff, A. Streit, W. Ziegler, M. Alexander, S. Childs (Eds.), Euro-Par 2007 Workshops: Parallel Processing. XVII, 236 pages. 2008.

Vol. 4851: S. Boztaş, H.-F.(F.) Lu (Eds.), Applied Algebra, Algebraic Algorithms and Error-Correcting Codes. XII, 368 pages. 2007.

Vol. 4848: M.H. Garzon, H. Yan (Eds.), DNA Computing. XI, 292 pages. 2008.

Vol. 4847: M. Xu, Y. Zhan, J. Cao, Y. Liu (Eds.), Advanced Parallel Processing Technologies. XIX, 767 pages. 2007.

Vol. 4846: I. Cervesato (Ed.), Advances in Computer Science – ASIAN 2007. XI, 313 pages. 2007.

Vol. 4838: T. Masuzawa, S. Tixeuil (Eds.), Stabilization, Safety, and Security of Distributed Systems. XIII, 409 pages. 2007.

Vol. 4835: T. Tokuyama (Ed.), Algorithms and Computation. XVII, 929 pages. 2007.

Vol. 4818: I. Lirkov, S. Margenov, J. Waśniewski (Eds.), Large-Scale Scientific Computing. XIV, 755 pages. 2008.

Vol. 4800: A. Avron, N. Dershowitz, A. Rabinovich (Eds.), Pillars of Computer Science. XXI, 683 pages. 2008.

Vol. 4783: J. Holub, J. Žďárek (Eds.), Implementation and Application of Automata. XIII, 324 pages. 2007.